Lecture Notes in Computer Sci

T0238830

Commenced Publication in 1973
Founding and Former Series Editors:
Gerhard Goos, Juris Hartmanis, and Jan van Leeuwen

Carlos Martín-Vide   Friedrich Otto
Henning Fernau (Eds.)

Language and Automata Theory and Applications

Second International Conference, LATA 2008
Tarragona, Spain, March 13-19, 2008
Revised Papers

 Springer

Volume Editors

Carlos Martín-Vide
Rovira i Virgili University, Research Group on Mathematical Linguistics
Plaza Imperial Tàrraco 1, 43005 Tarragona, Spain
E-mail: carlos.martin@urv.cat

Friedrich Otto
Universität Kassel, Fachbereich Elektrotechnik/Informatik
Wilhelmshöher Allee 73, 34121 Kassel, Germany
E-mail: otto@theory.informatik.uni-kassel.de

Henning Fernau
Universität Trier, Fachbereich 4, Abteilung Informatik/Wirtschaftsinformatik
Campus II, Gebäude H, 54286 Trier, Germany
E-mail: fernau@uni-trier.de

Library of Congress Control Number: 2008936681

CR Subject Classification (1998): F.4, I.1, I.5, F.1

LNCS Sublibrary: SL 1 – Theoretical Computer Science and General Issues

ISSN   0302-9743
ISBN-10   3-540-88281-2 Springer Berlin Heidelberg New York
ISBN-13   978-3-540-88281-7 Springer Berlin Heidelberg New York

Springer is a part of Springer Science+Business Media

springer.com

© Springer-Verlag Berlin Heidelberg 2008
Printed in Germany

Typesetting: Camera-ready by author, data conversion by Scientific Publishing Services, Chennai, India
Printed on acid-free paper   SPIN: 12524310   06/3180   5 4 3 2 1 0

Preface

These proceedings contain most of the papers that were presented at the Second International Conference on Language and Automata Theory and Applications (LATA 2008), held in Tarragona, Spain, during March 13–19, 2008.

The scope of LATA is rather broad, including: words, languages and automata; grammars (Chomsky hierarchy, contextual, multidimensional, unification, categorial, etc.); grammars and automata architectures; extended automata; combinatorics on words; language varieties and semigroups; algebraic language theory; computability; computational, descriptional, and parameterized complexity; decidability questions on words and languages; patterns and codes; symbolic dynamics; regulated rewriting; trees, tree languages and tree machines; term rewriting; graphs and graph transformation; power series; fuzzy and rough languages; cellular automata; DNA and other models of bio-inspired computing; quantum, chemical and optical computing; biomolecular nanotechnology; automata and logic; algorithms on automata and words; automata for system analysis and program verification; automata, concurrency and Petri nets; parsing; weighted machines; transducers; foundations of finite state technology; grammatical inference and algorithmic learning; text retrieval, pattern matching and pattern recognition; text algorithms; string and combinatorial issues in computational biology and bioinformatics; mathematical evolutionary genomics; language-based cryptography; data and image compression; circuits and networks; language-theoretic foundations of artificial intelligence and artificial life; digital libraries; and document engineering.

LATA 2008 received 134 submissions. Each of them was reviewed by at least three Program Committee members plus, in most cases, by additional external referees. After a thorough and vivid evaluation phase the committee decided to accept 40 papers (which means an acceptance rate of 29.85%). The conference programme also included three invited talks and two invited tutorials. Part of the success in the management of such a large number of submissions is due to the facilities provided by the EasyChair conference management system.

After the conference, the authors of the accepted and of the invited papers were asked to prepare revised versions of their papers (the former still satisfying the originally imposed 12-page limit). However, having post-proceedings offers the advantage of incorporating comments both of the reviewers and of the audience (received upon giving the talks). The present volume contains the 40 contributed papers plus extended abstracts of four of the invited papers.

June 2008

Carlos Martín-Vide
Friedrich Otto
Henning Fernau

Organization

LATA 2008 was hosted by the Research Group on Mathematical Linguistics (GRLMC) at Rovira i Virgili University, Tarragona, Spain.

Program Committee

Jorge Almeida	Porto, Portugal
Stefano Crespi-Reghizzi	Milan, Italy
Erzsébet Csuhaj-Varjú	Budapest, Hungary
Carsten Damm	Göttingen, Germany
Volker Diekert	Stuttgart, Germany
Frank Drewes	Umeå, Sweden
Manfred Droste	Leipzig, Germany
Zoltán Ésik	Tarragona, Spain
Henning Fernau	Trier, Germany
Jörg Flum	Freiburg, Germany
Rusins Freivalds	Riga, Latvia
Christiane Frougny	Paris, France
Max Garzon	Memphis, USA
Tero Harju	Turku, Finland
Lane Hemaspaandra	Rochester, USA
Markus Holzer	München, Germany
Hendrik Jan Hoogeboom	Leiden, The Netherlands
Kevin Knight	Marina del Rey, USA
Hans-Jörg Kreowski	Bremen, Germany
Dietrich Kuske	Leipzig, Germany
Thierry Lecroq	Rouen, France
Carlos Martín-Vide (Chair)	Tarragona, Spain
Victor Mitrana	Tarragona, Spain
Mark-Jan Nederhof	St. Andrews, UK
Mitsunori Ogihara	Rochester, USA
Friedrich Otto	Kassel, Germany
Jean-Éric Pin	Paris, France
Kai Salomaa	Kingston, Canada
Jacobo Torán	Ulm, Germany
Alfonso Valencia	Madrid, Spain
Hsu-Chun Yen	Taipei, Taiwan
Sheng Yu	London, Canada

Organizing Committee

Madalina Barbaiani
Gemma Bel-Enguix
Carlos Cruz Reyes
Adrian Horia Dediu
Szilárd Zsolt Fazekas
Mihai Ionescu
M. Dolores Jiménez-López
Alexander Krassovitskiy
Guangwu Liu

Remco Loos
Carlos Martín-Vide (Chair)
Zoltán-Pál Mecsei
Cătălin-Ionuţ Tîrnăucă
Cristina Tîrnăucă
Bianca Truthe
Sherzod Turaev
Florentina-Lilica Voicu

Additional Referees

Saïd Abdeddaïm
Giovanni Agosta
Cyril Allauzen
José J. Almeida
Dmitry Ananichev
Sergei V. Avgustinovich
Marie-Pierre Béal
Stefano Berardi
Achim Blumensath
Benedikt Bollig
Pierre Boullier
Luca Breveglieri
Robert Brijder
Alexander Burnstein
Olivier Carton
Giuseppa Castiglione
Julien Cervelle
Stephan Chalup
Jean-Marc
    Champarnaud
Yijia Chen
Alessandra Cherubini
Maxime Crochemore
Elena Czeizler
Eugen Czeizler
Jürgen Dassow
Michael Domaratzki
Andrew Duncan
Irène Anne Durand
Mahmoud El-Sakka
Mario Florido

Enrico Formenti
David de Frutos-Escrig
Yuan Gao
Paul Gastin
Zsolt Gazdag
Wouter Gelade
Jürgen Giesl
Daniel Gildea
Stefan Göller
Erich Grädel
Hermann Gruber
Stefan Gulan
Peter Habermehl
Vesa Halava
Yo-Sub Han
Kevin G. Hare
Ulrich Hertrampf
Mika Hirvensalo
Dieter Hofbauer
Benjamin Hoffmann
Johanna Högberg
Johanna Hörg
Hans Hüttel
Oscar Ibarra
Lucian Ilie
Aravind K. Joshi
Tomi Kärki
Oliver Keller
Bakhadyr Khoussainov
Renate
    Klempien-Hinrichs

Stavros Konstantinidis
Walter Kosters
Manfred Kufleitner
Sabine Kuske
Martin Kutrib
Peep Küngas
Arnaud Lefebvre
Hing Leung
Shiguo Lian
Markus Lohrey
Sylvain Lombardy
Michael Luttenberger
Alejandro Maass
Andreas Malcher
Andreas Maletti
Florin Manea
Wolfgang May
Ingmar Meinecke
Paul-André Melliès
Mark Mercer
Brink van der Merwe
Hartmut Messerschmidt
Antoine Meyer
Nelma Moreira
Rémi Morin
Peter Mosses
Moritz Müller
Loránd Muzamel
Marius Nagy
Paliath Narendran
Andrew J. Neel

Frank Neven
Dirk Nowotka
Enno Ohlebusch
Satoshi Okawa
Alexander Okhotin
Martin Otto
Michio Oyamaguchi
Edita Pelantová
Mati Pentus
Holger Petersen
Vinhthuy Phan
Erhard Plödereder
Natacha Portier
Matteo Pradella
Andreas P. Priesnitz
Mathieu Raffinot
George Rahonis
Antoine Rauzy
Wolfgang Reisig
Jacques Sakarovitch

Davide Sangiorgi
Pierluigi San Pietro
Nicolae Santean
Andrea Sattler-Klein
Gilles Schaeffer
Vincent Schmitt
Ilka Schnorr
Hinrich Schütze
Stefan Schwoon
Helmut Seidl
Carla Selmi
Géraud Sénizergues
Pedro V. Silva
David Soloveichik
Jirí Srba
Ludwig Staiger
Heiko Stamer
Ralf Stiebe
Wolfgang Thomas
Krisztián Tichler

Caroline von Totth
Yoshihito Toyama
Ralf Treinen
Sándor Vágvölgyi
Stefano Varricchio
György Vaszil
Kumar Neeraj Verma
Rudy van Vliet
Laurent Vuillon
Fabian Wagner
Shu Wang
John Watrous
Mark Weyer
Reinhard Winkler
Karsten Wolf
Thomas Worsch
Ryo Yoshinaka
Marc Zeitoun

Sponsoring Institutions

Thanks are to be given to the sponsor of the conference, namely, Fundació Caixa Tarragona.

Table of Contents

Tree-Walking Automata

Mikołaj Bojańczyk

Warsaw University

Abstract. A survey of tree-walking automata. The main focus is on how the expressive power is changed by adding features such as pebbles or non-determinism.

1 Introduction

A tree-walking automaton is a sequential device that can recognize properties of trees. The control of the automaton is always located in a single node of the tree; based on local properties of this node, the automaton chooses a new state and moves to a neighboring node. Tree-walking automata have been introduced already in a 1971 paper of Aho and Ullman [1]. The purpose of this talk is to survey the different types of tree-walking automata, with a special focus on expressive power.

A tree-walking automaton can be easily simulated by a branching bottom-up tree automaton, therefore tree-walking automata recognize only regular tree languages. However, the converse inclusion has been a notorious open problem for many years[1]; only recently did [2] establish that tree-walking automata are strictly less expressive than branching automata. Other fundamental properties have also been shown but recently: deterministic tree-walking automata are closed under complement [10], and recognize fewer languages than nondeterministic ones [3]. The proofs for the negative results—which show that tree-walking automata cannot recognize something—require involved combinatorics, and some algebra.

These are the main results on "bare" tree-walking automata. Things become even more interesting with extensions of the model. The problem with tree-walking automata, and also the reason why they are less expressive than branching automata, is that they easily get lost in a tree. One solution to this problem, due to Engelfriet and Hoogeboom [7], is to allow the automaton to mark tree nodes with pebbles.

There are several ways of adding pebbles to the automaton. One common property in all the pebble models is the use of stack discipline—where only the most recently placed pebble can be lifted—without which the model becomes undecidable. But beyond the stack discipline, there are several design choices: are the automata deterministic? does the head of the automaton need to be over a pebble when it is lifted? how many pebbles are there? which pebbles does the

---

[1] A footnote in the original paper [1] on tree-walking automata states that Michael Rabin has shown that tree-walking automata *do* recognize all regular tree languages.

C. Martín-Vide, F. Otto, and H. Fernau (Eds.): LATA 2008, LNCS 5196, pp. 1–2, 2008.

automaton see? For each of these choices, one gets an interesting and robust class of languages; these classes have been investigated in [4,6,7,8,9,10,12,5]. In all cases except one, the automata are weaker than branching automata, furthermore adding pebbles increases the expressive power.

Tree-walking automata, but even more so pebble automata, have a close relationship with logic. For each variant of the automaton, there is an equivalent logical description. For tree-walking automata and pebble automata with bounded numbers of pebbles, the logics are restricted fragments of transitive closure first-order logic (see [11] and [8], respectively). Automata with invisible pebbles, a form of pebble automata with an unbounded number of pebbles, capture all regular tree languages [9], and therefore correspond to monadic second-order logic. As far as logic is concerned, probably the most interesting class are the automata of [5], which extend pebble automata with a form of negation: these automata have the same expressive power as transitive closure first-order logic. In particular, since the extended pebble automata are still weaker than branching automata, it follows that transitive closure first-order logic is weaker than monadic second-order logic on trees (the two logics have the same expressive power over words).

References

1. Aho, A.V., Ullman, J.D.: Translations on a context-free grammar. Information and Control 19, 439–475 (1971)
2. Bojańczyk, M., Colcombet, T.: Tree-walking automata do not recognize all regular languages. In: ACM Symposium on the Theory of Computing, pp. 234–243 (2005)
3. Bojańczyk, M., Colcombet, T.: Tree-walking automata cannot be determinized. Theoretical Computer Science 350(2-3), 255–272 (2006)
4. Bojańczyk, M., Samuelides, M., Schwentick, T., Segoufin, L.: Expressive power of pebble automata. In: Bugliesi, M., Preneel, B., Sassone, V., Wegener, I. (eds.) ICALP 2006. LNCS, vol. 4051, pp. 157–168. Springer, Heidelberg (2006)
5. ten Cate, B., Segoufin, L.: XPath, transitive closure logic, and nested tree walking automata. In: Principles of Database Systems (2007)
6. Engelfriet, J., Hoogeboom, H., Van Best, J.: Trips on trees. Acta Cybernetica 14(1), 51–64 (1999)
7. Engelfriet, J., Hoogeboom, H.J.: Tree-walking pebble automata. In: Paum, G., Karhumaki, J., Maurer, H., Rozenberg, G. (eds.) Jewels Are Forever, Contributions to Theoretical Computer Science in Honor of Arto Salomaa, pp. 72–83. Springer, Heidelberg (1999)
8. Engelfriet, J., Hoogeboom, H.J.: Automata with nested pebbles capture first-order logic with transitive closure. Logical Methods in Computer Science, 3(2:3) (2007)
9. Engelfriet, J., Hoogeboom, H.J., Samwel, B.: XML transformation by tree-walking transducers with invisible pebbles. In: Principles of Database Systems, pp. 63–72 (2007)
10. Muscholl, A., Samuelides, M., Segoufin, L.: Complementing deterministic tree-walking automata. Information Processing Letters 99(1), 33–39 (2006)
11. Neven, F., Schwentick, T.: On the power of tree-walking automata. In: Welzl, E., Montanari, U., Rolim, J. (eds.) ICALP 2000. LNCS, vol. 1853. Springer, Heidelberg (2000)
12. Samuelides, M., Segoufin, L.: Complexity of pebble tree-walking automata. In: Fundamentals of Computation Theory, pp. 458–469 (2007)

Formal Language Tools for Template-Guided DNA Recombination

Michael Domaratzki

Department of Computer Science
University of Manitoba
Winnipeg, MB R3T 2N2 Canada
mdomarat@cs.umanitoba.ca

Certain stichotrichous ciliates, single-celled organisms with hair-like structures called cilia, have a well-studied ability to rearrange their DNA during a form of asexual reproduction called conjugation. Ciliates also demonstrate nuclear dualism: each ciliate has both a micronucleus and a macronucleus. The unscrambling of DNA during conjugation occurs when the scrambled version, contained in the micronucleus, is rearranged in a precise way to produce an unscrambled equivalent which forms the macronucleus.

The use of ciliates for natural computing has been one motivation for the study of computational aspects of several different models for ciliate DNA rearrangement (see, e.g., Ehrenfeucht *et al.* [8] for description of one model of DNA rearrangement in ciliates). If the mechanism by which ciliates rearrange their DNA can be understood, then it is conceivable that this mechanism could be modified to perform computation by rearrangement. For example, this line of research has been examined for one model of ciliate DNA rearrangement by Alhazov *et al.* [1], who have described a theoretical model for solving the Hamiltonian Path Problem using ciliate rearrangement.

Template-guided recombination (TGR) is one formal model for the rearrangement which occurs in stichotrichous ciliates. The theoretical model, proposed by Prescott *et al.* [11] and refined by Angeleska *et al.* [2] has been the subject of much research. Recent experimental evidence [10] suggests that TGR is an appropriate model of the rearrangement in stichotrichous ciliates. In the formal model of TGR, the unscrambling action is controlled by a set of templates; the experimental research suggests that existing genetic material forms these sets of templates which guide the rearrangement process.

TGR is easily interpreted in formal language-theoretic terms, including both an iterated and single-application variant. The iterated version gives a model which is more accurate in biological terms, as it represents the effect of repeated applications of the rearrangement process which is necessary for complete unscrambling. Much of the research on the computational aspects of TGR has focused on closure properties. Daley and McQuillan [4,5,6] have extensively studied the closure properties of TGR. In some of the iterated cases, the closure properties given were not effective; McQuillan et al. [9] presented effective constructions for these closure properties.

C. Martín-Vide, F. Otto, and H. Fernau (Eds.): LATA 2008, LNCS 5196, pp. 3–5, 2008.
© Springer-Verlag Berlin Heidelberg 2008

In all of the above results, TGR is considered as an operation on words with two operands. In this view of TGR as an inter-molecular operation, the operands represent separate DNA strands which are recombined using a template. Another aspect of the computational examinations of TGR is the consideration of unary (intra-molecular) operations, where the template acts on different regions of a single strand of DNA. Daley *et al.* [3] have examined the closure properties of the intra-molecular TGR operation. Angeleska *et al.* [2] also provide results on intra-molecular TGR in terms of their modified biological definition for TGR.

However, there is another approach to computational and language-theoretic questions related to TGR. In particular, we have recently examined the concept of equivalence for TGR [7]: given two sets of templates, do they define the same rearrangement process? This question can be asked for both the inter- and intra-molecular operations.

Results on equivalence provide tools for examining TGR as a model independently of its computational ability, and give tools for determining what changes should be made to the set of templates in order to affect the rearrangement process. The characterization of equivalence in formal language-theoretic terms yields decidability results for regular and context-free sets of trajectories.

Several questions related to equivalence remain open. In particular, we do not know if the condition which characterizes equivalence for TGR also applies to iterated TGR.

References

1. Alhazov, A., Petre, I., Rogojin, V.: Solutions to computational problems through gene assembly. In: Garzon, M.H., Yan, H. (eds.) DNA 2007. LNCS, vol. 4848, pp. 36–45. Springer, Heidelberg (2008)
2. Angeleska, A., Jonoska, N., Saito, M., Landweber, L.: RNA-Guided DNA assembly. Journal of Theoretical Biology 248, 706–720 (2007)
3. Daley, M., Domaratzki, M., Morris, A.: Intra-molecular template-guided recombination. International Journal of Foundations of Computer Science 18, 1177–1186 (2007)
4. Daley, M., McQuillan, I.: Template-guided DNA recombination. Theoretical Computer Science 330, 237–250 (2005)
5. Daley, M., McQuillan, I.: On computational properties of template-guided DNA recombination in ciliates. In: Carbone, A., Pierce, N.A. (eds.) DNA 2005. LNCS, vol. 3892, pp. 27–37. Springer, Heidelberg (2006)
6. Daley, M., McQuillan, I.: Useful templates and iterated template-guided DNA recombination in ciliates. Theory of Computing Systems 39, 619–633 (2006)
7. Domaratzki, M.: Equivalence in template-guided recombination. Natural Computing (to appear, 2008)
8. Ehrenfeucht, A., Harju, T., Petre, I., Prescott, D., Rozenberg, G.: Computation in Living Cells: Gene Assembly in Ciliates. Springer, Heidelberg (2004)
9. McQuillan, I., Salomaa, K., Daley, M.: Iterated TGR languages: Membership problem and effective closure properties. In: Chen, D.Z., Lee, D.T. (eds.) COCOON 2006. LNCS, vol. 4112, pp. 94–103. Springer, Heidelberg (2006)

10. Nowacki, M., Vijayan, V., Zhou, Y., Schotanus, K., Doak, T., Landweber, L.: RNA-mediated epigenetic programming of a genome-rearrangement pathway. Nature 451, 153–159 (2008)
11. Prescott, D., Ehrenfeucht, A., Rozenberg, G.: Template-guided recombination for IES elimination and unscrambling of genes in stichotrichous ciliates. Journal of Theoretical Biology 222, 323–330 (2003)

Subsequence Counting, Matrix Representations and a Theorem of Eilenberg

Benjamin Steinberg[*]

School of Mathematics and Statistics
Carleton University
Ottawa, ON, Canada
bsteinbg@math.carleton.ca

Introduction

Recently, Almeida, Margolis, Volkov and I have applied matrix representation theory [1] to give a simpler proof of results of Péladeau [4] and Weil [5] concerning marked products with counter. Eilenberg's theorem characterizing languages recognized by p-groups [2] is a special case of these results. In these proceedings I will give a simple proof of Eilenberg's Theorem based on representation theory that I came up with for a graduate course. The ideas are similar to those used in [1], which I presented during the conference.

1 Free Monoids, Algebras and Subsequences

A good reference for the material in this section is [3]. Let A be a finite alphabet. We use A^* for the free monoid and write 1 for the empty string. Recall that a language $L \subseteq A^*$ is recognized by a monoid M if there is a homomorphism $\varphi\colon A^* \to M$ such that $L = \varphi^{-1}\varphi(L)$. A language is regular if and only if it can be recognized by a finite monoid. Let p be a prime; then a finite p-group is a group of order p^n, some $n \geq 0$. Notice that the collection of finite p-groups is closed under direct product, subgroups and homomorphic images. Hence the set of languages recognized by a finite p-group is a Boolean algebra [2].

For words $u, v \in A^*$ define $\binom{u}{v}$ to be the number of ways to choose $|v|$ positions in u that spell the word v, i.e. the number of occurrences of v as a subsequence of u. More precisely, if $v = a_1 \cdots a_n$, then $\binom{u}{v}$ is the number of factorizations $u = v_0 a_1 v_1 \cdots a_n v_n$ with each $v_i \in A^*$. For example $\binom{abab}{ab} = 3$. Notice that $\binom{a^n}{a^m} = \binom{n}{m}$, whence the notation. Also one has a sort of Pascal's triangle rule:

$$\binom{ua}{vb} = \begin{cases} \binom{u}{vb} + \binom{u}{v} & a = b \\ \binom{u}{vb} & a \neq b. \end{cases}$$

The reader is referred to Lothaire [3, Sec. 6.3] for more on binomial coefficients.

---

[*] The author would like to acknowledge the support of an NSERC grant.

C. Martín-Vide, F. Otto, and H. Fernau (Eds.): LATA 2008, LNCS 5196, pp. 6–10, 2008.
© Springer-Verlag Berlin Heidelberg 2008

Let K be a field and A a finite alphabet. Then $K\langle A \rangle$ denotes the ring of polynomials over K in non-commuting variables A, that is, the free algebra over A. Elements of $K\langle A \rangle$ are of the form

$$f = \sum_{w \in A^*} c_w \cdot w \tag{1}$$

with $c_w \in K$ and all but finitely many $c_w = 0$. Multiplication is given by

$$\sum_{w \in A^*} c_w w \sum_{w \in A^*} d_w w = \sum_{w \in A^*} \left(\sum_{uv=w} c_u d_v \right) \cdot w.$$

For instance, in $K\langle x, y \rangle$ one has $(x + y)(x + 2y) = x^2 + 2xy + yx + 2y^2$. We identify K as a subring of $K\langle A \rangle$ via $c \mapsto c \cdot 1$.

If $f \in K\langle A \rangle$ is as per (1) and $w \in A^*$, we define $\langle f, w \rangle = c_w$; so $\langle f, w \rangle$ is the coefficient of w in f. Define, for $f \in K\langle A \rangle$,

$$v(f) = \min\{|w| : \langle f, w \rangle \neq 0\}$$

where we take $v(0) = \infty$. So, for instance, $v(xy^2 + yx) = 2$. One easily verifies that v is a discrete valuation, that is:

$$v(fg) = v(f) + v(g)$$
$$v(f + g) \geq \min\{v(f), v(g)\}.$$

It is immediate that $I_m = \{f \in K\langle A \rangle : v(f) \geq m\}$ is an ideal of $K\langle A \rangle$ for $m \geq 1$.

Define $\mu \colon A^* \to K\langle A \rangle$ by $\mu(a) = 1 + a$ for $a \in A$. That is $\mu(1) = 1$ and $\mu(a_1 \cdots a_n) = (1 + a_1) \cdots (1 + a_n)$. The map μ is known as the *Magnus transform* [3]. The classical binomial theorem admits the following generalization [3, Prop. 6.3.6].

Proposition 1. *Let $u \in A^*$. Then $\mu(u) = \displaystyle\sum_{v \in A^*} \binom{u}{v} v$.*

2 Eilenberg's Theorem

Fix a prime p. Suppose that $0 \leq r < p$ and $u \in A^*$. Define

$$L[u; r] = \left\{ w \in A^* : \binom{w}{u} \equiv r \bmod p \right\}.$$

Eilenberg proved the following theorem [2].

Theorem 2 (Eilenberg). *Let p be a prime. A regular language $L \subseteq A^*$ is recognized by a finite p-group if and only if it belongs to the Boolean algebra generated by languages of the form $L[u; r]$ with $u \in A^*$ and $0 \leq r < p$.*

Our goal is to use matrix representations to prove this theorem. From now on K will be $\mathbb{Z}_p$. Set $R = \mathbb{Z}_p\langle A\rangle$ and, for $m \geq 1$, put $R_m = R/I_m = \mathbb{Z}_p\langle A\rangle/I_m$; R_m is a truncated polynomial algebra. Notice that R_m is finite since each coset can be uniquely represented by a polynomial of degree at most $m - 1$. Define a truncated Magnus transform $\mu_m \colon A^* \to R_m$ by composing μ with the projection $R \to R_m$. We immediately obtain from Proposition 1

$$\mu_m(u) = \sum_{|v|<m} \binom{u}{v} v + I_m. \tag{2}$$

Let $G_m = \{f + I_m : \langle f, 1\rangle = 1\}$; evidently G_m is a submonoid of R_m containing the image of μ_m. In fact it turns out to be a p-group.

Proposition 3. *Let $f \in R$ be such that $v(f) \geq 1$. Then $(1 + f)^{p^k} \equiv 1 \bmod I_m$ whenever $p^k \geq m$.*

Proof. Since R has characteristic p, we have $(1 + f)^{p^k} = 1 + f^{p^k} \equiv 1 \bmod I_m$ since $v(f^{p^k}) = p^k v(f) \geq m$. □

Corollary 4. *The submonoid G_m of R_m is a finite p-group.*

Proof. Since f has constant term 1, it follows $v(f - 1) \geq 1$. Applying Proposition 3 to $f = 1 + (f - 1)$, we have $f^{p^k} + I_m = 1 + I_m$ for $p^k \geq m$. We conclude G_m is a p-group. □

We may now deduce that $L[u; r]$ is recognized by a p-group. Indeed, choose $m > |u|$. Viewing μ_m as a homomorphism $\mu_m \colon A^* \to G_m$, we have $L[u; r] = \mu_m^{-1}(T)$ where $T = \{f + I_m \in G_m : \langle f, u\rangle = r\}$. It follows that $\mu_m^{-1}\mu_m(L[u;r]) = L[u;r]$ and so the p-group G_m recognizes $L[u; r]$. We have thus proved:

Proposition 5. *If $m > |u|$, then the finite p-group G_m recognizes $L[u;r]$.*

In fact it is easy to see that any language recognized via μ_m is a Boolean combination of languages of the form $L[u; r]$ with $|u| < m$.

Proposition 6. *Suppose that $\mu_m^{-1}\mu_m(L) = L$. Then L is in the Boolean algebra $\mathscr{A}$ generated by languages of the form $L[u; r]$ with $|u| < m$.*

Proof. It suffices to show that if $T \subseteq \mu_m(A^*)$, then $\mu_m^{-1}(T)$ is in $\mathscr{A}$. Since $\mu_m^{-1}(T) = \bigcup_{f \in T} \mu_m^{-1}(\{f\})$, without loss of generality we may assume $T = \{f\}$. But (2) immediately yields

$$\mu_m^{-1}(\{f\}) = \bigcap_{|v|<m} L[v; \langle f, v\rangle]$$

completing the proof. □

In light of Proposition 6, to complete the proof it suffices to show that any onto homomorphism from A^* to a finite p-group factors through μ_m for some $m > 0$. Up until this point, the proof has been more or less the same as the one presented in [2]. Eilenberg uses group algebras to complete the proof; we opt for matrix representations.

Denote by $GL(n, p)$ the group of $n \times n$ invertible matrices over $\mathbb{Z}_p$. A matrix is called *unitriangular* if it is upper triangular with diagonal entries all equal to 1; the subgroup of unitriangular $n \times n$ matrices is denoted $UT(n, p)$. Notice that $|UT(n, p)| = p^{\binom{n}{2}}$ and hence $UT(n, p)$ is a p-group. In fact, since

$$|GL(n, p)| = \prod_{k=0}^{n-1} (p^n - p^k) = p^{\binom{n}{2}} \prod_{i=1}^{n} (p^i - 1),$$

it follows that $UT(n, p)$ is a p-Sylow subgroup of $GL(n, p)$ (using that $p \nmid p^i - 1$ for $i \geq 1$). It is now easy to prove that every finite p-group is isomorphic to a group of unitriangular matrices (I learned of this proof from Margolis).

Theorem 7. *Let G be a finite p-group of order $n = p^k$. Then G is isomorphic to a subgroup of $UT(n, p)$.*

Proof. By Cayley's Theorem, G is isomorphic to a subgroup of the symmetric group S_n. But S_n is isomorphic to the subgroup of $GL(n, p)$ consisting of the permutation matrices (zero-one matrices with a single one in each row and column). Thus without loss of generality we may assume G is a subgroup of $GL(n, p)$. But then Sylow's Theorem implies that some conjugate of G is contained in $UT(n, p)$, establishing the theorem. □

Denote by $M(n, p)$ the ring of $n \times n$ matrices over $\mathbb{Z}_p$. Let $N(n, p)$ be the ring of $n \times n$ upper triangular matrices with all diagonal entries equal to 0 The following fact is well known.

Lemma 8. *The ring $N(n, p)$ is nilpotent of index n, that is, $N(n, p)^n = 0$.*

Proof. Suppose that $A^{(1)}, \ldots, A^{(n)} \in N(n, p)$. Then since these matrices are upper triangular:

$$(A^{(1)} \cdots A^{(n)})_{ij} = \sum_{1 \leq i_1 \leq i_2 \leq \cdots \leq i_{n+1} \leq n} A^{(1)}_{i_1 i_2} A^{(2)}_{i_2 i_3} \cdots A^{(n)}_{i_n i_{n+1}}. \tag{3}$$

It follows that $i_k = i_{k+1}$ for some $1 \leq k \leq n$ and hence the right hand side of (3) is 0 as the diagonal entries of the $A^{(m)}$ are 0. □

From the lemma we deduce the desired property of the family $\{\mu_m\}$.

Corollary 9. *Let $\varphi \colon A^* \to G$ be an onto homomorphism with G a p-group of order $n = p^k$. Then there exists a unique homomorphism $\psi \colon \mu_n(A^*) \to G$ such that the diagram*

commutes.

Proof. Without loss of generality we may assume that $G \subseteq UT(n,p)$. Define $\alpha\colon A^* \to M(n,p)$ by $\alpha(a) = \varphi(a) - I$ (where I is the identity matrix). We may extend α uniquely to $\mathbb{Z}_p\langle A \rangle$ by setting

$$\alpha\left(\sum_{w \in A^*} c_w \cdot w\right) = \sum_{w \in A^*} c_w \alpha(w).$$

Notice that $\alpha(\mathbb{Z}_p\langle A \rangle) \subseteq N(n,p)$ since $\varphi(a) \in U(n,p)$ implies $\varphi(a) - I \in N(n,p)$. Lemma 8 then yields $I_n \subseteq \ker \alpha$. Thus there is an induced homomorphism $\psi\colon \mathbb{Z}_p\langle A \rangle / I_n \to M(n,p)$ given by $\psi(f + I_n) = \alpha(f)$. Observe that

$$\psi\mu_n(a) = \psi(1 + a + I_m) = \alpha(1 + a) = I + \alpha(a) = I + \varphi(a) - I = \varphi(a)$$

and so $\psi\mu_n = \varphi$. Uniqueness of ψ is clear since all monoids in question are generated by A.   □

There's not much else left to do to finish the proof of Eilenberg's theorem.

Proof (Eilenberg's Theorem). Proposition 5 shows that the Boolean algebra generated by the languages $L[u;r]$ consists of languages recognized by p-groups. For the converse, suppose L is recognized by a p-group G via a homomorphism $\varphi\colon A^* \to G$ (so $\varphi^{-1}\varphi(L) = L$). Factor $\varphi = \psi\mu_n$ as per Corollary 9. Then

$$L \subseteq \mu_n^{-1}\mu_n(L) \subseteq \mu_n^{-1}\psi^{-1}\psi\mu_n(L) = \varphi^{-1}\varphi(L) = L$$

and so L is a Boolean combination of the desired form by Proposition 6. This completes the proof.   □

References

1. Almeida, J., Margolis, S.W., Steinberg, B., Volkov, M.V.: Representation theory of finite semigroups, semigroup radicals and formal language theory. Trans. Amer. Math. Soc. (to appear)
2. Eilenberg, S.: Automata, languages, and machines. Vol. B. Academic Press, New York (1976); Tilson, B.: Depth decomposition theorem, Complexity of semigroups and morphisms. In: Pure and Applied Mathematics, Vol. 59 (1976)
3. Lothaire, M.: Combinatorics on words. Cambridge Mathematical Library. Cambridge University Press, Cambridge (1997)
4. Péladeau, P.: Sur le produit avec compteur modulo un nombre premier. RAIRO Inform. Théor. Appl. 26(6), 553–564 (1992)
5. Weil, P.: Closure of varieties of languages under products with counter. J. Comput. System Sci. 45(3), 316–339 (1992)

Synchronizing Automata
and the Černý Conjecture

Mikhail V. Volkov

Department of Mathematics and Mechanics,
Ural State University, 620083 Ekaterinburg, Russia
Mikhail.Volkov@usu.ru

Abstract. We survey several results and open problems related to synchronizing automata. In particular, we discuss some recent advances towards a solution of the Černý conjecture.

1 History and Motivations

Let $\mathscr{A} = \langle Q, \Sigma, \delta \rangle$ be a deterministic finite automaton (DFA), where Q denotes the state set, Σ stands for the input alphabet, and $\delta : Q \times \Sigma \to Q$ is the transition function defining an action of the letters in Σ on Q. The action extends in a unique way to an action $Q \times \Sigma^* \to Q$ of the free monoid Σ^* over Σ; the latter action is still denoted by δ. The automaton $\mathscr{A}$ is called *synchronizing* if there exists a word $w \in \Sigma^*$ whose action resets $\mathscr{A}$, that is to leave the automaton in one particular state no matter which state in Q it started at: $\delta(q, w) = \delta(q', w)$ for all $q, q' \in Q$. Any word w with this property is said to be a *reset* word for the automaton.

Fig. 1 shows an example of a synchronizing automaton with 4 states. The reader can easily verify that the word ab^3ab^3a resets the automaton leaving it in the state 1. With somewhat more effort one can also check that ab^3ab^3a is the shortest reset word for this automaton. The example in Fig. 1 is due to Černý, a Slovak computer scientist, in whose pioneering paper (1964) the notion of a synchronizing automaton explicitly appeared for the first time. (Černý called such automata *directable*. The word *synchronising* in this context was probably introduced by Hennie (1964).) Implicitly, however, this concept has

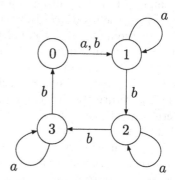

Fig. 1. A synchronizing automaton

been around since the earliest days of automata theory. The very first synchronizing automaton that we were able to trace back in the literature appeared in Ashby's classic book (1956, pp. 60–61). There Ashby presents a puzzle dealing with taming two ghostly noises, Singing and Laughter, in a haunted mansion. Each of the noises can be either on or off, and their behaviour depends on combinations of two possible actions, playing the organ or burning incense. Under

C. Martín-Vide, F. Otto, and H. Fernau (Eds.): LATA 2008, LNCS 5196, pp. 11–27, 2008.
© Springer-Verlag Berlin Heidelberg 2008

a suitable encoding, this leads to the following automaton with 4 states and 4 input letters:

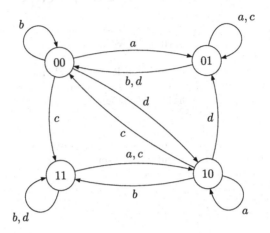

Fig. 2. Ashby's "ghost taming" automaton

Here 00 encodes the state when both Singing and Laughter are silent, 01 stands for the state when Singing is of but Laughter is on, etc. Similarly, a stands for the transition that happens when neither the organ is played nor incense is burned, b encodes the transition caused by organ-playing in the absence of incense-burning, etc. The problem is to ensure silence, in other words, to bring the automaton in Fig. 2 to the state 00. Ashby only solves the problem under the assumption that the automaton is in the state 11 and his suggested solution is encoded by the word acb. However, it is easy to check that acb is in fact a reset word for the automaton so applying the corresponding sequence of actions will get the house quiet from any initial configuration. It is not clear whether or not Ashby realized this nice feature of his automaton, and moreover, the fact that Ashby's automaton is synchronizing seems to be overlooked for many years.

Let us return to the genesis of the concept of synchronizing automata. In (Černý, 1964) this notion arose within the classic framework of Moore's "Gedanken-experiments" (1956). For Moore and his followers finite automata served as a mathematical model of devices working in discrete mode, such as computers or relay control systems. This leads to the following natural problem: how can we restore control over such a device if we do not know its current state but can observe outputs produced by the device under various actions? Moore (1956) has shown that under certain conditions one can uniquely determine the state at which the automaton arrives after a suitable sequence of actions (called an *experiment*). Moore's experiments were adaptive, that is, each next action was selected on the basis of the outputs caused by the previous actions. Ginsburg (1958) considered more restricted experiments that he called *uniform*. A uniform experiment[1] is just a fixed sequence of actions, that is, a word over

---

[1] After (Gill, 1961), the name *homing sequence* has become standard for the notion.

the input alphabet; thus, in Ginsburg's experiments outputs were only used for calculating the resulting state at the end of an experiment. From this, just one further step was needed to come to the setting in which outputs were not used at all. It should be noted that this setting is by no means artificial—there exist many practical situations when it is technically impossible to observe output signals. (Think of a satellite which loops around the Moon and cannot be controlled from the Earth while "behind" the Moon.)

It is not surprising that synchronizing automata were re-invented a number of times. First of all, the notion was very natural by itself and fitted fairly well in what was considered as the mainstream of automata theory in the 1960s. Second, Černý's paper (1964) published in Slovak language remained unknown in the English-speaking world for some time. As examples, we mention here the report (Laemmel & Rudner, 1969) and the paper (Fischler & Tannenbaum, 1970) both rediscovering results from (Černý, 1964). The books (Booth, 1967; Hennie, 1968; Kohavi, 1970) also present some information about synchronizing automata but do not refer to (Černý, 1964). It seems that the situation begun to change only in 1972 when the English translation of the book (Starke, 1969) appeared.

The original "Gedanken-experiments" motivation for studying synchronizing automata is still of importance, and reset words are frequently applied in model-based testing of reactive systems[2]. Rather unexpectedly, an additional source of problems related to synchronizing automata has come from *robotics* or, more precisely, from part handling problems in industrial automation such as part feeding, fixturing, loading, assembly and packing. Within this framework, the concept of a synchronizing automaton was again rediscovered in the mid-1980s by Natarajan (1986, 1989). We explain how abstract automata arise in part handling problems by means of a simple illustrative example from (Ananichev & Volkov, 2004).

Suppose that a part of a certain device has the shape shown in Fig. 3. Such parts arrive at manufacturing sites in boxes and they need to be sorted and oriented before assembly. For simplicity, assume that only four initial orientations of the part shown in Fig. 3 are possible, namely, the four shown in in Fig. 4.

Fig. 3. A part

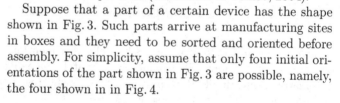

Fig. 4. Four possible orientations

Further, suppose that prior the assembly the part should take the "bump-left" orientation (the second one in Fig 4). Thus, one has to construct an orienter

---

[2] See (Cho et al, 1993; Boppana et al, 1999) as typical samples of technical contributions to the area and (Sandberg, 2005) for a recent survey.

which action will put the part in the prescribed position independently of its initial orientation.

Of course, there are many ways to design such an orienter but practical considerations favor methods which require little or no sensing, employ simple devices, and are as robust as possible. For our particular case, these goals can be achieved as follows. We put parts to be oriented on a conveyer belt which takes them to the assembly point and let the stream of the parts encounter a series of passive obstacles placed along the belt. We need two type of obstacles: high and low. A high obstacle should be high enough in order that any part on the belt encounters this obstacle by its rightmost low angle (we assume that the belt is moving from left to right). Being curried by the belt, the part then is forced to turn 90° clockwise. A low obstacle has the same effect whenever the part is in the "bump-dow" orientation (the first one in Fig. 4); otherwise it does not touch the part which therefore passes by without changing the orientation.

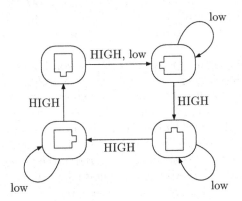

Fig. 5. The action of the obstacles

The scheme in Fig. 5 summarizes how the aforementioned obstacles effect the orientation of the part. The reader immediately recognizes the synchronizing automaton from Fig. 1. Remembering that its shortest reset word is the word ab^3ab^3a, we conclude that the series of obstacles

$$\text{low–HIGH–HIGH–HIGH–low–HIGH–HIGH–HIGH–low}$$

yields the desired sensor-free orienter.

Since the 1990s synchronizing automata usage in the area of robotic manipulation has grown into a prolific research direction but it is fair to say that publications in this direction deal mostly with implementation technicalities. However, amongst them there are papers of theoretical importance such as (Eppstein, 1990; Goldberg, 1993; Chen & Ierardi, 1994).

Speculating about further possible applications of synchronizing automata, one can think of *biocomputing*. Here we refer to recent experiments (Benenson et al, 2001, 2003) in which DNA molecules have been used as both hardware and software for finite automata of nanoscaling size. For instance,

Benenson et al (2003) produced a "soup of automata", that is, a solution containing 3×10^{12} identical automata per μl. All these molecular automata can work in parallel on different inputs, thus ending up in different and unpredictable states. In contrast to an electronic computer, one cannot reset such a system by just pressing a button; instead, in order to synchronously bring each automaton to its "ready-to-restart" state, one should spice the soup with (sufficiently many copies of) a DNA molecule whose nucleotide sequence encodes a reset word.

Clearly, from the viewpoint of applications, real or yet imaginary, algorithmic and complexity issues are of crucial importance. We discuss them in Section 2.

Putting applications aside, mathematicians since the 1960s have intensively studied synchronizing automata *per se*, as an interesting combinatorial object. These studies are mainly motivated by the *Černý conjecture*. Černý (1964) constructed for each $n > 1$ a synchronizing automaton $\mathscr{C}_n$ with n states which shortest reset word has length $(n-1)^2$ (the automaton in Fig. 1 is $\mathscr{C}_4$). Soon after that he conjectured that those automata represent the worst possible case, that is, every synchronizing automaton with n states can be reset by a word of length $(n-1)^2$. By now this simply looking conjecture is arguably the most longstanding open problem in the combinatorial theory of finite automata. We will discuss the Černý conjecture and some related partial results in Section 3.

Other mathematical motivations for studying synchronizing automata come from semigroup theory (see Ananichev & Volkov, 2004), multiple-valued logic and symbolic dynamics (see Mateescu & Salomaa, 1999). The latter connection is especially interesting in view of a recent breakthrough in the area—a (positive) solution to the Road Coloring Problem found by Trahtman (2008), but it clearly deserves a separate survey.

2 Algorithms and Complexity

It should be clear that not every DFA is synchronizing. Therefore, the very first question that we should address is the following one: *given an automaton $\mathscr{A}$, how to determine whether or not $\mathscr{A}$ is synchronizing?*

This question is in fact quite easy, and the most straightforward solution to it can be achieved via the classic power automaton construction. Recall that the *power automaton* $\mathcal{P}(\mathscr{A})$ of a given DFA $\mathscr{A} = \langle Q, \Sigma, \delta \rangle$ has the collection $\mathcal{P}'(Q)$ of the non-empty subsets of Q as the state set and the natural extension of δ to the set $\mathcal{P}'(Q) \times \Sigma$ as the transition function (still denoted by δ). In other words, for P being a non-empty subset of Q and $a \in \Sigma$, one sets $\delta(P, a) = \{\delta(p, a) \mid p \in P\}$. Fig. 6 presents the power automaton for the DFA $\mathscr{C}_4$ shown in Fig. 1.

Now it is obvious that a word $w \in \Sigma^*$ is a reset word for the DFA $\mathscr{A}$ if and only if w labels a path in $\mathcal{P}(\mathscr{A})$ starting at Q and ending at a singleton. (For instance, the bold path in Fig. 6 represents the shortest reset word ab^3ab^3a of the automaton $\mathscr{C}_4$.) Thus, the question of whether or not a given DFA $\mathscr{A}$ is synchronizing reduces to the following reachability question in the underlying digraph of the power automaton $\mathcal{P}(\mathscr{A})$: is there a path from Q to a singleton? The latter question can be easily answered by breadth-first search (see, e.g., Corman et al, 2001, Section 22.2).

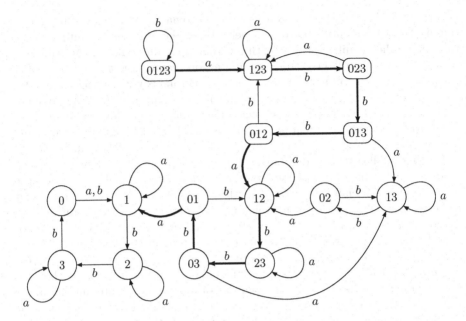

Fig. 6. The power automaton $\mathcal{P}(\mathscr{C}_4)$

The described procedure is conceptually very simple but rather inefficient because the power automaton $\mathcal{P}(\mathscr{A})$ is exponentially larger than $\mathscr{A}$. However, the following criterion of synchronizability (Černý, 1964, Theorem 2) gives rise to a polynomial algorithm.

Proposition 1. *A DFA $\mathscr{A} = \langle Q, \Sigma, \delta \rangle$ is synchronizing if and only if for every $q, q' \in Q$ there exists a word $w \in \Sigma^*$ such that $\delta(q, w) = \delta(q', w)$.*

One can treat Proposition 1 as a reduction of the synchronizability problem to a reachability problem in the subautomaton $\mathcal{P}^{[2]}(\mathscr{A})$ of $\mathcal{P}(\mathscr{A})$ whose states are 2-element and 1-element subsets of Q. Since the subautomaton has $\dfrac{|Q|(|Q|+1)}{2}$ states, breadth-first search solves this problem in $O(|Q|^2 \cdot |\Sigma|)$ time. This complexity bound assumes that no reset word is explicitly calculated. If one requires that, whenever $\mathscr{A}$ turns out to be synchronizing, a reset word is produced, then the best of the known algorithms (which is due to (Eppstein, 1990, Theorem 6), see also (Sandberg, 2005, Theorem 1.15)) has an implementation that consumes $O(|Q|^3 + |Q|^2 \cdot |\Sigma|)$ time and $O(|Q|^2 + |Q| \cdot |\Sigma|)$ working space, not counting the space for the output which is $O(|Q|^3)$.

For a synchronizing automaton, the power automaton can be used to construct shortest reset words which correspond to shortest paths from the whole state set to a singleton. Of course, this requires exponential (of $|Q|$) time in the worst case. Nevertheless, there were attempts to implement this approach (see, e.g., Rho et al, 1993; Trahtman, 2006a). One may hope that, as above, a suitable calculation in the "polynomial" subautomaton $\mathcal{P}^{[2]}(\mathscr{A})$ may yield a polynomial algorithm. However, it is not the case, and moreover, as we will see, it is very unlikely that any reasonable algorithm may exist for finding shortest reset words in general synchronizing automata. In the following discussion we assume the reader's acquaintance with some basics of computational complexity (such as the definitions of the complexity classes NP, coNP and PSPACE) that can be found, e.g., in (Garey & Johnson, 1979; Papadimitriou, 1994).

Consider the following decision problems:

SHORT-RESET-WORD: *Given a synchronizing automaton $\mathscr{A}$ and a positive integer ℓ, is it true that $\mathscr{A}$ has a reset word of length ℓ?*

SHORTEST-RESET-WORD: *Given a synchronizing automaton $\mathscr{A}$ and a positive integer ℓ, is it true that the minimum length of a reset word for $\mathscr{A}$ is equal to ℓ?*

Clearly, SHORT-RESET-WORD belongs to the complexity class NP: one can non-deterministically guess a word $w \in \Sigma^*$ of length ℓ and then check if w is a reset word for $\mathscr{A}$ in time $\ell|Q|$. Eppstein (1990) has proved that SHORT-RESET-WORD is NP-hard by a polynomial reduction from 3-SAT. Thus, SHORT-RESET-WORD is NP-complete. Other proofs for the same result (all via reductions from 3-SAT) have been suggested in (Goralčik & Koubek, 1995; Salomaa, 2003; Samotij, 2007). From the proofs it follows easily that SHORTEST-RESET-WORD is NP-hard; recently Samotij (2007) has shown that the negation of 3-SAT can be polynomially reduced to SHORTEST-RESET-WORD whence the latter problem is also coNP-hard. As a corollary, SHORTEST-RESET-WORD cannot belong to NP unless NP = coNP which is commonly considered to be very unlikely. In other words, even non-deterministic algorithms cannot find the minimum length of a reset word for a given synchronizing automaton in polynomial time.

On the other hand, the exhaustive search for reset words through all words over Σ of length $\leq \ell$ can be performed in polynomial (in fact, linear) space since one can reuse space. Thus, the problem SHORTEST-RESET-WORD belongs to the complexity class PSPACE; the question of the precise location of this problem with respect to the polynomial hierarchy still remains open. An upper bound has been recently found by Martjugin (unpublished) who has shown that the problem lies in the complexity class $\Sigma^2 \cap \Pi^2$.

By a standard argument, the hardness of the decision problem SHORT-RESET-WORD implies that its optimization version, in which one seeks a reset word of minimum length for a given synchronizing automaton, is hard as well. This did not exclude however that the optimization problem might admit a polynomial-time approximation algorithm, and moreover, all existing proofs for NP-hardness of SHORT-RESET-WORD were consistent with the conjecture that such an algorithm exists. However, recently Berlinkov (unpublished) has shown (assuming $P \neq NP$) that, for any given positive integer k, no polynomial algorithm can find for each synchronizing automaton $\mathscr{A}$ a reset word whose length would be bounded by k times the minimum length of reset words for $\mathscr{A}$. Thus, approximating the minimum length of reset words is hard.

We mention that Pixley et al (1992) suggested an heuristic algorithm for finding short reset words in synchronizing automata that was reported to perform rather satisfactory on a number of benchmarks from (Yang, 1991); further algorithms yielding short (though not necessarily shortest) reset words were implemented by Trahtman (2006a).

3   The Černý Conjecture

In this section we discuss results and open problems related to the following natural question: *given a positive integer n, how long can be reset words for synchronizing automata with n states?*

First of all, we recall Černý's construction (1964). For $n > 1$, let $\mathscr{C}_n$ stand for the DFA whose states are the residues modulo n and whose input letters a and b act as follows:

$$\delta(0,a) = 1, \ \delta(m,a) = m \text{ for } 0 < m < n, \ \delta(m,b) = m+1 \pmod{n}.$$

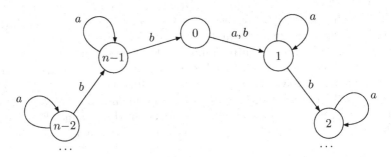

Fig. 7. The automaton $\mathscr{C}_n$

Černý (1964) has proved that $\mathscr{C}_n$ is a synchronizing automaton and that its shortest reset word is $(ab^{n-1})^{n-2}a$ of length $(n-1)^2$. (This series of automata was rediscovered many times, see, e.g., (Laemmel & Rudner, 1969; Fischler & Tannenbaum, 1970; Eppstein, 1990).) Thus, if we define the *Černý function* $C(n)$ as the maximum length of shortest reset words for synchronizing automata with n states, the above property of the series $\{\mathscr{C}_n\}$, $n = 2, 3, \ldots$, yields the inequality $C(n) \geq (n-1)^2$. The *Černý conjecture* is the claim that the equality $C(n) = (n-1)^2$ holds true.

In the literature, one often refers to (Černý, 1964) as the source of the Černý conjecture. In fact, the conjecture was not yet formulated in that paper. There Černý only observed that

$$(n-1)^2 \leq C(n) \leq 2^n - n - 1 \tag{1}$$

and concluded the paper with the following remark:

"The difference between the bounds increases rapidly and it is necessary to sharpen them. One can expect an improvement mainly for the upper bound."

The conjecture in its present-day form was formulated a bit later, after the expectation in the above quotation was confirmed by Starke (1966). (Namely, Starke improved the upper bound in (1) to $1 + \frac{n(n-1)(n-2)}{2}$, which was the first polynomial upper bound for $C(n)$.) Černý explicitly stated the conjecture $C(n) = (n-1)^2$ in his talk at the Bratislava Cybernetics Conference held in 1969; in print the conjecture first appeared in (Černý et al, 1971).

The best upper bound for the Černý function $C(n)$ achieved so far guarantees that for every synchronizing automaton with n states there exists a reset word of length $\frac{n^3-n}{6}$. Such a reset word arises as the output of the following greedy algorithm.

Algorithm 1.

---

input $\mathscr{A} = \langle Q, \Sigma, \delta \rangle$ (a DFA)

initialization $w \leftarrow 1$ (the empty word)
$\qquad\qquad\quad P \leftarrow Q$

while $|P| > 1$ find a word $v \in \Sigma^*$ of minimum length with $|\delta(P,v)| < |P|$; if none exists, **return** Failure

$\qquad w \leftarrow wv$
$\qquad P \leftarrow \delta(P,v)$

return w

If $|Q| = n$, then clearly the main loop of Algorithm 1 is executed at most $n - 1$ times. In order to evaluate the length of the output word w, we estimate the length of each word v produced by the main loop.

Consider a generic step at which $|P| = k > 1$ and let $v = a_1 \cdots a_\ell$ with $a_i \in \Sigma$, $i = 1, \dots, \ell$. Then it is easy to see that the sets

$$P_1 = P, \ P_2 = \delta(P_1, a_1), \ \dots, \ P_\ell = \delta(P_{\ell-1}, a_{\ell-1})$$

are k-element subsets of Q. Furthermore, since $|\delta(P_\ell, a_\ell)| < |P_\ell|$, there exist two distinct states $q_\ell, q'_\ell \in P_\ell$ such that $\delta(q_\ell, a_\ell) = \delta(q'_\ell, a_\ell)$. Now define 2-element subsets $R_i = \{q_i, q'_i\} \subseteq P_i$, $i = 1, \dots, \ell$, such that $\delta(q_i, a_i) = q_{i+1}$, $\delta(q'_i, a_i) = q'_{i+1}$ for $i = 1, \dots, \ell - 1$. Then the condition that v is a word of minimum length with $|\delta(P,v)| < |P|$ implies that $R_i \not\subseteq P_j$ for $1 \le j < i \le \ell$. Altogether, we arrive at the following purely combinatorial problem:

Question 1. Let Q be an n-element set, $P_1, \dots, P_\ell$ a sequence of its k-element subsets ($k > 1$) and $R_1, \dots, R_\ell$ a sequence of its 2-element subsets. Suppose that $R_i \subseteq P_i$ for each $i = 1, \dots, \ell$ but $R_i \not\subseteq P_j$ for $1 \le j < i \le \ell$. How big the number ℓ can be?

Question 1 was solved by Frankl (1982) who found the tight bound $\ell \leq \binom{n-k+2}{2}$. Summing up these inequalities from $k = n$ to $k = 2$, one arrives at the aforementioned bound

$$C(n) \leq \frac{n^3 - n}{6}. \tag{2}$$

In the literature the bound (2) is usually attributed to Pin who explained the above connection between Algorithm 3.1 and Question 1 and conjectured the estimation $\ell \leq \binom{n-k+2}{2}$ in his talk at the Colloquium on Graph Theory and Combinatorics held in Marseille in 1981; Frankl learned Question 1 from that talk. Accordingly, the usual reference for (2) is the paper (Pin, 1983) based on the talk. The full story is however more complicated. Actually, the bound (2) first appeared in (Fischler & Tannenbaum, 1970) where it was deduced from a combinatorial conjecture equivalent to Pin's one. Fischler & Tannenbaum presented their paper at the 11th FOCS conference but that time there was no Frankl in the audience so that the conjecture remained unproved and the paper eventually got lost in limbo. The bound (2) then reoccurred in Kohavi & Winograd (1971, 1973) but the argument justifying it in these papers was insufficient. Later both (2) and Frankl's solution to Question 1 were independently rediscovered in (Klyachko et al, 1987).

If one executes Algorithm 1 on the Černý automaton $\mathscr{C}_4$ (Fig. 6 is quite helpful here), one sees that the algorithm returns the word ab^2abab^3a of length 10 which is not the shortest reset word for $\mathscr{C}_4$. This reveals one of the main intrinsic difficulties of the synchronization problem: the standard optimality principle does not apply here since it is not true that the optimal solution behaves optimally also in all intermediate steps. In our example, the optimal solution is the word ab^3ab^3a but it cannot be found by the greedy algorithm because the algorithm chooses $v = b^2a$ rather than $v = b^3a$ on the second execution of the main loop.

Another difficulty behind the scene is that there are only very few examples of *extreme* synchronizing automata, that is n-state synchronizing automata whose shortest reset words have lengths $(n - 1)^2$. In fact, the Černý series $\mathscr{C}_n$, $n = 2, 3, \ldots$, is the only known infinite series of extreme synchronizing automata. Besides that, we know only a few isolated examples of such automata: up to isomorphism and adding/removing non-essential letters, there are three extreme automata with 3 states, three extreme automata with 4 states (see Fig. 8), one extreme automaton with 5 states recently found by Roman, see Fig. 9, and one extreme automaton with 6 states found by Kari (2001), see Fig. 10.

Moreover, even synchronizing automata whose shortest reset words have lengths close to the Černý bound are very rare. For instance, we are not aware of any 5-state synchronizing automaton whose shortest reset word has length 24, nor of any 6-state synchronizing automaton whose shortest reset word has length 33 or 34 or 35, etc. As for regular constructions of "almost-extreme" automata, we know just one series of n-state synchronizing automata with odd $n \geq 5$ such that the minimum length of reset words for the nth automaton in the series is equal to $(n - 1)(n - 2)$, see (Ananichev et al, 2007).

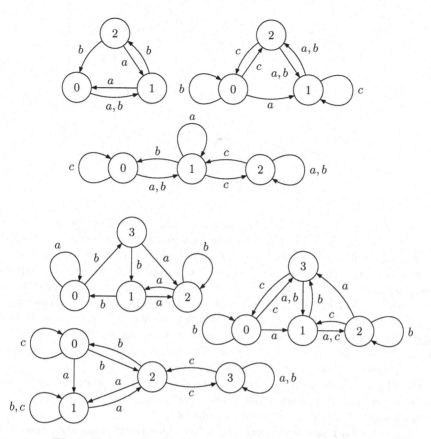

Fig. 8. Extreme synchronizing automata with 3 and 4 states

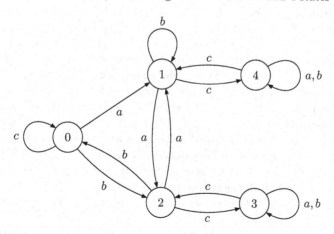

Fig. 9. Roman's automaton

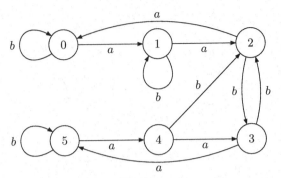

Fig. 10. Kari's automaton

In general, "slowly" synchronizing automata turn out to be rather exceptional, and this observation is supported also by probabilistic arguments. Indeed, if Q is an n-element set (with n large enough), then, on average, any product of $2n$ randomly chosen transformations of Q is known to be a constant map, see (Higgins, 1988). Being retold in automata-theoretic terms, this fact implies that a randomly chosen DFA with n states and a sufficiently large input alphabet tends to be synchronizing, and moreover, the length of its shortest reset word does not exceed $2n$. This means, in particular, that there is no hope to find new examples of "slowly" synchronizing automata, to say nothing of a counterexample to the Černý conjecture, via a random sampling experiment.

The Černý conjecture has been confirmed for various classes of synchronizing automata satisfying some additional restrictions. We conclude with a (non-complete) list of the most important results of this sort.

We begin with Eppstein's result (1990) in which restrictions are imposed on the action of the letters on the state set. Consider the set $\{0, 1, \ldots, n-1\}$ equipped with the natural cyclic order $0 \prec 1 \prec 2 \prec \cdots \prec n-1 \prec 1$ (here $k \prec \ell$ means that ℓ immediately follows k). If $i_1, i_2, \ldots, i_m$ are numbers in $\{0, 1, 2, \ldots, n-1\}$, we call the sequence $i_1, i_2, \ldots, i_m$ *cyclic* if, after removal of possible adjacent duplicate numbers, it is a subsequence of a cyclic shift of the sequence $0, 1, 2, \ldots, n-1$. In a slightly more formal language, we may say that $i_1, i_2, \ldots, i_m$ is a cyclic sequence if there exists no more than one index $t \in \{1, \ldots, m\}$ such that $i_{t+1} < i_t$ where i_{m+1} is understood as i_1 and $<$ stands for the usual strict linear order on $\{0, 1, 2, \ldots, n-1\}$. A transformation α of the set $\{0, 1, 2, \ldots, n-1\}$ is said to be *orientation preserving* if the numbers $0\alpha, 1\alpha, \ldots, (n-1)\alpha$ form a cyclic sequence. Now let $\mathscr{A} = \langle Q, \Sigma, \delta \rangle$ be a DFA with n states. We say that $\mathscr{A}$ is *orientable* if its states can be indexed by $0, 1, 2, \ldots, n-1$ so that all the transformations $\delta(\sqcup, a) : Q \to Q$ induced by the letters $a \in \Sigma$ are orientation preserving. For instance, Černý's automata $\mathscr{C}_n$, $n = 2, 3, \ldots$, are orientable.

Eppstein's interest in orientable automata (which he called *monotonic*) was motivated by the robotics applications of synchronizing automata. Indeed, in the problem of sensor-free orientation of polygonal parts one deals with solid bodies whence only those transformations of polygons are physically meaningful that preserve relative location of the faces of these polygons. It was observed already by Natarajan (1986) that in the "flat" case (when the polygonal parts do not leave a certain plane, say, the surface of a conveyer belt) this physical requirement leads precisely to orientation preserving transformations. In (Eppstein, 1990, Theorem 2) it is proved that, in accordance with the Černý conjecture, every orientable synchronizing automaton with n states has a reset word of length $(n-1)^2$. An extension of this result to a larger class of synchronizing automata whose letter actions mimic certain "spatial" transformations of solid polygons was obtained by Ananichev & Volkov (2004).

Dubuc (1998) has proved the Černý conjecture for yet another natural class of automata also containing the Černý series: automata in which a letter acts on the state set Q as a cyclic permutation of order $|Q|$.

In Kari's elegant paper (2003) the restriction has been imposed on the underlying digraphs of automata in question, namely, the Černý's conjecture has been verified for synchronizing automata with Eulerian digraphs. Moreover, it has been proved that if the underlying digraph of an n-state synchronizing automaton is Eulerian then there exists a reset word of length $(n-2)(n-1)+1$ (Kari, 2003, Theorem 2). It is unknown whether or not this bound is tight.

Recall that the *transition monoid* of a DFA $\mathscr{A} = \langle Q, \Sigma, \delta \rangle$ is the monoid consisting of all transformations $\delta(\sqcup, w) : Q \to Q$ induced by the words $w \in \Sigma^*$. Several authors have studied synchronization issues for automata whose transition monoids satisfy certain abstract properties. An important example of a property of automata expressed in this language is aperiodicity. A monoid is said to be *aperiodic* if all its subgroups are singletons; a DFA is called *aperiodic* (or *counter-free*) if its transition monoid is aperiodic. Aperiodic automata play a distinguished role in many aspects of formal language theory and its connections to logic, see the classic monograph by McNaughton & Papert (1971). Thus, studying synchronization within this important subclass of automata appears to be well justified, especially if one takes into account that the problem of finding short reset words is known to remain difficult when restricted to aperiodic automata. Indeed, inspecting the reductions from 3-SAT used in (Eppstein, 1990) or (Goralčik & Koubek, 1995) or (Salomaa, 2003), one can observe that in each case the construction results in an aperiodic automaton, and therefore, the question of whether or not a given aperiodic automaton admits a reset word whose length does not exceed a given positive integer is NP-complete.

Recently Trahtman (2007) has proved that every synchronizing aperiodic automaton with n states admits a reset word of length at most $\frac{n(n-1)}{2}$. Thus, the Černý conjecture holds true for synchronizing aperiodic automata. However, the problem of establishing a precise bound for the minimum length of reset words for synchronizing aperiodic automata with n states still remains open, and moreover, we do not even have a reasonably justified conjecture for this case. Denote by $C_A(n)$ the minimum length of reset words for synchronizing aperiodic automata with n states, that is, the restriction of the Černý function to the class of aperiodic automata. Then Trahtman's result can be expressed by the inequality $C_A(n) \leq \dfrac{n(n-1)}{2}$. However, the only non-trivial lower bound for

$C_A(n)$, which has been established so far, is linear, namely, $C_A(n) \geq n + \lfloor \frac{n}{2} \rfloor - 2$ for $n \geq 7$. (This bound comes from Ananichev's paper (2005).) One sees that the gap between the two bounds is fairly large. We believe that the actual value of $C_A(n)$ is closer to the lower bound.

In (Volkov, 2007) the results from (Trahtman, 2007) have been extended to a larger class of automata and improved. In particular, it has been proved that if the underlying digraph of an n-state aperiodic automaton is strongly connected, then the automaton has a reset word of length $\lfloor \frac{n(n+1)}{6} \rfloor$ (Volkov, 2007, Corollary 1).

Another large class of finite monoids which is of importance for formal language theory is known under the name **DS** and can be described as follows: a finite monoid M belongs to **DS** if and only if for all $x, y, z, t \in N$ the following condition holds:

$$MxM = MyM = MzM = MtM = MxyM \quad \text{implies} \quad MxyM = MztM.$$

(For the reader acquainted with some basics of semigroup theory, we recall an equivalent but more standard description of **DS**: a finite monoid M belongs to **DS** if and only if each regular $\mathcal{D}$-class of M is a subsemigroup in M.) Recently Almeida et al (2008) have proved the Černý conjecture for synchronizing automata with transition monoids in **DS**. Again, the problem of establishing a precise bound for the minimum length of reset words for synchronizing automata in this class still remains open.

References

Almeida, J., Margolis, S., Steinberg, B., Volkov, M.V.: Representation theory of finite semigroups, semigroup radicals and formal language theory. Trans. Amer. Math. Soc. (to appear, 2008), http://arxiv.org/abs/math/0702400v1

Ananichev, D.S.: The mortality threshold for partially monotonic automata. In: De Felice, C., Restivo, A. (eds.) DLT 2005. LNCS, vol. 3572, pp. 112–121. Springer, Heidelberg (2005)

Ananichev, D.S., Volkov, M.V.: Some results on Černý type problems for transformation semigroups. In: Araujo, I., Branco, M., Fernandes, V.H., Gomes, G.M.S. (eds.) Semigroups and Languages, pp. 23–42. World Scientific, Singapore (2004)

Ananichev, D.S., Volkov, M.V., Zaks, Y.I.: Synchronizing automata with a letter of deficiency 2. Theoret. Comput. Sci. 376(1-2), 30–41 (2007)

Ashby, W.R.: An Introduction to Cybernetics. Chapman & Hall, London (1956), http://pcp.vub.ac.be/books/IntroCyb.pdf

Benenson, Y., Paz-Elizur, T., Adar, R., Keinan, E., Livneh, Z., Shapiro, E.: Programmable and autonomous computing machine made of biomolecules. Nature 414(1), 430–434 (2001)

Benenson, Y., Adar, R., Paz-Elizur, T., Livneh, Z., Shapiro, E.: DNA molecule provides a computing machine with both data and fuel. Proc. National Acad. Sci. USA 100, 2191–2196 (2003)

Booth, T.L.: Sequential Machines and Automata Theory. J. Wiley & Sons, New York (1967)

Boppana, V., Rajan, S.P., Takayama, K., Fujita, M.: Model checking based on sequential ATPG. In: Computer Aided Verification. LNCS, vol. 1622, pp. 418–430. Springer, Berlin (1999)

Černý, J.: Poznámka k homogénnym eksperimentom s konečnými automatami. Matematicko-fyzikalny Časopis Slovensk. Akad. Vied 14(3), 208–216 (1964) (in Slovak)

Černý, J., Pirická, A., Rosenauerová, B.: On directable automata. Kybernetika 7(4), 289–298 (1971)

Chen, Y.-B., Ierardi, D.J.: The complexity of oblivious plans for orienting and distinguishing polygonal parts. Algorithmica 14, 367–397 (1995)

Cho, H., Jeong, S.-W., Somenzi, F., Pixley, C.: Synchronizing sequences and symbolic traversal techniques in test generation. J. Electronic Testing 4, 19–31 (1993)

Cormen, T.H., Leiserson, C.E., Rivest, R.L., Stein, C.: Introduction to Algorithms. MIT Press, Cambridge (2001)

Dubuc, L.: Sur le automates circulaires et la conjecture de Černý. RAIRO Inform. Theor. Appl. 32, 21–34 (1998) (in French)

Eppstein, D.: Reset sequences for monotonic automata. SIAM J. Comput. 19, 500–510 (1990)

Fischler, M.A., Tannenbaum, M.: Synchronizing and representation problems for sequential machines with masked outputs. In: Proc. 11th Annual Symp. Foundations Comput. Sci., pp. 97–103. IEEE, Los Alamitos (1970)

Frankl, P.: An extremal problem for two families of sets. Eur. J. Comb. 3, 125–127 (1982)

Garey, M.R., Johnson, D.S.: Computers and Intractability: A Guide to the Theory of NP-completeness. Freeman, San Francisco (1979)

Gill, A.: State-identification experiments in finite automata. Information and Control 4(2-3), 132–154 (1961)

Ginsburg, S.: On the length of the smallest uniform experiment which distinguishes the terminal states of a machine. J. Assoc. Comput. Mach. 5, 266–280 (1958)

Goldberg, K.: Orienting polygonal parts without sensors. Algorithmica 10, 201–225 (1993)

Goralčik, P., Koubek, V.: Rank problems for composite transformations. Int. J. Algebra and Computation 5, 309–316 (1995)

Hennie, F.C.: Fault detecting experiments for sequential circuits. In: Proc. 5th Annual Symp. Switching Circuit Theory and Logical Design, pp. 95–110. IEEE, Los Alamitos (1964)

Hennie, F.C.: Finite-state Models for Logical Machines. J. Wiley & Sons, New York (1968)

Higgins, P.M.: The range order of a product of i transformations from a finite full transformation semigroup. Semigroup Forum 37, 31–36 (1988)

Kari, J.: A counter example to a conjecture concerning synchronizing words in finite automata. EATCS Bull. 73, 146 (2001)

Kari, J.: Synchronizing finite automata on Eulerian digraphs. Theoret. Comput. Sci. 295, 223–232 (2003)

Klyachko, A.A., Rystsov, I.K., Spivak, M.A.: An extremal combinatorial problem associated with the bound of the length of a synchronizing word in an automaton. Kibernetika 25(2), 16–20 (1987) (in Russian); English translation: Cybernetics and Systems Analysis 23(2), 165–171

Kohavi, Z.: Switching and Finite Automata Theory. McGraw-Hill, New York (1970)

Kohavi, Z., Winograd, J.: Bounds on the length of synchronizing sequences and the order of information losslessness. In: Kohavi, Z., Paz, A. (eds.) Theory of Machines and Computations, pp. 197–206. Academic Press, New York (1971)

Kohavi, Z., Winograd, J.: Establishing certain bounds concerning finite automata. J. Comput. Syst. Sci. 7(3), 288–299 (1973)

Laemmel, A.E., Rudner, B.: Study of the application of coding theory, Report PIBEP-69-034, Polytechnic Inst. Brooklyn, Dept. Electrophysics, Farmingdale (1969)

Mateescu, A., Salomaa, A.: Many-valued truth functions, Černý's conjecture and road coloring. EATCS Bull. 68, 134–150 (1999)

McNaughton, R., Papert, S.A.: Counter-free Automata. MIT Press, Cambridge (1971)

Moore, E.F.: Gedanken-experiments on sequential machines. In: Shannon, C.E., McCarthy, J. (eds.) Automata Studies, Annals of Mathematics Studies, vol. 34, pp. 129–153. Princeton University Press, Princeton (1956)

Natarajan, B.K.: An algorithmic approach to the automated design of parts orienters. In: Proc. 27th Annual Symp. Foundations Comput. Sci., pp. 132–142. IEEE, Los Alamitos (1986)

Natarajan, B.K.: Some paradigms for the automated design of parts feeders. Internat. J. Robotics Research 8(6), 89–109 (1989)

Papadimitriou, C.H.: Computational Complexity. Addison-Wesley, Reading (1994)

Pin, J.-E.: On two combinatorial problems arising from automata theory. Ann. Discrete Math. 17, 535–548 (1983)

Pixley, C., Jeong, S.-W., Hachtel, G.D.: Exact calculation of synchronization sequences based on binary decision diagrams. In: Proc. 29th Design Automation Conf., pp. 620–623. IEEE, Los Alamitos (1992)

Rho, J.-K., Somenzi, F., Pixley, C.: Minimum length synchronizing sequences of finite state machine. In: Proc. 30th Design Automation Conf., pp. 463–468. ACM, New York (1993)

Salomaa, A.: Composition sequences for functions over a finite domain. Theoret. Comput. Sci. 292, 263–281 (2003)

Samotij, W.: A note on the complexity of the problem of finding shortest synchronizing words. In: Proc. AutoMathA 2007, Automata: from Mathematics to Applications, Univ. Palermo (CD) (2007)

Sandberg, S.: Homing and synchronizing sequences. In: Broy, M., Jonsson, B., Katoen, J.-P., Leucker, M., Pretschner, A. (eds.) Model-Based Testing of Reactive Systems. LNCS, vol. 3472, pp. 5–33. Springer, Heidelberg (2005)

Starke, P.H.: Eine Bemerkung über homogene Experimente. Elektronische Informationverarbeitung und Kybernetik 2, 257–259 (1966) (in German)

Starke, P. H.: Abstrakte Automaten, Deutscher Verlag der Wissenschaften, Berlin (1969) (in German); English translation: Abstract Automata, North-Holland, Amsterdam, American. Elsevier, New York (1972)

Trahtman, A.: An efficient algorithm finds noticeable trends and examples concerning the Černý conjecture. In: Královič, R., Urzyczyn, P. (eds.) MFCS 2006. LNCS, vol. 4162, pp. 789–800. Springer, Heidelberg (2006)

Trahtman, A.: The Černý conjecture for aperiodic automata. Discrete Math. Theoret. Comp. Sci. 9(2), 3–10 (2007)

Trahtman, A.: The Road Coloring Problem. Israel J. Math. (to appear, 2008), http://arxiv.org/abs/0709.0099

Volkov, M.V.: Synchronizing automata preserving a chain of partial orders. In: Holub, J., Žďárek, J. (eds.) CIAA 2007. LNCS, vol. 4783, pp. 27–37. Springer, Heidelberg (2007)

Yang, S.: Logic Synthesis and Optimization Benchmarks User Guide Version 3.0, Microelectronics Center of North Carolina, Research Triangle Park, NC (1991)

About Universal Hybrid Networks of Evolutionary Processors of Small Size*

Artiom Alhazov[1,2], Erzsébet Csuhaj-Varjú[3,4], Carlos Martín-Vide[5],
and Yurii Rogozhin[5,2]

[1] Åbo Akademi University, Department of Information Technologies,
Turku Center for Computer Science, FIN-20520 Turku, Finland
aalhazov@abo.fi
[2] Academy of Sciences of Moldova,
Institute of Mathematics and Computer Science,
Academiei 5, MD-2028, Chişinău, Moldova
{artiom,rogozhin}@math.md
[3] Computer and Automation Research Institute,
Hungarian Academy of Sciences,
Kende u. 13-17, 1111 Budapest, Hungary
csuhaj@sztaki.hu
[4] Eötvös Loránd University,
Faculty of Informatics, Department of Algorithms and Their Applications,
Pázmány Péter sétány 1/c, H-1117 Budapest, Hungary
[5] Rovira i Virgili University,
Research Group on Mathematical Linguistics,
Pl. Imperial Tàrraco 1, 43005 Tarragona, Spain
carlos.martin@urv.cat

Abstract. A hybrid network of evolutionary processors (an HNEP) is a graph with a language processor, input and output filters associated to each node. A language processor performs one type of point mutations (insertion, deletion or substitution) on the words in that node. The filters are defined by certain variants of random-context conditions. In this paper, we present a universal complete HNEP with 10 nodes simulating circular Post machines and show that every recursively enumerable language can be generated by a complete HNEP with 10 nodes. Thus, we positively answer the question posed in [5] about the possibility to generate an arbitrary recursively enumerable language over an alphabet V with a complete HNEP of a size smaller than $27 + 3 \cdot card(V)$.

Keywords: Bio-inspired computing, Hybrid networks of evolutionary processors, Small universal systems, Descriptional complexity, Circular Post machines.

* The first author gratefully acknowledges the support by Academy of Finland, project 203667. The fourth author gratefully acknowledges the support of European Commission, project MolCIP, MIF1-CT-2006-021666. The first and the fourth authors acknowledge the Science and Technology Center in Ukraine, project 4032.

C. Martín-Vide, F. Otto, and H. Fernau (Eds.): LATA 2008, LNCS 5196, pp. 28–39, 2008.

1 Introduction

Networks of language processors are finite collections of rewriting systems (language processors) organized in a communicating system [6]. The language processors are located at nodes of a virtual graph and operate over sets or multisets of words. During the functioning of the network, they rewrite the corresponding collections of words and then re-distribute the resulting strings according to a communication protocol assigned to the system. The language determined by the system is usually defined as the set of words which appear at some distinguished node in the course of the computation. One of the main questions related to networks of language processors is how much extent their generative power depends on the used operations and the size of the system. Particularly important variants are those ones where the language processors are based on elementary string manipulating rules, since these constructs give *insight into the limits of the power of the simplicity of basic language theoretic operations and that of distributed architectures.*

Networks of evolutionary processors (NEPs), introduced in [4], and also inspired by cell biology, are proper examples for these types of constructs. In this case, each processor represents a cell performing point mutations of DNA and controlling its passage inside and outside it through a filtering mechanism. The language processor corresponds to the cell, the generated word to a DNA strand, and operations insertion, deletion, or substitution of a symbol to the point mutations. It is known that a computationally universal behaviour emerges as a result of interaction of such simple components (see, for example [1,2]).

In the case of so-called *hybrid networks of evolutionary processors* (HNEPs), each language processor performs only one of the above operations on a certain position of the words in that node. The filters are defined by some variants of random-context conditions. The concept was introduced in [9], and proved computationally complete in [5], with $27 + 3 \cdot card(V)$ nodes for alphabet V.

In this paper, we present a *universal complete HNEP with 10 nodes* and prove that *every recursively enumerable language can be generated by a complete NHEP with the same number of nodes.* Although these bounds are not shown sharp, we significantly improve the previous result. The constructions demonstrate that distributed architectures of very small size, with uniform structure and with components based on very simple language theoretic operations are sufficient to obtain computational completeness.

2 Preliminaries

We recall some notions we shall use throughout the paper. An *alphabet* is a finite and nonempty set of symbols. The cardinality of a finite set A is written as $card(A)$. A sequence of symbols from an alphabet V is called a word over V. The set of all words over V is denoted by V^* and the empty word is denoted by ε; we use $V^+ = V^* \setminus \{\varepsilon\}$. The length of a word x is denoted by $|x|$, while we denote the number of occurrences of a letter a in a word x by $|x|_a$. For each nonempty word x, $alph(x)$ is the minimal alphabet W such that $x \in W^*$.

In our constructions, HNEPs simulate type-0 grammars in Kuroda normal form and Circular Post Machines.

A type-0 grammar in *Kuroda normal form* is a construct $\Gamma = (N, T, S, P)$, where N is the set of nonterminal symbols, T is the set of terminal symbols, N and T are disjoint sets, $S \in N$ is the start symbol, and P is the set of rules of the forms $A \longrightarrow a$, $A \longrightarrow BC$, $A \longrightarrow \varepsilon$, $AB \longrightarrow CD$, where $A, B, C, D \in N$ and $a \in T$. These grammars are known to generate all recursively enumerable languages.

Circular Post Machines (CPMs) were introduced in [7], where it was shown that all introduced variants of CPMs are computationally complete, and moreover, the same statement holds for CPMs with two symbols. In [8,3] several universal CPMs of variant 0 (CPM0) having small size were constructed, among them in [3] a universal CPM0 with 34 states and 2 symbols. In this article we use the deterministic variant of CPM0s.

A *Circular Post Machine* is a quintuple $(\Sigma, Q, \mathbf{q}_0, \mathbf{q}_f, P)$ with a finite alphabet Σ where 0 is the blank, a finite set of states Q, an initial state $\mathbf{q}_0 \in Q$, a terminal state $\mathbf{q}_f \in Q$, and a finite set of instructions of P with all instructions having one of the forms $\mathbf{p}x \to \mathbf{q}$ (erasing the symbol read), $\mathbf{p}x \to y\mathbf{q}$ (overwriting and moving to the right), $\mathbf{p}0 \to y\mathbf{q}0$ (overwriting and creation of a blank), where $x, y \in \Sigma$ and $\mathbf{p}, \mathbf{q} \in Q$.

The storage of this machine is a circular tape, the read and write head move only in one direction (to the right), and with the possibility to cut off a cell or to create and insert a new cell with a blank.

In the following, we summarize the necessary notions concerning so-called *evolutionary operations*, simple rewriting operations abstract local gene mutations.

For an alphabet V, we say that a rule $a \to b$, with $a, b \in V \cup \{\varepsilon\}$ is a *substitution rule* if both a and b are different from ε; it is a *deletion rule* if $a \neq \varepsilon$ and $b = \varepsilon$; and, it is an *insertion rule* if $a = \varepsilon$ and $b \neq \varepsilon$. The set of all substitution, deletion, and insertion rules over an alphabet V are denoted by Sub_V, Del_V, and Ins_V, respectively. Given such rules π, ρ, σ, and a word $w \in V^*$, we define the following *actions* of σ on w: If $\pi \equiv a \to b \in Sub_V$, $\rho \equiv a \to \varepsilon \in Del_V$, and $\sigma \equiv \varepsilon \to a \in Ins_V$, then

$$\pi^*(w) = \begin{cases} \{ubv : \exists u, v \in V^*(w = uav)\}, \\ \{w\}, \quad \text{otherwise} \end{cases} \tag{1}$$

$$\rho^*(w) = \begin{cases} \{uv : \exists u, v \in V^*(w = uav)\}, \\ \{w\}, \text{ otherwise} \end{cases} \tag{2}$$

$$\rho^r(w) = \begin{cases} \{u : \ w = ua\}, \\ \{w\}, \text{ otherwise} \end{cases} \tag{3}$$

$$\rho^l(w) = \begin{cases} \{v : \ w = av\}, \\ \{w\}, \text{ otherwise} \end{cases} \tag{4}$$

$$\sigma^*(w) = \{uav : \exists u, v, \in V^*(w = uv)\}, \tag{5}$$

$$\sigma^r(w) = \{wa\}, \ \sigma^l(w) = \{aw\}. \tag{6}$$

Symbol $\alpha \in \{*, l, r\}$ denotes the way of applying an insertion or a deletion rule to a word, namely, at any position ($a = *$), in the left-hand end ($a = l$), or in the right-hand end ($a = r$) of the word, respectively. Note that a substitution rule can be applied at any position. For every rule σ, action $\alpha \in \{*, l, r\}$, and $L \subseteq V^*$, we define the $\alpha - action$ of σ on L by $\sigma^\alpha(L) = \bigcup_{w \in L} \sigma^\alpha(w)$. For a given finite set of rules M, we define the $\alpha - action$ of M on a word w and on a language L by $M^\alpha(w) = \bigcup_{\sigma \in M} \sigma^\alpha(w)$ and $M^\alpha(L) = \bigcup_{w \in L} M^\alpha(w)$, respectively.

Before turning to the notion of an evolutionary processor, we define the filtering mechanism.

For disjoint subsets $P, F \subseteq V$ and a word $w \in V^*$, we define the predicate φ ($\varphi^{(2)}$ in terminology of [5]) as $\varphi(w; P, F) \equiv alph(w) \cap P \neq \emptyset \wedge F \cap alph(w) = \emptyset$. The construction of this predicate is based on *random-context conditions* defined by the two sets P *(permitting contexts)* and F *(forbidding contexts)*. For every language $L \subseteq V^*$ we define $\varphi(L, P, F) = \{w \in L \mid \varphi(w; P, F)\}$.

An *evolutionary processor over* V is a 5-tuple (M, PI, FI, PO, FO) where:

- Either $M \subseteq Sub_V$ or $M \subseteq Del_V$ or $M \subseteq Ins_V$. The set M represents the set of evolutionary rules of the processor. Note that every processor is dedicated to only one type of the above evolutionary operations.
- $PI, FI \subseteq V$ are the *input* permitting/forbidding contexts of the processor, while $PO, FO \subseteq V$ are the *output* permitting/forbidding contexts of the processor.

We denote the set of evolutionary processors over V by EP_V.

Definition 1. *A hybrid network of evolutionary processors (an HNEP, shortly) is a 7-tuple $\Gamma = (V, G, N, C_0, \alpha, \beta, i_0)$, where the following conditions hold:*

- *V is an alphabet.*
- *$G = (X_G, E_G)$ is an undirected graph with set of vertices X_G and set of edges E_G. G is called the* underlying graph *of the network.*
- *$N : X_G \longrightarrow EP_V$ is a mapping which associates with each node $x \in X_G$ the evolutionary processor $N(x) = (M_x, PI_x, FI_x, PO_x, FO_x)$.*
- *$C_0 : X_G \longrightarrow 2^{V^*}$ is a mapping which identifies the initial configuration of the network. It associates a finite set of words with each node of the graph G.*
- *$\alpha : X_G \longrightarrow \{*, l, r\}$; $\alpha(x)$ defines the action mode of the rules performed in node x on the words occurring in that node.*
- *$\beta : X_G \longrightarrow \{(1), (2)\}$ defines the type of the input/output filters of a node. More precisely, for every node, $x \in X_G$, we define the following filters: the input filter is given as $\rho_x(\cdot) = \varphi^{\beta(x)}(\cdot; PI_x, FI_x)$, and the output filter is defined as $\tau_x(\cdot) = \varphi^{\beta(x)}(\cdot, PO_x, FO_x)$. That is, $\rho_x(w)$ (resp.τ_x) indicates whether or not the word w can pass the input (resp. output) filter of x. More generally, $\rho_x(L)$ (resp. $\tau_x(L)$) is the set of words of L that can pass the input (resp. output) filter of x.*
- *$i_0 \in X_G$ is the* output node *of the HNEP.*

We say that $card(X_G)$ is the size of Γ. An HNEP is said to be a *complete* HNEP, if its underlying graph is a complete graph.

A configuration of an HNEP Γ, as above, is a mapping $C : X_G \longrightarrow 2^{V^*}$ which associates a set of words with each node of the graph. A component $C(x)$ of a configuration C is the set of words that can be found in the node x in this configuration, hence a configuration can be considered as the sets of words which are present in the nodes of the network at a given moment. A configuration can change either by an *evolutionary step* or by a *communication step*. When it changes by an evolutionary step, then each component $C(x)$ of the configuration C is changed in accordance with the set of evolutionary rules M_x associated with the node x and the way of applying these rules $\alpha(x)$. Formally, the configuration C' is obtained in *one evolutionary step* from the configuration C, written as $C \Longrightarrow C'$, iff $C'(x) = M_x^{\alpha(x)}(C(x))$ for all $x \in X_G$.

When it changes by a communication step, then each language processor $N(x)$, where $x \in X_G$, sends a copy of each of its words to every node processor where the node is connected with x, provided that this word is able to pass the output filter of x, and receives all the words which are sent by processors of nodes connected with x, providing that these words are able to pass the input filter of x. Formally, we say that configuration C' is obtained in *one communication step* from configuration C, written as $C \vdash C'$, iff $C'(x) = (C(x) - \tau_x(C(x))) \cup \bigcup_{\{x,y\} \in E_G} (\tau_y(C(y) \cap \rho_x(C(y))))$ for all $x \in X_G$.

For an HNEP Γ, a computation in Γ is a sequence of configurations C_0, $C_1, C_2, \ldots$, where C_0 is the initial configuration of Γ, $C_{2i} \Longrightarrow C_{2i+1}$ and $C_{2i+1} \vdash C_{2i+2}$, for all $i > 0$. If we use HNEPs as language generating devices, then the generated language is the set of all words which appear in the output node at some step of the computation. Formally, the language generated by Γ is $L(\Gamma) = \bigcup_{s \geq 0} C_s(i_0)$.

3   Main Results

3.1   Universality

Theorem 1. *Any CPM0 P with 2 symbols can be simulated by an HNEP P' with 10 nodes.*

Proof. Let us consider a CPM0 P with two symbols, 0 and 1, and f states, $q_i \in Q$, $i \in I = \{1, 2, \ldots, f\}$, where q_1 is the initial state and the only terminal state is $q_f \in Q$. Suppose that P stops in the terminal state q_f on every symbol, i.e., there are two instructions $q_f 0 \to Halt$ and $q_f 1 \to Halt$. (Notice, that it is easy to transform any CPM0 with n states into a CPM0 with $n + 1$ states that stops on every symbol in terminal state.)

So, we consider CPM0 P with instructions of the forms $q_i x \longrightarrow q_j$, $q_i x \longrightarrow yq_j$, $q_i 0 \longrightarrow yq_j 0$, $q_f 0 \longrightarrow Halt$, $q_f 1 \longrightarrow Halt$, where $q_i, q_j \in Q$, $x, y \in \{0, 1\}$. A configuration $w = xWq_i$ of CPM0 P describes that P in state $q_i \in Q$ considers symbol $x \in \{0, 1\}$ on the left-hand end of $W \in \{0, 1\}^*$. Let $I' = I \setminus \{f\}$ and $x, y \in \{0, 1\}$. In the following, we construct an HNEP P' simulating P. Starting with the initial configuration W_0 of CPM0 P in node 1 of HNEP P', we simulate every computation step performed by P with a sequence of computation steps in P'. If the computation in P is finite, then the final configuration W_f of P will be found at node 10 of P', moreover, any string that can be found at node 10 is a final configuration of P. In the case of an infinite computation in P, no string will appear in node 10 of P' and the computation in P' will never stop. In the Table 1 below a complete HNEP $P' = (V, G, N, C_0, \alpha, \beta, 10)$ with 10 nodes is described, where (for $i \in I', y \in X$)

$$V = Q \cup Q' \cup T \cup T' \cup S \cup S' \cup R \cup R' \cup X \cup X' \cup X'' \cup \{\hat{0}\} \text{ and}$$
$$Q' = \{q'_i\}, \ T = \{t_{i,y}\}, \ T' = \{t'_{i,y}\}, \ S = \{s_{i,y}\}, \ S' = \{s'_{i,y}\},$$
$$R = \{r_i \mid i \in I' \cup \{0\}\}, \ R' = \{r'_i\}, X = \{0, 1\}, \ X' = \{0', 1'\}, \ X'' = \{0'', 1''\}.$$

G is a complete graph with 10 nodes, N, C_0, α, β are presented in the Table 1 and node 10 is the output node of HNEP P'. We explain how P' simulates the instructions of CPM0 P. Due to the lack of space, we present only the necessary details.

Instruction $q_i x \longrightarrow q_j$: $xWq_i \xrightarrow{P} Wq_j$.
The simulation starts with xWq_i in node 1 of P'. By performing evolution steps on this string at node 1, we obtain $xWq_i \xrightarrow{1.1} xWq'_i \xrightarrow{1.2} \{x'Wq'_i, \ xW'q'_i\}$, where $W \in \{0,1\}^*$ and $W' \in \{0, 1, 0', 1'\}$. In the following communication step, only strings with q'_i and x' can leave node 1. Notice that strings $xW'q'_i$ do not contain symbols x' on the left-hand end. It is easy to see that during the next transformations it is not possible to delete x' if it is not on the left-hand end of the strings, so these strings will stay forever in node 8. Thus, we will not further consider strings that contain symbols x' not in the correct position. String $x'Wq'_i$ can enter nodes 2 or 3. Let us consider, for example, node 2 (the case for node 3 can be treated analogously). If the string enters node 2, then there exists an instruction $q_i 0 \longrightarrow q_j$ in CPM0 P and $x' = 0'$, so $0'Wq'_i \xrightarrow{2.1} 0'Wq_j$. In the following communication step, string $0'Wq_j$ can enter only node 8, where $0'Wq_j \xrightarrow{8.1} Wq_j$, and then the obtained string, Wq_j, can enter only node 1. So, we simulated instruction $q_i x \longrightarrow q_j$ of P in a correct manner.

In the Table 1 below $i, j \in I'$, $x, y \in X$, $x', y' \in X'$, $y'' \in X''$.

Table 1.

$N, C_0,$ α, β	M	PI, FI, PO, FO
1, $\{W_0\},$ $*, (2)$	$\{1.1 : q_i \to q'_i\} \cup$ $\{1.2 : x \to x'\} \cup$ $\{1.3 : y'' \to y\} \cup$ $\{1.4 : \hat{0} \to 0\}$	$PI = \emptyset,$ $FI = Q' \cup X'' \cup X' \cup \{q_f\},$ $PO = X',$ $FO = Q \cup X'' \cup \{\hat{0}\}$
2, $\emptyset,$ $*, (2)$	$\{2.1 : q'_i \to q_j \mid q_i 0 \to q_j\} \cup$ $\{2.2 : q'_i \to t_{j,y} \mid q_i 0 \to yq_j\} \cup$ $\{2.3 : q'_i \to s_{j,y} \mid q_i 0 \to yq_j 0\} \cup$ $\{2.4 : q'_i \to q_f \mid q_i 0 \to q_f\}$	$PI = \{q'_i \mid q_i 0 \to q_j\} \cup$ $\{q'_i \mid q_i 0 \to yq_j\} \cup$ $\{q'_i \mid q_i 0 \to yq_j 0\} \cup$ $\{q'_i \mid q_i 0 \to q_f\}$ $FI = \{1'\}, PO = \{0'\}, FO = Q'$
3, $\emptyset,$ $*, (2)$	$\{3.1 : q'_i \to q_j \mid q_i 1 \to q_j\} \cup$ $\{3.2 : q'_i \to t_{j,y} \mid q_i 1 \to yq_j\} \cup$ $\{3.3 : q'_i \to q_f \mid q_i 1 \to q_f\}$	$PI = \{q'_i \mid q_i 1 \to q_j\} \cup$ $\{q'_i \mid q_i 1 \to yq_j\} \cup$ $\{q'_i \mid q_i 1 \to q_f\}$ $FI = \{0'\}, PO = \{1'\}, FO = Q'$

Table 1. (*continued*)

$N, C_0,$ α, β	M	PI, FI, PO, FO
$4, \emptyset,$ $r, (2)$	$\{4.1 : \varepsilon \to r_0\}$	$PI = T \cup S, FI = R \cup R',$ $PO = \{r_0\}, FO = \emptyset$
$5, \emptyset,$ $*, (2)$	$\{5.1 : t_{i,y} \to t'_{i-1,y},$ $5.2 : s_{i,y} \to s'_{i-1,y} \mid 2 \leq i \leq f - 1\} \cup$ $\{5.3 : r_i \to r'_{i+1} \mid 0 \leq i \leq f - 2\} \cup$ $\{5.4 : t_{1,y} \to y\} \cup$ $\{5.5 : s_{1,y} \to y''\}$	$PI = T,$ $FI = T' \cup S' \cup R',$ $PO = R',$ $FO = T \cup S \cup R$
$6, \emptyset,$ $*, (2)$	$\{6.1 : t'_{i,y} \to t_{i-1,y},$ $6.2 : s'_{i,y} \to s_{i-1,y} \mid 2 \leq i \leq f - 2\} \cup$ $\{6.3 : r'_i \to r_{i+1} \mid 1 \leq i \leq f - 2\} \cup$ $\{6.4 : t'_{1,y} \to y\} \cup$ $\{6.5 : s'_{1,y} \to y''\}$	$PI = T',$ $FI = T \cup S \cup R,$ $PO = R,$ $FO = T' \cup S' \cup R'$
$7, \emptyset,$ $*, (2)$	$\{7.1 : r_i \to q_i,$ $7.2 : r'_i \to q_i \mid i \in I'\}$	$PI = R \cup R', FI = Q' \cup T \cup T' \cup$ $S \cup S', PO = X', FO = R \cup R'$
$8, \emptyset,$ $l, (2)$	$\{8.1 : x' \to \varepsilon\}$	$PI = X', FI = Q' \cup T \cup T' \cup$ $S \cup S' \cup R \cup R', PO = \emptyset, FO = X'$
$9, \emptyset,$ $l, (2)$	$\{9.1 : \varepsilon \to \hat{0}\}$	$PI = X'', FI = R \cup R' \cup X' \cup \{\hat{0}\},$ $PO = \{\hat{0}\}, FO = \emptyset$
$10, \emptyset,$ $*, (2)$	$\emptyset$	$PI = \{q_f\}, FI = V \setminus \{X \cup \{q_f\}\},$ $PO = \emptyset, FO = \{q_f\}$

Instruction $q_i x \longrightarrow y q_j : xW q_i \overset{P}{\longrightarrow} W y q_j.$
HNEP P' starts the simulation with $xW q_i$ in node 1. Then, two evolution steps follow, $xW q_i \overset{1.1}{\longrightarrow} xW q'_i \overset{1.2}{\longrightarrow} \{x'W q'_i, xW' q'_i\}$, where $W \in \{0, 1\}^*$ and $W' \in \{0, 1, 0', 1'\}$. Similarly to the previous case, we will consider only string of the form $x'W q'_i$. This string can enter nodes 2 or 3. Consider, for example, node 2 (the case for node 3 can be treated analogously). If the string enters node 2, then there is an instruction $q_i 0 \longrightarrow y q_j$ in CPM0 P and $x' = 0'$. So, $0'W q'_i \overset{2.2}{\longrightarrow} 0'W t_{j,y}$. String $0'W t_{j,y}$ can enter nodes 4 and 5. In the latter case the string will stay in node 5 forever, as it does not contain any symbol from R'. Suppose that the string enters node 4. Then, an evolution step, $0'W t_{j,y} \overset{4.1}{\longrightarrow} 0'W t_{j,y} r_0$, follows. Now, string $0'W t_{j,y} r_0$ can successfully be communicated only to node 5. Then, in nodes 5 and 6, the string is involved in the following evolution steps: For $t \in I'$,

$$0'W t_{j-t,y} r_t \overset{5.1}{\longrightarrow} 0'W t'_{j-(t+1),y} r_t \overset{5.3}{\longrightarrow} 0'W t'_{j-(t+1),y} r'_{t+1}$$

$$0'W t'_{j-(t+1),y} r'_{t+1} \overset{6.1}{\longrightarrow} 0'W t_{j-(t+2),y} r'_{t+1} \overset{6.3}{\longrightarrow} 0'W t_{j-(t+2),y} r_{t+2}.$$

The string enters in node 5 and 6 in circle, until the first index of t or t' will be decreased to 1. At that moment in node 5 (node 6) index of r (r') will be exactly $j - 1$, and it becomes j by rule 5.3 (6.3), i.e. the same, as the first index of t in string $0'W t_{j,y} r_0$ before entering node 5. After that, $0'W t_{1,y} r'_j \overset{5.4}{\longrightarrow} 0'W y r'_j$ or $0'W t'_{1,y} r_j \overset{6.4}{\longrightarrow} 0'W y r_j$. In the following communication step, string $0'W y r'_j$ or

$0'Wyr_j$ can enter only node 7, where the following evolution steps are performed: $0'Wyr_j \xrightarrow{7.1} 0'Wyq_j$, $0'Wyr'_j \xrightarrow{7.2} 0'Wyq_j$. String $0'Wyq_j$ is communicated to node 8, and this is the only node the string is able to enter. At node 8, evolution step $0'Wyq_j \xrightarrow{8.1} Wyq_j$ can be performed. Now, string Wyq_j can enter only node 1. So, instruction $q_i x \longrightarrow yq_j$ of P is correctly simulated.

Instruction $q_i 0 \longrightarrow yq_j 0$: $0Wq_i \xrightarrow{P} 0Wyq_j$.
The beginning of the simulation of instruction $q_i 0 \longrightarrow yq_j 0$ is the same as that of instruction $q_i 0 \longrightarrow yq_j$. The difference appears when rule 2.3 : $q'_i \to s_{j,y}$ is applied in node 2 instead of rule 2.2 : $q'_i \to t_{j,y}$ and at the end of the circle process in nodes 5 and 6, $s_{1,y}$ or $s'_{1,y}$ becomes y'' (rules 5.5 or 6.5) instead of y (rules 5.4 or 6.4). Strings $0'Wy''r'_j$ or $0'Wy''r_j$ can enter only node 7. Then, either evolution step $0'Wy''r_j \xrightarrow{7.1} 0'Wy''q_j$ or evolution step $0'Wy''r'_j \xrightarrow{7.2} 0'Wy''q_j$ follows. String $0'Wy''q_j$ can enter only node 8, where evolution step $0'Wy''q_j \xrightarrow{8.1} Wy''q_j$ is performed. The new string, $Wy''q_j$, can enter only node 9, where evolution step $Wy''q_j \xrightarrow{9.1} \hat{0}Wy''q_j$ follows. Then string $\hat{0}Wy''q_j$ can enter only node 1, where evolution steps $\hat{0}Wy''q_j \xrightarrow{1.3} \hat{0}Wyq_j \xrightarrow{1.4} 0Wyq_j$ are performed. Thus, instruction $q_i 0 \longrightarrow yq_j 0$ of P is correctly simulated.

Instruction $q_i x \longrightarrow q_f$: $xWq_i \xrightarrow{P} Wq_f$.
In nodes 2 or 3 we have rules $q'_i \to q_f$ (rules 2.4 or 3.3) and string $x'Wq'_i$ will be transformed to string $x'Wq_f$. After that it enters node 8 and changes to Wq_f. Now it enters node 10 as a result. So, CPM0 P is correctly modeled. We have demonstrated that the rules of P are simulated in P'. The proof that P' simulates only P comes from the construction of the rules in P', we leave the details to the reader.

Corollary 1. *There exists a universal HNEP with 10 nodes.*

3.2 Computational Completeness

Theorem 2. *Any recursively enumerable language can be generated by a complete HNEP of size 10.*

Proof. Let $\Gamma = (N, T, S, R)$ be a type-0 grammar in Kuroda normal form.
We construct a complete HNEP $\Gamma' = (V, G, N, C_0, \alpha, \beta, 10)$ of size 10 that simulates the derivations in Γ by the so-called *rotate-and-simulate* method. The rotate-and-simulate method means that the words found in the nodes are involved into either the rotation of the leftmost symbol (the leftmost symbol of the word is moved to the end of the word) or the simulation of a rule of R. To guarantee the correct simulation, a marker symbol, #, is introduced for indicating the end of the simulated word under the rotation. Assume that the symbols $N \cup T \cup \{\#\}$ are labeled in a one-to-one manner by $1, 2, \ldots, n$. More precisely let $N \cup T \cup \{\#\} = A = \{A_1, A_2, \ldots A_n\}$, $I = \{1, 2, \ldots, n\}$, $I' = \{1, 2, \ldots, n-1\}$, $I'' = \{2, 3 \ldots, n\}$, $I_0 = \{0, 1, 2, \ldots, n\}$, $I'_0 = \{0, 1, 2, \ldots, n-1\}$, $B_0 = \{B_{j,0} \mid j \in I\}$, $\# = A_n$, $T' = T \cup \#$. The alphabet V of the network is defined as follows:

$$V = A \cup B \cup B' \cup C \cup C' \cup D \cup D' \cup E \cup E' \cup \{\varepsilon'\}, \text{ where}$$
$$B = \{B_{i,j} \mid i \in I, \ j \in I_0\}, B' = \{B'_{i,j} \mid i,j \in I\}, C = \{C_i \mid i \in I\},$$
$$C' = \{C'_i \mid i \in I'\}, D = \{D_i \mid i \in I_0\}, D' = \{D'_i \mid i \in I\},$$
$$E = \{E_{i,j} \mid i,j \in I\}, E' = \{E'_{i,j} \mid i,j \in I\}.$$

G is a complete graph with 10 nodes, N, C_0, α, β are presented in Table 2 below and node 10 is the output node of HNEP Γ'.

A configuration of grammar Γ is a word $w \in \{N \cup T\}^*$. Each configuration w of Γ corresponds to a configuration $wB_{n,0}$ and configurations $w''A_n w'B_{i,0}$ of HNEP Γ', where $A_n = \#$, $w, w', w'' \in (N \cup T)^*$ and $w = w'A_i w''$.

The axiom $S = A_1$ of Γ corresponds to an initial word $A_1 \#$, represented as $A_1 B_{n,0}$ in node 1 of HNEP Γ'. Now we describe the how the rotation of a symbol and the application of an arbitrary rule of grammar Γ are simulated in Γ'. As above, due to the lack of space, we present only the necessary details.

Rotation

Let $A_{i_1} A_{i_2} \ldots A_{i_{k-1}} B_{i_k,0}$ be found at node 1, and let $w, w', w'' \in A^*$. Then, by evolution, $A_{i_1} A_{i_2} \ldots A_{i_{k-1}} B_{i_k,0} = A_{i_1} w B_{i_k,0} \xrightarrow{1.1} \{C_{i_1} w B_{i_k,0}, A_{i_1} w' C_{i_t} w'' B_{i_k,0}\}$ follows. Notice that during the simulation symbols C_i should be transformed to ε', and this symbol should be deleted from the left-hand end of the string (node 9). So, transformation of string $A_{i_1} w' C_{i_t} w'' B_{i_k,0}$ leads to a string that will stay in node 9 forever; thus, in the sequel, we will not consider strings with C_i not in the leftmost position. In the following communication step, string $C_{i_1} w B_{i_k,0}$ can enter only node 2. Then, in nodes 2 and 3 the string is involved in evolution steps followed by communication as follows:

$$C_{i_1-t} w B_{i_k,t} \xrightarrow{2.1} C'_{i_1-(t+1)} w B_{i_k,t} \xrightarrow{2.2} C'_{i_1-(t+1)} w B'_{i_k,t+1} \text{ (in node 2)},$$
$$C'_{i_1-t} w B'_{i_k,t} \xrightarrow{3.1} C_{i_1-(t+1)} w B'_{i_k,t} \xrightarrow{3.2} C_{i_1-(t+1)} w B_{i_k,t+1} \text{(in node 3)}.$$

The process continues in nodes 2 and 3 until index of C_i or C'_i will be decreased to 1. In this case rule $2.3 : C_1 \to \varepsilon'$ in node 2 or $3.3 : C'_1 \to \varepsilon'$ in node 3 will be applied and string $\varepsilon' w B'_{i_k,i_1}$ or $\varepsilon' w B_{i_k,i_1}$ appears in node 4. Then, in node 4, either evolution step $\varepsilon' w B'_{i_k,i_1} \xrightarrow{4.1} \varepsilon' w B'_{i_k,i_1} D_0$ or evolution step $\varepsilon' w B_{i_k,i_1} \xrightarrow{4.1} \varepsilon' w B_{i_k,i_1} D_0$ is performed. Strings $w B'_{i_k,i_1} D_0$ or $w B_{i_k,i_1} D_0$ can enter only node 5, where either evolution step $\varepsilon' w B'_{i_k,i_1} D_0 \xrightarrow{5.1} \varepsilon' w E_{i_k,i_1} D_0$ or evolution step $\varepsilon' w B_{i_k,i_1} D_0 \xrightarrow{5.2} \varepsilon' w E_{i_k,i_1} D_0$ follows. String $\varepsilon' w E_{i_k,i_1} D_0$ can enter only node 6. Then, in nodes 6 and 7 the string is involved in evolution steps followed by communication as follows:

$$\varepsilon' w E_{i_k,i_1-t} D_t \xrightarrow{6.1} \varepsilon' w E'_{i_k,i_1-(t+1)} D_t \xrightarrow{6.2} \varepsilon' w E'_{i_k,i_1-(t+1)} D'_{t+1} \text{ (in node 6)},$$
$$\varepsilon' w E'_{i_k,i_1-t} D'_t \xrightarrow{7.1} \varepsilon' w E_{i_k,i_1-(t+1)} D'_t \xrightarrow{7.2} \varepsilon' w E_{i_k,i_1-(t+1)} D_{t+1} \text{ (in node 7)}.$$

The process continues in nodes 6 and 7 until second index of $E_{i,j}$ or that of $E'_{i,j}$ will be decreased to 1. In this case, rule $6.3 : E_{i_k,1} \to A_{i_k}$ in node 6 or $7.3 : E'_{i_k,1} \to A_{i_k}$ in node 7 will be applied and string $\varepsilon' w A_{i_k} D'_{i_1}$ or $\varepsilon' w A_{i_k} D_{i_1}$ appears in node 8.

Notice, that rule 6.4: $A_n \to \varepsilon'$ can be applied. This case is discussed below. The next evolution step, performed in node 8, can either be $\varepsilon' w A_{i_k} D_{i_1} \xrightarrow{8.1} \varepsilon' w A_{i_k} B_{i_1,0}$ or $\varepsilon' w A_{i_k} D'_{i_1} \xrightarrow{8.2} \varepsilon' w A_{i_k} B_{i_1,0}$. In the following communication step, string $\varepsilon' w A_{i_k} B_{i_1,0}$ can enter node 9 or node 6.

1 Consider the last case (in this case $A_{i_1} \in T$).

At nodes 6, 9 and 10 the following evolution and communication steps are performed:

- Suppose that word $w A_{i_k} B_{i_1,0}$ does not contain nonterminal symbols (except A_n). Let $w A_{i_k} B_{i_1,0} = A_n w' A_{i_k} B_{i_1,0}$, where $w = A_n w'$. So, $w' A_{i_k} A_{i_1}$ is a result and it has appear in node 10. Notice, that if $w = w' A_n w''$ and $w' \neq \varepsilon$, then word $\varepsilon' w' A_n w'' A_{i_k} B_{i_1,0}$ leads to a word which will stay in node 9 forever (if rule 6.4 was applied) or will leave node 9 as word $w' A_n w'' A_{i_k} A_{i_1}$ and enter node 1, and will remain there forever. So, we will consider the following evolution of the word $\varepsilon' w A_{i_k} B_{i_1,0} = \varepsilon' A_n w' A_{i_k} B_{i_1,0}$: $\varepsilon' A_n w' A_{i_k} B_{i_1,0} \xrightarrow{6.5} \varepsilon' A_n w' A_{i_k} A_{i_1} \xrightarrow{6.4} \varepsilon' \varepsilon' w' A_{i_k} A_{i_1}$. Further, string $\varepsilon' \varepsilon' w' A_{i_k} A_{i_1}$ appears in node 9, where symbols ε' will be eliminated by rule 9.1 and, finally, word $w' A_{i_k} A_{i_1}$ enters node 10. This is a result.

  In the case of applying only rule 6.5, the resulting word $\varepsilon' A_n w' A_{i_k} A_{i_1}$ appears in node 9, where it becomes $A_n w' A_{i_k} A_{i_1}$, leaves node 9, enters node 1 and stays there forever.

- Suppose that word $w A_{i_k} B_{i_1,0}$ contains at least one nonterminal symbol (except A_n). In node 6 symbol $B_{i_1,0}$ is changed to A_{i_1}, after that the resulting word appears in node 1, where it will stay forever, since the output filter requires symbols from B_0.

2 Now consider the evolution of the word $\varepsilon' w A_{i_k} B_{i_1,0}$ in node 9. By applying the corresponding rules, we obtain $\varepsilon' w A_{i_k} B_{i_1,0} \xrightarrow{9.1} w A_{i_k} B_{i_1,0}$. Then, string $w A_{i_k} B_{i_1,0}$ enters node 1 and the rotation of a symbol is over. If $A_{i_1} \in T$, then the string can enter node 6. This case was considered above.

Table 2.

$N, \alpha, \beta, C_0,$	M	PI, FI, PO, FO
$1, *, (2),$ $\{A_1 B_{n,0}\}$	$\{\mathbf{1.1}: A_i \to C_i \mid i \in I,\ rotation\} \cup$ $\{\mathbf{1.2}: A_i \to \varepsilon' \mid i \in I',\ A_i \to \varepsilon\} \cup$ $\{\mathbf{1.3}: B_{j,0} \to B_{s,0} \mid A_j \to A_s,\ j, s \in I'\}$	$PI = \{A_n, B_{n,0}\},$ $FI = C \cup C' \cup \{\varepsilon'\},$ $PO = B_0, FO = \emptyset$
$2, *, (2), \emptyset$	$\{\mathbf{2.1}: C_i \to C'_{i-1},$ $\mathbf{2.2}: B_{j,k} \to B'_{j,k+1} \mid$ $i \in I'',\ j \in I,\ k \in I'_0\} \cup$ $\{\mathbf{2.3}: C_1 \to \varepsilon'\}$	$PI = C,$ $FI = C' \cup B' \cup \{\varepsilon'\},$ $PO = C' \cup \{\varepsilon'\},$ $FO = C \cup B$
$3, *, (2), \emptyset$	$\{\mathbf{3.1}: C'_i \to C_{i-1},$ $\mathbf{3.2}: B'_{j,k} \to B_{j,k+1} \mid$ $i \in I'',\ j \in I,\ k \in I'_0\} \cup$ $\{\mathbf{3.3}: C'_1 \to \varepsilon'\}$	$PI = C',$ $FI = C \cup B \cup \{\varepsilon'\},$ $PO = C \cup \{\varepsilon'\},$ $FO = C' \cup B'$
$4, r, (2), \emptyset$	$\{\mathbf{4.1}: \varepsilon \to D_0\}$	$PI = B \setminus B_0 \cup B',$ $FI = C \cup C' \cup B_0 \cup \{D_0\},$ $PO = \{D_0\}, FO = \emptyset$

Table 2. (*continued*)

$N, \alpha, \beta, C_0,$	M	PI, FI, PO, FO
$5, *, (2), \emptyset$	$\{\mathbf{5.1}: B_{j,k} \to E_{j,k},$ $\mathbf{5.2}: B'_{j,k} \to E_{j,k} \mid j, k \in I, rotation\} \cup$ $\{\mathbf{5.3}: B_{j,k} \to E_{s,t},$ $\mathbf{5.4}: B'_{j,k} \to E_{s,t} \mid$ $j, k, s, t \in I', A_j A_k \to A_s A_t\}$	$PI = \{D_0\},$ $FI = \emptyset,$ $PO = E,$ $FO = B \cup B'$
$6, *, (2), \emptyset$	$\{\mathbf{6.1}: E_{j,k} \to E'_{j,k-1},$ $\mathbf{6.2}: D_i \to D'_{i+1},$ $\mathbf{6.3}: E_{j,1} \to A_j \mid i \in I'_0, j \in I, k \in I''\} \cup$ $\{\mathbf{6.4}: A_n \to \varepsilon'\} \cup$ $\{\mathbf{6.5}: B_{j,0} \to A_j \mid A_j \in T\}$	$PI = E \cup \{B_{j,0} \mid A_j \in T\},$ $FI = E' \cup D' \cup C,$ $PO = D' \cup \{\varepsilon'\},$ $FO = E \cup D \cup$ $\{B_{j,0} \mid A_j \in T\}$
$7, *, (2), \emptyset$	$\{\mathbf{7.1}: E'_{j,k} \to E_{j,k-1},$ $\mathbf{7.2}: D'_i \to D_{i+1},$ $\mathbf{7.3}: E'_{j,1} \to A_j \mid i \in I', j \in I, k \in I''\}$	$PI = E', FI = E \cup D,$ $PO = D, FO = E' \cup D'$
$8, *, (2), \emptyset$	$\{\mathbf{8.1}: D_j \to B_{j,0},$ $\mathbf{8.2}: D'_j \to B_{j,0} \mid j \in I\} \cup$ $\{\mathbf{8.3}: D_j \to B_{s,t},$ $\mathbf{8.4}: D'_j \to B_{s,t} \mid A_j \to A_s A_t, j, s, t \in I'\}$	$PI = D \setminus \{D_0\} \cup D',$ $FI = E \cup E' \cup \{D_0\},$ $PO = \emptyset,$ $FO = D \cup D'$
$9, l, (2), \emptyset$	$\{\mathbf{9.1}: \varepsilon' \to \varepsilon\}$	$PI = \{\varepsilon'\},$ $FI = B \setminus B_0 \cup$ $B' \cup D \cup D',$ $PO = \emptyset, FO = \{\varepsilon'\}$
$10, *, (2), \emptyset$	$\emptyset$	$PI = T, FI = V \setminus T,$ $PO = \emptyset, FO = T$

Rule $A_i \longrightarrow \varepsilon$. Suppose that $A_i w B_{j,0}$ can be found at node 1 and let $w, w', w'' \in A^*$. Then, by evolution, either $A_i w B_{j,0} \xrightarrow{1.2} \varepsilon' w B_{j,0}$ or $A_t w' A_i w'' B_{j,0} \xrightarrow{1.2} A_i w' \varepsilon' w'' B_{j,0}$. String $\varepsilon' w B_{j,0}$ or $A_i w' \varepsilon' w'' B_{j,0}$ can enter node 9 or node 6 (considered above). String $A_i w' \varepsilon' w'' B_{j,0}$ will stay in node 9 forever. So, we will consider the transformation of only string $\varepsilon' w B_{j,0}$. At node 9, evolution step $\varepsilon' w B_{j,0} \xrightarrow{9.1} w B_{j,0}$ follows. Now, string $w B_{j,0}$ enters node 1. Thus, we correctly simulated rule $A_i \longrightarrow \varepsilon$ of grammar Γ.

Rule $A_i \longrightarrow A_j$. The evolution step performed at node 1 is $w B_{i,0} \xrightarrow{1.3} w B_{j,0}$. Since string $w B_{j,0}$ now is in node 1, we simulated the rule $A_i \longrightarrow A_j$ of grammar Γ in a correct manner.

Rule $A_j \longrightarrow A_s A_t$. At the end of the simulation of the rotation of a symbol in node 8 instead of applying rule $D_j \to B_{j,0}$ ($D'_j \to B_{j,0}$) rule $D_j \to B_{s,t}$ ($D'_j \to B_{s,t}$) will be applied. Then, at node 8 either evolution step $\varepsilon' w D_j \xrightarrow{8.3} \varepsilon' w B_{s,t}$ or evolution step $\varepsilon' w D'_j \xrightarrow{8.4} \varepsilon' w B_{s,t}$ is performed. Then, string $\varepsilon' w B_{s,t}$ can enter only node 4, where, by evolution, $\varepsilon' w B_{s,t} \xrightarrow{4.1} \varepsilon' w B_{s,t} D_0$. The process continues as above, in the case of simulating rotation, so, in several computation steps string $w A_s B_{t,0}$ will be obtained in node 9 which then successfully is communicated to node 1. So, we correctly simulated rule $A_j \longrightarrow A_s A_t$ of grammar Γ.

Rule $A_i A_j \longrightarrow A_s A_t$. In node 5 there are rules $\mathbf{5.3} : B_{i,j} \rightarrow E_{s,t}$ or $\mathbf{5.4} : B'_{i,j} \rightarrow E_{s,t}$. As in the case of simulating rotation, above, we will obtain string $w A_s B_{t,0}$ in node 9.

We have demonstrated how the rotation of a symbol and the application of rules of Γ are simulated by Γ'. By the constructions, the reader can easily verify that Γ and Γ' generate the same language.

Corollary 2. *The class of complete HNEPs with 10 nodes is computationally complete.*

4 Conclusions

We have presented a universal complete HNEP with 10 nodes and proved that complete HNEPs with 10 nodes generate all recursively enumerable languages. Thus, we positively answered question 1 from [5] and significantly improved the results of that paper.

References

1. Alhazov, A., Martín-Vide, C., Rogozhin, Y.: On the number of nodes in universal networks of evolutionary processors. Acta Informatica 43(5), 331–339 (2006)
2. Alhazov, A., Martín-Vide, C., Rogozhin, Y.: Networks of Evolutionary Processors with Two Nodes Are Unpredictable. In: Pre-Proceedings of the 1st International Conference on Language and Automata Theory and Applications, LATA 2007, GRLMC report 35/07, Rovira i Virgili University, Tarragona, Spain, pp.521–528 (2007)
3. Alhazov, A., Kudlek, M., Rogozhin, Y.: Nine Universal Circular Post Machines. Computer Science Journal of Moldova 10(3), 247–262 (2002)
4. Castellanos, J., Martín-Vide, C., Mitrana, V., Sempere, J.: Solving NP-complete problems with networks of evolutionary processors. In: Mira, J., Prieto, A. (eds.) IWANN 2001. LNCS, vol. 2084. Springer, Heidelberg (2001)
5. Csuhaj-Varjú, E., Martín-Vide, C., Mitrana, V.: Hybrid networks of evolutionary processors are computationally complete. Acta Informatica 41(4-5), 257–272 (2005)
6. Csuhaj-Varjú, E., Salomaa, A.: Networks of Parallel Language Processors. In: Păun, G., Salomaa, A. (eds.) New Trends in Formal Languages. Control, Cooperation, and Combinatorics. LNCS, vol. 1218, pp. 299–318. Springer, Heidelberg (1997)
7. Kudlek, M., Rogozhin, Y.: Small Universal Circular Post Machines. Computer Science Journal of Moldova 9(1), 34–52 (2001)
8. Kudlek, M., Rogozhin, Y.: New Small Universal Circular Post Machines. In: Freivalds, R. (ed.) FCT 2001. LNCS, vol. 2138, pp. 217–227. Springer, Heidelberg (2001)
9. Martín-Vide, C., Mitrana, V., Perez-Jimenez, M., Sancho-Caparrini, F.: Hybrid networks of evolutionary processors. In: Proc. of GECCO 2003. LNCS, vol. 2723, pp. 401–412. Springer, Heidelberg (2003)

On Bifix Systems and Generalizations

Jan-Henrik Altenbernd

RWTH Aachen University

Abstract. Motivated by problems in infinite-state verification, we study word rewriting systems that extend mixed prefix/suffix rewriting (short: bifix rewriting). We introduce several types of infix rewriting where infix replacements are subject to the condition that they have to occur next to tag symbols within a given word. Bifix rewriting is covered by the case where tags occur only as end markers. We show results on the reachability relation (or: derivation relation) of such systems depending on the possibility of removing or adding tags. Where possible we strengthen decidability of the derivation relation to the condition that regularity of sets is preserved, resp. that the derivation relation is even rational. Finally, we compare our model to ground tree rewriting systems and exhibit some differences.

1 Introduction

The algorithmic theory of prefix (respectively suffix) rewriting systems on finite words has long been well established, and a number of decision problems over such systems have been proven to be decidable. Such rewriting systems are a general view of pushdown systems, where symbols are pushed onto and removed from the top of a stack.

Büchi showed in [2] that the language derivable from a given word by prefix rewriting is regular (and that a corresponding automaton can be computed). In the theory of infinite-state system verification, the "saturation method" (for the transformation of finite automata) has been applied for this purpose (see e. g. [14,5,6]). Caucal [4] showed the stronger result that the derivation relation induced by a prefix rewriting system is a rational relation.

The extension to combined prefix and suffix rewriting goes back to Büchi and Hosken [3]. Karhumäki, Kunc, and Okhotin showed in [9] that when combining prefix and suffix rewriting, the corresponding derivation relation is still rational, and therefore preserves regularity of languages. They extended their work in [8] to rewriting systems with a center marker, simulating two stacks communicating with each other. They singled out a number of cases where universal computation power could already be achieved with very limited communication.

In a more restricted framework, Bouajjani, Müller-Olm and Touili studied dynamic networks of pushdown systems in [1]. Here, a collection of pushdown processes is treated as a word in which a special marker is used to separate the processes. Rewriting of such words is restricted to performing pushdown operations and to creating new processes, where the latter increases the number of markers. It was shown that reachability in this setting is decidable.

C. Martín-Vide, F. Otto, and H. Fernau (Eds.): LATA 2008, LNCS 5196, pp. 40–51, 2008.
© Springer-Verlag Berlin Heidelberg 2008

In the present paper, we develop a generalised framework of "tagged infix rewriting" which extends some of the cases mentioned above. We clarify the status of the word-to-word reachability relation (or derivation relation) for several types of tagged infix rewriting. More precisely, we determine whether this relation is undecidable, or decidable, or even decidable in two stronger senses: that the relation preserves effectively the regularity of a language, or that the derivation relation itself is rational. (By "effective" preservation of regular languages we mean that from a presentation of L by a finite automaton and from the rewriting system defining the relation R we obtain algorithmically a finite automaton for the image of L under the derivation relation of R.) So the motivation (and contribution) of the paper is twofold: first to push the frontier of decidability further for reachability problems over rewriting systems, and secondly to differentiate clearly between the three levels of decidability proofs mentioned above.

We define a generalisation of mixed prefix/suffix rewriting systems on words by introducing special symbols (*tags* or *markers*) to mark positions in words where rewriting can occur. Typically, a rewriting rule can transform a word $w = w_0 \#_1 w_1 \cdots \#_n w_n$ into a word $w' = w_0' \#_1 w_1' \cdots \#_n w_n'$ with $w_i = w_i'$ for all i except for some i_0 where w_{i_0}' is obtained from w_{i_0} by a prefix, suffix, or complete rewriting rule $U \hookrightarrow V$ with regular sets U, V (to be applied to the whole word $u \in U$ between two successive markers, replacing it by some $v \in V$). Thus, arbitrary words in finite sequences can be rewritten independently, extending a case studied in [9]. The variants we consider in this paper deal with the options that markers may be removed or added in the rewriting process. We show that the derivation relation is rational in the basic case mentioned above, where markers are always preserved, and that this fails in general for the other cases. However, we still obtain decidability of the reachability problem in all cases. For applications, our systems are close to models of concurrent processes where states are presented by words between tags, state transitions by local rewriting rules, and e. g. spawning of new processes by the insertion of tags.

The paper is structured as follows: In the subsequent section we summarise technical preliminaries. Section 3 introduces the basic models of bifix systems and its extension tagged infix rewriting, and we show that one obtains different levels of decidability of the derivation relation: We present cases where the derivation relation is not rational but effectively preserves regularity of languages, and where the latter condition fails but the word-to-word reachability problem is still decidable. This refined analysis also exhibits a substantial difference between the two cases of tag insertion and tag removal. The next section is devoted to a comparison of bifix systems and ground tree rewriting systems (and the closely related multi-stack systems).

2   Terminology

Automata and Languages. We use the standard terminology from automata theory and formal language theory (see e. g. [7]). We present nondeterministic finite automata (NFA) in the format $\mathcal{A} = (Q, \Sigma, q_0, \Delta, F)$, where Q is a finite set of states, Σ is a finite alphabet, $q_0 \in Q$ is the initial state, $F \subseteq Q$ is the set of final states, and $\Delta \subseteq Q \times (\Sigma \cup \{\varepsilon\}) \times Q$ is a finite set of transitions. We write $\mathcal{A} : p \xrightarrow{w} q$ to denote that there is a w-labelled path from state p to state q in $\mathcal{A}$. $\mathrm{Reg}(\Sigma)$ denotes the class of all regular languages over Σ. We will refer to *normalised NFAs* which have exactly one final state, and in which no incoming respectively outgoing transitions are allowed for the initial respectively final state. A (finite) transducer is an NFA $\mathcal{A} = (Q, \Gamma, q_0, \Delta, F)$, where $\Gamma \subseteq \Sigma^* \times \Sigma^*$ is a finite set of pairs of words over a finite alphabet Σ.

Relations. Let Σ be a finite alphabet. A relation $R \subseteq \Sigma^* \times \Sigma^*$ is *recognisable* if it is a finite union of products of regular languages over Σ, that is, $R = \bigcup_{i=1}^n L_i \times M_i$ for some $n \in \mathbb{N}$ and regular L_i, M_i; when using R as a rewriting system, we write rules in the form $L_i \hookrightarrow M_i$. R is *rational* if it is recognisable by a transducer, i. e. an NFA with transitions labelled by finite subsets of $\Sigma^* \times \Sigma^*$. We then write $R \in \mathrm{Rat}(\Sigma^* \times \Sigma^*)$.

For relations $R, S \subseteq \Sigma^* \times \Sigma^*$, we call $\mathrm{Dom}(R) = \{u \mid \exists v : (u, v) \in R\}$ the *domain* of R, and $\mathrm{Im}(R) = \{v \mid \exists u : (u, v) \in R\}$ the *image* of R. For $L \subseteq \Sigma^*$, we call $R(L) = \{v \mid \exists u \in L : (u, v) \in R\}$ the set derivable from L according to R. We define the concatenation of R and S as $R \cdot S = \{(ux, vy) \mid (u, v) \in R \wedge (x, y) \in S\}$, which we also shorten to RS, if no ambiguity arises, and their composition as $R \circ S = \{(u, w) \mid \exists v : (u, v) \in R \wedge (v, w) \in S\}$.

We call $I = \{(w, w) \mid w \in \Sigma^*\}$ the identity relation on Σ^*. Note that I is rational, but not recognisable. When considering iteration, we have to distinguish two cases. Let $R^* = \bigcup_{n \geq 0} R^n$, where $R^0 = \{(\varepsilon, \varepsilon)\}$, and $R^{n+1} = R^n \cdot R$, and let $R^{\circledast} = \bigcup_{n \geq 0} R^{(n)}$, where $R^{(0)} = I$, and $R^{(n+1)} = R^{(n)} \circ R$.

We recall some basic results about rational relations: $\mathrm{Rat}(\Sigma^* \times \Sigma^*)$ is closed under union, concatenation and the concatenation iteration *. Furthermore, if R is a rational relation, then $R(L)$ is regular for regular L, hence $\mathrm{Dom}(R)$ and $\mathrm{Im}(R)$ are regular. Finally, if R is a rational relation, and S is a recognisable relation, then $R \cap S$ is rational.

Mixed Prefix / Suffix Rewriting Systems. A *mixed prefix/suffix rewriting system* is a tuple $\mathcal{R} = (\Sigma, R, S)$, where Σ is a finite alphabet, and $R, S \subseteq \mathrm{Reg}(\Sigma) \times \mathrm{Reg}(\Sigma)$ are recognisable relations of rewriting rules. We write $w \xrightarrow[\mathcal{R}]{} w'$ if $(w, w') \in (RI \cup IS)$, i. e. R and S are used for prefix respectively suffix rewriting. We denote the derivation relation $\xrightarrow[\mathcal{R}]{}{}^{\circledast} = (RI \cup IS)^{\circledast}$ by $\mathcal{R}^{\circledast}$.

Proposition 1 ([9]). *The derivation relation $\mathcal{R}^{\circledast}$ of a mixed prefix/suffix rewriting system $\mathcal{R}$ is rational.*

3   Bifix Rewriting Systems and Extensions

As a first and minor extension of mixed prefix/suffix rewriting systems, we introduce *bifix rewriting systems*, which will serve as a basis for further extensions. A bifix rewriting system is a tuple $\mathcal{R} = (\Sigma, R, S, T)$, with Σ, R, S as in the case of mixed prefix/suffix rewriting systems, and where $T \subseteq \text{Reg}(\Sigma) \times \text{Reg}(\Sigma)$ is also a recognisable relation. We write $w \xrightarrow{\mathcal{R}} w'$ if $(w, w') \in (RI \cup IS \cup T)$, that is, R and S are used as before, and T is used to rewrite complete words. The other notions carry over.

As a first result, it is easy to see that Proposition 1 holds again:

Proposition 2. *The derivation relation of a bifix rewriting system is rational.*

Proof. We have to show that $W = (RI \cup IS \cup T)^\circledast$ is rational. For this, introduce $\# \notin \Sigma$, and consider $U = \#R \cup \#T\#$ and $V = S\#$. Then $\#W\# = (UI \cup IV)^\circledast \cap (\#\Sigma^*\# \times \#\Sigma^*\#)$, that is, we use rewriting of complete words with T for prefix rewriting, and we restrict the corresponding derivation relation to pairs of words with $\#$ at the beginning and end only. Since U, V, and $(\#\Sigma^*\# \times \#\Sigma^*\#)$ are recognisable, $(UI \cup IV)^\circledast$ is rational by Proposition 1, and it follows that $\{(\#u\#, \#v\#) \mid (u, v) \in W\}$ is rational. Removing the symbols $\#$ preserves this rationality, so W is rational.

3.1 Tagged Infix Rewriting Systems

Let Σ be a finite alphabet. We will use a finite set M of *tags* (or *markers*) with $M \cap \Sigma = \emptyset$ to mark positions in a finite word where rewriting can occur. Given Σ and M, let $P_{\Sigma,M} := M\Sigma^* \cup \Sigma^*M \cup M\Sigma^*M$ denote the set of all words over $\Sigma \cup M$ that contain at least one marker from M, but only at the beginning and/or end.

A *tagged infix rewriting system* (TIRS) is a structure $\mathcal{R} = (\Sigma, M, R)$ with disjoint finite alphabets Σ and M and a relation $R \subseteq P_{\Sigma,M} \times P_{\Sigma,M}$ which is a finite union of

$$
\begin{aligned}
&\textit{prefix rules of the form } \#U \hookrightarrow \#V \text{ (denoting } \#U \times \#V),\\
&\textit{suffix rules of the form } U\$ \hookrightarrow V\$, \text{ and} \qquad\qquad (1)\\
&\textit{bifix rules of the form } \#U\$ \hookrightarrow \#V\$,
\end{aligned}
$$

where $U, V \in \text{Reg}(\Sigma)$ and $\#, \$ \in M$. Note that when using R to rewrite a word w over $\Sigma \cup M$, all tags in w are preserved, and none are added. We write $xuy \xrightarrow{\mathcal{R}} xvy$ if $(u, v) \in R$ and $x, y \in (\Sigma \cup M)^*$, and we denote $\xrightarrow{\mathcal{R}}^\circledast$ by $\mathcal{R}^\circledast$.

As a first example, consider $\mathcal{R} = (\{a, b, c\}, \{\#\}, R)$ with the following set R of rules:

$$
\begin{aligned}
&\#\# \hookrightarrow \#acb\# && &&\text{(bifix rule)},\\
&\#a \hookrightarrow \#aa && \#a^+cb \hookrightarrow \#b &&\text{(prefix rules)},\\
&b\# \hookrightarrow bb\# && acb^+\# \hookrightarrow a\# &&\text{(suffix rules)}.
\end{aligned}
$$

Then $\mathcal{R}^\circledast(\{\#\#\}) = \#a^+cb^+\# \cup \#a^*\# \cup \#b^*\#$.

As a second example, note that the infinite grid can be generated with the simple TIRS $(\{a, b\}, \{\#\}, \{\# \hookrightarrow a\#, \# \hookrightarrow \#b\})$, starting with marker $\#$:

$$
\begin{array}{ccccc}
\# & \to & \#b & \to & \#bb & \to \cdots\\
\downarrow & & \downarrow & & \downarrow\\
a\# & \to & a\#b & \to & a\#bb & \to \cdots\\
\downarrow & & \downarrow & & \downarrow\\
aa\# & \to & aa\#b & \to & aa\#bb & \to \cdots\\
\downarrow & & \downarrow & & \downarrow
\end{array}
$$

Since the monadic second-order logic (MSO) of the infinite grid is undecidable (see e. g. [15]), we can immediately conclude the following.

Proposition 3. *The MSO theory of graphs generated by TIRSs is undecidable.*

It is well known that prefix (resp. suffix) and mixed prefix/suffix rewriting systems preserve regularity ([4,9]), that is, given such a system $\mathcal{R}$ and a regular set L, the set derivable from L according to $\mathcal{R}$ is again regular. It has also been shown that the derivation relation $\mathcal{R}^{\circledast}$ of such systems is rational. We show in the following that these results carry over to tagged infix rewriting systems.

Theorem 1. *The derivation relations of TIRSs are rational.*

Proof. Let $\mathcal{R} = (\Sigma, M, R)$ be a TIRS. We construct an NFA $\mathcal{A}_{\mathcal{R}} = (Q, \Gamma, q_0, \Delta, \{q_f\})$ whose edges are labelled with rational relations (i. e. Γ is a finite set of rational relations), such that $L(\mathcal{A}_{\mathcal{R}}) = \mathcal{R}^{\circledast}$. Since we know that every finite concatenation of rational relations is again rational, every path in $\mathcal{A}_{\mathcal{R}}$ from q_0 to q_f is labelled with a rational relation.

It is important to note that markers are preserved in the derivation process. Thus, the derivation relation is a concatenation of derivation relations of rewriting that occurs before the first marker (see (i) below), after the last marker (ii), or between two markers (iii), which are basically mixed prefix/suffix rewriting derivations.

We can therefore construct $\mathcal{A}$ as follows: For $\#, \$ \in M$, let $R_\# = \{(u,v) \mid (u\#, v\#) \in R\}$, $_\#R = \{(u,v) \mid (\#u, \#v) \in R\}$, and $_\#R_\$ = \{(u,v) \mid (\#u\$, \#v\$) \in R\}$. We choose $Q = \{q_0, q_f\} \cup \{s_m, t_m \mid m \in M\}$, that is, we take one source state s_m and one target state t_m for every marker m, and we set Δ to be the following set of edges labelled with relations:

$$\Delta = \{(s_m, \{m\} \times \{m\}, t_m) \mid m \in M\} \ \cup \ \{(q_0, I, q_f)\}$$
$$\cup \{(q_0, (IR_m)^{\circledast}, s_m) \mid m \in M\} \tag{i}$$
$$\cup \{(t_m, (_mRI)^{\circledast}, q_f) \mid m \in M\} \tag{ii}$$
$$\cup \{(t_m, (_mRI \cup IR_{m'} \cup {}_mR_{m'})^{\circledast}, s_{m'}) \mid m, m' \in M\} \ . \tag{iii}$$

We know that $\{m\} \times \{m\}$, $(IR_m)^{\circledast}$, $(_mRI)^{\circledast}$, and I are rational, and by Proposition 2 the same holds for $(_mRI \cup IR_{m'} \cup {}_mR_{m'})^{\circledast}$. $\qquad\square$

We can immediately deduce that TIRSs effectively preserve regularity.

3.2 Extending TIRSs by Removing Tags

We consider an extension of TIRSs where removing tags is allowed, thereby breaking up the preservation of markers. We will see that in this case some effective reachability analysis is still possible.

A *TIRS with tag-removing rules* is a structure $\mathcal{R} = (\Sigma, M, R)$ with disjoint finite alphabets Σ and M as before and a relation $R \subseteq P_{\Sigma,M} \times (P_{\Sigma,M} \cup \Sigma^*)$ containing rules of the basic form (1) and also rules of the forms $\#U \hookrightarrow V$,

$U\$ \hookrightarrow V$, $\#U\$ \hookrightarrow \#V$, $\#U\$ \hookrightarrow V\$$, and $\#U\$ \hookrightarrow V$, where $U, V \in \text{Reg}(\Sigma)$ and $\#, \$ \in M$. We show that in this case the derivation relation is not rational in general, but that regularity is still preserved (the latter result involving a nontrivial saturation construction).

Proposition 4. *Derivation relations of TIRSs with tag-removing rules are not rational in general.*

Proof. Consider $\mathcal{R} = (\{a, b\}, \{\#\}, R)$, where R contains only the rules $\#a \hookrightarrow b$ and $b\# \hookrightarrow a$. Then $\text{Dom}(\mathcal{R}^\circledast \cap (\#^*a\#^* \times \{a\})) = \{\#^n a\#^n \mid n \geq 0\}$ is not regular, and so $\mathcal{R}^\circledast$ is not rational. □

Before showing that such systems still preserve regularity, we need to introduce some more terminology. We call an NFA $\mathcal{A} = (Q, \Sigma \cup M, q_0, \Delta, F)$ *unravelled* if it satisfies the following conditions:

1. for every $q \in Q$: $|\{(q, m, p) \in \Delta \mid m \in M\}| \cdot |\{(p, m, q) \in \Delta \mid m \in M\}| = 0$; that is, every state is the source or the target state of transitions labelled with markers (or none of the above), but not both at the same time;
2. for every $m \in M$ and $(q, m, q') \in \Delta$: $|\{(q, a, r) \in \Delta \mid a \in \Sigma \cup M \cup \{\varepsilon\}\}| = 1$ and $|\{(r, a, q') \in \Delta \mid a \in \Sigma \cup M \cup \{\varepsilon\}\}| = 1$; that is, every source state of a marker transition has no other outgoing transitions, and every target state of a marker transition has no other incoming transitions.

Lemma 1. *For every NFA $\mathcal{A}$ over an alphabet $\Sigma \cup M$ one can effectively construct an unravelled NFA $\mathcal{A}'$ with $L(\mathcal{A}) = L(\mathcal{A}')$.*

Proof. Let $\mathcal{A} = (Q, \Sigma \cup M, q_0, \Delta, F)$ be an NFA. Construct $\mathcal{A}' = (Q', \Sigma \cup M, q'_0, \Delta', F')$ with

- $Q' := \{q'_0\} \cup \{(\overline{p}, a, q), (p, a, \overline{q}) \mid (p, a, q) \in \Delta\}$,
- $F' := \{(p, a, \overline{q}) \mid (p, a, q) \in \Delta, q \in F\} \cup \{q'_0 \mid q_0 \in F\}$, and
- $\Delta' := \{(q'_0, \varepsilon, (\overline{q_0}, a, q)) \mid (q_0, a, q) \in \Delta\}$
  $\cup \{((\overline{p}, a, q), a, (p, a, \overline{q})) \mid (p, a, q) \in \Delta\}$
  $\cup \{((p, a, \overline{q}), \varepsilon, (\overline{q}, b, r)) \mid (p, a, q), (q, b, r) \in \Delta\}$.

Then $L(\mathcal{A}') = L(\mathcal{A})$, and $\mathcal{A}'$ is unravelled.

A state $(\overline{p}, a, q)$ in $\mathcal{A}'$ symbolizes that p is the current state and (p, a, q) the next transition to be taken in a run of $\mathcal{A}$; $(p, a, \overline{q})$ denotes that q is the current state and (p, a, q) is the last transition used in a run of $\mathcal{A}$. After every such step, a transition of the form $((p, a, \overline{q}), \varepsilon, (\overline{q}, b, r))$ allows us to guess the next transition taken in a run of $\mathcal{A}$ (in this case (q, b, r)). We omit the details of the correctness proof due to space restrictions. □

The notion of unravelled NFA is important for the following theorem.

Theorem 2. *TIRSs with tag-removing rules effectively preserve regularity.*

Proof. Let $\mathcal{R} = (\Sigma, M, R)$ be a TIRS with tag-removing rules, and let $\mathcal{A} = (Q, \Sigma \cup M, q_0, \Delta, F)$ be an unravelled NFA with $L(\mathcal{A}) = L$. We provide an algorithm that constructs an NFA $\mathcal{A}'$ from $\mathcal{A}$ such that $L(\mathcal{A}') = \mathcal{R}^{\circledast}(L)$. For this, we first extend an initial automaton $\mathcal{A}_0 = (Q_0, \Sigma \cup M, q_0, \Delta_0, F)$ with $Q_0 := Q$ and $\Delta_0 := \Delta$ as follows.

We have to capture derivation at and between all possible combinations of markers, possibly involving the deletion of markers. If, for instance, there is a rule $\#U \hookrightarrow \#V$ in R, then it may be applied at different positions of the marker $\#$ in $\mathcal{A}$, and we thus have to distinguish between these applications to avoid side effects. Therefore, we add normalised NFAs for all $(p, m, q), (p', m', q') \in \Delta$ with $m, m' \in M$, taking disjoint copies for different applications of rules inside the given automaton:

- for every prefix rule of the form $mU \hookrightarrow mV$ or $mU \hookrightarrow V$ in R, we add
  $\mathcal{A}_{(p,q,V)} = (Q_{(p,q,V)}, \Sigma, s_{(p,q,V)}, \Delta_{(p,q,V)}, \{t_{(p,q,V)}\})$ with $L(\mathcal{A}_{(p,q,V)}) = V$;
  we set $Q_0 := Q_0 \cup Q_{(p,q,V)}$ and $\Delta_0 := \Delta_0 \cup \Delta_{(p,q,V)}$, and we add $(q, \varepsilon, s_{(p,q,V)})$
  (resp. $(p, \varepsilon, s_{(p,q,V)})$) to Δ_0;
- for every suffix rule of the form $Um' \hookrightarrow Vm'$ or $Um' \hookrightarrow V$ in R, we add
  $\mathcal{A}_{[p',q',V]} = (Q_{[p',q',V]}, \Sigma, s_{[p',q',V]}, \Delta_{[p',q',V]}, \{t_{[p',q',V]}\})$ with $L(\mathcal{A}_{[p',q',V]}) = V$; we set $Q_0 := Q_0 \cup Q_{[p',q',V]}$ and $\Delta_0 := \Delta_0 \cup \Delta_{[p',q',V]}$, and we add $(t_{[p',q',V]}, \varepsilon, p')$ (resp. $(t_{[p',q',V]}, \varepsilon, q')$) to Δ_0;
- for every bifix rule of the form $mUm' \hookrightarrow mVm'$, $mUm' \hookrightarrow mV$, $mUm' \hookrightarrow Vm'$, or $mUm' \hookrightarrow V$ in R, we add $\mathcal{A}_{(p,q,p',q',V)} = (Q_{(p,q,p',q',V)}, \Sigma, s_{(p,q,p',q',V)}, \Delta_{(p,q,p',q',V)}, \{t_{(p,q,p',q',V)}\})$ with $L(\mathcal{A}_{(p,q,p',q',V)}) = V$; we set $Q_0 := Q_0 \cup Q_{(p,q,p',q',V)}$ and $\Delta_0 := \Delta_0 \cup \Delta_{(p,q,p',q',V)}$, and we add $(q, \varepsilon, s_{(p,q,p',q',V)})$ in the first two cases resp. $(p, \varepsilon, s_{(p,q,p',q',V)})$ in the last two cases to Δ_0.

For the automaton $\mathcal{A}_0$ generated this way, we have $L(\mathcal{A}_0) = L(\mathcal{A})$.

For the sketch of the correctness proof later on, let Q_i denote the set of all initial states of the NFAs added for suffix rules, and let Q_f denote the set of all final states of the NFAs added for prefix and bifix rules.

After these preparatory steps, we now repeat the following saturation steps until no more transitions can be added, starting with $k = 0$:

1. If there are $(p, m, q) \in \Delta$, $r \in Q_0$, a prefix rule of the form $mU \hookrightarrow mV$ or $mU \hookrightarrow V$ in R, and a path $\mathcal{A}_k : q \xrightarrow{u} r$ for some $u \in U$, then we add the transition $(t_{(p,q,V)}, \varepsilon, r)$ to Δ_k to obtain $\mathcal{A}_{k+1}$, and we set $k := k + 1$.

   The following illustrates this for rules $mU_1 \hookrightarrow mV_1$ and $mU_2 \hookrightarrow V_2$ and a path $p \xrightarrow{m} q \xrightarrow{u} r$. The dotted lines denote the transitions added in the preparatory steps, while the dashed lines show the ε-transitions added in the saturation steps.

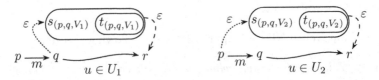

2. If there are $(p', m', q') \in \Delta$, $r \in Q_0$, a suffix rule of the form $Um' \hookrightarrow Vm'$ or $Um' \hookrightarrow V$ in R, and a path $\mathcal{A}_k : r \xrightarrow{u} p'$ for some $u \in U$, then we add the transition $(r, \varepsilon, s_{[p',q',V]})$ to Δ_k to obtain $\mathcal{A}_{k+1}$, and we set $k := k + 1$.

The following illustrates this for rules $U_3m' \hookrightarrow V_3m'$ and $U_4m' \hookrightarrow V_4$ and a path $r \xrightarrow{u} p' \xrightarrow{m'} q'$.

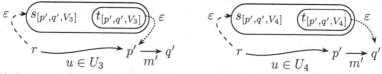

3. If there are $(p, m, q), (p', m', q') \in \Delta$ and a path $\mathcal{A}_k : q \xrightarrow{u} p'$ for some $u \in U$ for a bifix rule of the form

   (a) $mUm' \hookrightarrow mVm'$ or $mUm' \hookrightarrow Vm'$ in R, then we add the transition $(t_{(p,q,p',q',V)}, \varepsilon, p')$ to Δ_k;

   (b) $mUm' \hookrightarrow mV$ or $mUm' \hookrightarrow V$ in R, then we add the transition $(t_{(p,q,p',q',V)}, \varepsilon, q')$ to Δ_k;

   we obtain $\mathcal{A}_{k+1}$, and we set $k := k + 1$.

The case of bifix rules of the form $mU_5m' \hookrightarrow mV_5m'$, $mU_6m' \hookrightarrow mV_6$, $mU_7m' \hookrightarrow V_7m'$, and $mU_8m' \hookrightarrow V_8$ is basically a combination of cases 1. and 2. above.

After saturating $\mathcal{A}_0$ this way, we set $\mathcal{A}' := \mathcal{A}_k$, thereby obtaining the desired automaton with $L(\mathcal{A}') = \mathcal{R}^{\circledast}(L)$. Since only finitely many transitions can be added in the saturation steps, the algorithm terminates.

For the completeness of the algorithm, we can show by induction on n that if $z \xrightarrow[\mathcal{R}]{}^{(n)} w$ for some $z \in L(\mathcal{A})$, then there is a path $\mathcal{A}' : q_0 \xrightarrow{w} F$. For the soundness, we can show that if there is a path $\mathcal{A}' : q_0 \xrightarrow{w} F$, then $w \in \mathcal{R}^{\circledast}(L(\mathcal{A}))$. This follows directly from the more general claim

$$\mathcal{A}' : p \xrightarrow{w} q \text{ with } p \in Q \cup Q_i \wedge q \in Q \cup Q_f \Rightarrow \exists w' : w' \xrightarrow[\mathcal{R}]{}^{\circledast} w \wedge \mathcal{A}_0 : p \xrightarrow{w'} q .$$

For $p = q_0$ and $q \in F$ this yields the original claim. Note that we are using Q (states of the original automaton $\mathcal{A}$) in the claim, not Q_0. We omit the proof details due to space restrictions. $\square$

3.3   Extending TIRSs by Adding Tags

We extend our basic model such that R allows rules of the forms $\#U \hookrightarrow \#V$, $U\# \hookrightarrow V\#$, and $\#U\$ \hookrightarrow \#V\$$, where $U \subseteq \Sigma^*$ and $V \subseteq (\Sigma \cup M)^*$ are regular sets. This means that the right hand sides of rules may contain new tags, thereby allowing tags to be added when rewriting words.

It turns out that regularity is not preserved with this extension, and thus also the derivation relation is not rational in general. In view of Theorem 2, this illustrates well that the two cases of removing and of adding tags behave differently with respect to preservation of regularity.

Proposition 5. *TIRSs with tag-adding rules do not preserve regularity.*

Proof. Consider $\mathcal{R} = (\{a\}, \{\#\}, R)$, where R contains only the rule $\#a \hookrightarrow \#\#a\#$. Then $\mathcal{R}^\circledast(\{\#a\#\}) = \{\#^n a \#^n \mid n > 0\}$ is not regular. $\square$

However, we still keep decidability of the word-to-word reachability problem.

Theorem 3. *The word-to-word reachability problem for TIRSs with tag-adding rules is decidable.*

Proof. Let $\mathcal{R} = (\Sigma, M, R)$ be a TIRS with tag-adding rules, and let $u, v \in (\Sigma \cup M)^*$. Let $|w|_M$ denote the number of markers of M in w. If $|u|_M > |v|_M$, then clearly v is not reachable from u. Otherwise, a maximum of $n := |v|_M - |u|_M$ rewriting steps that add tags will suffice to obtain v from u, if at all possible. Let R_0 denote the set of rules of R that do not add tags, and let $R_1 = R \setminus R_0$. Similarly, let $\mathcal{R}_0 = (\Sigma, M, R_0)$ and $\mathcal{R}_1 = (\Sigma, M, R_1)$. Then we have to iterate the following at most n times to decide whether v is reachable from u, starting with $i = 0$ and $U_0 = \{u\}$:

1. Set $i := i + 1$, and compute $U_i' := \xrightarrow[\mathcal{R}_0]{}^\circledast (U_{i-1})$ and $U_i := \xrightarrow[\mathcal{R}_1]{} (U_i')$;
2. If $v \in U_i$, then v is reachable from u, else if $i = n$, then v is not reachable from u.

With the algorithm of Theorem 2, we can compute an NFA recognizing U_i' in every step, starting from an unravelled NFA recognizing U_{i-1}. Then, since $\xrightarrow[\mathcal{R}_1]{}$ is rational, U_i is also effectively regular. $\square$

3.4   Remarks on Further Extensions

There are several natural ways how the basic model of TIRS may be extended further. For instance, one may allow tag-removing and tag-adding rules at the same time, or rules might be allowed to rename the tags that are involved in a rewriting step. It is not difficult to see that these models allow to transfer information across tags in either direction, which makes it possible to move markers arbitrarily and thus to apply rewriting rules at any position within a word. Therefore, these models are Turing powerful, and all interesting properties over such systems are undecidable.

Another interesting extension is to allow information transfer across tags in only one direction, e.g. by allowing rules of the form $u\# \hookrightarrow \#v$. In [8], Karhumäki et al. distinguished the cases of *controlled* or *uncontrolled* transfer. In the controlled case, a connection of the u's and v's is allowed, that is, the word to be removed to the left of the marker $\#$ can determine the word to be added to the right of $\#$. In the uncontrolled case, no such connection is allowed, that is, the words to be removed and added are chosen independently. Karhumäki et al. showed that the language derivable from a regular initial set $L \subseteq \Sigma^* \# \Sigma^*$ is context-free in the case of uncontrolled transfer. For the controlled case however, they showed that even finite relations for the transfer suffice to obtain computational universality.

4   Comparison with Ground Tree Rewriting

Ground tree rewriting systems (GTRSs) have been studied intensively in [11]. They allow to substitute subtrees of finite ranked trees by other finite trees according to given rules. In this section we give a comparison with bifix rewriting systems.

Ranked trees are finite ordered trees over some ranked alphabet A which determines the labels and numbers of successors of nodes in a tree. T_A denotes the set of all finite trees over a given ranked alphabet A. A GTRS is a structure $\mathcal{R} = (A, \Sigma, R, t_{\mathrm{in}})$, where A is a ranked alphabet, Σ is an alphabet to label rewriting rules, R is a finite set of rewriting rules of the form $s \xrightarrow{\sigma} s'$, where $\sigma \in \Sigma$ and $s, s' \in T_A$, and $t_{\mathrm{in}} \in T_A$ is the initial tree.

Intuitively, a rule $s \xrightarrow{\sigma} s'$ may be applied to a tree $t \in T_A$ if s is a proper subtree of t. Applying the rule yields a tree that is obtained from t by replacing one occurrence of the subtree s by s'.

It is easy to realize the infinite $\mathbb{N} \times \mathbb{N}$ grid by a GTRS (using a tree of two unary branches of lengths i, j to represent vertex (i, j)). Hence the MSO theory of a GTRS graph is in general undecidable. As shown in [12], even the "universal reachability problem" ("Does every path from v reach a vertex in a regular tree set T?") is undecidable. On the other hand, as also shown in [12], the first-order theory with reachability (short: FO(R) theory) of a GTRS graph is decidable. In the FO(R) theory, the graph signature is extended by a symbol for the closure E^* of the edge relation E.

For bifix rewriting systems, the undecidability result on universal reachability is easily transferred from GTRSs. The proof for GTRSs only uses trees with two unary branches (for the representation of the left and right inscriptions of a Turing tape); in bifix rewriting systems, one simply combines the two branches into a single word with a separator between the left and right part.

It is remarkable that a converse simulation cannot work. This is clarified by the following result:

Theorem 4. *The FO(R) theory of a mixed prefix/suffix rewriting system is in general undecidable.*

For the proof, we remark that for the bifix rewriting system with rules $\Sigma \hookrightarrow \varepsilon$ for both prefix and suffix rewriting, the transitive closure gives the infix relation. As proved by Kuske [10], the first-order theory of Σ^* with the infix relation is undecidable.

This result shows that there is an essential difference between

- the "multiple stack" model that is inherent in ground tree rewriting (when a collection of unary branches is used as a list of stacks, with leaves as the top symbols of stacks), and
- the bifix rewriting model, where two stacks are easily simulated, but where an internal information flow between the two sides is possible.

5 Conclusion

We have introduced a general form of "tagged" rewriting system which extends the mixed prefix/suffix rewriting as studied in [3,9], and where reachability (or the derivation relation) is decidable. We studied systematically the effects of removing and adding tags and showed that these cases are not dual. At the same time, we exhibited examples which separate decidability proofs by preservation of regularity, by rationality, or just by recursiveness of the derivation relation.

Many questions arise from these results in infinite-state system verification, where the universe of words with the tagged infix rewriting relation is considered as an infinite transition graph. For example, it should be investigated which logics admit an algorithmic solution of the model-checking problem over tagged infix rewriting graphs (see e. g. [13]). Another field of study is the definition of natural extended models where the derivation relation is no more rational, but still decidable.

Acknowledgement. I thank Didier Caucal, Christof Löding, and Wolfgang Thomas for their support and fruitful discussions, and anonymous reviewers for their helpful remarks.

References

1. Bouajjani, A., Mueller-Olm, M., Touili, T.: Regular symbolic analysis of dynamic networks of pushdown systems. In: Abadi, M., de Alfaro, L. (eds.) CONCUR 2005. LNCS, vol. 3653, pp. 473–487. Springer, Heidelberg (2005)
2. Büchi, R.: Regular canonical systems. Archiv für Mathematische Logik und Grundlagenforschung 6, 91–111 (1964)
3. Büchi, R., Hosken, W.H.: Canonical systems which produce periodic sets. Mathematical Systems Theory 4(1), 81–90 (1970)
4. Caucal, D.: On the regular structure of prefix rewriting. In: CAAP 1990. LNCS, vol. 431, pp. 87–102. Springer, Heidelberg (1990)
5. Coquidé, J.-L., Dauchet, M., Gilleron, R., Vágvölgyi, S.: Bottom-up tree pushdown automata: classification and connection with rewrite systems. Theoretical Computer Science 127, 69–98 (1994)

6. Esparza, J., Hansel, D., Rossmanith, P., Schwoon, S.: Efficient algorithms for model checking pushdown systems. Technical Report TUM-I0002, Techn. Universität München, Institut für Informatik (2000)
7. Hopcroft, J., Motwani, R., Ullman, J.: Introduction to Automata Theory, Languages, and Computation. Addison-Wesley, Reading (2000)
8. Karhumäki, J., Kunc, M., Okhotin, A.: Communication of two stacks and rewriting. In: Bugliesi, M., Preneel, B., Sassone, V., Wegener, I. (eds.) ICALP 2006. LNCS, vol. 4052, pp. 468–479. Springer, Heidelberg (2006)
9. Karhumäki, J., Kunc, M., Okhotin, A.: Computing by commuting. Theoretical Computer Science 356(1-2), 200–211 (2006)
10. Kuske, D.: Theories of orders on the set of words. Theoretical Informatics and Applications 40, 53–74 (2006)
11. Löding, C.: Infinite Graphs Generated by Tree Rewriting. Doctoral thesis, RWTH Aachen University (2003)
12. Löding, C.: Reachability problems on regular ground tree rewriting graphs. Theory of Computing Systems 39(2), 347–383 (2006)
13. Mayr, R.: Process rewrite systems. Information and Computation 156(1-2), 264–286 (2000)
14. Salomaa, K.: Deterministic tree pushdown automata and monadic tree rewriting systems. Journal of Computer and System Sciences 37, 367–394 (1988)
15. Thomas, W.: Automata on infinite objects. In: van Leeuwen, J. (ed.) Handbook of Theoretical Computer Science, vol. B: Formal Models and Semantics, pp. 133–192. Elsevier, Amsterdam (1990)

Finite Automata, Palindromes, Powers, and Patterns

Terry Anderson, Narad Rampersad, Nicolae Santean[*], and Jeffrey Shallit

David R. Cheriton School of Computer Science
University of Waterloo
Waterloo, Ontario N2L 3G1, Canada
tanderson@uwaterloo.ca, nrampersad@cs.uwaterloo.ca
nsantean@iusb.edu, shallit@graceland.uwaterloo.ca

Abstract. Given a language L and a nondeterministic finite automaton M, we consider whether we can determine efficiently (in the size of M) if M accepts at least one word in L, or infinitely many words. Given that M accepts at least one word in L, we consider how long the shortest word can be. The languages L that we examine include the palindromes, the non-palindromes, the k-powers, the non-k-powers, the powers, the non-powers (also called primitive words), and words matching a general pattern.

1 Introduction

Let $L \subseteq \Sigma^*$ be a fixed language, and let M be a deterministic finite automaton (DFA) or nondeterministic finite automaton (NFA) with input alphabet Σ. In this paper we are interested in three questions:

1. Whether we can efficiently decide (in terms of the size of M) if $L(M)$ contains at least one element of L, that is, if $L(M) \cap L \neq \emptyset$;
2. Whether we can efficiently decide if $L(M)$ contains infinitely many elements of L, that is, if $L(M) \cap L$ is infinite;
3. Given that $L(M)$ contains at least one element of L, what is a good upper bound on the shortest element of $L(M) \cap L$?

As an example, consider the case where $\Sigma = \{a\}$, L is the set of primes written in unary, that is, $\{a^i : i \text{ is prime }\}$, and M is a NFA with n states.

To answer questions (1) and (2), we first rewrite M in Chrobak normal form [5]. Chrobak normal form consists of an NFA M' with a "tail" of $O(n^2)$ states, followed by a single nondeterministic choice to a set of disjoint cycles containing at most n states. Computing this normal form can be achieved in $O(n^5)$ steps by a result of Martinez [17].

Now we examine each of the cycles produced by this transformation. Each cycle accepts a finite union of sets of the form $(a^t)^* a^c$, where t is the size of

---

[*] Author's current address: Department of Computer and Information Sciences, Indiana University South Bend, 1700 Mishawaka Ave., P.O. Box 7111, South Bend, IN 46634, USA.

C. Martín-Vide, F. Otto, and H. Fernau (Eds.): LATA 2008, LNCS 5196, pp. 52–63, 2008.
© Springer-Verlag Berlin Heidelberg 2008

the cycle and $c \leq n^2 + n$; both t and c are given explicitly from M'. Now, by Dirichlet's theorem on primes in arithmetic progressions, $\gcd(t, c) = 1$ for at least one pair (t, c) induced by M' if and only if M accepts infinitely many elements of L. This can be checked in $O(n^2)$ steps, and so we get a solution to question (2) in polynomial time.

Question (1) requires a little more work. From our answer to question (2), we may assume that $\gcd(t, c) > 1$ for all pairs (t, c), for otherwise M accepts infinitely many elements of L and hence at least one element. Each element in such a set is of length $kt + c$ for some $k \geq 0$. Let $d = \gcd(t, c) \geq 2$. Then $kt + c = (kt/d + c/d)d$. If $k > 1$, this quantity is at least $2d$ and hence composite. Thus it suffices to check the primality of c and $t + c$, both of which are at most $n^2 + 2n$. We can precompute the primes $< n^2 + 2n$ in linear time using a modification of the sieve of Eratosthenes [18], and check if any of them are accepted. This gives a solution to question (1) in polynomial time.

On the other hand, answering question (3) essentially amounts to estimating the size of the least prime in an arithmetic progression, an extremely difficult question that is still not fully resolved [9], although it is known that there is a polynomial upper bound.

Thus we see that asking these questions, even for relatively simple languages L, can quickly take us to the limits of what is known in formal languages and number theory.

In this paper we examine questions (1)-(3) in the case where M is an NFA and L is either the set of palindromes, the set of k-powers, the set of powers, the set of words matching a general pattern, or their complements.

In some of these cases, there is previous work. For example, Ito et al. [12] studied several circumstances in which primitive words (non-powers) may appear in regular languages. As a typical result in [12], we mention: "A DFA over an alphabet of 2 or more letters accepts a primitive word iff it accepts one of length $\leq 3n - 3$, where n is the number of states of the DFA". Horváth, Karhumäki and Kleijn [11] addressed the decidability problem of whether a language accepted by an NFA is palindromic (i.e., every element is a palindrome). They showed that the language accepted by an NFA with n states is palindromic if and only if all its words of length shorter than $3n$ are palindromes.

A preliminary version of the full version of this paper is available online [2].

2   Notions and Notation

Let Σ be an alphabet, i.e., a nonempty, finite set of symbols (letters). By Σ^* we denote the set of all finite words over Σ, and by ε, the empty word. For $w \in \Sigma^*$, we denote by w^R the word obtained by reversing the order of symbols in w. A *palindrome* is a word w such that $w = w^R$. If L is a language over Σ, i.e., $L \subseteq \Sigma^*$, we say that L is *palindromic* if every word $w \in L$ is a palindrome.

Let $k \geq 2$ be an integer. A word y is a *k-power* if y can be written as $y = x^k$ for some non-empty word x. If y cannot be so written for any $k \geq 2$, then y is *primitive*. A 2-power is typically referred to as a *square*, and a 3-power as a *cube*.

Patterns are a generalization of powers. A *pattern* is a non-empty word p over a *pattern alphabet* Δ. The letters of Δ are called *variables*. A pattern p *matches* a word $w \in \Sigma^*$ if there exists a non-erasing morphism $h : \Delta^* \to \Sigma^*$ such that $h(p) = w$. Thus, a word w is a k-power if it matches the pattern a^k.

We define an NFA (or DFA) as the usual 5-tuple $M = (Q, \Sigma, \delta, q_0, F)$. The size of M is the total number N of its states and transitions. When we want to emphasize the components of M, we say M has n states and t transitions, and define $N := n + t$.

We note that if M is an NFA or NFA-ϵ, we can remove all states that either cannot be reached from the start state or cannot reach a final state (the latter are called *dead states*) in linear time (in the number of states and transitions) using depth-first search. We observe that $L(M) \neq \emptyset$ if and only if any states remain after this process, which can be tested in linear time. Similarly, if M is a NFA, then $L(M)$ is infinite if and only if the corresponding digraph has a directed cycle. This can also be tested in linear time.

We will also need the following well-known results [10]:

Theorem 1. *Let M be an NFA with n states. Then*

(a) $L(M) \neq \emptyset$ if and only if M accepts a word of length $< n$.
(b) $L(M)$ is infinite if and only if M accepts a word of length ℓ, $n \leq \ell < 2n$.

A language L is called *slender* if there is a constant C such that, for all $n \geq 0$, the number of words of length n in L is less than C. The following characterization of slender regular languages has been independently rediscovered several times in the past [14,24,19].

Theorem 2. *Let $L \subseteq \Sigma^*$ be a regular language. Then L is slender if and only if it can be written as a finite union of languages of the form uv^*w, where $u, v, w \in \Sigma^*$.*

For further background on finite automata and regular languages we refer the reader to Yu [26].

3   Testing If an NFA Accepts at Least One Palindrome

Over a unary alphabet, every string is a palindrome, so problems (1)-(3) become trivial. Let us assume, then, that the alphabet Σ contains at least two letters. Although the palindromes over such an alphabet are not regular, the language

$$L' = \{x \in \Sigma^* \ : \ xx^R \in L(M) \text{ or there exists } a \in \Sigma \text{ such that } xax^R \in L(M)\}$$

is, in fact, regular, as often shown in a beginning course in formal languages [10, p. 72, Exercise 3.4 (h)]. We can take advantage of this as follows:

Lemma 1. *Let M be an NFA with n states and t transitions. Then there exists an NFA M' with $n^2 + 1$ states and $\leq 2t^2$ transitions such that $L(M') = L'$.*

Corollary 1. *Given an NFA M with n states and t transitions, we can determine if M accepts a palindrome in $O(n^2 + t^2)$ time.*

Corollary 2. *Given an NFA M, we can determine if $L(M)$ contains infinitely many palindromes in quadratic time.*

Corollary 3. *If an NFA M accepts at least one palindrome, it accepts a palindrome of length $\leq 2n^2 - 1$.*

Rosaz [21] also gave a proof of this last corollary. The quadratic bound is tight, up to a multiplicative constant, in the case of alphabets with at least two letters, and even for DFAs:

Proposition 1. *For infinitely many n there exists a DFA M_n with n states over a 2-letter alphabet such that the shortest palindrome accepted by M_n is of length $\geq n^2/2 - 3n + 5$.*

4 Testing If an NFA Accepts at Least One Non-palindrome

In this section we consider the problem of deciding if an NFA accepts at least one non-palindrome. Equivalently, we consider the problem: *Given an NFA M, is $L(M)$ palindromic?*

Again, the problem is trivial for a unary alphabet, so we assume $|\Sigma| \geq 2$. Horváth, Karhumäki, and Kleijn [11] proved that the question is recursively solvable. In particular, they proved the following theorem:

Theorem 3. *$L(M)$ is palindromic if and only if $\{x \in L(M) : |x| < 3n\}$ is palindromic, where n is the number of states of M.*

For an NFA over an alphabet of at least 2 symbols, the $3n$ bound is easily seen to be optimal; for a DFA, however, the bound of $3n$ can be improved to $3n - 3$, and this is optimal.

While a naive implementation of Theorem 3 would take exponential time, in this section we show how to test palindromicity in polynomial time.

The main idea is to construct a "small" NFA M_t', for some integer $t > 1$, where no word in $L(M_t')$ is a palindrome, and M_t' accepts all non-palindromes of length $< t$ (in addition to some other non-palindromes). We omit the details of the construction (a similar construction appears in [25]).

Given an NFA M with n states, we now construct the cross-product with M_{3n}', and obtain an NFA A that accepts $L(M) \cap L(M_{3n}')$. By Theorem 3, $L(A) = \emptyset$ if and only if $L(M)$ is palindromic. We can determine if $L(A) = \emptyset$ in linear time. If M has n states and t transitions, then A has $O(n^2)$ states and $O(tn)$ transitions. Hence we have proved the following theorem.

Theorem 4. *Let M be an NFA with n states and t transitions. The algorithm sketched above determines whether M accepts a palindromic language in $O(n^2 + tn)$ time.*

In analogy with Corollary 2 and using a different construction than that of Theorem 4, we also have the following proposition.

Proposition 2. *Given an NFA M with n states and t transitions, we can determine in $O(n^2 + t^2)$ time if M accepts infinitely many non-palindromes.*

5   Testing If an NFA Accepts a Word Matching a Pattern

In this section we consider the computational complexity of the decision problem:

NFA PATTERN ACCEPTANCE
INSTANCE: An NFA M over the alphabet Σ and a pattern p over some alphabet Δ.
QUESTION: Does there exist $x \in \Sigma^+$ such that $x \in L(M)$ and x matches p?

Since the pattern p is given as part of the input, this problem is actually somewhat more general than the sort of problem formulated as Question 1 of the introduction, where the language L was fixed.

The following result was proved by Restivo and Salemi [20] (a more detailed proof appears in [4]).

Theorem 5 (Restivo and Salemi). *Let L be a regular language and let Δ be an alphabet. The set P_Δ of all non-empty patterns $p \in \Delta^*$ such that p matches a word in L is effectively regular.*

Observe that Theorem 5 implies the decidability of the **NFA PATTERN ACCEPTANCE** problem. It is possible to give a boolean matrix based proof of Theorem 5 (see Zhang [27] for a study of this boolean matrix approach to automata theory) that provides an explicit description of an NFA accepting P_Δ, but due to space constraints we omit this proof. However, the reader may perhaps deduce the argument from the proof of the following algorithmic result, which uses similar ideas.

Theorem 6. *The* **NFA PATTERN ACCEPTANCE** *problem is PSPACE-complete.*

Proof (sketch). We first show that the problem is in PSPACE. By Savitch's theorem [23] it suffices to give an NPSPACE algorithm. Let $M = (Q, \Sigma, \delta, q_0, F)$, where $Q = \{0, 1, \ldots, n-1\}$. For $a \in \Sigma$, let B_a be the $n \times n$ boolean matrix whose (i,j) entry is 1 if $j \in \delta(i, a)$ and 0 otherwise. Let $\mathcal{B}$ denote the semigroup generated by the B_a's. For $w = w_0 w_1 \cdots w_s \in \Sigma^*$, we write B_w to denote the matrix product $B_{w_0} B_{w_1} \cdots B_{w_s}$.

Let Δ be the set of letters occuring in p. We may suppose that $\Delta = \{1, 2, \ldots, k\}$. First, non-deterministically guess k boolean matrices $B_1, \ldots, B_k$. Next, for each i, verify that B_i is in the semigroup $\mathcal{B}$ by non-deterministically guessing a word w of length at most 2^{n^2} such that $B_i = B_w$. We guess w symbol-by-symbol and reuse space after perfoming each matrix multiplication while computing B_w. Then, if $p = p_0 p_1 \cdots p_r$, compute the matrix product $B = B_{p_0} B_{p_1} \cdots B_{p_r}$ and accept if and only if B describes an accepting computation of M.

To show hardness we reduce from the following PSPACE-complete problem [7, Problem AL6]. We leave the details to the reader.

DFA INTERSECTION

INSTANCE: An integer $k \geq 1$ and k DFAs $A_1, A_2, \ldots, A_k$, each over the alphabet Σ.

QUESTION: Does there exist $x \in \Sigma^*$ such that x is accepted by each A_i, $1 \leq i \leq k$? □

We may define various variations or special cases of the **NFA PATTERN ACCEPTANCE** problem, such as: **NFA ACCEPTS A k-POWER, NFA ACCEPTS INFINITELY MANY k-POWERS**, where each of these problems is defined in the obvious way. When k is part of the input (i.e., k is not fixed), these problems can be shown to be PSPACE-complete by a variation on the proof of Theorem 6. However, if k is fixed, both of these problems can be solved in polynomial time, as we now demonstrate.

Proposition 3. *Let M be an NFA with n states and t transitions, and set $N = n + t$, the size of M. For any fixed integer $k \geq 2$, there is an algorithm running in $O(n^{2k-1} t^k) = O(N^{2k-1})$ time to determine if M accepts a k-power.*

Proof (sketch). For a language $L \subseteq \Sigma^*$, we define $L^{1/k} = \{x \in \Sigma^* : x^k \in L\}$. It is well-known that if L is regular, then so is $L^{1/k}$. We leave it to the reader to verify that an NFA-ϵ M' accepting $L^{1/k}$ can be constructed with $n^{2k-1} + 1$ states and at most t^k distinct transitions. Testing whether or not $L(M')$ accepts a non-empty word can be done in linear time, so the running time of our algorithm is $O(n^{2k-1} t^k)$. □

Corollary 4. *We can decide if an NFA M with n states and t transitions accepts infinitely many k-powers in $O(n^{2k-1} t^k)$ time.*

We may also consider the problems **NFA ACCEPTS A $\geq k$-POWER** and **NFA ACCEPTS INFINITELY MANY $\geq k$-POWERS**, again defined in the obvious way. Here, even for fixed k, these problems are both PSPACE-complete. Setting $k = 2$ corresponds to the problems **NFA ACCEPTS A POWER** and **NFA ACCEPTS INFINITELY MANY POWERS**, so we see that both these problems are PSPACE-complete as well.

To show PSPACE-hardness for the "infinitely many" problems, we reduce from the **DFA INTERSECTION INFINITENESS** problem, which is defined similarly to the **DFA INTERSECTION** problem, except that we now ask if there are infinitely many words x such that x is accepted by each A_i. This problem is easily seen to be PSPACE-complete as well.

6 Testing If an NFA Accepts a Non-k-Power

In the previous section we showed that it is computationally hard to test if an NFA accepts a k-power (when k is not fixed). In this section we show how to efficiently test if an NFA accepts a non-k-power. Again, we find it more congenial to discuss the opposite problem, which is whether an NFA accepts nothing but k-powers.

First, we need some classical results from combinatorics on words.

Theorem 7 (Lyndon and Schützenberger [15]). *If x, y, and z are words satisfying an equation $x^i y^j = z^k$, where $i, j, k \geq 2$, then they are all powers of a common word.*

Theorem 8 (Lyndon and Schützenberger [15]). *Let u and v be non-empty words. If $uv = vu$, then there exists a word x and integers $i, j \geq 1$, such that $u = x^i$ and $v = x^j$. In other words, u and v are powers of a common word.*

We include here the following combinatorial result, which, when applied to words in a regular language, gives a sort of "pumping lemma" for powers in a regular language.

Proposition 4. *Let u, v, and w be words, $v \neq \varepsilon$, and let $f, g \geq 1$ be integers, $f \neq g$. If $uv^f w$ and $uv^g w$ are non-primitive, then $uv^n w$ is non-primitive for all integers $n \geq 1$. Further, if uvw and $uv^2 w$ are k-powers for some integer $k \geq 2$, then v and $uv^n w$ are k-powers for all integers $n \geq 1$.*

The following result is an analogue of Theorem 3, from which we will derive an efficient algorithm for testing if a finite automaton accepts only k-powers.

Theorem 9. *Let L be accepted by an n-state NFA M and let $k \geq 2$ be an integer.*

1. *Every word in L is a k-power if and only if every word in the set $\{x \in L : |x| \leq 3n\}$ is a k-power.*
2. *All but finitely many words in L are k-powers if and only if every word in the set $\{x \in L : n \leq |x| \leq 3n\}$ is a k-power.*

Further, if M is a DFA over an alphabet of size ≥ 2, then the bound $3n$ may be replaced by $3n - 3$.

Ito et al. [12] proved a similar result for primitive words: namely, that if L is accepted by an n-state DFA over an alphabet of two or more letters and contains a primitive word, then it contains a primitive word of length $\leq 3n - 3$. In other words, every word in L is a power if and only if every word in the set $\{x \in L : |x| \leq 3n - 3\}$ is a power.

The proof of Theorem 9 is similar to that of [12, Proposition 7], albeit with some additional complications. We shall give a complete proof in the full version of this paper.

The characterization due to Ito et al. [12, Proposition 10] (see also Dömösi, Horváth, and Ito [6, Theorem 3]) of the regular languages consisting only of powers, along with Theorem 2, implies that any such language is slender. A simple application of the Myhill–Nerode Theorem gives the following weaker result.

Proposition 5. *Let L be a regular language and let $k \geq 2$ be an integer. If all but finitely many words of L are k-powers, then L is slender. In particular, if L is accepted by an n-state DFA and all words in L of length $\geq \ell$ are k-powers, then for all $r \geq \ell$, the number of words in L of length r is at most n.*

The following characterization is analogous to the characterization of palindromic regular languages given in [11, Theorem 8], and follows from Proposition 5, Theorem 2, and the (omitted) proof of Proposition 4.

Theorem 10. *Let $L \subseteq \Sigma^*$ be a regular language and let $k \geq 2$ be an integer. The language L consists only of k-powers if and only if it can be written as a finite union of languages of the form uv^*w, where $u, v, w \in \Sigma^*$ satisfy the following: there exists a primitive word $x \in \Sigma^*$ and integers $i, j \geq 0$ such that $v = x^{ik}$ and $wu = x^{jk}$.*

Next we apply Theorem 9 to deduce the following algorithmic result.

Theorem 11. *Let $k \geq 2$ be an integer. Given an NFA M with n states and t transitions, it is possible to determine if every word in $L(M)$ is a k-power in $O(n^3 + tn^2)$ time.*

Proof (sketch). We create an NFA, M'_r, for $r = 3n$, such that no word in $L(M'_r)$ is a k-power, and M'_r accepts all non-k-powers of length $\leq r$ (and perhaps some other non-k-powers).

Note that we may assume that $k \leq r$. If $k > r$, then no word of length $\leq r$ is a k-power. In this case, to obtain the desired answer it suffices to test if the set $\{x \in L(M) : |x| \leq r\}$ is empty. However, this set is empty if and only if $L(M)$ is empty, and this is easily verified in linear time.

We now form a new NFA A as the cross product of M'_r with M. From Theorem 9, it follows that $L(A) = \emptyset$ iff every word in $L(M)$ is a k-power. Again, we can determine if $L(A) = \emptyset$ in linear time.

We omit the details of the construction of M'_r, noting only that M'_r can be constructed to have at most $O(r^2)$ states and $O(r^2)$ transitions. After constructing the cross-product, this gives a $O(n^3 + tn^2)$ bound on the time required to determine if every word in $L(M)$ is a k-power.   □

Theorem 9 suggests the following question: if M is an NFA with n states that accepts at least one non-k-power, how long can the shortest non-k-power be? Theorem 9 proves an upper bound of $3n$. A lower bound of $2n - 1$ for infinitely many n follows easily from the obvious $(n + 1)$-state NFA accepting $\mathsf{a}^n(\mathsf{a}^{n+1})^*$, where n is divisible by k. However, Ito et al. [12] gave a very interesting example that improves this lower bound: if $x = ((ab)^n a)^2$ and $y = baxab$, then x and xyx

are squares, but $xyxyx$ is not a power. Hence, the obvious $(8n + 8)$-state NFA that accepts $x(yx)^*$ has the property that the shortest non-k-power accepted is of length $20n + 18$. We generalize this lower bound by defining x and y as follows: let $u = (ab)^n a$, $x = u^k$, and $y = x^{-1}(xbau^{-1}x)^k x^{-1}$. We leave it to the reader to deduce the following result.

Proposition 6. *Let $k \geq 2$ be fixed. There exist infinitely many NFAs M with the property that if M has r states, then the shortest non-k-power accepted is of length $(2 + \frac{1}{2k-2})r - O(1)$.*

We may also apply part (2) of Theorem 9 to obtain an algorithm to check if an NFA accepts infinitely many non-k-powers.

Theorem 12. *Let $k \geq 2$ be an integer. Given an NFA M with n states and t transitions, it is possible to determine if all but finitely many words in $L(M)$ are k-powers in $O(n^3 + tn^2)$ time.*

7 Automata Accepting Only Powers

We now move from the problem of testing if an automaton accepts only k-powers to that of testing if it accepts only powers (of any kind). Just as Theorem 9 was the starting point for our algorithmic results in Section 6, the following theorem of Ito et al. [12] (stated here in a slightly stronger form than in the original) is the starting point for our algorithmic results in this section.

Theorem 13. *Let L be accepted by an n-state NFA M.*

1. *Every word in L is a power if and only if every word in the set $\{x \in L : |x| \leq 3n\}$ is a power.*
2. *All but finitely many words in L are powers if and only if every word in the set $\{x \in L : n \leq |x| \leq 3n\}$ is a power.*

Further, if M is a DFA over an alphabet of size ≥ 2, then the bound $3n$ may be replaced by $3n - 3$.

We next prove an analogue of Proposition 5. We need the following result, first proved by Birget [3], and later, independently, in a weaker form, by Glaister and Shallit [8].

Theorem 14. *Let $L \subseteq \Sigma^*$ be a regular language. Suppose there exists a set of pairs $S = \{(x_i, y_i) \in \Sigma^* \times \Sigma^* : 1 \leq i \leq n\}$ such that: (a) $x_i y_i \in L$ for $1 \leq i \leq n$; and (b) either $x_i y_j \notin L$ or $x_j y_i \notin L$ for $1 \leq i,j \leq n$, $i \neq j$. Then any NFA accepting L has at least n states.*

Proposition 7. *Let M be an n-state NFA and let ℓ be a non-negative integer such that every word in $L(M)$ of length $\geq \ell$ is a power. For all $r \geq \ell$, the number of words in $L(M)$ of length r is at most $7n$.*

Proof. We give the proof in full, as it illustrates an unusual and unexpected combination of techniques from both the theory of non-deterministic state complexity as well as the theory of combinatorics on words.

Let $r \geq \ell$ be an arbitrary integer. The proof consists of three steps.

Step 1. We consider the set A of words w in $L(M)$ such that $|w| = r$ and w is a k-power for some $k \geq 4$. For each such w, write $w = x^i$, where x is a primitive word, and define a pair (x^2, x^{i-2}). Let S_A denote the set of such pairs. Consider two pairs in S_A: (x^2, x^{i-2}) and (y^2, y^{j-2}). The word $x^2 y^{j-2}$ is primitive by Theorem 7 and hence is not in $L(M)$. The set S_A thus satisfies the conditions of Theorem 14. Since $L(M)$ is accepted by an n-state NFA, we must have $|S_A| \leq n$ and thus $|A| \leq n$.

Step 2. Next we consider the set B of cubes of length r in $L(M)$. For each such cube $w = x^3$, we define a pair (x, x^2). Let S_B denote the set of such pairs. Consider two pairs in S_B: (x, x^2) and (y, y^2). Suppose that xy^2 and yx^2 are both in $L(M)$. The word xy^2 is certainly not a cube; we claim that it cannot be a square. Suppose it were. Then $|x|$ and $|y|$ are even, so we can write $x = x_1 x_2$ and $y = y_1 y_2$ where $|x_1| = |x_2| = |y_1| = |y_2|$. Now if $xy^2 = x_1 x_2 y_1 y_2 y_1 y_2$ is a square, then $x_1 x_2 y_1 = y_2 y_1 y_2$, and so $y_1 = y_2$. Thus y is a square; write $y = z^2$. By Theorem 7, $yx^2 = z^2 x^2$ is primitive, contradicting our assumption that $yx^2 \in L(M)$. It must be the case then that xy^2 is a k-power for some $k \geq 4$. Thus, $xy^2 = u^k$ for some primitive u uniquely determined by x and y. With each pair of cubes x^3 and y^3 such that both xy^2 and yx^2 are in $L(M)$ we may therefore associate a k-power $u^k \in L(M)$, where $k \geq 4$. We have already established in Step 1 that the number of such k-powers is at most n. It follows that by deleting at most n pairs from the set S_B we obtain a set of pairs satisfying the conditions of Theorem 14. We must therefore have $|S_B| \leq 2n$ and thus $|B| \leq 2n$.

Step 3. Finally we consider the set C of squares of length r in $L(M)$. For each such square $w = x^2$, we define a pair (x, x). Let S_C denote the set of such pairs. Consider two pairs in S_C: (x, x) and (y, y). Suppose that xy and yx are both in $L(M)$. The word xy is not a square and must therefore be a k-power for some $k \geq 3$. We write $xy = u^k$ for some primitive u uniquely determined by x and y. In Steps 1 and 2 we established that the number of k-powers of length r, $k \geq 3$, is $|A| + |B| \leq 3n$. It follows that by deleting at most $3n$ pairs from the set S_C we obtain a set of pairs satisfying the conditions of Theorem 14. We must therefore have $|S_C| \leq 4n$ and thus $|C| \leq 4n$.

Putting everything together, we see that there are $|A| + |B| + |C| \leq 7n$ words of length r in $L(M)$, as required.   □

The bound of $7n$ in Proposition 7 is almost certainly not optimal. We now prove the following algorithmic result.

Theorem 15. *Given an NFA M with n states, it is possible to determine if every word in $L(M)$ is a power in $O(n^5)$ time.*

Proof (sketch). Checking if a word is a power can be done in linear time using the Knuth-Morris-Pratt algorithm [13]. By Theorem 13 and Proposition 7 it suffices to enumerate the words in $L(M)$ of lengths $1, 2, \ldots, 3n$, stopping if the number of such words in any length exceeds $7n$. If all these words are powers, then every word is a power. Otherwise, if we find a non-power, or if the number of words in any length exceeds $7n$, then not every word is a power. By the work of Mäkinen [16] or Ackerman & Shallit [1], we can enumerate these words in $O(n^5)$ time.   □

Using part (2) of Theorem 13 along with Proposition 7, one obtains the following in a similar manner.

Theorem 16. *Given an NFA M with n states, we can decide if all but finitely many words in $L(M)$ are non-powers in $O(n^5)$ time.*

8   Final Remarks

In this paper we examined the complexity of checking various properties of regular languages, such as consisting only of palindromes, containing at least one palindrome, consisting only of powers, or containing at least one power. In each case, we were able to provide an efficient algorithm or show that the problem is likely to be hard. Our results are summarized in the following table. We also report some upper and lower bounds on the length of a shortest palindrome, k-power, etc., accepted by an NFA; due to space constraints we must omit the proofs of these bounds. Here M is an NFA with n states and t transitions.

L	decide if $L(M) \cap L = \emptyset$	decide if $L(M) \cap L$ infinite	upper bound on shortest element of $L(M) \cap L$	worst-case lower bound known
palindromes	$O(n^2 + t^2)$	$O(n^2 + t^2)$	$2n^2 - 1$	$\frac{n^2}{2} - 3n + 5$
non-palindromes	$O(n^2 + tn)$	$O(n^2 + t^2)$	$3n - 1$	$3n - 1$
k-powers (k fixed)	$O(n^{2k-1}t^k)$	$O(n^{2k-1}t^k)$	kn^k	$\Omega(n^k)$
k-powers (k part of input)	PSPACE-complete	PSPACE-complete		
non-k-powers	$O(n^3 + tn^2)$	$O(n^3 + tn^2)$	$3n$	$(2 + \frac{1}{2k-2})n - O(1)$
powers	PSPACE-complete	PSPACE-complete	$(n+1)n^{n+1}$	$e^{\Omega(\sqrt{n \log n})}$
non-powers	$O(n^5)$	$O(n^5)$	$3n$	$\frac{5}{2}n - 2$

References

1. Ackerman, M., Shallit, J.: Efficient enumeration of regular languages. In: Holub, J., Žďárek, J. (eds.) CIAA 2007. LNCS, vol. 4783, pp. 226–242. Springer, Heidelberg (2007)
2. Anderson, T., Rampersad, N., Santean, N., Shallit, J.: Finite automata, palindromes, patterns, and borders, http://www.arxiv.org/abs/0711.3183

3. Birget, J.-C.: Intersection and union of regular languages and state complexity. Inform. Process. Lett. 43, 185–190 (1992)
4. Castiglione, G., Restivo, A., Salemi, S.: Patterns in words and languages. Disc. Appl. Math. 144, 237–246 (2004)
5. Chrobak, M.: Finite automata and unary languages. Theoret. Comput. Sci. 47, 149–158 (1986); Errata 302, 497–498 (2003)
6. Dömösi, P., Horváth, G., Ito, M.: A small hierarchy of languages consisting of non-primitive words. Publ. Math (Debrecen) 64, 261–267 (2004)
7. Garey, M., Johnson, D.: Computers and Intractability. Freeman, New York (1979)
8. Glaister, I., Shallit, J.: A lower bound technique for the size of nondeterministic finite automata. Inform. Process. Lett. 59, 75–77 (1996)
9. Heath-Brown, D.R.: Zero-free regions for Dirichlet L-functions, and the least prime in an arithmetic progression. Proc. Lond. Math. Soc. 64, 265–338 (1992)
10. Hopcroft, J.E., Ullman, J.D.: Introduction to Automata Theory, Languages, and Computation. Addison-Wesley, Reading (1979)
11. Horváth, S., Karhumäki, J., Kleijn, J.: Results concerning palindromicity. J. Inf. Process. Cybern. EIK 23, 441–451 (1987)
12. Ito, M., Katsura, M., Shyr, H.J., Yu, S.S.: Automata accepting primitive words. Semigroup Forum 37, 45–52 (1988)
13. Knuth, D., Morris Jr., J., Pratt, V.: Fast pattern matching in strings. SIAM J. Computing 6, 323–350 (1977)
14. Kunze, M., Shyr, H.J., Thierrin, G.: h-bounded and semi-discrete languages. Information and Control 51, 147–187 (1981)
15. Lyndon, R.C., Schützenberger, M.-P.: The equation $a^m = b^n c^p$ in a free group. Michigan Math. J. 9, 289–298 (1962)
16. Mäkinen, E.: On lexicographic enumeration of regular and context-free languages. Acta Cybernetica 13, 55–61 (1997)
17. Martinez, A.: Efficient computation of regular expressions from unary NFAs. In: DCFS 2002, pp. 174–187 (2002)
18. Pritchard, P.: Linear prime-number sieves: a family tree. Sci. Comput. Programming 9, 17–35 (1987)
19. Păun, G., Salomaa, A.: Thin and slender languages. Disc. Appl. Math. 61, 257–270 (1995)
20. Restivo, A., Salemi, S.: Words and patterns. In: Kuich, W., Rozenberg, G., Salomaa, A. (eds.) DLT 2001. LNCS, vol. 2295, pp. 215–218. Springer, Heidelberg (2002)
21. Rosaz, L.: Puzzle corner, #50. Bull. European Assoc. Theor. Comput. Sci. 76, 234 (February 2002); Solution 77, 261 (June 2002)
22. Rozenberg, G., Salomaa, A.: Handbook of Formal Languages. Springer, Berlin (1997)
23. Savitch, W.: Relationships between nondeterministic and deterministic tape complexities. J. Comput. System Sci. 4, 177–192 (1970)
24. Shallit, J.: Numeration systems, linear recurrences, and regular sets. Inform. Comput. 113, 331–347 (1994)
25. Shallit, J., Breitbart, Y.: Automaticity I: Properties of a measure of descriptional complexity. J. Comput. System Sci. 53, 10–25 (1996)
26. Yu, S.: Regular languages. In: Handbook of Formal Languages, Ch. 2, pp. 41–110 (1997)
27. Zhang, G.-Q.: Automata, Boolean matrices, and ultimate periodicity. Inform. Comput. 152, 138–154 (1999)

One-Dimensional Quantum Cellular Automata over Finite, Unbounded Configurations

Pablo Arrighi[1], Vincent Nesme[2], and Reinhard Werner[2]

[1] Université de Grenoble, LIG, 46 Avenue Félix Viallet,
38031 Grenoble Cedex, France
[2] Technische Universität Braunschweig, IMAPH, Mendelssohnstr. 3, 38106
Braunschweig, Germany

Abstract. One-dimensional quantum cellular automata (QCA) consist in a line of identical, finite dimensional quantum systems. These evolve in discrete time steps according to a causal, shift-invariant unitary evolution. By causal we mean that no instantaneous long-range communication can occur. In order to define these over a Hilbert space we must restrict to a base of finite, yet unbounded configurations. We show that QCA always admit a two-layered block representation, and hence the inverse QCA is again a QCA. This is a striking result since the property does not hold for classical one-dimensional cellular automata as defined over such finite configurations. As an example we discuss a bijective cellular automata which becomes non-causal as a QCA, in a rare case of reversible computation which does not admit a straightforward quantization. We argue that a whole class of bijective cellular automata should no longer be considered to be reversible in a physical sense. Note that the same two-layered block representation result applies also over infinite configurations, as was previously shown for one-dimensional systems in the more elaborate formalism of operators algebras [13]. Here the proof is simpler and self-contained, moreover we discuss a counterexample QCA in higher dimensions.

One-dimensional cellular automata (CA) consist in a line of cells, each of which may take one in a finite number of possible states. These evolve in discrete time steps according to a causal, shift-invariant function. When defined over infinite configurations, the inverse of a bijective CA is then itself a CA, and this structural reversibility leads to a natural block decomposition of the CA. None of this holds over finite, yet possibly unbounded, configurations.

Because CA are a physics-like model of computation it seems very natural to study their quantum extensions. The flourishing research in quantum information and quantum computer science provides us with appropriate context for doing so, both in terms of the potential implementation and the theoretical framework. Right from the very birth of the field with Feynman's 1986 paper, it was hoped that QCA may prove an important path to realistic implementations of quantum computers [8] – mainly because they eliminate the need for an external, classical control and hence the principal source of decoherence.

C. Martín-Vide, F. Otto, and H. Fernau (Eds.): LATA 2008, LNCS 5196, pp. 64–75, 2008.

Other possible aims include providing models of distributed quantum computation, providing bridges between computer science notions and modern theoretical physics, or anything like understanding the dynamics of some quantum physical system in discrete spacetime, i.e. from an idealized viewpoint. Studying QCA rather than quantum Turing machines for instance means we bother about the spatial structure or the spatial parallelism of things [2], for the purpose of describing a quantum protocol, or to model a quantum physical phenomena [12].

One-dimensional quantum cellular automata (QCA) consist in a line of identical, finite dimensional quantum systems. These evolve in discrete time steps according to a causal, shift-invariant unitary evolution. By causal we mean that information cannot be transmitted faster than a fixed number of cells per time step. Because the standard mathematical setting for quantum mechanics is the theory of Hilbert spaces, we must exhibit and work with a countable basis for our vectorial space. This is the reason why we only consider finite, unbounded configurations. An elegant alternative to this restriction is to abandon Hilbert spaces altogether and use the more abstract mathematical setting of C^*-algebras [4] – but here we want our proofs to be self-contained and accessible to the Computer Science community. Our main result is that QCA can always be expressed as two layers of an infinitely repeating unitary gate even over such finite configurations. The existence of such a two-layered block representation implies of course that the inverse QCA is again a QCA. Our proof method is mainly a drastic simplification of that of the same theorem over infinite configurations, adapted to finite unbounded configurations. Moreover in its present form the theorem over infinite configurations is stated for n-dimensions [13], which we prove is incorrect by presenting a two-dimensional QCA which does not admit a two-layered block representation.

It is a rather striking fact however that QCA admit the two-layered block representation in spite of their being defined over finite, unbounded configurations. For most purposes this saves us from complicated unitary tests such as [6,7,1]. But more importantly notice how this is clearly not akin to the classical case, where a CA may be bijective over such finite configurations, and yet not structurally reversible. In order to clarify this situation we consider a perfectly valid, bijective CA but whose inverse function is not a CA. It then turns out that its quantum version is no longer valid, as it allows superluminal signalling. Hence whilst we are used to think that any reversible computation admits a trivial quantization, this turns out not to be the case in the realm of cellular automata. Curiously the nonlocality of quantum states (entanglement) induces more structure upon the cellular automata – so that its evolution may remain causal as an operation (no superluminal signalling). Based upon these remarks we prove that an important, well-studied class of bijective CA may be dismissed as not physically reversible.

Outline. We provide a simple axiomatic presentation of QCA (Section 1). We reorganize a number of known mathematical results around the notion of subsystems in quantum theory (Section 2). Thanks to this small theory we prove the reversibility/block structure theorem in an elementary manner (Section 3).

In the discussion we show why the theorem does not hold as such in further dimensions; we exhibit superluminal signalling in the XOR quantum automata, and end with a general theorem discarding all injective, non surjective CA over infinite configurations as unphysical (Section 4).

Note that all proofs are omitted in this version of the paper, but all of them are available in full the longer version of the paper [3].

1   Axiomatics of QCA

We will now introduce the basic definitions of one-dimensional QCA.

In what follows Σ will be a fixed finite set of symbols (i.e. 'the alphabet', describing the possible basic states each cell may take) and q is a symbol such that $q \notin \Sigma$, which will be known as 'the quiescent symbol', which represents an empty cells. We write $q + \Sigma = \{q\} \cup \Sigma$ for short.

Definition 1 (finite configurations).
A (finite) configuration c over $q + \Sigma$ is a function $c : \mathbb{Z} \longrightarrow q + \Sigma$, with $i \longmapsto c(i) = c_i$, such that there exists a (possibly empty) interval I verifying $i \in I \Rightarrow c_i \in q + \Sigma$ and $i \notin I \Rightarrow c_i = q$. The set of all finite configurations over $\{q\} \cup \Sigma$ will be denoted $\mathcal{C}_f$.

Whilst configurations hold the basic states of an entire line of cells, and hence denote the possible basic states of the entire QCA, the global state of a QCA may well turn out to be a superposition of these. The following definition works because $\mathcal{C}_f$ is a countably infinite set.

Definition 2 (superpositions of configurations).
Let $\mathcal{H}_{\mathcal{C}_f}$ be the Hilbert space of configurations, defined as follows. To each finite configuration c is associated a unit vector $|c\rangle$, such that the family $(|c\rangle)_{c \in \mathcal{C}_f}$ is an orthonormal basis of $\mathcal{H}_{\mathcal{C}_f}$. A superposition of configurations is then a unit vector in $\mathcal{H}_{\mathcal{C}_f}$.

This space of QCA configurations is the same one as in [16,6,7,1]. It is isomorphic to the cyclic one considered in [11], but fundamentally different from the finite, bounded periodic space of [15] and the infinite setting of [13].

Definition 3 (Unitarity).
A linear operator $G : \mathcal{H}_{\mathcal{C}_f} \longrightarrow \mathcal{H}_{\mathcal{C}_f}$ is unitary if and only if $\{G|c\rangle \mid c \in \mathcal{C}_f\}$ is an orthonormal basis of $\mathcal{H}_{\mathcal{C}_f}$.

Definition 4 (Shift-invariance).
Consider the shift operation which takes configuration $c = \ldots c_{i-1} c_i c_{i+1} \ldots$ to $c' = \ldots c'_{i-1} c'_i c'_{i+1} \ldots$ where for all i $c'_i = c_{i+1}$. Let $\sigma : \mathcal{H}_{\mathcal{C}_f} \longrightarrow \mathcal{H}_{\mathcal{C}_f}$ be its linear extension to superpositions of configurations. A linear operator $G : \mathcal{H}_{\mathcal{C}_f} \longrightarrow \mathcal{H}_{\mathcal{C}_f}$ is said to be shift invariant if and only if $G\sigma = \sigma G$.

Definition 5 (Causality).

A linear operator $G : \mathcal{H}_{\mathcal{C}_f} \longrightarrow \mathcal{H}_{\mathcal{C}_f}$ *is said to be* causal *with radius* $\frac{1}{2}$ *if and only if for any* ρ, ρ' *two states over* $\mathcal{H}_{\mathcal{C}_f}$, *and for any* $i \in \mathbb{Z}$, *we have*

$$\rho|_{i,i+1} = \rho'|_{i,i+1} \quad \Rightarrow G\rho G^{\dagger}|_i = G\rho' G^{\dagger}|_i. \tag{1}$$

where we have written $A|_S$ *for the matrix* $Tr_{\bar{S}}(A)$, *i.e. the partial trace obtained from* A *once all of systems that are not in* S *have been traced out.*

In the classical case, the definition would be that the letter to be read in some given cell i at time $t + 1$ depends only on the state of the cells i and $i + 1$ at time t. This seemingly restrictive definition of causality is known in the classical case as a $\frac{1}{2}$-neighborhood cellular automaton. This is because the most natural way to represent such an automaton is to shift the cells by $\frac{1}{2}$ at each step, so that the state of a cell depends on the state of the two cells under it, as shown in figure 1. This definition of causality is actually not so restrictive, since by grouping cells into 'supercells' one can construct a $\frac{1}{2}$-neighborhood CA simulating the first one. The same thing can easily be done for QCA, so that this definition of causality is essentially done without loss of generality. Transposed to a quantum setting, we get the above definition: to know the state of cell number i, we only need to know the state of cells i and $i + 1$ before the evolution.

We are now set to give the formal definition of one-dimensional quantum cellular automata.

Definition 6 (QCA).

A one-dimensional quantum cellular automaton (QCA) is an operator $G : \mathcal{H}_{\mathcal{C}_f} \longrightarrow \mathcal{H}_{\mathcal{C}_f}$ *which is unitary, shift-invariant and causal.*

This is clearly the natural axiomatic quantization of the notion of cellular automata. An almost equivalent definition in the litterature is phrased in terms of homomorphism of a C^*-algebra [13]. On the other hand the definitions in [16,6,7,11,15,1,5] are not axiomatic, in the sense that they all make particular assumptions about the form of the local action of G, and G is then defined as a composition of these actions. The present work justifies some of these assumptions [5] to some extent.

The next theorem provides us with another characterization of causality, more helpful in the proofs. But more importantly it entails structural reversibility, i.e. the fact that the inverse function of a QCA is also a QCA. This theorem works for n-dimensional QCA as well as one. We are not aware of a rigorous proof of this fact for n-dimensional QCA in the previous litterature.

Theorem 1 (Structural reversibility)

Let G *be a unitary operator of* $\mathcal{H}_{\mathcal{C}_f}$ *and* $\mathcal{N}$ *a finite subset of* $\mathbb{Z}$. *The four properties are equivalent:*

(i) *For every states* ρ *and* ρ' *over the finite configurations, if* $\rho|_{\mathcal{N}} = \rho'|_{\mathcal{N}}$ *then* $\left(G\rho G^{\dagger}\right)|_0 = \left(G\rho' G^{\dagger}\right)|_0$.

(ii) *For every operator* A *localized on cell* 0, *then* $G^{\dagger}AG$ *is localized on the cells in* $\mathcal{N}$.

(iii) For every states ρ and ρ' over the finite configurations, if $\rho|_{-\mathcal{N}} = \rho'|_{-\mathcal{N}}$ then $\left(G^\dagger \rho G\right)|_0 = \left(G^\dagger \rho' G\right)|_0$.

(iv) For every operator A localized on cell 0, then $GAG^\dagger$ is localized on the cells in $-\mathcal{N}$.

When G satisfies these properties, we say that G is causal at 0 with neighbourhood $\mathcal{N}$. Here $-\mathcal{N}$ take the opposite of each of the elements of $\mathcal{N}$.

2   A Small Theory of Subsystems

The purpose of this section is to provide a series of mathematical results about 'When can something be considered a subsystem in quantum theory?'. Let us work towards making this sentence more precise. The 'something' will be an matrix algebra (a C^*-algebra over a finite-dimensional system):

Definition 7 (Algebras).
Consider $\mathcal{A} \subseteq M_n(\mathbb{C})$. We say that $\mathcal{A}$ is an algebra of $M_n(\mathbb{C})$ if and only if it is closed under weighting by a scalar (.), addition (+), matrix multiplication (∗), adjoint (†). Moreover for any S a subset of $M_n(\mathbb{C})$, we denote by curly S its closure under the above-mentioned operations.

The key issue here is that the notion of subsystem is usually a base-dependent one, i.e. one tends to say that $\mathcal{A}$ is a subsystem if $\mathcal{A} = M_p(\mathbb{C}) \otimes \mathbb{I}_q$, but this depends on a particular choice of basis/tensor decomposition. Let us make the definition base-independent, artificially at first.

Definition 8 (Subsystem algebras).
Consider $\mathcal{A}$ an algebra of $M_n(\mathbb{C})$. We say that $\mathcal{A}$ is a subsystem algebra of $M_n(\mathbb{C})$ if and only if there exists $p, q \in \mathbb{N}\,/\,pq = n$ and $U \in M_n(\mathbb{C})\,/\,U^\dagger U = UU^\dagger = \mathbb{I}_n$ such that $U\mathcal{A}U^\dagger = M_p(\mathbb{C}) \otimes \mathbb{I}_q$.

We now work our way towards simple characterizations of subsystem algebras.

Definition 9 (Center algebras).
For $\mathcal{A}$ an algebra of $M_n(\mathbb{C})$, we note $\mathcal{C}_\mathcal{A} = \{A \in \mathcal{A} \mid \forall B \in \mathcal{A}\; BA = AB\}$. $\mathcal{C}_\mathcal{A}$ is also an algebra of $M_n(\mathbb{C})$, which is called the center algebra of $\mathcal{A}$.

Theorem 2 (Characterizing one subsystem)
Let $\mathcal{A}$ be an algebra of $M_n(\mathbb{C})$ and $\mathcal{C}_\mathcal{A} = \{A \in \mathcal{A} \mid \forall B \in \mathcal{A}\; BA = AB\}$ its center algebra. Then $\mathcal{A}$ is a subsystem algebra if and only if $\mathcal{C}_\mathcal{A} = \mathbb{C}\mathbb{I}_n$.

Next we give two simple conditions for some algebras $\mathcal{A}$ and $\mathcal{B}$ to be splitted as a tensor product, namely commutation and generacy.

Theorem 3 (Characterizing several subsystems)
Let $\mathcal{A}$ and $\mathcal{B}$ be commuting algebras of $M_n(\mathbb{C})$ such that $\mathcal{A}\mathcal{B} = M_n(\mathbb{C})$. Then there exists a unitary matrix U such that, $U\mathcal{A}U^\dagger$ is $M_p(\mathbb{C}) \otimes \mathbb{I}_q$ and $U\mathcal{B}U^\dagger$ is $\mathbb{I}_p \otimes M_q(\mathbb{C})$, with $pq = n$.

Often however we want to split some algebras $\mathcal{A}$ and $\mathcal{B}$ as a tensor product, not over the union of the subsystems upon which they act, but over the intersection of the subsystems upon which they act. The next definition and two lemmas will place us in a position to do so.

Definition 10 (Restriction Algebras).
Consider $\mathcal{A}$ an algebra of $M_p(\mathbb{C}) \otimes M_q(\mathbb{C}) \otimes M_r(\mathbb{C})$. For A an element of $\mathcal{A}$, we write $A|_1$ for the matrix $Tr_{02}(A)$, i.e. the partial trace obtained from A once systems 0 and 2 have been traced out. Similarly so we call $\mathcal{A}|_1$ the restriction of $\mathcal{A}$ to the middle subsystem, i.e. the algebra generated by $\{\, Tr_{02}(A) \mid A \in \mathcal{A}\}$.

Indeed when we restrict our commuting algebras to the subsystem they have in common, their restrictions still commute.

Lemma 1 (Restriction of commuting algebras).
Consider $\mathcal{A}$ an algebra of $M_p(\mathbb{C}) \otimes M_q(\mathbb{C}) \otimes \mathbb{I}_r$ and $\mathcal{B}$ an algebra of $\mathbb{I}_p \otimes M_q(\mathbb{C}) \otimes M_r(\mathbb{C})$. Suppose $\mathcal{A}$ and $\mathcal{B}$ commute. Then so do $\mathcal{A}|_1$ and $\mathcal{B}|_1$.

Moreover when we restrict our generating algebras to the subsystem they have in common, theirs restrictions generate the subsystem.

Lemma 2 (Restriction of generating algebras).
Consider $\mathcal{A}$ an algebra of $M_p(\mathbb{C}) \otimes M_q(\mathbb{C}) \otimes \mathbb{I}_r$ and $\mathcal{B}$ an algebra of $\mathbb{I}_p \otimes M_q(\mathbb{C}) \otimes M_r(\mathbb{C})$. Suppose $\mathcal{A}\mathcal{B}|_1 = M_p(\mathbb{C})$. Then we have that $\mathcal{A}|_1 \mathcal{B}|_1 = M_p(\mathbb{C})$.

3   Block Structure

Now this is done we proceed to prove the structure theorem for QCA over finite, unbounded configurations. This is a simplification of [13]. The basic idea of the proof is that in a cell at time t we can separate what information will be sent to the left at time $t+1$ and which information will be sent to the right at time $t+1$. But first of all we shall need two lemmas. These are better understood by referring to Figure 1.

Lemma 3. *Let $\mathcal{A}$ be the image of the algebra of the cell 1 under the global evolution G. It is localized upon cells 0 and 1, and we call $\mathcal{A}|_1$ the restriction of $\mathcal{A}$ to cell 1.*

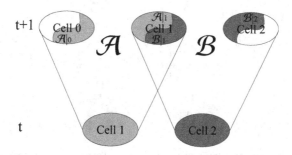

Fig. 1. Definitions of the algebras for the proof of the structure theorem

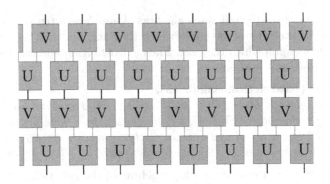

Fig. 2. QCA with two-layered block representation (U, V). Each line represents a cell, which is a quantum system. Each square represents a unitary U/V which gets applied upon the quantum systems. Time flows upwards.

Let $\mathcal{B}$ be the image of the algebra of the cell 2 under the global evolution G. It is localized upon cells 1 and 2, and we call $\mathcal{B}|_1$ the restriction of $\mathcal{B}$ to cell 1.

There exists a unitary U acting upon cell 1 such that $U\mathcal{A}|_0 U^\dagger$ is of the form $M_p(\mathbb{C}) \otimes \mathbb{I}_q$ and $U\mathcal{B}|_1 U^\dagger$ is of the form $\mathbb{I}_p \otimes M_q(\mathbb{C})$, with $pq = d$.

Lemma 4. Let $\mathcal{B}$ be the image of the algebra of the cell 2 under the global evolution G. It is localized upon cells 1 and 2, and we call $\mathcal{B}|_1$ the restriction of $\mathcal{B}$ to cell 1 and $\mathcal{B}|_2$ the restriction of $\mathcal{B}$ to cell 2. We have that $\mathcal{B} = \mathcal{B}|_1 \otimes \mathcal{B}|_2$.

Theorem 4 (Structure theorem)
Any QCA G is of the form described by Figures 2 and 3.

Note that this structure could be further simplified if we were to allow ancillary cells [1]. Therefore we have shown that one-dimensional QCA over finite, unbounded configurations admit a two-layered block representation. As we shall see n-dimensional QCA do not admit such a two-layered block representation,

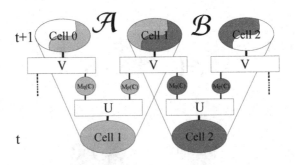

Fig. 3. Zooming into the two-layered block representation. The unitary interactions U and V are alternated repeatedly as shown.

contrary to what was stated in [13]. Whilst the proof remains similar in spirit, it has the advantage of being remarkably simpler and self-contained, phrased in the standard setting of quantum theory, understandable without heavy prerequisites in C^*-algebras. The proof technique is rather different from that of [15], for whom G is essentially a finite-dimensional matrix and hence can necessarily be approximated by a quantum circuit.

4   Quantizations and Consequences

The structure theorem for QCA departs in several important ways from the classical situation, giving rise to a number of apparent paradoxes. We begin this section by discussing some of these concerns in turns. Each of them is introduced via an example, which we then use to derive further consequences or draw the limits of the structure theorem. This will lead us to three original propositions.

Bijective CA and superluminal signalling. First of all, it is a well-known fact that not all bijective CA are structurally reversible. The modified XOR CA is a standard example of that.

Definition 11 (mXOR CA).
Let C_f be the set of finite configurations over the alphabet $q + \Sigma = \{q, 0, 1\}$. For all x, y in $q + \Sigma$ Let $\delta(qx) = q$, $\delta(xq) = x$, and $\delta(xy) = x \oplus y$ otherwise. We call $F : C_f \longrightarrow C_f$ the function mapping $c = \ldots c_{i-1} c_i c_{i+1} \ldots$ to $c' = \ldots \delta(c_{i-1} c_i) \delta(c_i c_{i+1}) \ldots$.

The mXOR CA is clearly shift-invariant, and causal in the sense that the state of a cell at $t + 1$ only depends from its state and that of its right neighbour at t. It is also bijective. Indeed for any $c' = \ldots qq c'_k c'_{k+1} \ldots$ with c'_k the first non quiescent cell, we have $c_k = q$, $c_{k+1} = c'_k$, and thereon for $l \geq k+1$ we have either $c_{l+1} = c_l \oplus c'_l$ if $c'_l \neq q$, or once again $c_{l+1} = q$ otherwise, etc. In other words the antecedent always exists (surjectivity) and is uniquely derived (injectivity) from left till right. But the mXOR CA is not structurally reversible. Indeed for some $c' = \ldots 000000000 \ldots$ we cannot know whether the antecedent of this large zone of zeroes is another large zone of zeroes or a large zone of ones – unless we deduce this from the left border as was previously described... but the left border may lie arbitrary far.

So classically there are bijective CA whose inverse is not a CA, and thus who do not admit any n-layered block representation at all. Yet surely, just by defining F over $\mathcal{H}_{C_f}$ by linear extension (e.g. $F(\alpha.|\ldots 01 \ldots\rangle + \beta.|\ldots 11 \ldots\rangle) = \alpha.F|\ldots 01 \ldots\rangle + \beta.F|\ldots 11 \ldots\rangle)$ we ought to have a QCA, together with its block representation, hence the apparent paradox.

In order to lift this concern let us look at the properties of this quantized F : $\mathcal{H}_{C_f} \longrightarrow \mathcal{H}_{C_f}$. It is indeed unitary as a linear extension of a bijective function, and it is shift-invariant for the same reason. Yet counter-intuitively it is non-causal. Indeed consider configurations $c_{\pm} = 1/\sqrt{2}.|\ldots qq\rangle(|00 \ldots 00\rangle \pm |11 \ldots 11\rangle)|qq \ldots\rangle$. We have $Fc_{\pm} = |\ldots qq00 \ldots 0\rangle|\pm\rangle|qq \ldots\rangle$, where we have used the usual notation

$|\pm\rangle = 1/\sqrt{2}.(|0\rangle\pm|1\rangle)$. Let i be the position of this last non quiescent cell. Clearly $(Fc_\pm)|_i = |\pm\rangle\langle\pm|$ is not just a function of $c|_{i,i+1} = (|0q\rangle\langle 0q| + |1q\rangle\langle 1q|)/2$, but instead depends upon this global $\pm$ phase. Another way to put it is that the quantized XOR may be used to transmit information faster than light. Say the first non quiescent cell is with Alice in Paris and the last non quiescent cell is with Bob in New York. Just by applying a phase gate Z upon her cell Alice can change c_+ into c_- at time t, leading to a perfectly measurable change from $|+\rangle$ to $|-\rangle$ for Bob. Again another way to say it is that operators localized upon cell 1 are not taken to operators localized upon cells 0 and 1, as was the case for QCA. For instance take $\mathbb{I}\otimes Z\otimes\mathbb{I}$ localized upon cell 1. This is taken to $F(\mathbb{I}\otimes Z\otimes\mathbb{I})F^\dagger$. But this operation is not localized upon cells 0 and 1, as it takes $|\dots qq00\dots0\rangle|+\rangle|qq\dots\rangle$ to $|\dots qq00\dots0\rangle|-\rangle|qq\dots\rangle$, whatever the position i of the varying $|\pm\rangle$. Note that because the effect is arbitrarily remote, this cannot be reconciled with just a cell grouping. Notice also the curious asymmetry of the scenario, which communicates towards the right.

Such a behaviour is clearly not acceptable. Although it seemed like a valid QCA, F must bi discarded as non-physical. A phenomenon which seems causal classically may turn out non-causal in its quantum extension. Clearly this is due to the possibility of having entangled states, which allow for more 'non-local' states, and hence strengthens the consequences of no-signalling. This is the deep reason why QCA, even on finite configurations, do admit a block representation. Now let us take a step back. If a CA is not structurally reversible, there is no chance that its QCA will be. Moreover according the current state of modern physics, quantum mechanics is the theory for describing all closed systems. Therefore we reach the following proposition, where the class B stands for the class of bijective but not structurally reversible CA upon finite configurations is known to coincide with the class of surjective but non injective CA upon infinite configurations, well-known to be quivalent to the class of bijective CA upon finite configurations but not upon infinite configuration.

Proposition 1 (Class B is not causally quantizable). *The quantization of a class B automata is not causal. It cannot be implemented by a series of finite quantum systems, isolated from the outside world.*

As far as CA are concerned this result removes much of the motivation of several papers which focus upon class B, since they become illegal physically in the formal sense above. As regards QCA the structure theorem also removes much of the motivation of the papers [6,7,11,1], which contain unitary decision procedures for possibly non-structurally reversible QCA.

Faster quantum signalling. Second, it is a well-known fact that there exists some 1/2-neighbourhood, structurally reversible CA, whose inverse is also of 1/2-neighbourhood, and yet which do not admit a two-layered block representation unless the cells are grouped into supercells. The Toffoli CA is a good example of that.

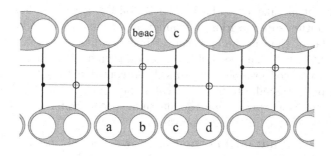

Fig. 4. The Toffoli CA

Definition 12 (Toffoli CA). *Let C_f be the set of finite configurations over the alphabet $\{00, 01, 01, 11\}$, with 00 now taken as the quiescent symbol. For all ab and cd taken in the alphabet let $\delta(abcd) = (b \oplus a.c)c$. We call $F : C_f \longrightarrow C_f$ the function mapping $c = \ldots c_{i-1}c_ic_{i+1}\ldots$ to $c' = \ldots \delta(c_{i-1}c_i)\delta(c_ic_{i+1})\ldots$. This is best described by Figure 4.*

The Toffoli CA is clearly shift-invariant, and of 1/2-neighbourhood. Let us check that its inverse is also of 1/2-neighbourhood. For instance say we seek to retrieve (c,d). c is easy of course. By shift-invariance retrieving d is like retrieving b. But since we have a and c in cleartext we can easily substract $a.c$ from $b \oplus a.c$. Now why does it not have a two-layered block representation without cell grouping? Remember the toffoli gate is the controlled-controlled-NOT gate. Here b is NOTed depending upon a and c, which pass through unchanged, same for d with the left and right neighbouring subcells, etc. So actually the Toffoli CA is just two layers of the toffoli gate, as we have shown in Figure 4. But we know that the toffoli gate cannot be obtained from two bit gates in classical reversible electronics, hence there cannot be a two-layered block representation without cell grouping.

So classically there exists some structurally reversible CA, of 1/2-neighbourhood, whose inverse is also of 1/2-neighbourhood, but do not admit a two-layered block representation without cell grouping. Yet surely, just by defining F over $\mathcal{H}_{C_f}$ by linear extension we ought to have a QCA, together with its block representation, and that construction does not need any cell grouping, hence again the apparent paradox.

Again in order to lift this concern let us look at the properties of this quantized $F : \mathcal{H}_{C_f} \longrightarrow \mathcal{H}_{C_f}$. It is indeed unitary and shift-invariant of course. This time it is also causal, but counter-intuitively it turns out not to be of 1/2-neighbourhood. Indeed from the formulation in terms of Toffoli gates as in Figure 4 one can show that the radius is 3/2 in a quantum mechanical setting. For instance one can check that putting $|+\rangle$ in the a-subcell, $|-\rangle$ in the b-subcell, and either $|0\rangle$ or $|1\rangle$ in the c-subcell of Fig. 4 at time t will yield either $|+\rangle$ or $|-\rangle$ in the a-subcell at time $t+1$. Basically this arises because unlike in the classical case where the control bit always emerges unchanges of a Toffoli gate, when a control bit is in a superposition (like a in the example given) it may emerge from the Toffoli gate modified.

Once more let us take a step back. The Toffoli CA is yet another case where exploiting quantum superpositions of configurations enables us to have information flowing faster than in the classical setting, just like for the XOR CA. But unlike the mXOR CA, the speed of information remains bounded in the Toffoli CA, and so up to cell grouping it can still be considered a QCA. The Toffoli CA is hence perfectly valid from a physical point of view, and causal, so long as we are willing to reinterprete what the maximal speed of information should be. Therefore we reach the following proposition.

Proposition 2 (Quantum information flows faster). *Let $F : C_f \longrightarrow C_f$ be a CA and $F : \mathcal{H}_{C_f} \longrightarrow \mathcal{H}_{C_f}$ the corresponding QCA, as obtained by linear extension of F. Information may flow faster in the the quantized version of F.*

This result is certainly intriguing, and one may wonder whether it might contain the seed of a novel development quantum information theory, as opposed to its classical counterpart.

No-go for n-dimensions. Finally, it is again well-known that in two-dimensions there exists some structurally reversible CA which do not admit a two-layered block representation, even after a cell-grouping. The standard example is that of Kari [10]:

Definition 13 (Kari CA). *Let C_f be the set of finite configurations over the alphabet $\{0, 1\}^9$, with 0^9 is now taken as the quiescent symbol. So each cell is made of 8 bits, one for each cardinal direction (North, North-East...) plus one bit in the center. At each time step, the North bit of a cell undergoes a NOT only if the cell lying North has center bit equal to 1, the North-East bit of a cell undergoes a NOT only if the cell lying North-East has center bit equal to 1, and so on. Call F this CA.*

The proof can easily be ported to the quantum case, as discussed in the longer version of the paper [3]. Hence we have a counterexample to the higher-dimensional case of the Theorem in [13]. We reach the following proposition.

Proposition 3 (No-go for n-dimensions). *There exists some 2-dimensional QCA which do not admit a two-layered block representation.*

Acknowledgements

We would like to thank Jacques Mazoyer, Torsten Franz, Holger Vogts, Jarkko Kari, Jérôme Durand-Lose, Renan Fargetton, Philippe Jorrand for a number of helpful conversations.

References

1. Arrighi, P.: An algebraic study of one-dimensional quantum cellular automata. In: Proceedings of MFCS 2006. LNCS, vol. 4162, pp. 122–133. Springer, Heidelberg (2006)
2. Arrighi, P., Fargetton, R.: Intrinsically universal one-dimensional quantum cellular automata. In: DCM 2007 (2007)

3. Arrighi, P., Nesme, V., Werner, R.: One-dimensional quantum cellular automata over finite, unbounded configurations, longer version of this paper, arXiv:0711.3517 (2007)
4. Bratteli, O., Robinson, D.: Operators algebras and quantum statistical mechanics 1. Springer, Heidelberg (1987)
5. Cheung, D., Perez-Delgado, C.A.: Local Unitary Quantum Cellular Automata, arXiv:/0709.0006
6. Dürr, C., LêThanh, H., Santha, M.: A decision procedure for well formed quantum cellular automata. Random Structures and Algorithms 11, 381–394 (1997)
7. Dürr, C., Santha, M.: A decision procedure for unitary quantum linear cellular automata. SIAM J. of Computing 31(4), 1076–1089 (2002)
8. Feynman, R.P.: Quantum mechanical computers. Found. Phys. 16, 507–531 (1986)
9. Gijswijt, D.: Matrix algebras and semidefinite programming techniques for codes, Ph.D. thesis, University of Amsterdam (2005)
10. Kari, J.: On the circuit depth of structurally reversible cellular automata. Fudamenta Informaticae 34, 1–15 (1999)
11. Meyer, D.: Unitarity in one dimensional nonlinear quantum cellular automata, arXiv:quant-ph/9605023 (1995)
12. Meyer, D.: From quantum cellular automata to quantum lattice gases. J. Stat. Phys. 85, 551–574 (1996)
13. Schumacher, B., Werner, R.F.: Reversible quantum cellular automata, arXiv:quant-ph/0405174
14. Shepherd, D.J., Franz, T., Werner, R.F.: Universally programmable quantum cellular automata. Phys. Rev. Lett. 97, 020502 (2006)
15. van Dam, W.: A Universal Quantum Cellular Automaton. In: Proc. of Phys. Comp. 1996, New England Complex Systems Institute, pp. 323–331 (1996); InterJournal manuscript 91(1996)
16. Watrous, J.: On one dimensional quantum cellular automata. In: Proc. of the 36th IEEE Symposium on Foundations of Computer Science, pp. 528–537 (1995)

The Three-Color and Two-Color Tantrix^{TM} Rotation Puzzle Problems Are NP-Complete Via Parsimonious Reductions*

Dorothea Baumeister and Jörg Rothe

Institut für Informatik, Universität Düsseldorf, 40225 Düsseldorf, Germany

Abstract. Holzer and Holzer [7] proved that the Tantrix^{TM} rotation puzzle problem with four colors is NP-complete, and they showed that the infinite variant of this problem is undecidable. In this paper, we study the three-color and two-color Tantrix^{TM} rotation puzzle problems (3-TRP and 2-TRP) and their variants. Restricting the number of allowed colors to three (respectively, to two) reduces the set of available Tantrix^{TM} tiles from 56 to 14 (respectively, to 8). We prove that 3-TRP and 2-TRP are NP-complete, which answers a question raised by Holzer and Holzer [7] in the affirmative. Since our reductions are parsimonious, it follows that the problems Unique-3-TRP and Unique-2-TRP are DP-complete under randomized reductions. Finally, we prove that the infinite variants of 3-TRP and 2-TRP are undecidable.

1 Introduction

The puzzle game Tantrix^{TM}, invented by Mike McManaway in 1991, is a domino-like strategy game played with hexagonal tiles in the plane. Each tile contains three colored lines in different patterns (see Figure 1). We are here interested in the variant of the Tantrix^{TM} rotation puzzle game whose aim it is to match the line colors of the joint edges for each pair of adjacent tiles, just by rotating the tiles around their axes while their locations remain fixed. This paper continues the complexity-theoretic study of such problems that was initiated by Holzer and Holzer [7]. Other results on the complexity of domino-like strategy games can be found, e.g., in Grädel's work [6]. Tantrix^{TM} puzzles have also been studied with regard to evolutionary computation [4].

Holzer and Holzer [7] defined two decision problems associated with four-color Tantrix^{TM} rotation puzzles. The first problem's instances are restricted to a finite number of tiles, and the second problem's instances are allowed to have infinitely many tiles. They proved that the finite variant of this problem is NP-complete and that the infinite problem variant is undecidable. The constructions in [7] use tiles with four colors, just as the original Tantrix^{TM} tile set. Holzer and Holzer posed the question of whether the Tantrix^{TM} rotation puzzle problem remains NP-complete if restricted to only three colors, or if restricted to otherwise reduced tile sets.

* Full version: [2]; see also Baumeister's Master Thesis "Complexity of the Tantrix^{TM} Rotation Puzzle Problem," Universität Düsseldorf, September 2007. Supported in part by DFG grants RO 1202/9-3 and RO 1202/11-1 and the Humboldt Foundation's TransCoop program.

C. Martín-Vide, F. Otto, and H. Fernau (Eds.): LATA 2008, LNCS 5196, pp. 76–87, 2008.

Table 1. Overview of complexity and decidability results for k-TRP and its variants

k	k-TRP is	Parsimonious?	Unique-k-TRP is	Inf-k-TRP is
1	in P (trivial)		in P (trivial)	decidable (trivial)
2	NP-compl., Cor. 3	yes, Thm. 2	DP-$\leq_{ran}^{p}$-compl., Cor. 4	undecidable, Thm. 3
3	NP-compl., Cor. 1	yes, Thm. 1	DP-$\leq_{ran}^{p}$-compl., Cor. 4	undecidable, Thm. 3
4	NP-compl., see [7]	yes, see [1]	DP-$\leq_{ran}^{p}$-compl., see [1]	undecidable, see [7]

In this paper, we answer this question in the affirmative for the three-color and the two-color version of this problem. For $1 \leq k \leq 4$, Table 1 summarizes the results for k-TRP, the k-color Tantrix™ rotation puzzle problem, and its variants. (All problems are formally defined in Section 2.)

Since the four-color Tantrix™ tile set contains the three-color Tantrix™ tile set, our new complexity results for 3-TRP imply the previous results for 4-TRP (both its NP-completeness [7] and that satisfiability *parsimoniously* reduces to 4-TRP [1]). In contrast, the three-color Tantrix™ tile set does not contain the two-color Tantrix™ tile set (see Figure 2 in Section 2). Thus, 3-TRP does not straightforwardly inherit its hardness results from those of 2-TRP, which is why both reductions, the one to 3-TRP and the one to 2-TRP, have to be presented. Note that they each substantially differ—both regarding the subpuzzles constructed and regarding the arguments showing that the constructions are correct—from the previously known reductions [7,1], and we will explicitly illustrate the differences between our new and the original subpuzzles.

Since we provide *parsimonious* reductions from the satisfiability problem to 3-TRP and to 2-TRP, our reductions preserve the uniqueness of the solution. Thus, the unique variants of both 3-TRP and 2-TRP are DP-complete under polynomial-time randomized reductions, where DP is the class of differences of NP sets. We also prove that the infinite variants of 3-TRP and 2-TRP are undecidable, via a circuit construction similar to the one Holzer and Holzer [7] used to show that the infinite 4-TRP problem is undecidable.

2   Definitions and Notation

Complexity-Theoretic Notions and Notation: We assume that the reader is familiar with the standard notions of complexity theory, such as the complexity classes P (deterministic polynomial time) and NP (nondeterministic polynomial time). DP denotes the class of differences of any two NP sets [9].

Let Σ^* denote the set of strings over the alphabet $\Sigma = \{0, 1\}$. Given any language $L \subseteq \Sigma^*$, $\|L\|$ denotes the number of elements in L. We consider both decision problems and function problems. The former are formalized as languages whose elements are those strings in Σ^* that encode the yes-instances of the problem at hand. Regarding the latter, we focus on the counting problems related to sets in NP. The counting version #A of an NP set A maps each instance x of A to the number of solutions of x. That is, counting problems are functions from Σ^* to $\mathbb{N}$. As an example, the counting version #SAT of SAT, the NP-complete satisfiability problem, asks how many satisfying assignments a given boolean formula has. Solutions of NP sets can be viewed as accepting paths of

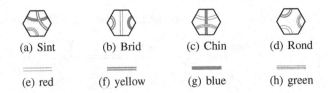

(a) Sint   (b) Brid   (c) Chin   (d) Rond

(e) red   (f) yellow   (g) blue   (h) green

Fig. 1. Tantrix^{TM} tile types and the encoding of Tantrix^{TM} line colors

NP machines. Valiant [10] defined the function class #P to contain the functions that give the number of accepting paths of some NP machine. In particular, #SAT is in #P.

The complexity of two decision problems, A and B, will here be compared via the *polynomial-time many-one reducibility*: $A \leq_m^p B$ if there is a polynomial-time computable function f such that for each $x \in \Sigma^*$, $x \in A$ if and only if $f(x) \in B$. A set B is said to be NP-complete if B is in NP and every NP set $\leq_m^p$-reduces to B.

Many-one reductions do not always preserve the number of solutions. A reduction that does preserve the number of solutions is said to be *parsimonious*. Formally, if A and B are any two sets in NP, we say A *parsimoniously reduces to* B if there exists a polynomial-time computable function f such that for each $x \in \Sigma^*$, #$A(x) = $ #$B(f(x))$.

Valiant and Vazirani [11] introduced the following type of *randomized polynomial-time many-one reducibility*: $A \leq_{ran}^p B$ if there exists a polynomial-time randomized algorithm F and a polynomial p such that for each $x \in \Sigma^*$, if $x \in A$ then $F(x) \in B$ with probability at least $1/p(|x|)$, and if $x \notin A$ then $F(x) \notin B$ with certainty. In particular, they proved that the unique version of the satisfiability problem, Unique-SAT, is DP-complete under randomized reductions.

Tile Sets, Color Sequences, and Orientations: The Tantrix^{TM} rotation puzzle consists of four different kinds of hexagonal tiles, named *Sint*, *Brid*, *Chin*, and *Rond*. Each tile has three lines colored differently, where the three colors of a tile are chosen among four possible colors, see Figures 1(a)–(d). The original Tantrix^{TM} colors are *red*, *yellow*, *blue*, and *green*, which we encode here as shown in Figures 1(e)–(h). The combination of four kinds of tiles having three out of four colors each gives a total of 56 different tiles.

Let C be the set that contains the four colors *red*, *yellow*, *blue*, and *green*. For each $i \in \{1,2,3,4\}$, let $C_i \subseteq C$ be some fixed subset of size i, and let T_i denote the set of Tantrix^{TM} tiles available when the line colors for each tile are restricted to C_i. For example, T_4 is the original Tantrix^{TM} tile set containing 56 tiles, and if C_3 contains, say, the three colors *red*, *yellow*, and *blue*, then tile set T_3 contains the 14 tiles shown in Figure 2(b).

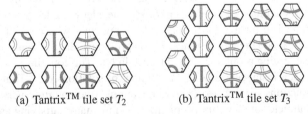

(a) Tantrix^{TM} tile set T_2 (b) Tantrix^{TM} tile set T_3

Fig. 2. Tantrix^{TM} tile sets T_2 (for *red* and *blue*) and T_3 (for *red*, *yellow*, and *blue*)

For T_3 and T_4, we require the three lines on each tile to have distinct colors, as in the original Tantrix^{TM} tile set. For T_1 and T_2, however, this is not possible, so we allow the same color being used for more than one of the three lines of any tile. Note that we care only about the sequence of colors on a tile, where we always use the clockwise direction to represent color sequences. However, since different types of tiles can yield the same color sequence, we will use just one such tile to represent the corresponding color sequence. For example, if C_2 contains, say, the two colors *red* and *blue*, then the color sequence *red-red-blue-blue-blue-blue* (which we abbreviate as rrbbbb) can be represented by a *Sint*, a *Brid*, or a *Rond* each having one short *red* arc and two *blue* additional lines, and we add only one such tile (say, the *Rond*) to the tile set T_2. That is, though there is some freedom in choosing a particular set of tiles, to be specific we fix the tile set T_2 shown in Figure 2(a). Thus, we have $\|T_1\| = 1$, $\|T_2\| = 8$, $\|T_3\| = 14$, and $\|T_4\| = 56$, regardless of which colors are chosen to be in C_i, $1 \le i \le 4$.

The six possible orientations for each tile in T_2 and in T_3, respectively, can be described by permuting the color sequences cyclically, and we omit the repetitions of color sequences (see the full version [2] for more details). For example, tile t_7 from T_2 has the same color sequence (namely, bbbbbb) in each of its six orientations. In Section 3, we will consider the counting versions of Tantrix^{TM} rotation puzzle problems and will construct parsimonious reductions. When counting the solutions of Tantrix^{TM} rotation puzzles, we will focus on color sequences only. That is, whenever some tile (such as t_7 from T_2) has distinct orientations with identical color sequences, we will count this as just one solution (and disregard such repetitions). In this sense, our reduction in the proof of Theorem 2 (which is presented in the full version [2]) will be parsimonious.

Definition of the Problems: We now recall some useful notation that Holzer and Holzer [7] introduced in order to formalize problems related to the Tantrix^{TM} rotation puzzle. The instances of such problems are Tantrix^{TM} tiles firmly arranged in the plane. To represent their positions, we use a two-dimensional hexagonal coordinate system, see [7] and also [2]. Let $T \in \{T_1, T_2, T_3, T_4\}$ be some tile set as defined above. Let $\mathscr{A} : \mathbb{Z}^2 \to T$ be a function mapping points in $\mathbb{Z}^2$ to tiles in T, i.e., $\mathscr{A}(x)$ is the type of the tile located at position x. Note that $\mathscr{A}$ is a partial function; throughout this paper (except in Theorem 3 and its proof), we restrict our problem instances to finitely many given tiles, and the regions of $\mathbb{Z}^2$ they cover may have holes (which is a difference to the original Tantrix^{TM} game).

Define $shape(\mathscr{A})$ to be the set of points $x \in \mathbb{Z}^2$ for which $\mathscr{A}(x)$ is defined. For any two distinct points $x = (a, b)$ and $y = (c, d)$ in $\mathbb{Z}^2$, x and y are neighbors if and only if $(a = c$ and $|b - d| = 1)$ or $(|a - c| = 1$ and $b = d)$ or $(a - c = 1$ and $b - d = 1)$ or $(a - c = -1$ and $b - d = -1)$. For any two points x and y in $shape(\mathscr{A})$, $\mathscr{A}(x)$ and $\mathscr{A}(y)$ are said to be neighbors exactly if x and y are neighbors. For k chosen from $\{1, 2, 3, 4\}$, define the following problem:

Name: k-Color Tantrix^{TM} Rotation Puzzle (k-TRP, for short).
Instance: A finite shape function $\mathscr{A} : \mathbb{Z}^2 \to T_k$, encoded as a string in Σ^*.
Question: Is there a solution to the rotation puzzle defined by $\mathscr{A}$, i.e., does there exist a rotation of the given tiles in $shape(\mathscr{A})$ such that the colors of the lines of any two adjacent tiles match at their joint edge?

Clearly, 1-TRP can be solved trivially, so 1-TRP is in P. On the other hand, Holzer and Holzer [7] showed that 4-TRP is NP-complete and that the infinite variant of 4-TRP is undecidable. Baumeister and Rothe [1] investigated the counting and the unique variant of 4-TRP and, in particular, provided a parsimonious reduction from SAT to 4-TRP. In this paper, we study the three-color and two-color versions of this problem, 3-TRP and 2-TRP, and their counting, unique, and infinite variants.

Definition 1. *A solution to a k-TRP instance $\mathscr{A}$ specifies an orientation of each tile in shape($\mathscr{A}$) such that the colors of the lines of any two adjacent tiles match at their joint edge. Let* $\mathrm{SOL}_{k\text{-}\mathrm{TRP}}(\mathscr{A})$ *denote the set of solutions of $\mathscr{A}$. Define the counting version of k-TRP to be the function #k-TRP mapping from Σ^* to $\mathbb{N}$ such that* $\#\text{-}\mathrm{TRP}(\mathscr{A}) = \|\mathrm{SOL}_{k\text{-}\mathrm{TRP}}(\mathscr{A})\|$. *Define the unique version of k-TRP as* Unique-k-TRP $= \{\mathscr{A} \mid \#k\text{-}\mathrm{TRP}(\mathscr{A}) = 1\}$.

The above problems are defined for the case of finite problem instances. The infinite Tantrix^{TM} rotation puzzle problem with k colors (Inf-k-TRP, for short) is defined exactly as k-TRP, the only difference being that the shape function $\mathscr{A}$ is not required to be finite and is represented by the encoding of a Turing machine computing $\mathscr{A} : \mathbb{Z}^2 \rightarrow T_k$.

3   Results

3.1   Parsimonious Reduction from SAT to 3-TRP

Theorem 1 below is the main result of this section. Notwithstanding that our proof follows the general approach of Holzer and Holzer [7], our specific construction and our proof of correctness will differ substantially from theirs. We will give a parsimonious reduction from SAT to 3-TRP. Let Circuit$_{\wedge,\neg}$-SAT denote the problem of deciding, given a boolean circuit c with AND and NOT gates only, whether or not there is a satisfying truth assignment to the input variables of c. The NP-completeness of Circuit$_{\wedge,\neg}$-SAT was shown by Cook [3], and it is easy to see that SAT parsimoniously reduces to Circuit$_{\wedge,\neg}$-SAT (see, e.g., [1]).

Theorem 1. SAT *parsimoniously reduces to* 3-TRP.

It is enough to show that Circuit$_{\wedge,\neg}$-SAT parsimoniously reduces to 3-TRP. The resulting 3-TRP instance simulates a boolean circuit with AND and NOT gates such that the number of solutions of the rotation puzzle equals the number of satisfying truth assignments to the variables of the circuit.

General remarks on our proof approach: The rotation puzzle to be constructed from a given circuit consists of different subpuzzles each using only three colors. The color *green* was employed by Holzer and Holzer [7] only to exclude certain rotations, so we choose to eliminate this color in our three-color rotation puzzle. Thus, letting C_3 contain the colors *blue, red,* and *yellow*, we have the tile set $T_3 = \{t_1, t_2, \ldots, t_{14}\}$, where the enumeration of tiles corresponds to Figure 2(b). Furthermore, our construction will be parsimonious, i.e., there will be a one-to-one correspondence between the solutions of the given Circuit$_{\wedge,\neg}$-SAT instance and the solutions of the resulting rotation puzzle instance. Note that part of our work is already done, since some subpuzzles constructed in [1] use only three colors and they each have unique solutions. However, the remaining subpuzzles have to be either modified substantially or to be constructed completely differently, and the arguments of why our modified construction is correct differs considerably from previous work [7,1].

Since it is not so easy to exclude undesired rotations without having the color *green* available, it is useful to first analyze the 14 tiles in T_3. In the remainder of this proof, when showing that our construction is correct, our arguments will often be based on which substrings do or do not occur in the color sequences of certain tiles from T_3. (Note that the full version of this paper [2] has a table that shows which substrings of the form uv, where $u,v \in C_3$, occur in the color sequence of t_i in T_3, and this table may be looked up for convenience.)

Holzer and Holzer [7] consider a boolean circuit c on input variables $x_1, x_2, \ldots, x_n$ as a sequence $(\alpha_1, \alpha_2, \ldots, \alpha_m)$ of computation steps (or "instructions"), and we adopt this approach here. For the ith instruction, α_i, we have $\alpha_i = x_i$ if $1 \leq i \leq n$, and if $n + 1 \leq i \leq m$ then we have either $\alpha_i = \text{NOT}(j)$ or $\alpha_i = \text{AND}(j,k)$, where $j \leq k < i$. Circuits are evaluated in the standard way. We will represent the truth value *true* by the color *blue* and the truth value *false* by the color *red* in our rotation puzzle. A technical difficulty in the construction results from the wire crossings that circuits can have. To construct rotation puzzles from *planar* circuits, Holzer and Holzer use McColl's planar "cross-over" circuit with AND and NOT gates to simulate such wire crossings [8], and in particular they employ Goldschlager's log-space transformation from general to planar circuits [5]. For the details of this transformation, we refer to [7].

Holzer and Holzer's original subpuzzles [7] should be compared with those in our construction. To illustrate the differences between our new and these original subpuzzles, modified or inserted tiles in our new subpuzzles presented in this section will always be highlighted by having a grey background.

Wire subpuzzles: Wires of the circuit are simulated by the subpuzzles WIRE, MOVE, and COPY. We present only the WIRE here; see [2] for MOVE and COPY.

A vertical wire is represented by a WIRE subpuzzle, which is shown in Figure 3. The original WIRE subpuzzle from [7] does not contain *green* but it does not have a unique solution, while the WIRE subpuzzle from [1] ensures the uniqueness of the solution but is using a tile with a *green* line. In the original WIRE subpuzzle, both tiles, a and b, have two possible orientations for each input color. Inserting two new tiles at positions x and y (see Figure 3) makes the solution unique. If the input color is *blue*, tile x must contain one of the following color-sequence substrings for the edges joint with tiles b and a: ry, rr, yy, or yr. If the input color is *red*, x must contain one of these substrings: bb, yb, yy, or by. Tile t_{12} satisfies the conditions yy and ry for the input color *blue*, and the conditions yb and yy for the input color *red*.

The solution must now be fixed with tile y. The possible color-sequence substrings of y at the edges joint with a and b are rr and ry for the input color *blue*, and yb and bb for the input color *red*. Tile t_{13} has exactly one of these sequences for each input color. Thus, the solution for this subpuzzle contains only three colors and is unique.

Gate subpuzzles: The boolean gates AND and NOT are represented by the AND and NOT subpuzzles. Both the original four-color NOT subpuzzle from [7] and the modified four-color NOT subpuzzle from [1] use tiles with *green* lines to exclude certain

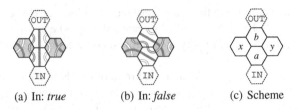

(a) In: *true*   (b) In: *false*   (c) Scheme

Fig. 3. Three-color WIRE subpuzzle

rotations. Our three-color NOT subpuzzle is shown in Figure 4. Tiles a, b, c, and d from the original NOT subpuzzle [7] remain unchanged. Tiles e, f, and g in this original NOT subpuzzle ensure that the output color will be correct, since the joint edge of e and b is always *red*. So for our new NOT subpuzzle in Figure 4, we have to show that the edge between tiles x and b is always *red*, and that we have unique solutions for both input colors.

First, let the input color be *blue* and suppose for a contradiction that the joint edge of tiles b and x were *blue*. Then the joint edge of tiles b and c would be *yellow*. Since x is a tile of type t_{13} and so does not contain the color-sequence substring bb, the edge between tiles c and x must be *yellow*. But then the edges of tile w joint with tiles c and x must both be *blue*. This is not possible, however, because w (which is of type t_{10}) does not contain the color-sequence substring bb. So if the input color is *blue*, the orientation of tile b is fixed with *yellow* at the edge of b joint with tile y, and with *red* at the edges of b joint with tiles c and x. This already ensures that the output color will be *red*, because tiles c and d behave like a WIRE subpuzzle. Tile x does not contain the color-sequence substring br, so the orientation of tile c is also fixed with *blue* at the joint edge of tiles c and w. As a consequence, the joint edge of tiles w and d is *yellow*, and due to the fact that the joint edge of tiles w and x is also *yellow*, the orientation of w and d is fixed as well. Regarding tile a, the edge joint with tile y can be *yellow* or *red*, but tile x has *blue* at the edge joint with tile y, so the joint edge of tiles y and a is *yellow*, and the orientation of all tiles is fixed for the input color *blue*. The case of *red* being the input color can be handled analogously.

The most complicated figure is the AND subpuzzle. The original four-color version from [7] uses four tiles with *green* lines and the modified four-color AND subpuzzle

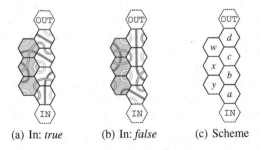

(a) In: *true*   (b) In: *false*   (c) Scheme

Fig. 4. Three-color NOT subpuzzle

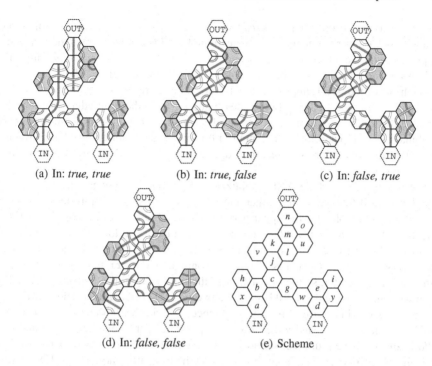

(a) In: *true, true*   (b) In: *true, false*   (c) In: *false, true*

(d) In: *false, false*   (e) Scheme

Fig. 5. Three-color AND subpuzzle

from [1] uses seven tiles with *green* lines. Figure 5 shows our new AND subpuzzle using only three colors and having unique solutions for all four possible combinations of input colors. To analyze this subpuzzle, we subdivide it into a lower and an upper part. The lower part ends with tile c and has four possible solutions (one for each combination of input colors), while the upper part, which begins with tile j, has only two possible solutions (one for each possible output color). The lower part can again be subdivided into three different parts.

The lower left part contains the tiles a, b, x, and h. If the input color to this part is *blue* (see Figures 5(a) and 5(b)), the joint edge of tiles b and x is always *red*, and since tile x (which is of type t_{11}) does not contain the color-sequence substring rr, the orientation of tiles a and x is fixed. The orientation of tiles b and h is also fixed, since h (which is of type t_2) does not contain the color-sequence substring by but the color-sequence substring yy for the edges joint with tiles b and x. By similar arguments we obtain a unique solution for these tiles if the left input color is *red* (see Figures 5(c) and 5(d)). The connecting edge to the rest of the subpuzzle is the joint edge between tiles b and c, and tile b will have the same color at this edge as the left input color.

Tiles d, e, i, w, and y form the lower right part. If the input color to this part is *blue* (see Figures 5(a) and 5(c)), the joint edge of tiles d and y must be *yellow*, since tile y (which is of type t_9) does not contain the color-sequence substrings rr nor ry for the edges joint with tiles d and e. Thus the joint edge of tiles y and e must be *yellow*, since i (which is of type t_6) does not contain the color-sequence substring bb for the edges joint with tiles y and e. This implies that the tiles i and w also have a fixed orientation. If the input color to the lower right part is *red* (see Figures 5(b) and 5(d)), a unique solution is obtained by similar arguments. The connection of the lower right part to the rest of the subpuzzle is the edge between tiles w and g. If the right input color is *blue*, this edge will also be *blue*, and if the right input color is *red*, this edge will be *yellow*.

The heart of the AND subpuzzle is its lower middle part, formed by the tiles c and g. The colors at the joint edge between tiles b and c and at the joint edge between tiles w and g determine the orientation of the tiles c and g uniquely for all four possible combinations of input colors. The output of this part is the color at the edge between c and j. If both input colors are *blue*, this edge will also be *blue*, and otherwise this edge will always be *yellow*.

The output of the whole AND subpuzzle will be *red* if the edge between c and j is *yellow*, and if this edge is *blue* then the output of the whole subpuzzle will also be *blue*. If the input color for the upper part is *blue* (see Figure 5(a)), each of the tiles j, k, l, m, and n has a vertical *blue* line. Note that since the colors *red* and *yellow* are symmetrical in these tiles, we would have several possible solutions without tiles o, u, and v. However, tile v (which is of type t_9) contains neither rr nor ry for the edges joint with tiles k and j, so the orientation of the tiles j through n is fixed, except that tile n without tiles o and u would still have two possible orientations. Tile u (which is of type t_2) is fixed because of its color-sequence substring yy at the edges joint with l and m, so due to tiles o and u the only color possible at the edge between n and o is *yellow*, and we have a unique solution. If the input color for the upper part is *yellow* (see Figures 5(b)–(d)), we obtain unique solutions by similar arguments. Hence, this new AND subpuzzle uses only three colors and has unique solutions for each of the four possible combinations of input colors.

Input and output subpuzzles: The input variables of the boolean circuit are represented by the subpuzzle BOOL. Our new three-color BOOL subpuzzle is presented in Figure 6, and since it is completely different from the original four-color BOOL subpuzzle from [7], no tiles are marked here. The subpuzzle in Figure 6 has only two possible solutions, one with the output color *blue* (if the corresponding variable is *true*), and one with the output color *red* (if the corresponding variable is *false*). The original four-color BOOL subpuzzle from [7] contains tiles with *green* lines to exclude certain rotations. Our three-color BOOL subpuzzle does not contain any *green* lines, but it might not be that obvious that there are only two possible solutions, one for each output color. The proof can be found in the full version [2].

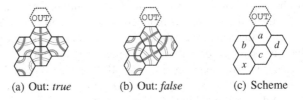

(a) Out: *true*   (b) Out: *false*   (c) Scheme

Fig. 6. Three-color BOOL subpuzzle

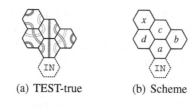

(a) TEST-true   (b) Scheme

Fig. 7. Three-color TEST subpuzzle

Finally, a subpuzzle is needed to check whether or not the circuit evaluates to *true*. This is achieved by the subpuzzle TEST-true shown in Figure 7(a). It has only one valid solution, namely that its input color is *blue*. Just like the subpuzzle BOOL, the original four-color TEST-true subpuzzle from [7], which was not modified in [1], uses *green* lines to exclude certain rotations. Again, since the new TEST-true subpuzzle is completely different from the original subpuzzle, no tiles are marked here. Our argument of why this subpuzzle is correct can be found in the full version [2].

The shapes of the subpuzzles constructed above have changed slightly. However, by Holzer and Holzer's argument [7] about the minimal horizontal distance between two wires and/or gates being at least four, unintended interactions between the subpuzzles do not occur. This concludes the proof of Theorem 1. □

Corollary 1. 3-TRP *is* NP-*complete.*

Since the tile set T_3 is a subset of the tileset T_4, we have 3-TRP $\leq_m^p$ 4-TRP. Thus, the hardness results for 3-TRP and its variants proven in this paper immediately are inherited by 4-TRP and its variants, which provides an alternative proof of these hardness results for 4-TRP and its variants established in [7,1]. In particular, Corollary 2 follows from Theorem 1 and Corollary 1.

Corollary 2 ([7,1]). 4-TRP *is* NP-*complete, via a parsimonious reduction from* SAT.

3.2 Parsimonious Reduction from SAT to 2-TRP

In contrast to the above-mentioned fact that 3-TRP $\leq_m^p$ 4-TRP holds trivially, the reduction 2-TRP $\leq_m^p$ 3-TRP (which we will show to hold due to both problems being NP-complete, see Corollaries 1 and 3) is not immediatedly straightforward, since the tile set T_2 is not a subset of the tile set T_3 (recall Figure 2 in Section 2). In this section, we study 2-TRP and its variants. Our main result here is Theorem 2 below the proof of which can be found in the full version [2].

Theorem 2. SAT *parsimoniously reduces to* 2-TRP.

Corollary 3. 2-TRP *is* NP-*complete.*

3.3 Unique and Infinite Variants of 3-TRP and 2-TRP

Parsimonious reductions preserve the number of solutions and, in particular, the uniqueness of solutions. Thus, Theorems 1 and 2 imply Corollary 4 below that also employs

Valiant and Vazirani's results on the DP-hardness of Unique-SAT under $\leq_{ran}^p$-reductions (which were defined in Section 2). The proof of Corollary 4 follows the lines of the proof of [1, Theorem 6], which states the analogous result for Unique-4-TRP in place of Unique-3-TRP and Unique-2-TRP.

Corollary 4

1. Unique-SAT *parsimoniously reduces to* Unique-3-TRP *and* Unique-2-TRP.
2. Unique-3-TRP *and* Unique-2-TRP *are* DP-*complete under* $\leq_{ran}^p$-*reductions.*

Holzer and Holzer [7] proved that Inf-4-TRP, the infinite Tantrix™ rotation puzzle problem with four colors, is undecidable, via a reduction from (the complement of) the empty-word problem for Turing machines. The proof of Theorem 3 below, which can be found in the full version [2], uses essentially the same argument but is based on our modified three-color and two-color constructions.

Theorem 3. *Both* Inf-2-TRP *and* Inf-3-TRP *are undecidable.*

4   Conclusions

This paper studied the three-color and two-color Tantrix™ rotation puzzle problems, 3-TRP and 2-TRP, and their unique and infinite variants. Our main contribution is that both 3-TRP and 2-TRP are NP-complete via a parsimonious reduction from SAT, which in particular solves a question raised by Holzer and Holzer [7]. Since restricting the number of colors to three and two, respectively, drastically reduces the number of Tantrix™ tiles available, our constructions as well as our correctness arguments substantially differ from those in [7,1]. Table 1 in Section 1 shows that our results give a complete picture of the complexity of k-TRP, $1 \leq k \leq 4$. An interesting question still remaining open is whether the analogs of k-TRP *without holes* still are NP-complete.

Acknowledgments. We are grateful to Markus Holzer and Piotr Faliszewski for inspiring discussions on Tantrix™ rotation puzzles, and we thank Thomas Baumeister for his help with writing a program for checking the correctness of our constructions and for producing reasonably small figures. We thank the anonymous LATA 2008 referees for helpful comments, and in particular the referee who let us know that he or she has also written a program for verifying the correctness of our constructions.

References

1. Baumeister, D., Rothe, J.: Satisfiability parsimoniously reduces to the Tantrix™ rotation puzzle problem. In: Durand-Lose, J., Margenstern, M. (eds.) MCU 2007. LNCS, vol. 4664, pp. 134–145. Springer, Heidelberg (2007)
2. Baumeister, D., Rothe, J.: The three-color and two-color Tantrix™ rotation puzzle problems are NP-complete via parsimonious reductions. Technical Report cs.CC/0711.1827, ACM Computing Research Repository (CoRR) (November 2007)
3. Cook, S.: The complexity of theorem-proving procedures. In: Proceedings of the 3rd ACM Symposium on Theory of Computing, pp. 151–158. ACM Press, New York (1971)

4. Downing, K.: Tantrix: A minute to learn, 100 (genetic algorithm) generations to master. Genetic Programming and Evolvable Machines 6(4), 381–406 (2005)

5. Goldschlager, L.: The monotone and planar circuit value problems are log space complete for P. SIGACT News 9(2), 25–29 (1977)

6. Grädel, E.: Domino games and complexity. SIAM Journal on Computing 19(5), 787–804 (1990)

7. Holzer, M., Holzer, W.: Tantrix^{TM} rotation puzzles are intractable. Discrete Applied Mathematics 144(3), 345–358 (2004)

8. McColl, W.: Planar crossovers. IEEE Transactions on Computers C-30(3), 223–225 (1981)

9. Papadimitriou, C., Yannakakis, M.: The complexity of facets (and some facets of complexity). Journal of Computer and System Sciences 28(2), 244–259 (1984)

10. Valiant, L.: The complexity of computing the permanent. Theoretical Computer Science 8(2), 189–201 (1979)

11. Valiant, L., Vazirani, V.: NP is as easy as detecting unique solutions. Theoretical Computer Science 47, 85–93 (1986)

Optional and Iterated Types for Pregroup Grammars

Denis Béchet[1], Alexander Dikovsky[1], Annie Foret[2], and Emmanuelle Garel[2]

[1] LINA CNRS – UMR 6241 – Université de Nantes
2, rue de la Houssiniére – BP 92208
44322 Nantes Cedex 03 – France
Denis.Bechet@univ-nantes.fr,
Alexandre.Dikovsky@univ-nantes.fr
[2] IRISA – Université de Rennes 1
Campus Universitaire de Beaulieu
Avenue du Général Leclerc
35042 Rennes Cedex – France
Annie.Foret@irisa.fr,
Emmanuelle.Garel@irisa.fr

Abstract. Pregroup grammars are a context-free grammar formalism which may be used to describe the syntax of natural languages. However, this formalism is not able to naturally define types corresponding to optional and iterated arguments such as optional complements of verbs or verbs' adverbial modifiers. This paper introduces two constructions that make up for this deficiency.

Keywords: Pregroups, Lambek Categorial Grammars, Categorial Dependency Grammar.

1 Introduction

Pregroup grammars (PG) [1] have been introduced as a simplification of Lambek calculus [2]. They have been used to model fragments of syntax of several natural languages: English [1], Italian [3], French [4], German [5,6], Japanese [7], Persian [8], etc. PG are based on the idea that the sentences are derived from words using only lexical rules. The syntactic properties of each word in the lexicon are defined as a finite set of its grammatical categories. These grammatical categories are types of a free pregroup generated by a set of basic types together with a partial order on the basic types. A sentence is correct with respect to a PG if for each word of the sentence, one can find in the lexicon such a type that the concatenation of the selected types can be proved in the pregroup calculus to be inferior than or equal to a particular basic type s. The PG are weakly equivalent to CF-grammars [9]. It doesn't mean that they are suitable to define the syntax of natural languages. In particular, any formalism desined for this

C. Martín-Vide, F. Otto, and H. Fernau (Eds.): LATA 2008, LNCS 5196, pp. 88–100, 2008.

purpose should naturally handle the optional and iterated constructions such as noun modifiers, attributes and relative clauses or verb's optional arguments, adverbials or location, manner and other circumstantial clauses. All of them are optional and their number is not bounded. Whereas the CF-grammars handle such constructions rather naturally, they are problematic for the conventional PG. Like the Lambek categorial grammars, the PG are resource sensitive. In particular, in PG proofs, every simple type (except s) should be linked and reduced with exactly one dual type. So a PG cannot define a simple type that is optional or linked to zero, one, or several duals. This effect can be simulated in PG using complex types, however this simulation has serious disadvantages (see a discussion below) .

In this paper, we propose a different solution to this problem adding new rules to the PG calculus. As a result, we obtain a class of PG with simple optional and iterated types, which express the optional and iterated constructions in the way the dependency grammars do ([10,11,12,13]). We prove that the new calculus is decidable.

2 Background

Definition 1 (Pregroup). *A* pregroup *is a structure* $(P, \leq, \cdot, l, r, 1)$ *such that* $(P, \leq, \cdot, 1)$ *is a partially ordered monoid*[1] *and* l, r *are two unary operations on* P *that satisfy for all element* $x \in P$, $x^l x \leq 1 \leq x x^l$ *and* $x x^r \leq 1 \leq x^r x$.

Definition 2 (Free Pregroup). *Let* $(P, \leq)$ *be an ordered set of basic types,* $P^{(\mathbb{Z})} = \{p^{(i)} \mid p \in P, i \in \mathbb{Z}\}$ *be the set of simple types and* $T_{(P,\leq)} = \left(P^{(\mathbb{Z})}\right)^* = \{p_1^{(i_1)} \cdots p_n^{(i_n)} \mid 0 \leq k \leq n, p_k \in P \text{ and } i_k \in \mathbb{Z}\}$ *be the set of types. The empty sequence in* $T_{(P,\leq)}$ *is denoted by 1. For* X *and* $Y \in T_{(P,\leq)}$, $X \leq Y$ *iff this relation is derivable in the following system where* $p, q \in P$, $n, k \in \mathbb{Z}$ *and* $X, Y, Z \in T_{(P,\leq)}$:

$$
\begin{array}{|cc|}
\hline
& \\
X \leq X \ \ (Id) & \dfrac{X \leq Y \quad Y \leq Z}{X \leq Z} \ (Cut) \\
& \\
\hline
& \\
\dfrac{XY \leq Z}{X p^{(n)} p^{(n+1)} Y \leq Z} \ (A_L) & \dfrac{X \leq YZ}{X \leq Y p^{(n+1)} p^{(n)} Z} \ (A_R) \\
& \\
\hline
& \\
\dfrac{X p^{(k)} Y \leq Z}{X q^{(k)} Y \leq Z} \ (IND_L) & \dfrac{X \leq Y q^{(k)} Z}{X \leq Y p^{(k)} Z} \ (IND_R) \\
\multicolumn{2}{|c|}{q \leq p \text{ if } k \text{ is even, and } p \leq q \text{ if } k \text{ is odd}} \\
& \\
\hline
\end{array}
$$

---

[1] We briefly recall that a *monoid* is a structure $< M, \cdot, 1 >$, such that $\cdot$ is associative and has a neutral element 1 ($\forall x \in M : 1 \cdot x = x \cdot 1 = x$). A partially ordered monoid is a monoid $< M, \cdot, 1 >$ with a partial order $\leq$ that satisfies $\forall a, b, c$: $a \leq b \Rightarrow c \cdot a \leq c \cdot b$ and $a \cdot c \leq b \cdot c$.

This construction, proposed by Buskowski [9], defines a pregroup that extends $\leq$ on basic types P to $T_{(P,\leq)}$[2,3].

The Cut Elimination. The cut rule in the Free Pregroup calculus can be eliminated: every derivable inequality has a cut-free derivation.

Definition 3 (Pregroup Grammar). *Let* $(P, \leq)$ *be a finite partially ordered set. A* pregroup grammar *based on* $(P, \leq)$ *is a lexicalized*[4] *grammar* $G = (\Sigma, I, s)$ *such that* $s \in T_{(P,\leq)}$. G *assigns a type* X *to a string* $v_1 \cdots v_n$ *of* Σ^* *iff for* $1 \leq i \leq n$, $\exists X_i \in I(v_i)$ *such that* $X_1 \cdots X_n \leq X$ *in the free pregroup* $T_{(P,\leq)}$. *The* language $\mathcal{L}(G)$ *is the set of strings in* Σ^* *that are assigned* s *by* G.

Example 1. Let us see the following sentence taken from "Un amour de Swann" by M. Proust: *Maintenant, tous les soirs, quand il l'avait ramenée chez elle, il fallait qu'il entrât.*[5] In Fig. 1 we show a proof of correctness of assignment of types to its fragment. The primitive types used in this proof are: π_3 and $\overline{\pi_3}$ = third person (subject) with $\pi_3 \leq \overline{\pi_3}$, p_2 = past participle, ω = object, s = sentence, s_5 = subjunctive clause, with $s_5 \leq s$, σ = complete subjunctive clause, τ = adverbial phrase. This grammar assigns s to the following sentence:

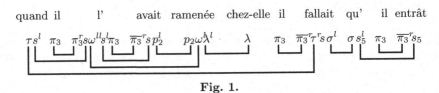

Fig. 1.

In more details, this grammar assigns σ to "qu'il entrât", due to:

$$\cfrac{\cfrac{\cfrac{\sigma = \sigma^{(0)} \leq \sigma^{(0)}}{\sigma s_5^l s_5 = \sigma^{(0)} s_5^{(-1)} s_5^{(0)} \leq \sigma^{(0)}}\,(A_L)}{\sigma s_5^l \pi_3 \pi_3^r s_5 = \sigma^{(0)} s_5^{(-1)} \pi_3^{(0)} \pi_3^{(1)} s_5^{(0)} \leq \sigma^{(0)}}\,(A_L)}{\sigma s_5^l \pi_3 \overline{\pi_3}^r s_5 = \sigma^{(0)} s_5^{(-1)} \pi_3^{(0)} \overline{\pi_3}^{(1)} s_5^{(0)} \leq \sigma^{(0)}}\,(IND_L)$$

figured as:

qu'   il entrât

$\sigma s_5^l \quad \pi_3 \quad \overline{\pi_3}^r s_5$

3   Optional and Iterated Primitive Types

In Fig. 2, we present a more traditional analysis of the sentence in Example 1 represented as a dependency tree.

---

[2] Left and right adjoints are defined by $(p^{(n)})^l = p^{(n-1)}$, $(p^{(n)})^r = p^{(n+1)}$, $(XY)^l = Y^l X^l$ and $(XY)^r = Y^r X^r$. We write p for $p^{(0)}$. We also iterate left and right adjoints for every $X \in T_{(P,\leq)}$: $X^{(0)} = X$, $X^{(n+1)} = (X^r)^{(n)}$ and $X^{(n-1)} = (X^l)^{(n)}$.

[3] $\leq$ is only a preorder. Thus, in fact, the pregroup is the quotient of $T_{(P,\leq)}$ by the equivalence relation $X \leq Y$ & $Y \leq X$.

[4] A lexicalized grammar is a triple (Σ, I, s): Σ is a finite alphabet, I assigns a finite set of categories (or types) to each $c \in \Sigma$, s is a category (or type) associated to correct sentences.

[5] [FR: *Now, every evening when he took back her to her home, he ought to enter*].

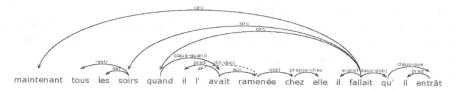

Fig. 2.

We see that the verb *fallait* governs three consecutive circumstantial phrases: *Maintenant, tous les soirs* and *quand il l'avait ramenée chez elle*. All the three are optional and there might be some more dependent circumstantial phrases in this sequence. We can also remark that the oblique object *chez elle* is optional for the verb *ramener*.

A Dependency Grammar Appoach

In rule-based dependency grammars (cf. [13]), as well as in the local environment dependency grammars of Sleator and Temperly [12], such optional and iterated constructions are defined using restricted regular expressions. In categorial dependency grammars (CDG) [14,15], which are a kind of categorial grammars with dependency types as categories, these constructions are defined using rather traditional reduction rules. For instance, the left iterated types are defined by the following two rules:

$$\mathbf{I^l.}\quad a[a^*\backslash\alpha] \vdash [a^*\backslash\alpha]$$
$$\mathbf{\Omega^l.}\quad [a^*\backslash\alpha] \vdash \alpha$$

Several Pregroup Approaches

Below, we denote the optional types by $a^?$ and the iterated types by a^*. Let us show how such types might be defined in the free pregroup. We see at least three approaches, among which only the last one is considered below in full detail.

A Simulation with Compound Types. The first way to define the optional types might be using the following definitions:

$$a^* \stackrel{\mathbf{Def}}{=} x_{a^*} x_{a^*}^r$$
$$a^? \stackrel{\mathbf{Def}}{=} x_{a^?} y_{a^?} y_{a^?}^r z_{a^?} z_{a^?}^r x_{a^?}^r$$

Here, the basic types x_{a^*}, $x_{a^?}$, $y_{a^?}$ and $z_{a^?}$ must not be used for any other purpose. This simulation is not perfect because the duals of the optional type $x_{a^*} x_{a^*}^r$ or of the iterated type $x_{a^?} y_{a^?} y_{a^?}^r z_{a^?} z_{a^?}^r x_{a^?}^r$ are not simple but compound types and that is problematic. In fact, we have to simulate a^*, a^r and a^l, such as we can obtain the composition of a^*, on the right with a^r and on the left with a^l. We have an other simulation to do with $a^?$, a^r and a^l.

A List-like Simulation. In order to simulate an iterated type $[\alpha/a^*]\, a^* \vdash \alpha$ we can distinguish two types, one type a for a first use in a sequence and one type $a^r a$ for next uses in a sequence of elements of type a (this encodes in fact one or more iterations of a). To fully encode a^*, we may add assignments b, whenever ba^* was intended. As in

$$
\begin{array}{cccc}
John & run & fast & yesterday \\
n & n^r sa^l & a & a^r a
\end{array}
$$

We have two assignments for run: in "John run", run $\mapsto n^r s$ but in "John run fast, yesterday", "run" $\mapsto n^r sa^l$. Unfortunately, this approach increases the number of types in the lexicon: if a type has k iterated simple types, the simulation associates 2^k types. The same problem occurs with a simulation of an optional simple type using two types, one with the optional simple type and the second without it.

Adding Rules to Pregroup Grammars. We propose another definition of the optional and iterated types adding to the PG calculus new rules. Our purpose is to ensure properties such as $a \leq a^?$, $a^* a \leq a^*$, $aa^* \leq a^*$, $1 \leq a^?$, $1 \leq a^*$ (see Corollary 1).

Definition 4 (PG with Optional and Iterated Types). *We add the following rules to a PG that define $p^?$ and p^* for p a basic type[6]:*

$$
\frac{XY \leq Z}{Xp^{?(2k+1)}Y \leq Z}\,(?-W_L) \qquad \frac{X \leq YZ}{X \leq Yp^{?(2k)}Z}\,(?-W_R)
$$

$$
\frac{Xp^{(2k+1)}Y \leq Z}{Xp^{?(2k+1)}Y \leq Z}\,(?-D_L) \qquad \frac{X \leq Yp^{(2k)}Z}{X \leq Yp^{?(2k)}Z}\,(?-D_R)
$$

$$
\frac{XY \leq Z}{Xp^{*(2k+1)}Y \leq Z}\,(*-W_L) \qquad \frac{X \leq YZ}{X \leq Yp^{*(2k)}Z}\,(*-W_R)
$$

$$
\frac{Xp^{*(2k+1)}p^{(2k+1)}Y \leq Z}{Xp^{*(2k+1)}Y \leq Z}\,(*-C_L) \qquad \frac{X \leq Yp^{(2k)}p^{*(2k)}Z}{X \leq Yp^{*(2k)}Z}\,(*-C_R)
$$

$$
\frac{Xp^{(2k+1)}p^{*(2k+1)}Y \leq Z}{Xp^{*(2k+1)}Y \leq Z}\,(*-C'_L) \qquad \frac{X \leq Yp^{*(2k)}p^{(2k)}Z}{X \leq Yp^{*(2k)}Z}\,(*-C'_R)
$$

As desired, this system enjoys the following property and corollary.

Proposition 1. *Let $U_i \leq 1$ for $0 \leq i \leq n$ and $C_j \leq a$ for $1 \leq j \leq n$. Then $U_0 C_1 U_1 \leq a^?$ and $\forall k, 1 \leq k \leq n : U_0 C_1 U_1 C_2 \cdots U_k a^* U_{k+1} \cdots C_n U_n \leq a^*$.*

---

[6] $p^?$ and p^* are considered as incomparable (with respect to the PG order) basic types: Rules (A_L) and (A_R) are valid. (IND_L) and (IND_R) are useless (if $p \neq q$ then $p^? \not\leq q$, $p^? \not\leq q^?$, $p^? \not\leq q^*$, etc.).

Proof. It is easy to check that if $X_i \leq Y_i$ for $1 \leq i \leq n$, $n \in \mathbb{N}$ then $X_1 \cdots X_n \leq Y_1 \cdots Y_n$. Thus $U_0 C_1 U_1 \leq a$ and using $(? - D_R)$, we find $U_0 C_1 U_1 \leq a^?$. Similarly, we have $U_0 C_1 U_1 C_2 \cdots U_k a^* U_{k+1} C_n U_n \leq a \cdots aa^* a \cdots a$. Using $(* - C_R)$ and $(* - C'_R)$, we find $U_0 C_1 U_1 C_2 \cdots U_k a^* U_{k+1} \cdots C_n U_n \leq a^*$.

Corollary 1 (Optional and Iterated Basic Types). *For a, a basic type:*

$$a^* a \leq a^*$$
$$a \leq a^? \qquad aa^* \leq a^*$$
$$1 \leq a^? \qquad 1 \leq a^*$$

Theorem 2. *This construction defines a pregroup that extends the free pregroup based on $(P, \leq)$.*

Proof. The structure is a monoid. It is partially ordered (the Cut rule implies transitivity) and moreover with a deduction of $X_1 \leq Y_1$ and a deduction of $X_2 \leq Y_2$, we can build a deduction of $X_1 X_2 \leq Y_1 Y_2$: the structure is a partially ordered monoid. Finally l and r define the left and right adjoints: the proofs use (A_L), (A_R), (Id).

Example 2. In Fig. 3, we show an analysis of the sentence of Proust in the calculus. The primitive types used below are: $\pi_3 =$ third person (subject), $p_2 =$ past participle, $\omega =$ object, $s =$ sentence, $s_5 =$ subjunctive clause, $\sigma =$ complete subjunctive clause, $d =$ determinant, $\rho =$ restrictive adjective, $\tau =$ adverbial phrase.

$\rho^{?^r}$ corresponds to a left optional restrictive adjective argument, $\lambda^{?^l}$ to a right locative argument and τ^{*^l} (or τ^{*^r}) to right (or left) iterated adverbial phrase arguments.

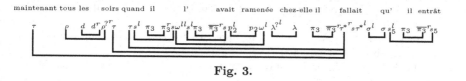

Fig. 3.

Theorem 3 (The Cut Elimination). *The cut rule can be eliminated in the extended calculus: every derivable inequality has a cut-free derivation.*

Proof. The proof is given in Appendix A.

Corollary 2 (Decidability). *The provability of an inequality in this system is decidable*

Proof. The provability of $X \leq Y$ for the free pregroup based on $(P, \leq)$ (without iterated or optional basic type) is a direct consequence of the elimination of the cut rule because the set of possible premises appearing in a cut-free deduction of $X \leq Y$ is finite. The proof is also correct with the addition of optional basic types. However, for a free pregroup with iterated basic types, this argument in not valid because $(*-C_L)$, $(*-C_R)$, $(*-C'_L)$, $(*-C'_R)$ introduce new occurrences of basic types in a cut-free derivation. However, we can limit the number of uses of these rules by the number of basic types in X and Y (this is a consequence of the parity condition on the exponent of the iterated or optional simple type). Thus, even if the search space is infinite, we can limit it to a finite subset.

4   Conclusion

This paper introduces in pregroups two new type constructors ? (option) and $*$ (iteration) allowing to handle in a natural way the optional and iterated constructions such as optional noun modifiers and complements of verbs or their circumstantials (adverbs, time or location clauses etc.). The extended sequent calculus for pregroups formalising the two constructors has natural properties and is decidable. The future work will concern the complexity of this calculus (see [16]) and the development of a parser for this new class of PG.

References

1. Lambek, J.: Type grammars revisited. In: Lecomte, A., Lamarche, F., Perrier, G. (eds.) LACL 1997. LNCS (LNAI), vol. 1582. Springer, Heidelberg (1999)
2. Lambek, J.: The mathematics of sentence structure. American Mathematical Monthly 65, 154–170 (1958)
3. Casadio, C., Lambek, J.: An algebraic analysis of clitic pronouns in italian. In: de Groote, P., Morrill, G., Retoré, C. (eds.) LACL 2001. LNCS (LNAI), vol. 2099. Springer, Heidelberg (2001)
4. Bargelli, D., Lambek, J.: An algebraic approach to french sentence structure. In: de Groote, P., Morrill, G., Retoré, C. (eds.) LACL 2001. LNCS (LNAI), vol. 2099. Springer, Heidelberg (2001)
5. Lambek, J.: Type grammar meets german word order. Theoretical Linguistics 26, 19–30 (2000)
6. Lambek, J., Preller, A.: An algebraic approach to the german noun phrase. Linguistic Analysis 31, 3–4 (2003)
7. Cardinal, K.: An algebraic study of Japanese grammar. Master's thesis, McGill University, Montreal (2002)
8. Sadrzadeh, M.: Pregroup analysis of persian sentences (2007)
9. Buszkowski, W.: Lambek grammars based on pregroups. In: de Groote, P., Morrill, G., Retoré, C. (eds.) LACL 2001. LNCS (LNAI), vol. 2099, pp. 95–109. Springer, Heidelberg (2001)
10. Tesnière, L.: Éléments de syntaxe structurale. Librairie C. Klincksiek, Paris (1959)
11. Hays, D.: Dependency theory: A formalism and some observations. Language 40, 511–525 (1964)
12. Sleator, D.D., Temperly, D.: Parsing English with a Link Grammar. In: Proc. IWPT 1993, pp. 277–291 (1993)
13. Kahane, S. (ed.): Les grammaires de dépendance, Paris, Hermes. Traitement automatique des langues, vol. 41(1) (2000)
14. Dikovsky, A.: Dependencies as categories. In: Kruiff, G.J., Duchier, D. (eds.) Proc. of Workshop Recent Advances in Dependency Grammars, In conjunction with COLING 2004, Geneva, Switzerland, August, 28th 2004, pp. 90–97 (2004)

15. Dekhtyar, M., Dikovsky, A.: Categorial dependency grammars. In: Moortgat, M., Prince, V. (eds.) Proc. of Int. Conf. on Categorial Grammars, Montpellier (2004)
16. Béchet, D., Dikovsky, A., Foret, A., Garel, E.: Introduction of option and iteration into pregroup grammars. In: Casadio, C., Lambek, J. (eds.) Computational Algebric Approaches to Morphology and Syntax, Polimetrica, Monza (Milan), Italy (2008)
17. Došen, K.: Cut Elimination in Categories. Kluwer Academic publishers, Dordrecht (1999)
18. Buszkowski, W.: Cut elimination for the lambek calculus of adjoints. In: Abrusci, V., Casadio, C. (eds.) New Perspectives in Logic and Formal Linguisitics, Proceedings Vth ROMA Workshop, Bulzoni Editore (2001)

Appendix A – Cut Elimination in S' : Proof Details

This proof assumes lemmas detailed in the Appendix B. Take again the systems precedently introduced and consider S the system without **Cut** and S' the system with the rule **Cut** Clearly a proof in S is also a proof in S'.

To show the converse, we proceed by induction on the number of **Cut** and on the length of a derivation γ_l, ending in **Cut** in S':

$$\gamma_l \left\{ \begin{array}{cc} \cdots & \cdots \\ \dfrac{}{X \le Y} R_l & \dfrac{}{Y \le Z} R_r \end{array} \right\} \gamma_r$$
$$\dfrac{\qquad\qquad\qquad\qquad}{X \le Z} \textbf{Cut}$$

- If R_l is the axiom rule, the last rule (cut) can be suppressed since R_r has the same conclusion as $\mathcal{D}$. If R_r is the axiom rule, the last rule (cut) can also be suppressed since R_l has the same conclusion as $\mathcal{D}$. We now assume that neither R_l nor R_r is the axiom rule.
- If R_l is the **Cut** rule, the induction hypothesis applies to γ_l, this **Cut** can be suppressed, in a proof γ'_l, and the final **Cut** can be suppressed in this deduction. If R_r is the **Cut** rule, we proceed similarly.

  We now assume that neither γ_l or γ_r has a **Cut** rule.

- We consider the remaining possibilities for R_l (left part) and R_r (right part): these cases are detailed below.
  - If R_l is a left rule, Y remains in the antecedent, we can easily permute R_l with **Cut** .
  - If R_l is a right rule, we apply a lemma or a rule on the right then **Cut** with the antecedent of R_l

R_l	R_r	method
$A_L, ..._L$	–	permute R_l with **Cut**
A_R	–	**Cut** and lemma (B1)
IND_R	–	**Cut** and IND_L
$? - W_R$	–	**Cut** and lemma (B2)
$? - D_R$	–	**Cut** and lemma (B3)
$* - W_R$	–	**Cut** and lemma (B4)
$* - C_R$	–	**Cut** and lemma (B5)
$* - C'_R$	–	**Cut** and lemma (B6)

- A typical case for a left rule is : $[\ R_l = ? - W_L\ ]$

$$\frac{\dfrac{X_1 Y_1 \leq Y}{X = X_1 p^{?(2k+1)} Y_1 \leq Y} \, ? - W_L \qquad \overset{\cdots}{\underline{}} R_r \Big\} \gamma_r \quad Y \leq Z}{X \leq Z} \text{Cut} \quad \mapsto$$

$$\frac{\dfrac{X_1 Y_1 \leq Y \qquad \overset{\cdots}{\underline{}} R_r \Big\} \gamma_r}{\dfrac{X_1 Y_1 \leq Z}{X_1 p^{?(2k+1)} Y_1 \leq Z} \, ? - W_L}}{}$$

– A typical case for a right rule is : [$R_l = ? - W_R$]

$$\frac{\dfrac{X_1 \leq Y_1 Z_1}{X_1 \leq Y_1 p^{?(2k)} Z_1 = Y} \, ? - W_R \qquad \overset{\cdots}{\underline{}} R_r \Big\} \gamma_r \quad Y \leq Z}{X \leq Z} \text{Cut} \quad \mapsto$$

$$\frac{X_1 \leq Y_1 Z_1 \qquad \dfrac{Y_1 p^{?(2k)} Z_1 \leq Z}{Y_1 Z_1 \leq Z} (lemma B2)}{X_1 \leq Z} \text{Cut}$$

The other cases are similar ∎

Appendix B - Lemmas for Cut elimination in S'

(1) if $U p^{(n+1)} p'^{(n)} V \leq Z$ with $p'^{(n)} \leq p^{(n)}$ and p, p' are primitive not an iterated or an optional type, then $UV \leq Z$

(1') if $U p^{(n+1)} p^{(n)} V \leq Z$ with p an iterated or an optional type, then $UV \leq Z$

(1") if $U P_{(2k+1)} p^{*(2k)} V \leq Z$ or $U p^{*(2k+2)} P_{(2k+1)} V \leq Z$
    where $P_{(2k+1)}$ has the form $\underbrace{p^{(2k+1)}}_{n_1 \, times} \underbrace{p^{*(2k+1)}}_{0 \, or \, 1 occ.} \underbrace{p^{(2k+1)}}_{n_2 \, times}$, then $UV \leq Z$

(2) if $U p^{?(2k)} V \leq Z$ then $UV \leq Z$ (3) if $U p^{?(2k)} V \leq Z$ then $U p^{(2k)} V \leq Z$

(4) if $U p^{*(2k)} V \leq Z$ then $UV \leq Z$

(5) if $U p^{*(2k)} V \leq Z$ then $U p^{(2k)} p^{*(2k)} V \leq Z$

(6) if $U p^{*(2k)} V \leq Z$ then $U p^{*(2k)} p^{(2k)} V \leq Z$

Proof. These properties are shown *for the system without* **Cut** by induction on the premise of the inequality, according to the last applied rule.

We show (1') and (1") separately after.

– The axiom cases are gathered below, *including those for (1') and (1")*.

(1), (1'), (1") case $U p^{(n+1)} p'^{(n)} V = Z$ with $p'^{(n)} \leq p^{(n)}$ or $p = p'$ is an iterated or an optional type

$$\text{then} \quad \frac{\dfrac{UV \leq UV}{UV \leq U p^{(n+1)} p^{(n)} V} A_R}{UV \leq U p^{(n+1)} p'^{(n)} V} IND_R \quad \text{or} \quad \frac{UV \leq UV}{UV \leq U p^{(n+1)} p^{(n)} V} A_R$$
$$\text{(for } p = p' \text{ iterated or optional)}$$

(1") Axiom case with $p = q^*$ let $Z = U Q_{(2k+1)} q^{*(2k)} V$ (first form) or $Z = U q^{*(2k+2)}$
    $Q_{(2k+1)} V$ (second form)
    where $Q_{(2k+1)}$ has the form $\underbrace{q^{(2k+1)}}_{n_1 \, times} \underbrace{q^{*(2k+1)}}_{n_3} \underbrace{q^{(2k+1)}}_{n_2 \, times}$

We proceed by induction on n_1, n_2, to show that in the first form of Z, if $Z = U \underbrace{q^{(2k+1)}}_{n_1 \text{ times}} \underbrace{q^{*(2k+1)}}_{n_3} \underbrace{q^{(2k+1)}}_{n_2 \text{ times}} q^{*(2k)} V$, then $UV \leq Z$

* for $n_3 = 0; n_1 = n_2 = 0$, we get $UV \leq Z = Uq^{*(2k)}V$, as conclusion of rule $* - W_R$ on $UV \leq UV$

* for $n_3 = 0; n_1 + n_2 > 0$,

$$\cfrac{\cfrac{\cfrac{UV \leq UV}{UV \leq Uq^{(2k+1)}q^{(2k)}V} A_R}{UV \leq Uq^{(2k+1)}q^{(2k)}q^{*(2k)}V} *W_R}{UV \leq Uq^{(2k+1)}q^{*(2k)}V} *C_R$$

then for $n_1 + n_2 > 1$:

$$\cfrac{\cfrac{UV \leq U \underbrace{q^{(2k+1)}}_{n_1+n_2-1} q^{*(2k)}V}{UV \leq U \underbrace{q^{(2k+1)}}_{n_1+n_2-1} \mathbf{q^{(2k+1)}q^{(2k)}}q^{*(2k)}V} A_R}{UV \leq U \underbrace{q^{(2k+1)}}_{n_1+n_2} q^{*(2k)}V} *C_R$$

* for $n_3 = 1; n_1 = n_2 = 0$, it is shown above, applying A_R
* for $n_3 = 1; n_1 > 0$ or $n_2 > 0$, we start from above when $n_3 = 0$

$$\cfrac{\cfrac{\cfrac{UV \leq U \underbrace{q^{(2k+1)}}_{n_1} q^{*(2k)}V}{UV \leq U \underbrace{q^{(2k+1)}}_{n_1} q^{*(2k+1)}q^{*(2k)}V} *W_L}{UV \leq U \underbrace{q^{(2k+1)}}_{n_1} q^{*(2k+1)}\mathbf{q^{(2k+1)}q^{(2k)}}q^{*(2k)}V} A_R}{UV \leq U \underbrace{q^{(2k+1)}}_{n_1} q^{*(2k+1)}\mathbf{q^{(2k+1)}}q^{*(2k)}V} *C_R$$

we then repeat these last two steps if $n_2 > 1$.

The second form is similar.

(2) case $Up^{?(2k)}V = Z$ then
$$\cfrac{UV \leq UV}{UV \leq Up^{?(2k)}V = Z} ? - W_R$$

(3) case $Up^{?(2k)}V = Z$ then
$$\cfrac{Up^{(2k)}V \leq Up^{(2k)}V}{Up^{(2k)}V \leq Up^{?(2k)}V = Z} ? - D_R$$

(4) case $Up^{*(2k)}V = Z$ then
$$\cfrac{UV \leq UV}{UV \leq Up^{*(2k)}V = Z} * - W_R$$

(5) case $Up^{*(2k)}V = Z$ then
$$\cfrac{Up^{(2k)}p^{*(2k)}V \leq Up^{(2k)}p^{*(2k)}V}{Up^{(2k)}p^{*(2k)}V \leq Up^{?(2k)}V = Z} * - C_R$$

(6) case $Up^{*(2k)}V = Z$ then
$$\cfrac{Up^{*(2k)}p^{(2k)}V \leq Up^{*(2k)}p^{(2k)}V}{Up^{*(2k)}p^{(2k)}V \leq Up^{?(2k)}V = Z} * - C'_R$$

- If *the last rule is a right rule*, it is easy to permute the induction hypothesis with this rule.
- In *all cases distinct from (1), if the last rule is a left rule distinct from A_L*, it cannot create the type $p^{?(2k)}$ or $p^{*(2k)}$ involved in the lemma. We can then permute the induction hypothesis with this rule. The same remark holds in *all cases distinct from (1), if the last rule is A_L, but does not create the type $p^{?(2k)}$ or $p^{*(2k)}$* involved in the lemma.
- In *all cases distinct from (1), if the last rule is A_L, and it creates $p^{?(2k)}$ or $p^{*(2k)}$*, let $p' = p^{?}$ according to case (2)(3), or $p' = p^{*}$ for (4)(5)(6) such that:

$$\frac{U p'^{(2k)} V = U' V' \leq Z}{U' p'^{(n)} p'^{(n+1)} V' \leq Z} \, A_L \quad \text{with } U p'^{(2k)} = U p'^{(n)} \text{ or } p'^{(2k)} V = p'^{(n+1)} V'$$

We then apply appropriate rules on $U'V' \leq Z$:

–in case(2), if $(n = 2k)$ we show $U' p^{?(2k+1)} V' \leq Z$, from $(? - W_L)$ on $U'V' \leq Z$

–in case(2), if $(n = 2k - 1)$ we show $U' p^{?(2k-1)} V' \leq Z$ similarly

–in case(3), if $(n = 2k)$ we show $U' p^{(2k)} p^{?(2k+1)} V' \leq Z$ by A_R then $(? - D_L)$

–in case(3), if $(n = 2k - 1)$ we show $U' p^{?(2k-1)} p^{(2k)} V' \leq Z$ by A_R then $(? - D_L)$

–in case(4), if $(n = 2k)$ we show $U' p^{*(2k+1)} V' \leq Z$, from $(* - W_L)$ on $U'V' \leq Z$

–in case(4), if $(n = 2k - 1)$ we show $U' p^{*(2k+1)} V' \leq Z$ similarly

–in case(5), if $(n = 2k)$ we show $U' p^{(2k)} p^{*(2k)} p^{*(2k+1)} V' \leq Z$, by A_L on $U'V' \leq Z$ we get : $U' p^{(2k)} p^{(2k+1)} V' \leq Z$,
then by A_L again : $U' p^{(2k)} p^{*(2k)} p^{*(2k+1)} p^{(2k+1)} V' \leq Z$, finally by $* - C_L$.

–in case(5), if $(n = 2k - 1)$ we show $U' p^{*(2k-1)} p^{(2k)} p^{*(2k)} V' \leq Z$, similarly : by A_L on $U'V' \leq Z$ we get : $U' p^{*(2k-1)} p^{*(2k)} V' \leq Z$,
then by A_L again : $U' p^{*(2k-1)} p^{(2k-1)} p^{(2k)} p^{*(2k)} V' \leq Z$, finally by $* - C_L$.

–in case(6), if $(n = 2k)$ we show $U' p^{*(2k)} p^{(2k)} p^{*(2k+1)} V' \leq Z$, by A_L on $U'V' \leq Z$ we get : $U' p^{*(2k)} p^{*(2k+1)} V' \leq Z$,
then by A_L again : $U' p^{*(2k)} p^{(2k)} p^{(2k+1)} p^{*(2k+1)} V' \leq Z$, finally by $* - C'_L$.

–in case(6), if $(n = 2k - 1)$ we show $U' p^{*(2k-1)} p^{*(2k)} p^{(2k)} V' \leq Z$, similarly : by A_L on $U'V' \leq Z$ we get : $U' p^{(2k-1)} p^{(2k)} V' \leq Z$,
then by A_L again : $U' p^{(2k-1)} p^{*(2k-1)} p^{*(2k)} p^{(2k)} V' \leq Z$, finally by $* - C'_L$.

- In *the inductive case for Lemma(1)*, we consider applications of the rule that can interfere with $p^{(n+1)} p'^{(n)}$ (in the other cases we can then permute the induction hypothesis with this rule) :

- A_L case
$$\frac{U' p'^{(n)} V \leq Z}{U \, p^{(n+1)} p'^{(n)} V = U' p^{(n)} \, p^{(n+1)} p'^{(n)} V \leq Z} \, A_L$$

$p^{(n)} \leq p'^{(n)}$, if $p = p'$, the premise is the desired inequality, otherwise we apply IND_L

- A_L case (second possibility) the premise is the desired inequality :
$$\frac{U p'^{(n+1)} V' \leq Z}{U p^{(n+1)} p'^{(n)} \, V = U p^{(n+1)} p'^{(n)} \, p'^{(n+1)} V' \leq Z} \, A_L$$

- IND_L case if $p'^{(n)} \leq p''^{(n)}$
$$\frac{U p^{(n+1)} p''^{(n)} V \leq Z}{U p^{(n+1)} p'^{(n)} V \leq Z} \, IND_L$$

we have $p^{(n)} \leq p'^{(n)} \leq p''^{(n)}$ and apply the induction hypothesis on the premise using $p^{(n)} \leq p''^{(n)}$.

- IND_L case if $p^{(n+1)} \leq q^{(n+1)}$

$$\frac{Uq^{(n+1)}p'^{(n)}V \leq Z}{Up^{(n+1)}p'^{(n)}V \leq Z} \; IND_L$$

we have $q^{(n)} \leq p^{(n)} \leq p'^{(n)}$ and apply the induction hypothesis on the premise using $q^{(n)} \leq p'^{(n)}$.

- *Separate proof for* (1') *and* (1"). We proceed similarly, with all other cases of the lemma already proved. The axiom cases are already shown. We consider below the case of a left rule that interferes with the formula involved in the lemma (otherwise we can permute induction and the rule):

  - $? - W_L$ case, we can write $p = q^?$,

    subcase $n = 2k$ subcase $n = 2k - 1$

    $$\frac{Uq^{?(2k)}V \leq Z}{Uq^{?(2k+1)}q^{?(2k)}V \leq Z} \; ? - W_L \qquad \frac{Uq^{?(2k)}V \leq Z}{Uq^{?(2k)}q^{?(2k-1)}V \leq Z} \; ? - W_L$$

    we then apply lemma B(2) on the premise.

  - $? - D_L$ case, we can write $p = q^?$,

    subcase $n = 2k$ subcase $n = 2k - 1$

    $$\frac{Uq^{(2k+1)}q^{?(2k)}V \leq Z}{Uq^{?(2k+1)}q^{?(2k)}V \leq Z} \; ? - D_L \qquad \frac{Uq^{?(2k)}q^{(2k-1)}V \leq Z}{Uq^{?(2k)}q^{?(2k-1)}V \leq Z} \; ? - D_L$$

    we then apply lemma B(3) on the premise then B(1) and get the result.

  - rules $* - W_L$

    Let $Q_{(2k+1)}$ has the form $\underbrace{q^{(2k+1)}}_{n_1 \text{ times}} \underbrace{q^{*(2k+1)}}_{n_3} \underbrace{q^{(2k+1)}}_{n_2 \text{ times}}$, where $n_3 \leq 1$, and non empty (if empty, we apply lemma (4))

    $$\frac{U'q^{*(2k)}V \leq Z}{UQ_{(2k+1)} \, q^{*(2k)}V = U'\mathbf{q}^{*(2k+1)} \, q^{*(2k)}V \leq Z} \; * - W_L \qquad \begin{array}{l} \text{we have} \\ UQ_{(2k+1)} = U'q^{*(2k+1)}, \\ n_2 = 0 \\ U' = U \, \underbrace{q^{(2k+1)}}_{n_1 \text{ times}} \end{array} \quad \text{we}$$

    then apply the induction on the premise, that has a similar form.
    The case $Uq^{*(2k+2)}Q_{(2k+1)}V \leq Z$ is similar.

  - $* - C_L$ case, subcase $n = 2k$.

    Let $Q_{(2k+1)}$ has the form $\underbrace{q^{(2k+1)}}_{n_1 \text{ times}} \underbrace{q^{*(2k+1)}}_{n_3} \underbrace{q^{(2k+1)}}_{n_2 \text{ times}}$, where $n_3 \leq 1$, and non empty (if empty, we apply lemma (4))

    $$\frac{UQ_{(2k+1)}q^{(2k+1)} \, q^{*(2k)}V = U'\mathbf{q}^{*(2k+1)}\mathbf{q}^{(2k+1)}q^{*(2k)}V \leq Z}{UQ_{(2k+1)} \, q^{*(2k)}V = U'\mathbf{q}^{*(2k+1)} \, q^{*(2k)}V \leq Z} \; * - C_L$$

    we have $UQ_{(2k+1)} = U'q^{*(2k+1)}$, $U' = U \, \underbrace{q^{(2k+1)}}_{n_1 \text{ times}}$,

    we then apply the induction on the premise, that has a similar form.

- $* - C_L$ case, subcase $n = 2k - 1$.
  Let $Q_{(2k-1)}$ has the form $\underbrace{q^{(2k-1)}}_{n_1 \ times} \underbrace{q^{*(2k-1)}}_{n_3} \underbrace{q^{(2k-1)}}_{n_2 \ times}$, where $n_3 \leq 1$ and non empty
  (if empty, we apply lemma (4))

$$\frac{U q^{*(2k)} \mathbf{q}^{*(2k-1)} \mathbf{q}^{(2k-1)} V' \leq Z}{U q^{*(2k)} Q_{(2k-1)} V = U q^{*(2k)} \mathbf{q}^{*(2k-1)} V' \leq Z} * - C_L$$

  we have $Q_{(2k-1)} V = q^{*(2k-1)} V'$, $V' = \underbrace{q^{(2k-1)}}_{n_2 \ times} V$,

  we then apply the induction on the premise, that has a similar form.
- rules $* - C_L'$ can be treated similarly to $* - C_L$

Transformations and Preservation of Self-assembly Dynamics through Homotheties

Florent Becker

Laboratoire d'Informatique du Parallélisme
UMR 5668 CNRS, INRIA, Université Lyon 1, ÉNS Lyon
46 Allée d'Italie – 69364 Lyon Cedex 07 – France
Florent.Becker@ens-lyon.fr

Abstract. We introduce a new notion in self-assembly, that of transforming the dynamics of assembly. This notion allows us to have transformation of the plane computed within the assembly process. Then we apply this notion to zooming. The possibility of zooming depends on the *order condition*. This shows that this condition, which arose from engineering concerns (how to design understandable tile systems) is indeed an important condition of regularity for the assembly process.

1 Introduction and Definitions

Self-assembly, a concept introduced by Winfree in [6] and studied by Winfree, Rothemund, Adelmann and others ([5],[1],[4]) is a model of universal computation by DNA, and of other physical mechanism of accretion and crystallization. This model, consists of a soup with floating Wang tiles with glue on their sides which can stick to each other whenever there is enough glue.

It is a kind of dynamic version of Wang tiling with interesting properties as well as a quite realistic model of some physical or biological construction processes, which has been used experimentally in [4]. In particular, the scale at which the assembled patterned are observable is an important parameter. It can for example be linked to time-complexity[5].

In this article, we are interested in transforming self-assembled tilings by changing the set of tiles. We want these transformations to preserve the dynamics of the assembly, that is the different ways in which a shape can be assembled. This allows us to say that the transformation has been computed *by the tile-set itself* rather than computing the image of a shape by the transformation and then finding a tile-set that assembles into that image. More precisely, we are interested in homotheties (or zooming), which allows to change the scale of assembly.

These homotheties have been used in the literature ([8],[7],...) as tools to enhance the properties of the assemblies, making it robust to error and even self-healing. Yet, their interaction with complex assembly processes were not rigorously defined, as all the tilesets were deterministic. Our notion of homotheties has two uses. First, it is an algorithmic notion, allowing us to scale some

C. Martín-Vide, F. Otto, and H. Fernau (Eds.): LATA 2008, LNCS 5196, pp. 101–112, 2008.

tilesets, and a framework for applying a geometrical transformation to a tileset. But it is also natural enough to be used as a tool for comparison of tilesets: if the "macrotiles" are observed rather than designed, they characterize the relationship between two tilesets operating in the same manner, but at a different scale.

We first explain these notions on an example, then we define formally what is a *dynamics* and what it means to *scale* it, which gives a formal framework for the study of the assembly process itself and its properties such as speed, robustness and so on. This formal definition leads to an impossibility result: the dynamics of some self-assembling tiling can depend on its precise size, with locality effects such as information coming from opposite sides of a tile. Though, we are able to show that given Rothemund's RC-condition [4], or even the slightly more generic order condition, tilesets can indeed be scaled, as these effect are then impossible. This shows that this order condition represents an important measure of regularity for the assembly process.

Our constructions, which can be extended to other transformations than homotheties, allow to define a notion of computation which is much more representative of what happens during the assembly.

1.1   The Model of Self-assembly

The model we study is based on Wang tiles, with some glue added to their sides. The model of self assembly is the following : at a given time, a tile can be added to a finite pattern if and only if its colors match these of its neighbors (like Wang tiles), and if the total strength of the bounds linking the tile to the pattern is more than a given parameter, called the *temperature*.

A Wang tile t on an alphabet Σ is an element of Σ^4. We will note them as follows: $t = (c_N(t), c_S(t), c_E(t), c_W(t))$.

We will use the usual direction functions on $\mathbb{Z}^2$: $N(x,y) = (x, y+1), S(x,y) = (x, y-1), E(x,y) = (x+1, y), W(x,y) = (x-1,y)$, and will note $-S = N, -N = S, -E = W, -W = E$. Given T a set of Wang tiles, and $p : \mathbb{Z}^2 \to T$ a pattern, we will say that $N(x,y)$ is the northern neighbor of (x,y), and that $c_N(p(x,y))$ and $c_S(p(N(x,y)))$ are adjacent colors (or sides), and similarly for other directions. We use the usual notations for intervals to represent intervals of integers: $[a,b] = \{a, \ldots, b\}, [a,b) = \{a, \ldots, b-1\}$, and so on.

Let J be a set of Wang tiles. A *finite pattern* is a partial mapping from $\mathbb{Z}^2$ to J whose domain is finite and 4-connected. We call the domain of a pattern its *shape*.

If a pattern is compatible with the colors on the edges of the tiles (that is two adjacent edges always have the same color), we will say that it is a (finite) *configuration* of J.

Definition 1. *A self-assembly system is a 5-tuplet $s = (\Sigma, g, T, t, seed)$ where*

- *Σ is a finite alphabet, the set of colors;*
- *$g : (\Sigma \to \mathbb{N})$ is called the strength function; for $c \in \Sigma$, we will say that $g(c)$ is the strength of the glue $(c, g(c))$*

- *T is a set of Wang tiles on the alphabet Σ, the* tile set*;*
- *$t \in \mathbb{N}$ is the* temperature*;*
- *seed $\in T$ is the* seed*.*

For graphical representations, we use the following conventions (see figure 1): each tile is represented by a square with symbols representing the glues on their side. The number of symbols on a side corresponds to the strength of the glue. The seed (marked by a star) will be the first finite pattern, which will initiate the growth process. The steps of this growth pattern are called transitions. A transition is the addition of a tile whose colors match those of a free slot in the pattern and whose bond are stronger than the temperature.

Definition 2 (transition). *Given a system $s = (\Sigma, G, T, t, seed)$ and two configurations c and c', there is a* transition *between c and c', which we will note $c \mapsto c'$ if*

$$\exists (x*, y*), \begin{cases} shape(c') = shape(c) \cup \{(x*, y*)\} \\ \forall (x, y) \in shape(c), c(x, y) = c'(x, y) \\ \sum_{\{d \in \{N, S, E, W\} | d(x*, y*) \in shape(c)\}} g(c_d(c'(x*, y*))) \geq t \end{cases}$$

The configurations c such that there is a sequence of transitions from the initial pattern ($\{(0, 0) \mapsto seed\}$) to them are called productions.

Note that we will limit ourselves to self-assembly systems with a temperature of 2, as is the case in the literature[1]. Most of the time, when there is no possible confusion, we use the terms tileset and system interchangeably, and leave the alphabet and the temperature (2) unspecified when their values are obvious from the context. We call direction of a transition the set $\{-d | d(x*, y*) \in c$ before the transition occurs.$\}$

Definition 3. *The dynamics of a self-assembled system is defined as the poset of the production, where the order relation is the transitive closure of transitions.*

Thus we have $c > c'$ (for c and c' two configurations) if and only if c is obtained from c' by adding one or several tiles according to the rules above.

Let us note that for any self-assembled system, its dynamics is an semi-lattice (that is, two elements always have a lower bound), and that for any production c, the set $\{x \leq c\}$ is a lattice. The proof of these two facts is stated in [3].

This dynamics captures the way the assembly takes place, and allows us to reason on the parallelism and synchronization phenomena which take place during the assembly. The properties of the growth process are reflected by those of the dynamics. For example, the speed of the growth in a given configuration is the outgoing degree of the vertex corresponding to that configuration. Confluent branches represent parallelism, and so on. For a detailed reference on (semi) lattices, see [2].

---

[1] Temperature 1 self-assembly is rather trivial, as it corresponds to assembling a Wang tiling with a greedy algorithm and higher temperatures do not seem to be very different from temperature 2.

A *final production* is a maximal element of the dynamics, that is a production to which no tile can be added. Given a tileset T, we say that T *assembles* the set L_T of all its final productions.

2 Scaling While Preserving Dynamics

2.1 An Example

When observing a tileset at larger scale, it is natural to consider each $s \times s$ square as a single tile, and to look at their interactions. We will call these squares *macrotiles*, and compare their interactions with those of the small scale tileset. Before any formal definition, we give an example of scaling, which will help to get the intuition of the kind of phenomenon we are going to capture.

In this section, we consider three tilesets. The first tileset, T, assembles a set of shapes $\mathcal{L}$. This set is the set of squares. The two other tilesets, F and D (for *same F*inal productions and *same D*ynamics) both assemble another set of shapes $\mathcal{L}'$, which is a scaled version of $\mathcal{L}$. We contrast F and D as D has an assembly process which is related to that of T, and not F.

The dynamics of T. Our reference tileset T, taken from [9] assembles the set of all squares. That is, from the configuration $\{(0,0) \mapsto seed\}$, T eventually reaches a configuration whose shape is $[0, n] \times [0, n]$, where n is determined during the assembly.

The assembly can be decomposed in two concurrent processes: the construction of the diagonal, and the filling of the square. The filling of the square is conditioned by the construction of the diagonal: if the diagonal has been built

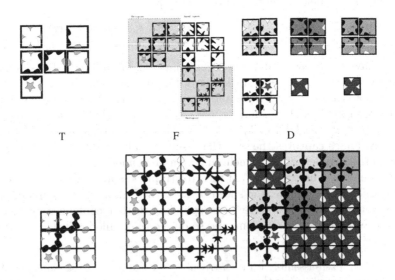

Fig. 1. The three tilesets T,F and D, and an example final production for each of them

up to the point (k, k), filling tiles can only be added in $[0, k] \times [0, k]$. This condition ensures that the shape we get when the diagonal process is stopped by a *stop* tile and the filling process is completed is a square.

The dynamics of F. The tileset F assembles the set of all squares of even size. This set is exactly the image of L by an homothety of factor 2. The assembly process is the following : in a first phase, a square is built as by T, then another of the same size n, then a $n \times 2n$ rectangle. Thus, the final shape is a $2n \times 2n$ rectangle. This assembly process is not at all related to the assembly process of T and the fact that the final productions of F are the images of those of T is somewhat a "happy coincidence".

The dynamics of D. The set of shapes assembled by D is also the set of all squares of even size, but the way the assembly works resembles much more to T: there is also a diagonal being built, and completed into a square. More precisely, let us consider the set of all 2×2 patterns which appear in squares whose lower-left corner is in $\{2k, 2k'|k, k' \in \mathbb{Z}^2\}$. We call these patterns *macrotiles*. Each of these macrotiles can be identified to a tile of T, and the interactions between these macrotiles "look like" the interactions between the corresponding tiles of T.

If we take a (non-terminal) production p of T, and replace each tile with the matching macrotile, the result is a (non-terminal) production p_D of D. The addition of a tile in p corresponds to the addition of the four tiles of the macrotile in D, and the addition of a tile in p is possible if and only if it is possible to add the tiles of the macrotile in p_D.

We say that the dynamics of a tileset D is the s-scaling of that of T whenever one can find a set of macrotiles of size s such that the interactions between the macrotiles of D correspond to the interactions between the tiles of T, and any production of D can be cut in macrotiles.

2.2   Formal Definitions

We now define more formally the notion of scaling the assembly of a tileset T. In this section, T, U are two tilesets, and the question is : "is the dynamics of U a scaling of that of T at scale s ?" We first define our transformations on shapes then extend them to patterns and dynamics.

Definition 4. *Let σ be a shape. Then $h_s(\sigma)$ is $\{(x, y)|(\lfloor x/s \rfloor, \lfloor y/s \rfloor) \in \sigma\}$.*

In order to be able to extend this definition to patterns, where each point is either empty or contains a tile, we need to know what to do with this information. For this, we will consider that $s \times s$ covered by tiles of U form "macrotiles".

Definition 5 (s-macrotile, interpretation). *A s-macrotile on U is a function $[0, s[^2 \to U$. U_s is the set of s-macrotiles on U. For a pattern p, the macrotile at (sx, sy) in p is the function from $[0, s[^2$ to U defined by $(i, j) \mapsto p(xs+i, ys+j)$. An interpretation i is a function from U_s to T.*

Now that we have grouping functions we can map patterns onto patterns. $i_s(p)$ is the pattern p shrinked by a factor s, and reinterpreted in T (when this definition makes sense).

Definition 6 (shrinking). *Any pattern p on U such that there is a shape $\sigma \subset \mathbb{Z}^2$ such that $shape(p) = h_s(\sigma)$ is said to be s-shrinkable.*

Given an interpretation i, for a s-shrinkable pattern p, $i_s(p)$ is the pattern on T defined by:

- *$shape(i_s(p))$ is the σ from above,*
- *for all $(x, y) \in \sigma$, if m is the s-macrotile at (xs, ys) in p, then $(i_s(p))(x, y) = i(m)$.*
- *otherwise, $(i_s(p))(x, y)$ is undefined.*

$D_{s,U}$ is the sublattice formed by the s-shrinkable productions of U.

Definition 7. *Let D_U be the lattice of productions of U, and D_T be the lattice of productions of T. We say that the dynamics of U is the image of the dynamics of T at scale s when*

1. *There is an interpretation i mapping the s-macrotiles of U to T, such that i_s is an isomorphism between $D_{s,U}$ and D_T.*
2. *for every production c of U, there is a $c' \geq c$ which is s-shrinkable.*

This definition formalizes the similarity we had between T and D in the example: we could attribute to each 2-macro-tile of D a tile of T such that this macro-tile behaved like the associated tile. The second condition ensures that we do not have out-of control cancerous growth patterns which just happen not to be shrinkable, and thus are not seen by $D_{s,U}$. Thus, as we can expect, we get the following corollary:

Corollary 1. *Let U and T be two tilesets, L_T and L_U the set of shapes they assemble, if the dynamics of U is the image of the dynamics of T with the interpretation i, then $L_U = h_s(L_T)$*

Proof. Let p be a final production of T. $i_s^{-1}(p)$ is soundly defined since i_s is an isomorphism between D_U and D_T, it is a final production of U, and its shape satisfies $shape(i_s^{-1}(p)) = h_s(shape(p))$.

Let p be a final production of U, p is shrinkable, and $i_s(p)$ is a final production of T, and its shape satisfies $shape(p) = h_s(shape(i_s(p)))$.   □

Our condition does not give a transition-to-transition matching between D_T and $D_{s,U}$, but only a path-to-path matching. This is because the parallelism in U can make several transitions from T take place at once.

Yet, it does allow us to link the transitions involved in both paths: to each path p through the dynamics of U, we can associate another path p' with the same transitions, from which one can extract a sequence of shrinkable patterns $(p*)$ such that the sequence $(i_s(p*))$ is a valid path in the dynamics of T. Let T and U be two tile-sets, s be an integer and i be an interpretation function of U_s in T, such that the dynamics of U is the image of that of T at scale s. Let $p = p(0) \ldots p(n)$ be a path in the dynamics of T. There is a path $p' = i_s^{-1}(p(0)) \ldots i_s^{-1}(p(1)) \ldots i_s(p(n))$ in the dynamics of U. This path is not unique, and there are paths going from $i_s^{-1} p(0)$ to $i_s^{-1}(n)$ without going through all the $i_s^{-1}(p(k))$, but as they have the same extremities as p', they all have the same transitions, but in a different order.

3   Zooming and Self-assembly

3.1   Zooming Sometimes Breaks Dynamics

Having defined a notion of preservation of dynamics, we would like to see if there is a way to scale the dynamics of any tileset. The answer is that it is not possible without some more conditions. What does it mean? It means that the notion of locality which is used by these self-assembling tilings is very local, and thus cannot be scaled. There is then an intrinsic scale in the process. By putting some more conditions on how the assembly takes place, we can scale the dynamics.

To show that zooming (or scaling) necessarily breaks the dynamics of an assembly, we use the fact that in our model, locality means simultaneity in reactions. Thus, wherever we break locality, we will need synchronization, which means putting tiles in advance. This is what breaks the dynamics.

Theorem 1. *There is a tileset T such that there is no Ts whose dynamics is the image of that of T at scale s for $s \geq 2$.*

Proof. Let T be the tileset of figure 2(a). T behaves thus: if the seed is at $(0,0)$, then when the assembly stops, there are two integers l_1 and l_{-1} such that there are tiles in the following places:

- in $\{-1\} \times \{0, \ldots, l_{-1}\}$; these tiles can each be either red or blue

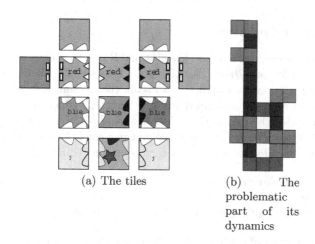

(a) The tiles

(b)   The problematic part of its dynamics

Fig. 2. Our counter-example

- in $\{1\} \times \{0, \ldots, l_1\}$, also in red or blue,
- in the subset S of $\{0\} \times \{0, \ldots, h_0\}$ defined by $\{(0, y) | \sigma_E(-1, y) = \sigma_W(1, y)\}$
  (where $h_0 = min(l_1, l_{-1})$).
- In $\{-2\} \times \{0, \ldots, l_{-1}\}$, wherever the tile in the same row at $x = -1$ is red
- In $\{2\} \times \{0, \ldots, l_1\}$, wherever the tile in the same row at $x = 1$ is red.

In the final productions, there are tiles in the central column whenever there are tiles at both $x = 2$ and $x = -2$ in the same row, or whenever there are neither.

This property cannot be ensured at greater scale. To show this, we do a proof *ad absurdum* by considering a system T_S which is a s-grouping of T. Let us consider the fragment of the dynamics of T represented on figure 2(b), in which no tile can be added at $(0, 1)$ or $(0, 2)$. This guarantee is given if the dependencies between tiles can be expressed by a poset.

Let $c'ij$ be the antecedent of cij by i_s, since i_s is an isomorphism, the $c'ij$ are in the same order as the cij. From $c'21$ and $c'23$, no tile can be added in the central column. Thus, there can be no strength 2 glue on the marked edges in $c'21$ and $c'23$ (represented in bold).

It is also impossible on the marked regions of $c'22$, because then they would have been added before either $c'11$ or $c'12$, and would be present in $c'21$ or c_s23. As there are no strength 2 glues on the marked areas, no tile can be added in the central column. This contradicts the fact that the dynamics of T_s is the image of that of T'.   □

Here, T was unscalable because the locality of the tiles could not be scaled: sometimes, adding a macrotile to a configuration of Ts would have involved two concurrent processes, none of which would have got the whole information on the macrotile to be added. So, in order to scale dynamics, we need a guarantee that we will always be able to have all the information on the tile to be added at one place.

3.2   The Order Condition

To avoid this problem, one can add the classical Row-Column condition [4], or an extension, the *order condition*. This condition removes all effects such as shown in the previous section, and allows us to scale the dynamics. It states that one can associate, to each tile in a production, a direction which corresponds to the way it has extended the production when it was added. Thus, as we know where each tile will extend the productions, we can put the information on colors and glues where it is needed, which allows us to scale the tiling.

Definition 8 (Order condition).. *Let c be a configuration. If there is a poset $<_c$ on the shape of c such that the shapes of the productions $p < c$ are the ideals of $<_c$, then c is said to satisfy the* order condition, *and $<_c$ is its* dependency order.

If all the productions of a tileset obey the order condition, we will say that it is an *ordered tileset*. Intuitively this condition states that the dependencies between the tiles can be presented as a poset, and that this poset depends only on the production and not on its history. $<_c$ represents the dependencies between tiles in c as a tile can be attached to a production p in order to give c if and only if all of its predecessors according to $<_c$ have already been attached. With this condition, the direction of each tile (as defined above) can be decided by looking only at $<_c$ and not at the history of the assembly. The direction of a tile is the opposite of the relative location of its predecessors for $<_c$. For example, if the predecessors of z are $S(z)$ and $W(z)$, the tile at z will have direction NE. This condition is slightly more general than Rothemund's RC condition[4], as it allows to build non-convex patterns. Yet, it is quite natural, and most if not all tilesets in the literature obey this order condition.

A corollary of this condition is that there is never "too much glue" when adding a tile: whenever a tile is added to a configuration, the sum of the glues on its adjacent sides is exactly 2

3.3   A Construction for the Order Case

Given this order condition, we are able to construct a scaled image of a given tile-set. We do this by cutting each of the tiles in s^2 pieces (where s is the scaling factor), and putting new glues on the new edges.

Theorem 2.. *Let T be an ordered-tileset, and s an integer. Then there exists a Ts whose dynamics is the image of that of T at scale s. Furthermore, Ts is also ordered.*

Proof. To each tile of T, one can associate, thanks to the order condition, a set of directions in which the tile can extend the productions. For example, the set of directions associated with a tile having a glue of strength 2 on its southern edge and glues of strength 1 on the other edges is $\{N, SE, SW\}$.

For each tile t and each direction d in which t can extend productions, t^d is the tile t with arrows on its edges showing that t has been added in direction d. These arrows indicate whether an edge is an input or output edge: for each edge e, the arrow points into the tile (in direction $-e$) if the angle between e and d is acute (either 0 or 45 degrees) , and out of the tile otherwise (in direction e). A side with an arrow pointing into the tile is an input side, and a side with an arrow pointing out of the tile is an output side. See figure 3.3.

We consider the tileset T' whose color set is $\Sigma \times \{N, S, E, W\}$, and whose tiles are the t_d for $t \in T$ with strength function s_T. Clearly T' has the same dynamics as T, because of the order condition. We then build a tileset T_s having which is a scaling of T', and thus of T.

Ts is defined as $Seed \cup Replicas$.

$Replicas$ is a set of pieces of tiles in T'. Each tile in T' is cut into s^2 squares of size $1/s$, and the internal edges are given new colors, each unique to the position of the edge inside the tile and to the tile. The details of these glues are shown on figure 3.3. Glues of strength 2 are only present once per edge, and are replaced by equivalent force 1 glues elsewhere on the edge. By convention, they only appear in the uppermost and leftmost parts of each macrotile.

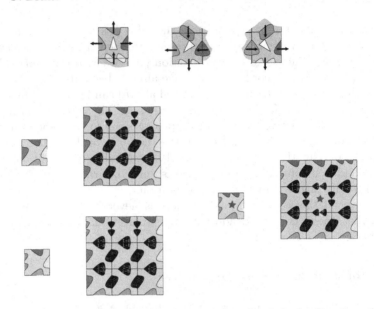

Fig. 3. Top, adding the arrows to the tiles of T: the triangle is the direction of the tile; bottom, cutting into pieces for *Seed* and *Replicas*; all internal edges are unique within each macrotile and between macrotiles

Seed is a set of tile which gives a square of size s with on its edges, the color of the edges of the seed of T.

The interpretation i that we use for this construction is simple: the macrotiles which correspond to the definition of *Replicas* are mapped to the matching tile of T. The macrotile corresponding to the definition of *Seed* is mapped to the seed of T. The other macrotiles are not actual.

Let us now look at the assembly of this tile set. Let us show that the dynamics of Ts is the image of that of T'.

We use the notion of local production. A *local production* is a configuration which appears when building an actual macrotile m. A local production for m is a configuration included in the square $[0, s)^2$ which can be reached from the empty square with the input colors for m on the sides adjacent to its input sides.

We need to show that if $i_s(c) \leq i_s(c')$ are productions of T', then $c \leq c'$. To show this, it is sufficient to show that if there is a transition between $i_s(c)$ and $i_s(c')$, then $c \leq c'$. This is easily seen, one only needs to reproduce the constructions of figure 3.3, which is always possible.

We add an element $\perp$ at the bottom of $D_{T'}$ and D_{Ts} (that is, for all $c \in D_{T'}, c \geq \perp$). We will say that $\perp$ is shrinkable, and $i_s(\perp) = \perp$. We prove the following lemma:

Lemma 1. *For any shrinkable production d and any $c \geq d$, there is a configuration e of T' such that:*

- *shape(e) is the smallest domain $\sigma \subset \mathbb{Z}^2$ such that $h_s(\sigma) \supset$ shape(c))*
- *$e \geq i_s(d)$*

- *For any square $S = [xs, (x+1)s[\times[ys, (y+1)s[$, c restricted to S is a local production of the macro-tile corresponding to $e(x,y)$*
- *and c verifies the order condition*

Proof. For any d, we prove the lemma by induction on c. When $c = d$, we take $e = i_s(d)$.

Let $c \geq d$, and τ be a transition from c to c' where one adds a tile t at $(sx + i, sy + j), 0 \leq i, j < s$. Let e be given by the induction hypothesis for c.

If $(x, y) \in shape(e)$, then e also verifies the lemma for c'. The condition on the shapes is clearly true, and also that $e \geq i_s(d)$. Since c is made of local productions and obeys the order condition, t can only be placed if it has its neighbors on its input sides (because of the arrows on the sides). Thus, it can only be added in accordance with the local productions, and cannot break the order condition.

If $(x, y) \notin shape(e)$, then let $t' \in T$ be the tile corresponding to the macrotile in which t appears. Let t_1 and (if needed) t_2 be the neighbors of t on its input sides, and the corresponding $t'_1, t'_2 \in T$. As t could be added next to t_1 and t_2, t' can be added between t'_1 and t'_2. Let $e' = e \cup \{(x, y) \to t'\}$. $e' \geq i_s(d)$, and it has the smallest shape such that $shape(c') \subset h_s(shape(e'))$. t is a local production of the macrotile corresponding to t', and c' has an order of dependency.

With this lemma, we get that if $c \leq c'$ and c, c' are shrinkable, then $i_s(c) \leq i_s(c')$, as when c is shrinkable, $i_s(e) = c$.

Let c be a production of Ts, then there is a corresponding production e of T', such that c is made of local productions of e. Let $(x_0, y_0, t_0) \ldots (x_k, y_k, t_k)$ be a chain of transitions from $\perp$ to e, and i be the smallest k such that $shape(c) \cap ([sx_k, s(x_k + 1)[\times[sy_k, s(y_k + 1)[]) \neq [sx_k, s(x_k + 1)[\times[sy_k, s(y_k + 1)[$, if it exists. Then, as $shape(c) \cap ([sx_k, s(x_k + 1)[\times[sy_k, s(y_k + 1)[])$ is a local production and its input neighbor(s) are complete, one can add a tile to c in $[sx_k, s(x_k + 1)[\times[sy_k, s(y_k + 1)[$. So if c is not shrinkable, there is a $c' > c$ shrinkable.

So the dynamics of Ts is the image of that of T'.

4   Conclusion

We have defined a notion of transformation of a tile-set that allows us to compute geometrical transformations of the plane within the tile-set, in parallel to whatever construction the tile-set does. This notion allows to pinpoint which aspects of locality a given transformation can break. We gave a demonstration framework for these questions. The fact that in all generality, this notion is not compatible with zooming can be interpreted as an acute sensibility of some self-assembly processes with locality.

Looking at dynamics for making transformation gives a new kind of algorithms on assemblies. Further investigation could lead to more uniform transformations (less dependence on the tileset): if we accept to introduce some inaccuracy in the tileset, then it is possible to have a set G of "growing tiles" such that for a tileset T, the dynamics of $G \cup T$ is the image of that of T by an homothety.

This construction can also be seen as a framework for implementing other geometrical transformations in self-assembly. The macrotiles construction is in fact a kind of tensor products between two tilesets, one assembling a finite pattern, and the other the image of the grid by a transformation. Whenever we have these two elements[2], we can implement the transformation. This expands in a very interesting manner the toolbox of the self-assemblist, allowing for new decompositions of shapes to be built.

References

1. Adleman, L., Cheng, Q., Goel, A., Huang, M.-D., Kempe, D., de Espanès, P.M., Rothemund, P.W.K.: Combinatorial optimization problems in self-assembly. In: STOC 2002: Proceedings of the thiry-fourth annual ACM symposium on Theory of computing, pp. 23–32 (2002)
2. Davey, B.A., Priestley, H.A.: Introduction to Lattices and Order. Cambridge University Press, Cambridge (2002)
3. Rothemund, P.W.K., Winfree, E.: The program-size complexity of self-assembled squares (extended abstract). In: STOC, pp. 459–468 (2000)
4. Rothemund, P.W.K.: Theory and Experiments in Algorithmic Self-Assembly. PhD thesis, University of Southern California (2001)
5. Soloveichik, D., Winfree, E.: Complexity of self-assembled shapes. In: Ferretti, C., Mauri, G., Zandron, C. (eds.) DNA 2004. LNCS, vol. 3384, pp. 344–354. Springer, Heidelberg (2005)
6. Winfree, E.: Algorithmic Self-Assembly of DNA. PhD thesis, Caltech (1998)
7. Winfree, E.: Nanotechnology: Science and computation. In: Chen, J., Jonoska, N., Rozenberg, G. (eds.) Nanotechnology: Science and Computation, Natural Computing, Chapter Self-healing tilesets. Springer, Heidelberg (2006)
8. Winfree, E., Bekbolatov, R.: Proofreading tile sets: Error correction for algorithmic self-assembly. In: Chen, J., Reif, J.H. (eds.) DNA 2003. LNCS, vol. 2943, pp. 126–144. Springer, Heidelberg (2003)
9. Éric Rémila et Ivan Rapaport. Self-assemblying (classes of) shapes with a constant number of tile. Technical report, LIP, ÉNS Lyon (2004)

---

[2] The grid part is not trivial for more involved transformations.

Deterministic Input-Reversal and Input-Revolving Finite Automata

Suna Bensch[1], Henning Bordihn[2], Markus Holzer[3], and Martin Kutrib[2]

[1] Institut für Informatik, Universität Potsdam,
August-Bebel-Straße 89, 14482 Potsdam, Germany
aydin@cs.uni-potsdam.de
[2] Institut für Informatik, Universität Giessen,
Arndtstraße 2, 35392 Giessen, Germany
{bordihn,kutrib}@informatik.uni-giessen.de
[3] Institut für Informatik, Technische Universität München,
Boltzmannstraße 3, 85748 Garching bei München, Germany
holzer@informatik.tu-muenchen.de

Abstract. Extended finite automata are finite state machines with the additional ability to manipulate the remaining part of the input. We investigate three types of deterministic extended automata, namely left-revolving, right-revolving, and input reversal finite automata. Concerning their computational capacity it is shown that nondeterminism is better than determinism, that is, for all three types of automata there is a language accepted by the nondeterministic versions but not accepted by any deterministic automaton of the same type. Concerning the closure properties most of the language families studied are not closed under standard operations. In particular, we show that the family of languages accepted by deterministic right-revolving finite automata is an anti-AFL which is not closed under reversal and intersection.

1 Introduction

In automata theory various classes of automata mainly differ in the resources of which they may make use during the computations. Typical resources are, for example, storages such as pushdown tapes [5], stack tapes [8], or Turing tapes, nondeterminism [11] or alternation [4]. For a more detailed discussion of machines and languages from an automata theoretical point of view see [7]. The investigations in [7] led to a rich theory of abstract families of automata, which is the equivalent to the theory of abstract families of languages. For the definition of an abstract family of languages (abbreviated AFL) we refer to [12].

In several recent papers, for example, see [1,2,3], extended finite automata have been considered. These models are (nondeterministic) finite state machines which are enriched with the ability to apply a string operation on the part of the input that has not been consumed yet. Extended finite automata are inspired by the model of flip pushdown automata [13] which can flip the contents of their pushdown stores in certain configurations. The authors in [9] showed

C. Martín-Vide, F. Otto, and H. Fernau (Eds.): LATA 2008, LNCS 5196, pp. 113–124, 2008.
© Springer-Verlag Berlin Heidelberg 2008

that $k + 1$ pushdown-flips are better than k, and established an interrelation between the pushdown-flips and reversal operations on the unprocessed input of a flip pushdown automaton. In [1], both pushdown and finite automata with input reversal operations have been studied. Moreover, in [2,3] further formal language string operations such as revolving and interchanging have been taken into consideration.

All the forerunner papers investigate nondeterministic automata as the most general case—except in [10], where deterministic flip-pushdown automata were considered. In the present paper, the additional resource of allowed input operations is traded against the resource of nondeterminism. That is, a (single) input operation remains permitted (its applicability depending on configurations), but the automata are restricted to be deterministic. The input reversal, left-revolving, and right-revolving operations will be allowed as prototypes of operations on the unconsumed input string.

After providing the definitions and notation, the computational power of the deterministic extended finite automata is investigated in Section 3. In particular, language families defined by those automata are related to the families of the Chomsky hierarchy and other well-known classes. Section 4 is devoted to compare the power of different input operations. In particular, the power of the deterministic machines is compared with the power of the corresponding nondeterministic ones, proving strict inclusion results. Finally, closure and non-closure properties of the families of deterministic extended finite automata languages under standard language operations are investigated. It turns out that right-revolving deterministic finite automata form a non-reversal and non-intersection closed anti-AFL, what is surprising for a language class defined by a deterministic automaton model. Although anti-AFLs are sometimes referred to an "unfortunate family of languages" there is linguistical evidence that such language families might be of crucial importance, since in [6] it was shown that the family of natural languages is an anti-AFL. Hence the question for uncommon automata models that induce anti-AFLs seem to be worth to consider.

2   Definitions and Preliminaries

We denote the powerset of a set S by 2^S. The empty word is denoted by λ, the reversal of a word w by w^R, and for the length of w we write $|w|$. For the number of occurrences of a symbol a in w we use the notation $|w|_a$. We use $\subseteq$ for *inclusions* and $\subset$ for *strict inclusions*.

In the following we consider finite automata that can reverse or shift the unread part of the input. We start with a uniform definition.

Definition 1. *A (nondeterministic) extended finite automaton is a 6-tuple* $A = (Q, \Sigma, \delta, \Delta, q_0, F)$, *where* Q *is a finite set of states,* Σ *is the input alphabet,* δ *and* Δ *are mappings from* $Q \times (\Sigma \cup \{\lambda\})$ *to* 2^Q, *where* δ *is called the transition function, and* Δ *is called the input operation function,* $q_0 \in Q$ *is the initial state, and* $F \subseteq Q$ *is the set of accepting states. Furthermore,* A *is said to be* λ*-free, if both* δ *and* Δ *are mappings from* $Q \times \Sigma$ *to* 2^Q.

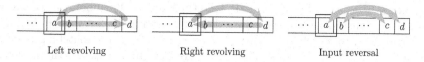

Left revolving   Right revolving   Input reversal

Fig. 1. Input operations

The different operations on the input are formally distinguished by different interpretations of the mapping Δ. To this end, we consider *configurations* of extended finite automata to be tuples (q, w), where $q \in Q$ is the current state, and $w \in \Sigma^*$ is the yet unread part of the input. If a is in $\Sigma \cup \{\lambda\}$ and w in Σ^*, then we write $(q, aw) \vdash_A (p, w)$, if p is in $\delta(q, a)$. Those transitions are referred to as ordinary transitions. An input operation is performed by applying the mapping Δ (cf. Fig. 1). For $a \in \Sigma \cup \{\lambda\}$, $b \in \Sigma$, $w \in \Sigma^*$, and p in $\Delta(q, a)$,

1. a *left-revolving transition* is defined by $(q, a) \vdash_A (p, a)$, $(q, awb) \vdash_A (p, baw)$,
2. a *right-revolving transition* is defined by $(q, aw) \vdash_A (p, wa)$, if $a \in \Sigma$, and $(q, bw) \vdash_A (p, wb)$ and $(q, \lambda) \vdash_A (p, \lambda)$, otherwise, and
3. an *input-reversal transition* is defined by $(q, aw) \vdash_A (p, w^R a)$.

The corresponding transitions are referred to as non-ordinary transitions. Note that, for any operation, if $p \in \Delta(q, \lambda)$, then $(q, \lambda) \vdash_A (p, \lambda)$.

Of particular interest are deterministic computations. A *deterministic* extended finite automaton is an extended finite automaton for which there is at most one choice of action for any possible configuration. A deterministic extended finite automaton $A = (Q, \Sigma, \delta, \Delta, q_0, F)$ with left-revolving, right-revolving, or input-reversal transitions is called a deterministic *left-revolving finite automaton* (lr-DFA), *right-revolving finite automaton* (rr-DFA), or *input-reversal finite automaton* (ir-DFA), respectively. If the automata are nondeterministic, then we use the notations lr-NFA, rr-NFA, or ir-NFA, respectively. The reflexive transitive closure of $\vdash_A$ is denoted by $\vdash_A^*$. The subscript A will be dropped whenever the meaning remains clear. The *language accepted* by A is $L(A) = \{ w \in \Sigma^* \mid (q_0, w) \vdash_A^* (q, \lambda), \text{ for } q \in F \}$. Unless stated otherwise, we denote the *family of languages accepted* by devices of type X by $\mathscr{L}(X)$, where $X \in \{\text{lr-DFA, ir-DFA, rr-DFA, lr-NFA, ir-NFA, lr-NFA}\}$.

In order to clarify our notation we give an example.

Example 2. The non-context-free language $\{ w \in \{a, b, c\}^* \mid |w|_a = |w|_b = |w|_c \}$ is accepted by the extended automaton $A = (Q, \{a, b, c\}, \delta, \Delta, q_0, \{q_0\})$, interpreted as either rr-DFA or lr-DFA, where $Q = \{q_0, q_{ab}, q_{ac}, q_{bc}, q_a, q_b, q_c\}$, and:

1. $\delta(q_0, a) = \{q_{bc}\}$	8. $\delta(q_{ab}, a) = \{q_b\}$	15. $\Delta(q_b, a) = \{q_b\}$
2. $\delta(q_{bc}, b) = \{q_c\}$	9. $\delta(q_{ab}, b) = \{q_a\}$	16. $\Delta(q_b, c) = \{q_b\}$
3. $\delta(q_{bc}, c) = \{q_b\}$	10. $\delta(q_a, a) = \{q_0\}$	17. $\Delta(q_c, a) = \{q_c\}$
4. $\delta(q_0, b) = \{q_{ac}\}$	11. $\delta(q_b, b) = \{q_0\}$	18. $\Delta(q_c, b) = \{q_c\}$
5. $\delta(q_{ac}, a) = \{q_c\}$	12. $\delta(q_c, c) = \{q_0\}$	19. $\Delta(q_{ab}, c) = \{q_{ab}\}$
6. $\delta(q_{ac}, c) = \{q_a\}$	13. $\Delta(q_a, b) = \{q_a\}$	20. $\Delta(q_{ac}, b) = \{q_{ac}\}$
7. $\delta(q_0, c) = \{q_{ab}\}$	14. $\Delta(q_a, c) = \{q_a\}$	21. $\Delta(q_{bc}, a) = \{q_{bc}\}$

From state q_0, automaton A tries to read three different symbols consecutively. It uses the transitions 1 to 12 to store the currently missing symbols in its finite control in order to search for it. Being in a search state, all non-matching symbols are shifted by the transitions 13 to 21. Thus, the input satisfies the property $|w|_a = |w|_b = |w|_c$ when the automaton reaches the accepting state.

It is straightforward to generalize the construction to an arbitrary number of symbols. That is, for any $i \geq 2$, the language

$$\{\, w \in \{a_1, a_2, \ldots, a_i\}^* \mid |w|_{a_1} = |w|_{a_2} = \cdots = |w|_{a_i} \,\}$$

is accepted by some rr-DFA and lr-DFA. $\square$

The situation is different for input-reversal finite automata. It is shown in [1] that nondeterministic ir-NFA accept exactly the linear context-free languages. So, clearly, ir-DFAs cannot accept non-context-free languages.

Example 3. The context-free language $\{\, wcw^R \mid w \in \{a, b\}^* \,\}$ is accepted by the ir-DFA $A = (\{q_0, q_a, q_b, q_a', q_b', q_f\}, \{a, b, c\}, \delta, \Delta, q_0, \{q_f\})$, where

1. $\delta(q_0, a) = \{q_a\}$ 4. $\delta(q_b', b) = \{q_0\}$ 7. $\delta(q_0, c) = \{q_f\}$
2. $\delta(q_0, b) = \{q_b\}$ 5. $\Delta(q_a, \lambda) = \{q_a'\}$
3. $\delta(q_a', a) = \{q_0\}$ 6. $\Delta(q_b, \lambda) = \{q_b'\}$

From state q_0 automaton A tries to read matching symbol pairs one symbol from each end of the input. The transitions 1 and 2 allow A to store the currently read input letter in the finite control in order to search for a corresponding mate letter, which must be at the end of the input. Then with transitions 5 through 8 the symbol at the end of the input is brought to the left, and with transitions 3 and 4 it is verified. Then the search process is repeated. Finally, with transition 9 the sole symbol c is read while A changes to the accepting state. It is straightforward to modify the construction such that the nondeterministic context-free language $\{\, ww^R \mid w \in \{a, b\}^* \,\}$ is accepted some ir-DFA. Similarly, the language $\{\, a^n b^n \mid n \geq 1 \,\}$ is an ir-DFA language. $\square$

The definition of deterministic extended finite automata allows λ-transitions of δ as well as of Δ. They have been included for the sake of compatibility and convenience, since often constructive proofs are much more readable if λ-transitions are used. In [3] it has been shown that λ-transitions do not increase the computational power of nondeterministic extended finite automata. The next theorem proves the same statement for the deterministic case.

Theorem 4. *For a deterministic extended finite automaton A of any type, one can construct a λ-free deterministic extended finite automaton B of the same type, such that $L(A) = L(B)$.*

Proof. Given a deterministic extended finite automaton $A = (Q, \Sigma, \delta, \Delta, q_0, F)$ we construct $B = (Q, \Sigma, \delta', \Delta', q_0, F')$ as follows. The non-λ-transitions of B are defined to be the non-λ-transitions of A, that is, $\delta'(p, a) = \delta(p, a)$ and $\Delta'(p, a) =$

$\Delta(p, a)$, for all $p \in Q$, $a \in \Sigma$. Next, we replace non-ordinary λ-transitions of A that appear on non-empty input. To this end, if $\Delta(p, \lambda)$ is defined, then we set $\Delta'(p, a) = \Delta(p, \lambda)$, for any $a \in \Sigma$. After all non-ordinary λ-transitions have been removed, we replace ordinary λ-transitions of A that appear on non-empty input. As A is deterministic, we can assume without loss of generality that no λ-cycles appear, that is, for all states p there is a unique state $\hat{p}$, such that $(p, \lambda) \vdash_A^* (\hat{p}, \lambda)$ with ordinary transitions only and $\delta(\hat{p}, \lambda)$ is undefined. Note that $\hat{p} = p$ holds if $\delta(p, \lambda)$ is undefined. Now, for all $p \in Q$ and all $a \in \Sigma$, we set $\delta'(p, a) = \delta(\hat{p}, a)$ if $\delta(\hat{p}, a)$ is defined and $\Delta'(p, a) = \Delta(\hat{p}, a)$ if $\Delta(\hat{p}, a)$ is defined. These transitions do not violate the determinism of B since, for any $a \in \Sigma$, neither $\delta(p, a)$ nor $\Delta(p, a)$ is defined if $\delta(p, \lambda)$ is defined. So far, B can simulate λ-transitions of A that appear on non-empty input. But there may be λ-transitions at the end of the computation when the whole input has been consumed. In order to retain deterministic computations, we can overcome the problem by adjusting the set of accepting states, since λ-transitions at the end of the computation can only change the finally reachable states. So, for any $p \in Q$, let $\Lambda_p = \{ q \in Q \mid (p, \lambda) \vdash_A^* (q, \lambda) \}$ be the set of all states that are reachable from some state p with ordinary and non-ordinary λ-transitions. Then we set $F' = F \cup \{ p \in Q \mid \Lambda_p \cap F \neq \emptyset \}$. □

3   Computational Capacity

In this section we investigate the computational power of deterministic extended finite automata. In particular, we compare the language families defined by those automata to well-known language families. Clearly, every regular language is accepted by any type of extended automaton in question. A straightforward Turing machine simulation yields the following upper bounds.

Theorem 5
1. Every language accepted by a right- or left-revolving finite automaton belongs to both complexity classes DTIME(n^2) *and* DSPACE(n).
2. Every language accepted by an input-reversal finite automaton belongs to both complexity classes DTIME(n) *and* DSPACE(n).

Obviously, *unary* languages accepted by extended finite automata are regular since neither left- and right-revolving nor input-reversal moves change the remaining part of the input. Therefore, non-ordinary moves can be omitted.

Theorem 6. *A unary language L is accepted by an extended finite automaton if and only if L is regular.*

An immediate consequence is that the inclusions of Theorem 5 are proper, since the non-regular language $\{ a^{n^2} \mid n \geq 1 \}$ belongs to the intersection DTIME(n) $\cap$ DSPACE(n).

Once it is known that all regular languages are accepted by deterministic extended finite automata, there is a natural question for better lower bounds in terms of known language families. A proper but still weak superclass of regular languages is the family of languages accepted by deterministic one-turn pushdown automata. We denote this family DLIN. Though deterministic extended finite automata accept rather complicated non-context-free languages, none of the deterministic extended finite automata under consideration can accept all languages from DLIN. Moreover, this will imply that nondeterminism is better than determinism for the cases of left-revolving and input-reversal automata.

Lemma 7. *Let* $L' = \{\, wcw^R \mid w \in \{a, b\}^* \,\}$. *The language* $L = L' \cup L'\{b\}\{a, b\}^*$ *is not accepted by any* lr-DFA.

Proof. In contrast to the assertion assume that L is accepted by some lr-DFA $A = (Q, \Sigma, \delta, \Delta, q_0, F)$ with n states. According to Theorem 4 we may assume A to be λ-free. The word $w_1 = a^{2n}ca^{2n}b^na^{2n}$ belongs to L. Due to the choice of n, there is an accepting computation such that some state q appears at least twice while A reads a's only, say

$$(q_0, w_1) \vdash^* (q, w_1') \vdash^+ (q, w_1'') \vdash^* (q_f, \lambda),$$

where $q \in Q$, $q_f \in F$. Moreover, we can derive $w_1' \neq w_1''$, and A has consumed at most n symbols a each while computing w_1' from w_1 and w_1'' from w_1', that is, $|w_1| - |w_1'| \leq n$ and $|w_1'| - |w_1''| \leq n$. We obtain $w_1' = a^{2n-i+j}ca^{2n}b^na^{2n-j}$, for some $0 \leq i+j \leq n$, where A consumes i symbols and performs j revolving steps. Similarly, we have $w_1'' = a^{2n-i+j-\ell+m}ca^{2n}b^na^{2n-j-m}$, for some $0 < \ell + m \leq n$. We conclude that there is an accepting computation

$$(q_0, a^{2n-\ell+m}ca^{2n}b^na^{2n-m}) \vdash^* (q, a^{2n-\ell+m-i+j}ca^{2n}b^na^{2n-m-j}) \vdash^* (q_f, \lambda)$$

which implies $\ell = m$.

Now we consider the word $w_2 = a^{2n}ca^{2n}$ that belongs to L. Since A is deterministic, the accepting computation on w_2 is

$$(q_0, w_2) \vdash^* (q, a^{2n-i+j}ca^{2n-j}) \vdash^* (q, a^{2n-i+j-\ell+m}ca^{2n-j-m}) \vdash^* (q_f', \lambda),$$

where $q_f' \in F$. Moreover, we obtain the computation

$$(q_0, a^{2n-\ell+m}ca^{2n-m}) \vdash^* (q, a^{2n-\ell+m-i+j}ca^{2n-m-j}) \vdash^* (q_f', \lambda).$$

Since $\ell = m$ the input $a^{2n}ca^{2n-m}$ is accepted. But $\ell + m > 0$ and $\ell = m$ imply $m > 0$, a contradiction. □

Theorem 8. *The families* DLIN *and* $\mathscr{L}(\text{lr-DFA})$ *are incomparable.*

Proof. Lemma 7 presents a deterministic one-turn pushdown automaton language not belonging to $\mathscr{L}(\text{lr-DFA})$. Example 2 shows that the non-context-free language $\{\, w \in \{a, b\}^* \mid |w|_a = |w|_b = |w|_c \,\}$ is accepted by an lr-DFA. □

Theorem 9. *The families* DLIN *and* $\mathscr{L}(\text{rr-DFA})$ *are incomparable.*

Proof. In [2] it is shown that the deterministic one-turn pushdown automaton language $\{\, a^n b^n \mid n \geq 1 \,\}$ is not accepted by any rr-NFA. On the other hand, by Example 2 the non-context-free language $\{\, w \in \{a, b\}^* \mid |w|_a = |w|_b = |w|_c \,\}$ is accepted by an rr-DFA. □

It remains to investigate whether $\mathscr{L}$(ir-DFA) is comparable with DLIN. It is known that nondeterministic ir-NFAs characterize the linear context-free languages [1]. Therefore the question arises whether this relation remains true for deterministic devices. The answer will be derived from the following lemma.

Lemma 10. *The family $\mathscr{L}$(ir-DFA) is properly included in $\mathscr{L}$(lr-DFA).*

Proof. As ir-NFAs characterize the linear context-free languages [1] and lr-DFAs can accept non-context-free languages, it is only left to prove the inclusion.

Given some ir-DFA $A = (Q, \Sigma, \delta, \Delta, q_0, F)$ we construct an equivalent lr-DFA $A' = (Q', \Sigma, \delta', \Delta', q_0', F')$. Basically, reversing the input means to read the input from right to left instead of left to right, and *vice versa*. The idea of the construction is to remember the direction (where l indicates from left to right and r from right to left), and to simulate right to left steps by reading a symbol which previously has been fetched from the back by a left-revolving step. To this end, it is convenient to store both the first and the last symbol of the remaining input as parts of the state. Without loss of generality, we assume that A is λ-free, and construct A' formally as follows.

$$Q' = (Q \cup \bar{Q}) \times (\Sigma \cup \{\sqcup\})^2 \times \{l, r\}, \quad q_0' = (q_0, \sqcup, \sqcup, l),$$

where $\bar{Q} = \{\bar{q} \mid q \in Q\}$ is a disjoint copy of Q, and $\sqcup$ is a blank.

For all $q \in Q$, $a \in \Sigma$, $x \in \Sigma \cup \{\sqcup\}$, $y \in \Sigma$, and $d \in \{l, r\}$, the transitions

$$\delta'((q, \sqcup, x, d), a) = (q, a, x, d),$$
$$\Delta'((q, y, \sqcup, d), \lambda) = (\bar{q}, y, \sqcup, d) \text{ and } \delta'((\bar{q}, y, \sqcup, d), a) = (q, y, a, d)$$

are applied to store the first symbol as the second component of the state and to fetch a symbol from the back and to store it as the third component of the state, whenever at least one of these components is blank.

If $\delta(p, a) = q$ is a transition of A, for $p, q \in Q$ and $a \in \Sigma$, then we simulate it for any $x \in \Sigma$ by

$$\delta'((p, a, x, l), \lambda) = (q, \sqcup, x, l) \text{ and } \delta'((p, x, a, r), \lambda) = (q, x, \sqcup, r).$$

If A reads the input from left to right or from right to left, then it applies the transition $\delta(p, a) = q$ to the first or the last input symbol, respectively. The left-revolving automaton A' has stored these symbols in the second and the third components of its state, which subsequently get blank, and a new symbol is fetched from the left or right end of the input, respectively.

If $\Delta(p, a) = q$ is an input-reversal transition of A, for $p, q \in Q$ and $a \in \Sigma$, then we simulate it for any $x \in \Sigma$ by

$$\delta'((p, a, x, l), \lambda) = (q, a, x, r) \text{ and } \delta'((p, x, a, r), \lambda) = (q, x, a, l).$$

These transitions change the first component of the state and l to r, or r to l.

A' is deterministic since A is so. Moreover, computations of A are simulated more or less directly until an input string of length 2 remains. In such a situation, the input of A' is already empty, since these symbols are always stored in the state. In order to cope with the missing last transitions, we can adjust the set of accepting states as follows. Let $p \in Q$, $x, y \in \Sigma$, and $d \in \{l, r\}$. Then

$$(p, x, y, d) \in F' \text{ if and only if } (p, xy) \vdash_A^+ (q_f, \lambda),$$

for some $q_f \in F$. If the input of A is of length less than 2, then we adjust the set of accepting states accordingly. □

Theorem 11. *The families* DLIN *and* $\mathscr{L}$(ir-DFA) *are incomparable.*

Proof. Lemma 7 shows that there is a language belonging to DLIN but not to $\mathscr{L}$(lr-DFA). By Lemma 10 it does not belong to $\mathscr{L}$(ir-DFA) either. On the other hand, it is easy to see that the language $\{ ww^R \mid w \in \{a, b\}^* \}$ is accepted by some ir-DFA but does not belong to DLIN. □

So far, we have related the computational power of deterministic extended finite automata to the expressive power of well-known language families. It turns out that all families in question properly include the regular languages and are incomparable with DLIN. Together with Theorem 5 it follows that all families are properly included in the deterministic context-sensitive languages. Together with the results from [1,2] we obtain that $\mathscr{L}$(ir-DFA) is properly included in the family of linear context-free languages, whereas the families $\mathscr{L}$(lr-DFA) and $\mathscr{L}$(rr-DFA) are incomparable to the class of context-free languages.

4   Comparing Modes

This section is devoted to the comparison of the different input operations. As mentioned before, we can separate nondeterministic classes from deterministic classes. Summarizing the investigations in the literature and the previous section, we have the proper inclusions $\mathscr{L}$(ir-DFA) $\subset$ $\mathscr{L}$(lr-DFA) (Lemma 10) and $\mathscr{L}$(ir-NFA) $\subset$ $\mathscr{L}$(lr-NFA) from [3]. Furthermore, $\mathscr{L}$(ir-NFA) is equal to the family of linear context-free languages [1], whereas all deterministic families are incomparable with DLIN which, in turn, is properly included in the linear context-free languages. Thus we immediately obtain the following theorem.

Theorem 12
1. *The family* $\mathscr{L}$(ir-DFA) *is properly included in* $\mathscr{L}$(ir-NFA).
2. *The family* $\mathscr{L}$(lr-DFA) *is properly included in* $\mathscr{L}$(lr-NFA). □

For the sake of completeness we now present the remaining separation result. It is a consequence of the different closure properties shown in the next section.

Theorem 13. *The family* $\mathscr{L}$(rr-DFA) *is properly included in* $\mathscr{L}$(rr-NFA).

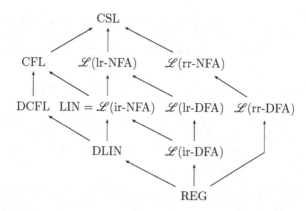

Fig. 2. Inclusion structure. All shown inclusions are strict and families that are not linked by a path are pairwise incomparable. CSL denotes the family of context-sensitive, CFL that of context-free, DCFL that of deterministic context-free, LIN that of linear context-free languages, and REG that of regular languages.

Proof. Clearly, the family $\mathscr{L}$(rr-NFA) is closed under union. In Lemma 18 it is shown that the deterministic family $\mathscr{L}$(rr-DFA) is not closed under union. $\square$

We continue to compare the power of the different input operations among themselves. The complete picture is shown in Figure 2.

Lemma 14. *There is a language* $L \in \mathscr{L}(\text{rr-DFA}) \setminus \mathscr{L}(\text{lr-NFA})$.

Proof. We use $L = \{ a^{2n}bv \mid n \geq 0, v \in \{a,b\}^*, n + |v|_a = 1 + |v|_b \}$ as witness language. Contrarily, assume that L is accepted by some lr-NFA $A = (Q, \Sigma, \delta, \Delta, q_0, F)$ with n states. Due to [2] A is assumed to be λ-free. We consider the word $w = a^{2n}b^{4n}a^{3n}$ which belongs to L. Due to the choice of n, there is an accepting computation such that some state q appears at least twice while A reads a's only, say

$$(q_0, w) \vdash^* (q, a^{2n-i+j}b^{4n}a^{3n-j}) \vdash^+ (q, a^{2n-i+j-\ell+m}b^{4n}a^{3n-j-m}) \vdash^* (q_f, \lambda),$$

where $q \in Q$, $q_f \in F$, and A consumes i symbols and performs j revolving steps until the first q appears, and it consumes further ℓ symbols and performs further m revolving steps until the second q appears. Furthermore, $i, j, \ell, m \leq n$ and $\ell + m > 0$. We conclude that there is an accepting computation

$$(q_0, a^{2n-\ell+m}b^{4n}a^{3n-m}) \vdash^* (q, a^{2n-\ell+m-i+j}b^{4n}a^{3n-m-j}) \vdash^* (q_f, \lambda),$$

which implies $a^{2n-\ell+m}b^{4n}a^{3n-m} \in L$. Therefore, $m - \ell$ is even and, thus, $l + m \geq 2$ since $\ell + m > 0$. On the other hand, since $n + \frac{m-\ell}{2} + 3n - m = 4n - \frac{\ell+m}{2} \neq 4n$, we have $a^{2n-\ell+m}b^{4n}a^{3n-m} \notin L$, a contradiction.

It remains to be shown that L is accepted by some rr-DFA A'. Basically, A' scans the leading a's, whereby every second symbol a is revolved until the first b appears. Subsequently, A' behaves as a known acceptor for the language $\{ w \in \{a,b\}^* \mid |w|_a = |w|_b \}$. $\square$

Corollary 15. *There is a language $L \in \mathscr{L}(\text{rr-DFA}) \setminus \mathscr{L}(\text{ir-NFA})$.*

Proof. The assertion follows by Lemma 14 and $\mathscr{L}(\text{ir-NFA}) \subset \mathscr{L}(\text{lr-NFA})$ shown in [3]. □

Lemma 16. *There is a language $L \in \mathscr{L}(\text{ir-DFA}) \setminus \mathscr{L}(\text{rr-NFA})$, hence we have $\mathscr{L}(\text{lr-DFA}) \setminus \mathscr{L}(\text{rr-NFA}) \neq \emptyset$.*

Proof. Example 3 revealed that the language $L = \{\, a^n b^n \mid n \geq 1 \,\}$ is accepted by some ir-DFA, hence also by some lr-DFA. On the other hand, it is shown in [2] that L is not accepted by any rr-NFA. □

Lemma 17. *There is a language $L \in \mathscr{L}(\text{lr-DFA}) \setminus \mathscr{L}(\text{ir-NFA})$.*

Proof. The witness language $\{\, w \in \{a, b, c\}^* \mid |w|_a = |w|_b = |w|_c \,\}$ does not belong to $\mathscr{L}(\text{ir-NFA})$ since it is not linear context free. But by Example 2 it is accepted by some lr-DFA. □

5   Closure Properties

We next discuss the closure properties of deterministic extended finite automata languages. Due to the lack of space we only give two lemmata exemplarily here. Further results are summarized in Table 1.

Table 1. Closure properties of families of deterministic extended automata languages; entry + means the the language family is closed under the corresponding operation, − means that it is *not* closed, and ? means that the answer is not known.

$\mathscr{L}(\cdot)$	Operation								
	$\cup$	$\cap$	$\sim$	$\cap_{reg}$	R	$\cdot$	$*$	h^{-1}	h_λ
lr-DFA	−	−	−	?	−	−	−	−	−
rr-DFA	−	−	−	−	−	−	−	−	−
ir-DFA	−	−	+	+	+	−	−	+	−

First, we show that $\mathscr{L}(\text{lr-DFA})$ and $\mathscr{L}(\text{rr-DFA})$ are neither closed under complementation nor under union, and then we proof that these families of languages are not closed under reversal.

Lemma 18. *The families $\mathscr{L}(\text{lr-DFA})$ and $\mathscr{L}(\text{rr-DFA})$ are neither closed under complementation nor under union.*

Proof. Let $L = \{\, w \in \{a, b\}^* \mid |w|_a = |w|_b \,\}$. We show that its complement $\bar{L}$ belongs neither to $\mathscr{L}(\text{lr-DFA})$ nor $\mathscr{L}(\text{rr-DFA})$. Contrarily assume a deterministic revolving finite automaton A accepts $\bar{L}$, and consider an accepting computation on input a^n, for n large enough. In order to accept the input, A has to read every symbol. So, it *must not* run into loops consisting of revolving transitions only.

This implies that after an initial part with some $i \geq 0$ ordinary transitions and $j \geq 0$ revolving transitions, the computation becomes cyclic, where some $k \geq 1$ ordinary transitions and $\ell \geq 0$ revolving transitions appear in a single loop. Let $n = i + j + c(k + \ell)$, for some constant number c which is large enough. For A being a rr-DFA we obtain the accepting computation

$$(q_0, a^n b^{n-k}) \vdash^* (q_i, a^{n-i-j} b^{n-k} a^j) \vdash^+$$
$$(q_i, a^{n-i-j-k-\ell} b^{n-k} a^{j+\ell}) \vdash^+ (q_i, b^{n-k} a^{j+c\cdot\ell}) \vdash^+ (q_f, \lambda),$$

for some state q_i. Due to the deterministic behavior, the computation on input $a^{n-k-\ell} b^{n-k} a^\ell$ is

$$(q_0, a^{n-k-\ell} b^{n-k} a^\ell) \vdash^* (q_i, a^{n-k-\ell-i-j} b^{n-k} a^{\ell+j}) \vdash^+$$
$$(q_i, b^{n-k} a^{\ell+j+(c-1)\cdot\ell}) \vdash^+ (q_f, \lambda),$$

which is a contradiction, since $a^{n-k-\ell} b^{n-k} a^\ell$ does not belong to $\bar{L}$.

If A is a left-revolving finite automaton, we obtain a contradiction with the inputs $a^{i+c\cdot k} b^{n-k} a^{j+c\cdot\ell} \in \bar{L}$ and $a^{i+(c-1)\cdot k+\ell} b^{n-k} a^{j+(c-1)\cdot\ell} \in L$.

Hence, both families are not closed under complementation. Almost the same reasoning can be used to show that the language $\{ w \in \{a,b\}^* \mid |w|_a = |w|_b \} \cup \{ a^n \mid n \geq 1 \}$ is not accepted by any rr-DFA and lr-DFA. Hence, both families are not closed under union either. □

Lemma 19. *Both $\mathscr{L}(\text{lr-DFA})$ and $\mathscr{L}(\text{rr-DFA})$ are not closed under reversal.*

Proof. By Lemma 14, the language $\{ a^{2n} bv \mid n \geq 0, v \in \{a,b\}^*, n+|v|_a = 1+|v|_b \}$ does not belong to $\mathscr{L}(\text{lr-NFA})$. On the other hand, its reversal is accepted by some lr-DFA A as follows. Automaton A starts to revolve a's from the back whereby every second a symbol is deleted (by a read transition). When the first b appears, A behaves as a known acceptor for the language $\{ w \in \{a,b\}^* \mid |w|_a = |w|_b \}$. So, $\mathscr{L}(\text{lr-DFA})$ is not closed under reversal.

Now, let L be any language from $\mathscr{L}(\text{rr-DFA}) \setminus \mathscr{L}(\text{lr-NFA})$. By Lemma 14 such languages exist. In contrast to the assertion assume $\mathscr{L}(\text{rr-DFA})$ were closed under reversal. Then $L^R \in \mathscr{L}(\text{rr-DFA})$ and, trivially, L^R is accepted by some nondeterministic right-revolving finite automaton, too. By a result in [2] we know, that if a language is accepted by a nondeterministic right-revolving finite automaton, then its reversal is accepted by a nondeterministic left-revolving finite automaton. We conclude $L \in \mathscr{L}(\text{lr-NFA})$, a contradiction. □

Finally, the closure of $\mathscr{L}(\text{ir-DFA})$ is trivial.

Lemma 20. *The family $\mathscr{L}(\text{ir-DFA})$ is closed under reversal.*

6 Conclusions

We have investigated left-revolving,- right-revolving and input reversal finite automata. The main interest was on deterministic computations. For all models we have separated deterministic from nondeterministic automata and have considered the relationships between the distinguished deterministic classes. The inclusion relation between the language families considered are depicted in Figure 2. Concerning the closure properties of the investigated language families we refer to Table 1, where we summarize our results. It was shown that most of the classes are not closed under standard language operations like union, concatenation or Kleene closure. Although it is not known whether deterministic left-revolving finite automata are closed under intersection with regular sets, these automata might be of interest in the framework of mathematical linguistics, since natural languages are an anti-AFL (see [6]). A model for natural languages should include the three non-context-free languages $L_1 = \{\, a^n b^n c^n \mid n \geq 1 \,\}$, $L_2 = \{\, a^n b^m c^n d^m \mid n, m \geq 1 \,\}$, and $L_3 = \{\, ww \mid w \in \{a, b\}^+ \,\}$. We know that L_1 and L_2 can be accepted by deterministic left-revolving automata and that L_3 can be accepted if it is marked appropriately.

References

1. Bordihn, H., Holzer, M., Kutrib, M.: Input reversals and iterated pushdown automata: A new characterization of Khabbaz geometric hierarchy of languages. In: Calude, C.S., Calude, E., Dinneen, M.J. (eds.) DLT 2004. LNCS, vol. 3340, pp. 102–113. Springer, Heidelberg (2004)
2. Bordihn, H., Holzer, M., Kutrib, M.: Revolving-input finite automata. In: De Felice, C., Restivo, A. (eds.) DLT 2005. LNCS, vol. 3572, pp. 168–179. Springer, Heidelberg (2005)
3. Bordihn, H., Holzer, M., Kutrib, M.: Hybrid extended finite automata. In: H. Ibarra, O., Yen, H.-C. (eds.) CIAA 2006. LNCS, vol. 4094, pp. 34–45. Springer, Heidelberg (2006)
4. Chandra, A.K., Kozen, D.C., Stockmeyer, L.J.: Alternation. Journal of the ACM 28, 114–133 (1981)
5. Chomsky, N.: Formal Properties of Grammars.In: Handbook of Mathematic Psychology, vol. 2, pp. 323–418. Wiley & Sons, New York (1962)
6. Culy, C.: Formal properties of natural language and linguistic theories. Linguistics and Philosophy 19, 599–617 (1996)
7. Ginsburg, S.: Algebraic and Automata-Theoretic Properties of Formal Languages. North-Holland, Amsterdam (1975)
8. Ginsburg, S., Greibach, S.A., Harrison, M.A.: One-way stack automata. Journal of the ACM 14, 389–418 (1967)
9. Holzer, M., Kutrib, M.: Flip-pushdown automata: $k+1$ pushdown reversals are better than k. In: International Colloquium on Automata, Languages and Programming (ICALP 2003). LNCS, vol. 2719, pp. 490–501. Springer, Heidelberg (2003)
10. Holzer, M., Kutrib, M.: Flip-pushdown automata: Nondeterminism is better than determinism. In: Developments in Language Theory (DLT 2003). LNCS, vol. 2710, pp. 361–372. Springer, Heidelberg (2003)
11. Rabin, M.O., Scott, D.: Finite automata and their decision problems. IBM Journal of Research and Development 3, 114–125 (1959)
12. Salomaa, A.: Formal Languages. Academic Press, London (1973)
13. Sarkar, P.: Pushdown automaton with the ability to flip its stack. Report TR01-081, Electronic Colloquium on Computational Complexity (ECCC) (2001)

Random Context in Regulated Rewriting *Versus* Cooperating Distributed Grammar Systems

Henning Bordihn[1] and Markus Holzer[2]

[1] Institut für Informatik, Universität Giessen,
Arndtstraße 2, D-35392 Giessen, Germany
`bordihn@informatik.uni-giessen.de`
[2] Institut für Informatik, Technische Universität München,
Boltzmannstraße 3, D-85748 Garching bei München, Germany
`holzer@informatik.tu-muenchen.de`

Abstract. It is well known that certain language families generated by cooperating distributed (CD) grammar systems can be characterized in terms of context-free random context grammars. In particular, the language families generated by CD grammar systems working in the t- and sf-modes of derivation obey a characterization in terms of ET0L systems, or equivalently by context-free disjoint forbidding random context grammars, and of context-free random context grammars with appearance checking, respectively. Now the question arises whether or not other random context like language families can be characterized in terms of CD grammar systems. We positively answer this question, proving that there are derivation modes for CD grammar systems, namely the negated versions of the aforementioned modes, which precisely characterize the family of context-free disjoint forbidding random context languages and that of languages generated by context-free random context grammars without appearance checking. In passing we show that every language generated by a context-free random context grammar without appearance checking can also be generated by a context-free recurrent programmed grammar without appearance checking, and *vice versa*.

1 Introduction

Random context is viewed as one of the prototype mechanisms in regulated rewriting [8]. The basic idea is, like in matrix or programmed grammars, to restrict the applicability of the rules in order to enhance the generative capacity of the underlying grammars. In the case of context-free grammars this yields languages families which are strict supersets of the family of all context-free languages. Random context grammars have been introduced by van der Walt [20] as string rewriting mechanisms, and are recently treated in the framework of picture and tree grammars, e.g., see [9,10]. In context-free random context string grammars, every production consists of an ordinary context-free (core) rewriting rule to which two sets of nonterminal symbols are associated, namely the sets of permitting and forbidding random context. The core rule of a production is

C. Martín-Vide, F. Otto, and H. Fernau (Eds.): LATA 2008, LNCS 5196, pp. 125–136, 2008.

applicable to a sentential form α only if all symbols from the permitting random context and no symbols from the forbidding random context appear in α, more precisely, in the context of the nonterminal to be replaced. One can distinguish three natural cases:

1. *Both permitting and forbidding random context can be used.* Then it is known that all recursively enumerable languages can be generated.
2. *There is only permitting random context.* Then still a strict superset of the family of context-free languages is obtained, which is included in the family of languages generated by context-free programmed (or, equivalently, context-free matrix) grammars without appearance checking. Whether or not the latter inclusion is strict is a longstanding open problem in regulated rewriting (see Open problem 1.2.2 in [8]). Permitting random context grammars are also referred to as random context grammars *without appearance checking.*
3. *There is only forbidding random context.* Again, a strict superset of the family of context-free languages (even of all languages generated by extended tabled context-free Lindenmayer systems, for short ET0L systems) is obtained. Moreover, these grammars are less powerful than type-0 Chomsky grammars. A precise characterization of the family of all ET0L languages is obtained in terms of context-free disjoint forbidding random context grammars, where in each production, the forbidding random context does not contain any symbol occurring in the context-free core rule, neither on its left-hand nor on its right-hand side [18,22].

If erasing rules are prohibited, then one is led to subclasses, in most cases strict. For a survey the reader may confer [8].

In the present paper, new characterizations of the families of context-free permitting and disjoint forbidding random context grammars in terms of cooperating distributed grammar systems will be given. Such systems have been introduced in [5] as models of distributed problem solving, after a forerunner paper [16] has treated a similar device in order to generalize two-level substitution grammars to a multi-level concept. Moreover they can be viewed as sequential counterparts of ET0L systems [3]. Context-free cooperating distributed grammar systems consist of a finite number of context-free grammars which jointly work on a common sentential form in turns. In what follows, we will restrict ourselves to context-free components without further mentioning. The conditions under which the grammar components may start and stop rewriting are determined by the cooperation protocol. For instance, the components may be required to perform, for some positive integer k, exactly, at least or at most k derivation steps. Besides these and the free mode of derivation, where no constraints have to be obeyed, two "competence based" cooperation protocols are treated as standard derivation modes:

1. In the t-mode a component, once started, has to remain active as long as it can apply one of its rules to the sentential form (terminating mode);
2. in the sf-mode a component, once started, has to remain active as long as it is able to rewrite every nonterminal occurring in the sentential form and, moreover, it can only start on sentential forms like this (full competence mode).

The latter cooperation protocol has been used in [16].

In subsequent papers, also hybrid cooperating distributed grammar systems have been considered, where several of the standard modes are combined. Two kinds of hybridization can be defined: the *external* hybrid modes, where different components can work according to distinct derivation modes [17], and *internal* hybrid modes, where all components rewrite according to one and the same derivation mode (of the system) but this mode is the Boolean combination of one or two of the standard modes [13,14]. Concerning the internal hybrid modes, only little has been done with respect to the negation of the "competence based" modes. In [4] the non-t-mode has been treated as one derivation mode in external hybrid grammar systems, as otherwise the system could never derive a terminal string. To the knowledge of the authors, the non-sf-mode has never been considered. In the present paper, the focus is set on these negated modes, where the external hybridization is circumvented by allowing derivations leading to a terminal string, as exceptions to the non-t condition. It turns out, that the language families corresponding to the non-t- and non-sf-modes are precisely the families of context-free permitting and context-free disjoint forbidding random context languages.

In the literature, random context has also been used in order to add to the power of ET0L systems, where a table is applicable only if the random context constraint associated to the table is obeyed. It is unknown whether or not any recursively enumerable language can be generated by some random context ET0L system. A characterization of this language family is given in terms of restricted context-free programmed grammars (with appearance checking), namely by *recurrent programmed grammars*. Here it is required that if a rewriting rule can be used according to the program of the grammar, then it may be used arbitrarily often in consecutive steps. Clearly, the class of context-free recurrent programmed grammars without appearance checking determines a subfamily of the family of context-free programmed grammars without appearance checking. The question of whether or not this inclusion is strict forms another open problem. In passing, we prove that this natural subfamily of the family of languages generated by context-free programmed grammars without appearance checking coincides with the family of context-free permitting random context grammars, thus with the family of languages generated by cooperating distributed grammar systems working in the non-t-mode of derivation. After all, the new characterizations of the regulated rewriting classes under consideration may shed new light on some problems in the field which are longstandingly open.

2   Definitions and Preliminaries

We assume the reader to be familiar with the standard notions of formal language theory as contained in [8]. In particular, for some alphabet V, let V^* be the set of all words over V, including the empty word λ. For $a \in V$ and $w \in V^*$, let $|w|_a$ denote the number of occurrences of a in w. The cardinality of a set M is denoted by $\#M$.

Further, the families of languages generated by context-free, context-sensitive, general type-0 Chomsky grammars, and ET0L systems are denoted by $\mathcal{L}(\mathrm{CF})$, $\mathcal{L}(\mathrm{CS})$, $\mathcal{L}(\mathrm{RE})$, and $\mathcal{L}(\mathrm{ET0L})$, respectively. We attach $-\lambda$ in our notation if erasing rules are not permitted. In what follows, we will consider two languages to be equal if they differ at most by the empty word λ.

A *context-free random context grammar* is a quadruple $G = (N, T, P, S)$, where N, T, and $S \in N$ are the set of nonterminals, the set of terminals, and the start symbol, respectively. Moreover, P is a finite set of context-free random context rules, i.e., triples of the form $(A \to \alpha, Q, R)$, where $A \to \alpha$ is a context-free production and $Q, R \subseteq N$ are its permitting and forbidding random context, respectively. For $x, y \in (N \cup T)^*$ we write $x \Rightarrow y$ if and only if $x = x_1 A x_2$, $y = x_1 \alpha x_2$, all symbols of Q appear in $x_1 x_2$, and no symbol of R appears in $x_1 x_2$. If either Q and/or R is empty, then the corresponding context check is omitted. The language generated by the random context grammar G is defined as $L(G) = \{\, w \in T^* \mid S \Rightarrow^* w \,\}$, where $\Rightarrow^*$ is the reflexive transitive closure of $\Rightarrow$. The family of languages generated by context-free random context grammars is denoted by $\mathcal{L}(\mathrm{RC}, \mathrm{CF}, \mathrm{ac})$. We replace CF by CF–$\lambda$ in that notation if erasing rules are forbidden. If no appearance checking features are involved, i.e., all forbidding random contexts are empty, then G is said to be a *context-free permitting random context grammar* or equivalently *context-free random context grammar without appearance checking*, and we are led to the language families $\mathcal{L}(\mathrm{RC}, \mathrm{CF})$ and $\mathcal{L}(\mathrm{RC}, \mathrm{CF}\text{–}\lambda)$. It is known that[1]

$$\mathcal{L}(\mathrm{CF}) \subset \mathcal{L}(\mathrm{RC}, \mathrm{CF}[-\lambda]) \subseteq \mathcal{L}(\mathrm{RC}, \mathrm{CF}[-\lambda], \mathrm{ac}) \subseteq \mathcal{L}(\mathrm{RE}),$$

and in particular $\mathcal{L}(\mathrm{RC}, \mathrm{CF}, \mathrm{ac}) = \mathcal{L}(\mathrm{RE})$ and that

$$\mathcal{L}(\mathrm{ET0L}) \subset \mathcal{L}(\mathrm{RC}, \mathrm{CF}\text{–}\lambda, \mathrm{ac}) \subset \mathcal{L}(\mathrm{CS}),$$

see, e.g., [8]. If all permitting random contexts are empty, then G is called *context-free forbidding random context grammar*; the corresponding families of languages are denoted by $\mathcal{L}(\mathrm{fRC}, \mathrm{CF})$ and $\mathcal{L}(\mathrm{fRC}, \mathrm{CF}\text{–}\lambda)$. A production $(A \to \alpha, \emptyset, R)$ of a context-free forbidding random context grammar is referred to as *disjoint* if the intersection of R with the set $\{A\} \cup \{\, B \mid |\alpha|_B > 0 \,\}$ of symbols occurring in its core rule $A \to \alpha$ is empty. A context-free forbidding random context grammar G is called *disjoint* if all productions of G are disjoint. We denote the language families generated by context-free disjoint forbidding random context grammars with and without erasing rules by $\mathcal{L}(\mathrm{dfRC}, \mathrm{CF})$ and $\mathcal{L}(\mathrm{dfRC}, \mathrm{CF}\text{–}\lambda)$, respectively, which are shown in [21] to be identical with the family $\mathcal{L}(\mathrm{ET0L})$. We are led to the following inclusion chain:

$$\mathcal{L}(\mathrm{ET0L}) = \mathcal{L}(\mathrm{dfRC}, \mathrm{CF}[-\lambda]) \subset \mathcal{L}(\mathrm{fRC}, \mathrm{CF}[-\lambda]) \subset \mathcal{L}(\mathrm{RC}, \mathrm{CF}[-\lambda], \mathrm{ac}).$$

The first strict inclusion follows from [8], the latter one from [11], see also [12]. The characterization of the family of ET0L languages in terms of context-free disjoint forbidding random context grammars is given in [21] and in [18] using a different approach.

---

[1] Whenever we use bracket notations like these, the statement is true both in case of ignoring the brackets and when neglecting the bracket contents.

A *context-free cooperating distributed grammar system* (CD grammar system, for short) with n components, $n \geq 1$, is a construct $\Gamma = (N, T, P_1, P_2, \ldots, P_n, S)$, where each $G_i = (N, T, P_i, S)$ is a context-free grammar. For $1 \leq i \leq n$, P_i is called a *component* of Γ. The *domain* of P_i, in symbols $\mathrm{dom}(P_i)$, is the set of nonterminals which can be rewritten by some production in P_i. Furthermore, P_i is said to be *sentential form competent* (*sf*-competent, for short) on a word $x \in (N \cup T)^*$ if, for any nonterminal A, $|x|_A > 0$ implies $A \in \mathrm{dom}(P_i)$; we write $P_i \models_{sf} x$ in this case and $P_i \not\models_{sf} x$ otherwise.

For $1 \leq i \leq n$, let $\Rightarrow_i$ be the usual yield relation of the context-free grammar $G_i = (N, T, P_i, S)$, and $\Rightarrow_i^*$ its reflexive and transitive closure. A t-mode (*sf*-mode) derivation step of Γ is defined by

$$x \Rightarrow_i^t y \text{ iff } x \Rightarrow_i^* y \text{ and there is no } z \text{ with } y \Rightarrow_i z,$$

and

$$x \Rightarrow_i^{sf} y \text{ iff } x \Rightarrow_i^* x' \Rightarrow_i y \text{ and } P_i \models_{sf} x' \text{ but } P_i \not\models_{sf} y,$$

for some words x, x', and y over $N \cup T$ and $1 \leq i \leq n$. Note that, consequently, component P_i is *sf*-competent on x and all intermediate sentential forms in $x \Rightarrow_i^* x'$, either. Therefore, if P_i is applied in the t-mode, then it has to continue rewriting until there is no nonterminal from $\mathrm{dom}(P_i)$ left in the sentential form. In the *sf*-mode it is active until and unless there appears a nonterminal in the sentential form which is not in $\mathrm{dom}(P_i)$, that is, until and unless it is *sf*-competent on the sentential forms.

The two negated derivation modes we will treat in the present paper shall model the following intuition. In the non-t-mode, each component has to perform an arbitrary number of derivations steps as long as at least one rule will remain applicable to the sentential form, except when the derivation produces a terminal word. That is the derivation process can be stopped whenever the sentential form contains at least one nonterminal of the domain of the component (hence, it is not a completed t-mode derivation) or the sentential form is terminal. In the non-*sf*-mode, a component can start and stop rewriting, if and only if it is *sf*-competent on the current sentential form (hence, it is not a completed *sf*-mode derivation). Therefore, they are defined as follows:

$$x \Rightarrow_i^{\bar{t}} y \text{ iff } x \Rightarrow_i^* y \text{ and there is a } z \text{ with } y \Rightarrow_i z,$$

and

$$x \Rightarrow_i^{\bar{sf}} y \text{ iff } x \Rightarrow_i^* y \text{ and } P_i \models_{sf} y,$$

for some words x and y over $N \cup T$ and $1 \leq i \leq n$.

Let f be some derivation mode, then the language generated by Γ working in the f-mode is the set

$$L(\Gamma) = \{\, w \in T^* \mid S = w_0 \Rightarrow_{i_1}^f w_1 \Rightarrow_{i_2}^f \cdots \Rightarrow_{i_m}^f w_m = w, \, m \geq 0,$$
$$1 \leq i_j \leq n, \text{ and } 1 \leq j \leq m \,\},$$

where $\Rightarrow_i^f$ denotes the f-mode derivation relation of the ith component.

For an overview about the generative capacity of CD grammar systems we refer to [6] and [7]. CD grammar systems working in the t-mode have been investigated in [5], where it was shown that context-free [λ-free] CD grammar systems working in the t-mode precisely characterize the family $\mathcal{L}(\text{ET0L})$ of languages generated by ET0L systems, i.e., $\mathcal{L}(\text{ET0L}) = \mathcal{L}(\text{CD}, \text{CF}[-\lambda], t)$. Moreover, CD grammar systems working in the sf-mode have been investigated, for example, in [1,2]. In [16] the equalities $\mathcal{L}(\text{P}, \text{CF}[-\lambda], \text{ac}) = \mathcal{L}(\text{CD}, \text{CF}[-\lambda], sf)$ have been shown. In other words, the family of languages generated by context-free [λ-free] CD grammar systems working in the sf-mode precisely characterize the family $\mathcal{L}(\text{P}, \text{CF}[-\lambda], \text{ac})$ of languages generated by programmed context-free [λ-free] grammars with appearance checking. The definition of a context-free programmed grammar is briefly recalled in the next section. We note that it is known that $\mathcal{L}(\text{P}, \text{CF}[-\lambda], \text{ac}) = \mathcal{L}(\text{RC}, \text{CF}[-\lambda], \text{ac})$ and hence $\mathcal{L}(\text{P}, \text{CF}, \text{ac})$ equals the family of recursively enumerable languages, while $\mathcal{L}(\text{P}, \text{CF-}\lambda, \text{ac})$ is a proper subset of the family of context-sensitive languages.

In order to clarify our definitions, we give two examples.

Example 1. Let Γ be the CD grammar system $\Gamma = (N, T, P_1, P_2, \ldots, P_6, S)$ with nonterminals $N = \{S, A, B, A', B'\}$, terminals $T = \{a, b, c\}$, and the production sets

$$P_1 = \{S \to AB, A \to A\} \qquad P_4 = \{A' \to A, B' \to B'\}$$
$$P_2 = \{A \to aA'b, B \to B\} \qquad P_5 = \{A \to A, B' \to B\}$$
$$P_3 = \{A' \to A', B \to B'c\} \qquad P_6 = \{A \to ab, B \to c\}.$$

It is easy to see that running Γ in the non-t-mode results in the non-context-free language $L_1 = \{a^n b^n c^n \mid n \geq 1\}$. The only way to start the derivation is to use production set P_1 leading to the sentential form AB. Then, for all natural numbers $n \geq 0$, we find $a^n A b^n B c^n \Rightarrow_2^{\bar{t}} a^{n+1} A' b^{n+1} B c^n \Rightarrow_3^{\bar{t}} a^{n+1} A' b^{n+1} B' c^{n+1} \Rightarrow_4^{\bar{t}} a^{n+1} A b^{n+1} B' c^{n+1} \Rightarrow_5^{\bar{t}} a^{n+1} A b^{n+1} B c^{n+1}$ or the terminating derivation $a^n A b^n B c^n \Rightarrow_6^{\bar{t}} a^{n+1} b^{n+1} c^{n+1}$. This shows the stated claim. Observe, that the rules of the form $X \to X$, for $X \in \{A, B, A', B'\}$, are needed to enforce that the derivation is non-blocking in the sense of a non-t-mode derivation.

When comparing the t-mode and the non-t-mode we find that in the first mode the CD grammar system Γ generates the empty set $\emptyset$. This fact is obvious, since rules of the form $X \to X$ in production sets enforce that the derivation is not terminating when X is already present in the sentential form or derived by the component under consideration. Therefore, the termination cannot start from axiom S since the only production set that can be applied on S is P_1 and it contains the rules $S \to AB$ and $A \to A$, which immediately leads to a non-terminating derivation.

Let us turn to our second example.

Example 2. The language $L_2 = \{a^n b^n c^n \mid n \geq 1\}$ is also generated by the CD grammar system $\Gamma = (\{S, A, B, A', B'\}, \{a, b\}, P_1, P_2, \ldots, P_6, S)$ with the production sets

$$P_1 = \{S \to AB, A \to A, B \to B\}$$
$$P_2 = \{A \to aA'b, A' \to A', B \to B\}$$
$$P_3 = \{A' \to A', B \to B'c, B' \to B'\}$$
$$P_4 = \{A \to A, A' \to A, B' \to B'\}$$
$$P_5 = \{A \to A, B \to B, B' \to B\}$$
$$P_6 = \{A \to ab, B \to c\}$$

if it is driven in the non-*sf*-mode of derivation. In fact, the successful derivation sequences are exactly those, which are already presented in the previous example for the non-*t*-mode of derivation. Again, the recurrent rules $X \to X$, for $X \in \{A, B, A', B'\}$ are of special interest. Here the purpose of these rules is to force the component to remain competent on the sentential form.

Finally, let us mention that CD grammar system Γ generates the empty set $\emptyset$ when run in conventional *sf*-mode of derivation. Observe, that one can modify the components $P_1, P_2, \ldots, P_6$ such that the CD grammar system generates L_2 when run in *sf*-mode. To this end, one has to erase the rules $A \to A$ and $B \to B$ from P_1, rule $A' \to A'$ from P_2, rule $B' \to B'$ from P_3, rule $A \to A$ from P_4, and rule $B \to B$ from P_5. In this way one obtains nearly the CD grammar system from the previous example (expect for the production set P_1).

3   Simulation Results

We consider context-free CD grammar systems working in the non-*t*- and non-*sf*-modes, showing equivalences to the context-free permitting and disjoint forbidding random context grammars, respectively. First, the non-*t*-mode is treated. In passing, another equivalence to a particular variant of programmed grammars, namely that of recurrent programmed grammars, is proved.

Theorem 3. $\mathcal{L}(\mathrm{RC}, \mathrm{CF}[-\lambda]) \subseteq \mathcal{L}(\mathrm{CD}, \mathrm{CF}[-\lambda], \bar{t})$.

Proof. Let $G = (N, T, P, S)$ be a context-free random context grammar (without appearance checking). With each random context production we associate a unique label p_j, $1 \leq j \leq \#P$. If $(A \to v, \{B_1, B_2, \ldots, B_{k_j}\}, \emptyset)$ is the production labeled p_j, for $1 \leq j \leq \#P$, then we set

$$N' = N \cup \{ A_{p_j}^{(i)} \mid A \in N, 1 \leq i \leq k_j, \text{ and } 1 \leq j \leq \#P \},$$

the union be disjoint, and construct a context-free CD grammar system $\Gamma = (N', T, P_1, P_2, \ldots, P_n, S)$ equivalent to G as follows. For $1 \leq j \leq \#P$, the following components are introduced:

$$\{A \to A_{p_j}^{(1)}, B_1 \to B_1\},$$
$$\{A_{p_j}^{(i)} \to A_{p_j}^{(i+1)}, B_{i+1} \to B_{i+1}\} \quad \text{for } 1 \leq i < k_j,$$
$$\{A_{p_j}^{(k_j)} \to v\} \cup \{B \to B \mid B \in N\}.$$

The application of the production labeled p_j can be successfully simulated if and only if all components associated with it are applied in sequence as listed in the construction. As Γ works in the non-*t*-mode, this is possible if and only if each symbol from the permitting context is present. Since Γ is λ-free if G is so, the proof is finished.   □

For the next theorem we need the definition of context-free programmed grammars. A *context-free programmed grammar* (see, e.g., [8]) is a septuple $G = (N, T, P, S, \Lambda, \sigma, \phi)$, where N, T, P, and S, $S \in N$, are as in the definition of context-free random context grammars; Λ is a finite set of labels (for the productions in P), such that Λ can be interpreted as a function which outputs a production when being given a label; σ and ϕ are functions from Λ into the set of subsets of Λ. Usually, the productions are written in the form $(r : A \to \alpha, \sigma(r), \phi(r))$, where r is the label of $A \to \alpha$. For (x, r_1) and (y, r_2) in $(N \cup T)^* \times \Lambda$ and $\Lambda(r_1) = A \to \alpha$, we write $(x, r_1) \Rightarrow (y, r_2)$ if and only if either $x = x_1 A x_2$, $y = x_1 \alpha x_2$ and $r_2 \in \sigma(r_1)$, or $x = y$ and rule $A \to \alpha$ is not applicable to x, and $r_2 \in \phi(r_1)$. In the latter case, the derivation step is done in *appearance checking mode*. The set $\sigma(r_1)$ is called *success field* and the set $\phi(r_1)$ *failure field* of r_1. As usual, the reflexive transitive closure of $\Rightarrow$ is denoted by $\Rightarrow^*$. The language generated by G is defined as $L(G) = \{ w \in T^* \mid (S, r_1) \Rightarrow^* (w, r_2) \text{ for some } r_1, r_2 \in \Lambda \}$. The family of languages generated by programmed grammars containing only context-free core rules is denoted by $\mathcal{L}(\mathrm{P}, \mathrm{CF}, \mathrm{ac})$. We replace CF by CF–$\lambda$ in that notation if erasing rules are forbidden. When no appearance checking features are involved, i.e., $\phi(r) = \emptyset$ for each label $r \in \Lambda$, we are led to the families $\mathcal{L}(\mathrm{P}, \mathrm{CF})$ and $\mathcal{L}(\mathrm{P}, \mathrm{CF}\text{–}\lambda)$. Obviously, $\mathcal{L}(\mathrm{P}, \mathrm{CF}[\text{–}\lambda]) \subseteq \mathcal{L}(\mathrm{P}, \mathrm{CF}[\text{–}\lambda], \mathrm{ac})$. For the relation between languages generated by context-free programmed grammars and languages generated by context-free random context grammars the following relations are well known, see, e.g., [8]: $\mathcal{L}(\mathrm{RC}, \mathrm{CF}[\text{–}\lambda]) \subseteq \mathcal{L}(\mathrm{P}, \mathrm{CF}[\text{–}\lambda])$ while $\mathcal{L}(\mathrm{RC}, \mathrm{CF}[\text{–}\lambda], \mathrm{ac}) = \mathcal{L}(\mathrm{P}, \mathrm{CF}[\text{–}\lambda], \mathrm{ac})$. Whether the former inclusion is strict or not is a long openstanding problem in regulated rewriting. Hence $\mathcal{L}(\mathrm{P}, \mathrm{CF}, \mathrm{ac})$ equals $\mathcal{L}(\mathrm{RE})$ and $\mathcal{L}(\mathrm{P}, \mathrm{CF}\text{–}\lambda, \mathrm{ac})$ is a strict subset of $\mathcal{L}(\mathrm{CS})$.

A special variant of programmed grammars are recurrent programmed grammars introduced in [22]. A context-free programmed grammar G is a *context-free recurrent programmed grammar* if for every $r \in \Lambda$ of G, we have $r \in \sigma(r)$, and if $\phi(r) \neq \emptyset$, then $\sigma(r) = \phi(r)$. The corresponding language families are denoted by $\mathcal{L}(\mathrm{RP}, \mathrm{CF}, \mathrm{ac})$ and $\mathcal{L}(\mathrm{RP}, \mathrm{CF}\text{–}\lambda, \mathrm{ac})$. When no appearance checking features are involved, i.e., $\phi(r) = \emptyset$ for each label $r \in \Lambda$, we omit ac in that notation, again. Obviously, by definition $\mathcal{L}(\mathrm{RP}, \mathrm{CF}[\text{–}\lambda]) \subseteq \mathcal{L}(\mathrm{P}, \mathrm{CF}[\text{–}\lambda])$ and $\mathcal{L}(\mathrm{RP}, \mathrm{CF}[\text{–}\lambda], \mathrm{ac}) \subseteq \mathcal{L}(\mathrm{P}, \mathrm{CF}[\text{–}\lambda], \mathrm{ac})$. Moreover, in [15] it was shown that

$$\mathcal{L}(\mathrm{RC}, \mathrm{CF}[\text{–}\lambda]) \subseteq \mathcal{L}(\mathrm{RP}, \mathrm{CF}[\text{–}\lambda]) \subset \mathcal{L}(\mathrm{RP}, \mathrm{CF}[\text{–}\lambda], \mathrm{ac}).$$

Note that the latter inclusion is a strict one. The following theorem shows how CD grammar systems working in non-t-mode of derivation can be simulated by recurrent programmed grammars without appearance checking.

Theorem 4. $\mathcal{L}(\mathrm{CD}, \mathrm{CF}[\text{–}\lambda], \bar{t}) \subseteq \mathcal{L}(\mathrm{RP}, \mathrm{CF}[\text{–}\lambda])$.

Proof. Assume the given CD grammar system Γ has n components $P_1, P_2 \ldots, P_n$, with

$$P_i = \{A_{i,1} \to v_{i,1}, A_{i,2} \to v_{i,2}, \ldots, A_{i,k_i} \to v_{i,k_i}\},$$

for $1 \leq i \leq n$. Then we consider the recurrent programmed grammar which possesses, for $1 \leq j \leq k_i$ and $1 \leq i \leq n$, the productions

$$([i,j] : A_{i,j} \to v_{i,j}, \{ [i,\ell] \mid 1 \leq \ell \leq k_i \} \cup \{ c_{i.\ell} \mid 1 \leq \ell \leq k_i \}, \emptyset)$$

and

$$(c_{i,j} : A_{i,j} \to A_{i,j}, \{c_{i,j}\} \cup \{ [i',\ell] \mid 1 \leq i' \leq n, 1 \leq \ell \leq k_{i'} \}, \emptyset).$$

Therefore, any number of consecutive steps performed by any component of Γ can be mimicked, but a change from one component P_i to another one can only successfully be done if a production labeled $c_{i,j}$ has been applied, for some j, testing the appearance of $A_j \in \mathrm{dom}(P_i)$ in the current sentential form. Thus, only non-t derivation steps can be simulated. In conclusion, the constructed context-free recurrent programmed grammar is equivalent to Γ and is λ-free if Γ has no erasing productions. This completes the proof. $\square$

Finally, we show how to simulate a context-free recurrent programmed grammar without appearance checking by a context-free random context grammar, also without appearance checking.

Theorem 5. $\mathcal{L}(\mathrm{RP}, \mathrm{CF}[-\lambda]) \subseteq \mathcal{L}(\mathrm{RC}, \mathrm{CF}[-\lambda])$.

Proof. For a context-free recurrent programmed grammar $G = (N, T, P, S, \Lambda, \sigma, \phi)$, we construct the context-free random context grammar $G' = (N', T, P', S')$ with $N' = N \cup \{ A_p, p, p' \mid p \in \Lambda \} \cup \{S'\}$, the unions being disjoint, where P' contains, for any $(p : A \to v, \sigma(p), \emptyset)$ in P, the following productions:

$$\begin{aligned}
&(S' \to Sp, \emptyset, \emptyset), \\
&(A \to A_p, \{p\}, \emptyset), \\
&(p \to p', \{A_p\}, \emptyset), \\
&(A_p \to v, \{p'\}, \emptyset), \\
&(p' \to q, \emptyset, \emptyset) \quad \text{for any } q \in \sigma(p), \\
&(p \to \lambda, \emptyset, \emptyset).
\end{aligned}$$

After initiating a derivation according to G' with the help of $(S' \to Sp, \emptyset, \emptyset)$, every application of a production from G is simulated by applying the corresponding productions from G' in the sequence they are listed above. Thus, every derivation of G can be mimicked by G'. Hence, $L(G) \subseteq L(G')$. On the other hand, the grammar G' can skip applying $A_p \to v$ in those simulation cycles. Then, symbols A_p can be resolved whenever p' is present in the sentential form, completing a former simulation of applying the production of G labeled p. As all productions are context-free this only yields words in $L(G)$. If the label symbol p, that is, the rightmost symbol in sentential forms αp, is erased before α has become terminal, then no further simulation cycles can be performed and the derivation is blocked. The fact that several occurrences of A can be replaced if p is present does not violate the simulation as G is a context-free *recurrent* programmed grammar. Therefore, $L(G') \subseteq L(G)$.

If G is λ-free, the only erasing productions of G' are used exactly once for deleting the rightmost symbol p in the sentential forms. By a standard technique

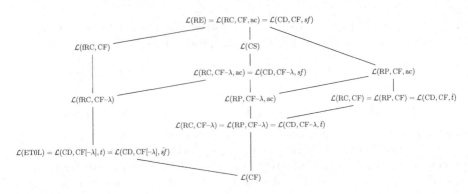

Fig. 1. Inclusion diagram—language families that are not linked by a path are not necessarily pairwise incomparable

in formal language theory the erasing productions can be eliminated from G' without affecting the generated language (for example, by encoding the label as a component of the rightmost nonterminal from G instead of as a distinct symbol); the details are omitted here.   □

Hence we have shown the following equalities.

Corollary 6. $\mathcal{L}(\text{CD}, \text{CF}[-\lambda], \bar{t}) = \mathcal{L}(\text{RC}, \text{CF}[-\lambda]) = \mathcal{L}(\text{RP}, \text{CF}[-\lambda])$.   □

Next, we turn to the non-sf-mode in relation to forbidding random context.

Theorem 7. $\mathcal{L}(\text{CD}, \text{CF}[-\lambda], \bar{sf}) = \mathcal{L}(\text{dfRC}, \text{CF}[-\lambda])$.

Proof. First we show $\mathcal{L}(\text{CD}, \text{CF}, \bar{sf}) \subseteq \mathcal{L}(\text{dfRC}, \text{CF})$. Let Γ be a CD grammar system with n components $P_1, P_2, \ldots, P_n$ and nonterminal set N. A production $A \rightarrow \alpha$ in P_i, for $1 \leq i \leq n$, can only be used in a non-sf-mode derivation, if every nonterminal occurring in α belongs to the domain of P_i. Thus we can assume without loss of generality that only rules with this property are present. Now, an arbitrary number of derivation steps can be performed by any component which is sf-competent whenever it starts rewriting the sentential form. This can be tested with the help of the forbidding random context: With each rule $A \rightarrow \alpha$ in P_i we associate the random context production $(A \rightarrow \alpha, \emptyset, R)$ with $R = N \setminus \text{dom}(P_i)$. By the explanation given above, the productions obtained are disjoint. This verifies the equivalence of the constructed context-free disjoint forbidding random context grammar and the CD grammar system Γ.

For the converse inclusion, we argue as follows. Let a context-free disjoint forbidding random context grammar $G = (N, T, P, S)$ be given. For the construction of a CD grammar system which is equivalent to G when working in the non-sf-mode, with each disjoint forbidding random context production $(A \rightarrow \alpha, \emptyset, R)$, we associate a component $\{A \rightarrow \alpha\} \cup \{B \rightarrow B \mid B \in N \setminus R\}$. As the forbidding random context production is disjoint, the set R does not contain A nor any nonterminal occurring in α. Therefore, $A \rightarrow \alpha$ is applicable if and only if no symbols from R are present in the current sentential form.

Note that, in both constructions, no new λ-rules are introduced. Hence the inclusions, and therefore also the equality, are also valid if λ-rules are not permitted.   □

Hence we have shown the following equalities.

Corollary 8. $\mathcal{L}(\mathrm{CD}, \mathrm{CF}[-\lambda], \bar{sf}) = \mathcal{L}(\mathrm{CD}, \mathrm{CF}[-\lambda], t) = \mathcal{L}(\mathrm{ET0L})$.   □

4   Conclusions

We have investigated whether context-free random context like language families can be characterized in terms of CD grammars systems. The inclusion relation between the language families considered are depicted in Figure 1. One of the most interesting results obtained are the equivalence of context-free random context grammars without appearance checking and of context-free recurrent programmed grammars also without appearance checking, with respect to their generative capacity.

Finally, it is worth mentioning, that also the language family $\mathcal{L}(\mathrm{RP}, \mathrm{CF}[-\lambda], \mathrm{ac})$ can be characterized in terms of ET0L random context grammars, namely

$$\mathcal{L}(\mathrm{RC}, \mathrm{E[P]T0L}) = \mathcal{L}(\mathrm{RC}, \mathrm{E[P]T0L}, \mathrm{ac}) = \mathcal{L}(\mathrm{RP}, \mathrm{CF}[-\lambda], \mathrm{ac})$$

and $\mathcal{L}(\mathrm{ET0L}) = \mathcal{L}(\mathrm{fRC}, \mathrm{E[P]T0L})$, where we refer to [22] for a definition of ET0L random context grammars. Moreover, in, e.g., [4,13,19], also a characterization of $\mathcal{L}(\mathrm{RP}, \mathrm{CF}[-\lambda], \mathrm{ac})$ in terms of CD grammar systems is given.

References

1. Bordihn, H.: On the number of components in cooperating distributed grammar systems. Theoretical Computer Science 330(2), 195–204 (2005)
2. Bordihn, H., Csuhaj-Varjú, E.: On competence and completeness in CD grammar systems. Acta Cybernetica 12, 347–360 (1996)
3. Bordihn, H., Csuhaj-Varjú, E., Dassow, J.: CD grammar systems versus L systems. In: Păun, G., Salomaa, A. (eds.) Grammatical Model of Multi-Agent Systems, Gordon and Breach, pp. 18–32 (1999)
4. Bordihn, H., Holzer, M.: Grammar systems with negated conditions in their cooperation protocols. Journal of Universal Computer Science 6(12), 1165–1184 (2000)
5. Csuhaj-Varjú, E., Dassow, J.: On cooperating/distributed grammar systems. Journal of Information Processing and Cybernetics (formerly: EIK) 26(1/2), 49–63 (1990)
6. Csuhaj-Varjú, E., Dassow, J., Kelemen, J., Păun, G.: Grammar Systems: A Grammatical Approach to Distribution and Cooperation, Grodon and Breach (1994)
7. Dassow, J., Păun, G., Rozenberg, G.: Grammar systems. In: Rozenberg, G., Salomaa, A. (eds.) Handbook of Formal Languages, vol. 2, pp. 155–213. Springer, Heidelberg (1997)
8. Dassow, J., Păun, G.: Regulated Rewriting in Formal Language Theory. In: EATCS Monographs in Theoretical Computer Science, vol. 18. Springer, Heidelberg (1989)

9. Ewert, S., van der Walt, A.P.J.: Generating pictures using random forbidding context. International Journal of Pattern Recognition and Ariticial Intelligence 12(7), 939–950 (1998)

10. Ewert, S., van der Walt, A.P.J.: Generating pictures using random permitting context. International Journal of Pattern Recognition and Ariticial Intelligence 13(3), 339–355 (1999)

11. Fernau, H.: Membership for 1-limited ET0L languages is not decidable. Journal of Information Processing and Cybernetics (formerly: EIK) 30(4), 191–211 (1994)

12. Fernau, H.: A predicate for separating language classes. Bulletin of the European Association for Theoretical Computer Science 56, 96–97 (1995)

13. Fernau, H., Freund, R., Holzer, M.: Hybrid modes in cooperating distributed grammar systems: internal versus external hybridization. Theoretical Computer Science 259(1–2), 405–426 (2001)

14. Fernau, H., Holzer, M., Freund, R.: Bounding resources in cooperating distributed grammar systems. In: Bozapalidis, S. (ed.) Proceedings of the 3rd International Conference Developments in Language Theory, Aristotle University of Thessaloniki, Thessalomiki, Greece, July 1997, pp. 261–272 (1997)

15. Fernau, H., Wätjen, D.: Remarks on regulated limited ET0L systems and regulated context-free grammars. Theoretical Computer Science 194, 35–55 (1998)

16. Meersman, R., Rozenberg, G.: Cooperating grammar systems. In: Winkowski, J. (ed.) Proceedings of the 7th Sympmosium Mathematical Foundations of Computer Science, Zakopane, Poland. LNCS, vol. 64, pp. 364–374. Springer, Heidelberg (1978)

17. Mitrana, V.: Hybrid cooperating/distributed grammar systems. Computers and Artificial Intelligence 12(1), 83–88 (1993)

18. Penttonen, M.: ET0L-grammars and N-grammars. Information Processing Letters 4(1), 11–13 (1975)

19. ter Beek, M.H., Csuhaj-Varjú, E., Holzer, M., Vaszil, G.: On competence in CD grammar systems. In: Calude, C.S., Calude, E., Dinneen, M.J. (eds.) DLT 2004. LNCS, vol. 3340, pp. 76–88. Springer, Heidelberg (2004)

20. van der Walt, A.P.J.: Random context languages. In: Freiman, C.V., Griffith, J.E., Rosenfeld, J.L. (eds.) Proceedings of the IFIP Congress 71, Ljubljana, Yugoslavia, August 1971, vol. 1, pp. 66–68. North-Holland, Amsterdam (1971)

21. von Solms, S.H.: On T0L languages over terminals. Information Processing Letters 3(3), 69–70 (1975)

22. von Solms, S.H.: Some notes on ET0L-languages. International Journal on Computer Mathematics 5, 285–296 (1976)

Extending the Overlap Graph for Gene Assembly in Ciliates*

Robert Brijder and Hendrik Jan Hoogeboom

Leiden Institute of Advanced Computer Science, Universiteit Leiden,
Niels Bohrweg 1, 2333 CA Leiden, The Netherlands
rbrijder@liacs.nl

Abstract. Gene assembly is an intricate biological process that has been studied formally and modeled through string and graph rewriting systems. Recently, a restriction of the general (intramolecular) model, called simple gene assembly, has been introduced. This restriction has subsequently been defined as a string rewriting system. We show that by extending the notion of overlap graph it is possible to define a graph rewriting system for two of the three types of rules that make up simple gene assembly. It turns out that this graph rewriting system is less involved than its corresponding string rewriting system. Finally, we give characterizations of the 'power' of both types of graph rewriting rules. Because of the equivalence of these string and graph rewriting systems, the given characterizations can be carried over to the string rewriting system.

1 Introduction

Gene assembly is a highly involved process occurring in one-cellular organisms called ciliates. Ciliates have two both functionally and physically different nuclei called the micronucleus and the macronucleus. Gene assembly occurs during sexual reproduction of ciliates, and transforms a micronucleus into a macronucleus. This process is highly parallel and involves a lot of splicing and recombination operations – this is true for the stichotrichs group of ciliates in particular. During gene assembly, each gene is transformed from its micronuclear form to its macronuclear form.

Gene assembly has been extensively studied formally, see [6]. The process has been modeled as either a string or a graph rewriting system [5,7]. Both systems are 'almost equivalent', and we refer to these as the general model. In [8] a restriction of this general model has been proposed. While this model is less powerful than the general model, it is powerful enough to allow each known gene [4] in its micronuclear form to be transformed into its macronuclear form. Moreover this model is less involved and therefore called the simple model. The simple model was first defined using signed permutations [8], and later proved

---

* This research was supported by the Netherlands Organization for Scientific Research (NWO) project 635.100.006 'VIEWS'.

C. Martín-Vide, F. Otto, and H. Fernau (Eds.): LATA 2008, LNCS 5196, pp. 137–148, 2008.

equivalent to a string rewriting system [3]. The graph rewriting system of the general model is based an overlap graphs. This system is an abstraction from the string rewriting system in the sense that certain local properties within the strings are lost in the overlap graph. Therefore overlap graphs are not suited for the simple gene assembly model. In this paper we show that by naturally extending the notion of overlap graph we can partially define simple gene assembly as a graph rewriting system. These extended overlap graphs form an abstraction of the string model, and is some way easier to deal with. This is illustrated by characterizing the power of two of the three types of recombination operations that make up simple gene assembly. While this characterization is based on extended overlap graphs, due to its equivalence, it can be carried over to the string rewriting system for simple gene assembly.

2   Background: Gene Assembly in Ciliates

In this section we very briefly describe the process of gene assembly. For a detailed account of this process we refer to [6]. Gene assembly occurs in a group of one-cellular organisms called ciliates. A characterizing property of ciliates is that they have two both functionally and physically different nuclei called the micronucleus (MIC) and the macronucleus (MAC). All the genes occur in both the MIC and the MAC, but in very different forms. For each gene however, one can distinguish a number of segments $M_1, \ldots, M_\kappa$, called MDSs (macronuclear destined segments), appearing in both the MIC and MAC form of that gene. In the MAC form the MDSs appear as in Figure 1: each two consecutive MDSs overlap in the MAC gene. The gray areas in the figure where the MDSs overlap are called *pointers*. Moreover, there are two sequences denoted by b and e, which occur on M_1 and M_κ respectively, that indicate the beginning and ending of the gene. The sequences b and e are called *markers*. In the MIC form the MDSs appear scrambled and inverted with non-coding segments, called IESs (internal eliminated segments), in between. As an example, Figure 2 shows the MIC form of the gene that encodes for the actin protein in a ciliate called sterkiella nova (see [10,4]). Notice that that the gene consists of nine segments, and that MDS M_2 occurs inverted, i.e. rotated 180 degrees, in the gene. The process of gene assembly transforms the MIC into the MAC, thereby transforming each gene in the MIC form to the MAC form. Hence, for each gene the MDSs are 'sorted' and put in the right orientation (i.e., they do not occur inverted). This links gene assembly to the well-known theory of sorting by reversal [2].

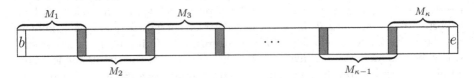

Fig. 1. The structure of a MAC gene consisting of κ MDSs

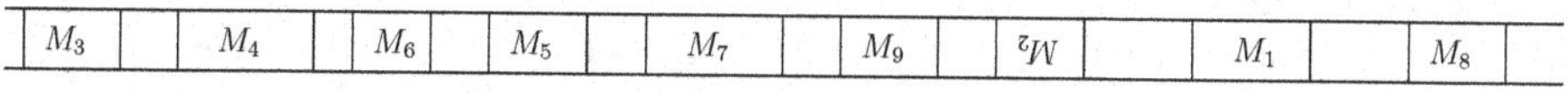

M_3		M_4		M_6	M_5		M_7		M_9		$^z\!/\!\!V$		M_1		M_8	

Fig. 2. The structure of the MIC gene encoding for the actin protein in sterkiella nova

It is postulated that there are three types of recombination operations that cut-and-paste the DNA to transform the gene from the MIC form to the MAC form. These operations are defined on pointers, so one can abstract from the notion of MDSs by simply considering the MIC gene as a sequence of pointers and markers, see Figure 3 corresponding to the gene in MIC form of Figure 2. The pointers are numbered according to the MDS they represent: the pointer on the left (right, resp.) of MDS M_i is denoted by i ($i + 1$, resp.). Pointers or markers that appear inverted are indicated by a bar: hence pointers 2 and 3 corresponding to MDS M_2 appear inverted and are therefore denoted by $\bar{2}$ and $\bar{3}$ respectively. In the general model the markers are irrelevant, so in that case only the sequence of pointers is used.

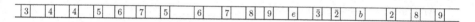

Fig. 3. Sequence of pointers and markers representing the gene in MIC form

3 Legal Strings with Markers

For an arbitrary finite alphabet A, we let $\bar{A} = \{\bar{a} \mid a \in A\}$ with $A \cap \bar{A} = \emptyset$. We use the 'bar operator' to move from A to $\bar{A}$ and back from $\bar{A}$ to A. Hence, for $p \in A \cup \bar{A}$, $\bar{\bar{p}} = p$. For a string $u = x_1 x_2 \cdots x_n$ with $x_i \in A$, the *inverse* of u is the string $\bar{u} = \bar{x}_n \bar{x}_{n-1} \cdots \bar{x}_1$. We denote the empty string by λ.

We fix $\kappa \geq 2$, and define the alphabet $\Delta = \{2, 3, \ldots, \kappa\}$ and the alphabet $\Pi = \Delta \cup \bar{\Delta}$. The elements of Π are called *pointers*. For $p \in \Pi$, we define $\|p\|$ to be p if $p \in \Delta$, and $\bar{p}$ if $p \in \bar{\Delta}$, i.e., $\|p\|$ is the 'unbarred' variant of p. A *legal string* is a string $u \in \Pi^*$ such that for each $p \in \Pi$ that occurs in u, u contains exactly two occurrences from $\{p, \bar{p}\}$.

Let $M = \{b, e\}$ with $\Delta \cap \{b, e\} = \emptyset$. The elements of M are called *markers*. We let $\Xi = \Delta \cup \{b, e\}$, and let $\Psi = \Xi \cup \bar{\Xi}$. We define the morphism $\mathrm{rm} : \Psi^* \to \Pi^*$ as follows: $\mathrm{rm}(a) = a$, for all $a \in \Pi$, and $\mathrm{rm}(m) = \lambda$, for all $m \in M \cup \bar{M}$. We say that a string $u \in \Psi^*$ is an *extended legal string* if $\mathrm{rm}(u)$ is a legal string and u has one occurrence from $\{b, \bar{b}\}$ and one occurrence from $\{e, \bar{e}\}$. We fix $m \notin \Psi$ and define for each $q \in M \cup \bar{M}$, $\|q\| = m$.

An extended legal string represents the sequence of pointers and markers of a gene in MIC form. Hence, the extended legal string corresponding to Figure 3 is $34456756789e\bar{3}\bar{2}b289$. The legal string corresponding to this figure is $3445675678 9\bar{3}\bar{2}289$ (without the markers). Legal strings are considered in the general model since markers are irrelevant there.

The *domain* of a string $u \in \Psi^*$ is $\mathrm{dom}(u) = \{\|p\| \mid p \text{ occurs in } u\}$. Note that $m \in \mathrm{dom}(v)$ for each extended legal string v. Let $q \in \mathrm{dom}(u)$ and let q_1 and q_2 be the two occurrences of u with $\|q_1\| = \|q_2\| = q$. Then q is *positive* in u if exactly one of q_1 and q_2 is in Ξ (the other is therefore in $\bar{\Xi}$). Otherwise, q is *negative* in u.

Example 1. String $u = 24b4e\bar{2}$ is an extended legal string since $\mathrm{rm}(u) = 244\bar{2}$ is a legal string. The domain of u is $\mathrm{dom}(u) = \{m, 2, 4\}$. Now, m and 2 are positive in u, and 4 is negative in u.

Let $u = x_1 x_2 \cdots x_n$ be an (extended) legal string with $x_i \in \Xi$ for $1 \leq i \leq n$, and let $p \in \mathrm{dom}(u)$. The *p-interval* of u is the substring $x_i x_{i+1} \cdots x_j$ where i and j with $i < j$ are such that $\|x_i\| = \|x_j\| = p$.

Next we consider graphs. A signed graph is a graph $G = (V, E, \sigma)$, where V is a finite set of *vertices*, $E \subseteq \{\{x, y\} \mid x, y \in V, x \neq y\}$ is a set of (undirected) *edges*, and $\sigma : V \to \{+, -\}$ is a *signing*, and for a vertex $v \in V$, $\sigma(v)$ is the *sign* of v. We say that v is *negative* in G if $\sigma(v) = -$, and v is *positive* in G if $\sigma(v) = +$. A signed directed graph is a graph $G = (V, E, \sigma)$, where the set of edges are directed $E \subseteq V \times V$. For $e = (v_1, v_2) \in E$, we call v_1 and v_2 *endpoints* of e. Also, e is an edge *from v_1 to v_2*.

4   Simple and General String Pointer Rules

Gene Assembly has been modeled using three types of string rewriting rules on legal strings. These types of rules correspond to the types of recombination operations that perform gene assembly. We will recall the string rewriting rules now – together they form the string pointer reduction system, see [5,6]. The string pointer reduction system consists of three types of reduction rules operating on legal strings. For all $p, q \in \Pi$ with $\|p\| \neq \|q\|$:

- the *string negative rule* for p is defined by $\mathbf{snr}_p(u_1 pp u_2) = u_1 u_2$,
- the *string positive rule* for p is defined by $\mathbf{spr}_p(u_1 p u_2 \bar{p} u_3) = u_1 \bar{u}_2 u_3$,
- the *string double rule* for p, q is defined by $\mathbf{sdr}_{p,q}(u_1 p u_2 q u_3 p u_4 q u_5) = u_1 u_4 u_3 u_2 u_5$,

where $u_1, u_2, \ldots, u_5$ are arbitrary (possibly empty) strings over Π.

We now recall a restriction to the above defined model. The motivation for this restricted model is that it is less involved but still general enough to allow for the successful assembling of all known experimental obtained micronuclear genes [4]. The restricted model, called *simple* gene assembly, is originally defined on signed permutations, see [8,9]. In [3], the model is consequently defined (in an equivalent way) as a string rewriting system – similar to the string pointer reduction system described above for the general model. We recall it here. It turns out that it is necessary to use extended legal strings adding symbols b and e to legal strings.

The simple string pointer reduction system consists of three types of reduction rules operating on *extended* legal strings. For all $p, q \in \Pi$ with $\|p\| \neq \|q\|$:

- the *string negative rule* for p is defined by $\mathbf{snr}_p(u_1ppu_2) = u_1u_2$ as before,
- the *simple string positive rule* for p is defined by $\mathbf{sspr}_p(u_1pu_2\bar{p}u_3) = u_1\bar{u}_2u_3$, where $|u_2| = 1$, and
- the *simple string double rule* for p, q is defined by $\mathbf{ssdr}_{p,q}(u_1pqu_2pqu_3) = u_1u_2u_3$,

where u_1, u_2, and u_3 are arbitrary (possibly empty) strings over Ψ. Note that the string negative rule is not changed, and that the simple version of the string positive rule requires $|u_2| = 1$, while the simple version of the string double rule requires $u_2 = u_4 = \lambda$ (in the string double rule definition).

Example 2. Let $u = 5\bar{2}44\bar{5}3\bar{6}26b3\bar{e}$ be an extended legal string. Then within the simple string pointer reduction system only $\mathbf{snr}_4$ and $\mathbf{sspr}_{\bar{6}}$ are applicable to u. We have $\mathbf{sspr}_{\bar{6}}(u) = 5\bar{2}44\bar{5}3\bar{2}b3\bar{e}$. Within the string pointer reduction system also $\mathbf{spr}_5$ and $\mathbf{spr}_{\bar{2}}$ are applicable to u. We will use u (in addition to a extended legal string v, which is defined later) as a running example.

A composition $\varphi = \rho_n \cdots \rho_2 \rho_1$ of string pointer rules ρ_i is a *reduction* of (extended) legal string u, if φ is applicable to (i.e., defined on) u. A reduction φ of legal string u is *successful* if $\varphi(u) = \lambda$, and a reduction φ of extended legal string u is *successful* if $\varphi(u) \in \{be, eb, \bar{e}\bar{b}, \bar{b}\bar{e}\}$. A successful reduction corresponds to the transformation using recombination operations of a gene in MIC form to MAC form. It turns out that not every extended legal string has a successful reduction using only simple rules – take e.g. $b234\bar{2}\bar{3}\bar{4}e$.

Example 3. In our running example, $\varphi = \mathbf{sspr}_{\bar{3}} \mathbf{sspr}_2 \mathbf{sspr}_5 \mathbf{snr}_4 \mathbf{sspr}_{\bar{6}}$ is a successful reduction of u, since $\varphi(u) = \bar{b}\bar{e}$. All rules in φ are simple.

5 Extended Overlap Graph

The general string pointer reduction system has been made more abstract by replacing legal strings by so-called overlap graphs, and replacing string rewriting rules by graph rewriting rules. The obtained model is called the graph pointer reduction system. Unfortunately, this model is not fully equivalent to the string pointer reduction system since the string negative rule is not faithfully simulated. Also, overlap graphs are not suited for a graph model for simple gene assembly. We propose an extension to overlap graphs that allows one to faithfully model the string negative rule and the simple string positive rule using graphs and graph rewriting rules. First we recall the definition of overlap graph.

Definition 1. The *overlap graph* for (extended) legal string u is the signed graph (V, E, σ), where $V = \mathrm{dom}(u)$ and for all $p, q \in \mathrm{dom}(u)$, $\{p, q\} \in E$ iff $q \in \mathrm{dom}(p')$ and $p \in \mathrm{dom}(q')$ where p' (q', resp.) is the p-interval (q-interval) of u. Finally, for $p \in \mathrm{dom}(u)$, $\sigma(p) = +$ iff p is positive in u.

Example 4. Consider again extended legal string $u = 5\bar{2}44\bar{5}3\bar{6}26b3\bar{e}$. Then the overlap graph $\mathcal{G}_u$ of u is given in Figure 4.

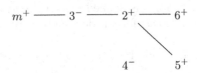

Fig. 4. The overlap graph of u from Example 4

We say that $p, q \in \text{dom}(u)$ *overlap* if there is an edge between p and q in the overlap graph of u. We now define the extended overlap graph.

Definition 2. The *extended overlap graph* for (extended) legal string u is the signed directed graph (V, E, σ), denoted by $\mathcal{G}_u$, where $V = \text{dom}(u)$ and for all $p, q \in \text{dom}(u)$, there is an edge (q, p) iff q or $\bar{q}$ occurs in the p-interval of u. Finally, for $p \in \text{dom}(u)$, $\sigma(p) = +$ iff p is positive in u.

Notice first that between any two (different) vertices p and q we can have the following possibilities:

1. There is no edge between them. This corresponds to $u = u_1 p u_2 p u_3 q u_4 q u_5$ or $u = u_1 q u_2 q u_3 p u_4 p u_5$ for some (possibly empty) strings $u_1, \ldots, u_5$ and possibly inversions of the occurrences of p and q in u.
2. There are exactly two edges between them, which are in opposite direction. This corresponds to the case where p and q overlap in u.
3. There is exactly one edge between them. If there is an edge from p to q, then this corresponds to the case where $u = u_1 q u_2 p u_3 p u_4 q u_5$ for some (possibly empty) strings $u_1, \ldots, u_5$ and possibly inversions of the occurrences of p and q in u.

As usual, we represent two directed edges in opposite direction (corresponding to case number two above) by one undirected edge. In the remaining we will use this notation and consider the extended overlap graph as having two sets of edges: undirected edges and directed edges. In general, we will call graphs with a special vertex m and having both undirected edges and directed edges *simple marked graphs*.

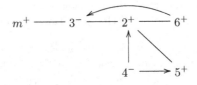

Fig. 5. The extended overlap graph of u from Example 5

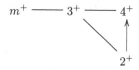

Fig. 6. The extended overlap graph of v from Example 5

Example 5. Consider again extended legal string $u = 5\bar{2}44\bar{5}3\bar{6}26b3\bar{e}$. Then the extended overlap graph $\mathcal{G}_u$ of u is given in Figure 5. Also, the extended overlap graph of $v = \bar{4}23\bar{2}4\bar{e}\bar{3}b$ is given in Figure 6.

The undirected graph obtained by removing the directed edges is denoted by $[\mathcal{G}_u]$. This is the 'classical' overlap graph of u, cf. Figures 4 and 5. On the other hand, the directed graph obtained by removing the undirected edges is denoted by $[[\mathcal{G}_u]]$. This graph represents the proper nesting of the p-intervals in the legal string.

6 Simple Graph Rules

We will now define two types of rules for simple marked graphs γ. Each of these rules transform simple marked graph of a certain form into another simple marked graph. We will subsequently show that in case γ is the extended overlap graph of a legal string, then these rules faithfully simulate the effect of the **snr** and **sspr** rules on the underlying legal string.

Definition 3. Let γ be a simple marked graph. Let p be any vertex of γ not equal to m.

- The *graph negative rule* for p, denoted by $\mathbf{gnr}_p$, is applicable to γ if p is negative, there is no undirected edge e with p as an endpoint, and there is no directed edge from a vertex *to* p in γ. The result is the simple marked graph $\mathbf{gnr}_p(\gamma)$ obtained from γ by removing vertex p and removing all edges connected to p. The set of all graph negative rules is denoted by Gnr.
- The *simple graph positive rule* for p, denoted by $\mathbf{sgpr}_p$, is applicable if p is positive, there is exactly one undirected edge e with p as an endpoint, and there is no directed edge from a vertex *to* p in γ. The result is the simple marked graph $\mathbf{sgpr}_p(\gamma)$ obtained from γ by removing vertex p, removing all edges connected to p, and flipping the sign of the other vertex q of e (i.e. changing the sign of q to $+$ if it is $-$ and to $-$ if it is $+$). The set of all simple graph positive rules is denoted by sGpr.

These rules are called simple graph pointer rules.

Remark 1. The **sgpr** rule is much simpler than the **gpr** for 'classical' overlap graphs. One does not need to compute the 'local complement' of the set of adjacent vertices. Obviously, this is because the simple rule allows only a single pointer/marker in the p-interval.   □

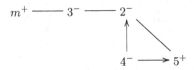

Fig. 7. The simple marked graph $\mathbf{sgpr}_6(\mathcal{G}_u)$

Example 6. Rules $\mathbf{gnr}_4$ and $\mathbf{sgpr}_6$ are the only applicable rules on the simple marked graph $\gamma = \mathcal{G}_u$ of Figure 5. Simple marked graph $\mathbf{sgpr}_6(\gamma)$ is depicted in Figure 7.

Similar as for strings, a composition $\varphi = \rho_n \cdots \rho_2 \rho_1$ of graph pointer rules ρ_i is a *reduction* of simple marked graph γ, if φ is applicable to (i.e., defined on) γ. A reduction φ of γ is *successful* if $\varphi(\gamma)$ is the graph having only vertex m where m is negative. For $S \subseteq \{\text{Gnr}, \text{sGpr}\}$, we say that γ is *successful in S* if there is a successful reduction of γ using only graph pointer rules from S.

Example 7. In our running example, $\varphi = \mathbf{sgpr}_3 \, \mathbf{sgpr}_2 \, \mathbf{sgpr}_5 \, \mathbf{gnr}_4 \, \mathbf{sgpr}_6$ is a successful reduction of $\mathcal{G}_u$.

We now show that these two types of rules faithfully simulate the string negative rule and the simple string positive rule.

Lemma 1. *Let u be a legal string and let $p \in \Pi$. Then $\mathbf{snr}_p$ is applicable to u iff $\mathbf{gnr}_{\|p\|}$ is applicable to $\mathcal{G}_u$. In this case, $\mathcal{G}_{\mathbf{snr}_p(u)} = \mathbf{gnr}_{\|p\|}(\mathcal{G}_u)$.*

Proof. We have $\mathbf{snr}_p$ is applicable to u iff $u = u_1 p p u_2$ for some strings u_1 and u_2 iff $\|p\|$ is negative in u and the $\|p\|$-interval is empty iff $\|p\|$ is negative in $\mathcal{G}_u$ and there is no undirected edge with $\|p\|$ as endpoint and there is no directed edge to $\|p\|$ iff $\mathbf{gnr}_{\|p\|}$ is applicable to $\mathcal{G}_u$.

In this case, $\mathcal{G}_{\mathbf{snr}_p(u)}$ is obtained from $\mathcal{G}_u$ by removing vertex $\|p\|$ and the edges connected to $\|p\|$, hence $\mathcal{G}_{\mathbf{snr}_p(u)}$ is equal to $\mathbf{gnr}_{\|p\|}(\mathcal{G}_u)$. □

Fig. 8. A commutative diagram illustrating Lemma 1

The previous lemma is illustrated as a commutative diagram in Figure 8. The next lemma shows that a similar diagram can be made for the simple string positive rule.

Lemma 2. *Let u be a legal string and let $p \in \Pi$. Then $\mathbf{sspr}_p$ is applicable to u iff $\mathbf{sgpr}_{\|p\|}$ is applicable to $\mathcal{G}_u$. In this case, $\mathcal{G}_{\mathbf{sspr}_p(u)} = \mathbf{sgpr}_{\|p\|}(\mathcal{G}_u)$.*

Proof. We have $\mathbf{sspr}_p$ is applicable to u iff $u = u_1 p u_2 \bar{p} u_3$ for some strings u_1, u_2, and u_3 with $|u_2| = 1$ iff $\|p\|$ is positive in u (or equivalently in $\mathcal{G}_u$) and there is exactly one undirected edge e with $\|p\|$ as endpoint and there is no directed edge with $\|p\|$ as endpoint iff $\mathbf{sgpr}_{\|p\|}$ is applicable to $\mathcal{G}_u$.

In this case, $\mathcal{G}_{\mathbf{sspr}_p(u)}$ is obtained from $\mathcal{G}_u$ by removing vertex $\|p\|$, removing all edges connected to $\|p\|$, and flipping the sign of the other vertex of e. Hence $\mathcal{G}_{\mathbf{sspr}_p(u)}$ is equal to $\mathbf{sgpr}_{\|p\|}(\mathcal{G}_u)$. □

Example 8. In our running example, one can easily verify that the extended overlap graph of $\mathbf{sspr}_{\bar{6}}(u) = 5\bar{2}44\bar{5}3\bar{2}b3\bar{e}$ is equal to graph $\mathbf{sgpr}_6(\mathcal{G}_u)$ given in Figure 7.

Fig. 9. The extended overlap graph of $w = b234234e$

One may be wondering at this point why we have not defined the simple graph double rule. To this aim, consider extended legal string $w = b234234e$. Note that $\mathbf{ssdr}_{2,3}$ and $\mathbf{ssdr}_{3,4}$ are applicable to w, but $\mathbf{ssdr}_{2,4}$ is not applicable to w. However, this information is lost in $\mathcal{G}_w$ – applying the isomorphism that interchanges vertices 2 and 3 in $\mathcal{G}_w$ obtains us $\mathcal{G}_w$ again, see Figure 9. Thus, given only $\mathcal{G}_w$ it is impossible to deduce applicability of the simple graph double rule.

To successfully define a simple graph double rule, one needs to retain information on which pointers are next to each other, and therefore different concepts are required. However, this concept would require that the linear representation of the pointers in an extended legal string is retained. Hence, string representations are more natural compared to graph representations.

The next lemma shows that simple marked graphs that are extended overlap graphs are quite restricted in form. We will use this restriction in the next section.

Lemma 3. *Let u be a legal string. Then $[[\mathcal{G}_u]]$ is acyclic and transitively closed.*

Proof. There is a (directed) edge from p to q in $[[\mathcal{G}_u]]$ iff the p-interval is completely contained in the q-interval of u. A nesting relation of intervals is acyclic and transitive. □

Remark 2. We have seen that $[\mathcal{G}_u]$ is the overlap graph of u. Not every graph is an overlap graph – a characterization of which graphs are overlap graphs is shown in [1]. Hence, both $[[\mathcal{G}_u]]$ and $[\mathcal{G}_u]$ are restricted in form compared to graphs in general. □

7 Characterizing Successfulness

In this section we characterize successfulness of simple marked graphs in $S \subseteq \{\text{Gnr}, \text{sGpr}\}$. First we consider the case $S = \{\text{Gnr}\}$.

Remark 3. In the general (not simple) model, which has different graph pointer rules and is based on overlap graphs, successfulness in S has been characterized for those S which include the graph negative rules (note that these rules are different from the graph negative rules defined here) – the cases where S does not contain the graph negative rules remain open. □

Theorem 1. *Let γ be a simple marked graph. Then γ is successful in $\{\text{Gnr}\}$ iff each vertex of γ is negative, γ has no undirected edges, and γ is acyclic.*

Proof. Since $[[\gamma]] = \gamma$ is acyclic, there is a linear ordering $(p_1, p_2, \ldots, p_n)$ of the vertices of γ such that if there is an edge from p_i to p_j, then $i < j$. The result now follows by the definition of **gnr**. In this case, linear ordering $(p_1, p_2, \ldots, p_n)$ corresponds to a successful reduction $\varphi = \mathbf{gnr}_{p_{n-1}} \cdots \mathbf{gnr}_{p_2} \mathbf{gnr}_{p_1}$ of γ. □

Using Lemma 3, more can be said if $\gamma = \mathcal{G}_u$ for some legal string u.

Corollary 1. *Let $\gamma = \mathcal{G}_u$ for some legal string u. Then γ is successful in $\{\text{Gnr}\}$ iff each vertex of γ is negative and γ has no undirected edges. In this case, since there are no overlapping pointers, γ is the transitive closure of a forest, where edges in the forest are directed from children to their parents.*

Next we turn to the case $S = \{\text{Gnr}, \text{sGpr}\}$.

Theorem 2. *Let γ be a simple marked graph. Then γ is successful in $\{\text{Gnr}, \text{sGpr}\}$ iff the following conditions hold:*

1. *$[\gamma]$ is a (undirected) forest,*
2. *for each vertex v of γ, the degree of v in $[\gamma]$ is even iff v is negative in γ, and*
3. *for each tree in the forest we can identify a root, where m is one such root, such that the graph obtained by replacing each undirected edge e in γ by a directed edge from the child to the parent in the tree to which e belongs, is acyclic.*

Proof. Proof sketch. It can be verified that each of both statements hold iff there is an linear ordering $L = (p_1, p_2, \ldots, p_n)$ of the vertices of γ such that $p_n = m$, and for each p_i with $i \in \{1, \ldots, n\}$ the following holds:

1. if $i < n$, then there is at most one undirected edge between p_i and another vertex p_j with $j > i$,
2. the number of undirected edges connected to p_i is even iff p_i is negative in γ, and
3. there is no directed edge from a vertex p_j to p_i with $j > i$.

In this case, linear ordering L corresponds to a successful reduction φ of γ in $\{\text{Gnr}, \text{sGpr}\}$ where the graph rules are applied in the order described by L and the vertices corresponding to roots in forest $[\gamma]$ (except m) are used in **gnr** rules, while the other vertices are used in **sgpr** rules. Hence $\varphi = \rho_{p_{n-1}} \cdots \rho_{p_2} \rho_{p_1}$ where ρ_{p_i} is a **gnr** rule precisely when p_i is a root.   □

Example 9. Consider again extended legal string u of Example 5 with its extended overlap graph $\mathcal{G}_u$ given in Figure 5. Notice that $[\mathcal{G}_u]$ (see Figure 4) is a forest, fulfilling condition 1) of Theorem 2. The forest consists of two trees, one of which is the single vertex 4. Therefore, 4 is necessarily the root of the tree, and by condition 3) m is the root of the other tree. All conditions of Theorem 2 hold, and therefore $\mathcal{G}_u$ is successful in $\{\text{Gnr}, \text{sGpr}\}$. According to the proof of Theorem 2, $(6, 4, 5, 2, 3, m)$, $(4, 6, 5, 2, 3, m)$, and $(4, 5, 6, 2, 3, m)$ are the linear orderings of the vertices that correspond to successful reductions of $\mathcal{G}_u$ in $\{\text{Gnr}, \text{sGpr}\}$. Moreover, in each case vertex 4 corresponds to the gnr_4 rule while the other pointers correspond to **sgpr** rules. Thus, e.g., $\varphi = \text{sgpr}_3\, \text{sgpr}_2\, \text{sgpr}_5\, \text{gnr}_4\, \text{sgpr}_6$ is a successful (graph) reduction of $\mathcal{G}_u$. By Lemma 2, this in turn corresponds to a successful (string) reduction φ' of u – one can verify that we can take $\varphi' = \text{sspr}_{\bar{3}}\, \text{sspr}_2\, \text{sspr}_5\, \text{snr}_4\, \text{sspr}_{\bar{6}}$.

The case $S = \{\text{sGpr}\}$ is now an easy corollary.

Corollary 2. *Let γ be a simple marked graph. Then γ is successful in $\{\text{sGpr}\}$ iff the all of the conditions of Theorem 2 hold, and moreover $[\gamma]$ is a connected graph. Or, equivalently, $[\gamma]$ is a (undirected) tree.*

Note that for the case $S = \{\text{sGpr}\}$, condition 3) Theorem 2 can be stated more succinctly (since we cannot choose any root): "the graph obtained by replacing each undirected edge in γ by a directed edge from the child to the parent in tree $[\gamma]$ with root m is acyclic".

Example 10. Consider again extended legal string u of Example 5 with its extended overlap graph $\mathcal{G}_u$ given in Figure 5. Then by Corollary 2, $\mathcal{G}_u$ is *not* successful in $\{\text{sGpr}\}$, since condition 1 is violated – $[\mathcal{G}_u]$ is not a tree as it has two connected components.

Reconsider now extended legal string v of Example 5 with its extended overlap graph $\mathcal{G}_v$ given in Figure 6. By Corollary 2, $\mathcal{G}_v$ is successful in $\{\text{sGpr}\}$. By the proof of Theorem 2, $(2, 4, 3, m)$ is a linear ordering of the vertices corresponding to a successful reduction $\varphi = \text{sgpr}_3\, \text{sgpr}_4\, \text{sgpr}_2$ of $\mathcal{G}_v$. Moreover, by the proof of Theorem 2, linear ordering $(4, 2, 3, m)$ does *not* correspond to a successful reduction of $\mathcal{G}_v$ (or of v).

8   Discussion

We have shown that we can partially model simple gene assembly based on a natural extension of the well-known concept of overlap graph. The model is partial in the sense that the simple double string rule does not have graph rule counterpart. Within this partial model we characterize which micronuclear genes can be successfully assembled using 1) only graph negative rules, 2) only simple graph positive rules, and 3) both of these types of rules. These results carry over to the corresponding simple string pointer rules.

What remains to be found is a graph rule counterpart of the simple double string rule. However such a counterpart would require different concepts since the overlap graph or any natural extension does not capture the requirement that pointers p and q (in the rule) are next to each other in the string.

References

1. Bouchet, A.: Circle graph obstructions. J. Comb. Theory, Ser. B 60(1), 107–144 (1994)
2. Brijder, R., Hoogeboom, H.J., Rozenberg, G.: Reducibility of gene patterns in ciliates using the breakpoint graph. Theoretical Computer Science 356, 26–45 (2006)
3. Brijder, R., Langille, M., Petre, I.: A string-based model for simple gene assembly. In: Csuhaj-Varjú, E., Ésik, Z. (eds.) FCT 2007. LNCS, vol. 4639, pp. 161–172. Springer, Heidelberg (2007)
4. Cavalcanti, A.R.O., Clarke, T.H., Landweber, L.F.: MDS_IES_DB: a database of macronuclear and micronuclear genes in spirotrichous ciliates. Nucleic Acids Res. 33, D396–D398 (2005)
5. Ehrenfeucht, A., Harju, T., Petre, I., Prescott, D.M., Rozenberg, G.: Formal systems for gene assembly in ciliates. Theoretical Computer Science 292, 199–219 (2003)
6. Ehrenfeucht, A., Harju, T., Petre, I., Prescott, D.M., Rozenberg, G.: Computation in Living Cells – Gene Assembly in Ciliates. Springer, Heidelberg (2004)
7. Ehrenfeucht, A., Petre, I., Prescott, D.M., Rozenberg, G.: String and graph reduction systems for gene assembly in ciliates. Mathematical Structures in Computer Science 12, 113–134 (2002)
8. Harju, T., Petre, I., Rozenberg, G.: Modelling Simple Operations for Gene Assembly. In: Nanotechnology: Science and Computation, pp. 361–373. Springer, Heidelberg (2006)
9. Langille, M., Petre, I.: Simple gene assembly is deterministic. Fundam. Inform. 73(1-2), 179–190 (2006)
10. Prescott, D.M., DuBois, M.: Internal eliminated segments (IESs) of oxytrichidae. J. Euk. Microbiol. 43, 432–441 (1996)

Automatic Presentations for Cancellative Semigroups

Alan J. Cain[1], Graham Oliver[2], Nik Ruškuc[1], and Richard M. Thomas[2]

[1] School of Mathematics and Statistics, University of St Andrews, North Haugh,
St Andrews, Fife KY16 9SS, United Kingdom
alanc@mcs.st-andrews.ac.uk, nik@mcs.st-andrews.ac.uk
[2] Department of Computer Science, University of Leicester, University Road,
Leicester, LE1 7RH, United Kingdom
G.Oliver@mcs.le.ac.uk, rmt@mcs.le.ac.uk

Abstract. This paper studies FA-presentable structures and gives a complete classification of the finitely generated FA-presentable cancellative semigroups. We show that a finitely generated cancellative semigroup is FA-presentable if and only if it is a subsemigroup of a virtually abelian group.

1 Introduction

This paper is concerned with structures with automatic presentations. Recall that a *structure* $\mathcal{A}$ is a tuple $(A, R_1, \ldots, R_n)$ where:

- A is a set called the *domain* of $\mathcal{A}$;
- for each i with $1 \leqslant i \leqslant n$, there is an integer $r_i \geqslant 1$ such that R_i is a subset of A^{r_i}; r_i is called the *arity* of R_i.

An obvious instance of a structure is a relational database. However, there are many other natural examples; for instance, a semigroup is a structure $(S, \circ)$, where $\circ$ has arity 3, and a group is a structure $(G, \circ, e, ^{-1})$, where $\circ$ has arity 3, e has arity 1, and $^{-1}$ has arity 2.

Khoussainov and Nerode introduced [12] the notion of *FA-presentable* structures (or structures with *automatic presentations*); see Definition 3 below. These are interesting from a computer science perspective, in that they have some nice algorithmic and logical properties (such as a decidable model checking problem).

One important field of research has been the attempt to classify classes of FA-presentable structures. As any finite structure is FA-presentable, we are really only interested in infinite structures here. In some cases this means that we have no real examples (for example, any FA-presentable integral domain is finite [13]). Essentially the only cases where we have a complete classification are those of:

- Boolean algebras [13];
- ordinals [5];
- finitely generated groups [17].

C. Martín-Vide, F. Otto, and H. Fernau (Eds.): LATA 2008, LNCS 5196, pp. 149–159, 2008.
© Springer-Verlag Berlin Heidelberg 2008

For a number of further results for FA-presentable groups, see [16]; for some necessary conditions for trees and linear orders to be FA-presentable, see [14].

As far as groups are concerned, we also have the notion of an "automatic group" in the sense of [6]. This has been generalized to semigroups (as in [3,18,11]). The considerable success of the theory of automatic groups was another motivation to have a general notion of FA-presentable structures; see also [19,20]. We note that a structure admitting an automatic presentation is often called an "automatic structure"; although we will avoid that term, the reader should be aware of the terminological clash with the different notion of an automatic structure for a group or semigroup in the sense of [6,3].

In this paper we will be particularly concerned with FA-presentable semigroups. When one moves from groups to semigroups, it appears that the problem becomes significantly more difficult. For example, if one has an undirected graph Γ with vertices V and edges E, then we have a semigroup with elements $S = V \cup \{e, 0\}$, where we have the following products:

$$uv = \begin{cases} e & \text{if } u, v \in V \text{ and } \{u, v\} \in E; \\ 0 & \text{if } u, v \in V \text{ and } \{u, v\} \notin E; \end{cases}$$

$$ue = eu = u0 = 0u = 0 \text{ for } u \in V \cup \{e, 0\}.$$

Moreover, if we form the semigroup S from the graph Γ in this way, then S is FA-presentable if and only if Γ is FA-presentable. It is known [13] that the isomorphism problem for FA-presentable graphs is Σ_1^1-complete; hence the isomorphism problem for FA-presentable semigroups is also Σ_1^1-complete.

Given this, it seems sensible to restrict oneself to some naturally occurring classes of semigroups. Given the classification of the FA-presentable finitely generated groups referred to above, a natural class to consider is that of the FA-presentable finitely generated cancellative semigroups. In this paper we give a complete classification of these structures:

Theorem 1. *A finitely generated cancellative semigroup is FA-presentable if and only if it embeds into a virtually abelian group.*

We remark that there are many examples of *non-cancellative* finitely generated FA-presentable semigroups. For example, it is easy to see that adjoining a zero to a semigroup always preserves FA-presentability and destroys cancellativity. In addition, all finite semigroups, whether cancellative or not, are FA-presentable.

2 Automatic Presentations

A semigroup is a set equipped with an associative binary operation $\circ$, although the operation symbol is often suppressed, so that $s \circ t$ is denoted st. We recall the idea of a "convolution mapping" which we will need throughout this paper:

Definition 1. *Let L be a regular language over a finite alphabet A. Define, for $n \in \mathbb{N}$,*

$$L^n = \{(w_1, \ldots, w_n) : w_i \in L \text{ for } i = 1, \ldots, n\}.$$

Let $ be a new symbol not in A. The mapping $\mathrm{conv} : (A^*)^n \to ((A \cup \{\$\})^n)^*$ is defined as follows. Suppose

$$w_1 = w_{1,1}w_{1,2} \cdots w_{1,m_1}, \; w_2 = w_{2,1}w_{2,2} \cdots w_{2,m_2}, \; \ldots, \; w_n = w_{n,1}w_{n,2} \cdots w_{n,m_n},$$

where $w_{i,j} \in A$. Then $\mathrm{conv}(w_1, \ldots, w_n)$ is defined to be

$$(w_{1,1}, w_{2,1}, \ldots, w_{n,1})(w_{1,2}, w_{2,2}, \ldots, w_{n,2}) \cdots (w_{1,m}, w_{2,m}, \ldots, w_{n,m}),$$

where $m = \max\{m_i : i = 1, \ldots, n\}$ and with $w_{i,j} = \$$ whenever $j > m_i$.

Observe that the conv maps an n-tuple of words to a word of n-tuples. We then have:

Definition 2. Let A be a finite alphabet, and let $R \subseteq (A^*)^n$ be a relation on A^*. Then R is said to be regular if

$$\{\mathrm{conv}(w_1, \ldots, w_n) : (w_1, \ldots, w_n) \in R\}$$

is a regular language over $(A \cup \{\$\})^n$.

Having done this, we can now define the concept of an "automatic presentation" for a structure:

Definition 3. Let $\mathcal{S} = (S, R_1, \ldots, R_n)$ be a relational structure. Let L be a regular language over a finite alphabet A, and let $\phi : L \to S$ be a surjective mapping. Then (L, ϕ) is an automatic presentation for $\mathcal{S}$ if:

1. the relation $L_= = \{(w_1, w_2) \in L^2 : \phi(w_1) = \phi(w_2)\}$ is regular, and
2. for each relation R_i of arity r_i, the relation

$$L_{R_i} = \{(w_1, w_2, \ldots, w_{r_i}) \in L^{r_i} : (\phi(w_1), \ldots, \phi(w_{r_i})) \in R_i\}$$

is regular. A structure with an automatic presentation is said to be FA-presentable.

As noted in Section 1, a semigroup can be viewed as a relational structure with the binary operation $\circ$ becoming a ternary relation. The following definition simply restates the preceding one in the special case where the structure is a semigroup:

Definition 4. Let S be a semigroup. Let L be a regular language over a finite alphabet A, and let $\phi : L \to S$ be a surjective mapping. Then (L, ϕ) is an automatic presentation for S if the relations

$$L_= = \{(w_1, w_2) \in L^2 : \phi(w_1) = \phi(w_2)\},$$
$$L_\circ = \{(w_1, w_2, w_3) \in L^3 : \phi(w_1)\phi(w_2) = \phi(w_3)\}$$

are both regular.

An *interpretation* of one structure inside another is, loosely speaking, a copy of the former inside the latter. The following definition is restricted to an interpretation of one semigroup inside another.

Definition 5. *Let S and T be semigroups. Let $n \in \mathbb{N}$. An (n-dimensional) interpretation of T in S consists of the following:*

- *a first-order formula $\psi(x_1, \ldots, x_n)$, called the* domain formula, *which specified those n-tuples of elements of S used in the interpretation;*
- *a surjective map $f : \psi(S^n) \to T$, called the* co-ordinate map *(where $\psi(S^n)$ denotes the set of n-tuples of elements of S satisfying the formula ψ);*
- *a first-order formula $\theta_=(x_1, \ldots, x_n; y_1, \ldots, y_n)$ that is satisfied by*

$$(a_1, \ldots, a_n; b_1, \ldots, b_n)$$

  if and only if $f(a_1, \ldots, a_n) = f(b_1, \ldots, b_n)$ in the semigroup T;
- *a first-order formula $\theta_\circ(x_1, \ldots, x_n; y_1, \ldots, y_n; z_1, \ldots, z_n)$ that is satisfied by*

$$(a_1, \ldots, a_n; b_1, \ldots, b_n; c_1, \ldots, c_n)$$

  if and only if $f(a_1, \ldots, a_n)f(b_1, \ldots, b_n) = f(c_1, \ldots, c_n)$ in the semigroup T.

The following result, although here stated only for semigroups, is true for structures generally:

Proposition 1 ([1, Proposition 3.13]). *Let S and T be semigroups. If S has an automatic presentation and there is an interpretation of T in S, then T has an automatic presentation.*

The fact that a tuple of elements $(a_1, \ldots, a_n)$ of a structure S satisfies a first-order formula $\theta(x_1, \ldots, x_n)$ is denoted $S \models \theta(a_1, \ldots, a_n)$. We then have:

Proposition 2 ([12]). *Let S be a structure with an automatic presentation. For every first-order formula $\theta(x_1, \ldots, x_n)$ over the structure there is an automaton which accepts $(w_1, \ldots, w_n)$ if and only if $S \models \theta(\phi(w_1), \ldots, \phi(w_n))$.*

(Proposition 1 is actually a consequence of Proposition 2.)

As mentioned in Section 1, a classification of the finitely generated groups with an automatic presentation was given in [17]. For convenience, we state the result here (along with some extra details from [17] that we will need later). Recall that a group G is said to be *virtually abelian* if it has an abelian subgroup A of finite index. If G is finitely generated, then the subgroup A is finitely generated as well. Using the fact that any finitely generated abelian group is the direct sum of finitely many cyclic groups, we may assume that A is of the form $\mathbb{Z}^n$ for some $n \geqslant 0$.

Theorem 2 ([17]). *A finitely generated group admits an automatic presentation if and only if it is virtually abelian. In particular, a group G with a subgroup*

$\mathbb{Z}^n$ of index ℓ admits an automatic presentation (L, ϕ), where L is the language of words

$$g_i \mathrm{conv}(\varepsilon_1 z_1, \ldots, \varepsilon_n z_n),$$

where $\varepsilon_i \in \{+, -\}$, z_i is a natural number in reverse binary notation, $g_1, \ldots, g_\ell$ are representatives of the cosets of $\mathbb{Z}^n$ in G, with $\phi : L \to G$ being defined in the natural way:

$$\phi(g_i \mathrm{conv}(\varepsilon_1 z_1, \ldots, \varepsilon_n z_n)) = g_i(\varepsilon_1 z_1, \ldots, \varepsilon_n z_n).$$

3 Growth

In the proof of Theorem 2 above in [17], one essential ingredient was the notion of growth:

Definition 6. *Let S be a semigroup generated by a finite set X. Define $\delta(s)$ to be the length of the shortest product of elements of X that equals s, i.e.*

$$\delta(s) = \min\{n \in \mathbb{N} : s = x_1 \cdots x_n \text{ for some } x_i \in X\}.$$

The growth function $\gamma : \mathbb{N} \to \mathbb{N}$ *of S is given by*

$$\gamma(n) = |\{s \in S : \delta(s) \leqslant n\}|.$$

If the function γ is asymptotically bounded above by a polynomial function (that is, if there exists a polynomial function β and some $N \in \mathbb{N}$ such that $\beta(n) > \gamma(n)$ for $n > N$), then S is said to have polynomial growth.

Note that whether a semigroup has polynomial growth or not is independent of the choice of finite generating set [8]. We now have the following result:

Theorem 3. *Any finitely generated semigroup admitting an automatic presentation has polynomial growth.*

For a proof of the Theorem 3 for finitely generated groups, see [17]; the proof given there immediately generalizes to semigroups. We remark that polynomial growth is dependent on the structures in question being semigroups: general algebras admitting automatic presentations are only guaranteed to have exponential growth [12, Lemma 4.5].

4 The Characterization

Recall that a semigroup S has a group of left (respectively, right) quotients G if S embeds into G and every element of G is of the form $t^{-1}s$ (respectively, st^{-1}) for $s, t \in S$. If a semigroup S has a group of left (respectively, right) quotients, then this group is unique up to isomorphism. For further information on groups of left and right quotients, see [4, Section 1.10].

The following result, due to Grigorchuk, generalizes the result of Gromov [9] that a finitely generated group of polynomial growth is virtually nilpotent (i.e. it has a nilpotent subgroup of finite index):

Theorem 4 ([7]). *A finitely generated cancellative semigroup has polynomial growth if and only if it has a virtually nilpotent group of left quotients.*

We then have the following immediate consequence of Theorems 4 and 3:

Corollary 1. *Let S be a finitely generated cancellative semigroup that admits an automatic presentation. Then the group of left quotients of S exists and is virtually nilpotent.*

Note that the groups of left and right quotients of subsemigroups of virtually nilpotent groups coincide (see [15] or [2, Sections 5.2–5.3]). We now have:

Proposition 3. *Let S be a finitely generated cancellative semigroup that admits an automatic presentation. Then the [necessarily virtually nilpotent] group of left (and right) quotients of S admits an automatic presentation.*

Proof. Let G be the group of left (and right) quotients of S. The strategy is to show that G has a 2-dimensional interpretation in S.

- The domain formula is tautological: $\phi(x_1, x_2) := x_1 = x_1$. Thus all pairs of elements of S are used.
- The co-ordinate map is $f(x_1, x_2) = x_1^{-1} x_2$. Since G is the group of left quotients of S, the mapping f is surjective as required.
- The formula $\theta_=$ is given by

$$\theta_=(x_1, x_2; y_1, y_2) := (\exists a, b)(x_1 a = x_2 b \wedge y_1 a = y_2 b),$$

since

$$f(x_1, x_2) = f(y_1, y_2)$$
$$\iff (\exists a, b)(f(x_1, x_2) = ab^{-1} \wedge f(y_1, y_2) = ab^{-1})$$
$$\iff (\exists a, b)(x_1^{-1} x_2 = ab^{-1} \wedge y_1^{-1} y_2 = ab^{-1})$$
$$\iff (\exists a, b)(x_1 a = x_2 b \wedge y_1 a = y_2 b).$$

- The formula $\theta_\circ$ is given by

$$\theta_\circ(x_1, x_2; y_1, y_2; z_1, z_2) :=$$
$$(\exists a, b, c, d)(cx_1 a = dy_2 b \wedge cx_2 = dy_1 \wedge z_2 b = z_1 a),$$

since

$$f(x_1, x_2) f(y_1, y_2) = f(z_1, z_2)$$
$$\iff (\exists a, b)(f(x_1, x_2) f(y_1, y_2) = ab^{-1} \wedge f(z_1, z_2) = ab^{-1})$$
$$\iff (\exists a, b)(x_1^{-1} x_2 y_1^{-1} y_2 = ab^{-1} \wedge z_1^{-1} z_2 = ab^{-1})$$
$$\iff (\exists a, b, c, d)(c^{-1} d = x_2 y_1^{-1} \wedge x_1^{-1} c^{-1} dy_2 = ab^{-1} \wedge z_1^{-1} z_2 = ab^{-1})$$
$$\iff (\exists a, b, c, d)(cx_2 = dy_1 \wedge dy_2 b = cx_1 a \wedge z_2 b = z_1 a). \qquad \square$$

We are now in a position to prove one direction of Theorem 1:

Proposition 4. *A finitely generated cancellative semigroup admitting an automatic presentation embeds into a finitely generated virtually abelian group.*

Proof. Let S be a finitely generated cancellative semigroup with an automatic presentation. By Proposition 3, its group of left quotients G has an automatic presentation. Since S is finitely generated, G is also. Theorem 2 then shows that G is virtually abelian. ◻

The other direction is provided by:

Proposition 5. *Every finitely generated subsemigroup of a virtually abelian group admits an automatic presentation.*

Proof. Let G be a virtually abelian group. Let $\mathbb{Z}^n$ be a finite-index abelian subgroup of G. By replacing $\mathbb{Z}^n$ by its core (the maximal normal subgroup of G contained in $\mathbb{Z}^n$) if necessary, we may assume that $\mathbb{Z}^n$ is normal in G. Let k be the index of $\mathbb{Z}^n$ in G. Let A be a finite alphabet representing a subset of G, and let S be the semigroup generated by this subset. Throughout this proof, denote by $\overline{w}$ the element of S represented by the word w over an alphabet representing a generating set. This notational distinction is necessary to avoid confusion when there are several representatives for the same element.

Let $B = \{a \in A : \overline{a} \in \mathbb{Z}^n\}$ and let $C = A - B$. So B consists of all letters in A representing elements of the abelian subgroup $\mathbb{Z}^n$ and C consists of letters representing elements of other cosets of $\mathbb{Z}^n$.

Introduce a new alphabet D representing the set

$$\{\overline{w} : w \in C^{\leqslant k}, \overline{w} \in \mathbb{Z}^n\}.$$

Notice that since the set $C^{\leqslant k}$ is finite, so is D. Furthermore, the semigroup S is generated by $\overline{B} \cup \overline{C} \cup \overline{D}$. We next observe the following lemma:

Lemma 1. *Every element of the semigroup S is represented by a word over $B \cup C \cup D$ that contains at most $k^2 - 1$ letters from C.*

Proof. Let $s \in S$, and let $w \in (B \cup C \cup D)^+$ with $\overline{w} = s$. Then w is of the form

$$u_0 c_1 u_1 c_2 \cdots u_{m-1} c_m u_m, \tag{1}$$

where each u_i lies in $(B \cup D)^*$ and each c_i in C. The aim is to show that such a word w can be transformed into one that still represents $s \in S$ but contains at most $k^2 - 1$ letters from C

First stage. For any word w of the form (1) and for $i = 0, \ldots, m - 1$, let $\psi_w(i)$ be maximal such that $\overline{c_{i+1} u_{i+1} \cdots c_m u_m}$ and $\overline{c_{\psi_w(i)+1} u_{\psi_w(i)+1} \cdots c_m u_m}$ lie in the same coset of $\mathbb{Z}^n$ in G. It is clear that $\psi_w(i)$ is always defined and is not less than i. Notice that since there are k distinct cosets of $\mathbb{Z}^n$ in G, $\psi_w(i)$ can take at most k distinct values as i ranges from 0 to $m - 1$. Furthermore, for each i, $\overline{c_{i+1} u_{i+1} \cdots c_{\psi_w(i)} u_{\psi_w(i)}}$ lies in $\mathbb{Z}^n$ and so commutes with $\overline{u_i}$.

Define a mapping $\beta' : (B \cup C \cup D)^+ \to (B \cup C \cup D)^+$ as follows: for w of the form (1), $\beta'(w)$ is defined to be

$$u_0 c_1 u_1 c_2 \cdots c_i c_{i+1} u_{i+1} \cdots c_{\psi_w(i)} u_{\psi_w(i)} u_i c_{\psi_w(i)+1} \cdots u_{m-1} c_m u_m,$$

where i is minimal with $\psi_w(i) \neq i$, and $\beta'(w) = w$ if $\psi_w(i) = i$ for all i. By the remark at the end of the last paragraph, $\overline{w} = \overline{\beta'(w)}$.

The mapping $\beta : (B \cup C \cup D)^+ \to (B \cup C \cup D)^+$ is defined by $\beta(w) = (\beta')^p(w)$, where p is minimal with $(\beta')^p(w) = (\beta')^{p+1}(w)$. Again, $\overline{w} = \overline{\beta(w)}$.

So $\beta(w)$ is the word obtained from w by shifting each u_i rightwards to one of at most k distinct positions between the various letters c_j. Thus $\beta(w)$ has the form (1) with at most k of the words u_i being non-empty.

Second stage. Define a mapping $\gamma' : (B \cup C \cup D)^+ \to (B \cup C \cup D)^+$ as follows: if $w \in (B \cup C \cup D)^+$ has a subword $v \in C^{\leqslant k}$ with $\overline{v} \in \mathbb{Z}^n$, then choose the leftmost, shortest such subword and replace it with the letter of D representing the same element of S. (Such a letter exists by the definition of D.)

The mapping $\gamma : (B \cup C \cup D)^+ \to (B \cup C \cup D)^+$ is defined by $\gamma(w) = (\gamma')^p(w)$, where p is minimal with $(\gamma')^p(w) = (\gamma')^{p+1}(w)$. Since each application of γ' that results in a different word decreases the number of letters from C present, such a p must exist. Observe that $\overline{w} = \overline{\gamma(w)}$ and that $\gamma(w)$ cannot contain a subword of k letters from C, for such a string must contain a subword representing an element of $\mathbb{Z}^n$.

Third stage. The final mapping $\delta : (B \cup C \cup D)^+ \to (B \cup C \cup D)^+$ is given by $\delta(w) = (\gamma\beta)^p(w)$, where p is minimal with $(\gamma\beta)^p(w) = (\gamma\beta)^{p+1}(w)$. Observe that $\overline{w} = \overline{\delta(w)}$. Now, $\delta(w)$ is of the form (1) with at most k words u_i being nonempty and does not contain k consecutive letters from C. So separated by the k nonempty words u_i are strings of at most $k-1$ letters from C. So the total number of letters from C in $\delta(w)$ is at most $(k-1) \times (k+1) = k^2 - 1$. $\square$

We now return to the proof of Proposition 5. Choose a set of representatives $g_1, \ldots, g_k$ for the cosets of $\mathbb{Z}^n$ in G. Suppose $B \cup D = \{b_1, \ldots, b_q\}$.

For $c_1, \ldots, c_m \in C$ with $0 \leqslant m \leqslant k^2 - 1$, define

$$P_{c_1 \cdots c_m} = \{u_0 c_1 u_1 c_2 \cdots u_{m-1} c_m u_m : u_i = b_1^{\alpha_{i,1}} \cdots b_q^{\alpha_{i,q}}, \alpha_{i,j} \in \mathbb{N} \cup \{0\}\}.$$

By Lemma 1 and the fact that the elements $\overline{b_j}$ commute, every element of S is represented by an element in at least one of the sets $P_{c_1 \cdots c_m}$. That is,

$$S = \bigcup_{\substack{c_1, \ldots, c_m \in C \\ 0 \leqslant m \leqslant k^2 - 1}} \overline{P_{c_1 \cdots c_m}}. \tag{2}$$

By Theorem 2, the virtually abelian group G has an automatic presentation (L, ϕ), where L is the language of words

$$g_h \mathrm{conv}(\varepsilon_1 z_1, \ldots, \varepsilon_n z_n), \tag{3}$$

where $\varepsilon_i \in \{+,-\}$ and z_i is a natural number in reverse binary notation. (In L, the coset representative g_h functions simply as a *symbol*.) The aim is now to show that the subset of L representing elements of S is regular. To do so, it suffices to show that the set of words in L representing elements of $\overline{P_{c_1 \cdots c_m}}$ is regular, since (2) is a finite union.

To this end, fix $c_1, \ldots, c_m$ and write P for $P_{c_1 \cdots c_m}$. Let $z_{i,j} \in \mathbb{Z}^n$ be such that $\overline{b_j c_{i+1} \cdots c_m} = \overline{c_{i+1} \cdots c_m} z_{i,j}$. Let $u_0 c_1 u_1 \cdots c_m u_m \in P$ with $u_i = b_1^{\alpha_{i,1}} \cdots b_q^{\alpha_{i,q}}$. Then

$$\overline{u_0 c_1 u_1 \cdots c_m u_m} = \overline{c_1 \cdots c_m} \prod_{i=0}^{m} \prod_{j=1}^{q} z_{i,j}^{\alpha_{i,j}},$$

or, switching to additive notation and supposing $\overline{c_1 \cdots c_m} = g_h(z_1', \ldots, z_n')$ and $z_{i,j} = (z_{i,j,1}, \ldots, z_{i,j,n})$ for all i, j:

$$\overline{u_0 c_1 u_1 \cdots c_m u_m} = g_h(z_1', \ldots, z_n') \sum_{i=0}^{m} \sum_{j=1}^{q} \alpha_{i,j}(z_{i,j,1}, \ldots, z_{i,j,n}).$$

Therefore define $\theta(z_1, \ldots, z_n)$ to be

$$(\exists \alpha_{0,1}, \ldots, \alpha_{m,q})\Big((\alpha_{0,1} \geqslant 0) \wedge \ldots \wedge (\alpha_{m,q} \geqslant 0)$$

$$\wedge \Big(z_1 = z_1' + \sum_{i=0}^{m} \sum_{j=1}^{m} \alpha_{i,j} z_{i,j,1}\Big)$$

$$\wedge \Big(z_2 = z_2' + \sum_{i=0}^{m} \sum_{j=1}^{m} \alpha_{i,j} z_{i,j,2}\Big)$$

$$\vdots$$

$$\wedge \Big(z_n = z_n' + \sum_{i=0}^{m} \sum_{j=1}^{m} \alpha_{i,j} z_{i,j,n}\Big)\Big),$$

where $\alpha_{i,j} z_{i,j,k}$ is understood to be shorthand for

$$\underbrace{\alpha_{i,j} + \ldots + \alpha_{i,j}}_{z_{i,j,k} \text{ times}}.$$

By a special case of Theorem 2, the structure $(\mathbb{Z}, +)$ admits an automatic presentation (M, ψ), where M is the set of words ϵz, where $\epsilon \in \{+,-\}$ and z is in reverse binary notation. Furthermore, it is clear that, in this presentation, the relation $\geqslant$ is regular. That is, (M, ψ) is an automatic presentation for $(\mathbb{Z}, +, \geqslant)$.

The set of words in L representing elements of $\overline{P}$ is then

$$\{g_h \text{conv}(z_1, \ldots, z_n) : (\mathbb{Z}, +, \geqslant) \models \theta(\psi(z_1), \ldots, \psi(z_n))\}.$$

(Recall that g_h is the representative of the coset in which $\overline{c_1 \cdots c_m}$ lies.) By Proposition 2, this set is a regular subset of L.

Union together the [finitely many] regular subsets of L obtained for the various $c_1, \ldots, c_m$ to see that the set L_S consisting of those words in L representing elements of S is regular. So S admits the automatic presentation $(L_S, \phi|_{L_S})$. $\square$

Propositions 5 and 4 together yield Theorem 1.

Acknowledgements. The constructive comments of the referees were much appreciated. Part of the work described in this paper was undertaken whilst the fourth author was on study leave from the University of Leicester and the support of the University in this regard is appreciated. The fourth author would also like to thank Hilary Craig for all her help and encouragement.

References

1. Blumensath, A.: Automatic Structures (Diploma Thesis, RWTH Aachen) (1999)
2. Cain, A.J.: Presentations for Subsemigroups of Groups (Ph.D. thesis, University of St Andrews) (2005)
3. Campbell, C.M., Robertson, E.F., Ruškuc, N., Thomas, R.M.: Automatic semigroups. Theoret. Comput. Sci. 250, 365–391 (2001)
4. Clifford, A.H., Preston, G.B.: The Algebraic Theory of Semigroups. Mathematical Surveys 7, vol. 1. American Mathematical Society (1961)
5. Delhommé, C.: Automaticité des ordinaux et des graphes homogènes. C. R. Math. Acad. Sci. Paris 339, 5–10 (2004)
6. Epstein, D.B.A., Cannon, J.W., Holt, D.F., Levy, S.V.F., Paterson, M.S., Thurston, W.P.: Word Processing in Groups. Jones & Bartlett (1992)
7. Grigorchuk, R.I.: Semigroups with cancellations of degree growth. Mat. Zametki 43, 305–319, 428 (1988) (in Russian); Grigorchuk, R. I.: Cancellative semigroups of power growth. Math. Notes 43, 175–183 (1988) (translation in)
8. Grigorchuk, R.I.: On growth in group theory. In: Proceedings of the International Congress of Mathematicians, Kyoto 1990, vol. I, II, pp. 325–338. Math. Soc. Japan (1991)
9. Gromov, M.: Groups of polynomial growth and expanding maps. Inst. Hautes Études Sci. Publ. Math. 53, 53–78 (1981)
10. Howie, J.M.: Fundamentals of Semigroup Theory. London Mathematical Society Monographs 12. Oxford University Press, Oxford (1995)
11. Hudson, J.F.P.: Regular rewrite systems and automatic structures. In: Almeida, J., Gomes, G.M.S., Silva, P.V. (eds.) Semigroups, Automata and Languages, pp. 145–152. World Scientific, Singapore (1998)
12. Khoussainov, B., Nerode, A.: Automatic presentations of structures. In: Leivant, D. (ed.) Logic and Computational Complexity (Indianapolis, IN, 1994). LNCS, vol. 960, pp. 367–392. Springer, Heidelberg (1995)
13. Khoussainov, B., Nies, A., Rubin, S., Stephan, F.: Automatic structures: richness and limitations. In: Proceedings of the 19th IEEE Symposium on Logic in Computer Science, pp. 110–119. IEEE Computer Society, Los Alamitos (2004)
14. Khoussainov, B., Rubin, S., Stephan, F.: Automatic partial orders. In: Proceedings of the 18th IEEE Symposium on Logic in Computer Science, pp. 168–177. IEEE Computer Society, Los Alamitos (2003)

15. Neumann, B.H., Taylor, T.: Subsemigroups of nilpotent groups. Proc. Roy. Soc. Ser. A 274, 1–4 (1963)
16. Nies, A.: Describing groups. Bull. Symbolic Logic 13, 305–339 (2007)
17. Oliver, G.P., Thomas, R.M.: Automatic presentations for finitely generated groups. In: Diekert, V., Durand, B. (eds.) STACS 2005. LNCS, vol. 3404, pp. 693–704. Springer, Heidelberg (2005)
18. Otto, F., Sattler-Klein, A., Madlener, K.: Automatic monoids versus monoids with finite convergent presentations. In: Nipkow, T. (ed.) RTA 1998. LNCS, vol. 1379, pp. 32–46. Springer, Heidelberg (1998)
19. Pelecq, L.: Isomorphismes et automorphismes des graphes context-free, équationnels et automatiques. PhD Thesis, Bordeaux 1 University (1997)
20. Sénizergues, G.: Definability in weak monadic second-order logic of some infinite graphs. In: Compton, K., Pin, J.-E., Thomas, W. (eds.) Automata Theory: Infinite Computations, Wadern, Germany. Dagstuhl Seminar, vol. 9202, p. 16 (1992)

Induced Subshifts and Cellular Automata

Silvio Capobianco

School of Computer Science, Reykjavík University, Reykjavík, Iceland
silvio@ru.is

Abstract. Given a shift subspace over a finitely generated group, we define the subshift induced by it on a larger group. Then we do the same with cellular automata and, while observing that the new automaton can model a different abstract dynamics, we remark several properties that are shared with the old one. After that, we simulate the old automaton inside the new one, and discuss some consequences and restrictions.

Keywords: Dynamical system, shift subspace, cellular automaton. *Mathematics Subject Classification 2000:* 37B15, 68Q80.

1 Introduction

Cellular automata (briefly, CA) are presentations of global dynamics in local terms: the phase space is made of *configurations* on an underlying lattice structure, and the transition function is induced by a pointwise evolution rule, which changes the state at a node of the grid by only considering finitely many *neighbouring* nodes. Modern CA theory employs tools from group theory, symbolic dynamics, and topology (cf. [2,4,6]). The lattice structure is provided by a *Cayley graph* of a finitely generated group: the "frames" of this class generalize the "classical" hypercubic ones, allowing more complicated grid geometries. Such broadening, however, preserves the requirement for finite neighbourhoods, and this fact allows the definition of global evolution laws in local terms. Moreover, the phase space can be a *subshift*, leaving out some configurations, but allowing *translations* of single elements and *limits* of sequences. While this can be questionable when seeing CA as computation devices, we remark how the richer framework simplifies dealing with *simulations* between CA.

In this paper, we examine how subshifts and CA on a given group define other subshifts and CA on another group, the former being a *subgroup* of the latter. A lemma about mutual inclusion between images of shift subspaces via global CA functions, showing that it is preserved *either way* when switching between the smaller group and the larger one, ensures that our definitions are well posed. We then show how several properties are transferred from the old objects to the new ones, some even either way as well; this is of interest, because the new dynamics is usually *richer* than the old one. A simulation of the original automaton into the induced one is then explicitly constructed. This extends to the case of arbitrary, finitely generated groups the usual embedding of d-dimensional cellular automata into $(d + k)$-dimensional ones; a consequence of this fact will be the collapse of

C. Martín-Vide, F. Otto, and H. Fernau (Eds.): LATA 2008, LNCS 5196, pp. 160–171, 2008.
© Springer-Verlag Berlin Heidelberg 2008

the hierarchy of cellular automata dynamics over free non-abelian groups. Some remarks about *sofic shifts* are also made throughout the discussion.

2   Background

A **dynamical system** (briefly, d.s.) is a pair (X, F) where the **phase space** X is compact and metrizable and the **evolution function** $F : X \to X$ is continuous. If $Y \subseteq X$ is closed (equivalently, compact) and $F(Y) \subseteq Y$, then (Y, F) is a **subsystem** of (X, F). A **morphism** from a d.s. (X, F) to a d.s. (X', F') is a continuous $\vartheta : X \to X'$ such that $\vartheta \circ F = F' \circ \vartheta$; an **embedding** is an injective morphism, a **conjugacy** a bijective morphism.

Let G be a group. We write $H \leq G$ if H is a subgroup of G. If $H \leq G$ and $x\rho y$ iff $x^{-1}y \in H$, then ρ is an equivalence relation over G, whose classes are called the **left cosets** of H, one of them being H itself. If J is a **set of representatives** of the left cosets of H (one representative per coset) then $(j, h) \mapsto jh$ is a bijection between $J \times H$ and G. A **(right) action** of G over a set X is a collection $\phi = \{\phi_g\}_{g \in G} \subseteq X^X$ such that $\phi_{gh}(x) = \phi_h(\phi_g(x))$ for all $g, h \in G$, $x \in X$, and $\phi_{1_G}(x) = x$ for all $x \in X$. Observe that the ϕ_g's are invertible, with $(\phi_g)^{-1} = \phi_{(g^{-1})}$. When ϕ is clear from the context, $\phi_g(x)$ is often written x^g. Properties of functions (e.g., continuity) are extended to actions by saying that ϕ has property P iff each ϕ_g has property P.

If G is a group and $S \subseteq G$, the **subgroup generated by** S is the set $\langle S \rangle$ of all $g \in G$ such that

$$g = s_1 s_2 \cdots s_n \tag{1}$$

for some $n \geq 0$, with $s_i \in S$ or $s_i^{-1} \in S$ for all i. S is a **set of generators** for G if $\langle S \rangle = G$; a group is **finitely generated** (briefly, f.g.) if it has a finite set of generators (briefly, f.s.o.g.). The **length** of $g \in G$ with respect to S is the least $n \geq 0$ such that (1) holds, and is indicated by $\|g\|_S$. The **distance** of g and h w.r.t. S is the length $d_S^G(g, h)$ of $g^{-1}h$; the **disk** of center g and radius R w.r.t. S is $D_{R,S}^G(g) = \{h \in G \mid d_S^G(g, h) \leq R\}$. In all such writings, G and/or S will be omitted if irrelevant or clear from the context; g, if equal to 1_G.

An **alphabet** is a finite set with two or more elements; all alphabets are given the discrete topology. A **configuration** is a map $c \in A^G$ where A is an alphabet and G is a f.g. group. Observe that the product topology on A^G is induced by any of the distances d_S defined by putting $d_S(c_1, c_2) = 2^{-r}$, r being the minimum length w.r.t. S of a $g \in G$ s.t. $c_1(g) \neq c_2(g)$. The **natural action** σ^G of G over A^G is defined as

$$(\sigma_g^G(c))(h) = c(gh) \quad \forall c \in A^G \quad \forall g, h \in G . \tag{2}$$

Observe that σ^G is continuous. A closed subset X of A^G that is invariant by σ^G is called a **shift subspace**, or simply **subshift**. The restriction of σ^G to X is again called the natural action of G over X and indicated by σ^G. From now on, unless differently stated, we will write c^g for $\sigma_g^G(c)$.

Let $E \subseteq G$, $|E| < \infty$. A **pattern** on A with **support** E is a map $p : E \to A$; we write $E = \operatorname{supp} p$. A pattern p **occurs** in a configuration c if there exists

$g \in G$ such that $(c^g)_{|\text{supp}\,p} = p$; p is **forbidden** otherwise. Given a set $\mathcal{F}$ of patterns, the set of all the configurations $c \in A^G$ for which all the patterns in $\mathcal{F}$ are forbidden is indicated as $\mathsf{X}_{\mathcal{F}}^{A,G}$; A and/or G will be omitted if irrelevant or clear from the context. It is well known [4,6] that X is a subshift iff $X = \mathsf{X}_{\mathcal{F}}^{A,G}$ for some $\mathcal{F}$; X is a **shift of finite type** if $\mathcal{F}$ can be chosen finite. A pattern p is forbidden for $X \subseteq A^G$ if it is forbidden for all $c \in X$, i.e. $(c^g)_{|\text{supp}\,p} \neq p$ for all $c \in X$, $g \in G$; if X is a subshift, this is the same as $c_{|\text{supp}\,p} \neq p$ for all $c \in X$.

A map $F : A^G \to A^G$ is **uniformly locally definable** (UL-definable) if there exist $\mathcal{N} \subseteq G$, $|\mathcal{N}| < \infty$, and $f : A^{\mathcal{N}} \to A$ such that

$$(F(c))(g) = f\left((c^g)_{|\mathcal{N}}\right) \tag{3}$$

for all $c \in A^G$, $g \in G$; in this case, we write $F = F_f^{A,G}$. Observe that any UL-definable function F is continuous and commutes with the natural action of G on A^G; **Hedlund's theorem** [4,5] states that, if $X \subseteq A^G$ is a subshift and $F : X \to A^G$ is continuous and commutes with the natural action of G over X, then F is the restriction to X of a UL-definable function. Moreover, remark that, if X is a subshift and F is UL-definable, then $F(X)$ is a subshift too: if X is of finite type, we say that $F(X)$ is a **sofic shift**.

A **cellular automaton** (CA) with alphabet A and **tessellation group** G is a triple $\langle X, \mathcal{N}, f \rangle$ where the **support** $X \subseteq A^G$ is a subshift, the **neighbourhood index** $\mathcal{N} \subseteq G$ is finite, and the **local evolution function** $f : A^{\mathcal{N}} \to A$ satisfies $F_f^{A,G}(X) \subseteq X$; the restriction $F_{\mathcal{A}}$ of $F_f^{A,G}$ to X is the **global evolution function**, and $(X, F_{\mathcal{A}})$ is the **associate dynamical system**. Observe that $(X, F_{\mathcal{A}})$ is a subsystem of $(A^G, F_f^{A,G})$. When speaking of bijectivity, finiteness of type, etc. *of $\mathcal{A}$*, we simply "confuse" it and either $F_{\mathcal{A}}$ or X. $\mathcal{A}$ is a **presentation** of (X', F') if the latter and $(X, F_{\mathcal{A}})$ are conjugate. We call $CA(A, G)$ the class of d.s. having a presentation as CA with alphabet A and tessellation group G.

A pattern p is a **Garden of Eden** (briefly, GoE) for a CA $\mathcal{A} = \langle X, \mathcal{N}, f \rangle$ if it is allowed for X and forbidden for $F_{\mathcal{A}}(X)$. Any CA having a GoE pattern is nonsurjective; compactness of X and continuity of $F_{\mathcal{A}}$ ensure that the vice versa holds as well [4,7]. $\mathcal{A}$ is **preinjective** if $F_{\mathcal{A}}(c_1) \neq F_{\mathcal{A}}(c_2)$ for any two $c_1, c_2 \in X$ such that $\{g \in G \mid c_1(g) \neq c_2(g)\}$ is finite and nonempty. If G is *amenable* (cf. [2,3]; $\mathbb{Z}^d$ is amenable for all d) and $X = A^G$, then $\mathcal{A}$ is surjective iff it is preinjective [2]; this can be false [2,4] if G is not amenable or $X \neq A^G$.

3   Induced Subshifts

Let $X = \mathsf{X}_{\mathcal{F}}^{A,G}$. The idea of "induced subshift" that first comes to the mind is

Definition 1. *Let $X = \mathsf{X}_{\mathcal{F}}^{A,G}$ be a subshift, and let $G \leq \Gamma$. The subshift induced by X on A^{Γ} is $X' = \mathsf{X}_{\mathcal{F}}^{A,\Gamma}$.*

According to Definition 1, X' is what we obtain instead of X by interpreting $\mathcal{F}$ in the context provided by Γ instead of G. However, since different sets of patterns can define identical subshifts, we must ensure that Definition 1 is well posed and X' only depends on X rather than $\mathcal{F}$, i.e., $\mathsf{X}_{\mathcal{F}_1}^{A,G} = \mathsf{X}_{\mathcal{F}_2}^{A,G}$ must imply $\mathsf{X}_{\mathcal{F}_1}^{A,\Gamma} = \mathsf{X}_{\mathcal{F}_2}^{A,\Gamma}$. This actually follows from

Lemma 1. *Let A be an alphabet, and let G and Γ be f.g. groups with $G \leq \Gamma$. For $i = 1, 2$, let $\mathcal{F}_i$ be a set of patterns on A with supports contained in G, let $\mathcal{N}_i$ be a finite nonempty subset of G, and let $f_i : A^{\mathcal{N}_i} \to A$. Then*

$$F_{f_1}^{A,G}\left(\mathsf{X}_{\mathcal{F}_1}^{A,G}\right) \subseteq F_{f_2}^{A,G}\left(\mathsf{X}_{\mathcal{F}_2}^{A,G}\right) \text{ iff } F_{f_1}^{A,\Gamma}\left(\mathsf{X}_{\mathcal{F}_1}^{A,\Gamma}\right) \subseteq F_{f_2}^{A,\Gamma}\left(\mathsf{X}_{\mathcal{F}_2}^{A,\Gamma}\right) .$$

Proof. Let J be a set of representatives of the left cosets of G in Γ such that $1_G = 1_\Gamma \in J$. To simplify notation, we will write

$$X_i = \mathsf{X}_{\mathcal{F}_i}^{A,G} , \ \Xi_i = \mathsf{X}_{\mathcal{F}_i}^{A,\Gamma} , \ F_i = F_{f_i}^{A,G} , \ \Phi_i = F_{f_i}^{A,\Gamma} ,$$

so that the thesis becomes

$$F_1(X_1) \subseteq F_2(X_2) \text{ iff } \Phi_1(\Xi_1) \subseteq \Phi_2(\Xi_2) .$$

For the "if" part, let $c \in F_1(X_1)$, and let $x_1 \in X_1$ satisfy $F_1(x_1) = c$. Define $\xi_1 \in A^\Gamma$ by $\xi_1(jg) = x_1(g)$ for all $j \in J$, $g \in G$: then for all $j \in J$, $g \in G$, $p \in \mathcal{F}_1$

$$(\xi_1^{jg})_{|\text{supp}\,p} = (x_1^g)_{|\text{supp}\,p} \neq p ,$$

hence $\xi_1 \in \Xi_1$. Put $\chi = \Phi_1(\xi_1)$: by hypothesis, there exists $\xi_2 \in \Xi_2$ such that $\Phi_2(\xi_2) = \chi$, and by construction,

$$\chi(g) = f_1((\xi_1^g)_{|\mathcal{N}_1}) = f_1((x_1^g)_{|\mathcal{N}_1}) = c(g) \ \forall g \in G .$$

Let $x_2 = (\xi_2)_{|G}$: then $x_2 \in X_2$ by construction. But

$$f_2((x_2^g)_{|\mathcal{N}_2}) = f_2((\xi_2^g)_{|\mathcal{N}_2}) = \chi(g) = c(g) \ \forall g \in G ,$$

thus $c \in F_2(X_2)$.

For the "only if" part, let $\chi \in \Phi_1(\Xi_1)$, and let $\xi_1 \in \Xi_1$ satisfy $\Phi_1(\xi_1) = \chi$. For each $j \in J$, define $x_{1,j} \in A^G$ as $x_{1,j}(g) = \xi_1(jg)$ for all $g \in G$. It is straightforward to check that $x_{1,j} \in X_1$ for all $j \in J$: let $c_j = F_1(x_{1,j})$. By hypothesis, for all $j \in J$ there exists $x_{2,j} \in X_2$ such that $F_2(x_{2,j}) = c_j$: define $\xi_2 \in A^\Gamma$ by $\xi_2(jg) = x_{2,j}(g)$ for all $j \in J$, $g \in G$. It is straightforward to check that $\xi_2 \in \Xi_2$; but for all $j \in J$, $g \in G$

$$f_2((\xi_2^{jg})_{|\mathcal{N}_2}) = f_2((x_{2,j}^g)_{|\mathcal{N}_2}) = c_j(g) = f_1((x_{1,j}^g)_{|\mathcal{N}_1}) = f_1((\xi_1^{jg})_{|\mathcal{N}_1}) = \chi(jg) ,$$

thus $\chi \in \Phi_2(\Xi_2)$. $\square$

Corollary 1. *In the hypotheses of Lemma 1,*

1. $\mathsf{X}_{\mathcal{F}_1}^{A,G} \subseteq F_{f_2}^{A,G}(\mathsf{X}_{\mathcal{F}_2}^{A,G})$ *iff* $\mathsf{X}_{\mathcal{F}_1}^{A,\Gamma} \subseteq F_{f_2}^{A,\Gamma}(\mathsf{X}_{\mathcal{F}_2}^{A,\Gamma})$,
2. $F_{f_1}^{A,G}(\mathsf{X}_{\mathcal{F}_1}^{A,G}) \subseteq \mathsf{X}_{\mathcal{F}_2}^{A,G}$ *iff* $F_{f_1}^{A,\Gamma}(\mathsf{X}_{\mathcal{F}_1}^{A,\Gamma}) \subseteq \mathsf{X}_{\mathcal{F}_2}^{A,\Gamma}$, *and*
3. $\mathsf{X}_{\mathcal{F}_1}^{A,G} \subseteq \mathsf{X}_{\mathcal{F}_2}^{A,G}$ *iff* $\mathsf{X}_{\mathcal{F}_1}^{A,\Gamma} \subseteq \mathsf{X}_{\mathcal{F}_2}^{A,\Gamma}$.

Proof. Consider the neighbourhood index $\{1_G\}$ and the local evolution function $f(1_G \mapsto a) = a$. Apply Lemma 1. □

Corollary 2. *Let A be an alphabet, let G and Γ be f.g. groups with $G \leq \Gamma$, and let $\mathcal{F}$ be a set of patterns on A with supports contained in G. If $\mathsf{X}_{\mathcal{F}}^{A,G}$ is sofic then $\mathsf{X}_{\mathcal{F}}^{A,\Gamma}$ is sofic.*

Proof. By hypothesis, $\mathsf{X}_{\mathcal{F}}^{A,G} = F(\mathsf{X}_{\mathcal{F}'}^{A,G})$ for some UL-definable function F and finite set of patterns $\mathcal{F}'$. Apply points 1 and 2 of Corollary 1. □

4   Induced Cellular Automata

After having found a way to construct subshifts on large groups from subshifts of smaller groups, we work on doing the same with cellular automata.

Definition 2. *Let $\mathcal{A} = \langle X, \mathcal{N}, f \rangle$ be a CA with alphabet A and tessellation group G, and let Γ be a f.g. group such that $G \leq \Gamma$. The CA induced by $\mathcal{A}$ on Γ is the cellular automaton*

$$\mathcal{A}' = \langle X', \mathcal{N}, f \rangle \ , \tag{4}$$

where X' is the subshift induced by X on A^Γ.

Again, $\mathcal{A}'$ is what we obtain by interpreting $\mathcal{F}$, $\mathcal{N}$, and f in the context provided by Γ instead of G. Lemma 1 ensures that $\mathcal{A}'$ is well defined.

In general, $\mathcal{A}'$ is not conjugate to $\mathcal{A}$: just consider the case Γ finite, G proper, $\mathcal{F} = \emptyset$. However, some important properties—notably, surjectivity—are preserved in the passage from the original CA to the induced one; which is not surprising, because intuitively $F_f^{A,\Gamma}$ is going to operate "slice by slice" on A^Γ, each "slice" being "shaped" as G. The next statement extends a result in [2] from the case $X = A^G$ to the general case when X is an arbitrary subshift.

Theorem 1. *Let $\mathcal{A} = \langle X, \mathcal{N}, f \rangle$ be a CA with alphabet A and tessellation group G, let $G \leq \Gamma$, and let $\mathcal{A}'$ be the CA induced by $\mathcal{A}$ on Γ.*

1. *$\mathcal{A}$ is surjective iff $\mathcal{A}'$ is surjective.*
2. *$\mathcal{A}$ is preinjective iff $\mathcal{A}'$ is preinjective.*
3. *$\mathcal{A}$ is injective iff $\mathcal{A}'$ is injective.*

Proof. Let $\mathcal{F}$ satisfy $X = \mathsf{X}_{\mathcal{F}}^{A,G}$ (and $X' = \mathsf{X}_{\mathcal{F}}^{A,\Gamma}$). Take J as in proof of Lemma 1.

To prove the "if" part of point 1, suppose $\mathcal{A}$ has a GoE pattern p. By contradiction, assume that there exists $\chi \in X'$ such that $F_{\mathcal{A}'}(\chi)_{|\operatorname{supp} p} = p$. Let c be the restriction of χ to G. Then, since both $\mathcal{N}$ and $\operatorname{supp} p$ are subsets of G by hypothesis,

$$(F_{\mathcal{A}}(c))(x) = f\left((c^x)_{|\mathcal{N}}\right) = f\left((\chi^x)_{|\mathcal{N}}\right) = (F_{\mathcal{A}'}(\chi))(x) = p(x)$$

for every $x \in \operatorname{supp} p$: this is a contradiction.

To prove the "only if" part of point 1, suppose $\mathcal{A}'$ has a GoE pattern π. By hypothesis, there exists $\chi \in X'$ such that $\chi_{|\text{supp}\,\pi} = \pi$. For all $j \in J$ define $c_j \in A^G$ as

$$c_j(g) = \chi(jg) \ \forall g \in G \,,$$

and for all $j \in J$ such that $jG \cap \text{supp}\,\pi \neq \emptyset$ define the pattern p_j over G as

$$p_j(x) = \pi(jx) \ \forall x \text{ s.t. } jx \in \text{supp}\,\pi \,.$$

Observe that $c_j \in X$ for all j, and that $p_j = (c_j)_{|jG\cap\text{supp}\,\pi}$ when defined. But at least one of the patterns p_j must be a GoE for $\mathcal{A}$: otherwise, for all $j \in J$, either $jG \cap \text{supp}\,\pi = \emptyset$, or there would exist $k_j \in X'$ such that $F_\mathcal{A}(k_j)_{|\text{supp}\,p_j} = p_j$. In this case, however, $\kappa \in A^\Gamma$ defined by $\kappa(jg) = k_j(g)$ for all $j \in J$, $g \in G$ would satisfy $\kappa \in X'$ and $F_{\mathcal{A}'}(\kappa)_{|\text{supp}\,\pi} = \pi$, against π being a GoE for $\mathcal{A}'$.

For the "if" part of point 2, suppose $c_1, c_2 \in X$ differ on all and only the points of a finite nonempty $U \subseteq G$, but $F_\mathcal{A}(c_1) = F_\mathcal{A}(c_2)$. For all $j \in J$, $g \in G$, put $\chi_1(jg) = c_1(g)$, and set $\chi_2(jg)$ as $c_2(g)$ if $j = 1_\Gamma$, $c_1(g)$ otherwise. Then χ_1 and χ_2 belong to X' and differ precisely on U. Moreover, for every $\gamma \in \Gamma$, either $\gamma \in G$ or $\gamma\mathcal{N} \cap G = \emptyset$, so either $(F_{\mathcal{A}'}(\chi_i))(\gamma) = (F_\mathcal{A}(c_i))(\gamma)$ or $(F_{\mathcal{A}'}(\chi_1))(\gamma) = (F_{\mathcal{A}'}(\chi_2))(\gamma)$.

For the "only if" part of point 2, suppose $\mathcal{A}$ is preinjective. Let $\chi_1, \chi_2 \in X'$ differ on all and only the points of a finite nonempty $U' \subseteq \Gamma$. For $i \in \{1, 2\}$, $\gamma \in \Gamma$, let $c_{i,\gamma}$ be the restriction of χ_i^γ to G: these are all in X, because a pattern occurring in $c_{i,\gamma}$ also occurs in χ_i, and cannot belong to $\mathcal{F}$. Let $U_\gamma = \{g \in G \mid c_{1,\gamma}(g) \neq c_{2,\gamma}(g)\}$: then $|U_\gamma| \leq |U|$ for all $\gamma \in \Gamma$, plus $U_\gamma \neq \emptyset$ for at least one γ. For such γ, there exists $g \in G$ such that $(F_\mathcal{A}(c_{1,\gamma}))(g) \neq (F_\mathcal{A}(c_{2,\gamma}))(g)$: then by construction $(F_{\mathcal{A}'}(\chi_1))(\gamma g) \neq (F_{\mathcal{A}'}(\chi_2))(\gamma g)$ as well.

The proof of point 3 is straightforward to see. For the "if" part, let $c_1 \neq c_2$, $F_\mathcal{A}(c_1) = F_\mathcal{A}(c_2)$, and consider $\chi_i(\gamma) = c_i(g)$ iff $\gamma = jg$. For the "only if" part, given $\chi_1 \neq \chi_2$, consider $c_{i,j}(g) = \chi_i(jg)$, and observe that $F_\mathcal{A}(c_{1,j}) \neq F_\mathcal{A}(c_{2,j})$ for at least one $j \in J$. $\qquad\square$

Surjectivity and preinjectivity are always shared by $\mathcal{A}$ and $\mathcal{A}'$, even when these two properties are not equivalent. Moreover, even if $\mathcal{A}$ and $\mathcal{A}'$ are non-conjugate, there always exists an *embedding* of the former into the latter.

Lemma 2. *Let A be an alphabet, and let G and Γ be f.g. groups with $G \leq \Gamma$; let $\mathcal{A} = \langle X, \mathcal{N}, f \rangle$ be a CA with alphabet A and tessellation group G, and let $\mathcal{A}' = \langle X', \mathcal{N}, f \rangle$ be the CA induced by $\mathcal{A}$ over Γ. Let J be a set of representatives of the left cosets of G in Γ, and let $\iota_J : A^G \to A^\Gamma$ be defined by*

$$(\iota_J(c))(\gamma) = c(g) \ \text{iff} \ \exists j \in J : \gamma = jg \,. \tag{5}$$

Then ι_J is an embedding of $\mathcal{A}$ into $\mathcal{A}'$, so that

$$\iota_J(\mathcal{A}) = \langle \iota_J(X), \mathcal{N}, f \rangle \tag{6}$$

is a CA conjugate to $\mathcal{A}$. In particular, $CA(A, G) \subseteq CA(A, \Gamma)$.

Proof. First, we observe that ι_J is injective and $\iota_J(X) \subseteq X'$. In fact, if $c_1(g) \neq c_2(g)$, then $(\iota_J(c_1))(jg) \neq (\iota_J(c_2))(jg)$ for all $j \in J$. Moreover, should a pattern p exist such that $(\iota_J(c))(\gamma x) = p(x)$ for all $x \in \operatorname{supp} p \subseteq G$, by writing $\gamma = jg$ and applying (5) we would find $c(gx) = p(x)$ for all $x \in \operatorname{supp} p$, a contradiction.

Next, we show that ι_J is continuous. Let S be a f.s.o.g. for G, Σ a f.s.o.g. for Γ. Let $R \geq 0$, and let

$$E_R = \{g \in G \mid \exists j \in J \mid jg \in D_{R,\Sigma}^{\Gamma}\} .$$

Since the writings $\gamma = jg$ are unique and $D_{R,\Sigma}^{\Gamma}$ is finite, E_R is finite too. Let $E_R \subseteq D_{r,S}^G$: if $(c_1)_{|D_{r,S}^G} = (c_2)_{|D_{r,S}^G}$, then $(\iota_J(c_1))_{|D_{R,\Sigma}^{\Gamma}} = (\iota_J(c_2))_{|D_{R,\Sigma}^{\Gamma}}$.

Next, we show that ι_J is a morphism of d.s. For every $c \in A^G$, $\gamma = jg \in \Gamma$, $x \in \mathcal{N}$ we have $\gamma x \in jG$ and $(\iota_J(c))(\gamma x) = (\iota_J(c))(jgx) = c(gx)$. Thus,

$$((F_{A'} \circ \iota_J)(c))(\gamma) = f(\iota_J(c)^{\gamma}|_{\mathcal{N}}) = f(c^g|_{\mathcal{N}}) = (F_A(c))(g) = ((\iota_J \circ F_A)(c))(\gamma) ,$$

so that $F_{A'} \circ \iota_J = \iota_J \circ F_A$. Moreover, $F_{A'}(\iota_J(X)) = \iota_J(F_A(X)) \subseteq \iota_J(X)$ because $F_A(X) \subseteq X$.

Finally, we observe that $\iota_J(X)$ is a subshift. In fact, if $X = \mathsf{X}_{\mathcal{F}}^{A,G}$, then $\iota_J(X) = \mathsf{X}_{\mathcal{F} \cup \mathcal{F}'}^{A,\Gamma}$, where

$$\mathcal{F}' = \left\{ p \in A^{\{j_1 g, j_2 g\}} \mid j_1, j_2 \in J, g \in G, j_1 \neq j_2, p(j_1 g) \neq p(j_2 g) \right\} . \tag{7}$$

It is straightforward that $\iota_J(X) \subseteq \mathsf{X}_{\mathcal{F} \cup \mathcal{F}'}^{A,\Gamma}$. Let $\chi \in \mathsf{X}_{\mathcal{F} \cup \mathcal{F}'}^{A,\Gamma}$: then $c(g) = \chi(jg)$ is well defined, and $\chi = \iota_J(c)$ by construction. Moreover, for every $g \in G$, $p \in F$, and any $j \in G$ $(c^g)_{\operatorname{supp} p} = (\chi^{jg})_{\operatorname{supp} p} \neq p$, so $c \in X$ and $\chi \in \iota_J(X)$. $\square$

Observe that, in the hypotheses of Lemma 2, we are *not* assuming $1_\Gamma \in J$. Hence, in general, $E_R \not\subseteq D_{R,S}^G$, even if $S \subseteq \Sigma$. As a counterexample, let $\Gamma = \mathbb{Z}^2$, $G = \{(x,0) \mid x \in \mathbb{Z}\}$, $J = \{(1,0)\} \cup \{(0,y) \mid y \in \mathbb{Z}, y \neq 0\}$, $S = \{(1,0)\}$, $\Sigma = \{(1,0), (0,1)\}$: then $E_1 = \{(0,0), (-1,0), (-2,0)\} \not\subseteq D_{1,S}^G$.

Lemma 2 says that growing the tessellation group does not shrink the class of presentable dynamics. This is also true for growing the alphabet, and holds up to alphabet *bijections* and group *isomorphisms*. Before proving this fact, we state a definition, on which these and other results will be based.

Definition 3. *Let X be a set, A an alphabet, G a group, ϕ an action of G over X. X is discernible on A by ϕ if there exists a continuous function $\pi : X \to A$ such that, for any two distinct $x_1, x_2 \in X$, there exists $g \in G$ such that $\pi(\phi_g(x_1)) \neq \pi(\phi_g(x_2))$.*

Observe, in Definition 3, the continuity requirement, which demands that $\pi(x) = \pi(y)$ if $x, y \in X$ are "near enough". We now state and prove the following result from [1].

Theorem 2. *Let A be an alphabet, G a f.g. group, (X, F) a d.s. The following are equivalent:*

1. $(X, F) \in CA(A, G)$;
2. there exists a continuous action ϕ of G over X such that F commutes with ϕ and X is discernible on A by ϕ.

Proof. We start with supposing that $\mathcal{A} = \langle Y, \mathcal{N}, f \rangle$ is a presentation of (X, F). Let $\theta : X \to Y$ be a conjugacy from (X, F) to $(Y, F_{\mathcal{A}})$; put

$$\phi_g = \theta^{-1} \circ \sigma_g^G \circ \theta$$

for all $g \in G$, and

$$\pi(x) = (\theta(x))(1_G) .$$

Remark that $\phi = \{\phi_g\}_{g \in G}$ is an action of G over X and that $(\theta(x))(g) = (\theta(x))^g(1_G)$ for all x and g. Continuity of ϕ and commutation with F are straightforward to verify. If $x_1 \neq x_2$, then $(\theta(x_1))(g) \neq (\theta(x_2))(g)$ for some $g \in G$, thus

$$\pi(\phi_g(x_1)) = (\sigma_g^G(\theta(x_1)))(1_G) \neq (\sigma_g^G(\theta(x_2)))(1_G) = \pi(\phi_g(x_2)) .$$

For the reverse implication, let π as in Definition 3: then $\tau : X \to A^G$ defined by

$$(\tau(x))(g) = \pi(\phi_g(x))$$

is injective. Moreover, $(\tau(\phi_g(x))(h) = \pi(\phi_h(\phi_g(x))) = \pi(\phi_{gh}(x)) = (\tau(x))(gh)$ for every $x \in X$, $g, h \in G$: thus, $\tau \circ \phi_g = \sigma_g^G \circ \tau$ for all $g \in G$, and $X' = \tau(X)$ is invariant under σ^G.

We now prove that τ is continuous. Let $\lim_{n \in \mathbb{N}} x_n = x$ in X: by continuity of π and ϕ, $\lim_{n \in \mathbb{N}} (\tau(x_n))(g) = (\tau(x))(g)$ in A for all G. Since A is discrete, for each $g \in G$ there exists n_g such that $\pi(\phi_g(x_n)) = \pi(\phi_g(x))$ for every $n > n_g$. Hence, for every finite $E \subseteq G$, if $n > n_E = \max_{g \in E} n_g$, then $\tau(x_n)$ and $\tau(x)$ coincide on E: this is the same as saying that $\lim_{n \in \mathbb{N}} \tau(x_n) = \tau(x)$ in the product topology of A^G.

Since X and A^G are compact and Hausdorff, X' is closed in A^G and a subshift, while τ is a homeomorphism between X and X'. Define $F' : X' \to X'$ by $F' = \tau \circ F \circ \tau^{-1}$: then (X', F') is a d.s. and τ is a conjugacy between (X, F) and (X', F'). But for every $g \in G$

$$\phi_g \circ \tau^{-1} = (\tau \circ \phi_{g^{-1}})^{-1} = (\sigma_{g^{-1}}^G \circ \tau)^{-1} = \tau^{-1} \circ \sigma_g^G ,$$

thus

$$\sigma_g^G \circ F' = \tau \circ \phi_g \circ F \circ \tau^{-1} = \tau \circ F \circ \phi_g \circ \tau^{-1} = F' \circ \sigma_g^G ;$$

hence, F' commutes with σ^G. By Hedlund's theorem, there exist a finite $\mathcal{N}' \subseteq G$ and a map $f' : A^{\mathcal{N}'} \to A$ such that $(F'(c))_g = f'(c^g|_{\mathcal{N}'})$ for all $c \in X$, $g \in G$: then $\langle X', \mathcal{N}', f' \rangle$ is a presentation of (X, F) as a cellular automaton.   □

Theorem 2 has two immediate consequences, the first one being *Richardson's lemma* [8]: if $(X, F) \in CA(A, G)$ is invertible and ϕ is as in Theorem 2, then it is straightforward to check that F^{-1} commutes with ϕ, so that $(X, F^{-1}) \in CA(A, G)$. The second one is

Lemma 3. *Let A and B be alphabets, and let G and Γ be f.g. groups.*

1. *If $|A| \leq |B|$ then $CA(A, G) \subseteq CA(B, G)$.*
2. *If G is isomorphic to Γ then $CA(A, G) = CA(A, \Gamma)$.*

Proof. To prove point 1, let $\iota : A \to B$ be injective. Let $(X, F) \in CA(A, G)$, and let ϕ satisfy point 2 of Theorem 2, π being the discerning map. Then X is discernible over B by ϕ, $\iota \circ \pi$ being the discerning map.

To prove point 2, let $\psi : G \to \Gamma$ be a group isomorphism. Let $(X, F) \in CA(A, G)$ and let ϕ satisfy point 2 of Theorem 2, π being the discerning map. Define $\phi' = \{\phi'_\gamma\}_{\gamma \in \Gamma}$ as

$$\phi'_\gamma = \phi_{\psi^{-1}(\gamma)} .$$

It is straightforward to check that ϕ' is an action which commutes with F. Let $x_1 \neq x_2$: if $g \in G$ is such that $\pi(\phi_g(x_1)) \neq \pi(\phi_g(x_2))$, then $\pi(\phi'_{\psi(g)}(x_1)) \neq \pi(\phi'_{\psi(g)}(x_2))$ as well. Thus ϕ' satisfies condition 2 of Theorem 2, and $(X, F) \in CA(A, \Gamma)$. From the arbitrariness of (X, F) follows $CA(A, G) \subseteq CA(A, \Gamma)$: by swapping the roles of G and Γ and repeating the argument with ψ^{-1} in place of ψ we obtain the reverse inclusion. $\square$

Lemma 2 and Lemma 3 together yield

Theorem 3. *Let A, B be alphabets and G, Γ be f.g. groups. If $|A| \leq |B|$ and G is isomorphic to a subgroup of Γ, then $CA(A, G) \subseteq CA(B, \Gamma)$.*

Proof. Let $G \cong H \leq \Gamma$. Then $CA(A, G) = CA(A, H) \subseteq CA(A, \Gamma) \subseteq CA(B, \Gamma)$. $\square$

Corollary 3. *Let $\mathbb{F}_n$ be the free group on $n < \infty$ generators. For every alphabet A and every $n > 1$, $CA(A, \mathbb{F}_n) = CA(A, \mathbb{F}_2)$.*

Proof. Follows from Theorem 3 and the fact that $\mathbb{F}_2$ has a free subgroup on infinitely many generators. $\square$

Remark that it is not possible to replace $\mathcal{F}'$ (7) with the smaller set

$$\mathcal{F}'' = \left\{ p \in A^{\{j_1, j_2\}} \mid j_1, j_2 \in J, p(j_1) \neq p(j_2) \right\} .$$

Indeed, for $\iota_J(X)$ to be of finite type, it would then suffice X being of finite type and G being of finite index in Γ. Instead, we have

Theorem 4. *Let Γ be the group of ordered pairs (i, k), $i \in \{0, 1\}$, $k \in \mathbb{Z}$ with the product*

$$(i_1, k_1)(i_2, k_2) = (i_1 + i_2 - 2i_1 i_2, (-1)^{i_2} k_1 + k_2) .$$

Let $A = \{a, b\}$, $G = \{(0, k), k \in \mathbb{Z}\} \leq \Gamma$, and $J = \{(0, 0), (1, 0)\}$. Then $\iota_J(A^G)$ is not a shift of finite type.

Proof. Let $S = \{(1,0),(0,1)\}$: it is straightforward to check that $\langle S \rangle = \Gamma$.

By contradiction, assume that $\iota_J(A^G) = \mathsf{X}_{\mathcal{F}}^{A,\Gamma}$ with $|\mathcal{F}| < \infty$; it is not restrictive to choose $\mathcal{F}$ so that $\mathrm{supp}\, p = D_{M,S}^{\Gamma}$ for all $p \in \mathcal{F}$. Let $\delta \in A^{\Gamma}$ satisfy $\delta(x) = b$ iff $x = (0,0)$: then $\delta \notin \iota_J(A^G)$, so there must exist $p \in \mathcal{F}$, $\eta \in \Gamma$ such that $\delta^{\eta}|_{\mathrm{supp}\, p} = p$. It is straightforward to check that there exists exactly one $y \in D_M^{\Gamma}$ such that $p(y) = b$, and that $y = \eta^{-1} = (i, (-1)^{1-i}x)$ if $\eta = (i, x)$.

Now, for all $k \in \mathbb{Z}$ we have $d_S^{\Gamma}((0,k),(1,k)) = \|(1,2k)\|_S^{\Gamma} = 2|k|+1$. This can be checked by observing the following two facts. Firstly, $(1,2k) = (1,0)(0,t)\ldots(0,t)$, with $2|k|$ factors $(0,t)$, and $t = 1$ or $t = -1$ according to $k > 0$ or $k < 0$. Secondly, multiplying (i,x) on the right by $(0,1)$ or $(0,-1)$ does not change the value of i, while multiplying (i,x) on the right by $(1,0)$ does not change $|x|$: hence, at least one multiplication by $(1,0)$ and $2|k|$ multiplications by either $(0,1)$ or $(0,-1)$ are necessary to reach $(1,2k)$ from $(0,0)$.

For $i \in \{0,1\}$ let $\gamma_i = (i, 2M+1)$. Let $\chi \in A^{\Gamma}$ be such that $\chi(\gamma) = b$ iff $\gamma = \gamma_0$ or $\gamma = \gamma_1$: then $\chi \in \iota_J(A^G)$. However, since $\eta^{-1} \in D_{M,S}^{\Gamma}$, for all $x \in D_M^{\Gamma}(\eta^{-1})$ we have $\gamma_0\eta x \in D_{2M}^{\Gamma}(\gamma_0)$. Hence, either $x = \eta^{-1}$, $\gamma_0\eta x = \gamma_0$, and $\chi^{\gamma_0\eta}(x) = b$; or $x \neq \eta^{-1}$, $0 < d_S(\gamma_0, \gamma_0\eta x) \leq 2M < 4M + 3 = d_S(\gamma_0, \gamma_1)$, and $\chi^{\gamma_0\eta}(x) = a$. Thus, $(\chi^{\gamma_0\eta})|_{\mathrm{supp}\, p} = p$: this is a contradiction. $\square$

Corollary 4. *For cellular automata on arbitrary f.g. groups, finiteness of type is not invariant by conjugacy. In particular, for subshifts on arbitrary f.g. groups, finiteness of type is not a topological property.*

The first statement in Corollary 4 seems to collide with Theorem 2.1.10 of [6], stating that any two conjugate subshifts of $A^{\mathbb{Z}}$ are either both of finite type or both not of finite type. Actually, in the cited result, conjugacies are always intended as being between *shift dynamical systems*, which is a much more specialized situation than ours. Moreover, the tessellation group is always $\mathbb{Z}$, so that *the action is also the same*, while we have different groups and different actions. Last but not least, translations are UL-definable if and only if the translating factor is *central*, i.e. commutes with every element in the tessellation group: thus, the only groups where *all* the translations are UL-definable are the abelian groups. On the other hand, the second statement remarks the well-known phenomenon that homeomorphisms do not preserve finiteness of type, not even in "classical" symbolic dynamics. For instance, the even shift (there is always an even number of 0's between any two 1's) is not of finite type, but it is homeomorphic to the Cantor set, thus also to the full shift.

Things are better for direct products.

Theorem 5. *Let H and K be f.g. groups; let S be a finite set of generators for H such that $1_H \notin S$ and $H = \langle S \rangle$; let $\Gamma = H \times K$, $G = \{1_H\} \times K$, $J = H \times \{1_K\}$. Let A be an alphabet and let*

$$\mathcal{F}_S = \left\{ p \in A^{\{(1_H,1_K),(s,1_K)\}} \mid s \in S \cup S^{-1} \setminus \{1_H\}, p((1_H,1_K)) \neq p((s,1_K)) \right\}.$$

For every set $\mathcal{F}$ of patterns on A with supports contained in G, $\iota_J(\mathsf{X}_{\mathcal{F}}^{A,G}) = \mathsf{X}_{\mathcal{F} \cup \mathcal{F}_S}^{A,\Gamma}$. In particular, if $X \subseteq A^G$ is a shift of finite type, then $\iota_J(X)$ is also a shift of finite type.

Proof. First, observe that $\mathcal{F}_S \subseteq \mathcal{F}'$, where $\mathcal{F}'$ is given by (7), so that $\iota_J(\mathsf{X}_{\mathcal{F}}^{A,G}) = \mathsf{X}_{\mathcal{F}\cup\mathcal{F}'}^{A,\Gamma} \subseteq \mathsf{X}_{\mathcal{F}\cup\mathcal{F}_S}^{A,\Gamma}$.

Let now $\chi \in A^\Gamma \setminus \iota_J(X)$; suppose that no $p \in \mathcal{F}$ occurs in χ. Let $h_1, h_2 \in H$, $k \in K$ satisfy $\chi((h_1, k)) \neq \chi((h_2, k))$, and let $h_1^{-1}h_2 = s_1 s_2 \cdots s_N$ be a writing of minimal length of the form (1). For $i \in \{0, \ldots, N\}$ let $a_i = \chi(h_1 s_1 \ldots s_i, k)$; for $i \in \{1, \ldots, N\}$ define $p_i : \{(1_H, 1_K), (s_i, 1_K)\} \to A$ by $p_i(1_H, 1_K) = a_{i-1}$ and $p_i(s_i, 1_K) = a_i$. Since $a_0 \neq a_N$, $a_{i-1} \neq a_i$ for some i: then $p_i \in \mathcal{F}_S$ and $(\chi^{(h_1 s_1 \cdots s_{i-1}, k)})_{|\mathrm{supp}\, p_i} = p_i$. Since χ is arbitrary, $\mathsf{X}_{\mathcal{F}\cup\mathcal{F}_S}^{A,\Gamma} \subseteq \iota_J(\mathsf{X}_{\mathcal{F}}^{A,G})$. □

Observe that G needs not to be of finite index in Γ. We conclude with

Theorem 6. *Let A, G, Γ, and J be as in Lemma 2. Suppose $\iota_J(X)$ is a shift of finite type for every shift of finite type $X \subseteq A^G$. Then $\iota_J(X)$ is a sofic shift for every sofic shift $X \subseteq A^G$.*

Proof. Let $X = F(Y)$ for some shift of finite type $Y \subseteq A^G$ and UL-definable function $F : A^G \to A^G$. Let $\mathcal{N} \subseteq G$, $|\mathcal{N}| < \infty$, and $f : A^{\mathcal{N}} \to A$ be such that $(F(c))_g = f(c^g|_{\mathcal{N}})$ for all $c \in A^G$, $g \in G$; let $\mathcal{A} = \langle A^G, \mathcal{N}, f \rangle$ and let F' be the global evolution function of $\iota_J(\mathcal{A})$. By Lemma 2, $F' \circ \iota_J = \iota_J \circ F$, so that $\iota_J(X) = \iota_J(F(Y)) = F'(\iota_J(Y))$ is the image of a shift of finite type via a UL-definable function. □

5 Conclusions

We have shown how to construct new shift subspaces and cellular automata by enlarging their underlying groups; we have then remarked the properties of old objects inherited by the new ones, while taking note of some exceptions; finally, we have observed how enlarging the group makes the class of presentable dynamics grow. However, there is surely much work to do; in particular, the problem whether the reverse of Corollary 2—namely, that $\mathsf{X}_{\mathcal{F}}^{A,G}$ is sofic if $\mathsf{X}_{\mathcal{F}}^{A,\Gamma}$ is—does or does not hold, has not yet, at the best of our knowledge, found a solution. Aside of looking ourselves for the answers to such questions, our hope is that our work can be interesting, or even useful, to researchers in the field.

Acknowledgements

The author was partially supported by the project "The Equational Logic of Parallel Processes" (nr. 060013021) of The Icelandic Research Fund. We also thank the anonymous referees for their many insightful suggestions.

References

1. Capobianco, S.: Structure and Invertibility in Cellular Automata. PhD thesis, University of Rome "La Sapienza" (2004)
2. Ceccherini-Silberstein, T.G., Machì, A., Scarabotti, F.: Amenable groups and cellular automata. Ann. Inst. Fourier, Grenoble 42, 673–685 (1999)

3. de la Harpe, P.: Topics in Geometric Group Theory. Chicago Lectures in Mathematics. University of Chicago Press (2000)
4. Fiorenzi, F.: Cellular automata and strongly irreducible shifts of finite type. Theor. Comp. Sci. 299, 477–493 (2003)
5. Hedlund, G.A.: Endomorphisms and automorphisms of the shift dynamical system. Math. Syst. Th. 3, 320–375 (1969)
6. Lind, D., Marcus, B.: An introduction to symbolic dynamics and coding. Cambridge University Press, Cambridge (1995)
7. Machí, A., Mignosi, F.: Garden of Eden Configurations for Cellular Automata on Cayley Graphs on Groups. SIAM J. Disc. Math. 6, 44–56 (1993)
8. Richardson, D.: Tessellations with local transformations. J. Comp. Syst. Sci. 6, 373–388 (1972)

Hopcroft's Algorithm and Cyclic Automata

Giusi Castiglione, Antonio Restivo, and Marinella Sciortino

Università di Palermo, Dipartimento di Matematica e Applicazioni,
via Archirafi, 34 - 90123 Palermo, Italy
{giusi,restivo,mari}@math.unipa.it

1 Introduction

Minimization of deterministic finite automata is a largely studied problem of the
Theory of Automata and Formal Languages. It consists in finding the unique (up
to isomorphism) minimal deterministic automaton recognizing a set of words.
The first approaches to this topic can be traced back to the 1950's with the works
of Huffman and Moore (cf. [12,15]). Over the years several methods to solve this
problem have been proposed but the most efficient algorithm in the worst case
was given by Hopcroft in [11]. Such an algorithm computes in $O(n \log n)$ the minimal automaton equivalent to a given automaton with n states. The Hopcroft's
algorithm has been widely studied, described and implemented by many authors
(cf. [13,4,16,2]). In particular, in [4] the worst case of the algorithm is considered.
The authors of [4] introduce an infinite family of automata associated to circular
words. The circular words taken into account are the de Bruijn words, and, by
using the associated automata, it is shown that the complexity of the algorithm
is tight. More precisely, the Hopcroft's algorithm has some degrees of freedom
(see Section 3) in the sense that there can be several executions of the algorithm
on a given deterministic automaton. The running time of such executions can be
different. With regard to the family of automata proposed in [4] it is shown that
there exist some "unlucky" sequences of choices that slow down the computation
to achieve the bound $\Omega(nlogn)$. However, there are also executions that run in
linear time for the same automata. The authors of [4] leave the open problem
whether there are automata on which all the executions of Hopcroft's algorithm
do not run in linear time. In [16] the author proves that the exact worst case
(i.e. in terms of the exact number of operations) of the algorithm for unary languages is reached only for the family of automata given in [4] when a queue is
used in the implementation. He remarks that a stack implementation is more
convenient for such automata. At the same time in [16] the author conjectures
that there is a strategy for processing the stack used in the implementation such
that the minimization of all unary languages will be realized in linear time by
the Hopcroft's algorithm.

In the present paper we provide an answer to both the above questions by
giving an infinite family of automata for which the running time is $\Theta(n \log n)$,
whatever implementation strategy is used. Such automata, presented in Section
6, are obtained by developing the idea of Berstel and Carton proposed in [4].
We reach our aim by investigating, in a more general context, the connections

C. Martín-Vide, F. Otto, and H. Fernau (Eds.): LATA 2008, LNCS 5196, pp. 172–183, 2008.

between combinatorial properties of circular words and minimality conditions of the associated automata (see Section 4).

We show in Section 5 that, in the case of automata associated to circular standard words, the Hopcroft's algorithm has a unique execution. Moreover, we prove in Section 6 that, in case of automata associated to circular Fibonacci words, the unique execution of the Hopcroft's algorithm runs in time $\Theta(n \log n)$. The result is obtained by the exact computation of the running time of the algorithm, expressed in terms of the Fibonacci convolution sequence.

2   Minimization of Finite State Automata

In this section we give some basics about minimization of finite automata.

Let $\mathcal{A} = (Q, \Sigma, \delta, q_0, F)$ be a *deterministic finite automaton* (*DFA*) over the finite alphabet Σ, where Q is a finite state set, δ is a *transition function* $Q \times \Sigma \rightarrow Q$, $q_0 \in Q$ is the *initial state* and $F \subseteq Q$ the set of *final states*. If C is a subset of Q and $a \in \Sigma$, with $\delta_a^{-1}(C)$ we denote the set $\{q \in Q | \delta(q, a) \in C\}$.

Two finite automata are *equivalent* if they recognize the same language.

A *DFA* is *minimal* if it has the minimum number of states among all its equivalent deterministic automata. For each regular language there exists a unique (up to isomorphism) minimal automaton recognizing it. It is computed by using the Nerode equivalence, as described below.

For any state $p \in Q$, it is considered the language

$$\mathcal{L}_p(\mathcal{A}) = \{v \in \Sigma^* | \delta(p, v) \in F\}.$$

The *Nerode equivalence* on Q, denoted by $\sim$, is defined as follows: for $p, q \in Q$, $p \sim q$ if $\mathcal{L}_p(\mathcal{A}) = \mathcal{L}_q(\mathcal{A})$.

We say that an equivalence relation $\sim$ defined on the set Q of the states of $\mathcal{A}$ is a *congruence* of $\mathcal{A}$ if it is compatible with the transitions of $\mathcal{A}$, i.e. for any $a \in \Sigma$, $p \sim q$ implies $\delta(p, a) \sim \delta(q, a)$.

It is also known (cf. [9]) that the Nerode equivalence is the coarsest congruence of $\mathcal{A}$ that saturates F, i.e. such that F is union of classes of the congruence.

The minimal automaton equivalent to a given *DFA* can be computed by merging states which are equivalent w.r.t. the Nerode equivalence. Let $\mathcal{A} = (Q, \Sigma, \delta, q_0, F)$ be a *DFA* that recognizes the language $\mathcal{L}$, and $\Pi = \{Q_1, Q_2, ..., Q_m\}$ the partition corresponding to Nerode equivalence. For $q \in Q_i$, the class Q_i is denoted by $[q]$. Then the minimal automaton that recognizes $\mathcal{L}$ is $\mathcal{MA} = (\overline{Q}, \Sigma, \overline{\delta}, \overline{q_0}, \overline{F})$, where:

- $\overline{Q} = \{Q_1, Q_2, ..., Q_m\}$

- $\overline{q_0} = [q_0]$

- $\overline{\delta}([q], a) = [\delta(q, a)]$, $\forall q \in Q$ and $\forall a \in \Sigma$

- $\overline{F} = \{[q] | q \in F\}$

The Nerode equivalence is commonly computed by the Moore construction (cf. [15]) as follows.

For any integer $k \geq 0$, we define

$$\mathcal{L}_p^k = \{v \in \mathcal{L}_p \mid |v| \leq k\}.$$

Then the equivalence $\sim_k$ on Q is defined as follows:

$$p \sim_k q \Leftrightarrow \mathcal{L}_p^k = \mathcal{L}_q^k.$$

Such a relation means that in order to distinguish the two states p and q, a word of length at least $k + 1$ is needed.

Theorem 1. *(Moore) The Nerode equivalence is equal to $\sim_{|Q|-2}$.*

There exist several methods which can be used to compute the Nerode equivalence in order to minimize a finite automaton $\mathcal{A}$. Some of them operate by successive refinements of a partition of the states of a given *DFA* (cf. [15,11]). Moore's algorithm compute in time $O(|\Sigma||Q|^2)$ the minimal automaton, the Hopcroft's algorithm is the most efficient in the worst case and its running time is $O(|\Sigma||Q| \log |Q|)$. The Brzozowski's method (cf. [6]) operates by reversal and determinization repeated twice and it can be applied also to a non deterministic finite automata. The time complexity is exponential in the worst case, but it has good performance in practice (cf. [7]). Other methods work only for a restricted class of automata, for instance for acyclic automata (cf. [17,8]) and local automata (cf. [3]).

A taxonomy of finite automata minimization algorithms is given in [21]. Very recently many authors have worked on experimental comparison of minimization algorithms (cf. [19,1]).

Here we are interested in the study of the Hopcroft's algorithm.

3   Hopcroft's Algorithm

In 1971 Hopcroft gave an algorithm for minimizing a finite state automaton with n states, over an alphabet Σ, in $O(|\Sigma| n \log n)$ time (c.f. [11]). This algorithm has been widely studied and described by many authors (see for example [10,13,14,21]) cause of the difficult to give its theoretical justification, to prove correctness and to compute running time.

In this section we give a brief description of the Hopcroft's algorithm.

Given an automaton $\mathcal{A} = (Q, \Sigma, \delta, q_0, F)$, it computes the coarsest congruence that saturates F. Let us observe that the partition $\{F, Q \setminus F\}$, trivially, saturates F.

Given a partition $\Pi = \{Q_1, Q_2, ..., Q_m\}$ of Q, we say that the pair (Q_i, a), with $a \in \Sigma$, *splits* the class Q_j if $\delta_a^{-1}(Q_i) \cap Q_j \neq \emptyset$ and $Q_j \not\subseteq \delta_a^{-1}(Q_i)$. In this case the class Q_j is split into $Q_j' = \delta_a^{-1}(Q_i) \cap Q_j$ and $Q_j'' = Q_j \setminus \delta_a^{-1}(Q_i)$. Let us note that if Π saturates F then $\Pi \setminus \{Q_j\} \cup \{Q_j', Q_j''\}$ saturates F and it is coarser than Π. Furthermore, we have that a partition Π is a congruence if and only if for any $1 \leq i, j \leq m$ and any $a \in \Sigma$, the pair (Q_i, a) does not splits Q_j.

MINIMIZATION $(\mathcal{A} = (Q, \Sigma, \delta, q_0, F))$
1. $\Pi \leftarrow \{F, Q \setminus F\}$
2. **for all** $a \in \Sigma$ **do**
3. $\mathcal{W} \leftarrow \{(min(F, Q \setminus F), a)\}$
4. **while** $\mathcal{W} \neq \emptyset$ **do**
5.   *choose and delete any* (C, a) *from* $\mathcal{W}$
6.   **for all** $B \in \Pi$ **do**
7.   **if** B **is split from** (C, a) **then**
8. $B' \leftarrow \delta_a^{-1}(C) \cap B$
9. $B'' \leftarrow B \setminus \delta_a^{-1}(C)$
10. $\Pi \leftarrow \Pi \setminus \{B\} \cup \{B', B''\}$
11.   **for all** $b \in \Sigma$ **do**
12.   **if** $(B, b) \in \mathcal{W}$ **then**
13. $\mathcal{W} \leftarrow \mathcal{W} \setminus \{(B, b)\} \cup \{(B', b), (B'', b)\}$
14.   **else**
15. $\mathcal{W} \leftarrow \mathcal{W} \cup \{(min(B', B''), b)\}$

The main idea of the algorithm is the following. It starts from the partition $\{F, Q \setminus F\}$ and refines it by means of splitting operations until it obtains a congruence, i.e. until no split is possible. To do that it maintains the current partition Π and a set $\mathcal{W} \subseteq \Pi \times \Sigma$, called *waiting set*, that contains the pairs for which it has to check whether some classes of the current partition is split.

The main loop of the algorithm takes and deletes one pair (C, a) from $\mathcal{W}$ and, for each class B of Π, checks if it is split by (C, a). If it is the case, the class B in Π is replaced by the two sets B' and B'' obtained from the split. For each $b \in \Sigma$, if $(B, b) \in \mathcal{W}$, it is replaced by (B', b) and (B'', b), otherwise the pair $(min(B', B''), b)$ is added to $\mathcal{W}$ (with the notation $min(B', B'')$ we mean the set with minimum cardinality between B' and B''). Let us observe that a class is split by (B', b) if and only if it is split by B'', hence, the pair $(min(B', B''), b)$ is chosen for convenience.

We point out that the algorithm is not deterministic because the pair (C, a) to be processed at each step is freely chosen. This implies that for each automaton there can be many different executions that produce the same partition and, as consequence, different running time. The most common implementations of the set $\mathcal{W}$ use a FIFO strategy, but it is still unknown whether there exist more convenient data structures. In [2] an implementation using a LIFO strategy is given.

Another "nondeterministic" choice intervenes when a set B is split into B' and B'' and B is not present in $\mathcal{W}$. In this case the algorithm chooses which set B' or B'' to be added to $\mathcal{W}$ and such a choice is based on the minimal number of states in these two sets. When both B' and B'' have the same number of states, we could add indifferently B' or B''.

As regards the running time of the algorithm we can observe that the splitting of classes of the partition, with respect to the pair (C, a), takes a time proportional to the cardinality of the set C. Hence, the running time of the algorithm is proportional to the sum of the cardinality of all sets processed.

Hopcroft proved that the running time is bounded by $O(|\Sigma||Q|\log|Q|)$. In [4] the authors prove that this bound is tight, in the sense that they provide a class of unary automata for which there exist executions of the Hopcroft's algorithm that run in time $O(|\Sigma||Q|\log|Q|)$ by using a FIFO strategy for the implementation of $\mathcal{W}$. Such a bound is reached by using a non-splitting strategy in choosing the class to add at each step in $\mathcal{W}$. Other strategy could produce executions that run in linear time for the same automata. In [16] the author proves that the exact worst case of the algorithm (i.e. in terms of the exact number of states in the waiting set) for unary languages is reached when a strategy FIFO is used and only for the family of automata given in [4]. He remarks that a stack implementation is more convenient for such automata. In Section 6 we present a family of unary automata representing the worst case for the Hopcroft's algorithm, whatever strategies for implementing $\mathcal{W}$ and for choosing the class to add at each step are used.

4   Circular Words and Cyclic Automata

In this section we consider a class of automata over a unary alphabet. These automata have a very simple structure since they are just made of a single cycle. The final states of these automata are defined by a pattern given by a given binary circular word.

Let $A = \{0, 1\}$ be a binary alphabet and v, u be two words in A^*. We say that v and u are conjugate if for some words $z, w \in A^*$ one has that $v = zw$ and $u = wz$. It is easy to see that conjugation is an equivalence relation. Note that many combinatorial properties of words in A^* can be thought as properties of the respective conjugacy classes. So, in order to investigate some structural properties of conjugacy classes of words, in this section we consider a conjugacy class of words as a *circular word*. In particular, we denote by (w) the circular word corresponding to all the conjugates of the word w. Circular words has been widely studied, in particular in [5] such a notion is used in order to shed some light onto the circular structure of Christoffel words.

A word w in A^* is called primitive if all conjugates of its are distinct. In this case we say that the circular word (w) is *primitive*. For instance, it is easy to verify that the circular word $(bcabcabca)$ is not primitive, while $(abaab)$ is primitive.

We say that a word $v \in A^*$ is a *factor* of a circular word (w) if v is a factor of some conjugate of w. Equivalently, a factor of (w) is a factor of ww of length not greater that $|w|$. Not that, while each factor of w is also a factor of (w), there exist circular words (w) having factors that are not factor of w. For instance, ca is a factor of (abc) without being factor of abc.

The following proposition is a little improvement of a result proved in [5].

Proposition 1. *Let w a word of length $n \geq 2$. The following statements are equivalent:*

1. *(w) is primitive;*
2. *for $k = 0, \ldots, n-1$ the circular word (w) has at least $k+1$ factors of length k;*
3. *(w) has n factors of length $n - 1$.*

We can define analogously the notion of special factor of a circular binary word. Given a circular word (w) defined over the alphabet A, we say that u is a *special factor* of (w) if both $u0$ and $u1$ are factors of (w). For instance, 00 is a special factor of (01001100) because 001 and 000 are factors.

We recall the notion of standard word. Let $d_1, d_2, ..., d_n, ...$ be a sequence of natural integers, with $d_1 \geq 0$ and $d_i > 0$, for $i = 2, ..., n, ...$. Consider the following sequence of words $\{s_n\}_{n \geq 0}$ over the alphabet A: $s_0 = 1$, $s_1 = 0$, $s_{n+1} = s_n^{d_n} s_{n-1}$ for $n \geq 1$. Each finite words s_n in the sequence is called *standard word*. It is uniquely determined by the (finite) directive sequence $(d_0, d_1, ..., d_{n-1})$. In the special case the directive sequence is of the form $(1, 1, ..., 1, ...)$ we obtain the sequence of Fibonacci words.

We say that a circular word defined over a binary alphabet A is *standard* if some word in its conjugacy class is a standard word. For instance, the circular word $(baaba)$ is standard because the $abaab$ is a Fibonacci word.

The following proposition, proved in [5] provides some characterization of the standard circular words.

Proposition 2. *Let w a word of length $n \geq 2$. The following statements are equivalent:*

1. *(w) is a standard circular word;*
2. *for $k = 0, ..., n - 1$ the circular word (w) has exactly $k + 1$ factors of length k;*
3. *(w) has $n - 1$ factors of length $n - 2$ and w is primitive.*

From such a proposition we can derive the following

Corollary 1. *Let w be a word of length n. The circular word (w) is standard if and only if for each $k = 0, ..., n - 2$ there exists a unique special factor of (w) of length k.*

Proof. If (w) is standard then it has exactly $k+1$ factors of length k. In particular each factor of length k has a unique right extension into a factor of length $k + 1$ except one, which has two right extensions. Conversely, if for $k = 0, ..., n - 2$ there exists a unique special factor of (w) of length k then for each $k = 0, ..., n-1$ there exist exactly $k + 1$ factors of (w) of length k.   □

Following the idea introduced in [4], we associate to a circular word (w) an automaton $\mathcal{A}_w$.

Definition 1. *Let (w) be a circular word, where $w = a_1 a_2 ... a_n$ be a word of length n over the binary alphabet $A = \{0, 1\}$. The cyclic automaton associated to w, denoted by $\mathcal{A}_w$, is the automaton (Q, Σ, δ, F) such that:*

- $Q = \{1, 2, ..., n\}$
- $\Sigma = \{a\}$
- $\delta(i, a) = (i + 1), \forall\, i \in Q \setminus \{n\}$ and $\delta(n, a) = 1$
- $F = \{i \in Q|\ a_i = 1\}$

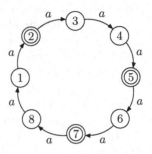

Fig. 1. Cyclic automaton $\mathcal{A}_w$ for $w = 01001010$

See Fig.1 for example. We do not specify the initial state because for our aim it does not matter.

We want to observe as combinatorial properties of the word w are closed to the properties of the state of the automaton $\mathcal{A}_w$.

Remark 1. For any $i \in Q$ and any k, $0 \leq k \leq n$, the path starting from i and having label a^k corresponds to the factor $a_i a_{i+1} ... a_{i+k}$ of (w), i.e. the factor of (w) of length $k + 1$ starting from the position i. From the definition of the equivalence $\sim_k$ one has that $i \sim_k j$ if the factors of w of length $k + 1$ starting from i and j, respectively, are equal.

Theorem 2. $\mathcal{A}_w$ *is minimal iff* (w) *is primitive.*

Proof. By the theorem of Moore, $\mathcal{A}_w$ is minimal if and only if each class of $\sim_{n-2}$ is a singleton. By the previous remark, this corresponds to the fact that all the factors of (w) of length $n - 1$ have a unique occurrence in (w). This means, by Proposition 1, that w is primitive. $\square$

We call *cyclic standard* the automaton $\mathcal{A}_w$ associated to a circular standard word.

5   Hopcroft's Algorithm on Cyclic Automata

In this section we deepen the connection between the refinements strategy of the Hopcroft's algorithm on unary cyclic automata and the combinatorial properties of circular words. By using such properties we analyze algorithm's behavior with respect to cyclic automata associated to circular standard words.

For a word $u \in A^*$, we define a subset Q_u of states of $\mathcal{A}_w$ as the set of positions of occurrences of the factor u in (w). Trivially, we have that $Q_\epsilon = Q$, $Q_1 = F$ and $Q_0 = Q \setminus F$. Let u be a factor of (w) such that Q_u is a class of the partition of Q. We say that Q_u is a *splitting* subset of states if there exists $v \in A^*$ such that Q_v is a class of the partition and it splits Q_u.

The following proposition establishes a close relation between the execution of the Hopcroft's algorithm on a cyclic automata and the notion of special factor of a circular word.

Proposition 3. *If Q_u is a splitting subset of states, then u is a prefix of a special factor of (w).*

Proof. Since Q_u is a splitting subset, then there exists a binary word v such that Q_v splits Q_u. It follows that there exists $b \in \{0,1\}$ such that some occurrences of u are occurrences of bv, too. If $|v| < |u|$, then bv should be a prefix of u, so each occurrence of u is an occurrence of bv and Q_v could not split Q_u. So, we have that $|u| \le |v|$. Let us denote by z the longest word whose prefix is u and $Q_u = Q_z$. If Q_v splits Q_z then there exist $c \in \{0,1\}$ and a word s in $\{0,1\}^+$ such that $cv = zs$. Moreover the resulting sets are Q_{zs} and Q_{zt} where $t \in \{0,1\}^+$ and $t \ne s$. So, it follows that u is prefix of a special factor of (w). □

Corollary 2. *Let Q_u and Q_v be classes of the partition. If Q_v splits Q_u and $|u| = |v|$ then u is a special factor of (w) and the resulting sets are Q_{u0} and Q_{u1}.*

The classes that appear during each execution of Hopcroft's algorithm on cyclic automaton $\mathcal{A}_w$ are all of the form Q_u for some factor u of (w). Then the nodes of the associated derivation tree are labeled with Q_u.

Let us denote by Π_k and $\mathcal{W}_k$ the partition and the waiting set at the k-th step of the algorithm.

Proposition 4. *The executions of the Hopcroft's algorithm on a cyclic standard automaton $\mathcal{A}_w$, where (w) is a circular standard word, are uniquely determined. In particular, at each step $1 \le k \le n - 2$, $\Pi_k = \{Q_v | v$ is a factor of length $k\}$ and $|\mathcal{W}_k| = 1$.*

Proof. We prove the statement by induction on k. Let us suppose that the number of occurrences of 1's in (w) is greater than 0's, so $|Q_0| < |Q_1|$. Let us consider $k = 1$. In this case the starting partition $\Pi_1 = \{Q_0, Q_1\}$ and $\mathcal{W}_1 = Q_0$. Let us suppose now that $\mathcal{P}_k = \{Q_v | v$ is a factor of length $k\}$ and $|\mathcal{W}_k| = 1$. In particular $\mathcal{W}_k$ contains Q_v where v is one of the factors of (w) of length k that were added during the $(k-1)$-th step. Let u be the k-length factor of (w) such that Q_v splits Q_u. Such a word exists and satisfies the equality $bv = ua$, for some $a, b \in \{0,1\}$. By Corollary 2, u is the unique special factor of length k. After the split the resulting sets are Q_{u0} and Q_{u1} and for each k-length factor $z \ne u$, there exists $b \in \{0,1\}$ such that $Q_z = Q_{zb}$. Then $\Pi_{k+1} = \{Q_v | v$ is a factor of length $k + 1\}$ and $|\mathcal{W}_{k+1}| = 1$. Note that whatever subset of states (Q_{u0} or Q_{u1}) is included in $\mathcal{W}_{k+1}$, then both Q_{u0} and Q_{u1} will split the same subset at the next step. In fact, if Q_{u0} splits Q_t for some $(k+1)$-length factor t of (w), then $td = cv0$, so $t = cv$. It is easy to see that Q_{u1} splits Q_t, too. Moreover the resulting sets are equal to the ones obtained by Q_{u0}. □

Remark 2. From Proposition 4 and its proof it follows that the execution of the Hopcroft's algorithm on cyclic standard automata is deterministic. Since $\mathcal{W}$ contains only one element at each step, both LIFO and FIFO strategies produce the same execution. Furthermore, if the class Q_u is split in Q_{u0} and Q_{u1}, the execution does not change whatever class resulting from the split is added to the waiting set $\mathcal{W}$.

6   Hopcroft's Algorithm and Fibonacci Words

In this section we recall definition of Fibonacci numbers and words and some of its properties. Furthermore we study Hopcroft's algorithm on automata associated to Fibonacci words and give the main result of the paper.

The infinite Fibonacci word f, over the alphabet $\{0,1\}$ is the limit of the infinite sequence $\{f_n\}_{n \geq 0}$ of binary words inductively defined as $f_0 = 1$, $f_1 = 0$, $f_{n+1} = f_n f_{n-1}$, $n \geq 2$. Words f_n are called finite Fibonacci words. We denote by f_n also each circular Fibonacci word. The numbers $F_n = |f_n|$, for $n \geq 1$, are the Fibonacci numbers defined by the recurrence equation $F_{n+1} = F_n + F_{n-1}$, with $F_0 = F_1 = 1$.

We want to analyze the execution of cyclic automata associated to circular Fibonacci words f_n, with $n \geq 2$, that we denote by $\mathcal{A}_{f_n} = (Q, \Sigma, \delta, F)$. Since finite Fibonacci words are standards words we will refer to results in Section 5.

First, we recall some combinatorial properties of finite Fibonacci words that will be useful.

For each $n > 1$, with Q_v^n we denote the set of states corresponding to the positions of the occurrences of v in the circular word f_n. Trivially, we have that $|Q_0^n| = F_{n-1}$ and $|Q_1^n| = F_{n-2}$.

For each n, the circular word f_n can be factorized in $\{0, 01\}$ or, equivalently, in $\{10, 100\}$. Hence, if $v \neq 1$ is a factor of f_n starting with 0 (resp. 1) then $v \in (0 + 01)^+$ (resp. $(100 + 10)^+$). Hence, if we consider the two morphism $\varphi, \psi : \Sigma \to \Sigma^*$ defined as follows:

$$\varphi(1) = 0, \ \varphi(0) = 01$$

and

$$\psi(1) = 10, \ \psi(0) = 100$$

we have the following remarks.

Remark 3. For each v factor of f_{n-1}, $\varphi(v)$ is a factor of f_n and

$$|Q_{\varphi(v)}^n| = |Q_v^{n-1}|.$$

Remark 4. For each v factor of f_{n-2}, $\psi(v)$ is a factor of f_n and

$$|Q_{\psi(v)}^n| = |Q_v^{n-2}|.$$

Previous remarks allows us to create a correspondence between occurrences of factors in the n-th finite Fibonacci words and occurrences of factors in the $n-1$-th and $n-2$-th ones. That is a useful tools to compute the running time of the execution on $\mathcal{A}_{f_n}$ by using a recursive method.

Lemma 1. *Let $v \in \Sigma^*$, for each n we have that:*

- *v is a special factor of f_{n-1} iff $\varphi(v)0$ is a special factor of f_n.*
- *v is a special factor of f_{n-2} iff $\psi(v)10$ is a special factor of f_n.*

Proof. If v is a special factor of f_{n-1} then both $v0$ and $v1$ are factors of f_{n-1}. Let ut consider $\varphi(v)0$. It is a factor of f_n because equal to $\varphi(v1)$. Furthermore, $\varphi(v0) = \varphi(v)01$ and $\varphi(v10) = \varphi(v)001$ are factors of f_n that is $\varphi(v)0$ is a special factor of f_n. One can analogously prove that v is a special factor of f_{n-2} then $\psi(v)10$ is a special factor of f_n. Now, let us observe that, since Fibonacci words does not contain the factor 11, special factors of any finite Fibonacci words ends with 0. Hence, if w is a special factors of f_n, either w starts with 0 (and then $w \in (0+01)^*0$) or w starts with 1 (and then $w \in (10+100)^*10$). In the first case $w = u0$ and let $v = \varphi^{-1}(u)$. Since $u0$ is special then $u01$ and $u00$ (and necessarily $u001$) exist. We have that both $\varphi^{-1}(u)0 = \varphi^{-1}(u01)$ and $\varphi^{-1}(u)10 = \varphi^{-1}(u001)$ are factors of f_{n-1}. One can proceed analogously if w starts with 1. $\square$

For each $n \geq 2$, with $\mathbf{c}(F_n)$ we denote the running time of the algorithm to minimize $\mathcal{A}_{f_n}$.

Proposition 5. *For each $n \geq 2$,*

$$\mathbf{c}(F_n) = \mathbf{c}(F_{n-1}) + \mathbf{c}(F_{n-2}) + F_{n-2},$$

with the position $\mathbf{c}(F_1) = \mathbf{c}(F_0) = 0$.

Proof. The starting partition of Q is $\Pi = \{Q_0^n, Q_1^n\}$ and the waiting set is $\mathcal{W} = Q_1^n$. Recall that, by Proposition 4, at each step k of the execution,

$$\mathcal{P}_k = \{Q_v | v \text{ is a } k\text{-length factor of } f_n\}$$

and only one class Q_v, with v special factor of f_n, is split in Q_{v0} and Q_{v1}. Then, at each step k, $|\mathcal{S}_k| = 1$ and

$$\mathcal{S}_k = \{min(Q_{v0}^n, Q_{v1}^n) | v \text{ is a } k\text{-length special factor of } f_n\}.$$

Our aim is to compute the running time of Hopcroft's algorithm on $\mathcal{A}_{f_n}$, that is, the sum of the sizes of the minimum classes that each time result from the splitting. Let $sp(f_n)$ be the set of special factors of f_n, $\mathcal{L}_0 = sp(f_n) \cap (0+01)^*0$ and $\mathcal{L}_1 = sp(f_n) \cap (10+100)^*10$.

$$\mathbf{c}(F_n) = |Q_1^n| + \sum_{w \in sp(f_n)} min(|Q_{w0}^n|, |Q_{w1}^n|) =$$

$$|Q_1^n| + \sum_{w \in \mathcal{L}_0} min(|Q_{w0}^n|, |Q_{w1}^n|) + \sum_{w \in \mathcal{L}_1} min(|Q_{w0}^n|, |Q_{w1}^n|) =$$

$$|Q_1^n| + \sum_{\varphi(v)0 \in \mathcal{L}_0} min(|Q_{\varphi(v)00}^n|, |Q_{\varphi(v)01}^n|) + \sum_{\psi(v)10 \in \mathcal{L}_1} min(|Q_{\psi(v)100}^n|, |Q_{\psi(v)101}^n|) =$$

$$|Q_1^n| + \sum_{v \in sp(f_{n-1})} min(|Q_{v0}^{n-1}|, |Q_{v1}^{n-1}|) + \sum_{v \in sp(f_{n-2})} min(|Q_{v0}^{n-2}|, |Q_{v1}^{n-2}|) =$$

$$F_{n-2} + \mathbf{c}(F_{n-1}) + \mathbf{c}(F_{n-2}). \qquad \square$$

Previous theorem states that the sequence $\{c(F_n)\}_{n \geq 0}$ of running time of Hopcroft's algorithm on automaton associated to finite Fibonacci words is the Fibonacci convolution sequence (sequence A001629 in [18]). It is a well-known sequence that involves Fibonacci and Lucas numbers (sequence A000204 in [18]) indeed

$$c(F_{n+1}) = \frac{1}{5}(nL_n - F_n),$$

Theorem 3. *Hopcroft's algorithm on cyclic automata associated to finite Fibonacci words has a unique execution that runs in time $\Theta(|Q|log|Q|)$.*

Proof. We know that $|Q| = F_n$. The n-th term of the sequence $\{c(F_n)\}_{n \geq 0}$ is

$$c(F_n) = \tfrac{1}{5}((n-1)F_n + 2nF_{n-1}) \text{ (cf. [20])}.$$

The theorem follows by the relation between n and F_n, $F_n = [\frac{\phi^n}{\sqrt{5}}]$, where $[x]$ is the nearest integer function and ϕ is the golden ratio $\frac{1+\sqrt{5}}{2}$. In fact, one can prove, by simple computations, that definitively we have

$$\frac{k}{\phi}F_n log F_n \leq c(F_n) \leq kF_n log F_n,$$

where $k = \frac{3}{5log\phi}$. $\qquad\qquad\qquad\square$

References

1. Almeida, M., Moreira, N., Reis, R.: On the performance of automata minimization algorithms. Technical Report DCC-2007-03, Universidade do Porto (2007)
2. Baclet, M., Pagetti, C.: Around Hopcroft's algorithm. In: H. Ibarra, O., Yen, H.-C. (eds.) CIAA 2006. LNCS, vol. 4094, pp. 114–125. Springer, Heidelberg (2006)
3. Béal, M.-P., Crochemore, M.: Minimizing local automata. In: ISIT 2007 (to appear, 2007)
4. Berstel, J., Carton, O.: On the complexity of Hopcroft's state minimization algorithm. In: Domaratzki, M., Okhotin, A., Salomaa, K., Yu, S. (eds.) CIAA 2004. LNCS, vol. 3317, pp. 35–44. Springer, Heidelberg (2005)
5. Borel, J.P., Reutenauer, C.: On Christoffel classes. RAIRO-Theoretical Informatics and Applications 450, 15–28 (2006)
6. Brzozowski, J.A.: Canonical regular expressions and minimal state graphs for definite events. Mathematical Theory of Automata
7. Champarnaud, J.-M., Khorsi, A., Paranthon, T.: Split and join for minimizing: Brzozowskis algorithm. In: Proceedings of PSC 2002 (Prague Stringology Conference), pp. 96–104 (2002)
8. Daciuk, J., Watson, R.E., Watson, B.W.: Incremental construction of acyclic finite-state automata and transducers. In: Finite State Methods in Natural Language Processing, Bilkent University, Ankara, Turkey (1998)
9. Eilenberg, S.: Automata, Languages, and Machines, vol. A (1974)
10. Gries, D.: Describing an algorithm by Hopcroft. Acta Inf. 2, 97–109 (1973)

11. Hopcroft, J.E.: An $n \log n$ algorithm for mimimizing the states in a finite automaton. In: Paz, A., Kohavi, Z. (eds.) Theory of machines and computations (Proc. Internat. Sympos. Technion, Haifa, 1971), pp. 189–196. Academic Press, New York (1971)
12. Huffman, D.A.: The synthesis of sequencial switching circuits. J. Franklin Institute 257, 161–190 (1954)
13. Knuutila, T.: Re-describing an algorithm by Hopcroft. Theoret.Comput. Sci. 250, 333–363 (2001)
14. Matz, O., Miller, A., Potthoff, A., Thomas, W., Valkema, E.: Report on the program AMoRE. Technical Report 9507, Inst. f. Informatik u.Prakt. Math., CAU Kiel (1995)
15. Moore, E.F.: Gedaken experiments on sequential machines. In: Automata Studies, pp. 129–153 (1956)
16. Paun, A.: On the Hopcroft's minimization algorithm. CoRR, abs/0705.1986 (2007)
17. Revuz, D.: Minimisation of acyclic deterministic automata in linear time. Theor. Comput. Sci. 92(1), 181–189 (1992)
18. Sloane, N.J.A.: The On-Line Encyclopedia of Integer Sequences, http://www.research.att.com/~njas/sequences/
19. Tabakov, D., Vardi, M.Y.: Experimental evaluation of classical automata constructions
20. Vajda, S.: Fibonacci and Lucas numbers, and the golden section. Technical Report 98/183, Ellis Horwood Ltd., Chichester (1989)
21. Watson, B.: A taxonomy of finite automata minimization algorithms. Technical Report 93/44, Eindhoven University of Technology, Faculty of Mathematics and Computing Science (1994)

Efficient Inclusion Checking for Deterministic Tree Automata and DTDs

Jérôme Champavère, Rémi Gilleron, Aurélien Lemay, and Joachim Niehren

INRIA Futurs and Lille University, LIFL, Mostrare project

Abstract. We present a new algorithm for testing language inclusion $L(A) \subseteq L(B)$ between tree automata in time $O(|A| * |B|)$ where B is deterministic. We extend this algorithm for testing inclusion between automata for unranked trees A and deterministic DTDs D in time $O(|A| * |\Sigma| * |D|)$. No previous algorithms with these complexities exist.

1 Introduction

Language inclusion for tree automata is a basic decision problem that is closely related to universality and equivalence [5,14,15]. Tree automata algorithms are generally relevant for XML document processing [11,17,7,13]. Regarding inclusion checking, a typical application is inverse type checking for tree transducers [10]. Another one is schema-guided query induction [4], the motivation for the present study. There, candidate queries produced by the learning process are to be checked for consistency with deterministic DTDs, such as for HTML.

We investigate language inclusion $L(A) \subseteq L(B)$ for tree automata A and B under the assumption that B is (bottom up) deterministic, not necessarily A. Without this assumption the problem becomes DEXPTIME complete [15]. Deterministic language inclusion still subsumes universality of deterministic tree automata $L(B) = T_\Sigma$ up to a linear time reduction, as well as equivalence of two deterministic automata $L(A) = L(B)$. The converse might be false, i.e., we cannot rely on polynomial time equivalence tests, as for instance, by comparing number of solutions [14] or minimal deterministic tree automata.

In the case of standard tree automata for ranked trees, the well-known naive algorithm for inclusion goes through complementation. It first computes an automaton B^c that recognizes the complement of the language of B, and then checks whether the intersection automaton for B^c and A has a nonempty language. The problematic step is to complete B before complementing its final states, since completion might require to add rules for all possible left-hand sides. The overall running time may thus become $O(|A| * |\Sigma| * |B|^n)$, which is exponential in the maximal rank n of function symbols in the signature Σ.

It seems folklore that one can bound the maximal arity of a signature to 2. This can be done by transforming ranked trees into binary trees, and then converting automata for ranked trees into automata for binary trees correspondingly. The problem is to preserve determinism in such a construction, while the size of automata should grow at most linearly. We show how to obtain such a transformation by stepwise tree automata [3,5]. Thereby we obtain an inclusion test in

C. Martín-Vide, F. Otto, and H. Fernau (Eds.): LATA 2008, LNCS 5196, pp. 184–195, 2008.

time $O(|A| * |\Sigma| * |B|^2)$. This is still too much in practice with DTDs, where A and B may be of size 500 and Σ of size 100.

Our first contribution is a more efficient algorithm for standard tree automata on binary trees that verifies inclusion in time $O(|A| * |B|)$ if B is deterministic. This bound is independent of the size of the signature, even though Σ is not fixed. As a second contribution, we show how to test inclusion between automata A for unranked trees and deterministic DTDs D in time $O(|A|*|\Sigma|*|D|)$. Determinism is required by the XML standards. Our algorithm first computes the Glushkov automata of all regular expressions of D in time $O(|\Sigma| * |D|)$, which is possible for deterministic DTDs [2]. The second step is more tedious. We would like to transform the collection of Glushkov automata to a deterministic stepwise tree automaton of the same size. Unfortunately, this seems difficult to achieve, since the usual construction of [9] eliminates ϵ-rules on the fly, which may lead to a quadratic blowup of the number of rules (not the number of states). We solve this problem by introducing factorized tree automata, which use ϵ-transitions for representing deterministic automata more compactly. We show how to adapt our inclusion test to factorized tree automata and thus to DTDs.

Related Work and Outline. Heuristic algorithms for inclusion between non-deterministic schemas that avoid the high worst-case complexity were proposed in [16]. The complexity of inclusion for various fragments of DTDs and extended DTDs was studied in [8]. The algorithms presented there assume the same types of regular expressions on both sides of the inclusion test. When applied to deterministic DTDs, the same complexity results may be obtainable. Our algorithm permits richer left-hand sides.

We reduce inclusion for the ranked case to the binary case in Section 2. The efficient algorithm for binary tree automata is given in Section 3. In Section 4, we introduce deterministic factorized tree automata and lift the algorithm for inclusion testing to them. Finally in Section 5, we apply them to testing inclusion of automata for unranked trees in deterministic DTDs.

2 Standard Tree Automata for Ranked Trees

We reduce the inclusion problem of tree automata for ranked trees [5] to the case of binary trees with a single binary function symbol.

A ranked signature Σ is a finite set of function symbols $f \in \Sigma$, each of which has an arity $n \geq 0$. A constant $a \in \Sigma$ is a function symbol of arity 0. A tree $t \in T_\Sigma$ is either a constant $a \in \Sigma$ or a tuple $f(t_1, \ldots, t_n)$ consisting of a function symbol f of arity n and n trees $t_1, \ldots, t_n \in T_\Sigma$.

A *tree automaton* (possibly with ϵ-rules) A over Σ consists of a finite set $\mathsf{s}(A)$ of states, a subset $\mathsf{final}(A) \subseteq \mathsf{s}(A)$ of final states, and a set $\mathsf{rules}(A)$ of rules of the form $f(p_1, \ldots, p_n) \to p$ or $p' \xrightarrow{\epsilon} p$ where $f \in \Sigma$ has arity n and $p_1, \ldots, p_n, p, p' \in \mathsf{s}(A)$. We write $p' \xrightarrow{\epsilon}_A p$ iff $p' \xrightarrow{\epsilon} p \in \mathsf{rules}(A)$, $\xrightarrow{\epsilon}{}^*_A$ for the reflexive transitive closure of $\xrightarrow{\epsilon}_A$, and $\xrightarrow{\epsilon}{}^{\leq 1}_A$ for the union of $\xrightarrow{\epsilon}_A$ and the identity relation on $\mathsf{s}(A)$.

The *size* $|A|$ of A is the sum of the cardinality of $\mathsf{s}(A)$ and the number of symbols in $\mathsf{rules}(A)$, i.e., $\sum_{f(p_1,\ldots,p_n)\to p \in \mathsf{rules}(A)}(n+2)$. The cardinality of the signature can be ignored, since our algorithms will not take unused function symbols into account. Every tree automaton A defines an evaluator $\mathsf{eval}_A :$ $T_{\Sigma \cup \mathsf{s}(A)} \to 2^{\mathsf{s}(A)}$ such that $\mathsf{eval}_A(f(t_1,\ldots,t_n)) = \{p \mid p_1 \in \mathsf{eval}_A(t_1), \ldots, p_n \in$ $\mathsf{eval}_A(t_n), f(p_1,\ldots,p_n) \to p' \in \mathsf{rules}(A), p' \xrightarrow{\epsilon}{}^*_A p\}$ and $\mathsf{eval}_A(p) = \{p\}$. A tree $t \in T_\Sigma$ is *accepted* by A if $\mathsf{final}(A) \cap \mathsf{eval}_A(t) \neq \emptyset$. The *language* $L(A)$ is the set of trees accepted by A.

A tree automaton is (bottom-up) *deterministic* if it has no ϵ-rules, and if no two rules have the same left-hand side. It is complete if there are rules for all potential left-hand sides. It is well-known that deterministic complete tree automata can be complemented in linear time, by switching the final states.

We will study the inclusion problem for tree automata, whose input consists of a ranked signature Σ, tree automata A with ϵ-rules and deterministic B, both over Σ. Its output is the truth value of $L(A) \subseteq L(B)$. We can deal with this problem by restriction to so-called stepwise signatures $\Sigma_@$, consisting of a single binary function symbol @ and a finite set of constants $a \in \Sigma$. A stepwise tree automaton [3] is a tree automaton over a stepwise signature.

Proposition 1. *The above inclusion problem for ranked trees can be reduced in linear time to the corresponding inclusion problem for stepwise tree automata over binary trees.*

We first encode ranked trees into binary trees via Currying. Given a ranked signature Σ we define the corresponding signature $\Sigma_@ = \{@\} \uplus \Sigma$ whereby all symbols of Σ become constants. Currying is defined by a function $\mathsf{curry} : T_\Sigma \to T_{\Sigma_@}$ which for all trees $t_1,\ldots,t_n \in T_\Sigma$ and $f \in \Sigma$ satisfies:

$$\mathsf{curry}(f(t_1,\ldots,t_n)) = f@\mathsf{curry}(t_1)@ \ldots @\mathsf{curry}(t_n)$$

For instance, $f(a,g(a,b),c)$ is mapped to $f@a@(g@a@b)@c$ which is infix notation with left-most parenthesis for the tree $@(@(@(f,a),@(@(g,a),b)),c)$. Now we encode tree automata A over Σ into stepwise tree automata $\mathsf{step}(A)$ over $\Sigma_@$, such that the language is preserved up to Currying, i.e., such that $L(\mathsf{step}(A)) = \mathsf{curry}(L(A))$. The states of $\mathsf{step}(A)$ are the prefixes of left-hand sides of rules in A, i.e., words in $\Sigma(\mathsf{s}(A))^*$:

$$\mathsf{s}(\mathsf{step}(A)) = \{fq_1 \ldots q_i \mid f(q_1,\ldots,q_n) \to q \in \mathsf{rules}(A),\ 0 \le i \le n\} \uplus \mathsf{s}(A)$$

Its rules extend prefixes step by step by states q_i according to the rule of A. Since constants do not need to be extended, we distinguish two cases in Fig. 1.

Lemma 1. *The encoding of tree automata A over Σ into stepwise tree automata $\mathsf{step}(A)$ over $\Sigma_@$ preserves determinism, the tree language modulo Currying, and the automata size up to a constant factor of 3.*

As a consequence, $L(A) \subseteq L(B)$ is equivalent to $L(\mathsf{step}(A)) \subseteq L(\mathsf{step}(B))$, and can be tested in this way modulo a linear time transformation. Most importantly, the determinism of B carries over to $\mathsf{step}(B)$.

$$\frac{f(q_1,\ldots,q_n) \to q \in \mathsf{rules}(A) \qquad 1 \leq i < n}{\begin{array}{c} f \to f \in \mathsf{rules}(\mathsf{step}(A)) \\ fq_1\ldots q_{i-1}@q_i \to fq_1\ldots q_i \in \mathsf{rules}(\mathsf{step}(A)) \\ fq_1\ldots q_{n-1}@q_n \to q \in \mathsf{rules}(\mathsf{step}(A)) \end{array}}$$

$$\frac{a \to q \in \mathsf{rules}(A)}{a \to q \in \mathsf{rules}(\mathsf{step}(A))}$$

Fig. 1. Transforming ranked tree automata into stepwise tree automata

3   Stepwise Tree Automata for Binary Trees

We present a new inclusion test that applies to stepwise tree automata over binary trees. We first characterize inclusion into deterministic tree automata, second, express the characterization in Datalog [6] and third, turn it into an efficient algorithm. While the two first steps are easy, the last step is nontrivial.

Characterization of Inclusion. We call a state $p \in \mathsf{s}(A)$ *accessible* if there exists a tree t such that $p \in \mathsf{eval}_A(t)$. We call p *co-accessible* if there exists a tree $t \in T_{\Sigma \cup \{p\}}$ with a unique occurrence of p such that $\mathsf{eval}_A(t) \cap \mathsf{final}(A) \neq \emptyset$. A tree automaton is *productive* if all its states are accessible and co-accessible. We denote the *product* of two automata A and B with the same signature by $A \times B$. The state set of $A \times B$ is $\mathsf{s}(A) \times \mathsf{s}(B)$. For inferring its rules, we assume that B does not have ϵ-rules:

$$\frac{a \to p \in \mathsf{rules}(A) \quad a \to q \in \mathsf{rules}(B)}{a \to (p,q)} \qquad \frac{p_1@p_2 \to p \in \mathsf{rules}(A) \quad q_1@q_2 \to q \in \mathsf{rules}(B)}{(p_1,q_1)@(p_2,q_2) \to (p,q)} \qquad \frac{p' \xrightarrow{\epsilon} p \in \mathsf{rules}(A) \quad q \in \mathsf{s}(B)}{(p',q) \xrightarrow{\epsilon} (p,q)}$$

We do not care about final states of $A \times B$ since these are useless in our characterization of inclusion.

Proposition 2. *Inclusion $L(A) \subseteq L(B)$ for a productive stepwise tree automaton A with ϵ-rules and a deterministic stepwise tree automaton B fails iff:*

fail$_0$: *there exists a rule $a \to p \in \mathsf{rules}(A)$ but no state $q \in \mathsf{s}(B)$ such that $a \to q \in \mathsf{rules}(B)$, or*

fail$_1$: *there exist accessible states (p_1,q_1) and (p_2,q_2) of $A \times B$ and a rule $p_1@p_2 \to p \in \mathsf{rules}(A)$ but no state $q \in \mathsf{s}(B)$ such that $q_1@q_2 \to q \in \mathsf{rules}(B)$, or*

fail$_2$: *some accessible state (p,q) of $A \times B$ satisfies $p \in \mathsf{final}(A)$ but $q \notin \mathsf{final}(B)$.*

Proof. If one of the failure conditions holds, then failure of inclusion follows from the hypotheses that A is productive and B deterministic.

For the converse, let us consider a tree t such that $t \in L(A)$ and $t \notin L(B)$. There are two cases to be considered, depending on $\mathsf{eval}_B(t)$.

(i)   Assume $\mathsf{eval}_B(t) = \emptyset$. There exists a minimal subtree t' of t such that $\mathsf{eval}_B(t') = \emptyset$, too. If $t' = a$ is a leaf then $\mathsf{eval}_A(a) \neq \emptyset$, since $t \in L(A)$, and $\mathsf{eval}_B(a) = \emptyset$, thus **fail$_0$** holds. If $t' = t_1@t_2$, then there exist $p_1 \in \mathsf{eval}_A(t_1)$, $p_2 \in \mathsf{eval}_A(t_2)$ and $p_1@p_2 \to p \in \mathsf{rules}(A)$, since $t \in L(A)$. Since t' is defined as a

$$(\text{acc}_{/1}) \; \frac{a \to p \in \text{rules}(A) \quad a \to q \in \text{rules}(B)}{\text{acc}(p,q).} \qquad (\text{acc}_{/2}) \; \frac{p' \xrightarrow{\epsilon}_A p \in \quad q \in \text{s}(B)}{\text{acc}(p,q):-\; \text{acc}(p',q).}$$

$$(\text{acc}_{/3}) \; \frac{p_1@p_2 \to p \in \text{rules}(A) \quad q_1@q_2 \to q \in \text{rules}(B)}{\text{acc}(p,q):-\; \text{acc}(p_1,q_1), \text{acc}(p_2,q_2).}$$

$$(\text{frb}) \; \frac{p_1@p_2 \to p \in \text{rules}(A) \quad \nexists q.q_1@q_2 \to q \in \text{rules}(B)}{\text{frb}(p_2,q_2):-\; \text{acc}(p_1,q_1).}$$
$$\text{frb}(p_1,q_1):-\; \text{acc}(p_2,q_2).$$

$$(\text{fail}_0) \; \frac{a \to p \in \text{rules}(A) \quad \nexists q.a \to q \in \text{rules}(B)}{\texttt{fail}_0.}$$

$$(\text{fail}_1) \; \frac{p \in \text{s}(A) \quad q \in \text{s}(B)}{\texttt{fail}_1:-\; \text{acc}(p,q), \text{frb}(p,q).} \qquad (\text{fail}_2) \; \frac{p \in \text{final}(A) \quad q \notin \text{final}(B)}{\texttt{fail}_2:-\; \text{acc}(p,q).}$$

Fig. 2. Transforming tree automata A and B into a Datalog program $D_1(A,B)$

minimal subtree and B is deterministic, $\text{eval}_B(t_1) = \{q_1\}$, $\text{eval}_B(t_2) = \{q_2\}$, and since $\text{eval}_B(t') = \emptyset$, there is no rule $q_1@q_2 \to q \in \text{rules}(B)$. This leads to $\texttt{fail}_1$.

(ii) If $\text{eval}_B(t) \neq \emptyset$ then there exists $q \in \text{eval}_B(t)$; B being deterministic this q is necessarily unique. Since $t \notin L(B)$, $q \notin \text{final}(B)$. Moreover, since $t \in L(A)$, there exists $p \in \text{eval}_A(t) \cap \text{final}(A)$. This leads to $\texttt{fail}_2$.   □

Testing the Characterization. The following efficiency theorem for ground Datalog will be fundamental to all what follows. Given a Datalog program P (without negation), we write $\text{lfp}(P)$ for its least fixed point semantics.

Theorem 1 (Efficiency of Ground Datalog [6]). *For every ground Datalog program P, the least fixed point semantics $\text{lfp}(P)$ can be computed in linear time $O(|P|)$ where the size $|P|$ is the number of symbols in P.*

This result holds even without any bound on the arity of the relation symbols of P, which will be very useful later on. If relation symbols of higher arities are used, the number of their arguments is accounted for by the size of P.

Fig. 2 presents a Datalog program $D_1(A,B)$ that verifies the characterization of $L(A) \subseteq L(B)$ in Proposition 2. Transformation rules $(\text{acc}_{/1})$, $(\text{acc}_{/2})$, and $(\text{acc}_{/3})$ define clauses for accessibility in $A \times B$ through predicate acc. The clauses produced by transformation rule (frb) define *forbidden states* of $A \times B$ through predicate frb. These are states that lead to $\texttt{fail}_1$ when accessed. Transformation rules (fail_0), (fail_1), and (fail_2) define clauses for failures. The characterization of inclusion from Proposition 2 is captured in the following sense:

Proposition 3. *Let A and B be stepwise tree automata for binary trees. If A is productive and B deterministic then:*

$$L(A) \subseteq L(B) \Leftrightarrow \text{lfp}(D_1(A,B)) \cap \{\texttt{fail}_0, \texttt{fail}_1, \texttt{fail}_2\} = \emptyset$$

The sum of the sizes of the clauses defined by transformation rules $(\text{acc}_{/1})$, $(\text{acc}_{/2})$, $(\text{acc}_{/3})$, (fail_0), (fail_1), and (fail_2) is $O(|A| * |B|)$. The sizes of the clauses

$(\text{frb}_{/2}^c)\ \dfrac{p_1@p_2 \to p \in \text{rules}(A)\quad q_1 \in \text{s}(B)\quad Q_2 = \{q_2 \mid q_1@q_2 \to q \in \text{rules}(B)\}}{\text{frb}^c(p_2, Q_2) :- \text{acc}(p_1, q_1).}$

$(\text{frb}_{/1}^c)\ \dfrac{p_1@p_2 \to p \in \text{rules}(A)\quad q_2 \in \text{s}(B)\quad Q_1 = \{q_1 \mid q_1@q_2 \to q \in \text{rules}(B)\}}{\text{frb}^c(p_1, Q_1) :- \text{acc}(p_2, q_2).}$

Fig. 3. Grouping (frb) transformations

defined by transformation rule (frb) sum up to $O(|A| * |\text{s}(B)|^2)$. The overall size of the ground Datalog program $D_1(A, B)$ is $O(|A| * (|B| + |\text{s}(B)|^2))$, which may be $O(|A| * |B|^2)$ in the worst case. Therefore, using Theorem 1, inclusion can be decided in time $O(|A| * |B|^2)$.

Efficient Algorithm. This running time is not better than that of the naive algorithm. The square factor is due to the computation of forbidden states for capturing fail_1. Since (frb) rules cannot be inferred efficiently enough with a Datalog program, we introduce a new predicate frb^c that will group (frb) rules. Using an appropriate data structure, the frb predicates can be induced efficiently from frb^c. The semantics of the latter is given below:

$$A, B \models \text{frb}^c(p, Q) \Leftrightarrow \forall q \in \text{s}(B) \setminus Q,\ A, B \models \text{frb}(p, q)$$

Formally, we impose an order $<$ on $\text{s}(B)$ and consider $\text{frb}^c(p, \{q_1, \ldots, q_n\})$ as $(n + 1)$-ary literals $\text{frb}^c(p, q_{i_1}, \ldots, q_{i_n})$ such that $\{q_{i_1}, \ldots, q_{i_n}\} = \{q_1, \ldots, q_n\}$ and $q_{i_1} < \ldots < q_{i_n}$.

In Fig. 3, we propose two transformations $(\text{frb}_{/1}^c)$ and $(\text{frb}_{/2}^c)$ for inferring frb^c clauses, both of which group (frb) transformations. Note that for every state p, there may be several sets Q such that $\text{frb}^c(p, Q)$ gets inferred. Therefore, we will have to test efficiently whether a state belongs to the union of complements of those state sets. This will be further detailed.

Let us consider the transformation of tree automata A and B into a ground Datalog program $D_2(A, B)$ defined by transformation rules $(\text{acc}_{/1})$, $(\text{acc}_{/2})$, $(\text{acc}_{/3})$, $(\text{frb}_{/1}^c)$, $(\text{frb}_{/2}^c)$, (fail_0), and (fail_2). The clauses producing acc, fail_0, and fail_2 of $D_1(A, B)$ and $D_2(A, B)$ are identical and their number is in $O(|A| * |B|)$.

$$
\dfrac{\left.\begin{array}{c} \\ p_1@p_2 \to p \in \text{rules}(A) \end{array}\right| \left.\begin{array}{c} q_1@q_2^1 \to q^1 \in \text{rules}(B) \\ \vdots \\ q_1@q_2^n \to q^n \in \text{rules}(B) \end{array}\right\} \text{all the rules for } q_1}{\begin{array}{l} \text{frb}^c(p_2, \{q_2^1, \ldots, q_2^n\}) :- \text{acc}(p_1, q_1). \\ \text{acc}(p, q^1) :- \text{acc}(p_1, q_1), \text{acc}(p_2, q_2^1). \\ \qquad\qquad \vdots \\ \text{acc}(p, q^n) :- \text{acc}(p_1, q_1), \text{acc}(p_2, q_2^n). \end{array}}
$$

Fig. 4. Rewriting grouping rules for complexity analysis of $(\text{frb}_{/2}^c)$ clauses

The number of frb^c clauses introduced by rule $(\mathrm{frb}^c_{/1})$ is in $O(|A| * |\mathsf{s}(B)|)$ but the size of each such clause is $n+1$ which in the worst case could be $|\mathsf{s}(B)| + 1$, and symmetrically for $(\mathrm{frb}^c_{/2})$. The overall size of all frb^c clauses, however, is bounded by the overall number of acc clauses, which in turn is bounded by $O(|A| * |B|)$, too! To see this, we can rewrite $(\mathrm{frb}^c_{/2})$ as shown in Fig. 4, such that the corresponding $(\mathsf{acc}_{/3})$ clauses are inferred simultaneously (and these don't overlap). Therefore the overall size of $D_2(A, B)$ is in $O(|A| * |B|)$.

Inclusion Test. First it computes $\mathsf{lfp}(D_2(A, B))$ in time $O(|A|*|B|)$. If fail_0 or fail_2 belong to $\mathsf{lfp}(D_2(A, B))$ then inclusion does not hold, so false is returned. Otherwise, we test for fail_1 in a second step, by checking for all states $\mathsf{acc}(p, q) \in \mathsf{lfp}(D_2(A, B))$ whether there is a $\mathsf{frb}^c(p, Q) \in \mathsf{lfp}(D_2(A, B))$ such that $q \in \mathsf{s}(B) \setminus Q$. If so, false is returned, otherwise true.

We have to prove that the second step can be done in time $O(|A| * |B|)$. For every state $p \in \mathsf{s}(A)$, there are some state sets $Q_1, \ldots, Q_m$ such that for $1 \leq j \leq m$, $\mathsf{frb}^c(p, Q_j) \in \mathsf{lfp}(D_2(A, B))$ and we have to check efficiently whether some state q is in $\bigcup_{j=1}^{j=m} \mathsf{s}(B) \setminus Q_j$. For this, we define a data structure $\mathsf{bad\_states}(p)$ as an array T of size $|\mathsf{s}(B)| + 1$. Counters are indexed by elements in $\mathsf{s}(B)$ and one counter is indexed by 0. All counter values are set to 0 initially. The initialization of all $\mathsf{bad\_states}(p)$ can be done in $O(|\mathsf{s}(A)| * |\mathsf{s}(B)|)$. For every p and every Q_j such that $\mathsf{frb}^c(p, Q_j) \in$

$\|0\|$	q_1	q_2	q_3	q_4	q_5	$\cdots$
$p\|2\|$	0	1	2	1	0	$\cdots$

Fig. 5. Data structure $\mathsf{bad\_state}$. Here, it has been set up $\mathsf{frb}^c(p, \{q_2, q_3\})$ and $\mathsf{frb}^c(p, \{q_3, q_4\})$. When $\mathsf{frb}^c(p, \{q_2, q_3\})$ is set up, $\mathsf{bad\_state}(p)(0)$, $\mathsf{bad\_state}(p)(q_2)$ and $\mathsf{bad\_state}(p)(q_3)$ are incremented. We have for instance $\mathsf{frb}(p, q_4)$ since $\mathsf{bad\_state}(p)(q_4) = 1$ is lower than $\mathsf{bad\_state}(p)(0) = 2$. In fact, the only *not forbidden* state is (p, q_3) because q_3 belongs to the intersection of $\{q_2, q_3\}$ and $\{q_3, q_4\}$.

$\mathsf{lfp}(D_2(A, B))$, counter $T[0]$ is incremented by 1 and counter $T[q]$ is incremented by 1 for all $q \in Q_j$, which can be done in time $O(|Q_j|)$. As the overall size of frb^c clauses in $\mathsf{lfp}(D_2(A, B))$ is in $O(|A| * |B|)$, the computation of all $\mathsf{bad\_states}(p)$ can be done in time $O(|A| * |B|)$.

It remains to test for all states $\mathsf{acc}(p, q) \in \mathsf{lfp}(D_2(A, B))$ whether there exists a $\mathsf{frb}^c(p, Q) \in \mathsf{lfp}(D_2(A, B))$ such that $q \in \mathsf{s}(B) \setminus Q$. This is done by checking whether $T[q] < T[0]$ in $\mathsf{bad\_states}(p)$ (see, e.g., Fig. 5). Indeed, if $\{Q \mid \mathsf{frb}^c(p, Q) \in \mathsf{lfp}(D_2(A, B))\} = \{Q_1, \ldots, Q_m\}$ then $\mathsf{bad\_states}(p)$ is defined such that $T[q] = T[0]$ iff $q \in \bigcap_{j=1}^{j=m} Q_j$, thus $T[q] < T[0]$ iff $q \in \bigcup_{j=1}^{j=m} \mathsf{s}(B) \setminus Q_j$. Each such test costs $O(1)$ so the overall time is bounded by $O(|A| * |B|)$.

This concludes the inclusion test for stepwise tree automata, for productive A and deterministic B. Every tree automaton can be made productive in linear time. Higher arities can be reduced to 2 by Proposition 1. This yields:

Theorem 2. *Let A and B be standard tree automata for ranked trees of some signature Σ possibly with ϵ-rules. If B is deterministic, inclusion $L(A) \subseteq L(B)$ can be decided in time $O(|A| * |B|)$ independently of the size of Σ.*

4   Factorized Tree Automata

We next relax the determinism assumption on B in a controlled manner, that will be crucial to deal with DTDs. We replace B by deterministic factorized automata, that we introduce. These are automata with ϵ-rules, that represent deterministic automata in a compact manner.

Definition 1. *A factorized tree automaton F over a stepwise signature Σ is a stepwise tree automaton with ϵ-rules and a partition $\mathsf{s}(F) = \mathsf{s}_1(F) \uplus \mathsf{s}_2(F)$ such that if $q_1 @ q_2 \to q$ in $\mathsf{rules}(F)$ then $q_1 \in \mathsf{s}_1(F)$ and $q_2 \in \mathsf{s}_2(F)$.*

We say that q is of sort i in F if $q \in \mathsf{s}_i(F)$. The sort determines which states may be used in the i-th position of the binary symbol @ in rules of F.

Every factorized automaton F defines a tree automaton $\mathsf{ta}(F)$ without ϵ-rules that recognizes the same language. Both automata have the same signature and states; the rules of $\mathsf{ta}(F)$ are inferred as follows from those of F:

$$(E_1) \ \frac{a \to q \in \mathsf{rules}(F)}{a \to q \in \mathsf{rules}(\mathsf{ta}(F))} \qquad (E_2) \ \frac{q_1 \xrightarrow{\epsilon}{}^*_F r_1 \quad q_2 \xrightarrow{\epsilon}{}^*_F r_2 \quad r_1 @ r_2 \to q \in \mathsf{rules}(F)}{q_1 @ q_2 \to q \in \mathsf{rules}(\mathsf{ta}(F))}$$

We set $\mathsf{final}(\mathsf{ta}(F)) = \{q \mid q \xrightarrow{\epsilon}{}^*_F r, \ r \in \mathsf{final}(F)\}$. Note that the size of $\mathsf{ta}(F)$ may be $O(|\mathsf{rules}(F)| * |\mathsf{s}(F)|^2)$ which is cubic in that of F in the worst case. Besides their succinctness, the truly interesting bit about factorized tree automata is their notion of determinism.

Definition 2. *A factorized tree automaton F is (bottom-up) deterministic if:*

d_0: *the ϵ-free part of F is (bottom-up) deterministic;*
d_1: *for all $q \in \mathsf{s}(F)$ and sorts $i \in \{1, 2\}$, there is at most one state r of sort i such that $q \xrightarrow{\epsilon}{}^*_F r$.*

Nonredundant ϵ-rules must change the sort: if $q \xrightarrow{\epsilon}_F r$ for two states of the same sort then $r = q$ by d_1 and $q \xrightarrow{\epsilon}{}^*_F q$. A similar argument shows that all proper chains of ϵ-rules are redundant so that $\xrightarrow{\epsilon}{}^*_F$ is equal to $\xrightarrow{\epsilon}{}^{\leq 1}_F$.

Proposition 4. *The tree automaton $\mathsf{ta}(F)$ represented by a deterministic factorized tree automaton F is deterministic.*

Proof. Let $B = \mathsf{ta}(F)$ which by construction is free of ϵ-rules. For every constant $a \in \Sigma$, the uniqueness of q such that $a \to q \in \mathsf{rules}(B)$ follows from d_0. For every $q_1 @ q_2 \to q$ in $\mathsf{rules}(B)$ we have to show that q is uniquely determined by q_1 and q_2. By d_1 there is at most one state r_1 of sort 1 such that $q_1 \xrightarrow{\epsilon}{}^*_F r_1$ at most one r_2 of sort 2 such that $q_2 \xrightarrow{\epsilon}{}^*_F r_2$. Condition d_0 implies that there exists at most one state q such that $r_1 @ r_2 \to q \in \mathsf{rules}(F)$. $\qquad\square$

We fix a stepwise tree automaton A and a deterministic factorized tree automaton F, and let us $B = \mathsf{ta}(F)$. We now show how to test language inclusion

$$(\text{acc}_{/3a}) \quad \frac{p_1@p_2 \to p \in \text{rules}(A) \quad q_1@q_2 \to q \in \text{rules}(F)}{\text{acc}(p,q) :- \text{ f.acc}(p_1,q_1), \text{f.acc}(p_2,q_2).}$$

$$(\text{f.acc}) \quad \frac{p \in \text{s}(A) \quad q \xrightarrow{\epsilon \leq 1}_F r}{\text{f.acc}(p,r) :- \text{ acc}(p,q).}$$

$$(\text{f.frb}_2^c) \quad \frac{p_1@p_2 \to p \in \text{rules}(A) \quad q_1 \in \text{s}_1(F)}{\text{f.frb}_2^c(p_2, Q_2^F(q_1)) :- \text{ f.acc}(p_1,q_1).}$$

$$(\text{f.frb}_1^c) \quad \frac{p_1@p_2 \to p \in \text{rules}(A) \quad q_2 \in \text{s}_2(F)}{\text{f.frb}_1^c(p_1, Q_1^F(q_2)) :- \text{ f.acc}(p_2,q_2).}$$

$$(\text{frb}_2^c) \quad \frac{p_1@p_2 \to p \in \text{rules}(A) \quad q_1 \in \text{s}(F)}{\text{frb}_2^c(p_2, R_2^F) :- \text{ acc}(p_1,q_1).}$$

$$(\text{frb}_1^c) \quad \frac{p_1@p_2 \to p \in \text{rules}(A) \quad q_2 \in \text{s}(F)}{\text{frb}_1^c(p_1, R_1^F) :- \text{ acc}(p_2,q_2).}$$

$$(\text{frb}_{/1a}^c) \quad \frac{p_1@p_2 \to p \in \text{rules}(A) \quad q_2 \notin R_2^F}{\text{frb}^c(p_1, \emptyset) :- \text{ acc}(p_2,q_2).}$$

$$(\text{frb}_{/2a}^c) \quad \frac{p_1@p_2 \to p \in \text{rules}(A) \quad q_1 \notin R_1^F}{\text{frb}^c(p_2, \emptyset) :- \text{ acc}(p_1,q_1).}$$

$$(\text{fail}_{2a}) \quad \frac{p \in \text{final}(A) \quad \forall r. \, q \xrightarrow{\epsilon \leq 1}_F r \Rightarrow r \notin \text{final}(F)}{\text{fail}_2 :- \text{ acc}(p,q).}$$

The clauses from $(\text{acc}_{/1})$, $(\text{acc}_{/2})$ and (fail_0) in $D_2(A,F)$ belong to $D_3(A,F)$, too. We use sets of states $Q_2^F(q_1) = \{q' \mid q_1@q' \to q'' \in \text{rules}(F)\}$, $Q_1^F(q_2)$ symmetrically, and sets of states reaching a sort $R_i^F = \{q \mid \exists r \in \text{s}_i(F). \, q \xrightarrow{\epsilon \leq 1}_F r\}$.

Fig. 6. Inferring clauses of Datalog program $D_3(A,F)$ simulating $D_2(A,B)$

$L(A) \subseteq L(B)$ from A and F without computing B. This is done by the ground Datalog program $D_3(A,F)$ of Fig. 6.

We need new predicates for properties of F in order to infer corresponding properties of B. The accessibility predicate f.acc for F subsumes the accessibility predicate acc for B. Subsumption may by proper as stated by the rule (f.acc) of $D_3(A,F)$. *Vice versa*, we infer accessibility in F from accessibility in B according to the rule $(\text{acc}_{/3a})$. Rules $(\text{acc}_{/1})$ and $(\text{acc}_{/2})$ of $D_2(A,F)$ remain valid for accessibility in B, too.

Lemma 2. $\text{acc}(p,q) \in \text{lfp}(D_3(A,F))$ *iff* $\text{acc}(p,q) \in \text{lfp}(D_2(A,B))$

We need to refine predicate frb into predicates frb_1 and frb_2 that take sorts into account, and corresponding predicates f.frb_1 and f.frb_2 in the factorized case. Their semantics can be defined as follows, where A, B are tree automata and F is a factorized tree automaton. The semantics of frb_1 and f.frb_1 are symmetric.

$$A, B \models \text{frb}_2(p_2, q_2) \Leftrightarrow \exists p, p_1, q_1. \, A, B \models \text{acc}(p_1,q_1), \, p_1@p_2 \to p \in \text{rules}(A), \, q_2 \notin Q_2^B(q_1)$$
$$A, F \models \text{f.frb}_2(p_2, r_2) \Leftrightarrow \exists p, p_1, r_1. \, A, F \models \text{f.acc}(p_1,r_1), \, p_1@p_2 \to p \in \text{rules}(A), \, r_2 \notin Q_2^F(r_1)$$

The relation to the previous predicate frb is that $A, B \models \text{frb}(p,q)$ if and only if $A, B \models \text{frb}_1(p,q) \vee \text{frb}_2(p,q)$.

The Datalog program $D_3(A,F)$ infers for sorts $i \in \{1,2\}$ literals with predicates f.frb_i^c that are to be understood by grouping of f.frb_i literals, and similarly frb_i^c by grouping of frb_i. These grouping mechanisms account for sorts, since complementation is with respect to sorts.

$$A, F \models \mathsf{f.frb}_i^c(p, Q) \Leftrightarrow \forall q \in \mathsf{s}_i(F) \setminus Q.\ A, F \models \mathsf{f.frb}_i(p, q)$$
$$A, F \models \mathsf{frb}_i^c(p, Q) \Leftrightarrow \forall q \in \mathsf{s}_i(F) \setminus Q.\ A, F \models \mathsf{frb}_i(p, q)$$

The clauses produced by ($\mathsf{f.frb}_i^c$) and (frb_i^c) are sound with respect to this semantics for deterministic F's. This is easier to see for ($\mathsf{f.frb}_i^c$) than for (frb_i^c). We prove it by the next lemma. For states p, q, r and sorts $i \in \{1, 2\}$ we define:

$$A, B \vdash \mathsf{frb}_i(p, q) \quad \text{iff} \quad \exists Q \subseteq \mathsf{s}(B) \setminus \{q\}.\ \mathsf{frb}^c(p, Q) \in \mathsf{lfp}(D_2(A, B)) \ via \ (\mathsf{frb}_{/i}^c)$$
$$A, F \vdash \mathsf{f.frb}_i(p, r) \quad \text{iff} \quad \exists R \subseteq \mathsf{s}(F) \setminus \{r\}.\ \mathsf{f.frb}_i^c(p, R) \in \mathsf{lfp}(D_3(A, F))$$
$$A, F \vdash \mathsf{frb}_i(p, r) \quad \text{iff} \quad \exists R \subseteq \mathsf{s}(F) \setminus \{r\}.\ \mathsf{frb}_i^c(p, R) \in \mathsf{lfp}(D_3(A, F))$$
$$\text{or } \mathsf{frb}^c(p, \emptyset) \in \mathsf{lfp}(D_3(A, F)) \ via \ (\mathsf{frb}_{/ia}^c)$$

Lemma 3 (Core). $A, B \vdash \mathsf{frb}_i(p, q)$ *iff* $A, F \vdash \mathsf{frb}_i(p, q)$ *or the unique state* r *of sort* i *with* $q \xrightarrow{\epsilon \ \leq 1}_F r$ *exists and satisfies* $A, F \vdash \mathsf{f.frb}_i(p, r)$.

Since the size of $D_3(A, F)$ is in $O(|A| * |F|)$ we can compute the set of all $\mathsf{f.frb}_1^c(p, R)$ and $\mathsf{f.frb}_2^c(p, R)$ literals in $\mathsf{lfp}(D_3(A, F))$ in time $O(|A| * |F|)$. It remains to infer the induced literals $\mathsf{f.frb}_i(p, r)$ literals in an efficient manner. Computing all $A, F \vdash \mathsf{f.frb}_i(p, r)$ from $\mathsf{lfp}(D_3(A, F))$ can be done by the same clever algorithm as before for deducing all $A, B \vdash \mathsf{frb}(p, q)$ given $\mathsf{lfp}(D_2(A, B))$. Note, however, that we now need two different data structures for the two sorts.

Theorem 3. *For a stepwise tree automaton with ϵ-rules A and a deterministic factorized tree automaton F over the same signature, inclusion $L(A) \subseteq L(F)$ can be decided in time $O(|A| * |F|)$.*

5   Automata for Unranked Trees and DTDs

We lift our results to deterministic tree automata for unranked trees possibly with factorization, so that they become applicable to deterministic DTDs.

An unranked signature Σ is a finite set of symbols (without arity restrictions). The set T_Σ^u of unranked trees over Σ is the least set that contains all pairs $a(t_1, \ldots, t_n)$ where $a \in \Sigma$ and $(t_1, \ldots, t_n)$ is a possibly empty sequence of unranked trees in T_Σ^u. Currying carries over literally from ranked to unranked trees. This yields a bijection $\mathsf{curry} : T_\Sigma^u \to T_{\Sigma_\otimes}$. Thus, we can reuse stepwise tree automata to recognize languages of unranked trees, and as before, we can factorize them. So, as a corollary of Theorem 3 we have:

Corollary 2. *For a stepwise tree automaton for unranked trees A and a deterministic factorized tree automaton for unranked trees F over the same signature Σ, $L(A) \subseteq L(F)$ can be decided in time $O(|A| * |F|)$ independently of $|\Sigma|$.*

Note that hedge automata [5] can be translated in linear time to stepwise tree automata with ϵ-rules [9]. The automata of [12] support factorization, too. We finally show how to convert deterministic DTDs D to deterministic factorized tree automata in time $O(|\Sigma| * |D|)$, so that we can reuse our algorithm for testing inclusion in deterministic DTDs. Factorization avoids the quadratic blowup from translating hedge to stepwise automata [9].

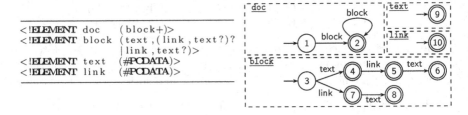

```
<!ELEMENT doc     (block+)>
<!ELEMENT block   (text,(link,text?)?
                  |link,text?)>
<!ELEMENT text    (#PCDATA)>
<!ELEMENT link    (#PCDATA)>
```

Fig. 7. An example DTD and the corresponding Glushkov automata

A DTD D with elements in a set Σ is a function mapping letters $a \in \Sigma$ to regular expressions e over Σ, what we write $a \to_D e$ in this case. One of these elements is the distinguished start symbol. The language $L_a(D) \subseteq T_\Sigma^u$ of elements a of a DTD D is the smallest set of unranked trees such that:

$$L_a(D) = \{a(t_1,\ldots,t_n) \mid a \to_D e, \; a_1 \ldots a_n \in L(e), t_i \in L_{a_i}(D) \text{ for } 1 \le i \le n\}$$

The language of a DTD D is $L(D) = L_a(D)$ where a is the start symbol of D. The size of D is the total number of symbols in the regular expressions of D. An example in XML syntax is given in Fig. 7. The set of elements of D is $\Sigma = \{\text{doc}, \text{block}, \text{text}, \text{link}\}$, of which the element doc is the start symbol. The regular expression for #PCDATA recognizes only the empty word.

A DTD is deterministic if all its regular expressions are one-unambiguous [2]. This is equivalent to say that all corresponding Glushkov automata are deterministic, which is a requirement by the W3C. See Fig. 7 for the example.

Theorem 4 (Brüggemann-Klein [1]). *The collection of Glushkov automata for a deterministic DTD D over Σ can be computed in time $O(|\Sigma| * |D|)$.*

We transform a collection of Glushkov automata for a deterministic DTD D into a single factorized tree automaton F as follows. The set of states of sort 1 of F is the disjoint union of the states of the Glushkov automata. The states of sort 2 of F are the elements of D. For every element a, we connect all final states q of its Glushkov automaton to the state a, i.e., $q \xrightarrow{\epsilon} a \in \text{rules}(F)$. The only final state of F is the start symbol of the DTD D. The result is a finite automaton, that represents a factorized

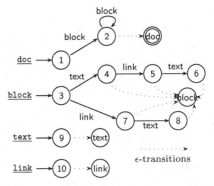

Fig. 8. The deterministic factorized tree automaton for the DTD in Fig. 7

tree automaton, as for instance in Fig. 8. This needs time of at most $O(|\Sigma| * |D|)$. Note that the size of the example automaton would grow quadratically, when eliminating ϵ-edges. For every $a \in \Sigma$, there is $a \to q \in \text{rules}(F)$ for the unique initial state q of the Glushkov automaton of a. For every transition $q \xrightarrow{a} q'$ of one of the Glushkov automata, we add a rule $q@a \to q' \in \text{rules}(F)$.

Note that F is deterministic as a factorized automaton. The ϵ-free part of F is deterministic since all Glushkov automata are: d_0. Let q be a state of the Glushkov automaton for some letter a. The only state of sort 1 q can reach by ϵ-edges is a and the only state of sort 2 is q itself. All other states of F are elements of $a \in \Sigma$, which have no outgoing ϵ-edges: d_1.

Theorem 5. *Deterministic DTDs D over Σ can be translated to deterministic factorized tree automata that recognize the same language in time $O(|\Sigma| * |D|)$.*

Corollary 3. *Language inclusion of hedge automata A over Σ in deterministic DTDs D with elements in Σ can be decided in time $O(|A| * |\Sigma| * |D|)$.*

References

1. Brüggemann-Klein, A.: Regular expressions to finite automata. Theoretical Computer Science 120(2), 197–213 (1993)
2. Brüggemann-Klein, A., Wood, D.: One-unambiguous regular languages. Information and Computation 142(2), 182–206 (1998)
3. Carme, J., Niehren, J., Tommasi, M.: Querying unranked trees with stepwise tree automata. In: van Oostrom, V. (ed.) RTA 2004. LNCS, vol. 3091, pp. 105–118. Springer, Heidelberg (2004)
4. Champavère, J., Gilleron, R., Lemay, A., Niehren, J.: Towards schema-guided XML query induction. In: ICML CAGI Workshop (2007)
5. Comon, H., et al.: Tree automata techniques and applications (2007), http://tata.gforge.inria.fr
6. Dantsin, E., Eiter, T., Gottlob, G., Voronkov, A.: Complexity and expressive power of logic programming. ACM computing surveys 33(3), 374–425 (2001)
7. Maneth, S., Berlea, A., Perst, T., Seidl, H.: XML type checking with macro tree transducers. In: 24th PODS, pp. 283–294 (2005)
8. Martens, W., Neven, F., Schwentick, T.: Complexity and decision problems for XML Schemas and chain regular expressions. Journal extension of MFCS 2004 (2008)
9. Martens, W., Niehren, J.: On the minimization of XML schemas and tree automata for unranked trees. J. of Comp. and Sys. Sci. 73(4), 550–583 (2007)
10. Milo, T., Suciu, D., Vianu, V.: Type checking XML transformers. J. of Comp. and Sys. Sci. 1(66), 66–97 (2003)
11. Neumann, A., Seidl, H.: Locating matches of tree patterns in forests. In: Arvind, V., Ramanujam, R. (eds.) FST TCS 1998. LNCS, vol. 1530, pp. 134–145. Springer, Heidelberg (1998)
12. Raeymaekers, S.: Information Extraction from Web Pages Based on Tree Automata Induction. PhD thesis, Katholieke Universiteit Leuven (2008)
13. Schwentick, T.: Automata for XML—a survey. J. of Comp. and Sys. Sci. 73(3), 289–315 (2007)
14. Seidl, H.: Deciding equivalence of finite tree automata. SIAM Journal on Computing 19(3), 424–437 (1990)
15. Seidl, H.: Haskell overloading is DEXPTIME-complete. Information Processing Letters 52(2), 57–60 (1994)
16. Tozawa, A., Hagiya, M.: XML schema containment checking based on semi-implicit techniques. In: Int. Conf. on Impl. and Appl. of Automata (2003)
17. Dal Zilio, S., Lugiez, D.: XML schema, tree logic and sheaves automata. In: 18th RTA. LNCS, vol. 2706, pp. 246–263. Springer, Heidelberg (2003)

Consensual Definition of Languages by Regular Sets*

Stefano Crespi Reghizzi and Pierluigi San Pietro

Dipartimento di Elettronica e Informazione, Politecnico di Milano
P.za Leonardo da Vinci 32, I–20133 Milano
crespi@elet.polimi.it,
sanpietr@elet.polimi.it

Abstract. A new language definition model is introduced and investigated, based on agreement or consensus between similar strings. Considering a regular set of strings over a bipartite alphabet made by pairs of unmarked/marked symbols, a match relation is introduced, in order to specify when such strings agree. Then a regular set over the bipartite alphabet can be interpreted as defining another language over the unmarked alphabet, called the consensual language. A string is in the consensual languages if a set of corresponding matching strings is in the original language. The family defined by this approach includes the regular languages and also interesting non-semilinear languages. The word problem can be solved in polynomial time, using a multi-counter machine. Closure properties of consensual languages are proved for intersection with regular sets and inverse alphabetical homomorphism.

1 Introduction

An ever present, common sense idea in language modelling research is that, for a string to to be a valid phrase, it should comply with several constraints at once. Theories of grammar have taken various approaches for expressing the constraints by different mechanisms, such as by superimposing semantic constraints to syntactic ones, or by using intersections of, say, context-free languages. Here we propose a very simple novel mechanism, where the constraints are expressed by an elementary character by character agreement between strings belonging to a regular language. The alphabet is bipartite, made by pairs of unmarked/marked characters. The agreement is formalized by a k-ary relation, called match, that is satisfied by a set of k equally long strings if, in each position, exactly one string has an unmarked character and the other strings have the same character but marked. In our metaphor we view such strings as providing consensus on the validity of the corresponding unmarked string. This justifies the name "consensual" proposed for the new family, which strictly includes the regular one.

Here some reader may prefer to jump to the definition (Def. 1, 2 and 3) of consensual language before reading the next discussion of the position of the new model from the perspective of language theory.

With respect to their storage, language recognition devices can be classified as using tapes (Turing machines, push-down machines, nested push-down machines) or counters. The latter case includes various models of counter machines and also Petri Nets.

* Partially supported by PRIN 2005015419, FIRB "Applicazioni della Teoria degli Automi all'Analisi, Compilazione e Verifica di Software Critico e in Tempo Reale", and CNR-IEIIT.

C. Martín-Vide, F. Otto, and H. Fernau (Eds.): LATA 2008, LNCS 5196, pp. 196–208, 2008.
© Springer-Verlag Berlin Heidelberg 2008

Consensual languages are recognized by real-time non-deterministic multi counter machines with a linear bound on the counter values.

Considering the complexity of the word recognition problem, consensual languages belong to the polynomial time class.

With respect to generative capacity, the new family shares little ground with the families of context-free and mildly context-sensitive [6] languages. For instance, the Dyck language over two letters can be defined but not the language of palindromes. On the other hand interesting non-semilinear languages (in the Parikh sense [5]) can be easily defined.

Next we compare the computation performed by a consensual recognizer versus a finite non-deterministic alternating automaton [1]. Although both machines perform simultaneous computations for recognizing a given string, they apply entirely different acceptance criteria. All possible computations must be successful for a string to be recognized by an alternating machine when using universal non-determinism, and their number may be exponential with respect to the string length. On a consensual device, the computations performed on the finite automaton, which can be assumed to be deterministic, are not labelled by the input string (except in the trivial case when the language is regular) but by matching strings over the marked/unmarked alphabet. The number of computations is bounded by the input length.

Recalling that certain Petri net language families [2] include non-semilinear languages and that their recognizers use counters, a vague resemblance between the two models may be mentioned. In fact C.A. Petri introduced his nets as a formal model of synchronization between computations performed by finite automata and our model too specifies a matching rule between the labels of different computations.

Notwithstanding the fact that the proposed approach has little to do with any classical model we know, we hope its simplicity and expressivity may attract some attention.

The paper is organized as follows. Section 2 lists the basic definitions, and provides an example giving evidence of the strict inclusion of regular languages. Section 3 shows the Parikh image may be not linear, and proves some closure properties. Section 4 presents the construction of a counter machine as recognizer. Then it shows that the word recognition problem is polynomial in time, and that the languages of palindromes and replicas exceed the power of consensual languages. The conclusion mentions some directions of continuation.

2   First Definitions

Let Σ be an alphabet called *terminal*, and let $\underline{\Sigma}$ be the disjoint alphabet obtained by marking each symbol $a \in \Sigma$ as $\underline{a}$, referred to as the *marked copy* of a. The set $\Sigma \cup \underline{\Sigma}$ is called *internal* alphabet.

The empty string is denoted by the letter ϵ. Given a string x, its length is denoted by $|x|$ and the i-th character is $x(i)$, $1 \leq i \leq |x|$.

A deterministic finite automaton DFA over the internal alphabet is specified as $A = (\Sigma \cup \underline{\Sigma}, Q, \delta, q_1, F)$ where Q is the state set, $\delta : Q \times (\Sigma \cup \underline{\Sigma})^* \to Q$ is the state-transition function, q_1 is the initial state, and F is the set of final states. For convenience, the set Q is considered to be ordered: $q_1, q_2, \ldots q_{|Q|}$, with the first state q_1 being the initial state. Furthermore, to simplify some proofs, the transition function δ is always assumed to be *total*. Hence, $\delta(q_1, y)$ is defined for every string y over the internal alphabet.

Definition 1. *The partial, symmetrical, and associative binary operator, called* match

$$@ : (\Sigma \cup \underline{\Sigma}) \times (\Sigma \cup \underline{\Sigma}) \to (\Sigma \cup \underline{\Sigma})$$

is defined as follows, for all $a \in \Sigma$:

$$a@\underline{a} \quad = \underline{a}@a = a$$
$$\underline{a}@\underline{a} \quad = \quad \underline{a}$$
$$\textit{in every other case} = \textit{undefined}$$

The operator can be naturally extended to strings of equal length, by assuming $\epsilon@\epsilon = \epsilon$. For any strings $w, w' \in (\Sigma \cup \underline{\Sigma})^*$, $|w| = |w'|$, for every $a, b \in \Sigma \cup \underline{\Sigma}$

$$aw@bw' = (a@b)(w@w')$$

where we assume the match yields precedence to concatenation.

Since the operation is associative and symmetrical, for any number $m > 1$ of component strings $w_1, w_2, \ldots, w_m$ we may write $w_1@w_2@ \ldots @w_m$ without parentheses and in any order.

The number m is called the *breadth* of the match.

If $w_1@w_2@ \ldots @w_m$ is defined, then the strings $w_1, w_2, \ldots, w_m$ are said to *(weakly) match*; furthermore, if $w_1@w_2@ \ldots @w_m \in \Sigma^*$, they are said to *strongly match*.

Notice the match is undefined on strings w, w' of unequal lengths, or else if there exists a position i such that $w(i)@w'(i)$ is undefined. The latter condition occurs in three cases: when both characters are in Σ, when both are in $\underline{\Sigma}$ and differ, and when either one is marked but is not the marked copy of the other.

Next we extend the match operator to languages over the internal alphabet. For two languages, $L', L'' \subseteq (\Sigma \cup \underline{\Sigma})^*$:

$$L'@L'' = \{w'@w'' \mid w' \in L', w'' \in L''\}$$

Proposition 1. *If $L', L'' \subseteq (\Sigma \cup \underline{\Sigma})^*$ are regular then $L'@L''$ is also regular.*

Proof. Let $A' = (\Sigma \cup \underline{\Sigma}, Q', \delta', q'_1, F')$, $A'' = (\Sigma \cup \underline{\Sigma}, Q'', \delta'', q''_1, F'')$ be two DFA for L', L'', respectively. Let $A'@A''$ be the (possibly nondeterministic) finite automaton $(\Sigma \cup \underline{\Sigma}, Q' \times Q'', \delta, (q'_1, q''_1), F' \times F'')$, with $\delta : (Q' \times Q'') \times (\Sigma \cup \underline{\Sigma}) \to 2^{Q' \times Q''}$ such that for every $q', p' \in Q', q'', p'' \in Q''$, for every $a \in \Sigma$:

$$\langle p', p'' \rangle \in \delta(\langle q', q'' \rangle, a) \text{ if } p' = \delta'(q', a), p'' = \delta''(q'', \underline{a})$$
$$\langle p', p'' \rangle \in \delta(\langle q', q'' \rangle, a) \text{ if } p' = \delta'(q', \underline{a}), p'' = \delta''(q'', a)$$
$$\langle p', p'' \rangle \in \delta(\langle q', q'' \rangle, \underline{a}) \text{ if } p' = \delta'(q', \underline{a}), p'' = \delta''(q'', \underline{a})$$

The construction is similar to the traditional Cartesian product machine of two DFAs' for recognizing their intersection. Here $A'@A''$ recognizes $L'@L''$, because the construction has been modified to match an a with $\underline{a}$ and an $\underline{a}$ with $\underline{a}$, but not to match a with a. □

The repeated application of the match operation to a language is formalized next.

Definition 2. *The* closure under match, *or* @-closure, *of a language* $L \subseteq (\Sigma \cup \underline{\Sigma})^*$ *is:*

$$L^@ = L \cup \{w_1 @ w_2 @ \ldots @ w_m \mid m > 1, w_1, w_2 \ldots, w_m \in L\}$$

An alternative definition can be given. Let $L^{1@} = L, L^{i@} = L @ L^{i-1@}, i > 1$.

$$L^@ = \bigcup_{i \geq 1} L^{i@}$$

Focusing on languages over the terminal alphabet Σ, the main definition comes next.

Definition 3. *The* consensual language with base $L \subseteq (\Sigma \cup \underline{\Sigma})^*$ *is the set*

$$\mathcal{C}(L) = L^@ \cap \Sigma^*$$

Let L be in a language family $\mathcal{F}$: then $\mathcal{C}(L)$ is called a consensual language based on $\mathcal{F}$, and the corresponding family is written $\mathcal{C}_{\mathcal{F}}$.

In this paper we study the family of consensual languages based on the family of regular languages, $\mathcal{C}_{\mathcal{REG}}$.

Example 1. Consider the regular expression

$$R = \underline{a}^* a \underline{a}^* \underline{b}^* b \underline{b}^*$$

Then $R^@$ is the set of strings of the form:

$$\underline{a}^* a_1 \underline{a}^* a_2 \underline{a}^* \ldots a_m \underline{b}^* b_1 \underline{b}^* b_2 \ldots b_m \underline{b}^*$$

where $m \geq 1$, and each a_i is a and each b_i is b. The consensual language with base R is $\mathcal{C}(R) = \{a^n b^n \mid n > 0\}$. For instance, the string $aaabbb$ can be obtained in different ways, matching together the strings of R in the first column, or matching those in the second column:

$a\underline{aabbb}$	$a\underline{aa}bb\underline{b}$
$a\underline{aa}bb\underline{b}$	$\underline{a}aa\underline{bb}b$
$\underline{aaa}bb\underline{b}$	$\underline{aaa}bb\underline{b}$

Hence, if the base L is regular, $\mathcal{C}(L)$ and $L^@$ may be not regular. However, from Prop. 1, for any finite i, $L^{i@}$ is regular. This corresponds, in Def. 2, to the case where at most i strings $w_1, \ldots, w_i$ are matched. Notice that in general $L^{i@} \not\subseteq L^{(i+1)@}$ even when $L^@$ is regular. For instance, consider the regular expression $R = \underline{a}^* a \underline{a}^*$: $R^@$ is $(\underline{a}^* a \underline{a}^*)^+$ while $R^{i@}$ is $(\underline{a}^* a \underline{a}^*)^i$.

3   First Properties

We introduce further useful terminology and make intuitive comments about previous definitions and concepts.

By Def. 1, 2, in a strong match the component strings and the result have the same length $n = |w_1| = |w|$ and for each position $1 \leq i \leq n$ exactly one component string, say w_k is unmarked, i.e. $w_k(i) \in \Sigma$ and $w_r(i) \in \underline{\Sigma}$ for all $r \neq k$. We say that string w_k *places* the character into position i and the other strings *consent* to it.

The match of identical strings containing at least one unmarked character is undefined. Therefore we may safely assume that in a match closure all component strings are distinct and that any phrase w in a consensual language is the result of a strong match with breadth at most $|w| + 1$, that is

$$\mathcal{C}(L) = (L \cap \Sigma^*) \cup \{w \in \Sigma^* \mid w = w_1 @ w_2 @ \ldots @ w_k, 1 \leq k \leq |w| + 1, w_i \in L\} \quad (1)$$

Consider a deterministic recognizer (DFA) of the base language R. A string w is in the consensual language $\mathcal{C}(R)$, if, and only if, the automaton performs $1 \leq k \leq |w| + 1$ successful computations, accepting a set of strings that strongly match to w. We may say that such computations strongly (or weakly) match.

The case $k = 1$ clearly corresponds to the usual recognition condition of a DFA. As the consensual language of Ex. 1 is not regular we have

Proposition 2. *The family $\mathcal{C}_{\mathcal{REG}}$ of consensual languages on a regular language base strictly includes the family of regular languages.*

The next examples show languages which are not semilinear (in Parikh's sense [5]).

Example 2

1. Series of identical unary integers.
   Choose the base:
   $$R_1 = (\underline{a}^* a \underline{a}^* \underline{b})^+ \cup (\underline{a}^+ b)^+$$
   Then the consensual language is:
   $$L_1 = \mathcal{C}(R_1) = \{a^n b a^n b a^n b \ldots a^n b \mid n > 0\}$$

2. Enumeration of unary integers.
   The language $L_2 = \{aba^2b \ldots ba^n b \mid n > 0\}$ is consensually defined by the regular base
   $$R_2 = \underbrace{(\underline{a}^* a \underline{a}^* \underline{b})^+}_{1} \cup \underbrace{(\underline{a}^+ \underline{b})^* \underline{a}^+ b (\underline{a}^* a \underline{a}^* \underline{b})^*}_{2}$$
   Call a run a minimal substring of L_2 delimited by b or by an edge. R_2 is the union of two clauses, numbered 1 and 2. Clause 1 places an a in each one of the n runs. Clause 2 places one b and one a in each run to its right. Therefore the j-th run, $1 \leq j \leq n$ will get exactly j letters a.

3. Series of exponential unary numbers.
   For $\Sigma = \{a, b, c\}$ let
   $$R_3 = \underline{\Sigma}^* a (\underline{a} \cup \underline{c})^* \underline{b} (\underline{a} \cup \underline{c})^* c \underline{a} c \underline{\Sigma}^* \cup \underline{a} c b \left((\underline{a} c)^+ b\right)^* (\underline{a} \underline{c})^+ b \cup acb$$
   The consensual language $\mathcal{C}(R_3)$ is
   $$L_3 = \{ac\, b(ac)^2 b(ac)^4 b(ac)^8 b \ldots (ac)^{2^m} b \mid m \geq 0\}$$
   The proof is in the Appendix.

Proposition 3. *The family C_{REG} is closed under:*

1. *intersection with regular languages;*
2. *inverse alphabetic homomorphism;*
3. *mirror operation;*
4. *marked concatenation [5] of consensual languages;*
5. *union of consensual languages with disjoint alphabets.*

Proof.

1. Let $R \subseteq (\Sigma \cup \underline{\Sigma})^*, S \subseteq \Sigma^*$ be two regular languages, and let $h : \Sigma \cup \underline{\Sigma} \to \Sigma$ be the alphabetic homomorphism defined by $h(a) = h(\underline{a}) = a$ for every $a \in \Sigma$. We claim that

$$C(R) \cap S = C\left(R \cap h^{-1}(S)\right)$$

   thus proving the statement.

   Let $x \in C(R) \cap S$. Hence, $\exists k, 1 \le k \le |x|, \exists x_1, \ldots x_k \in R$ such that $x_1 @ x_2 \ldots @ x_k = x$ and for every i, $1 \le i \le k$, $h(x_i) = x$. Hence, every $x_i \in h^{-1}(x) \subseteq h^{-1}(S)$ since $x \in S$. Hence, for every $i, 1 \le i \le k, x_i \in R \wedge x_i \in h^{-1}(S)$: if follows that $x \in C\left(R \cap h^{-1}(S)\right)$.

   Assume now $x \in C\left(R \cap h^{-1}(S)\right)$. Hence, $\exists k, 1 \le k \le |x|, \exists x_1, \ldots x_k$ such that $x_1 @ x_2 \ldots @ x_k = x$ and for every i, $1 \le i \le k$, $x_i \in R \cap h^{-1}S$, with $h(x_i) = x$. Then $x \in C(R)$ (since each $x_i \in R$). Also, $x \in h^{-1}(S)$ (since each $x_i \in h^{-1}(S)$). Therefore, $x \in S$, since $S = h^{-1}(S) \cap \Sigma^*$ and $x \in \Sigma^*$.

2. Let $R \subseteq (\Sigma \cup \underline{\Sigma})^*$ be a regular language, and let Δ be another finite alphabet. Let $h : \Delta \to \Sigma$ be a homomorphism. We need to prove that $h^{-1}(C(R))$ is a consensual language with regular base. Extend first h to the internal alphabet as follows: $\widehat{h} : \Delta \cup \underline{\Delta} \to \Sigma \cup \underline{\Sigma}$ is defined as $\widehat{h}(A) = h(A)$, $\widehat{h}(\underline{A}) = \underline{h(A)}$ for every $A \in \Delta$. We notice that $\widehat{h}^{-1}(\underline{a} @ \underline{a}) = \widehat{h}^{-1}(\underline{a}) = \widehat{h}^{-1}(\underline{a}) @ \widehat{h}^{-1}(\underline{a})$, and that $\widehat{h}^{-1}(a @ a) = \widehat{h}^{-1}(a) = \widehat{h}^{-1}(a) @ \widehat{h}^{-1}(a)$, while both $\widehat{h}^{-1}(a) @ \widehat{h}^{-1}(\underline{a})$ and $\widehat{h}^{-1}(\underline{a} @ a)$ are undefined. Hence, $\widehat{h}^{-1}(X @ Y) = \widehat{h}^{-1}(X) @ \widehat{h}^{-1}(Y)$ for every $X, Y \in \Sigma \cup \underline{\Sigma}$. Hence, if $u, u' \in (\Sigma \cup \underline{\Sigma})^*$ then $\widehat{h}^{-1}(u @ u') = \widehat{h}^{-1}(u) @ \widehat{h}^{-1}(u')$. We now claim that $\widehat{h}^{-1}(R^@) = \left(\widehat{h}^{-1}(R)\right)^@$. From here the thesis follows, since $\widehat{h}^{-1}(R)$ is regular and $h^{-1}(C(R)) = \widehat{h}^{-1}(R^@) \cap \Delta^* = \left(\widehat{h}^{-1}(R)\right)^@ \cap \Delta^*$. Let $x \in \widehat{h}^{-1}(R^@)$. Hence, there is $w \in R^@$ such that $x \in h^{-1}(w)$. By Def. 3, there exist $k > 0$ strings $w_1, \ldots, w_k \in R$, with $1 \le k \le |x|$, such that $w_1 @ \ldots @ w_k = w$. Hence, $x \in \widehat{h}^{-1}(w) = \widehat{h}^{-1}(w_1) @ \ldots @ \widehat{h}^{-1}(w_k) \subseteq \left(\widehat{h}^{-1}(R)\right)^@$. Let $x \in \left(\widehat{h}^{-1}(R)\right)^@$. Hence, there exist $k > 0$ strings $x_1, \ldots, x_k \in \widehat{h}^{-1}(R)$ such that $x_1 @ \ldots @ x_k = x$. Therefore, there exist $k > 0$ strings $w_1, \ldots, w_k \in R$ such that $x_1 \in \widehat{h}^{-1}(w_1), \ldots, x_k \in \widehat{h}^{-1}(w_k)$, and hence:
$$x = x_1 @ \ldots @ x_k \subseteq \widehat{h}^{-1}(w_1) @ \ldots @ \widehat{h}^{-1}(w_k) =$$
$$= \widehat{h}^{-1}(w_1 @ w_2 @ \ldots @ w_k) \subseteq \widehat{h}^{-1}(R^@).$$

3. For items 3, 4, and 5 the obvious proofs are based on simple transformations of the DFAs' recognizing the base languages. □

4   Consensual Languages Are in P

We have observed that the DFA accepting the base language L (on the internal alphabet) can be used to recognize the strings over Σ that strongly match to a phrase w of the consensual language $C(L)$ (on the alphabet Σ).

A recognition algorithm for consensual languages performs as many computations on the DFA, as the breadth of the match producing w, which following Eq. 1 is bounded by $|w| + 1$. All computations start in the initial state and at any time their labels are strings on the internal alphabet, such that they strongly match to a prefix of w. Notice that the algorithm is in general non-deterministic, although the base automaton is deterministic. Acceptance occurs when all the computations reach a final state. A configuration reached by the algorithm at some point can be naturally encoded by a multi-set of states of the DFA: the multiplicity of a state q_j in the multi-set encodes the number of computations that have reached state q_j. Concretely, a multi-set can be represented by multiplicity counters, and the recognition algorithm becomes a counter machine.

Let $\mathbb{N}$ be the set of natural numbers. Boldface capital letters, such as $\mathbf{V}, \mathbf{V}', \mathbf{V}'', \ldots$, denote vectors on $\mathbb{N}^{|Q|}$. V_k denotes the k-th component of $\mathbf{V} = [V_1 V_2 \ldots V_k \ldots V_{|Q|}]$. The Kronecker function is $\gamma_{k,k} = 1$ and $\gamma_{k,h} = 0$ for every $h \neq k$. It will be used to add or subtract one from a counter V_j, leaving all other counters unchanged. Another useful shorthand is the norm of a vector $\mathbf{V}$, $\|\mathbf{V}\| = \sum_{j=1}^{|Q|} V_j$.

Consider a base language $L = L(A)$ with $A = (\Sigma \cup \underline{\Sigma}, Q = \langle q_1, q_2, \ldots q_{|Q|} \rangle, \delta, q_1, F)$ a DFA defined as in Sect. 2.

Definition 4. *The* consensual transition relation *for A $\Rightarrow_A \subseteq \mathbb{N}^{|Q|} \times \Sigma \times \mathbb{N}^{|Q|}$ is defined for every $a \in \Sigma$, $\mathbf{V}, \mathbf{V}' \in \mathbb{N}^{|Q|}$, as: $\mathbf{V} \overset{a}{\Rightarrow}_A \mathbf{V}'$ holds if, an only if, $\exists h, k$, with $1 \leq h, k \leq |Q|$ such that:*

1. $V_h > 0 \wedge \delta(q_h, a) = q_k$
2.
$$\forall i, 1 \leq i \leq |Q|, \quad V_i' = \gamma_{k,i} + \sum_{j:q_i = \delta(q_j, \underline{a})} (V_j - \gamma_{h,j}\gamma_{k,i})$$

We say that character a is the label of the transition from $\mathbf{V}$ to $\mathbf{V}'$.

In general the consensual transition relation is not deterministic, in the sense that there may be $\mathbf{V}' \neq \mathbf{V}''$ such that $\mathbf{V} \overset{a}{\Rightarrow}_A \mathbf{V}'$, $\mathbf{V} \overset{a}{\Rightarrow}_A \mathbf{V}''$. In fact, more than one pair of states q_h, q_k may verify the conditions of the definition of $\overset{a}{\Rightarrow}_A$, even if the original automaton A is deterministic.

A nondeterministic (one-way) counter machine with $|V|$ counters may easily implement the consensual transition relation of a DFA, which only specifies linear relations among counters. In practice, a counter machine reads in a move the current (unmarked) input character and updates the counter values as specified in Def. 4.

The move simulates a set of strongly matching transitions performed by the DFA on marked and unmarked characters. More precisely, exactly one instance of a state (q_h of Eq. 1) places the current character a, while the remaining ($V_h - 1$) instances consent to it. For any other counter $i \neq h$, V_i instances of state q_i consent to character a.

To complete the simulation, the counter updating rule calculates how many times each state is entered by a transition originating in the current state instances.

For ease of reading, in the following, we will write $\Rightarrow$ instead of $\Rightarrow_A$ when no confusion can arise. For a string $y \in \Sigma^*$ the notation

$$\mathbf{V} \overset{y}{\Rightarrow} \mathbf{V}'$$

stands for

$$\mathbf{V} = \mathbf{V}_0 \overset{y(1)}{\Rightarrow} \mathbf{V}_1 \overset{y(2)}{\Rightarrow} \mathbf{V}_2 \ldots \mathbf{V}_{|y|-1} \overset{y(|y|)}{\Rightarrow} \mathbf{V}_{|y|} = \mathbf{V}'$$

Definition 5. *Let A as above be a DFA, and $\mathbf{V} \in \mathbb{N}^{|Q|}$ be a vector. If $V_1 > 0$ and $V_i = 0$ for $i > 1$ then $\mathbf{V}$ is called* initial.
If $\sum_{i:q_i \in F} V_i > 0$ and $\sum_{i:q_i \notin F} V_i = 0$ then $\mathbf{V}$ is called final.

Definition 6. *The consensual language of the automaton A, denoted $L^@(A)$, is the set of strings $w \in \Sigma^*$ such that there exist an initial vector $\mathbf{V}^0$, $1 \le \|\mathbf{V}^0\| \le |w| + 1$, and a final vector $\mathbf{V}$: $\mathbf{V}^0 \overset{w}{\Rightarrow} \mathbf{V}$.*

A counter automaton implementing relation $\Rightarrow$ accepts a string w if it starts with an initial vector and checks, upon reading the last character of w, whether in the last configuration some counters associated with the final states of the DFA differ from zero and all other counters are null.

We observe another cause of nondeterminism for a counter automaton implementing $\Rightarrow$ is that there are multiple initial configurations, since the value $\|\mathbf{V}^0\|$ of the initial counter is any integer up to $|w| + 1$.[1]

The sum of the counter values of a configuration is constant during a computation as next stated. In fact, the transition function of the DFA is assumed to be total, hence if $\mathbf{V} \overset{a}{\Rightarrow} \mathbf{V}'$ then

$$\sum_{1 \le i \le |Q|} (\gamma_{k,i}) + \sum_{j:q_i = \delta(q_j, \underline{a})} (V_j - \gamma_{h,j}\gamma_{k,i}) =$$
$$= 1 + \sum_{1 \le i \le |Q|} \sum_{j:q_i = \delta(q_j, \underline{a})} (V_j) - \sum_{1 \le i \le |Q|} \sum_{j:q_i = \delta(q_j, \underline{a})} \gamma_{h,j}\gamma_{k,i} =$$
$$= \sum_{1 \le i \le |Q|} \sum_{j:q_i = \delta(q_j, \underline{a})} (V_j) = \|\mathbf{V}\|.$$

Lemma 1. *Let $\mathbf{V}^0$ be a vector. For every word $w \in \Sigma^+$, if there exist $\mathbf{V}^1 \ldots \mathbf{V}^{|w|}$ such that*

$$\mathbf{V}^0 \overset{w(1)}{\Rightarrow} \mathbf{V}^1 \overset{w(2)}{\Rightarrow} \mathbf{V}^2 \ldots \overset{w(|w|)}{\Rightarrow} \mathbf{V}^{|w|}$$

then for every i, $1 \le i \le |w|$, $\|\mathbf{V}^i\| = \|\mathbf{V}^0\|$

Since a counter machine simulating $\Rightarrow$ has $|Q|$ counters (i.e., a fixed number) and the sum of its counter values is always bounded by the length of the input string, its computational complexity is NLOGSPACE:

Proposition 4. *Given a DFA A, the word problem for $L^@(A)$ can be computed in nondeterministic logspace.*

---

[1] This is not essential and is done to simplify definitions and proofs. It would be equivalent to define a counter machine that always starts with $V_1 = 1$.

Our goal is to prove that $C(L(A)) = L^@(A)$, thus proving that $C_{\mathcal{REG}}$ is in the complexity class NLOGSPACE and hence in P [4]. The proof requires a few technical definitions and lemmas

Definition 7. *Let $m \geq 1$ be an integer, and let A be a finite automaton. A m-vector $\underline{y} \in \left(2^{\mathbb{N}}\right)^{|Q|}$ is a vector of (possibly empty) $|Q|$ components $\mathcal{Y}_1, \ldots, \mathcal{Y}_{|Q|} \subseteq \{1, \ldots m\}$, such that $\mathcal{Y}_1, \ldots \mathcal{Y}_{|Q|}$ is a partition of $\{1, \ldots, m\}$.*

The intended meaning of an m-vector $\underline{Y}$ is to represent a partition of strings $y_1, \ldots y_m$ over the internal alphabet into subsets.

Definition 8. *Relation $\dashrightarrow \subseteq \left(2^{\mathbb{N}}\right)^{|Q|} \times \Sigma \times \left(2^{\mathbb{N}}\right)^{|Q|}$ is defined, for every $a \in \Sigma$ for every m-vectors $\underline{y}, \underline{y}'$ as $\underline{y} \overset{a}{\dashrightarrow} \underline{y}'$ if, and only if:*
$\exists r, 1 \leq r \leq m, \exists h, k, \text{ with } 1 \leq h, k \leq |Q|, \text{ such that } r \in \mathcal{Y}_h \text{ and } \delta(q_h, a) = q_k \text{ and for every } i, 1 \leq i \leq |Q|,$

$$\mathcal{Y}'_i = \bigcup_{1 \leq j \leq |Q|: \delta(q_j, \underline{a}) = q_i} \{z \mid z \in \mathcal{Y}_j, z \neq r\} \cup (\text{ if } i = k \text{ then } \{r\} \text{ else } \emptyset)$$

This definition of $\dashrightarrow$ is actually equivalent, as explained next, to relation $\Rightarrow$, but sets of integers are considered rather than integers. Their equivalence allows to use $\dashrightarrow$ instead of $\Rightarrow$, in order to simplify proofs. Notice that a Turing machine implementing $\Rightarrow$ only uses logarithmic space in the size of the input, while implementing a m-vector $\underline{y}$ requires a super-logarithmic memory.

Lemma 2. *Let $y \in \Sigma^+$, let $m \geq 1$ and let A be a DFA.*

1. *Given two m-vectors for A $\underline{y}, \underline{y}'$ such that $\underline{y} \overset{y}{\dashrightarrow} \underline{y}'$ then there exist two vectors for A: $\mathbf{V}, \mathbf{V}'$, such that for every i, $1 \leq i \leq |Q|$, $|\mathcal{Y}_i| = V_i, |\mathcal{Y}'_i| = V'_i$ and $\mathbf{V} \overset{y}{\Rightarrow} \mathbf{V}'$.*

2. *Given two vectors for $A, \mathbf{V}, \mathbf{V}'$, such that $\mathbf{V} \overset{y}{\Rightarrow} \mathbf{V}'$ there exist two m-vectors for A: $\underline{y}, \underline{y}'$ such that for every i, $1 \leq i \leq |Q|$, $|\mathcal{Y}_i| = V_i, |\mathcal{Y}'_i| = V'_i$ and $\underline{y} \overset{y}{\dashrightarrow} \underline{y}'$.*

Proof. 1) The proof obviously only has to show the case for $y = a \in \Sigma$. Assume $\underline{y} \overset{a}{\dashrightarrow} \underline{y}'$. Let r, k, h as in Def. 8. $\mathbf{V} \overset{a}{\Rightarrow} \mathbf{V}'$ follows from the fact that the update rules for $\dashrightarrow$ update each set $\mathcal{Y}'_i$ with a cardinality given by the sum of cardinalities of sets $\mathcal{Y}_j$, using the same update mechanism of $\Rightarrow$. In fact, for $i \neq k$,

$$\mathcal{Y}'_i = \bigcup_{1 \leq j \leq |Q|: \delta(q_j, \underline{a}) = q_i} \{z \mid z \in \mathcal{Y}_j, z \neq r\} =$$

$$\{z \mid z \in \mathcal{Y}_h, z \neq r\} \cup \bigcup_{1 \leq j \leq |Q|, j \neq h: \delta(q_j, \underline{a}) = q_i} \{z \mid z \in \mathcal{Y}_j\}$$

since $r \in \mathcal{Y}_h$.
Hence,
$$|\mathcal{Y}'_i| = V_h - 1 + \sum_{1 \leq j \leq |Q|, j \neq h: \delta(q_j, \underline{a}) = q_i} V_j = \sum_{1 \leq j \leq |Q|: \delta(q_j, \underline{a}) = q_i} (V_j - \gamma_{h,j} \gamma_{k,i})$$

$= \gamma_{k,i} + \sum_{1 \le j \le |Q| : \delta(q_j, \underline{a}) = q_i} (V_j - \gamma_{h,j}\gamma_{k,i})$ since $\gamma_{k,i} = 0$. If $i = k$, then the same argument finds the same expression for $|\mathcal{Y}'_i|$, but this time with $\gamma_{k,i} = 1$. Hence, since $V'_i = |\mathcal{Y}'_i|$ for every i, $\mathbf{V} \overset{a}{\rightrightarrows} \mathbf{V}'$.

2) Assume now $\mathbf{V} \overset{a}{\rightrightarrows} \mathbf{V}'$, and let k, h as in Def. 4. Hence, $|V_h| > 0$. Let $m = \|\mathbf{V}\|$. Let $\underline{y}$ be any m-vector satisfying $|\mathcal{Y}_i| = V_i$ for every i. Hence, $\mathcal{Y}_h \ne \emptyset$. Let r be any value in $\mathcal{Y}_h$. Then define $\underline{y}'$ as in Def. 8 (this is always possible when $\mathcal{Y}_h \ne \emptyset$). But this definition does not depend on the actual value r chosen among the elements of $\mathcal{Y}_h$. The same counting argument as for part (1) above shows that $|Y'_i| = V_i$ for every i.   □

The following lemma is easier to prove for $--\rightarrow$ rather than for $\rightrightarrows$:

Lemma 3. *Let $y \in \Sigma^*, m > 0$, let A be a DFA and let $\mathcal{Y}, \mathcal{Y}'$ be m-vectors such that $\mathcal{Y} \overset{y}{--\rightarrow} \mathcal{Y}'$ and $\mathcal{Y}_1 = \{1, \dots, m\}$, $Y_i = \emptyset$ for $i > 1$. Then there exist $y_1, \dots y_m$ strings of length $|y|$ over the internal alphabet such that $y = y_1@\dots@y_m$, if, and only, for every i, $1 \le i \le |Q|$,*

$$\mathcal{Y}'_i = \{z \mid 1 \le z \le m, \delta(q_1, y_z) = q_i\}$$

Proof. Let $y, m, \underline{\mathcal{Y}}, \underline{\mathcal{Y}}'$ be as in the hypothesis of the statement. Consider the case that there exist $y_1, \dots y_m$ such that $y = y_1@\dots@y_m$. The proof that for each i $\mathcal{Y}'_i = \{z \mid 1 \le z \le m, \delta(q_1, y_z) = q_i\}$ is by induction on $|y|$. Let $y = xa, a \in \Sigma, x \in \Sigma^*$. If $x = \epsilon$ then $m = 1$, $y_1 = a$. Let k be such that $\delta(q_1, a) = q_K$. Then $Y'_i = \emptyset$ for $i \ne k$, $\mathcal{Y}'_K = \{1\}$ (by Def. 8 with $h = 1, r = 1$). If $|x| > 0$, then there exist $x_1, \dots x_m$ such $x = x_1@\dots@x_m$, with each x_i simply being the prefix of y_i with length $|y| - 1$. Hence, there exists r in $\{1 \dots m\}$ such that $y_i = x_i\underline{a}$ for $i \ne r$, $y_r = x_r a$. By induction hypothesis there exists a m-vector $\underline{\mathcal{Y}}''$ such that $\underline{\mathcal{Y}} --\rightarrow \underline{\mathcal{Y}}''$ with $Y''_i = \{z \mid 1 \le z \le m, \delta(q_1, x_z) = q_i\}$. Let h, k be such that $r \in \mathcal{Y}_h, \delta(q_h, a) = q_k$. By applying Def. 8, the thesis follows: for each z, i with $z \ne r$, one has $z \in \mathcal{Y}'_i$ if, and only if, there exists $j: z \in \mathcal{Y}''_j$, and $\delta(q_j, \underline{a}) = q_i$. But $y_z = x_z\underline{a}$, hence $\delta(q_1, y_z) = q_i$. Similarly for the case $z = r: r \in \mathcal{Y}'_k, y_r = x_r a, \delta(q_1, y_r) = q_k$.

Consider the converse case $\mathcal{Y}'_i = \{z \mid 1 \le z \le m, \delta(q_1, y_z) = q_i\}$. The proof that $\exists y_1, \dots, y_m$ such that $y = y_1@\dots@y_m$ is also by induction on $|y|$. Let $y = xa$, $a \in \Sigma, ax \in \Sigma^*$. If $x = \epsilon$, then $m = 1$, $y_1 = a$. Then there exists k such that $\delta(q_1, a) = q_K, Y'_i = \emptyset$ for $i \ne k, \mathcal{Y}'_K = \{1\}$. Hence, $y = y_1$. For $|x| > 0$, let $\underline{Y}''$ be a m-vector such that $\underline{\mathcal{Y}} \overset{x}{--\rightarrow} \underline{\mathcal{Y}}'' \overset{a}{--\rightarrow} \underline{\mathcal{Y}}'$. By induction hypothesis there exist $x_1, \dots, x_m$, such that $\mathcal{Y}''_i = \{z \mid 1 \le z \le m, \delta(q_1, x_z) = q_i\}$ and $x = x_1@\dots@x_m$. Therefore, there exist r, h, k such that $\delta(q_1, x_r) = q_h, \delta(q_1, x_r a) = q_k$. By Def. 8, $r \in \mathcal{Y}'_k$. Consider now $z \ne r$. By Def. 8, $z \in \mathcal{Y}'_i$ if, and only if, there exists j such that $z \in \mathcal{Y}''_j$ and $\delta(q_j, \underline{a}) = q_i$. But $z \in \mathcal{Y}''_j$ if, and only if, $\delta(q_1, x_z) = q_j$. Hence, $z \in \mathcal{Y}'_i$ if, and only if, $\delta(q_1, x_z\underline{a}) = q_i$. Then $y_r = x_r a, y_z = x_z\underline{a}$ for $z \ne r$ are such that $\mathcal{Y}'_i = \{z \mid 1 \le z \le m, \delta(q_1, y_z) = q_i\}$. By induction hypothesis, $x = x_1@\dots@x_m$, and hence $y = xa = y_1@\dots@y_m$.

From Lemma 3 and Lemma 2, the equivalence with consensual languages follows almost immediately:

Proposition 5. *Let $R \subseteq (\Sigma \cup \underline{\Sigma})^*$ be a regular language, and A be its DFA. Then $L^@(A) = \mathcal{C}(R)$*

Corollary 1. *The word problem for the family C_{REG} is in the time complexity class P.*

5   Further Results

In order to compare consensual languages with some classical language families, we show that certain languages exceed the capacity of consensual languages based on regular sets.

Proposition 6. *The languages $\{ ucu^R \mid u \in \{a,b\}^* \}$ and $\{ ucu \mid u \in \{a,b\}^* \}$ are not in the family C_{REG}.*

Proof. Let $L = \{ ucu \mid u \in \{a,b\}^* \}$. The proof for $\{ ucu^R \mid u \in \{a,b\}^* \}$ is completely analogous. Assume by contradiction there is a DFA $A = (\{a,b,\underline{a},\underline{b}\}, Q, \delta, q_1, F)$ such that $L^@(A) = L$. Given in input a string of length $n > 0$, by Lemma 1, every vector $\mathbf{V}$ appearing in a configuration of $A^@$ is such that $\|\mathbf{V}\| \leq n + 1$. Hence, there are only $(n+1)^{|Q|}$ different vectors. However, there are 2^n different strings in $\{a,b\}^n$. For n large enough, the number of possible strings is much larger than the number of different vectors: there exist $u, w \in \{a,b\}^n$, $u \neq w$, such that there are initial vectors $\mathbf{V}^1, \mathbf{V}^2$, final vectors $\mathbf{V}', \mathbf{V}''$ and a vector $\mathbf{V}$: $\mathbf{V}^1 \overset{u}{\rightrightarrows} \mathbf{V} \overset{cu}{\rightrightarrows} \mathbf{V}'$, $\mathbf{V}^2 \overset{w}{\rightrightarrows} \mathbf{V} \overset{cw}{\rightrightarrows} \mathbf{V}''$. But then also $\mathbf{V}^1 \overset{u}{\rightrightarrows} \mathbf{V} \overset{cw}{\rightrightarrows} \mathbf{V}''$, a contradiction since $ucw \notin L$. $\square$

From Prop. 6, and from Ex. 2 it follows:

Corollary 2

1. *The family C_{REG} is not comparable with the families of context-free languages and of tree adjoining languages [3];*
2. *The commutative (or Parikh) image of a language in C_{REG} may be not semilinear.*

Among the typical context-free languages, the Dyck sets (see e.g. [5]) with two or more pairs of parentheses trespass the family C_{REG}. To see it it suffices to observe that the proof of Prop. 6 also applies to the case of a language

$$L = \{ uc\,h(u^R) \mid u \in \{a,b\}^* \} \text{ where } h \text{ is the morphism } h(a) = a', h(b) = b'$$

Let D_2 be the Dyck language with opening parentheses a, b and closing parentheses a', b' respectively. Let R be the regular language composed of all strings on $\{a, b, a', b'\}$ where there is no occurrence of the factors $a'a, b'b, a'b, b'a$. Hence, $D_2 \cap R = L$, and if D_2 were in C_{REG} then by closure of C_{REG} under intersection with regular languages also L would be in C_{REG}.

6   Conclusion

Our simple notion of consensus by strong matching is admittedly not the only one possible and sensible, yet it permits rather remarkable selectivity. For instance two variants would be to allow (or to oblige) a finite number k of component strings to place each character in each position of the match. Such variants would then model systems where stronger consensus between independent computations is possible (or is requested), in order for a string to be accepted. We believe our definitions though possibly the simplest, already capture a rather reach range of language paradigms.

Since the formalism is new, the research is in its early stages. There remain many problems open for investigation, for instance concerning minimality and decidability of equivalence or of ambiguity, as well as closure properties.

We also hope the consensual approach could be fruitfully studied for different families of base languages.

References

1. Chandra, A.K., Kozen, D., Stockmeyer, L.J.: Alternation. Journal of ACM 28, 114–133 (1981)
2. Jantzen, M.: On the hierarchy of Petri net languages. R.A.I.R.O. Informatique théorique/ Theoretical Informatics 13(1), 19–30 (1979)
3. Joshi, A., Schabes, Y.: Tree-adjoining grammars. In: Rozenberg, G., Salomaa, A. (eds.) Handbook of Formal Languages, vol. 3, pp. 69–124. Springer, Berlin (1997)
4. Kozen, D.: Theory of Computation. Springer, New York (2006)
5. Salomaa, A.: Theory of Automata. Pergamon Press, Oxford (1969)
6. Vijay-Shanker, K., Weir, D.J.: The equivalence of four extensions of context-free grammars. Mathematical Systems Theory 27(6), 511–546 (1994)

Appendix

The third language of Ex. 2.

Lemma 4. *Let* $\Sigma = \{a, b, c\}$ *and let*

$$R_3 = \underbrace{\Sigma^* a(\underline{a} \cup \underline{c})^* \underline{b}\,(\underline{a} \cup \underline{c})^* \, c\underline{a}c\Sigma^*}_{1} \cup \underbrace{\underline{a}cb((\underline{a}\underline{c})^+ b)^*(\underline{a}\underline{c})^+ b}_{2} \cup \underbrace{\underline{a}cb}_{3}$$

i.e., R_3 *is the union of three expressions, numbered 1, 2 and 3. We claim that the consensual language* $\mathcal{C}(R_3)$ *is*

$$L_3 = \{acb(ac)^2 b(ac)^4 b(ac)^8 b \dots (ac)^{2^m} b \mid m \geq 0\}$$

Proof. We show that $L_3 \subseteq \mathcal{C}(R_3)$. Any string in $\mathcal{C}(R_3)$, apart from acb, must be obtained by matching a string in Expression 2 with strings in Expression 1, since neither regular expression can generate alone a string in Σ^+. We now show, by induction on the number m for strings of the form $w_m = acb(ac)^2 b(ac)^4 b(ac)^8 b \dots (ac)^{2^m}$ (which exhaust all language L_3), that if $w_m \in L_3$ then $w_m \in \mathcal{C}(R_3)$. The base step is $m = 0$, corresponding to the case acb, both in $\mathcal{C}(R_3)$ and in L_3. Assume now that the induction hypothesis holds for $m - 1$. Hence, string $w_{m-1} = acb(ac)^2 b(ac)^4 b(ac)^8 b \dots (ac)^{2^{m-1}} b \in \mathcal{C}(R_3)$. But w_{m-1} must be obtained as a match of $h > 0$ strings $x_1, x_2, \dots x_h$ of Expression 1, with one string $y_{m-1} = \underline{a}cb(\underline{a}\underline{c})^2 b \dots (\underline{a}\underline{c})^{2^{m-2}} b(\underline{a}\underline{c})^{2^{m-1}} b$ of Expression 2. But also

$$y_m = \underline{a}cb(\underline{a}\underline{c})^2 b \dots \underline{b}(\underline{a}\underline{c})^{2^{m-1}} (\underline{a}\underline{c})^{2^m} b$$

is in Expression 2, and if x_i is in Expression 1 then also $x_i' = x_i(\underline{ac})^{2^m}\underline{b}$ is in Expression 1, since Expression 1 ends with $\underline{\Sigma}^*$. Therefore, by matching y_m with $x_1', \ldots x_m'$, one obtains:

$$w_m' = acb(ac)^2 b(ac)^4 b(ac)^8 b \ldots b(ac)^{2^{m-2}} b(\underline{ac})^{2^{m-1}} b(a\underline{c})^{2^m} b \in R_3^{\textcircled{a}}.$$

The strings

$$z_i = \underline{acb}(\underline{ac})^2\underline{b} \ldots \underline{b}(\underline{ac})^{2^{m-2}}\underline{b}(\underline{ac})^i a\underline{c}(\underline{ac})^{2^{m-1}-i-1}\underline{b}(\underline{ac})^{2i}a\underline{c}a\underline{c}(\underline{ac})^{2^m-2i-2}\underline{b}$$

are in Expression 1, for every i, $1 \leq i \leq 2^{m-1}$. Hence, for every a placed in group $m-1$ (the group with 2^{m-1} occurrences of ac), there must be two occurrences of c in group m. Hence, the number of c's (and therefore also of a's) in group m must be twice the number of c's in group $m-1$:

$$w_m = acb(ac)^2 b(ac)^4 b(ac)^8 b \ldots (ac)^{2^{m-1}} b(ac)^{2^m} b \in \mathcal{C}(R_3).$$

For the converse case, notice first that, since any w in $\mathcal{C}(R_3)$ must match a string in Expression 2, $\mathcal{C}(R_3) \subseteq acb((ac)^+ b)^*$. An induction on the number $m \geq 0$ of groups $b(ac)^+ b$ in strings of $\mathcal{C}(R_3)$ completes the proof. The base case is obvious (corresponding to the string acb). By induction hypothesis, all strings with $m-1 \geq 0$ groups are in L_3. Assume that there is a string x_m with m groups but which is not in L_3. Hence, there exists $i > 0$ such that the group in position i has a number of c which is not the double of the number of a of the group in position $i-1$. But $i = m$, otherwise one could define also a string which is not in L_3 while having less than m groups, contradicting the induction hypothesis. However, the only way to place a c in group m is by using strings in Expression 1, which place two occurrences of c in group m for an occurrence of a in group m-1. Since no other strings can place a in group $m-1$, then the number of c's in group m must be exactly the double of the number of a in group $m-1$ (2^{m-1} by induction hypothesis), that is there are 2^m occurrences of c in group m.

k-Petri Net Controlled Grammars

Jürgen Dassow[1] and Sherzod Turaev[2]

[1] Otto-von-Guericke-Universität Magdeburg
PSF 4120, D-39016 Magdeburg, Germany
dassow@iws.cs.uni-magdeburg.de
[2] GRLMC, Universitat Rovira i Virgili
Pl. Imperial Tàrraco 1, 43005 Tarragona, Spain
sherzod.turaev@estudiants.urv.cat

Abstract. We introduce a new type of regulated grammars using Petri nets as a control of the derivations of a context-free grammar. We investigate the generative power and give closure properties of the families of languages generated by such Petri net controlled grammars.

1 Introduction

A context-free grammar and its derivation process can be described by a Petri net, where places correspond to nonterminals and terminals, transitions are the counterpart of the productions, and the tokens reflect the occurrences of symbols in the sentential form, and there is a one-to-one correspondence between the application of (sequences of) rules and the firing of a (sequence of) transitions. Therefore it is a very natural and very easy idea to control the derivations in a context-free grammar by adding some features to the associated Petri net.

In [2] and [9] it has been shown that by adding some places and arcs which satisfy some structural requirements one can generate well-known families of languages as random context languages, vector languages and matrix languages. Thus the control by Petri nets can be considered as a unifying approach to different types of control (note that random context is a control by occurrence/non-occurrence of letters whereas matrices give a prescribed set of sequences in which the productions have to be applied).

In this paper we add additional places, called *counters*, and additional arcs associated with the new places. Adding k places leads to a control by k-Petri nets. The aim of this paper is the study of properties of the family of languages which can be generated by context-free grammars with a control by k-Petri nets. We present results on the generative power and we give some closure properties.

The paper is organized as follows. In Section 2 we give some notions and definitions from the theories of formal languages and Petri nets needed in the sequel. Moreover we introduce the Petri net associated with a context-free grammar.

In Section 3 we construct the new Petri net control mechanism and define the corresponding grammars. Furthermore, we give some examples. In Section 4 we show that context-free grammars with the simple control by one additional place can generate non-context-free languages. We also give relations to valence

C. Martín-Vide, F. Otto, and H. Fernau (Eds.): LATA 2008, LNCS 5196, pp. 209–220, 2008.
© Springer-Verlag Berlin Heidelberg 2008

grammars and vector grammars. Furthermore, we show that we get an infinite hierarchy with respect to the numbers of additional places. In Section 5 we investigate the fundamental closure properties of the families of languages generated by k-Petri net controlled grammars.

2 Preliminaries

The reader is assumed to be familiar with basic notions of formal language theory and Petri net theory as, e.g. contained in [4,5,7,8].

2.1 Grammars

Let $\Sigma = \{a_1, a_2, \ldots, a_k\}$ be an alphabet. A *string* over Σ is a sequence of symbols from the alphabet. The *length* of a string w is denoted by $|w|$, and the number of a symbol a in a string w by $|w|_a$. The *empty* string is denoted by λ which is of length 0. The number of the occurrences of all symbols of $U \subseteq \Sigma$ in a string w is denoted by $|w|_U$. The set of all strings over the alphabet Σ is denoted by Σ^*. A subset L of Σ^* is called a *language*. For languages $L, L' \subseteq \Sigma^*$, the operation *shuffle* is defined by $\text{Shuf}(L, L') = \{x_1 y_1 \cdots x_n y_n \mid x_1 \cdots x_n \in L, y_1 \cdots y_n \in L', x_i, y_i \in \Sigma^*, 1 \leq i \leq n\}$ and for $L \subseteq \Sigma^*$: $\text{Shuf}^1(L) = L$; $\text{Shuf}^i(L) = \text{Shuf}(\text{Shuf}^{i-1}(L), L)$ for $i \geq 2$; $\text{Shuf}^*(L) = \cup_{i \geq 1} \text{Shuf}^i(L)$.

A *context-free grammar* is a quadruple $G = (V, \Sigma, S, R)$ where V and Σ are the finite sets of *nonterminal* and *terminal* symbols, respectively, $S \in V$ is the *start* symbol and $R \in V \times (V \cup \Sigma)^*$ is the set of *(production) rules*. Usually, a rule (A, x) is written as $A \to x$. A rule of the form $A \to \lambda$ is called an *erasing rule*. $x \in (V \cup \Sigma)^+$ *directly derives* $y \in (V \cup \Sigma)^*$, written as $x \Rightarrow y$, iff there is a rule $r = A \to \alpha \in R$ such that $x = x_1 A x_2$ and $y = x_1 \alpha x_2$. The reflexive and transitive closure of $\Rightarrow$ is denoted by $\Rightarrow^*$. A derivation using the sequence of rules $\pi = r_1 r_2 \cdots r_n$ is denoted by $\overset{\pi}{\Rightarrow}$ or $\xrightarrow{r_1 r_2 \cdots r_n}$. The *language* generated by G is defined by $L(G) = \{w \in \Sigma^* \mid S \Rightarrow^* w\}$.

A *vector grammar* is a quadruple $G = (V, \Sigma, S, M)$ where V, Σ, S are defined as for context-free grammars, and M is a finite set of strings over a set R of context-free rules called *matrices*. The language generated by G is $L(G) = \{w \in \Sigma^* \mid S \overset{\pi}{\Rightarrow} w$ and $\pi \in \text{Shuf}^*(M)\}$.

An *additive valence grammar* is a quintuple $G = (V, \Sigma, S, R, v)$ where V, Σ, S, R are defined as for context-free grammars and v is a mapping from R into the set $\mathbb{Z}$ of integers. The language generated by G consists of all strings $w \in \Sigma^*$ such that there is a derivation $S \xrightarrow{r_1 r_2 \cdots r_k} w$ such that $\sum_{i=1}^{k} v(r_i) = 0$.

A *positive valence grammar* is a quintuple $G = (V, \Sigma, S, R, v)$ whose components are defined as for additive valence grammars. The language generated by G consists of all strings $w \in \Sigma^*$ such that there is a derivation $S \xrightarrow{r_1 r_2 \cdots r_k} w$ such that $\sum_{i=1}^{k} v(r_i) = 0$ and for any $1 \leq j < k$, $\sum_{i=1}^{j} v(r_i) \geq 0$.

The families of languages generated by context-free, vector, additive valence and positive valence grammars (with erasing rules) are denoted by **CF**, **V**, **aV** and **pV** (**CF**$^\lambda$, **V**$^\lambda$, **aV**$^\lambda$ and **pV**$^\lambda$), respectively.

2.2   Petri Nets

A *Petri net* is a construct $N = (P, T, F, \varphi)$ where P and T are disjoint finite sets of *places* and *transitions*, respectively, $F \subseteq (P \times T) \cup (T \times P)$ is the set of *directed arcs*, $\varphi : F \to \{1, 2, \dots\}$ is a *weight function*.

A Petri net can be represented by a bipartite directed graph with the node set $P \cup T$ where places are drawn as *circles*, transitions as *boxes* and arcs as *arrows* with labels $\varphi(p, t)$ or $\varphi(t, p)$. If $\varphi(p, t) = 1$ or $\varphi(t, p) = 1$, the label is omitted.

A mapping $\mu : P \to \{0, 1, 2, \dots\}$ is called a *marking*. For each place $p \in P$, $\mu(p)$ gives the number of *tokens* in p. Graphically, tokens are drawn as small solid *dots* inside circles. $^\bullet x = \{y \mid (y, x) \in F\}$ and $x^\bullet = \{y \mid (x, y) \in F\}$ are called *pre-* and *post-sets* of $x \in P \cup T$, respectively.

A transition $t \in T$ is *enabled* by marking μ iff $\mu(p) \geq \varphi(p, t)$ for all $p \in P$. In this case t can *occur* (*fire*). Its occurrence transforms the marking μ into the marking μ' defined for each place $p \in P$ by $\mu'(p) = \mu(p) - \varphi(p, t) + \varphi(t, p)$. A finite sequence $t_1, t_2, \dots, t_k$ of transitions is called *an occurrence sequence* enabled at a marking μ if there are markings $\mu_1, \mu_2, \dots, \mu_k$ such that $\mu \xrightarrow{t_1} \mu_1 \xrightarrow{t_2} \dots \xrightarrow{t_k} \mu_k$. In short this sequence can be written as $\mu \xrightarrow{t_1 t_2 \cdots t_k} \mu_k$ or $\mu \xrightarrow{\nu} \mu_k$ where $\nu = t_1 t_2 \cdots t_k$.

A *marked* Petri net is a system $N = (P, T, F, \varphi, \iota)$ where (P, T, F, φ) is a Petri net, ι is the *initial marking*. We also define the *final marking* τ at which we stop the execution of N under some conditions. An occurrence sequence ν of transitions is called *successful* if it is enabled at the initial marking ι and finished at a final marking τ.

2.3   cf Petri Nets

The construction of the following type of Petri nets is based on the idea of using similarity between the firing of a transition and the application of a production rule in a derivation in which places are terminal and nonterminal symbols and tokens are separate occurrences of symbols.

Definition 1. *Let $G = (V, \Sigma, S, R)$ be a context-free grammar. Let $\beta : P \to (V \cup \Sigma)$ and $\gamma : T \to R$ be bijections. The cf Petri net corresponding to the grammar G is a marked Petri net $N = (P, T, F, \varphi, \iota)$ where*

- *each place $p \in P$ is labeled by the corresponding symbol $x = \beta(p) \in V \cup \Sigma$ and each transition $t \in T$ is labeled by the corresponding rule $r = \gamma(t) \in R$;*
- *$(p, t) \in F$ iff $\gamma(t) = r$, $r : A \to w \in R$ where $\beta(p) = A$ and $\varphi(p, t) = 1$;*
- *$(t, p) \in F$ iff $\gamma(t) = r$, $r : A \to w \in R$ where $\beta(p) = x$, $|w|_x > 0$ and $\varphi(t, p) = |w|_x$;*
- *$\iota(\beta^{-1}(S)) = 1$ and $\iota(\beta^{-1}(x)) = 0$ for all $x \in (V \cup \Sigma) \setminus \{S\}$.*

Example 2. Let G_1 be a context-free grammar with the following rules: $r_0 : S \to AB$, $r_1 : A \to aAb$, $r_2 : A \to ab$, $r_3 : B \to cB$, $r_4 : B \to c$ (the other components of the grammar can be seen from these rules). Figure 1 illustrates the corresponding cf Petri net N. Obviously, $L(G_1) = \{a^n b^n c^m \mid n \geq 1, m \geq 1\}$.

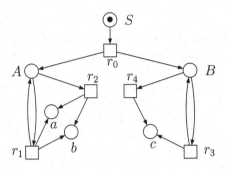

Fig. 1. A cf Petri net N

The following proposition, which directly follows from the definition, shows the similarity between terminal derivations in a context-free grammar and successful occurrences in the corresponding cf Petri net.

Proposition 3. *Let $N = (P, T, F, \varphi, \iota)$ be the cf Petri net corresponding to a context-free grammar $G = (V, \Sigma, S, R)$. Then $S \xrightarrow{r_1 r_2 \cdots r_n} w \in \Sigma^*$ is a derivation in G iff $\iota \xrightarrow{t_1 t_2 \cdots t_n} \tau$ is an occurrence sequence of transitions in N such that $t_i = \gamma^{-1}(r_i), 1 \leq i \leq n$, and $\tau(\beta^{-1}(x)) = 0$ for all $x \in V$.*

3   Petri Net Controlled Grammars

Now we consider cf Petri nets with $k \geq 1$ additional places, pre-sets and post-sets of which are disjoint, respectively. Let $N = (P, T, F, \varphi, \iota)$ be the corresponding cf Petri net corresponding to a context-free grammar $G = (V, \Sigma, S, R)$. For $k \geq 1$ we set $T_1^k = \{t_{11}^k, t_{12}^k, \ldots, t_{1m(1)}^k\} \subset T$ and $T_2^k = \{t_{21}^k, t_{22}^k, \ldots, t_{2m(2)}^k\} \subset T$ where $T_i^{k_1} \cap T_j^{k_2} = \emptyset$ for $1 \leq i < j \leq 2$ or $1 \leq k_1 < k_2 \leq k$ and $T_1^i = \emptyset$ iff $T_2^i = \emptyset$ for any $1 \leq i \leq k$.

Let $Q = \{q_1, q_2, \ldots, q_k\}$ be the set of new places called *counters*. Let $F_1 = \{(t, q_i) \mid t \in T_1^i, 1 \leq i \leq k\}$ and $F_2 = \{(q_i, t) \mid t \in T_2^i, 1 \leq i \leq k\}$.

Definition 4. *A k-Petri net (in short k-PN) with respect to the grammar G is a system $N_k = (P', T', F', \varphi', \iota', M')$ where*

- *$P' = P \cup Q$, $T' = T$ and $F' = F \cup F_1 \cup F_2$,*
- *$\varphi'(x, y) = \varphi(x, y)$ if $(x, y) \in F$ and $\varphi'(x, y) = 1$ if $(x, y) \in F_1 \cup F_2$,*
- *$\iota'(\beta^{-1}(S)) = 1$ and $\iota'(p) = 0$ for all $p \in P' \setminus \{\beta^{-1}(S)\}$,*
- *M' is the set of final markings and for any $\tau' \in M'$, $\tau'(p) = 0$ for all $p \in P' \setminus \{\beta^{-1}(x) \mid x \in \Sigma\}$.*

Definition 5. *Let $k \geq 1$. A k-PN controlled grammar is a quintuple $G = (V, \Sigma, S, R, N_k)$ where V, Σ, S, R are defined as for a context-free grammar and N_k is the k-PN with respect to the context-free grammar (V, Σ, S, R).*

Definition 6. *The language generated by a k-PN controlled grammar G, denoted by $L(G)$, consists of all strings $w \in \Sigma^*$ such that there is a derivation $S \overset{\pi}{\Rightarrow} w$, $\pi = r_1 r_2 \cdots r_n$ where $\nu = t_1 t_2 \cdots t_n$, $t_i = \gamma^{-1}(r_i) \in T$, $1 \leq i \leq n$ is an occurrence sequence of the transitions of N_k enabled at the initial marking ι and finished at a final marking τ.*

We denote the family of languages generated by k-PN controlled grammars (without erasing rules) by $\mathbf{PN}_k^\lambda$ ($\mathbf{PN_k}$), $k \geq 1$.

We give two examples which will be used in the sequel.

Example 7. Figure 2 illustrates a 1-PN N_1 which is constructed from the cf Petri net N in Figure 1 adding a single counter place q. Let $G_2 = (V, \Sigma, S, R, N_1)$ be the 1-PN controlled grammar where V, Σ, S, R are defined as for the grammar G_1 in Example 2. It is easy to see that $L(G_2) = \{a^n b^n c^n \mid n \geq 1\}$.

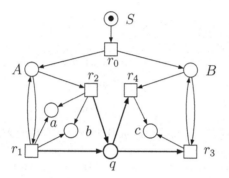

Fig. 2. A 1-PN N_1

Example 8. Let G_3 be a 2-PN controlled grammar with the production rules:
$r_0 : S \to A_1 B_1 A_2 B_2$, $r_1 : A_1 \to a_1 A_1 b_1$, $r_2 : A_1 \to a_1 b_1$, $r_3 : B_1 \to c_1 B_1$, $r_4 : B_1 \to c_1$, $r_5 : A_2 \to a_2 A_2 b_2$, $r_6 : A_2 \to a_2 b_2$, $r_7 : B_2 \to c_2 B_2$, $r_8 : B_2 \to c_2$. and the corresponding 2-PN N_2 is given in Figure 3. Then it is easy to see that G_3 generates the language $L(G_3) = \{a_1^n b_1^n c_1^n a_2^m b_2^m c_2^m \mid n, m \geq 1\}$.

Lemma 9. *The language $L' = \{a_1^n b_1^n c_1^n a_2^m b_2^m c_2^m \mid n, m \geq 1\}$ cannot be generated by a 1-PN controlled grammar.*

Proof. Suppose the contrary: there is a 1-PN controlled grammar $G = (V, \Sigma, S, R, N_1)$ where $\Sigma = \{a_1, b_1, c_1, a_2, b_2, c_2\}$ such that $L(G) = L'$. Since the set V is finite, for large n and m each derivation $S \Rightarrow^* w \in L'$ contains a subderivation of the form D: $A \Rightarrow^* xAy$ where $A \in V$ and $x, y \in \Sigma^*$. As L' is infinite, there are words with large enough length obtained by iterating such a derivation D arbitrarily many times. Suppose

$$S \Rightarrow^* uAv \Rightarrow^* uxAyv \Rightarrow^* \cdots \Rightarrow^* ux^n Ay^n v \Rightarrow^* w' \in \Sigma^* \qquad (1)$$

is also a derivation in G. Then x^n and y^n are subwords of w'. It follows that x and y can be only powers of two symbols from $\Sigma \cup \{\lambda\}$. Therefore, in order to

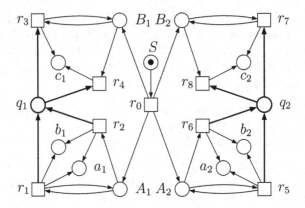

Fig. 3. A 2-PN N_2

generate a word $w = a_1^n b_1^n c_1^n a_2^m b_2^m c_2^m \in L'$ for large n and m, we need at least three subderivations of the form

$$D_1 : A_1 \Rightarrow^* x_1 A_1 y_1, \tag{2}$$

$$D_2 : A_2 \Rightarrow^* x_2 A_2 y_2, \tag{3}$$

$$D_3 : A_3 \Rightarrow^* x_3 A_3 y_3 \tag{4}$$

where $x_1, x_2, x_3, y_1, y_2, y_3$ are powers of the symbols from Σ, i.e. $x_i = \alpha_i^{k_i}$ and $y_i = \beta_i^{l_i}$ where $\alpha_i, \beta_i \in \Sigma$ and $k_i + l_i \geq 1, i = 1, 2, 3$.

Obviously, any of the subderivations (2)–(4) produces additional tokens at the counter or consumes tokens from the counter or does not change the number of tokens at the counter.

According to the production and consumption of tokens by the subderivations (2)–(4) we distinguish some cases and show a contradiction in all cases. By reasons of space we discuss here only two cases (the other ones can be handled analogously).

Case I. One of the derivations (2)–(4) does not produce and consume any token at the counter. Without loss of generality we can assume that this derivation is (2). If $S \Rightarrow^* u A_1 v \Rightarrow^* u w v \in L'$ then for any $k > 1$ we apply (2) k times and get a word which is not in L', i.e.,

$$S \Rightarrow^* u A_1 v \Rightarrow^* u x_1 A_1 y_1 v \Rightarrow^* u x_1^2 A_1 y_1^2 v \Rightarrow^* u x_1^k A_1 y_1^k v \Rightarrow^* u x_1^k w y_1^k v \notin L'$$

since (2) increases only the powers of at most two symbols.

Case II. The subderivation (2) produces $p \geq 1$ tokens, the subderivation (3) consumes $q \geq 1$ tokens, $\{\alpha_1, \beta_1, \alpha_2, \beta_2\} \cap \{a_i, b_i, c_i\} \neq \emptyset$ for $1 \leq i \leq 2$, and there is a derivation $S \Rightarrow^* u_1 A_1 u_2 A_2 u_3 \Rightarrow^* u_1 w_1 u_2 w_2 u_3 \in L'$. Then the derivation

$$S \Rightarrow^* u_1 A_1 u_2 A_2 u_3$$

$$\Rightarrow^* u_1 x_1 A_1 y_1 u_2 A_2 u_3 \Rightarrow^* u_1 x_1^k A_1 y_1^k u_2 A_2 u_3$$

$$\Rightarrow^* u_1 x_1^k A_1 y_1^k u_2 x_2 A_2 y_2 u_3 \Rightarrow^* u_1 x_1^k A_1 y_1^k u_2 x_2^l A_2 y_2^l u_3$$

$$\Rightarrow^* u_1 x_1^k w_1 y_1^k u_2 x_2^l w_2 y_2^l u_3$$

where $k, l \geq 1$, should be in G. It can be done by choosing the numbers k, l and the symbols $\alpha_1, \beta_1, \alpha_2, \beta_2 \in \Sigma \cup \{\lambda\}$.

If we choose $k = q$ and $l = p$, then we also come to the final marking $(kp - lq = 0)$. By the supposition $1 \leq |\{\alpha_1, \beta_1, \alpha_2, \beta_2\} \cap \{a_i, b_i, c_i\}| \leq 2$ for at least one i, $1 \leq i \leq 2$. Thus at most two letters from $\{a_i, b_i, c_i\}$ are changed in there power.

4 Hierarchy Results

We start with a simple fact.

Lemma 10. CF $\subsetneq$ PN$_1$.

Proof. It is clear that $\mathbf{CF} \subseteq \mathbf{PN_1}$ if we take $T_1 = T_2 = \emptyset$. From Example 7 it follows that $\mathbf{CF} \subsetneq \mathbf{PN_1}$. □

Now we present some relations to (positive) additive valence languages.

Lemma 11. PN$_1$ $\subseteq$ pV *and* **PN$_1^\lambda$ $\subseteq$ pV$^\lambda$.**

Proof. Let $G = (V, \Sigma, S, R, N_1)$ where $N_1 = (P, T, F, \varphi, \iota, M)$ is the corresponding 1-PN with the counter q (with the notions of Definition 4) be a 1-PN controlled grammar (with or without erasing rules). We define a positive valence grammar $G' = (V, \Sigma, S, R, v)$ where V, Σ, S, R are defined as for the grammar G and for each $r \in R$, the mapping v is defined by $v(r) = 1$ if $\gamma^{-1}(r) \in {}^\bullet q$, $v(r) = -1$ if $\gamma^{-1}(r) \in q^\bullet$ and $v(r) = 0$ otherwise.

Let $S \overset{\pi}{\Rightarrow} w \in \Sigma^*$, $\pi = r_1 r_2 \cdots r_k$ be a derivation in G. Then $\nu = t_1 t_2 \cdots t_k$, $t_i = \gamma^{-1}(r_i), 1 \leq i \leq k$ is an occurrence sequence of transitions in N_1 enabled at the initial marking ι and finished at a final marking τ. By definition if $|\nu|_t > 0$ for some $t \in {}^\bullet q$ then there is a transition $t' \in q^\bullet$ such that $|\nu|_{t'} > 0$. Let $U_1 = \{t_{11}, t_{12}, \ldots, t_{1k(1)}\} \subseteq {}^\bullet q$ where $|\nu|_{t_{1j}} > 0, 1 \leq j \leq k(1)$ and $U_2 = \{t_{21}, t_{22}, \ldots, t_{2k(2)}\} \subseteq q^\bullet$ where $|\nu|_{t_{2j}} > 0, 1 \leq j \leq k(2)$.

As $\mu_i(q) \geq 0$ for any occurrence step $1 \leq i \leq k$ we have $|\nu|_{U_1} \geq |\nu|_{U_2}$, consequently, $v(r_1) + v(r_2) + \ldots + v(r_j) \geq 0$ for any $1 \leq j < k$ and from $\iota(q) = \tau(q) = 0$, $\tau \in M$, it follows that $\sum_{t \in U_1} |\nu|_t - \sum_{t \in U_2} |\nu|_t \overset{\text{def}}{=} \sum_{i=1}^n v(r_i) = 0$. Hence, $L(G) \subseteq L(G')$.

Let $D : S \overset{\pi}{\Rightarrow} w \in \Sigma^*$, $\pi = r_1 r_2 \cdots r_k$ be a derivation in G' where $v(r_1) + v(r_2) + \ldots + v(r_k) = 0$ and $v(r_1) + v(r_2) + \ldots + v(r_j) \geq 0$ for any $1 \leq j < k$. By construction of G' D is a derivation in (V, Σ, S, R).

According to the bijection $\gamma : T \to R$, there is an occurrence sequence in N_1: $\mu_0 \overset{\nu}{\to} \mu_k$, $\nu = t_1 t_2 \cdots t_k$ such that $t_i = \gamma^{-1}(r_i), 1 \leq i \leq k$.

$\mu_0 = \iota$ as D starts from S, i.e. $\mu_0(\beta^{-1}(S)) = 1$ and $\mu_0(\beta^{-1}(x)) = 0$ for all $x \in (V \cup \Sigma) \setminus \{S\}$ as well as $\mu_0(q) = 0$.

Since $w \in \Sigma^*$ for the last marking, we have $\mu_k(\beta^{-1}(x)) = 0$ for all $x \in V$. From $\sum_{i=1}^j v(r_i) \geq 0$, it follows that $\mu_i(q) \geq 0$ for any $1 \leq j < k$. Also $\sum_{i=1}^k v(r_i) \overset{\text{def}}{=} \sum_{\gamma^{-1}(r) \in {}^\bullet q} v(r) + \sum_{\gamma^{-1}(r) \in q^\bullet} v(r) = 0$ shows that in k steps the counter place q has received and given the same number of tokens, i.e. $\mu_k(q) = 0$. Hence $\mu_k \in M$. Consequently, $L(G') \subseteq L(G)$. □

Lemma 12. $\mathbf{aV} \subsetneq \mathbf{PN_2}$ *and* $\mathbf{aV^\lambda} \subsetneq \mathbf{PN_2^\lambda}$.

Proof. Let $G = (V, \Sigma, S, R, v)$ be an additive valence grammar (with or without erasing rules). Without loss of generality we can assume that $v(r) \in \{1, 0, -1\}$ for each $r \in R$ (theorem 2.1.10 in [4]).

For each rule $r : A \rightarrow \alpha \in R$, $v(r) \neq 0$ we add a nonterminal symbol A_r and a pair of rules $r' : A \rightarrow A_r$, $r'' : A_r \rightarrow \alpha$ and set $V' = V \cup \{A_r \mid r : A \rightarrow \alpha \in R, v(r) \neq 0\}$, $R' = R \cup \{r' : A \rightarrow A_r, r'' : A_r \rightarrow \alpha \mid r : A \rightarrow \alpha \in R, v(r) \neq 0\}$.

Let $N = (P, T, F, \varphi, \iota)$ be the cf Petri net corresponding to (V', Σ, S, R'). We construct 2-PN $N'_2 = (P', T', F', \varphi', \iota, M)$ by adding the counter places q, q' and the arcs: (t, q), $t = \gamma^{-1}(r)$ for each $r \in R$, $v(r) = 1$ and (q, t), $t = \gamma^{-1}(r)$ for each $r \in R$, $v(r) = -1$, (t', q'), $t' = \gamma^{-1}(r')$ for each $r \in R$, $v(r) = -1$, (q', t'), $t' = \gamma^{-1}(r')$ for each $r \in R$, $v(r) = 1$. The value of the weight function φ' for each new arc is 1. Consider 2-PN controlled grammar $G' = (V', \Sigma, S, R', N'_2)$.

$L(G') \subseteq L(G)$ is obvious: according to the definitions of the initial and final markings for the counter places q and q', i.e. $\iota(q) = \iota(q') = 0$ and for any $\tau \in M$, $\tau(q) = \tau(q') = 0$, we have $\sum_{i=1}^{n} v(r_i) = 0$, for any derivation $S \xrightarrow{r_1 r_2 \cdots r_n} w \in \Sigma^*$.

Let $S \xrightarrow{r_1 r_2 \cdots r_n} w \in \Sigma^*$ be a derivation in G. For any $1 \leq k \leq n$

(1) if $\sum_{i=1}^{k} v(r_i) > 0$ then for the rule r_{k+1}, $v(r_{k+1}) \in \{1, 0, -1\}$ in G, we also choose the rule r_{k+1} in G';

(2) if $\sum_{i=1}^{k} v(r_i) \leq 0$ then (a) for the rule r_{k+1}, $v(r_{k+1}) \neq 0$ in G, we choose the rules r'_{k+1} and r''_{k+1} in G'; (b) for the rule r_{k+1}, $v(r_{k+1}) = 0$ in G then we also choose r_{k+1}. Hence, $w \in L(G')$.

The strict inclusion follows from the fact that $\{a_1^n b_1^n c_1^n a_2^m b_2^m c_2^m \mid n, m \geq 1\} \in \mathbf{PN_2}$ cannot be generated by an additive valence grammar (Example 2.1.7 in [4]). □

The following lemma shows that, for any $n \geq 1$, an n-PN controlled grammar generates a vector language.

Lemma 13. *For any* $n \geq 1$, $\mathbf{PN_n} \subseteq \mathbf{V}$ *and* $\mathbf{PN_n^\lambda} \subseteq \mathbf{V^\lambda}$.

Theorem 14. *For* $k \geq 1$, $\mathbf{PN_k} \subsetneq \mathbf{PN_{k+1}}$ *and* $\mathbf{PN_k^\lambda} \subsetneq \mathbf{PN_{k+1}^\lambda}$.

Proof. We first prove that $\mathbf{PN_1} \subseteq \mathbf{PN_2}$.

Let $G = (V, \Sigma, S, R, N_1)$ be a 1-PN controlled grammar (with or without erasing rules) where $N_1 = (P, T, F, \varphi, \iota, M)$ 1-PN with the counter place q. Let $^\bullet q = \{t_{11}, t_{12}, \ldots, t_{1k(1)}\}, k(1) \geq 1$ and $q^\bullet = \{t_{21}, t_{22}, \ldots, t_{2k(2)}\}, k(2) \geq 1$ where $t_{ij} = \gamma^{-1}(r_{ij})$, $r_{ij} : A_{ij} \rightarrow w_{ij}$, $1 \leq i \leq 2$, $1 \leq j \leq k(i)$ and by definition $^\bullet q \cap q^\bullet = \emptyset$.

We set $V' = V \cup \{A'_{ij} \mid 1 \leq i \leq 2, 1 \leq j \leq k(i)\}$ where A'_{ij}, $1 \leq i \leq 2$, $1 \leq j \leq k(i)$ are new nonterminal symbols. For each rule $r_{ij} : A_{ij} \rightarrow w_{ij}$, $1 \leq i \leq 2$, $1 \leq j \leq k(i)$ we add the new rules $r'_{ij} : A_{ij} \rightarrow A'_{ij}$, $r''_{ij} : A'_{ij} \rightarrow w_{ij}$ and we consider the 2-PN controlled grammar $G' = (V', \Sigma, S, R', N'_2)$ where R' consists of all rules of R and all rules constructed above.

Let $p_{ij} = \beta^{-1}(A_{ij})$ where $A_{ij} \rightarrow w_{ij}, 1 \leq i \leq 2, 1 \leq j \leq k(i)$. We add the new places and transitions: $p'_{ij}, t'_{ij}, t''_{ij}$, $1 \leq i \leq 2$, $1 \leq j \leq k(i)$ and define *extended* bijections $\widehat{\beta}$ and $\widehat{\gamma}$ by $\widehat{\beta}(p) = \beta(p)$ if $p \in P$ and $\widehat{\beta}(p) = A'_{ij}$ if $p = p'_{ij}$, $1 \leq i \leq 2, 1 \leq j \leq k(i)$, and $\widehat{\gamma}(t) = \gamma(t)$ if $t \in T$, $\widehat{\gamma}(t) = r'_{ij}$ if $t = t'_{ij}$ and $\widehat{\gamma}(t) = r''_{ij}$ if $t = t''_{ij}$, $1 \leq i \leq 2, 1 \leq j \leq k(i)$. Let q' be a new counter place.

We construct 2-PN $N'_2 = (P', T', F', \varphi', \iota', M')$ corresponding to the grammar (V', Σ, S, R') where

$$P' = P \cup \{p'_{ij} \mid 1 \leq i \leq 2, 1 \leq j \leq k(i)\} \cup \{q'\},$$

$$T' = T \cup \{t'_{ij} \mid 1 \leq i \leq 2, 1 \leq j \leq k(i)\} \cup \{t''_{ij} \mid 1 \leq i \leq 2, 1 \leq j \leq k(i)\},$$

$$F' = F \cup \{(p_{ij}, t'_{ij}) \mid 1 \leq i \leq 2, 1 \leq j \leq k(i)\}$$
$$\cup \{(t'_{ij}, p'_{ij}) \mid 1 \leq i \leq 2, 1 \leq j \leq k(i)\}$$
$$\cup \{(p'_{i,j}, t''_{i,j}) \mid 1 \leq i \leq 2, 1 \leq j \leq s(i)\}$$
$$\cup \{(t''_{ij}, p) \mid p = \beta^{-1}(x), |w_{ij}|_x > 0, 1 \leq i \leq 2, 1 \leq j \leq k(i)\}$$
$$\cup \{(t''_{ij}, q') \mid 1 \leq j \leq k(1)\}$$
$$\cup \{(q', t''_{2j}) \mid 1 \leq j \leq k(2)\}.$$

For the weight function we set $\varphi'(x, y) = \varphi(x, y)$ if $(x, y) \in F$ and $\varphi'(x, y) = 1$, otherwise.

The initial and final markings are defined by $\iota'(\widehat{\beta}^{-1}(S)) = 1$, $\iota'(p) = 0$ for all $p \in P' \backslash \{\widehat{\beta}^{-1}(S)\}$ and for any $\tau' \in M'$, $\tau'(p) = 0$ for all $p \in P' \backslash \{\widehat{\beta}^{-1}(x) \mid x \in \Sigma\}$.

The inclusion $L(G) \subseteq L(G')$ is obvious, it follows from the construction of G'. Let $S \overset{\pi}{\Rightarrow} w$, $\pi = r_1 r_2 \cdots r_n$ be a derivation in G' with the occurrence sequence $\nu = t_1 t_2 \cdots t_n$ of transitions N'_2 where $t_i = \widehat{\gamma}^{-1}(r_i), 1 \leq i \leq n$ enabled at ι' and finished at $\tau' \in M'$.

It is obvious that if for some $1 \leq i \leq 2$, $1 \leq j \leq k(i)$ a production rule $|\pi|_{r'_{ij}} > 0$, $r'_{ij} : A_{ij} \rightarrow A'_{ij}$ then $|\pi|_{r''_{ij}} > 0$, $r''_{ij} : A'_{ij} \rightarrow w_{ij}$, and $|\pi|_{r'_{ij}} = |\pi|_{r''_{ij}}$.

Without loss of generality we can assume that a rule r''_{ij} is the next to a rule r'_{ij} in π (as to the nonterminal A'_{ij} only the rule r''_{ij} is applicable and we can change the order in which the derivation π is used). Then we can replace any derivation steps of the form $x_1 A_{ij} x_2 \Rightarrow_{r'_{ij}} x_1 A'_{ij} x_2 \Rightarrow_{r''_{ij}} x_1 w_{ij} x_2$ by $x_1 A_{ij} x_2 \Rightarrow_{r_{ij}} x_1 w_{ij} x_2$.

Accordingly, the occurrence sequence $\mu \overset{t'_{ij}}{\longrightarrow} \mu' \overset{t''_{ij}}{\longrightarrow} \mu''$ is replaced by $\mu \overset{t_{ij}}{\longrightarrow} \mu''$ where $t_{ij} = \widehat{\gamma}^{-1}(r_{ij})$, $t'_{ij} = \widehat{\gamma}^{-1}(r'_{ij})$ and $t''_{ij} = \widehat{\gamma}^{-1}(r''_{ij})$, $1 \leq i \leq 2$, $1 \leq j \leq k(i)$. Clearly, $L(G') \subseteq L(G)$.

Let us consider the general case $k \geq 1$. Let $G = (V, \Sigma, S, R, N_k)$ be a *k*-PN controlled grammar (with or without erasing rules) where $N_k = (P, T, F, \varphi, \iota, M)$ is a *k*-PN with the counters $q_1, q_2, \ldots, q_k$. We can repeat the arguments of the proof for $k = 1$ considering q_k instead of q and adding the new counter place q_{k+1}.

For $k \geq 1$, let the language L_k be defined by

$$L_k = \{\prod_{i=1}^{k} a_i^{n_i} b_i^{n_i} c_i^{n_i} \mid n_i \geq 1, 1 \leq i \leq k\}.$$

Then we can show analogously to Example 8 and Lemma 9 that, for $k \geq 1$,

$$L_{k+1} \in \mathbf{PN_{k+1}} \text{ and } L_{k+1} \notin \mathbf{PN_k}.$$

Thus the inclusions are strict.   □

5   Closure Properties

We define the following binary form for k-PN controlled grammars, which will be used in some of the next proofs.

Definition 15. *A k-PN controlled grammar $G = (V, \Sigma, S, R, N_k)$ is said to be in a binary form if for each rule $A \to \alpha$ the length of the α is not greater than 2, i.e. $|\alpha| \leq 2$.*

Lemma 16 (Binary Form). *For each k-PN controlled grammar there exists an equivalent k-PN controlled grammar in the binary normal form.*

Lemma 17 (Union). *The family of languages $\mathbf{PN_k}$, $k \geq 1$ is closed under union.*

Proof. Let $G_1 = (V_1, \Sigma_1, S_1, R_1, N_{k1})$ and $G_2 = (V_2, \Sigma_2, S_2, R_2, N_{k2})$ be k-PN controlled grammars. We assume (without loss of generality) that $V_1 \cap V_2 = \emptyset$. Let $(P_i, T_i, F_i, \varphi_i, \iota_i)$ be the cf Petri net corresponding to the context-free grammar $(V_i, \Sigma_i, S_i, R_i)$, $i = 1, 2$ and let $Q_i = \{q_{ij} \mid i = 1, 2, 1 \leq j \leq k\}$, $i = 1, 2$ be the set of the counters. Then k-PN is defined by $N_{ki} = (P_i \cup Q_i, T_i, F_i \cup F_{i1} \cup F_{i2}, \varphi_i', \iota_i', M_i')$, $i = 1, 2$ (with the notions of Definition 4). We construct the k-PN controlled grammar $G = (V_1 \cup V_2 \cup \{S\}, \Sigma_1 \cup \Sigma_2, S, R_1 \cup R_2 \cup \{S \to S_1, S \to S_2\}, N_k)$ where $N_k = (P, T, F, \varphi, \iota, M)$ is defined by

- $P = P_1 \cup P_2 \cup Q_1 \cup \{q\}$ where q is labeled by S;
- $T = T_1 \cup T_2 \cup \{t_{01}, t_{02}\}$ where t_{01} and t_{02} are labeled by the rules $S \to S_1$ and $S \to S_2$, respectively;
- $F = F_1 \cup F_2 \cup F_{11} \cup F_{12} \cup \{(q, t_{0i}), (t_{0i}, q_{0i}) \mid i = 1, 2\} \cup \{(t, q_{1i}) \mid (t, q_{2i}) \in F_{21}, 1 \leq i \leq k\} \cup \{(q_{1i}, t) \mid (q_{2i}, t) \in F_{22}, 1 \leq i \leq k\}$ where q_{0i} is labeled by S_i, $i = 1, 2$;
- $\varphi(x, y) = \varphi_i'(x, y)$ if $(x, y) \in F_i$, $i = 1, 2$ and $\varphi(x, y) = 1$ otherwise;
- $\iota(q) = 1$, $\iota(q_{01}) = \iota(q_{02}) = 0$ and $\tau(q) = \tau(q_{01}) = \tau(q_{02}) = 0$ for any $\tau \in M$ $\iota(p) = \iota_i'(p)$ and for any $\tau \in M$, $\tau(p) = \tau_i'(p)$ if $p \in P_i \setminus \{q_{0i}\}$, $i = 1, 2$; $\iota(p) = 0$ and for any $\tau \in M$, $\tau(p) = 0$ if $p \in Q_1$.

By the construction of N_k any occurrence of its transitions can start by firing of t_{01} or t_{02} then transitions of T_1 or transitions of T_2 can occur, correspondingly we start a derivation with the rule $S \to S_1$ or $S \to S_2$ then we can use rules of R_1 or R_2.

A string w is in $L(G)$ iff there is a derivation $S \Rightarrow S_i \Rightarrow^* w \in L(G_i)$, $i = 1, 2$. On the other hand, we can initialize any derivation $S_i \Rightarrow^* w \in L(G_i)$ with the rule $S \to S_i$, $i = 1, 2$, i.e. $w \in L(G)$.   □

Lemma 18 (Concatenation). *The family of languages* $\mathbf{PN_k}$, $k \geq 1$ *is not closed under concatenation.*

Proof. Let L_k be the language given at the end of the proof of Theorem 14. Then $L_k \in \mathbf{PN_k}$ and $L_{2k} = L_k \cdot L_k \notin \mathbf{PN_k}$.

Using the standard proof we can show that

Lemma 19. *For* $L_1 \in \mathbf{PN_k}$, $k \geq 1$ *and* $L_2 \in \mathbf{PN_m}$, $m \geq 1$, $L_1 \cdot L_2 \in \mathbf{PN_{k+m}}$.

Lemma 20 (Substitution). *The family of languages* $\mathbf{PN_k}$, $k \geq 1$ *is closed under substitution by context-free languages.*

Proof. Let $G = (V, \Sigma, S, R, N_k)$ be a k-PN controlled grammar with k-Petri net $N_k = (P, T, F, \varphi, \iota, M)$. Let $P = P_V \cup P_\Sigma$ where $P_V = \{p \mid p = \beta^{-1}(A), A \in V\}$ and $P_\Sigma = \{p \mid p = \beta^{-1}(a), a \in \Sigma\}$. We consider a substitution $s : \Sigma^* \to 2^{U^*}$ with $s(a) \in \mathbf{CF}$ for each $a \in \Sigma$. Let $G_a = (V_a, \Sigma_a, S_a, R_a)$ be a context-free grammar for $s(a)$, $a \in \Sigma$. We can assume that $V \cap V_a = \emptyset$ for any $a \in \Sigma$ and $V_a \cap V_b = \emptyset$ for any $a, b \in \Sigma$, $a \neq b$.

Let $N_a = (P_a, T_a, F_a, \varphi_a, \iota_a)$ be the corresponding cf Petri net to $G_a, a \in \Sigma$ with the bijections β_a and γ_a. For each place $p \in P_\Sigma$ we replace its label $a = \beta(p)$ by S_a in N_k and we define the k-PN controlled grammar $G' = (V \cup \bigcup_{a \in \Sigma} V_a, U, S, R' \cup \bigcup_{a \in \Sigma} R_a, N'_k)$ where R' is the set of rules obtained by replacing each occurrence of $a \in \Sigma$ by S_a in R and N'_k is defined by $N'_k = (P \cup \bigcup_{a \in \Sigma} P_a, T \cup \bigcup_{a \in \Sigma} T_a, F \cup \bigcup_{a \in \Sigma} F_a, \varphi', \iota', M')$ where

- $\varphi'(x, y) = \varphi(x, y)$ if $(x, y) \in F$ and $\varphi'(x, y) = \varphi_a(x, y)$ if $(x, y) \in F_a$, $a \in \Sigma$;
- $\iota'(p) = \iota(p)$ if $p \in P$ and $\iota'(p) = 0$ if $p \in P_a$, $a \in \Sigma$;
- for any $\tau' \in M'$, $\tau'(p) = 0$ if $p \in P$ or $p \in P_a \setminus \{\beta_a^{-1}(b) \mid b \in \Sigma_a\}$, $a \in \Sigma$.

Obviously, $L(G') \in \mathbf{PN_k}$. $\square$

Lemma 21 (Mirror Image). *The family of languages* $\mathbf{PN_k}$, $k \geq 1$ *is closed under mirror image.*

Proof. Let $G = (V, \Sigma, S, R, N_k)$ be a k-PN controlled grammar. The context-free grammar (V, Σ, S, R) and its reversal (V, Σ, S, R^R) have the same corresponding cf Petri net $N = (P, T, F, \varphi, \iota)$ as N does not preserve the order of the positions of the output places for each transition. Thus we can also use the k-Petri net N_k as a control mechanism for the grammar (V, Σ, S, R^R), i.e. we define $G^R = (V, \Sigma, S, R^R, N_k)$. Clearly, $L(G^R) \in \mathbf{PN_k}$. $\square$

Lemma 22 (Intersection with Regular Languages). *The family of languages* $\mathbf{PN_k}$, $k \geq 1$ *is closed under intersection with regular languages.*

The results of the previous lemmas are summarized in the following theorem

Theorem 23. *The family of languages* $\mathbf{PN_k}$, $k \geq 1$ *is closed under union, substitution, mirror image, intersection with regular languages and it is not closed under concatenation.*

References

1. ter Beek, M.H., Kleijn, H.C.M.: Petri net control for grammar systems. In: Brauer, W., Ehrig, H., Karhumäki, J., Salomaa, A. (eds.) Formal and Natural Computing. LNCS, vol. 2300, pp. 220–243. Springer, Heidelberg (2002)
2. Ceska, M., Marek, V.: Petri nets and random-context grammars. In: Proc. 35th Spring Conference: Modeling and Simulation of Systems (MOSIS 2001), MARQ Ostrava Hardec nad Moravici, pp. 145–152 (2001)
3. Crespi-Reghizzi, S., Mandrioli, D.: Petri nets and commutative grammars. Internal Report No. 74-5, Laboratorio di Calcolatori, Instituto de Elettrotecnica ed Elettronica del Politecnico di Milano, Italy (1974)
4. Dassow, J., Păun, G.: Regulated rewriting in formal language theory. Springer, Berlin (1989)
5. Hack, M.: Petri net languages. Computation Structures Group Memo 124, p. 128. MIT, Cambridge (1975)
6. Hauschildt, D., Jantzen, M.: Petri nets algorithms in the theory of matrix grammars. Acta Informatica 31, 719–728 (1994)
7. Hopcroft, J.E., Ullman, J.D.: Introduction to automata theory, languages, and computation. Addison-Wesley Longman Publishing Co., Inc, Amsterdam (1990)
8. Reisig, W., Rozenberg, G. (eds.): Lectures on Petri nets I: Basic models. LNCS, vol. 1491. Springer, Heidelberg (1998)
9. Turaev, S.: Petri Net Controlled Grammars. In: Proc. 3rd Doctoral Workshop on MEMICS-2007, Znojmo, Czech Republic, pp. 233–240 (2007)

2-Synchronizing Words

Paweł Gawrychowski[1,*] and Andrzej Kisielewicz[2,**]

[1] Institute of Computer Science, University of Wrocław
ul. Joliot-Curie 15, 50-383 Wrocław, Poland
gawry1@gmail.com
[2] Institute of Mathematics, University of Wrocław
pl. Grunwaldzki 2, 50-384 Wrocław, Poland
kisiel@math.uni.wroc.pl

Abstract. A word w over an alphabet Σ is n-synchronizing if it resets every $(n + 1)$-state synchronizing automaton over this alphabet. For a fixed n and Σ, n-synchronizing words can be recognized in polynomial time, yet no practical algorithm is known. In this paper we show that one cannot expect to find such an algorithm. We prove that the problem of recognizing 2-synchronizing words, where the input consists of a word and an alphabet, is co-NP-complete. We also show that the length of a 2-synchronizing word is at least $2|\Sigma|^2$, which improves the lower bound known so far.

Keywords: Formal languages, synchronizing automata.

1 Introduction

In this paper we consider deterministic finite automata $\mathcal{A} = \langle Q, \Sigma, \delta \rangle$ (in short DFA) with the state set Q, the input alphabet Σ, and the total transition function $\delta : Q \times \Sigma \to Q$. The action of Σ on Q given by δ extends naturally on the action of the words over Σ on Q; this action, in this paper, will be denoted simply by concatenation: $qw = \delta(q, w)$. Similarly, Qw will stand for $\cup_{q \in Q} \delta(q, w)$.

The automaton $\mathcal{A}$ is called *synchronizing* if there exists a word $w \in \Sigma^*$ whose action *resets* $\mathcal{A}$ in the sense that $|Qw| = 1$. The problem of DFA synchronization is very natural and its various aspects were considered in the literature (see e.g., [2,3,7] and [8] for more references). The most famous is Černý's conjecture which states that for a synchronizing automaton with n states there exists a reset word of length not exceeding $(n-1)^2$ (see [5]).

In [1], a word w over a fixed alphabet Σ is called n-synchronizing if it resets every synchronizing DFA with $n + 1$ states (cf. also [3]). In [9] it is proved that each such a word is n-*full*, that is, it contains any word of length n among its subwords (in particular, the alphabet in question is determined by the letters occurring in the word). Since there are finitely many automata with $n + 1$ states

---

* Research supported by MNiSW grant number N206 024 31/3826, 2006-2008 and AutoMathA program of the ESF.
** Research supported by Polish KBN grant P03A 03430.

C. Martín-Vide, F. Otto, and H. Fernau (Eds.): LATA 2008, LNCS 5196, pp. 221–231, 2008.
© Springer-Verlag Berlin Heidelberg 2008

and a fixed finite alphabet, recognizing n-synchronizing words is decidable and may be done in polynomial time. (This is in contrast with the fact that recognizing 2-collapsing words is co-NP-complete; cf. [6]). In fact, the language of n-synchronizing words over a fixed alphabet is regular, as it is an intersection of finitely many regular languages (see [3]). Yet, so far no efficient algorithm of recognizing 2-synchronizing words is known. For example, algebraic characterization from [1] leads to an exponential time algorithm (with respect to the word length). In this paper we prove that the problem becomes clearly hard once we do not fix the alphabet. This is important, because it shows that certain types of characterizations and algorithms sought so far cannot exist.

Another question in this context is that concerning the minimum possible length of an n-synchronizing word. It has received much attention in a similar setting of n-collapsing words (recall that w is n-collapsing if $|Qw| \leq |Q| - n$ for each automaton for which such decrease is possible; each n-collapsing word is easily seen to be also n-synchronizing, but the converse is not true). Let $C(n,t)$ and $S(n,t)$ be the minimum possible lengths of an n-collapsing word and an n-synchronizing word over an alphabet of size t, respectively. Construction of [9] shows that

$$S(n,t) \leq C(n,t) \leq t^{\frac{1}{2}(n^2+n)} + o(t^{\frac{1}{2}(n^2+n)})$$

thus improving the work of [12]. On the other hand, as an immediate consequence of the fact that both n-collapsing and n-synchronizing words are n-full, we get the following lower bound (see [9]):

$$t^n + n - 1 \leq S(n,t) \leq C(n,t)$$

In [11] the above lower bound has been improved to $2t^2$ for the case of $C(2,t)$. We strengthen this result by proving the same for $S(2,t)$.

2   Computational Complexity

Definition 1. *A word w over a fixed alphabet Σ is called n-synchronizing if it resets every synchronizing DFA over Σ with $n + 1$ states (in the sense that $|Qw| = 1$).*

It is well known that each such word is n-full, meaning that it contains any word of length n as a factor (to see why, take automaton recognizing word s of length n; it has $n + 1$ states and is synchronized by w iff s is a subword of w). It justifies the following:

Definition 2. *A word w over any alphabet is n-synchronizing if it is n-synchronizing over the alphabet consisting of the letters occurring in w.*

In this section we will consider the following decision problem:

INSTANCE: A word w;

QUESTION: Is w 2-synchronizing?

where letters of w are reasonably encoded (for example as $1^i 0$).

The problem above can be viewed as one concerning automata with exactly 3 states. Given a word w, let $\Sigma(w)$ denote the letters occurring in w. To show that the above problem is co-NP-complete, we will focus on the following complementary problem:

INSTANCE: A word w;
QUESTION: Is there a synchronizing automaton $\mathcal{A} = \langle\{0,1,2\}, \Sigma(w), \delta\rangle$ such that $|\{0,1,2\}w| > 1$?

To prove that this problem is NP-hard we shall use a reduction from 3SAT. Our idea is to treat 3-state DFA's as objects encoding truth assignments. To make things easier (and working) we shall restrict ourselves to a certain type of synchronizing DFA, which we call *TA-automata*, and to a special kind of words designed to encode instances of 3-SAT. We assume that each such a word is of the form $w = pq$, where p is a fixed word resetting every synchronizing DFA except for TA. For TA-automata we will have $\{0,1,2\}p = \{1,2\}$. The remaining suffix q will depend on the instance $\mathbb{C}$ of 3-SAT so that for each TA-automaton we will have $|\{1,2\}q| > 1$ if and only if the corresponding truth assignment satisfies all the clauses of $\mathbb{C}$.

Given an automaton $\mathcal{A} = \langle Q, \Sigma, \delta\rangle$, transformations of Q determined by the letters of Σ are called transformations of $\mathcal{A}$ and are denoted by corresponding letters, if no confusion can arise. Following [6], we say that a transformation a is of type $\{x, y\}/z$ if $xa = ya$, while z is the unique state not in Qa.

To define TA-automata and construct p we will need some insight into the structure of synchronizing automata. It is clear that for a DFA to be synchronizing at least one letter has to act as a non-permutation (though this condition is clearly not sufficient). As we are interested in automata that are "hardest" to synchronize, we omit from our consideration those DFA that can be reset by a word of length not exceeding 3.

Lemma 1. *Let $\mathcal{A}$ be a synchronizing DFA with 3 states that cannot be reset by a word of length at most 3. Then the states may be labeled with numbers $0, 1, 2$ so that the following conditions hold:*

(i) *There is a transformation a of type $\{0,1\}/0$ in $\mathcal{A}$.*
(ii) *All transformations of $\mathcal{A}$ are either permutations or transformations of type $\{0,1\}/0$.*
(iii) *All permutations of $\mathcal{A}$ belong to $\{(0,1,2), (0)(1,2), (0,1)(2), (0)(1)(2)\}$.*
(iv) *Either $(0,1,2)$ is among the permutations of $\mathcal{A}$ or both $(0)(1,2)$ and $(0,1)(2)$ are among the permutations of $\mathcal{A}$.*

Proof. First note that since no single letter is resetting $\mathcal{A}$, all transformations of $\mathcal{A}$ are either permutations or transformations of type $\{x,y\}/z$. If a is of type $\{x,y\}/z$ with $z \notin \{x,y\}$, then aa is resetting $\mathcal{A}$. This yields (i).

Suppose that $w = a_1 a_2 \ldots a_n$, is a synchronizing word with the smallest $n \geq 4$. We may assume that a_1 is a non-permutation and that $a_1 = a$. By assumption, for all $i = 1, \ldots, n-1$,

$$|\{0,1,2\}a_1 \ldots a_i| = 2$$

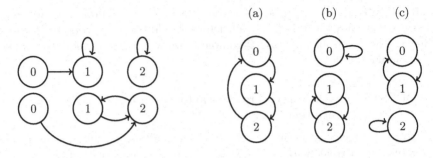

Fig. 1. Transformations of type $\{0,1\}/0$

Fig. 2. $\mathcal{A}$ from Lemma 1 contains (a) or both (b) and (c) among its permutations

and all $\{0,1,2\}a_1 \ldots a_i$ are pairwise distinct (otherwise there would be a shorter synchronizing word). Since there are only 3 different subsets of $\{0,1,2\}$ consisting of exactly 2 elements, we get that in fact $n = 4$ (which proves the well known fact that Černý's conjecture holds for $|Q| = 3$). We may consider now in detail the sequence of images under prefixes of w:

$$\{0,1,2\} \xrightarrow{a_1} \{1,2\} \xrightarrow{a_2} \{0,2\} \xrightarrow{a_3} \{0,1\} \xrightarrow{a_4} \{s\}$$

for some $s \in \{1,2\}$. Note that $\{0,1\}$ has to appear as the last 2-element image, since otherwise the synchronizing word would be shorter. For the same reason it is clear that each non-permutation is of the type $\{0,1\}/z$, and further that $z = 0$, which proves (ii). It follows also that a_3 and a_2 are permutations. Consequently, for each permutation c, $\{1,2\}c = \{0,2\}$ or $\{1,2\}$, which proves (iii); and if $c = (0,1,2)$ is among the permutations of $\mathcal{A}$, then $acca$ is a synchronizing word; otherwise, we must have $a_2 = (0,1)(2)$ and $a_3 = (0)(1,2)$, proving (iv).   □

Now we define *TA-automaton* to be one isomorphic to $\mathcal{A} = \langle Q, \Sigma, \delta \rangle$ with $Q = \{0,1,2\}$ and $\Sigma = \{a,b\} \cup V$, such that a acts as a non-permutation of type $\{0,1\}/0$, b acts as the permutation $(0,1,2)$, and all $v \in V$ act as permutations $(0)(1,2)$ or $(0)(1)(2)$ (the identity) (see Figure 3).

To construct the word p we will use the obvious fact that if w resets $\mathcal{A}$, then each word with a factor w resets $\mathcal{A}$ as well. We will need the following:

Lemma 2. *Let $a, b \in \Sigma$. If $p \in \Sigma^*$ contains as a factor every word:*

(i) *xyz, for each $x, y, z \in \Sigma$,*
(ii) *$xyzx$, for each $x, y, z \in \Sigma$ satisfying $x \neq a$ or $y \neq b$,*
(iii) *ab^3za, for each $z \in \Sigma$,*

then p is synchronizing for each synchronizing automaton $\mathcal{A}$ which is not TA.

Proof. Observe that the above conditions easily ensure that p resets each synchronizing automaton which can be synchronized by a word of length at most 3 or has a non-permutation among transformations corresponding to letters different than a. So, we may assume that the letter a acts as the transformation of type $\{0,1\}/0$ (and is the only non-permutation), and (iv) of Lemma 1 holds.

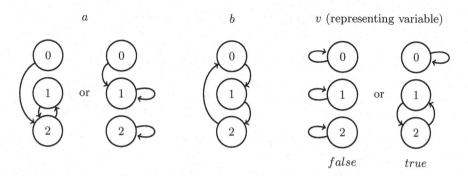

Fig. 3. Transformations of TA-automaton

Now, if there are $d, e \in \Sigma$ such that d acts as $(0,1)(2)$ and e acts as $(0)(1,2)$, then we have two possibilities:

(i) $d \neq b$, then $xyzx$ is synchronizing for $x = a, y = d$ and $z = e$,
(ii) $d = b$, then ab^3za is synchronizing for $z = e$ (since $d^3 = b^3 = b = d$).

Otherwise, there is $c \in \Sigma$ acting as $(0,1,2)$ and again we have two possibilities:

(i) $c \neq b$, then $xyzx$ is synchronizing with $x = a, y = z = c$,
(ii) $c = b$ is the only letter acting as $(0,1,2)$.

In the latter case, if there is $d \in \Sigma$ acting as $(0,1)(2)$, then $xyzx$ is synchronizing with $x = a, y = d, z = b$; otherwise, every letter different from a and b acts as $(0)(1,2)$ or as the identity, which means that $\mathcal{A}$ is TA.   □

Recall that we wish to make p synchronizing for exactly those synchronizing automata which are not TA. Using the lemma above it is easy to ensure that p will be synchronizing for required automata, but making sure that it is **not** synchronizing for TA-automata requires more work.

Lemma 3. *Let $a, b \in \Sigma$ and p be a concatenation of the following words:*

(i) $x^2(xyz)z^5a$, *for each* $x, y \in \Sigma$,
(ii) $y^2x^5(xyzx)x^5a$, *for each* $x, y, z \in \Sigma$ *satisfying* $x \neq a$ *or* $y \neq b$,
(iii) ab^3za, *for each* $z \in \Sigma$.

Then p is synchronizing exactly for those synchronizing automata that are not TA. For TA-automata we have $\{0,1,2\}p = \{1,2\}$.

Proof. Due to the previous lemma we only have to check whether $\{1,2\}w = \{1,2\}$ for every w being one of the above words when $\mathcal{A}$ is TA-automaton. It is an easy (although a little tedious) case inspection:

(i) $\{1,2\}x^2(xyz)z^5a = \{1,2\}x^3yz^6a = \{1,2\}yz^6a$

$$\{1,2\}yz^6a = \begin{cases} \{1,2\}z^6a = \{1,2\}a = \{1,2\} & \text{if } y \neq b \\ \{0,2\}z^6a = \{0,2\}a = \{1,2\} & \text{if } y = b, z \neq a \\ \{0,2\}z^6a = \{1,2\}a = \{1,2\} & \text{if } y = b, z = a \end{cases}$$

(ii) $y^2x^5(xyzx)x^5a \equiv y^2x^6yzx^6a$. We have two possibilities:
- $x = a, y \neq b$, so

$$\{1,2\}y^2x^6yzx^6a = \{1,2\}x^6yzx^6a = \{1,2\}yzx^6a = \{1,2\}zx^6a$$

$$\{1,2\}zx^6a = \begin{cases} \{1,2\}x^6a = \{1,2\}a = \{1,2\} & \text{if } z \neq b \\ \{0,2\}x^6a = \{1,2\}a = \{1,2\} & \text{if } z = b \end{cases}$$

- $x \neq a$, then $y^2x^6yzx^6a \equiv y^2yza \equiv y^3za$ and

$$\{1,2\}y^3za = \{1,2\}za = \begin{cases} \{0,2\}a = \{1,2\} & \text{if } z = b \\ \{1,2\}a = \{1,2\} & \text{if } z \neq b \end{cases}$$

(iii) $\{1,2\}ab^3za = \{1,2\}za = \{1,2\}$.

As p is just a concatenation of the words w for which $\{1,2\}w = \{1,2\}$, we know that $\{1,2\}p = \{1,2\}$. On the other hand, at least one of these words contains a, so in fact $\{0,1,2\}p = \{1,2\}$. □

Now we can proceed to construct the suffix q of the input word. Let $\mathbb{C} = \{C_1, \ldots, C_m\}$ be a 3-SAT instance, where C_i are 3-literal clauses, and let V be the set of variables occurring in $\mathbb{C}$. With each truth assignment T for $\mathbb{C}$ we associate a TA-automaton $\mathcal{A} = \langle \{0,1,2\}, \Sigma, \delta \rangle$, where $\Sigma = \{a,b\} \cup V$, a acts as a non-permutation of type $\{0,1\}/0$, b acts as $(0,1,2)$, and all $v \in V$ act as permutations $(0)(1)(2)$ or $(0)(1,2)$, depending on whether $T(v)$ is false or true, respectively. We put

$$q = c_1ac_2a\ldots ac_ma$$

where $c_i \in (\{b\} \cup V)^*$ will be an encoding of the clause C_i. Observe that $|\{1,2\}q| = 1$ if and only if there is $i \in \{1, \ldots, m\}$ such that $2 \notin \{1,2\}c_i$. So we wish to construct c_i so that $\{1,2\}c_i = \{0,1\}$ if and only if $\mathcal{A}$ corresponds to a truth assignment T that makes C_i false (note that at this point we consider only TA-automata).

Permutation	linear function
(0)(1)(2)	x
(012)	$x + 1$
(021)	$x + 2$
(0)(12)	$2x$
(01)(2)	$2x + 1$
(02)(1)	$2x + 2$

Fig. 4. Correspondence between elements of S_3 and nonconstant linear functions over $\mathbb{Z}_3$

At this step we make use of the fact that that the permutations of S_3 can be presented as nonconstant linear functions $\alpha x + \beta$ over $\mathbb{Z}_3$ (see Figure 4). For example, in terms of the linear functions $b \equiv x + 1$, and each $v \in V$ is equal either to x or to $2x$. This correspondence gives us a convenient way of dealing with permutation transformations of a TA-automaton (their composition can be viewed as a multilinear polynomial, which is easy to describe).

For a given TA-automaton $\mathcal{A}$ (associated with a truth assignment T as described above), we define

$$
val(v) = \begin{cases} 1 & \text{if } v \text{ is set to false in } T \\ 2 & \text{otherwise.} \end{cases}
$$

Then $v \equiv val(v)x$ for any truth assignment. Our aim is to construct c_i so that $c_i \equiv x + 2$ whenever the corresponding clause is valuated false, and $c_i \equiv x + 1$, otherwise (then $\{1,2\}c_i = \{0,1\}$ or $\{0,2\}$, respectively). To this aim we will need the following observation:

Lemma 4. *For each mapping $\phi : 2^V \to \mathbb{Z}_3$ there is a word $w \in (\{b\} \cup V)^*$ such that*

$$
w \equiv x + \sum_{U \subseteq V} \phi(U) \prod_{v \in U} val(v)
$$

for all truth assignments on V.

Proof. Obviously, it is enough to show that for each $U \subseteq V$ there is a word w such that

$$
w \equiv x + \prod_{v \in U} val(v)
$$

We apply induction on the size of U:

(i) $U = \emptyset$, then we can take $w = b$.
(ii) $|U| > 0$. Choose $v_i \in U$. By induction hypothesis there is a word w' such that

$$
w' \equiv x + \prod_{v \in U'} val(v)
$$

where $U' = U \setminus \{v_i\}$. We can simply take $w = v_i w' v_i$ because

$$
v_i w' v_i \equiv val(v_i) \left(val(v_i)x + \prod_{v \in U'} val(v) \right) \equiv x + \prod_{v \in U} val(v)
$$

(the last equality follows from the fact that $x^2 \equiv 1 \pmod 3$ for $x \in \{1,2\}$). $\quad\square$

Now, let $C_i = \{l_1, l_2, l_3\}$ be a clause in $\mathbb{C}$. To construct c_i with the required properties observe first that for any $r, s, t \in \{1,2\}$ the product $r(s+2)(t+2) \equiv 2 \pmod 3$ if and only if $r = s = t = 2$. By the last lemma there exists a word $c_i \in \{b, v_1, v_2, v_3\}^*$ such that:

$$
c_i \equiv x + \alpha_1 val(v_1)\,(\alpha_2 val(v_2) + 2)\,(\alpha_3 val(v_3) + 2), \quad \text{where } \alpha_i = \begin{cases} 1 & \text{if } l_i = \neg v_i \\ 2 & \text{if } l_i = v_i \end{cases}
$$

As c_i maps 0 to 2 if and only if all l_i are set to false, $\{1,2\}c_i a = \{1,2\}$ or $\{z\}$, depending on whether TA-automaton A corresponds to a truth assignment that makes C_i true or false, respectively.

For the whole word $w = pq$ we have that there exists a synchronizing automaton A such that w does not reset A if and only if there is a truth assignment T such that all the clauses in $\mathbb{C}$ are satisfied. Definitions in Lemma 3 and Lemma 4 show that the whole construction of w may be done in polynomial time. The original problem is easily seen to be in co-NP (as we only have to guess automata for which the given word is not synchronizing and check whether it is indeed the case), so we have completed the proof of the following theorem:

Theorem 1. *The problem of recognizing 2-synchronizing words is co-NP-complete.*

It seems that the above reasoning used to settle the computational complexity of recognizing n-synchronizing words for $n = 2$ could be used to prove co-NP-completeness of this problem for other values of n. It turns out, however, that choosing suitable encoding requires finding automata having shortest synchronizing word of greatest length for the given value of $|Q|$. It seems possible for $n = 3$ but the details become very tedious. For greater values of n this approach does not seem feasible as the number of different automata that should be considered becomes very large.

3   Lower Bound

Let $S(n, t)$ be the minimum possible length of a n-synchronizing word over $\Sigma = \{a_1, \ldots, a_t\}$. Knowing that n-synchronizing word must be n-full yields:

Lemma 5
$$S(n, t) \geq t^n + n - 1$$

Proof. Take any n-synchronizing word w over Σ. For each $s \in \Sigma^n$ mark the last letter of its first occurrence in w. As no letter was marked twice and the first $n - 1$ letters could not been marked $|w| \geq t^n + n - 1$. $\square$

Knowing that n-synchronizing word must be n-full is not enough to show better lower bound as we can easily construct a n-full word of length $t^n + n - 1$ using de Bruijn sequence. Using a stronger method we will prove

$$S(2, t) \geq 2t^2$$

which is much closer to the known values of $S(2, 2) = 8$ and $S(2, 3) = 20$.

The general approach is similar to the one of [11], where it was used to deal with collapsing words. Let w be a 2-synchronizing word over $\Sigma = \{a_1, \ldots, a_t\}$. We will show that each $a_k \in \Sigma$ must occur in w at least $2t$ times. Choose any $a_k \in \Sigma$ and rewrite w as

$$w = w_0 a_k^+ w_1 a_k^+ w_2 \ldots a_k^+ w_m a_k^+ w_{m+1}, \qquad \text{each } w_i \in (\Sigma \setminus \{a_k\})^*$$

To show that m cannot be small, we will focus on some subset of all synchronizing automata which is easy to deal with, yet complicated enough. Let a_k correspond to a transformation of type $\{0,1\}/0$ and the remaining $n = t - 1$ letters to permutations belonging to the set $\{x, x + 1, x + 2\}$ (where we again represent elements of S_3 as nonconstant linear functions over $\mathbb{Z}_3$).

As long as at least one permutation is different than x, defined automaton $\mathcal{A}$ is synchronizing. a_k is the only nonpermutation so the only way for w to synchronize $\mathcal{A}$ is that $|\{1,2\}w_i a_k| = 1$ holds for some $i \in \{1, \ldots, m\}$. This condition is equivalent to $0 w_i = 2$ for some $i \in \{1, \ldots, m\}$.

Thus from the assumption that w is 2-synchronizing follows that for some $v_1, \ldots, v_m \in \mathbb{Z}_3^n$

$$\forall_{x \in \mathbb{Z}_3^n \setminus \vec{0}} \exists_{i \in \{1, \ldots, m\}} \ v_i x \equiv 2 \pmod 3 \tag{1}$$

holds, where vector v_i is simply a compact description of w_i: its k-th coordinate is the number of times (modulo 3) the k-th element of $\Sigma \setminus \{a_k\}$ occurs in w_i. Now we can focus on this algebraic formulation, which turns out to be quite easy to deal with.

Define

$$A = \left\{ \begin{pmatrix} v_1 \\ \ldots \\ v_m \end{pmatrix} x \mid x \in \mathbb{Z}_3^n \right\}$$

and observe that $\dim A = n$ as otherwise the equation

$$\begin{pmatrix} v_1 \\ \ldots \\ v_m \end{pmatrix} x = \vec{0}$$

would have at least two different solutions, one of them being nonzero and contradicting (1). Hence $n \leq m$, but we can prove even more. W.l.o.g assume that

$$A = \{[y_1, \ldots, y_n, u_1 y, \ldots, u_{m-n} y] \mid y \in \mathbb{Z}_3^n\}$$

We can restate (1) in the following form:

$$\forall_{y \in \{0,1\}^n \setminus \vec{0}} \exists_{i \in \{1, \ldots, m-n\}} \ u_i y \equiv 2 \pmod 3 \tag{2}$$

It is very easy to find $u_1, \ldots, u_{m-n}$ for which the above condition is satisfied when $m - n \geq n$:

$$u_i^{(j)} = \begin{cases} 2 & \text{if } i = j \text{ and } i \leq n \\ 0 & \text{otherwise} \end{cases}$$

What is a little surprising, this simple construction is the best we can hope for. To prove its optimality, assume that $m - n < n$. From (2) we get that it is possible to find $m - n + 1$ vectors $v_1, \ldots, v_{m-n+1}$ such that sum of their each nonempty subset have at least one coordinate equal to 2 (modulo 3). Thus to get contradiction we need the following lemma:

Lemma 6. *Take any $v_1, \ldots, v_{k+1} \in \mathbb{Z}_3^k$. Then $\sum_{v \in V} v \in \{0,1\}^k$ (where coordinates are taken modulo 3) for some nonempty $V \subseteq \{v_1, \ldots, v_{k+1}\}$.*

Proof. We use the beautiful Combinatorial Nullstellensatz method of Noga Alon (see [10]). Assume that there are $v_1, \ldots, v_{k+1} \in \mathbb{Z}_3^k$ such that $\sum_{v \in V} v \notin \{0,1\}^k$ for each nonempty V. We can restate this condition by defining

$$P(x_1, \ldots, x_{k+1}) = \prod_{i=1}^{k} 1 + \sum_{j=1}^{k+1} v_j^{(i)} x_j$$

and saying that $P(x_1, \ldots, x_{k+1}) = 0$ for each $[x_1, \ldots, x_{k+1}] \in \{0,1\}^{k+1} \setminus \vec{0}$. Now take

$$P_{all}(x_1, \ldots, x_{k+1}) = P(x_1, \ldots, x_{k+1}) - \prod_{i=1}^{k+1} (1 - x_i)$$

which equals 0 for each $[x_1, \ldots, x_{k+1}] \in \{0,1\}^{k+1}$. As each x_i is 0 or 1, the same is true for the truncated version of P_{all}, $\overline{P_{all}}$, in which we replace each x_i^e by x_i (for example $x_1^2 x_2^3 + x_1^3 x_2$ becomes $2x_1 x_2$). $\overline{P_{all}}$ is a multilinear polynomial over $\mathbb{Z}_3$ so the following lemma applies:

Lemma 7. *Let $Q = Q(x_1, \ldots, x_m)$ be a multilinear polynomial over $\mathbb{Z}_3$. If $Q(x_1, \ldots, x_m) = 0$ for each $[x_1, \ldots, x_m] \in \{0,1\}^m$ then $Q \equiv 0$.*

Proof. Induction on m.

(i) $m = 1$, then $Q(x_1) = ax_1 + b$. As $Q(0) = Q(1) = 0$, $a = b = 0$.
(ii) $m > 1$, then $Q(x_1, \ldots, x_m) = Q_1(x_1, \ldots, x_{m-1}) + x_m Q_2(x_1, \ldots, x_{m-1})$. By induction hypothesis, $Q_1 \equiv 0$. But if $Q_1 \equiv 0$ then by induction hypothesis also $Q_2 \equiv 0$ so the whole Q vanishes. □

Now applying the above lemma we get that $\overline{P_{all}} \equiv 0$ which is a contradiction: P has degree at most k so the term $\prod_{i=1}^{k+1} x_i$ cannot vanish in $\overline{P_{all}}$. □

This argument shows that $m \geq 2n = 2(t-1)$. Combining with the fact that a_k^2 must be a factor of w, we get that a_k occurs in w at least $2t$ times. As a_k was arbitrarily chosen:

Theorem 2. *The length of a 2-synchronizing word over Σ must be at least $2|\Sigma|^2$.*

The improved lower bound still seems to be really far from being accurate (though it is much better that the previously known, especially for the case of $S(2,2)$ and $S(2,3)$). The best known explicit construction gives words that are actually n-collapsing so it should be possible to improve it by using the fact that n-synchronizing word does not have to be n-collapsing (but it seems that such modification is not easy).

References

1. Ananichev, D.S., Cherubini, A., Volkov, M.V.: Image reducing words and subgroups of free groups. Theor. Comput. Sci. 307, 77–92 (2003)
2. Ananichev, D.S., Volkov, M.V.: Synchronizing monotonic automata. Theor. Comput. Sci. 327(1), 225–239 (2004)

3. Ananichev, D.S., Volkov, M.V.: Collapsing words vs. synchronizing words. In: Kuich, W., Rozenberg, G., Salomaa, A. (eds.) DLT 2001. LNCS, vol. 2295, pp. 166–174. Springer, Heidelberg (2002)

4. Ananichev, D.S., Petrov, I.V., Volkov, M.V.: Collapsing words: A Progress Report. In: De Felice, C., Restivo, A. (eds.) DLT 2005. LNCS, vol. 3572, pp. 11–21. Springer, Heidelberg (2005)

5. Černý, J.: Poznámka k. homogénnym experimentom s konecnými automatmi. Mat. fyz. čas SAV 14, 208–215 (1964)

6. Cherubini, A., Gawrychowski, P., Kisielewicz, A., Piochi, B.: A combinatorial approach to collapsing words. In: Královič, R., Urzyczyn, P. (eds.) MFCS 2006. LNCS, vol. 4162, pp. 256–266. Springer, Heidelberg (2006)

7. Kari, J.: Synchronizing finite automata on Eulerian digraphs. In: Sgall, J., Pultr, A., Kolman, P. (eds.) MFCS 2001. LNCS, vol. 2136, pp. 432–438. Springer, Heidelberg (2001)

8. Mateescu, A., Salomaa, A.: Many-valued truth functions, Cerny'y's conjecture and road coloring. EATCS Bull. 68, 134–150 (1999)

9. Margolis, S.W., Pin, J.-E., Volkov, M.V.: Words guaranteeing minimum image. Internat. J. Foundations Comp. Sci. 15, 259–276 (2004)

10. Alon, N.: Combinatorial Nullstellensatz. Comb. Probab. Comput. 8, 7–29 (1999)

11. Pribavkina, E.V.: On Some Properties of the Language of 2-Collapsing Words. In: De Felice, C., Restivo, A. (eds.) DLT 2005. LNCS, vol. 3572. Springer, Heidelberg (2005)

12. Sauer, N., Stone, M.G.: Composing functions to reduce image size. Ars Combinatoria 31, 171–176 (1991)

Not So Many Runs in Strings

Mathieu Giraud[1,2]

[1] CNRS, LIFL, Université Lille 1, 59 655 Villeneuve d'Acsq cedex, France
[2] INRIA Lille Nord-Europe, 59 650 Villeneuve d'Ascq, France
mathieu.giraud@lifl.fr

Abstract. Since the work of Kolpakov and Kucherov in [5,6], it is known that $\rho(n)$, the maximal number of runs in a string, is linear in the length n of the string. A lower bound of $3/(1+\sqrt{5})n \sim 0.927n$ has been given by Franek and al. [3,4], and upper bounds have been recently provided by Rytter, Puglisi and al., and Crochemore and Ilie (1.6n) [8,7,1]. However, very few properties are known for the $\rho(n)/n$ function. We show here by a simple argument that $\lim_{n\mapsto\infty} \rho(n)/n$ exists and that this limit is never reached. Moreover, we further study the asymptotic behavior of $\rho_p(n)$, the maximal number of runs with period at most p. We provide a new bound for some microruns : we show that there is no more than $0.971n$ runs of period at most 9 in binary strings. Finally, this technique improves the previous best known upper bound, showing that the total number of runs in a binary string of length n is below 1.52n.

1 Introduction

The study of repetitions is an important field of research, both for word combinatorics theory and for practice, with applications in domains like computational biology or cryptanalysis. The notion of *run* (also called maximal repetition or m-repetition [5]) allows a compact representation of the set of all tandem periodicities, even fractional, in a string. The proper counting of those runs is important for all algorithms dealing with repetitions.

Since the work of Kolpakov and Kucherov in [5,6], it is known that $\rho(n)$, the maximal number of runs in a string, is linear in the length n of the string. They gave the first algorithm computing all runs in a linear time, but without an actual constant.

Upper bounds have been recently provided by Rytter ($5n$) [8] and Puglisi, Simpson, and Smyth ($3.48n$) [7]. The best upper bound known today, $1.6n$, was obtained by Crochemore and Ilie [1]. They count separately the *microruns*, that is the runs with short periods, and the runs with larger ones. Crochemore and Ilie show that the number of microruns with period at most 9 verifies $\rho_9(n) \leq n$. For larger runs, they prove that

$$\rho_{\geq p}(n) \leq \frac{2}{p}\left(\sum_{i=0}^{\infty}\left(\frac{2}{3}\right)^i\right)n = \frac{6}{p}\cdot n$$

C. Martín-Vide, F. Otto, and H. Fernau (Eds.): LATA 2008, LNCS 5196, pp. 232–239, 2008.
© Springer-Verlag Berlin Heidelberg 2008

Table 1. Values of $\rho(n)$ for small values of n for binary strings, from [5]

n	5	6	7	8	9	10	11	12	13	14	15	16	17	18	19	20	21	22	23	24	25	26	27	28	29	30	31
$\rho(n)$	2	3	4	5	5	6	7	8	8	10	10	11	12	13	14	15	15	16	17	18	19	20	21	22	23	24	25

A lower bound of αn, with $\alpha = 3/(1 + \sqrt{5}) = 0.927...$, has been given by [3] then [4]. In [3], Franek, Simpson and Smyth propose a sequence of strings (x_n) with increasing lengths such that $\lim_{n \mapsto \infty} r(x_n)/|x_n| = \alpha$, where $r(x)$ is the number of runs in the string x. In [4], Franek and Yang show that α is an asymptotic lower bound by showing that there exists a whole family of asymptotic lower bounds arbitrarily close to α.

In fact, very few properties are known for the $\rho(n)/n$ function [4,9]. In this paper, after giving some definitions (Section 2), we show by a simple rewriting argument that $\ell = \lim_{n \mapsto \infty} \rho(n)/n$ exists and that this limit is never reached (Section 3), proving that

$$\frac{\rho(n)}{n} \leq \ell - \frac{1}{4n}$$

Section 4 proves the convergence of $\rho(n)/n$ even in the case of a fixed alphabet, for example for binary strings. Moreover, we further study the asymptotic behavior of $\rho_p(n)$, the number of runs with short periods (Section 5), showing that $\ell_p = \lim_{n \mapsto \infty} \rho_p(n)/n$ exists and that, for some constant z_p,

$$\ell_p - \frac{z_p}{n} \leq \frac{\rho_p(n)}{n} \leq \ell_p \leq \ell$$

Practically, this inequality implies that the count of some microruns is *below* n, and this improves the upper bound of [1] to $1.52n$ for binary strings. Section 6 gives some concluding remarks.

2   Definitions

Let $x = x_1 x_2 \ldots x_n$ be a string over an alphabet. Let $p \geq 1$ be an integer. We say that x has a *period* p if for any i with $1 \leq i \leq n - p$, $x_{i+p} = x_i$. We denote by $x[i...j]$ the substring $x_i x_{i+1} \ldots x_j$. A *run* is a substring $x[i...j]$:

- which has a period $p \leq \lfloor (j - i + 1)/2 \rfloor$,
- that is maximal : if they exist, neither $x_{i-1} = x_{i-1+p}$, nor $x_{j+1} = x_{j+1-p}$,
- and such that $x[i...i+p-1]$ is primitive : it is not an integer power of another string.

We define by $r_p(x)$ the number of runs of period $\leq p$ in x, called *microruns* in [1], and by $r(x) = r_{\lfloor |x|/2 \rfloor}(x)$ the total number of runs in x. For example, the four runs of $x = $ atattatt are $x[4,5] = $ tt, $x[7,8] = $ tt, $x[1,4] = $ atat and $x[2,8] = $ tattatt, and thus $r_1(x) = 2$, $r_2(x) = 3$, and $r_3(x) = r(x) = 4$.

Given an integer $n \geq 2$, we now consider all strings of length n. We define as

$$\rho_p(n) = \max\{r_p(x) \mid |x| = n\}$$

the maximal number of runs of period $\leq p$ in a string of length n. Then we define as

$$\rho(n) = \max\{r(x) \mid |x| = n\} = \rho_{\lfloor n/2 \rfloor}(n)$$

the maximal total number of runs in a string of length n. Kolpakov and Kucherov gave in [6] some values for $\rho(n)$ (Table 1). Table 3, at the end of this paper, shows some values for $\rho_p(n)$. Note that $r(x) = \rho(|x|)$ does not imply that $r_p(x) = \rho_p(|x|)$ for all p : for example, $r(\text{aatat}) = 2 = \rho(5)$ but $r_1(\text{aatat}) = 1 < \rho_1(5) = 2$.

Finally, we can define values $r_{\geq p}(x)$ and $\rho_{\geq p}(n)$ for *macroruns*, that is runs with a period at least p. Again, $r(x) = \rho(|x|)$ does not imply that $r_{\geq p}(x) = \rho_{\geq p}(|x|)$. For example, $r_{\geq 2}(\text{aatt}) = 0 < \rho_{\geq 2}(4) = 1 = r_{\geq 2}(\text{atat})$.

3   Asymptotic Behavior of the Number of Runs

Franek and al. [3,4] list some known properties for $\rho(n)$:

- For any n, $\rho(n+2) \geq \rho(n) + 1$
- For any n, $\rho(n+1) \leq \rho(n) + \lfloor n/2 \rfloor$
- For some n, $\rho(n+1) = \rho(n)$
- For some n, $\rho(n+1) = \rho(n) + 2$

We add the following two simple properties.

Proposition 1. *The function ρ is superadditive : for any m and n, we have* $\rho(m+n) \geq \rho(m) + \rho(n)$.

Proof. Take two strings x et y of respective lengths m and n such that $r(x) = \rho(m)$ and $r(y) = \rho(n)$. Let $\bar{y}$ be a rewriting of y with characters not present in x. Then $x\bar{y}$ is a string of length $m + n$ containing exactly the runs of x and the rewritten runs of y. Thus $\rho(m + n) \geq r(x\bar{y}) = r(x) + r(y) = \rho(m) + \rho(n)$.

Proposition 2. *For any n, $\rho(4n) \geq 4\rho(n) + 1$*

Proof. Take a string x of length n with $r(x) = \rho(n)$. Let $\bar{x}$ be a rewriting of x with characters not present in x. Then $r(x\bar{x}x\bar{x}) \geq 4r(x) + 1$.

We have in particular $\rho(tn) \geq t\rho(n)$, giving our main result :

Theorem 1. *$\rho(n)/n$ converges to its upper limit ℓ. Moreover, the limit is never reached, as for any n we have*

$$\frac{\rho(n)}{n} \leq \ell - \frac{1}{4n}$$

Proof. Let ℓ be the upper limit of $\rho(n)/n$. This limit is finite because of [6]. Given ε, there is a n_0 such that $\rho(n_0)/n_0 \geq \ell - \varepsilon/2$. For any $n \geq n_0$, let be $t = \lfloor n/n_0 \rfloor$. Then we have $\rho(n)/n \geq \rho(tn_0)/n \geq t\rho(n_0)/n$ by Proposition 1, thus $\rho(n)/n \geq t/(t+1) \cdot \rho(n_0)/n_0$. Let be t_0 such that $t_0/(t_0 + 1) \cdot \rho(n_0)/n_0 \geq \rho(n_0)/n_0 - \varepsilon/2$. Then, for any $n \geq t_0 n_0$, we have $\rho(n)/n \geq \ell - \varepsilon$, thus $\ell = \lim_{n \mapsto \infty} \rho(n)/n$. Finally, Proposition 2 gives $\ell \geq \rho(4n)/4n \geq \rho(n)/n + \frac{1}{4n}$.

The proof of convergence of $f(n)/n$ when f is superadditive is known as Fekete's Lemma [2,10]. This convergence result was an open question of [4]. In fact, the motivation of [4] was the remark that *"the sequence $|x_i|$ (of [3]) is only "probing" the domain of the function $\rho(n)$ and $r(x_i)$ is "pushing" the value of $\rho(n)$ above αn in these "probing" points"*. Then Franek and Yang [4] prove that every $\alpha - \varepsilon$ is an actual asymptotic lower bound by building specific sequences. With Theorem 1, it is now sufficient to study bounds on any $(\rho(n_i)/n_i)$ sequence (for a growing sequence (n_i)) to give bounds on $\rho(n)/n$.

Note that this convergence does not imply monotonicity. In fact, if $\ell < 1$, then $\rho(n)/n$ is asymptotically non monotonic, as there will be in this case an infinity of n's such that $\rho(n+1) = \rho(n)$. Note also that, although Proposition 1 and 2 require to double the alphabet size, the alphabet remains finite : the proof of Theorem 1 only requires to double once this alphabet size. Moreover, it is possible to prove Proposition 1 without rewriting in a larger alphabet, thus proving the convergence of $\rho(n)/n$ when considering only binary strings. This second proof, more elaborated, is given in the next section.

The bound $\ell - \frac{1}{4n}$ can be improved. For example, with a rewriting similar to the one used in Proposition 2, it can be shown that $\rho(2n^2) \geq (2n+1)\rho(n)$, giving by successive iterations $\rho(n)/n \leq \ell - \frac{1}{2n}$. This has not been reported here to keep the proof simple.

Concerning microruns with period at most p, Proposition 1 still holds :

Proposition 3. *For any p, m, and n, we have $\rho_p(m+n) \geq \rho_p(m) + \rho_p(n)$. Thus for any p, $\rho_p(n)/n$ converges to its upper limit ℓ_p.*

The proof is the same as above. On the contrary, Proposition 2 may be not true for microruns. For example, for any n, $\rho_1(n) = \lfloor n/2 \rfloor$, and thus for any even n, we have $\rho_1(n)/n = \ell_1 = 1/2$.

Finally, Theorem 1 is fully valid for macroruns, and as a rewriting argument shows that $\rho_{\geq p}(n) \geq \rho(\lfloor n/p \rfloor)$, we have $\ell_{\geq p} \geq \ell/p$.

4   A Proof of Proposition 1 for Fixed Alphabets

Here we prove Proposition 1 without rewriting in a larger alphabet, thus proving the convergence of $\rho(n)/n$ when considering only binary strings. This proof is borrowed and simplified from one part of a proof of Franek and al. (Theorem 2 of [3]). A key observation is that some runs of x and y are merged in xy only when a word z^2 is both a suffix of x and a prefix of y (case a_2 on Figure 1). We first have this property :

Proposition 4. *Let Σ be an alphabet with $|\Sigma| \geq 2$, and let x and y be strings on Σ such that $|y| \geq 1$. Then there exists strings x' and y' on Σ such that $|x'| + |y'| = |x| + |y|$, $|y'| < |y|$ and $r(x') + r(y') \geq r(x) + r(y)$.*

Proof. Let w be the largest string, eventually empty, such that w is a suffix of x and a prefix of y. Thus $x = uw$ and $y = wv$ for some strings u and v. Let

$x' = uwv$ and $y' = w$. Clearly $|x'| + |y'| = |x| + |y|$ and $|y'| \leq |y|$. Without loss of generality, we assume that y is not a suffix of x. (If it is not the case, we rewrite y into $\bar{y}$ using an isomorphism of Σ onto itself.) Thus $|y'| < |y|$. Now we count the runs of x and y. The runs of period p that have $2p$ characters ("a square") completely included in w were once in x and once in y. Such runs can be found again once in x' and once in y'. By definition of w, all the others runs of x and y are found exactly once in x', without being merged.

To prove Proposition 1, we take two strings x_0 and y_0 of respective lengths m and n such that $r(x_0) = \rho(m)$ and $r(y_0) = \rho(n)$. Applying recursively Proposition 4 gives a finite sequence of pairs of strings $(x_0, y_0), (x_1, y_1), \ldots (x_t, y_t)$ with $r(x_i) + r(y_i) \geq r(x_{i-1}) + r(y_{i-1})$ and $|y_0| > |y_1| > \ldots > |y_t| = 0$ for some t. Finally $|x_t| = |x_0| + |y_0| = m + n$, and thus $\rho(m + n) \geq r(x_t) \geq r(x_0) + r(y_0) = \rho(m) + \rho(n)$, proving Proposition 1.

Note that the proof of Franek and al. in [3] was in a different context, and that no result leading to our Proposition 1 was stated as such in their paper.

5   On the Number of Microruns

In this section, we study the asymptotic behavior of the number of microruns beyond the result of Proposition 3. Additionally, we provide a new bound on the number of some microruns (see the end of the section).

The idea to bound the number of microruns is to count the *new runs* created by the concatenation of two strings. Let x and y be two strings, and s be a run of xy with period q. Then s is exactly in one of the following two cases (Figure 1) :

- a) s has a substring that is a run (with the same period q) completely included in x, or in y, or in both;
- b) s has strictly less than 2 periods in x and in y.

We call the runs in the case b) the *new runs* between x and y, and we denote by $z_p(x, y)$ the number of such runs. Then $r_p(xy) \leq r_p(x) + r_p(y) + z_p(x, y)$, the

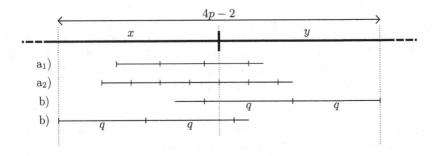

Fig. 1. a_1) Runs with a substring that is a run included in x. a_2) Runs with two substrings that are runs both in x and y. b) "New runs" between x and y. To count the new runs with period $q \leq p$, it is sufficient to consider words of length $4p - 2$.

inequality coming from the fact that a run from x can be merged with a run from y (case a_2 on Figure 1). We can bound the number of new runs, and thus have an upper bound on $r_p(xy)$:

Proposition 5. *Let* $z_p = \max\{z_p(x, y) \mid |x| = |y| = 2p-1\}$ *the maximal number of new runs between words of length* $2p - 1$. *Then, for every strings* x *and* y *of any length, we have* $z_p(x, y) \leq z_p$.

Proposition 6. *For any* p, m, *and* n, $\rho_p(m + n) \leq \rho_p(m) + \rho_p(n) + z_p$.

Proof. (Proposition 5.) Any new run with period $q \leq p$ has less than $2q - 1 \leq 2p - 1$ characters in x, and in y (Figure 1). (Proposition 6.) Let x and y be two strings such that $|x| = m$, $|y| = n$, and $r_p(xy) = \rho_p(m + n)$. Then $\rho_p(m + n) = r_p(xy) \leq r_p(x) + r_p(y) + z_p(x, y) \leq \rho_p(m) + \rho_p(n) + z_p$.

Table 2 provides actual values of z_p for small values of p. An immediate bound on z_p is $z_p \leq z_{p-1} + 2$. Knowing bounds on z_p helps to characterize the asymptotic behavior of the number of microruns :

Theorem 2. *For any* p *and* n, *we have* $\ell_p \leq \rho_p(n)/n + z_p/n$, *and thus*

$$\ell_p - \frac{z_p}{n} \leq \frac{\rho_p(n)}{n} \leq \ell_p \leq \ell$$

Proof. By Proposition 6, for any $t \geq 1$, we have $\rho_p(tn) \leq t\rho_p(n) + (t - 1)z_p$. Thus $\rho_p(tn)/tn \leq \rho_p(n)/n + \frac{t-1}{t}z_p/n$. Taking this inequality to the limit gives the result.

Thus we know that the convergence of $\rho_p(n)/n$ to ℓ_p is faster than z_p/n. Note that we do not have a similar result for $\rho(n)$, as we do not have a convenient way to bound $\rho(m + n)$ like in Proposition 5.

With Theorem 2, one can show that the number of some microruns is *below* n. For example, we have for binary strings $z_9 = 7$ and $\rho_9(34) = 26$, thus $\ell_9 \leq 33/34 = 0.970....$ Thus *there are less than* $0.971n$ *runs of period* ≤ 9 *in any binary string of length* n. This result is better than Lemma 2 of [1] which proved the n bound by the count of amortizing positions for centers of runs.

Using the result of Crochemore and Ilie's Proposition 1 [1] for large runs, we get an upper bound on $\rho(n)/n$. The best bound we obtain, with $z_{10} = 7$ and $\rho_{10}(34) = 26$, gives $\ell_{10} \leq 33/34$, and finally, with Crochemore and Ilie's Proposition 1 :

$$\ell \leq \ell_{10} + \frac{6}{11} = 1.516...$$

Thus *the number of runs in a binary string of length* n *is not more than* $1.52n$. This result could be further extended by choosing other periods for the count of microruns, by computation or by other techniques.

Table 2. Values for z_p for binary strings with worst-case examples of length $\leq 4p - 2$

p	z_p	example
1, 2	1	t t
3	2	ttat ta
4	4	ataaata attaat
5, 6, 7	5	ttatatta taatataa
8	6	ttttattattttat taattattaa
9, 10	7	ttatatatattatata taatatatataa

Table 3. Values of $\rho_p(n)$ for small values of n and p for binary strings. For each n, the value in bold shows the smallest p such that $\rho_p(n) = \rho(n)$.

n	5	6	7	8	9	10	11	12	13	14	15	16	17	18	19	20	21	22	23	24	25	26	27	28	29	30	31	32	33	34	35
$\rho(n)$	2	3	4	5	5	6	7	8	8	10	10	11	12	13	14	15	15	16	17	18	19	20	21	22	23	24	25	26	27	27	28
1, 2	**2**	**3**	3	4	4	5	5	6	6	7	7	8	8	9	9	10	10	11	11	12	12	13	13	14	14	15	15	16	16	17	17
3		3	**4**	4	**5**	**6**	6	7	**8**	9	9	10	11	12	12	13	14	15	15	16	17	18	18	19	20	21	21	22	23	24	24
4				**5**	5	6	**7**	**8**	8	9	9	10	11	12	13	13	14	15	16	17	18	18	19	20	21	21	22	23	24	24	25
5						6	7	8	8	9	**10**	**11**	11	12	13	14	14	15	16	17	18	18	19	20	21	21	22	23	24	25	25
6								8	8	9	10	11	11	12	13	14	14	15	16	17	18	18	19	20	21	21	22	23	24	25	25
7										**10**	10	11	**12**	**13**	**14**	14	**15**	15	16	**18**	18	19	20	21	21	22	23	24	25	26	26
8												11	12	13	14	**15**	15	**16**	**17**	18	18	19	20	21	22	23	24	25	26	26	27
9														13	14	15	15	16	17	18	18	19	20	21	22	23	24	25	26	26	27
10																15	15	16	17	18	18	19	20	21	22	23	24	25	26	26	27
11																		16	17	18	**19**	19	**21**	21	22	23	24	25	26	26	27
12																				18	19	19	21	21	22	23	24	25	26	26	27
13																						**20**	21	**22**	**23**	**24**	**25**	**26**	**27**	**27**	**28**
14																								22	23	24	25	26	27	27	28

6   Perspectives

The results on the asymptotic behavior of the functions ρ and ρ_p of Theorems 1 and 2 simplify the research on lower and upper bounds. We hope that these results will bring a better understanding of the number of runs and be a step towards proving the conjecture of [5] ($\ell \leq 1$) or the stronger conjecture of [3] ($\ell = 0.927...$).

A side result of Theorem 2 was a new upper bound for some microruns, and thus an upper bound for the total number of runs. This upper bound can be lowered again by doing a more precise analysis, theoretical or computational, of the z_p values. This would require large evaluations of some z_p and $\rho_p(n)$ values. Other techniques could provide better bounds for microruns. For example, it should be possible to push the idea of Crochemore and Ilie by finding more amortizing positions than the number of centers of runs in a given interval of positions. Again, when the number of possible positions grows, the complexity of their method increases.

For the lower bound, it remains to be shown if one can find strings with more runs than those of [3,4]. Although Theorem 1 also provides a way to have a lower bound on $\rho(n)/n$, all the computations we ran gave not better bounds than the 0.927 bound of [3,4].

Now an important question is if the actual value of ℓ can be found with such a separation between microruns and macroruns. The inequality $\ell \leq \ell_p + \ell_{\geq p+1}$

may be strict for some p, as the values $\ell_1 = 1/2$ and $\ell_{\geq 2} > 1/2$ may suggest. If this inequality is strict for several p's, the conjecture may be impossible to prove by this way if one choose a bad splitting period p.

Another open question is if one of the constants $\ell_p = \lim_{n \mapsto \infty} \rho_p(n)/n$ is equal to ℓ, or if, more probably, the limit ℓ is obtained by considering asymptotically runs with any period. Finally, it remains to be proven if strings on binary alphabets can always achieve the highest number of runs.

References

1. Crochemore, M., Ilie, L.: Maximal repetitions in strings. J. Comput. Systems Sci. 74(5), 796–807 (2008)
2. Fekete, M.: Über die Verteilung der Wurzeln bei gewissen algebraischen Gleichungen mit ganzzahligen Koeffizienten. Mathematische Zeitschrift 17, 228–249 (1923)
3. Franek, F., Simpson, R.J., Smyth, W.F.: The maximum number of runs in a string. In: Proceedings of the 2003 Australasian Workshop on Combinatorial Algorithms (AWOCA 2003), pp. 26–35 (2003)
4. Franek, F., Yang, Q.: An asymptotic lower bound for the maximal-number-of-runs function. In: Prague Stringology Conference 2006, pp. 3–8 (2006)
5. Kolpakov, R., Kucherov, G.: Maximal repetitions in words or how to find all squares in linear time. Technical Report 98-R-227, LORIA (1998)
6. Kolpakov, R., Kucherov, G.: On maximal repetitions in words. Journal on Discrete Algorithms 1(1), 159–186 (2000)
7. Puglisi, S.J., Simpson, J., Smyth, B.: How many runs can a string contain? Theoretical Computer Science 401(1-3), 165–171 (2008)
8. Rytter, W.: The number of runs in a string: improved analysis of the linear upper bound. Information and Computation 205(9), 1459–1469 (2007)
9. Smyth, B.: The maximum number of runs in a string. In: International Workshop on Combinatorial Algorithms (IWOCA 2007), Problems Session (2007)
10. van Lint, J.L., Wilson, R.M.: A course in combinatorics. Cambridge University Press, Cambridge (1992)

A Hybrid Approach to Word Segmentation of Vietnamese Texts

Lê Hông Phuong[1], Nguyên Thi Minh Huyên[2], Azim Roussanaly[1], and Hô Tuòng Vinh[3]

[1] LORIA, Nancy, France
[2] Vietnam National University, Hanoi, Vietnam
[3] IFI, Hanoi, Vietnam

Abstract. We present in this article a hybrid approach to automatically tokenize Vietnamese text. The approach combines both finite-state automata technique, regular expression parsing and the maximal-matching strategy which is augmented by statistical methods to resolve ambiguities of segmentation. The Vietnamese lexicon in use is compactly represented by a minimal finite-state automaton. A text to be tokenized is first parsed into lexical phrases and other patterns using pre-defined regular expressions. The automaton is then deployed to build linear graphs corresponding to the phrases to be segmented. The application of a maximal-matching strategy on a graph results in all candidate segmentations of a phrase. It is the responsibility of an ambiguity resolver, which uses a smoothed bigram language model, to choose the most probable segmentation of the phrase. The hybrid approach is implemented to create *vnTokenizer*, a highly accurate tokenizer for Vietnamese texts.

1 Introduction

As many occidental languages, Vietnamese is an alphabetic script. Alphabetic scripts usually separate words by blanks and a tokenizer which simply replaces blanks with word boundaries and cuts off punctuation marks, parentheses and quotation marks at both ends of a word, is already quite accurate [5]. However, unlike other languages, in Vietnamese blanks are not only used to separate words, but they are also used to separate syllables that make up words. Furthermore, many of Vietnamese syllables are words by themselves, but can also be part of multi-syllable words whose syllables are separated by blanks between them. In general, the Vietnamese language creates words of complex meaning by combining syllables that most of the time also possess a meaning when considered individually. This linguistic mechanism makes Vietnamese close to that of syllabic scripts, like Chinese. That creates problems for all natural language processing tasks, complicating the identification of what constitutes a word in an input text.

Many methods for word segmentation have been proposed. These methods can be roughly classified as either dictionary-based or statistical methods, while many state-of-the-art systems use hybrid approaches [6].

C. Martín-Vide, F. Otto, and H. Fernau (Eds.): LATA 2008, LNCS 5196, pp. 240–249, 2008.

We present in this paper an efficient hybrid approach for the segmentation of Vietnamese text. The approach combines both finite-state automata technique, regular expression parsing and the maximal-matching method which is augmented by statistical methods to deal with ambiguities of segmentation. The rest of the paper is organized as follows. The next section gives the construction of a minimal finite-state automaton that encodes the Vietnamese lexicon. Sect. 3 discusses the application of this automaton and the hybrid approach for word segmentation of Vietnamese texts. The developed tokenizer for Vietnamese and its experimental results are shown in Sect. 4. Finally, we conclude the paper with some discussions in Sect. 5.

2   Lexicon Representation

In this section, we first briefly describe the Vietnamese lexicon and then introduce the construction of a minimal deterministic, acyclic finite-state automaton that accepts it.

2.1   Vietnamese Lexicon

The Vietnamese lexicon edited by the Vietnam Lexicography Center (Vietlex[1]) contains $40,181$ words, which are widely used in contemporary spoken language, newspapers and literature. These words are made up of $7,729$ syllables. It is noted that Vietnamese is an inflexionless language, this means that every word has exactly one form.

There are some interesting statistics about lengths of words measured in syllables as shown in Table 1. Firstly, there are about 81.55% of syllables which are words by themselves, they are called single words; 15.69% of words are single ones. Secondly, there are 70.72% of compound words which are composed of two syllables. Finally, there are $13,59\%$ of compounds which are composed of at least three syllables; only $1,04\%$ of compounds having more than four syllables.

Table 1. Lengths of words measured in syllables

Length	#	%
1	6,303	15.69
2	28,416	70.72
3	2,259	5.62
4	2,784	6.93
≥ 5	419	1.04
Total	40,181	100

The high frequency of two-syllable compounds suggests us a simple but efficient method to resolve ambiguities of segmentation. The next paragraph presents the representation of the lexicon.

---

[1] http://www.vietlex.com/

2.2   Lexicon Representation

Minimal deterministic finite state automata (MDFA) have been known to be the best representation of a lexicon. They are not only compact but also give the optimal access time to data [1]. The Vietnamese lexicon is represented by an MDFA.

We implement an algorithm developed by J. Daciuk *et al.* [2] that incrementally builds a minimal automaton in a single phase by adding new strings one by one and minimizing the resulting automaton on-the-fly.

The minimal automaton that accepts the Vietnamese lexicon contains $42,672$ states in which $5,112$ states are final ones. It has $76,249$ transitions; the maximum number of outgoing transitions from a state is 85, and the maximum number of incoming transitions to a state is $4,615$. The automaton operates in optimal time in the sense that the time to recognize a word corresponds to the time required to follow a single path in the deterministic finite-state machine, and the length of the path is the length of the word measured in characters.

3   Vietnamese Word Segmentation

We present in this section an application of the lexicon automaton for the word segmentation of Vietnamese texts. We first give the specification of segmentation task.

3.1   Segmentation Specification

We have developed a set of segmentation rules based on the principles discussed in the document of the ISO/TC 37/SC 4 work group on word segmentation (2006) [3]. Notably, the segmentation of a corpus follows the following rules:

1. *Compounds:* word compounds are considered as words if their meaning is not compound from their sub parts, or if their usage frequency justifies it.
2. *Derivation:* when a bound morpheme is attached to a word, the result is considered as a word. The reduplication of a word (common phenomenon in Vietnamese) also gives a lexical unit.
3. *Multiword expressions:* expressions such as "because of" are considered as lexical units.
4. *Proper names:* name of people and locations are considered as lexical units.
5. *Regular patterns:* numbers, times and dates are recognized as lexical units.

3.2   Word Segmentation

An input text for segmentation is first analyzed by a regular expression recognizer for detection of regular patterns such as proper names, common abbreviations, numbers, dates, times, email addresses, URLs, punctuations, *etc.* The recognition of arbitrary compounds, derivation, and multiword expressions is committed to a regular expression that extracts phrases of the text.

The regular recognizer analyzes the text using a greedy strategy in that all patterns are scanned and the longest matched pattern is taken out. If a pattern is a phrase, that is a sequence of syllables and spaces, it is passed to a segmenter for detection of word composition. In general, a phrase usually has several different word compositions; nevertheless, there is typically one correct composition which the segmenter need to determine.

A simple segmenter could be implemented by the maximal matching strategy which selects the segmentation that contains the fewest words [8]. In this method, the segmenter determines the longest syllable sequence which starts at the current position and is listed in the lexicon. It takes the recognized pattern, moves the position pointer behind the pattern, and starts to scan the next one. Although this method works quite well since long words are more likely to be correct than short words. However, this is a too greedy method which sometimes leads to wrong segmentation because of a large number of overlapping candidate words in Vietnamese. Therefore, we need to list all possible segmentations and design a strategy to select the most probable correct segmentation from them.

A phrase can be formalized as a sequence of blank-separated syllables $s_1 s_2 \cdots s_n$. We ignore for the moment the possibility of seeing a new syllable or a new word in this sequence. Due to the fact that, as we showed in the previous section, most of Vietnamese compound words are composed of two syllables, the most frequent case of ambiguities involves three consecutive syllables $s_i s_{i+1} s_{i+2}$ in which both of the two segmentations $(s_i s_{i+1})(s_{i+2})$ and $(s_i)(s_{i+1} s_{i+2})$ may be correct, depending on context. This type of ambiguity is called *overlap ambiguity*, and the string $s_i s_{i+1} s_{i+2}$ is called an overlap ambiguity string.

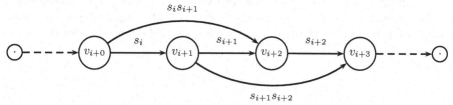

Fig. 1. Graph representation of a phrase

The phrase is represented by a linearly directed graph $G = (V, E)$, $V = \{v_0, v_1, \ldots, v_n, v_{n+1}\}$, as shown in Fig. 1. Vertices v_0 and v_{n+1} are respectively the start and the end vertex; n vertices $v_1, v_2, \ldots, v_n$ are aligned to n syllables of the phrase. There is an arc (v_i, v_j) if the consecutive syllables $s_{i+1}, s_{i+2}, \ldots, s_j$ compose a word, for all $i < j$. If we denote accept$(\mathcal{A}, s)$ the fact that the lexicon automaton $\mathcal{A}$ accepts the string s, the formal construction of the graph for a phrase is shown in Algorithm 1. We can then propose all segmentations of the phrase by listing all shortest paths on the graph from the start vertex to the end vertex.

As illustrated in Fig. 1, each overlap ambiguity string results in an ambiguity group, therefore, if a graph has k ambiguity groups, there are 2^k segmentations of the underlying phrase[2]. For example, the ambiguity group in Fig. 1 gives two segmentations $(s_i s_{i+1})s_{i+2}$ and $s_i(s_{i+1} s_{i+2})$.

---

[2] If these ambiguity groups do not overlap each other.

Algorithm 1. Construction of the graph for a phrase $s_1 s_2 \ldots s_n$

1: $V \leftarrow \emptyset$;
2: **for** $i = 0$ **to** $n + 1$ **do**
3:    $V \leftarrow V \cup \{v_i\}$;
4: **end for**
5: **for** $i = 0$ **to** n **do**
6:    **for** $j = i$ **to** n **do**
7:       **if** (accept($\mathcal{A}_W, s_i \cdots s_j$)) **then**
8:          $E \leftarrow E \cup \{(v_i, v_{j+1})\}$;
9:       **end if**
10:    **end for**
11: **end for**
12: **return** $G = (V, E)$;

We discuss in the next subsection the ambiguity resolver which we develop to choose the most probable segmentation of a phrase in the case it has overlap ambiguities.

3.3   Resolution of Ambiguities

The ambiguity resolver uses a bigram language model which is augmented by the linear interpolation smoothing technique.

In n-gram language modeling, the probability of a string $P(s)$ is expressed as the product of the probabilities of the words that compose the string, with each word probability conditional on the identity of the last $n - 1$ words, *i.e.*, if $s = w_1 \cdots w_m$ we have

$$P(s) = \prod_{i=1}^{m} P(w_i | w_1^{i-1}) \approx \prod_{i=1}^{m} P(w_i | w_{i-n+1}^{i-1}), \tag{1}$$

where w_i^j denotes the words $w_i \cdots w_j$. Typically, n is taken to be two or three, corresponding to a bigram or trigram model, respectively.[3]

In the case of a bigram model $n = 2$, to estimate the probabilities $P(w_i | w_{i-1})$ in (1), we can use training data, and take the maximum likelihood (ML) estimate for $P(w_i | w_{i-1})$ as follows

$$P_{ML}(w_i | w_{i-1}) = \frac{P(w_{i-1} w_i)}{P(w_{i-1})} = \frac{c(w_{i-1} w_i)/N}{c(w_{i-1})/N} = \frac{c(w_{i-1} w_i)}{c(w_{i-1})},$$

where $c(\alpha)$ denotes the number of times the string α occurs and N is the total number of words in the training data.

The maximum likelihood estimate is a poor one when the amount of training data is small compared to the size of the model being built, as is generally the

---

[3] To make the term $P(w_i | w_{i-n-1}^{i-1})$ meaningful for $i < n$, one can pad the beginning of the string with a distinguished token. We assume there are $n - 1$ such distinguished tokens preceding each phrase.

case in language modeling. A zero bigram probability can lead to errors of the modeling. Therefore, a variety of smoothing techniques have been developed to adjust the maximum likelihood estimate in order to produce more accurate probabilities. Not only do smoothing methods generally prevent zero probabilities, but they also attempt to improve the accuracy of the model as a whole. Whenever a probability is estimated from few counts, smoothing has the potential to significantly improve estimation [7].

We adopt the linear interpolation technique to smooth the model. This is a simple yet effective smoothing technique which is widely used in the domain of language modeling [4]. In this method, the bigram model is interpolated with a unigram model $P_{ML}(w_i) = c(w_i)/N$, a model that reflects how often each word occurs in the training data. We take our estimate $\widehat{P}(w_i|w_{i-1})$ to be

$$\widehat{P}(w_i|w_{i-1}) = \lambda_1 P_{ML}(w_i|w_{i-1}) + \lambda_2 P_{ML}(w_i), \tag{2}$$

where $\lambda_1 + \lambda_2 = 1$ and $\lambda_1, \lambda_2 \geq 0$.

The objective of smoothing techniques is to improve the performance of a language model, therefore the estimation of λ values in (2) is related to the evaluation of the language model. The most common metric for evaluating a language model is the probability that the model assigns to test data, or more conveniently, the derivative measured of entropy. For a smoothed bigram model that has probabilities $p(w_i|w_{i-1})$, we can calculate the probability of a sentence $P(s)$ using (1). For a test set T composed of n sentences $s_1, s_2, \ldots, s_n$, we can calculate the probability $P(T)$ of the test set as the product of the probabilities of all sentences in the set $P(T) = \prod_{i=1}^{n} P(s_i)$. The entropy $H_p(T)$ of the model on data T is defined by

$$H_p(T) = \frac{-\log_2 P(T)}{N_T} = -\frac{1}{N_T} \sum_{i=1}^{n} \log_2 P(s_i), \tag{3}$$

where N_T is the length of the text T measured in words. The entropy is inversely related to the average probability a model assigns to sentences in the test data, and it is generally assumed that lower entropy correlates with better performance in applications.

Starting from a part of the training set which is called the "validation" data, we define $C(w_{i-1}, w_i)$ to be the number of times the bigram (w_{i-1}, w_i) is seen in the validation set. We need to choose λ_1, λ_2 to maximize

$$L(\lambda_1, \lambda_2) = \sum_{w_{i-1}, w_i} C(w_{i-1}, w_i) \log_2 \widehat{P}(w_i|w_{i-1}) \tag{4}$$

such that $\lambda_1 + \lambda_2 = 1$, and $\lambda_1, \lambda_2 \geq 0$.

The λ_1 and λ_2 values can be estimated by an iterative process given in Algorithm 2. Once all the parameters of the bigram model have been estimated, the smoothed probabilities of bigrams can be easily computed by (2). These results are used by the resolver to choose the most probable segmentation of a

---

Algorithm 2. Estimation of values λ

1: $\lambda_1 \leftarrow 0.5, \lambda_2 \leftarrow 0.5$;
2: $\epsilon \leftarrow 0.01$;
3: **repeat**
4:     $\widehat{\lambda}_1 \leftarrow \lambda_1, \quad \widehat{\lambda}_2 \leftarrow \lambda_2$;
5:     $c_1 \leftarrow \sum_{w_{i-1}, w_i} \frac{C(w_{i-1}, w_i)\lambda_1 P_{ML}(w_i|w_{i-1})}{\lambda_1 P_{ML}(w_i|w_{i-1}) + \lambda_2 P_{ML}(w_i)}$;
6:     $c_2 \leftarrow \sum_{w_{i-1}, w_i} \frac{C(w_{i-1}, w_i)\lambda_2 P_{ML}(w_i)}{\lambda_1 P_{ML}(w_i|w_{i-1}) + \lambda_2 P_{ML}(w_i)}$;
7:     $\lambda_1 \leftarrow \frac{c_1}{c_1 + c_2}, \quad \lambda_2 \leftarrow 1 - \widehat{\lambda}_1$;
8:     $\widehat{\epsilon} \leftarrow \sqrt{(\widehat{\lambda}_1 - \lambda_1)^2 + (\widehat{\lambda}_2 - \lambda_2)^2}$;
9: **until** $(\widehat{\epsilon} \leq \epsilon)$;
10: **return** λ_1, λ_2;

---

phrase, say, s, by comparing probabilities $P(s)$ which is estimated using (1). The segmentation with the greatest probability will be chosen.

We present in the next section the experimental setup and obtained results.

4   Experiments

We present in this section the experimental setup and give a report on results of experiments with the hybrid approach presented in the previous sections. We also describe briefly vnTokenizer, an automatic software for segmentation of Vietnamese texts.

4.1   Corpus Constitution

The corpus upon which we evaluate the performance of the tokenizer is a collection of 1264 articles from the "Politics – Society" section of the Vietnamese newspaper *Tui tr* (The Youth), for a total of $507,358$ words that have been manually spell-checked and segmented by linguists from the Vietnam Lexicography Center. Although there can be multiple plausible segmentations of a given Vietnamese sentence, only a single correct segmentation of each sentence is kept. We assume a single correct segmentation of a sentence for two reasons. The first one is of its simplicity. The second one is due to the fact that we are not currently aware of any effective way of using multiple segmentations in typical applications concerning Vietnamese processing.

We perform a 10-fold cross validation on the test corpus. In each experiment, we take 90% of the gold test set ($\approx 456,600$ lexical units) as training set, and 10% as test set. We present in the next paragraph the training and results of the model.

4.2   Results

In an experiment, the bigram language model is trained on a training set. An estimation of parameters λs in the Algorithm 2 is given in Table 2. With a given error $\epsilon = 0.03$, the estimated parameters converge after four iterations.

Table 2. Estimation of lambda values

Step	λ_1	λ_2	ϵ
0	0.500	0.500	1.000
1	0.853	0.147	0.499
2	0.952	0.048	0.139
3	0.981	0.019	0.041
4	0.991	0.009	0.015

The above experimental results reveal a fact that the smoothing technique basing on the linear interpolation adjusts well bigram and unigram probabilities, it thus improves the estimation and the accuracy of the model as a whole. Table 3 presents the values of precisions, recalls and F-measures of the system on two versions with or without ambiguity resolution. Precision is computed as the count of common tokens over tokens of the automatically segmented files, recall as the count of common tokens over tokens of the manually segmented files, and F-measure is computed as usual from these two values.

Table 3. Precision, recall and F-measure of the system

Precision	Recall	F-measure
0.948	0.960	0.954
0.950	0.963	0.956

The system has good recall ratios, about 96%. However, the use of the resolver for resolution of ambiguities only slightly improves the overall accuracy. This can be explained by the fact that the bigram model exploits a small amount of training data compared to the size of the universal language model. It is hopeful that the resolver may improve further the accuracy if it is trained on larger corpora.

4.3   vnTokenizer

We have developed a software tool named *vnTokenzier* that implements the presented approach for automatic word segmentation of Vietnamese texts. The tool is written in Java and bundled as an Eclipse plug-in and it has already been integrated into *vnToolkit*, an Eclipse Rich Client[4] application which is intended to be a general framework integrating tools for processing of Vietnamese text. *vnTokenizer* plug-in, *vnToolkit* and related resources, include the lexicon and test corpus are freely available for download[5]. They are distributed under the GNU General Public License[6].

---

[4] http://www.eclipse.org/rcp/
[5] http://www.loria.fr/~lehong/projects.php
[6] http://www.gnu.org/copyleft/gpl.html

5   Conclusion

We have presented an efficient hybrid approach to word segmentation of Vietnamese texts that gives a relatively high accuracy. The approach has been implemented to produce *vnTokenizer*, an automatic tokenizer for Vietnamese texts.

By analyzing results of experiments, we found two types of ambiguity strings in word segmentation of Vietnamese texts: (1) *overlap ambiguity strings* and (2) *combination ambiguity strings*. A sequence of syllables $s_1 s_2 \ldots s_n$ is called a combination ambiguity string if it is a compound word by itself and there exists its sub sequences which are also words by themselves in some context. For instance, the word *b ba* (a kind of pajamas) may be segmented into two words *b* and *ba* (the third wife), and there exists contexts under which this segmentation is both syntactically and semantically correct. Being augmented with a bigram model, our tokenizer is able to resolve effectively overlap ambiguity strings, but combination ambiguity strings have not been discovered. There is a delicate reason, it is that combination ambiguities require a judgment of the syntactic and semantic sense of the segmentation – a task where an agreement cannot be reached easily among different human annotators. Furthermore, we observe that the relative frequency of combination ambiguity strings in Vietnamese is small. In a few ambiguity cases involving bigrams, we believe that a trigram model resolver would work better. These questions would be of interest for further research to improve the accuracy of the tokenizer.

Finally, we found that the majority of errors of segmentation are due to the presence in the texts of compounds absent from the lexicon. Unknown compounds are a much greater source of segmenting errors than segmentation ambiguities. Future efforts should therefore be geared in priority towards the automatic detection of new compounds, which can be performed by means either statistical in a large corpus or rule-based using linguistic knowledge about word composition.

Acknowledgements

The work reported in this article would not have been possible without the enthusiastic collaboration of all the linguists at the Vietnam Lexicography Center. We thank them for their help in data preparation.

References

1. Maurel, D.: Electronic Dictionaries and Acyclic Finite-State Automata: A State of The Art. Grammars and Automata for String Processing (2003)
2. Daciuk, J., Mihov, S., Watson, B.W., Watson, R.E.: Incremental Construction of Minimal Acyclic Finite-State Automata. Computational Linguistics 26(1) (2000)
3. ISO/TC 37/SC 4 AWI N309, Language Resource Management - Word Segmentation of Written Texts for Mono-lingual and Multi-lingual Information Processing - Part I: General Principles and Methods. Technical Report, ISO (2006)

4. Jelinke, F., Mercer, R.L.: Interpolated estimation of Markov source parameters from sparse data. In: Proceedings of the Workshop on Pattern Recognition in Practice, The Netherlands (1980)
5. Schmid, H.: Tokenizing. In: Lüdeling, A., Kytö, M. (eds.) Corpus Linguistics. An International Handbook. Mouton de Gruyter, Berlin (2007)
6. Gao, J., et al.: Chinese Word Segmentation and Named Entity Recognition: A Pragmatic Approach. Computational Linguistics (2006)
7. Chen, S.F., Goodman, J.: An Empirical Study of Smoothing Techniques for Language Modeling. In: Proceedings of the 34th Annual Meeting of the ACL (1996)
8. Wong, P., Chan, C.: Chinese Word Segmentation based on Maximum Matching and Word Binding Force. In: Proceedings of the 16th Conference on Computational Linguistics, Copenhagen, DK (1996)

On Linear Logic Planning and Concurrency

Ozan Kahramanoğulları

Imperial College London, Department of Computing
ozank@doc.ic.ac.uk

Abstract. We present an approach to linear logic planning where an explicit correspondence between partial order plans and multiplicative exponential linear logic proofs is established. This is performed by extracting partial order plans from sound and complete encodings of planning problems in multiplicative exponential linear logic in a way that exhibits a non-interleaving behavioral concurrency semantics. Relying on this fact, we argue that this work is a crucial step for establishing a common language for concurrency and planning that will allow to carry techniques and methods between these two fields.

1 Introduction

Planning[1] and concurrency are two fields of computer science that evolved independently, aiming at solving tasks that are similar in nature but different in perspective: while planning formalisms focus on finding a plan, if there *exists* such a plan, that solves a given planning problem; the focus in concurrency theory is on the global behaviour of a given concurrent system, resulting in *universally quantified* queries, e.g., deadlock freeness, verification of a security protocol. In contrast to approaches to planning, in order to be able to handle such queries, languages for concurrency are equipped with a rich arsenal of mathematical methods that allow for an analysis of equivalence of processes.

In concurrency theory, parallel and sequential composition are expressed at the same level of representation, since they are equivalently important notions for expressing concurrent processes. However, in planning, although parallel behaviour between actions have been studied in partial order planners, e.g., UCPOP [28], Graphplan [1], these investigations focused on increasing the efficiency of the planners. In these approaches, the independence and causality between partially ordered actions, which is crucial from a concurrency theoretic point of view, is often specified by means of linguistic constraints (see, e.g., [19,2]). Another line of research, which aims at capturing the concurrent behaviour of actions in the logical AI literature, e.g., in [30], defines concurrency over the parametrised time spans shared by the actions.[2]

Linear logic is widely recognised as a logic of concurrency (see, e.g., [26]) also because of its resource conscious features. In this paper, we propose linear

---

[1] A preliminary version of this paper has been presented as short paper at the 14th Int. Conference on Logic for Programming Artificial Intelligence and Reasoning.

[2] For a survey on reasoning about actions, planning and concurrency, see [13].

C. Martín-Vide, F. Otto, and H. Fernau (Eds.): LATA 2008, LNCS 5196, pp. 250–262, 2008.
© Springer-Verlag Berlin Heidelberg 2008

logic planning (see. e.g., [25,24,17,18]) as a platform for a common language for planning and concurrency, aiming at bringing these two fields closer and allowing the techniques and tools in both fields to be interchanged. We establish a strict correspondence between partial order plans and the proofs of multiplicative exponential linear logic encodings of planning problems. The partial order plans which we extract from the proofs by an algorithm exhibit a non-interleaving behavioural concurrency semantics. Our result also contributes to the field of petri nets because of the strict correspondence between the reachability problem in petri nets and linear logic planning problems (see, e.g., [4]).

As the underlying formalism we employ the proof theoretic formalism of the calculus of structures (see, e.g., [11,31]) instead of the sequent calculus. The distinguishing feature of this formalism is deep inference: the inference rules can be applied at arbitrary depths inside logical expressions. This brings about properties of proofs and deductive systems that are interesting from the point of view of computer science applications (see, e.g., [13]): in particular, in the complementary work in [15], the logic that we use heavily relies on a notion of deep inference as in the calculus of structures [11,14], and it cannot be given as a sequent calculus system, as it was shown by Tiu [33]. Furthermore, more possibilities in the permutability of the inference rules in the calculus of structures yield optimised presentations (decomposition) of proofs [31]. By using the formalism of calculus of structures instead of the sequent calculus, it becomes possible to profit from these properties and also combine the results of this paper and the results of [15] for a common language for planning and concurrency, e.g., in [16]. Space restrictions do not allow to give the proofs here, we refer to [13].

2   Linear Logic Planning and Concurrency

In the following, we review multiplicative exponential linear logic in its deep inference presentation, following [31].

2.1   MELL in the Calculus of Structures

There are countably many *atoms*, denoted by $a, b, c, \ldots$ The *formulae* P, Q, R, $S \ldots$ of multiplicative exponential linear logic are generated by

$$R ::= a \mid 1 \mid \bot \mid (R \,\text{⅋}\, R) \mid (R \otimes R) \mid\, !R \mid\, ?R \mid \bar{R},$$

where a stands for any atom, 1 and $\bot$, called *one* and *bottom*. A formula $(R_1 \,\text{⅋}\, R_h)$ is a *par formula*, $(R_1 \otimes R_h)$ is a *times formula*, $!R$ is an *of-course formula*, and $?R$ is a *why-not formula*; $\bar{R}$ is the *negation* of the formula R. Formulae are considered to be equivalent modulo the relation $\approx$, which is the smallest congruence relation induced by the equations for associativity and commutativity for par and times formulae together with the following equations.

$$(\bot \,\text{⅋}\, R) \approx R \quad ??R \approx ?R \quad !!R \approx !R \quad \overline{(R \,\text{⅋}\, T)} \approx (\bar{R} \otimes \bar{T}) \quad \overline{?R} \approx !\bar{R} \quad \bar{\bar{R}} \approx R$$
$$(1 \otimes R) \approx R \quad\quad \bot \approx ?\bot \quad\quad 1 \approx !1 \quad\quad \overline{(R \otimes T)} \approx (\bar{R} \,\text{⅋}\, \bar{T}) \quad \overline{!R} \approx ?\bar{R}$$

A *formula context*, denoted as in $S\{\ \}$, is a formula with a hole that does not appear in the scope of negation. The formula R is a *subformula* of $S\{R\}$ and $S\{\ \}$ is its *context*. Context braces are omitted if no ambiguity is possible.

An *inference rule* is a scheme of the kind $\rho\,\dfrac{T}{R}$, where ρ is the *name* of the rule, T is its *premise* and R is its *conclusion*. A typical deep inference rule has the shape $\rho\,\dfrac{S\{T\}}{S\{R\}}$ and specifies a step of rewriting, by the implication $T \Rightarrow R$ inside a generic context $S\{\ \}$, which is linear implication in our case. Rules with empty contexts correspond to the case of the sequent calculus.

The following rules give the multiplicative exponential linear logic system in the calculus of structures [31], or system ELS. The rules of system ELS are called *atomic interaction* ($\mathsf{ai}{\downarrow}$), *switch* (s), *promotion* ($\mathsf{p}{\downarrow}$), *weakening* ($\mathsf{w}{\downarrow}$), and *absorption* ($\mathsf{b}{\downarrow}$), respectively.

$$\mathsf{ai}{\downarrow}\,\frac{S\{1\}}{S(a\,\bindnasrepma\,\bar{a})} \qquad \mathsf{s}\,\frac{S((R\,\bindnasrepma\,T)\otimes U)}{S((R\otimes U)\,\bindnasrepma\,T)} \qquad \mathsf{p}{\downarrow}\,\frac{S\{!(R\,\bindnasrepma\,T)\}}{S(!R\,\bindnasrepma\,?T)} \qquad \mathsf{w}{\downarrow}\,\frac{S\{\bot\}}{S\{?R\}} \qquad \mathsf{b}{\downarrow}\,\frac{S(?R\,\bindnasrepma\,R)}{S\{?R\}}$$

A derivation Δ is a finite chain of instances of these inference rules. A derivation can consist of just one formula. The top-most formula in a derivation, if present, is called the *premise*, and the bottom-most formula is called its *conclusion*. A derivation Δ whose premise is T, conclusion is R, and inference rules are in $\mathscr{S}$ is written as $\Delta\,\Big\|\,\mathscr{S}$ with $\dfrac{T}{R}$. A *proof* Π is a finite derivation whose premise is the unit 1.

2.2   Linear Logic Planning

Following [25,9], a linear logic planning problem $\mathscr{P}$ is given by $\langle \mathcal{I}, \mathcal{G}, \mathscr{A} \rangle$ where $\mathcal{I} : \{r_1, \ldots, r_m\}$ is a multiset[3] of fluents called the *initial state*. The multiset $\mathcal{G} : \{g_1, \ldots, g_n\}$ of fluents is the *goal state*. $\mathscr{A}$ is a finite set of *actions* of the form $\mathsf{a} : \{c_1, \ldots, c_p\} \to \{e_1, \ldots, e_q\}$, where $\{c_1, \ldots, c_p\}$ and $\{e_1, \ldots, e_q\}$ are multisets of fluents called *conditions* and *effects*, respectively, and a is the *name* of the action. Action a is applicable to a state $\mathcal{S}$ iff $\{c_1, \ldots, c_p\} \dot{\subseteq} \mathcal{S}$. The application of such an action a to a state $\mathcal{S}$ is defined by the function Φ, where it is applicable, as $\Phi(\mathsf{a}, \mathcal{S}) = (\mathcal{S} \dot{-} \{c_1, \ldots, c_p\}) \dot{\cup} \{e_1, \ldots, e_q\}$.

A goal $\mathcal{G}$ is satisfied iff there is a *plan* P, i.e., a sequence of actions $\mathsf{P} = \langle \mathsf{a}_1; \ldots; \mathsf{a}_k \rangle$ such that $\Phi(\mathsf{a}_k, \ldots, \Phi(\mathsf{a}_1, \mathcal{I}) \ldots) = \mathcal{G}$. Then, we say P *transforms the initial state* $\mathcal{I}$ into the state $\mathcal{G}$. If there exists such a plan P then P is a solution for the planning problem $\mathscr{P}$. Then we say P *solves* $\mathscr{P}$. We denote the empty plan with $\circ$. If it is more convenient, $\Phi(\mathsf{a}_k, \ldots, \Phi(\mathsf{a}_1, \mathcal{I}) \ldots)$ is abbreviated with $\Phi(\mathsf{P}, \mathcal{I})$. The *length of a plan* is the number of actions in that plan.

---

[3] Multisets are denoted by the curly brackets "$\{\ \}$". $\dot{\cup}$, $\dot{-}$ and $\dot{\subseteq}$ denote the multiset operations corresponding to the usual set operations $\cup$, $-$ and $\subseteq$, respectively.

Example 1. Consider the following planning problem: the actions α and β, respectively, buy an apple for fifty cents and buy a banana for fifty cents, i.e., $\mathscr{A} = \{\alpha : \{f\} \rightarrow \{a\}, \ \beta : \{f\} \rightarrow \{b\}\}$. The initial state and goal state are $\mathcal{I} = \{f, f\}$ and $\mathcal{G} = \{a, b\}$, respectively. The solutions for this planning problem are the plans $\langle\alpha; \beta\rangle$ and $\langle\beta; \alpha\rangle$. A plan which is executable at the initial state, however not a solution for this problem is the plan $\langle\alpha; \alpha\rangle$.

It is important to observe that the explicit representation of resources, given by the multiset representation, demonstrates that there are no resource conflicts between the actions α and β in the solution plans above. This observation permits the parallel execution of these two actions when there is no hardware constraints. However, in the encodings of linear logic planning, e.g, in [25,9,18], plans are extracted by sequentially reading the proper axioms corresponding to actions from the leaves of the proof tree, constructed by using the cut-rule. This does not allow to observe such a parallel execution semantics and breaks the cut-elimination property (see, e.g., [8,31]).

Let us now present our encoding of the planning problems in multiplicative exponential linear logic, which allows the construction of cut-free proofs.

Definition 1. *Let* $\mathsf{a} = \{c_1, \ldots, c_p\} \rightarrow \{e_1, \ldots, e_q\}$ *be an action. An action formula for* a *(A_a) is of the form* $?(\bar{c}_1 \otimes \ldots \otimes \bar{c}_p \otimes (e_1 \,\invamp\, \ldots \,\invamp\, e_q))$ *. A problem formula is of the form* $!(r_1 \,\invamp\, \ldots \,\invamp\, r_m \,\invamp\, (\bar{t}_1 \otimes \ldots \otimes \bar{t}_n))$ *.*

Definition 2. *Given a planning problem* $\mathscr{P} = \langle \mathcal{I}, \mathcal{G}, \mathscr{A}\rangle$ *where* $\mathcal{I} = \{r_1, \ldots, r_m\}$ *is the initial state,* $\mathcal{G} = \{t_1, \ldots, t_n\}$ *is the goal state, and* $\mathscr{A}$ *is a set of actions.*

$$(?A_1 \,\invamp\, \ldots \,\invamp\, ?A_s \,\invamp\, !(r_1 \,\invamp\, \ldots \,\invamp\, r_m \,\invamp\, (\bar{t}_1 \otimes \ldots \otimes \bar{t}_n)))$$

is the planning problem formula (ppf) that corresponds to $\mathscr{P}$ *where* $A_1, \ldots, A_s$ *are action formulae for all the actions* $\mathsf{a} \in \mathscr{A}$ *.*

Example 2. When we consider the planning problem of Example 1, we obtain the planning problem formula $?(\bar{f} \otimes a) \,\invamp\, ?(\bar{f} \otimes a) \,\invamp\, !(f \,\invamp\, f \,\invamp\, (\bar{a} \otimes \bar{b}))$.

Lemma 1. *(i.) The rule action below is derivable (sound) in* ELS. *(ii.) Let* $\mathsf{a} : \{c_1, \ldots, c_p\} \rightarrow \{e_1, \ldots, e_q\}$ *be an action, and* $\mathcal{S} = \{r_1, \ldots, r_m\}$ *and* $\mathcal{S}' = \{t_1, \ldots, t_n\}$ *be states. For some formulae R, T, and E*

$$\Phi(a, \mathcal{S}) = \mathcal{S}' \quad iff \quad \text{action} \frac{(?(\bar{c}_1 \otimes \ldots \otimes \bar{c}_p \otimes E) \,\invamp\, !(t_1 \,\invamp\, \ldots \,\invamp\, t_n \,\invamp\, R) \,\invamp\, T)}{(?(\bar{c}_1 \otimes \ldots \otimes \bar{c}_p \otimes E) \,\invamp\, !(r_1 \,\invamp\, \ldots \,\invamp\, r_m \,\invamp\, R) \,\invamp\, T)}.$$

Lemma 2. *The following rule is derivable (sound) in* ELS.

$$\text{termination} \frac{1}{(?A_1 \,\invamp\, \ldots \,\invamp\, ?A_s \,\invamp\, !(g_1 \,\invamp\, \ldots \,\invamp\, g_m \,\invamp\, (\bar{g}_1 \otimes \ldots \otimes \bar{g}_m)))}$$

Theorem 3. *Let* $\mathcal{P} = (?A_1 \,\mathbin{\invamp}\, \ldots \,\mathbin{\invamp}\, ?A_s \,\mathbin{\invamp}\, !\,(r_1 \,\mathbin{\invamp}\, \ldots \,\mathbin{\invamp}\, r_m \,\mathbin{\invamp}\, (\bar{t}_1 \otimes \ldots \otimes \bar{t}_n)))$ *be a ppf that corresponds to a planning problem* $\mathcal{P}$. *There is a proof with* k *number of applications of the* $\mathsf{p}{\downarrow}$ *rule iff there is a plan* p *with length* k *that solves the planning problem* $\mathcal{P}$, *where* $\mathcal{I} = \{\, r_1, \ldots, r_m \,\}$ *and* $\mathcal{G} = \{\, t_1, \ldots, t_n \,\}$.

Theorem 3 states the equivalence of existence of a plan solving a planning problem with the existence of a proof of the encoding of this planning problem. However, by resorting to this theorem, and Lemma 1 and Lemma 2, we can use the inference rules action and termination as the operational semantics of a planner: these inference rules can be used as machine instructions in an implementation for searching for plans.

Example 3. A proof of the planning problem of Example 1 can be constructed bottom-up as follows. The shaded regions denote the action formula being used at every inference step and the name of the action is displayed on the right-hand side of the inference rules. Then the plan can be extracted by reading these names bottom-up. Thus, the proof below reads the plan $\langle \alpha; \beta \rangle$.

$$
\begin{array}{l}
\text{termination} \; \dfrac{1}{?(\bar{f} \otimes a) \,\mathbin{\invamp}\, ?(\bar{f} \otimes b) \,\mathbin{\invamp}\, !\,(a \,\mathbin{\invamp}\, b \,\mathbin{\invamp}\, (\bar{a} \otimes \bar{b}))} \\[2ex]
\text{action} \; \dfrac{}{?(\bar{f} \otimes a) \,\mathbin{\invamp}\, ?(\bar{f} \otimes b) \,\mathbin{\invamp}\, !\,(a \,\mathbin{\invamp}\, f \,\mathbin{\invamp}\, (\bar{a} \otimes \bar{b}))} \; \beta \\[2ex]
\text{action} \; \dfrac{}{?(\bar{f} \otimes a) \,\mathbin{\invamp}\, ?(\bar{f} \otimes b) \,\mathbin{\invamp}\, !\,(f \,\mathbin{\invamp}\, f \,\mathbin{\invamp}\, (\bar{a} \otimes \bar{b}))} \; \alpha
\end{array}
$$

3   Independence and Causality in Plans

Given that a planning problem has a solution, the ppf of this planning problem can be proved in many different ways. Although these different proofs are distinct syntactic objects, they can be considered equivalent, because they share the instances of the rule ai↓ applied to the same pairs of atoms. Such an equivalence can be observed, for example, when two multiplicative linear logic proofs are mapped to the same proof net [8] or they can be decomposed to the same proof by permutation of the inference rules [31].

Example 4. We can construct the following proof which is a different syntactic object from the proof in Example 3, however the ai↓ rule instances, hidden in the instances of the rules action and termination in the proof below (see Lemma 1 and Lemma 2), are identical in these two proofs.

$$
\begin{array}{l}
\text{termination} \; \dfrac{1}{?(\bar{f} \otimes a) \,\mathbin{\invamp}\, ?(\bar{f} \otimes b) \,\mathbin{\invamp}\, !\,(a \,\mathbin{\invamp}\, b \,\mathbin{\invamp}\, (\bar{a} \otimes \bar{b}))} \\[2ex]
\text{action} \; \dfrac{}{?(\bar{f} \otimes a) \,\mathbin{\invamp}\, ?(\bar{f} \otimes b) \,\mathbin{\invamp}\, !\,(f \,\mathbin{\invamp}\, b \,\mathbin{\invamp}\, (\bar{a} \otimes \bar{b}))} \; \alpha \\[2ex]
\text{action} \; \dfrac{}{?(\bar{f} \otimes a) \,\mathbin{\invamp}\, ?(\bar{f} \otimes b) \,\mathbin{\invamp}\, !\,(f \,\mathbin{\invamp}\, f \,\mathbin{\invamp}\, (\bar{a} \otimes \bar{b}))} \; \beta
\end{array}
$$

In this section, by using this idea, we present an algorithm for extracting partial order plans, which exhibit a concurrency semantics, from the proofs of the ppf.

Definition 4. *Let Π be a proof* $\rho \dfrac{\overset{\displaystyle \Pi' \| \text{ELS}}{S\{T\}}}{S\{R\}}$ *where the atoms in an action formula are labelled with the name of that action. Furthermore, whenever there is an application of the rule* b↓*, the labels of the atoms in the premise, which are copied, are extended with a natural number that does not occur with the same action name elsewhere in the proof. Similarly, in a problem formula, all the positive and negative atoms are labelled with* init *and* goal*, respectively. Let Label denote the set of all the labels occurring in Π. The function μ on Π is defined as follows. If $\Pi = 1$, then $\mu(\Pi) = 1$. Otherwise,*

- *if ρ is the application of a rule other than* ai↓ *then $\mu(\Pi) = \mu(\Pi')$.*
- *if ρ is the application of the rule* ai↓ *where R is the formula $(a_{\mathsf{l}} \,\mathbin{\rotatebox[origin=c]{180}{\&}}\, \bar{a}_{\mathsf{k}})$ for an atom a such that $\mathsf{l}, \mathsf{k} \in Label$, then*

$$\mu(\Pi) = \{\,(\mathsf{l}, \mathsf{k})\,\} \cup \mu\left(\overset{\displaystyle \Pi' \| \text{ELS}}{S\{1\}} \right) \,.$$

Given a proof Π of $\mathcal{P}$, a constraint set *of Π for $\mathcal{P}$ ($\mathcal{C}_{P,\Pi}$) is given with $\mu(\Pi)$.*

Proposition 1. *For any proof Π of a ppf, $\mu(\Pi)$ terminates in linear time in the number of atoms in Π.*

The constraint sets are obtained by recording the atoms that get annihilated by the ai↓ instances. This idea is very similar to using proof nets [8] as a means for identifying classes of equivalent proofs up to permutation of inference rules, or the ideas used to describe classes of proofs that are equivalent upto a geometric criterion similar to proof nets [21]. Because each atom that gets annihilated is produced and consumed by a specific action formula, a constraint set provides an explicit record of causality between the actions producing and consuming each atom in the execution of a plan. Thus, the actions that are partially ordered in a constraint set are independent events in the execution because they do not have any resource conflicts.

Example 5. A proof of Example 1 that is syntactically different from the proofs given in Example 3 and Example 4 is as follows.

$$\mathsf{p}{\downarrow}^2 \dfrac{\mathsf{s}^7 \dfrac{\mathsf{ai}{\downarrow} \dfrac{\mathsf{ai}{\downarrow} \dfrac{\mathsf{ai}{\downarrow} \dfrac{\mathsf{ai}{\downarrow} \dfrac{1}{!\,(b_\beta \,\mathbin{\rotatebox[origin=c]{180}{\&}}\, \bar{b}_{\mathsf{goal}})}}{!\,((a_\alpha \,\mathbin{\rotatebox[origin=c]{180}{\&}}\, \bar{a}_{\mathsf{goal}}) \otimes (b_\beta \,\mathbin{\rotatebox[origin=c]{180}{\&}}\, \bar{b}_{\mathsf{goal}}))}}{!\,((f_{\mathsf{init}} \,\mathbin{\rotatebox[origin=c]{180}{\&}}\, \bar{f}_\beta) \otimes (a_\alpha \,\mathbin{\rotatebox[origin=c]{180}{\&}}\, \bar{a}_{\mathsf{goal}}) \otimes (b_\beta \,\mathbin{\rotatebox[origin=c]{180}{\&}}\, \bar{b}_{\mathsf{goal}}))}}{!\,((f_{\mathsf{init}} \,\mathbin{\rotatebox[origin=c]{180}{\&}}\, \bar{f}_\alpha) \otimes (f_{\mathsf{init}} \,\mathbin{\rotatebox[origin=c]{180}{\&}}\, \bar{f}_\beta) \otimes (a_\alpha \,\mathbin{\rotatebox[origin=c]{180}{\&}}\, \bar{a}_{\mathsf{goal}}) \otimes (b_\beta \,\mathbin{\rotatebox[origin=c]{180}{\&}}\, \bar{b}_{\mathsf{goal}}))}}{!\,((\bar{f}_\alpha \otimes a_\alpha) \,\mathbin{\rotatebox[origin=c]{180}{\&}}\, ?(\bar{f}_\beta \otimes b_\beta) \,\mathbin{\rotatebox[origin=c]{180}{\&}}\, !\,(f_{\mathsf{init}} \,\mathbin{\rotatebox[origin=c]{180}{\&}}\, f_{\mathsf{init}} \,\mathbin{\rotatebox[origin=c]{180}{\&}}\, (\bar{a}_{\mathsf{goal}} \otimes \bar{b}_{\mathsf{goal}})))}}{?(\bar{f}_\alpha \otimes a_\alpha) \,\mathbin{\rotatebox[origin=c]{180}{\&}}\, ?(\bar{f}_\beta \otimes b_\beta) \,\mathbin{\rotatebox[origin=c]{180}{\&}}\, !\,(f_{\mathsf{init}} \,\mathbin{\rotatebox[origin=c]{180}{\&}}\, f_{\mathsf{init}} \,\mathbin{\rotatebox[origin=c]{180}{\&}}\, (\bar{a}_{\mathsf{goal}} \otimes \bar{b}_{\mathsf{goal}}))}$$

When we plug into the function μ of Definition 4 this proof, or any of the proofs in Example 3 or Example 4 expanded with respect to Lemma 1 and Lemma 2, we obtain the following partial order.

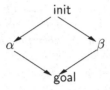

Proposition 5. *Let* $\mathcal{P}$ *be a ppf defined on the action set* $\mathscr{A}$ *and* $\mathcal{C}_{\mathcal{P},\Pi}$ *be a constraint set of a proof* Π *for* $\mathcal{P}$. (*i*) $\mathcal{C}_{\mathcal{P},\Pi}$ *is antisymmetric.* (*ii*) $\mathcal{C}_{\mathcal{P},\Pi}$ *is irreflexive.*

Definition 6. *Let* $\mathcal{P}$ *be a ppf defined on the action set* $\mathscr{A}$ *and* $\mathcal{C}_{\mathcal{P},\Pi}$ *be a constraint set of a proof* Π *for* $\mathcal{P}$. *The concurrent plan of* Π *for* $\mathcal{P}$ *is* $\Gamma_{\mathcal{P},\Pi}$ *is the transitive reduction (cover relation) of* $\mathcal{C}_{\mathcal{P},\Pi}$.

Definition 7. *A plan* P *is induced by a strict total order* $\prec$ *if for any pair* $(x,y) \in \prec$, *x appears to the left of y in* P.

Theorem 8. *Let* $\mathcal{P}$ *be a ppf of a planning problem* $\mathscr{P}$, Π *a proof of* $\mathcal{P}$ *and* $\Gamma_{\mathcal{P},\Pi}$ *the concurrent plan of* Π *for* $\mathcal{P}$. *For any strict total order* $\prec \supseteq \Gamma_{\mathcal{P},\Pi}$, *if* P *is a plan induced by* $\prec$ *then* P *solves* $\mathscr{P}$.

This theorem provides an interleaving semantics of the plans computed as proofs of ppf. Let us now give a non-interleaving semantics for these plans.

4   Partial Order Plans with a Concurrency Semantics

In linear logic planning, states are defined over the data structure multiset, actions are considered as multiset rewriting rules. Multiset rewriting is also complete for representing computations of place/transition petri nets [29] (see, e.g., [4,12]). In such an encoding, the multiset rewrite rules represent the possible firings of the transitions of a petri net. The places of the net are represented by elements of multisets. Such a view allows to consider a planning problem as the reachability problem of the corresponding petri net and vice versa.

4.1   Planning and Concurrency

When planning problems are considered from the point of view of concurrent computations, e.g., as those in petri nets, due to the explicit representation of resources, multiset rewriting planning allows to observe true concurrency in the computations: in a language with true concurrency, when two actions are partially ordered, the outcome of their execution in parallel is same as the outcome of their execution in either order. Because the explicit treatment of resources provides a representation of independence and causality, when two actions are partially ordered, in an execution that involves both of these actions, they are independent in terms of the resources that they require to be executed. Thus, their parallel composition results in an action that has the same effect as their execution in any order. It becomes possible to define the parallel composition of two actions by taking the multiset-union of their condition and effect multisets [15]. Here, we speak about concurrency, in contrast to only parallelism, because the common predecessors and successors of the two composed actions provide a synchronisation mechanism when they are considered as points in time.

Example 6. In the planning problem of Example 1, we can compose the two actions α and β and obtain the action $\{f, f\} \to \{a, b\}$, which corresponds to concurrent executions of these two actions. These two actions are then synchronised by their predecessor init and successor goal.

However, when planning problems are modelled in common AI planning formalisms, that are based on properties instead of resources, e.g., STRIPS [5], it is not always possible to observe true concurrency in the partial order plans of these languages, e.g., UCPOP [28] and Graphplan [1]. A simple modification of the famous dining philosophers problem is helpful to see the reason for this.

Example 7. There are two hungry philosophers, a and b, sitting at a dinner table. In order for a philosopher to eat, she must have a fork. However, there is only one fork on the table. The problem consists in finding a plan where both philosophers have eaten. The solution of this problem is a plan in which a and b eat in either order. A plan where a and b eat concurrently cannot be a solution for this problem, because a and b cannot have the fork at the same time. Because the fork is a resource, which cannot be shared, eating of one is dependent on the others finishing eating and leaving the fork. Hence, these two actions can be executed in either order but not in parallel. A simple encoding of this scenario as a planning problem allows to observe such a semantics: let $\mathcal{I} = \{h_a, h_b, f\}$, $\mathcal{G} = \{e_a, e_b, f\}$ and $\mathcal{A} = \{a : \{h_a, f\} \to \{e_a, f\}, \quad b : \{h_b, f\} \to \{e_b, f\}\}$. In the encoding above, for a philosopher x, the resource h_x denotes that x is hungry and e_x denotes that x has eaten. f denotes the resource fork. The actions a and b can be executed in either order. However, their parallel composition results in the action $[a, b] : \{h_a, h_b, f, f\} \to \{e_a, e_b, f, f\}$ which requires two instances of the resource f in order to be executed. Thus the parallel composition of these two actions cannot be executed in the initial state $\mathcal{I}$. An encoding of this problem by means of properties, in a propositional language, in a way which delivers such a semantics is not straight-forward, if not impossible.

Due to the explicit treatment of resources, in contrast to the partial order planners in the literature, such as, UCPOP or Graphplan, the approach of the present paper respects the dependency and causality between actions in a planning domain, and results in a non-interleaving, behavioural concurrency semantics, namely, labelled event structure semantics.

4.2 Labelled Event Structure Semantics of Planning Problems

Labelled event structures (LES) is a non-interleaving branching-time behavioural model of concurrency [34]. An interleaving model of concurrency is equipped with an expansion law that identifies parallel composition by means of choice and sequential composition. In an interleaving model, parallel composition of two events indicates that these events can take place in either order. A model for concurrency without such an expansion law is said to be a non-interleaving model: when two events are composed in parallel they can take place simultaneously or in either order. In such a view of the systems, the independence and causality between the events of the system is central. In a LES the causality between actions is captured in terms of their dependencies in a partial order.

Apart from the causality given by a partial order, in a LES, the nondeterminism in the computation is captured by a conflict relation, which is a symmetric irreflexive relation of events. In a planning perspective, this corresponds to actions that are applicable in the same state, but are in conflict. When two actions are in conflict with each other, execution of one of them instead of the other determines a different state space ahead. This provides a branching-time model of the possible computations.

By using the operational semantics given by the rule action, in [13], we provide a procedure for obtaining a LES from the specification of a planning problem: for each planning problem, we obtain a labelled transition system determined by the possible applications of the rule action. We then apply the standard techniques for obtaining a LES from these transition systems (see, e.g., [34,10]). The LES obtained takes all the possible computations in the planning domain into consideration and reflects the plans that are in conflict with each other. Because of the explicit treatment of resources in linear logic, the computations in the LES reflect the independence and causality between actions of the computation. The concurrent plans of Definition 6 reflect this semantics.

Example 8. Consider the planning problem of Example 1. The computations of this system are given as the LES depicted below, where # denotes the conflict relation. The four events are partially ordered because they are causally independent. The events α_1 and β_2 are in conflict because they cannot co-occur. Similarly, the events α_2 and β_1 are in conflict. The events that are not in conflict and causally independent can co-occur, e.g., α_1 and α_2 or α_1 and β_1.

$$\overset{\bullet}{\alpha_1}\cdots\#\cdots\overset{\bullet}{\beta_2}\qquad\overset{\bullet}{\alpha_2}\cdots\#\cdots\overset{\bullet}{\beta_1}$$

Example 9. There are two tables. On Table 1 there are four blocks which are stacked on top of each other as shown on the left-hand side of the Figure 1. The only available action takes a block from Table 1 and puts it on Table 2. The goal of the problem is moving three of the blocks from Table 1 to Table 2. Because block a is stacked on blocks c and d, blocks c and d cannot be moved before block a. Similarly, block d cannot be moved before block b.

The LES displaying the causality and independence of the planning problem is on the right-hand side of Figure 1. Each event there represents moving the corresponding block. Events a and b are independent, however b and c are also independent although event c causally requires event a in order to occur. The events c and d are independent but they cannot co-occur in an execution of the system because they are in conflict. The possible concurrent plans are the two maximal

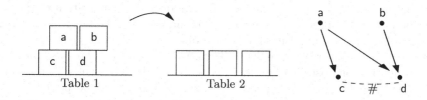

Fig. 1. A planning problem and the corresponding LES

conflict-free partial order in LES, i.e., { (init, a), (init, b), (b, goal), (a, c), (c, goal) } and { (init, a), (init, b), (a, d), (b, d), (d, goal) }. The interleavings of these concurrent plans are given by the plans $\langle a; c; b \rangle$, $\langle b; a; c \rangle$, $\langle a; b; c \rangle$, $\langle a; b; d \rangle$ and $\langle b; a; d \rangle$, which solve this problem.

5   Relation to Other Work

The reachability problem in petri nets is known to be EXPSPACE-hard [22]. Thus, the encoding of petri nets as multiset rewriting systems in multiplicative exponential linear logic delivers the lower bound of this logic to be EXPSPACE-hard [23]. When the complexity of a language is seen as a measure of expressive power, this also sets the scene for the expressive power of the propositional languages based on multiset rewriting in comparison to propositional languages based on STIRPS: given that planning in STRIPS is PSPACE-complete [3], because PSPACE is a strict subset of EXPSPACE multiset rewriting is strictly more expressive than propositional languages based on STRIPS. In order to achieve the same expressive power, these languages must be enriched with a constant-only first order language, i.e., DATALOGSTRIPS. However, a characterisation of STRIPS in multiset planning is possible (see, e.g., [20]).

If we consider the planning languages based on properties, e.g., ADL [27], we see that any planning problem expressed in these languages can be expressed as a multiset planning problem: [32] shows that multiset planning languages can be employed to encode the domain descriptions of the action description language $\mathcal{A}$ [6]. As it is stated in [7], because the action description language $\mathcal{A}$ is equivalent to the propositional fragment of the planning language ADL the result of [32] also implies that the multiset rewriting approach can be used for ADL domains.

For a more extensive survey on reasoning about actions, planning and concurrency we refer to [13].

6   Discussion

In [15,13], we have introduced a deductive language [4] for multiset planning within an extension of multiplicative exponential linear logic with a noncommutative

---

[4] Prototype implementations of planners based on this approach, mainly in Maude language, are available at http://www.doc.ic.ac.uk/~ozank/maude_cos.html

self-dual operator [11]. In this language, the sequential composition of the actions is represented by means of the non-commutative self-dual logical operator, whereas the parallel composition of the actions is naturally mapped to the commutative par operator of linear logic. Thus, by means of this language parallel and sequential composition of actions and plans can be represented at the same logical level as in process algebra and logical reasoning can be performed on these plan expressions. Ongoing work includes using the ideas of this paper to provide an event structure semantics to this deductive language with a proof theoretical operational semantics. This language should then benefit from a rich arsenal of tools and techniques that are imported from the both fields of planning and concurrency, and find applications, e.g., in modelling biological systems as complex reactive systems [16].

Acknowledgements. The author would like to thank Alessio Guglielmi, Steffen Hölldobler, Luca Cardelli, Max Kanovich and anonymous referees for valuable comments and improvements.

References

1. Blum, A., Furst, M.: Fast planning through planning graph analysis. Artificial Intelligence 90, 281–300 (1997)
2. Boutilier, C., Brafman, R.: Partial-order planning with concurrent interacting actions. Journal of Artificial Intelligence Research 14, 105–136 (2001)
3. Bylander, T.: Complexity results for serial decomposability. In: Proc. of the Tenth National Conf. on AI (AAAI-1992), San Jose, pp. 729–734. AAAI Press, Menlo Park (1992)
4. Cervesato, I.: Petri nets and linear logic: a case study for logic programming. In: Proceedings of the Joint Conference on Declarative Programming: GULP-PRODE 1995, Marina di Vietri, Ital (1995)
5. Fikes, R.E., Nilsson, H.J.: STRIPS: A new approach to the application of theorem proving to problem solving. Artificial Intelligence 2, 189–205 (1971)
6. Gelfond, M., Lifschitz, V.: Representing action and change by logic programs. Journal of Logic Programming 17(2/3–4), 301–321 (1993)
7. Gelfond, M., Lifschitz, V.: Action languages. Electronic Transactions on Artificial Intelligence 2 (3–4), 193–210 (1998)
8. Girard, J.-Y.: Linear logic. Theoretical Computer Science 50, 1–102 (1987)
9. Große, G., Hölldobler, S., Schneeberger, J.: Linear deductive planning. Journal of Logic and Computation 6 (2), 233–262 (1996)
10. Guglielmi, A.: Abstract Logic Programming in Linear Logic Independence and Causality in a First Order Calculus. PhD thesis, Universita di Pisa (1996)
11. Guglielmi, A.: A system of interaction and structure. ACM Transactions on Computational Logic 8(1), 1–64 (2007)
12. Ishihara, K., Hiraishi, K.: The completeness of linear logic for petri net models. Logic Journal of IGPL 9(4), 549–567 (2001)
13. Kahramanoğulları, O.: Nondeterminism and Language Design in Deep Inference. PhD thesis, Technische Universität Dresden (2006)
14. Kahramanoğulları, O.: System BV is NP-complete. Annals of Pure and Applied Logic (to appear, 2007)

15. Kahramanoğulları, O.: Towards planning as concurrency. In: Proceedings of the IASTED International Conference on Artificial Intellgence and Applications, AIA 2005, Innsbruck, Austria, pp. 387–393 (2005)

16. Kahramanoğulları, O.: A deductive compositional approach to petri nets for systems biology. Poster presentation at the Computational Methods in Systems Biology Conference (2007)

17. Kanovich, M.I., Vauzeilles, J.: The classical AI planning problems in the mirror of horn linear logic: semantics, expressibility, complexity. Mathematical Structures in Computer Science 11(6), 689–716 (2001)

18. Kanovich, M.I., Vauzeilles, J.: Strong planning under uncertainty in domains with numerous but identical elements (a generic approach). Theoretical Computer Science 379, 84–119 (2007)

19. Craig, A.: Knoblock. Generating parallel execution plans with a partial-order planner. In: Artificial Intelligence Planning Systems, pp. 98–103 (1994)

20. Küngas, P.: Linear logic for domain-independent AI planning (extended abstract). In: Proc. of Doctoral Consortium at 13th Int. Conf. on Automated Planning and Scheduling, ICAPS 2003, Trento, Italy, pp. 68–72 (2003)

21. Lamarche, F., Straßburger, L.: Naming proofs in classical propositional logic. In: Urzyczyn, P. (ed.) TLCA 2005. LNCS, vol. 3461, pp. 246–261. Springer, Heidelberg (2005)

22. Lipton, R.J.: The reachability problem requires exponential space. Technical Report 62, Yale University (1976)

23. Martí-Oliet, N., Meseguer, J.: From petri nets to linear logic. Mathematical Structures in Computer Science 1, 66–101 (1991)

24. Mart'i-Oliet, N., Meseguer, J.: Action and change in rewriting logic. In: Pareschi, R., Fronhofer, B. (eds.) Dynamic Worlds: From the Frame Problem to Knowledge Management, vol. 11–2, pp. 1–53. Kluwer Academic Publishers, Dordrecht (1999)

25. Masseron, M., Tollu, C., Vauzeilles, J.: Generating plans in linear logic I–II. In: Veni Madhavan, C.E., Nori, K.V. (eds.) FSTTCS 1990. LNCS, vol. 472, pp. 63–75. Springer, Heidelberg (1990)

26. Miller, D.: The π-calculus as a theory in linear logic: Preliminary results. In: Lamma, E., Mello, P. (eds.) ELP 1992. LNCS, vol. 660, pp. 242–265. Springer, Heidelberg (1993)

27. Pednault, E.P.D.: ADL: Exploring the middle ground between STRIPS and the situation calculus. In: Brachmann, R., Levesque, H.J., Reiter, R. (eds.) Principles of Knowledge Representation and Reasoning: Proc. of the First Int. Conf (KR-1989), Toronto, ON, pp. 324–332. Morgan Kaufmann, San Francisco (1989)

28. Penberthy, J., Weld, D.: UCPOP: A sound, complete, partial order planner for ADL. In: KR 1992. Principles of Knowledge Representation and Reasoning: Proceedings of the Third International Conference, pp. 103–114 (1992)

29. Petri, C.A.: Kommunikation mit Automaten. PhD thesis, Institut für Instrumentelle Mathematik, Bonn (1962)

30. Reiter, R.: Natural actions, concurrency and continuous time in the situation calculus. In: Proceedings of the International Conference on Principles of Knowledge Representation and Reasoning, pp. 2–13. Morgan Kaufmann, Cambridge (1996)

31. Straßburger, L.: MELL in the calculus of structures. Theoretical Computer Science 309, 213–285 (2003)

32. Thielscher, M.: Representing Actions in Equational Logic Programming. In: Van Hentenryck, P. (ed.) Proc. of the Int. Conf. on Logic Programming (ICLP), Santa Margherita Ligure, Italy, pp. 207–224. MIT Press, Cambridge (1994)
33. Tiu, A.: A system of interaction and structure II: The need for deep inference. Logical Methods in Computer Science 2(2),4: 1–24 (2006)
34. Winskel, G., Nielsen, M.: Models for concurrency. In: Handbook of Logic in Computer Science, vol. 4, pp. 1–148. Oxford University Press, Oxford (1995)

On the Relation between Multicomponent Tree Adjoining Grammars with Tree Tuples (TT-MCTAG) and Range Concatenation Grammars (RCG)

Laura Kallmeyer and Yannick Parmentier

Collaborative Research Center 441, University of Tübingen, Germany
lk@sfs.uni-tuebingen.de, parmenti@sfs.uni-tuebingen.de

Abstract. This paper investigates the relation between TT-MCTAG, a formalism used in computational linguistics, and RCG. RCGs are known to describe exactly the class PTIME; simple RCG even have been shown to be equivalent to linear context-free rewriting systems, i.e., to be mildly context-sensitive. TT-MCTAG has been proposed to model free word order languages. In general, it is NP-complete. In this paper, we will put an additional limitation on the derivations licensed in TT-MCTAG. We show that TT-MCTAG with this additional limitation can be transformed into equivalent simple RCGs. This result is interesting for theoretical reasons (since it shows that TT-MCTAG in this limited form is mildly context-sensitive) and, furthermore, even for practical reasons: We use the proposed transformation from TT-MCTAG to RCG in an actual parser that we have implemented.

1 Introduction

1.1 Tree Adjoining Grammars (TAG)

Tree Adjoining Grammar (TAG, [1]) is a tree-rewriting formalism. A TAG consists of a finite set of trees (elementary trees). The nodes of these trees are labelled with nonterminals and terminals (terminals only label leaf nodes). Starting from the elementary trees, larger trees are derived by substitution (replacing a leaf with a new tree) and adjunction (replacing an internal node with a new tree). In case of an adjunction, the tree being adjoined has exactly one leaf that is marked as the foot node (marked with an asterisk). Such a tree is called an *auxiliary* tree. When adjoining it to a node n, in the resulting tree, the subtree with root n from the old tree is attached to the foot node of the auxiliary tree. Non-auxiliary elementary trees are called *initial* trees. A derivation starts with an initial tree. In a final derived tree, all leaves must have terminal labels. For a sample derivation see Fig. 1.

C. Martín-Vide, F. Otto, and H. Fernau (Eds.): LATA 2008, LNCS 5196, pp. 263–274, 2008.

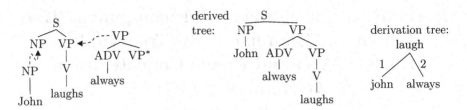

Fig. 1. TAG derivation for *John always laughs*

Definition 1 (Tree Adjoining Grammar)
A Tree Adjoining Grammar (TAG) is a tuple $G = \langle I, A, N, T \rangle$ with

- *N and T being disjoint finite sets, the nonterminals and terminals*
- *I being a finite set of initial trees with nonterminals N and terminals T, and*
- *A being a finite set of auxiliary trees with nonterminals N and terminals T.*

The internal nodes in $I \cup A$ can be marked as OA (obligatory adjunction) and NA (null adjunction, i.e., no adjunction allowed).

Definition 2 (TAG derivation and tree language). *Let $G = \langle I, A, N, T \rangle$ be a TAG. Let γ and γ' be finite trees.*

- *$\gamma \Rightarrow \gamma'$ in G iff there is a node position p and a tree γ_0' that is either elementary or derived from some elementary tree such that $\gamma' = \gamma[p, \gamma_0']^1$.*
  $\overset{*}{\Rightarrow}$ is the reflexive transitive closure of $\Rightarrow$.
- *The tree language of G is $L_T(G) = \{\gamma \mid \alpha \overset{*}{\Rightarrow} \gamma$ for some $\alpha \in I$, all leaves in γ have terminal labels and there are no remaining OA nodes in $\gamma\}$.*
  The string language $L(G)$ contains all yields of trees from the tree language.

TAG derivations are represented by derivation trees (unordered trees) that record the history of how the elementary trees are put together. A derived tree is the result of carrying out the substitutions and adjunctions, i.e., the derivation tree describes uniquely the derived tree. Each edge in a derivation tree stands for an adjunction or a substitution. The edges are labelled with Gorn addresses[2]. E.g., the derivation tree in Fig. 1 indicates that the elementary tree for *John* is substituted for the node at address 1 and *always* is adjoined at node address 2 (the fact that the former is an adjunction, the latter a substitution can be inferred from the fact that the node at address 1 is a leaf that is no foot node while the node at address 2 is an internal node).

Definition 3 (TAG derivation tree). *Let $G = \langle I, A, N, T \rangle$ be a TAG. Let γ be a tree derived as follows in G:*
  $\gamma = \gamma_0[p_1, \gamma_1] \dots [p_k, \gamma_k]$ where γ_0 is an instance of an elementary tree and the substitutions/adjunctions of the $\gamma_1, \dots, \gamma_k$ are all the substitutions/adjunctions to γ_0 that are performed to derive γ.

---

[1] For trees $\gamma, \gamma_1, \dots, \gamma_n$ and pairwise different node positions $p_1, \dots, p_n$ in γ, $\gamma[p_1, \gamma_1] \dots [p_n, \gamma_n]$ denotes the result of subsequently substituting/adjoining the $\gamma_1, \dots, \gamma_n$ to the nodes in γ with addresses $p_1, \dots, p_n$ respectively.

[2] The root address is ϵ, and the jth child of a node with address p has address pj.

Then the corresponding derivation tree has a root labelled with γ_0 that has k daughters. The edges from γ_0 to these daughters are labelled with $p_1, \ldots, p_k$, and the daughters are the derivation trees for the derivations of $\gamma_1, \ldots, \gamma_k$.

1.2   Range Concatenation Grammars (RCG)

This section defines RCGs [2,3].

Definition 4 (Range Concat-enation Grammar). *A positive Range Concatenation Grammar is a tuple $G = \langle N, T, V, S, P \rangle$ such that a) N is a finite set of predicates, each with a fixed arity; b) T and V are disjoint alphabets of terminals and of variables; c) $S \in N$ is the start predicate, a predicate of arity 1; d) P is a finite set of* clauses
$$A_0(x_{01}, \ldots, x_{0a_0}) \to \epsilon, \text{ or}$$
$$A_0(x_{01}, \ldots, x_{0a_0}) \to A_1(x_{11}, \ldots, x_{1a_1}) \ldots A_n(x_{n1}, \ldots, x_{na_n}) \text{ with } n \geq 1$$
where $A_i \in N, x_{ij} \in (T \cup V)^*$ and a_i the arity of A_i.

Since throughout the paper, we use only positive RCGs, whenever we say "RCG", we actually mean "positive RCG"[3]. An RCG with maximal predicate arity n is called an RCG of arity n.

When applying a clause with respect to a string $w = t_1 \ldots t_n$, the arguments of the predicates are instantiated with substrings of w, more precisely with the corresponding ranges. A range $\langle i, j \rangle$ with $0 \leq i < j \leq n$ corresponds to the substring between positions i and j, i.e., to the substring $t_{i+1} \ldots t_j$. If $i = j$, then $\langle i, j \rangle$ corresponds to the empty string ϵ. If $i > j$, then $\langle i, j \rangle$ is undefined.

Definition 5. *For a given clause, an instantiation with respect to a string $w = t_1 \ldots t_n$ consists of a function $f : \{t' \mid t' \text{ is an occurrence of some } t \in T \text{ in the clause}\} \cup V \to \{\langle i, j \rangle \mid i \leq j, i, j \in \mathbb{N}\}$ such that*

a) *for all occurrences t' of a $t \in T$ in the clause: $f(t') := \langle i, i+1 \rangle$ for some $i, 0 \leq i < n$ such that $t_i = t$,*
b) *for all $v \in V$: $f(v) = \langle j, k \rangle$ for some $0 \leq j \leq k \leq n$, and*
c) *if consecutive variables and occurrences of terminals in an argument in the clause are mapped to $\langle i_1, j_1 \rangle, \ldots, \langle i_k, j_k \rangle$ for some k, then $j_m = i_{m+1}$ for $1 \leq m < k$. By definition, we then state that f maps the whole argument to $\langle i_1, j_k \rangle$.*

The derivation relation is defined as follows:

Definition 6 (RCG derivation and string language)

- *For a predicate A of arity k, a clause $A(\ldots) \to \ldots$, and ranges $\langle i_1, j_1 \rangle, \ldots, \langle i_k, j_k \rangle$ with respect to a given w: if there is a instantiation of this clause with left-hand-side $A(\langle i_1, j_1 \rangle, \ldots, \langle i_k, j_k \rangle)$, then in one derivation step $(\ldots \Rightarrow \ldots) A(\langle i_1, j_1 \rangle, \ldots, \langle i_i, j_k \rangle)$ can be replaced with the right-hand side of this instantiation. $\overset{*}{\Rightarrow}$ is the reflexive transitive closure of $\Rightarrow$.*

---

[3] The negative variant allows for negative predicate calls of the form $\overline{A(\alpha_1, \ldots, \alpha_n)}$. Such a predicate is meant to recognize the complement language of its positive counterpart. See [3].

- *The language of an RCG G is*
  $L(G) = \{w \mid S(\langle 0, |w| \rangle) \overset{*}{\Rightarrow} \epsilon \text{ with respect to } w\}.$

For illustration, consider the RCG $G = \langle \{S, A, B\}, \{a, b\}, \{X, Y, Z\}, S, P \rangle$ with
$P = \{S(X\,Y\,Z) \to A(X, Z)\,B(Y),\, A(a\,X, a\,Y) \to A(X, Y),\, B(b\,X) \to B(X),\, A(\epsilon, \epsilon)$
$B(\epsilon) \to \epsilon\}.$

$L(G) = \{a^n b^k a^n \mid k, n \in \mathbb{N}\}$. Take $w = aabaa$. The derivation starts with
$S(\langle 0, 5 \rangle)$. First we apply the following clause instantiation:

$$
\begin{array}{cccccc}
S(X & Y & Z) & \to A(X, & Z) & B(Y) \\
\langle 0,2 \rangle & \langle 2,3 \rangle & \langle 3,5 \rangle & \langle 0,2 \rangle & \langle 3,5 \rangle & \langle 2,3 \rangle \\
aa & b & aa & aa & aa & b
\end{array}
$$

With this instantiation, $S(\langle 0,5 \rangle) \Rightarrow A(\langle 0,2 \rangle, \langle 3,5 \rangle) B(\langle 2,3 \rangle)$. Then

$$
\begin{array}{ccc}
B(b & X) & \to B(X) \\
\langle 2,3 \rangle & \langle 3,3 \rangle & \langle 3,3 \rangle \\
b & \epsilon & \epsilon
\end{array}
\quad \text{and } B(\epsilon) \to \epsilon
$$

lead to $A(\langle 0,2 \rangle, \langle 3,5 \rangle) B(\langle 2,3 \rangle) \Rightarrow A(\langle 0,2 \rangle, \langle 3,5 \rangle) B(\langle 3,3 \rangle) \Rightarrow A(\langle 0,2 \rangle, \langle 3,5 \rangle).$

$$
\begin{array}{cccccc}
A(a & X & a & Y) & \to A(X, & Y) \\
\langle 0,1 \rangle & \langle 1,2 \rangle & \langle 3,4 \rangle & \langle 4,5 \rangle & \langle 1,2 \rangle & \langle 4,5 \rangle \\
a & a & a & a & a & a
\end{array}
$$

leads to $A(\langle 0,2 \rangle, \langle 3,5 \rangle) \Rightarrow A(\langle 1,2 \rangle, \langle 4,5 \rangle)$. Then

$$
\begin{array}{cccccc}
A(a & X & a & Y) & \to A(X, & Y) \\
\langle 1,2 \rangle & \langle 2,2 \rangle & \langle 4,5 \rangle & \langle 5,5 \rangle & \langle 2,2 \rangle & \langle 5,5 \rangle \\
a & \epsilon & a & \epsilon & \epsilon & \epsilon
\end{array}
\quad \text{and } A(\epsilon, \epsilon) \to \epsilon
$$

lead to $A(\langle 1,2 \rangle, \langle 4,5 \rangle) \Rightarrow A(\langle 2,2 \rangle, \langle 5,5 \rangle) \Rightarrow \epsilon$

Definition 7 (Simple Range Concatenation Grammar). *An RCG is*

- non-combinatorial *if each of the arguments in the right-hand sides of the clauses are single variables.*
- linear *if no variable appears more than once in the left-hand sides of a clause or more than once in the right-hand side of a clause.*
- non-erasing *if each variable occurring in the left-hand side of a clause occurs also in its right-hand side and vice versa.*
- simple *if it is non-combinatorial, linear and non-erasing.*

Simple RCGs and linear context-free rewriting systems (LCFRS, [4]) are equivalent (see [5]). Consequently, simple RCGs are mildly context-sensitive [6].

1.3   From TAG to RCG

Now let us sketch the general idea of the transformation from TAG to RCG, following [7]: The RCG contains predicates $\langle \alpha \rangle(X)$ and $\langle \beta \rangle(L, R)$ for initial and auxiliary trees respectively. X covers the yield of α and all trees added to α, while L and R cover those parts of the yield of β (including all trees added to β) that are to the left and the right of the foot node of β. The clauses in the RCG reduce the argument(s) of these predicates by identifying those parts that come from the elementary tree α/β itself and those parts that come from one of the elementary trees added by substitution or adjunction. A sample TAG with an equivalent RCG is shown in Fig. 2.

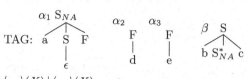

Equivalent RCG:

$S(X) \rightarrow \langle\alpha_1\rangle(X) \,|\, \langle\alpha_2\rangle(X) \,|\, \langle\alpha_3\rangle(X)$ (every word in the language is the yield of an $\alpha \in I$)

$\langle\alpha_1\rangle(aF) \rightarrow \langle\alpha_2\rangle(F) \,|\, \langle\alpha_3\rangle(F)$ (the yield of α_1 is a followed by the tree that substitutes at F)

$\langle\alpha_1\rangle(aB_1B_2F) \rightarrow \langle\beta\rangle(B_1, B_2)\langle\alpha_2\rangle(F) \,|\, \langle\beta\rangle(B_1, B_2)\langle\alpha_3\rangle(F)$ (or β adjoins to S in α; then the yield is a followed by the left part of β, the right part of β and the tree substituted at F)

$\langle\beta\rangle(B_1b, cB_2) \rightarrow \langle\beta\rangle(B_1, B_2)$ (β can adjoin to its root; then the left part is the left part of the adjoined β followed by b; the right part is c followed by the right part of the adjoined β)

$\langle\alpha_2\rangle(d) \rightarrow \epsilon$ $\langle\alpha_3\rangle(e) \rightarrow \epsilon$ $\langle\beta\rangle(b, c) \rightarrow \epsilon$ (the yields of α_2, α_3 and β can be d, e and the pair b (left) and c (right) resp.)

Fig. 2. A sample TAG and an equivalent RCG

2   TT-MCTAG

For a range of linguistic phenomena, multicomponent TAG (MCTAG, [4]) have been proposed. The motivation is the desire to split the contribution of a single lexical item (e.g., a verb and its arguments) into several elementary trees. An MCTAG consists of sets of elementary trees, so-called multicomponents. If a multicomponent is used in a derivation, all its members must be used.

Definition 8 (MCTAG). *A multicomponent TAG (MCTAG) is a tuple $G = \langle I, A, N, T, \mathcal{A} \rangle$ where $G_{TAG} := \langle I, A, N, T \rangle$ is a TAG, and $\mathcal{A}$ is a partition of $I \cup A$, the set of elementary tree sets.*

The particular type of MCTAG we are concerned with is Tree-Tuple MCTAG with Shared Nodes (TT-MCTAG, [8]). TT-MCTAG were introduced to deal with free word order phenomena in languages such as German. An example is (1) where the argument *es* of *reparieren* precedes the argument *der Mechaniker* of *verspricht* and is therefore not adjacent to the predicate it depends on:

(1) ... dass es der Mechaniker zu reparieren verspricht
    ... that it the mechanic   to repair   promises
    '... that the mechanic promises to repair it'

A TT-MCTAG is slightly different from standard MCTAG since the elementary tree sets contain two parts: 1. one lexicalized tree γ, marked as the unique *head tree*, and 2. a set of auxiliary trees, the *argument trees*. Such a pair is called a tree tuple. During derivation, the argument trees must either adjoin directly to their head tree or they must be linked by a chain of adjunctions at root nodes to a tree that attaches to the head tree. In other words, in the corresponding TAG derivation tree, the head tree must dominate the auxiliary trees such that all positions on the path between them, except the first one, must be ϵ. This captures the notion of adjunction under node sharing from [9][4].

Definition 9 (TT-MCTAG). *Let $G = \langle I, A, N, T, \mathcal{A} \rangle$ be an MCTAG. G is a TT-MCTAG iff*

1. *every $\Gamma \in \mathcal{A}$ has the form $\{\gamma, \beta_1, \ldots, \beta_n\}$ where γ contains at least one leaf with a terminal label, the head tree, and $\beta_1, \ldots, \beta_n$ are auxiliary trees, the argument trees. We write such a set as a tuple $\langle \gamma, \{\beta_1, \ldots, \beta_n\} \rangle$.*
2. *A derivation tree D for some $t \in L(\langle I, A, N, T \rangle)$ is licensed as a TAG derivation tree in G iff D satisfies the following conditions (MC) ("multicomponent condition") and (SN-TTL) ("tree-tuple locality with shared nodes"):*
   (a) *(MC) There are k pairwise disjoint instances $\Gamma_1, \ldots, \Gamma_k$ of elementary tree sets from $\mathcal{A}$ for some $k \geq 1$ such that $\bigcup_{i=1}^{k} \Gamma_i$ is the set of node labels in D.*
   (b) *(SN-TTL) for all nodes $n_0, n_1, \ldots, n_m$, $m > 1$, in D with labels from the same elementary tree tuple such that n_0 is labelled by the head tree: for all $1 \leq i \leq m$: either $\langle n_0, n_i \rangle \in \mathcal{P}_D$[5] or there are $n_{i,1}, \ldots, n_{i,k}$ with auxiliary tree labels such that $n_i = n_{i,k}$, $\langle n_0, n_{i,1} \rangle \in \mathcal{P}_D$ and for $1 \leq j \leq k-1$: $\langle n_{i,j}, n_{i,j+1} \rangle \in \mathcal{P}_D$ where this edge is labelled with ϵ.*

Fig. 3 shows a TT-MCTAG derivation for (1). Here, the NP_{nom} auxiliary tree adjoins directly to *verspricht* (its head) while the NP_{acc} tree adjoins to the root of a tree that adjoins to the root of a tree that adjoins to *reparieren*.

In the general case, the recognition problem for TT-MCTAG is NP-hard [10]. In the following, we define a limitation for TT-MCTAG based on a suggestion from [10]: TT-MCTAG are of rank k if, at any time during the derivation, at most k argument trees depending on higher head trees in the derivation tree are still waiting for adjunction.

Definition 10 (k-TT-MCTAG). *Let $G = \langle I, A, N, T, \mathcal{A} \rangle$ be a TT-MCTAG. G is of rank k (or a k-TT-MCTAG for short) iff for each derivation tree D licenced in G, the following holds:*
   (TT-k) There are no nodes $n, h_0, \ldots, h_k, a_0, \ldots, a_k$ in D such that the label of a_i is an argument tree of the label of h_i and $\langle h_i, n \rangle, \langle n, a_i \rangle \in \mathcal{P}_D^+$ for $0 \leq i \leq k$.

---

[4] The intuition is that if a tree γ' adjoins to some γ, its root in the resulting derived tree somehow belongs both to γ and γ', it is shared by them. A further tree β adjoining to this node can then be considered as adjoining to γ, not only to γ' as in standard TAG. Note that we assume that foot nodes do not allow adjunctions, otherwise node sharing would also apply to them.

[5] For a tree γ, $\mathcal{P}_\gamma$ is the parent relation on the nodes, i.e., $\langle x, y \rangle \in \mathcal{P}_\gamma$ for nodes x, y in γ iff x is the mother of y.

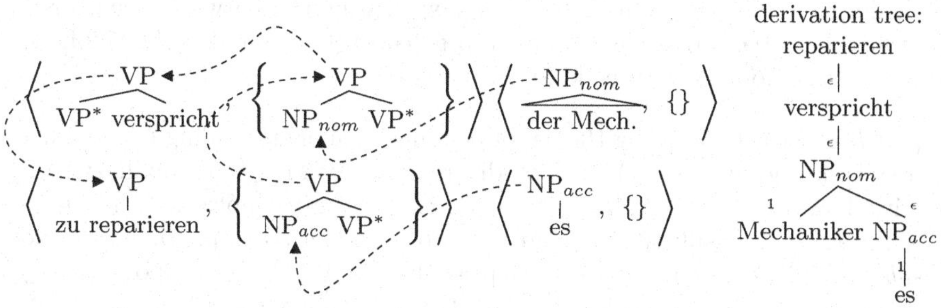

Fig. 3. TT-MCTAG derivation of (1)

With the analyses proposed in [8], that lead to a binary branching structure with a verbal projection line, the linguistic signification of this restriction is roughly that for every VP node on the verbal projection line, at most k NPs can be scrambled over this node. It is hard to say whether such a restriction is empirically valid. Note however, that the number of verbs that allow for non-local scrambling of their arguments is limited. Furthermore, the number of arguments of these verbs is fixed. This indicates that such a limit k actually exists, although it might be motivated rather by semantic and pragmatic reasons than by syntactic reasons.

3   From k-TT-MCTAG to RCG

We construct equivalent simple RCGs for k-TT-MCTAG in a way similar to the RCG construction for TAG. There are predicates $\langle \gamma \rangle$ for the elementary trees (not the tree sets) that characterize the contribution of γ. Recall that each TT-MCTAG is a TAG, a TT-MCTAG derivation is a derivation in the underlying TAG. (This is how we defined TT-MCTAG.) Consequently, we can construct the RCG for the underlying TAG, enrich the predicates in a way that allows to keep track of the "still to adjoin" argument trees and constrain thereby further the RCG clauses. In this case, the yield of a predicate corresponding to a tree γ contains not only γ and its arguments but also arguments of predicates that are higher in the derivation tree and that are adjoined below γ via node sharing[6].

Our construction leads to an RCG of arity 2 with complex predicate names. In order to keep the number of necessary predicates finite, the limit k is crucial.

A predicate $\langle \gamma \rangle$ must encode the set of argument trees that depend on higher head trees and that still need to be adjoined. We call this set the list of pending arguments (LPA). These trees need to either adjoin to the root or to be passed to the LPA of the root-adjoining tree. The LPA is a multiset since we allow for several occurrences of a single tree.

---

[6] An alternative possibility is to consider only γ and its arguments as the yield of γ. This leads to an RCG with simpler predicate names (the LPAs are not be needed) but with predicates of higher arity since the contribution of γ can be discontinuous: Every argument of a higher head adjoining below γ interrupts the contribution of γ. This construction is much more complex than the one we choose here.

In order to reduce the number of clauses, we distinguish between tree clauses (predicates $\langle \gamma... \rangle$) and branching clauses (predicates $\langle adj... \rangle$ and $\langle sub... \rangle$) following [2]. We therefore have three kinds of predicates:

1. $\langle \gamma, LPA \rangle$ with LPA being the list of pending arguments coming from higher trees (not arguments of γ). This predicate has arity 2 if γ is an auxiliary tree, arity 1 otherwise. $\langle \gamma, LPA \rangle$-clauses distribute the variables for the yields of the trees that substitute or adjoin into γ among corresponding adj and sub predicates. Furthermore, they pass the LPA to the root-position adj predicate and distribute the arguments of γ among the LPAs of all adj predicates.
2. $\langle adj, \gamma, dot, LPA \rangle$ as intermediate predicates (of arity 2). Here, LPA contains a) the list of higher args if $dot = \epsilon$, and b) arguments of γ. We assume as a condition that it contains only trees that can be adjoined to dot in γ. $\langle adj, \gamma, dot, LPA \rangle$-clauses adjoin a γ' to the dot in γ. If $\gamma' \in LPA$, then the new predicate receives $LPA \setminus \{\gamma'\}$. Otherwise, γ' must be a head and LPA is passed unchanged.
3. $\langle sub, \gamma, dot \rangle$ as intermediate predicates (arity 1). $\langle sub, \gamma, dot \rangle$-clauses substitute a γ' into dot in γ.

More precisely, the construction goes as follows:

We define the decoration string σ_γ of an elementary tree γ as in [2]: each internal node has two variables L and R and each substitution node has one variable X (L and R represent the left and right parts of the yield of the adjoined tree and X represents the yield of a substituted tree). In a top-down-left-to-right traversal the left variables are collected during the top-down traversal, the terminals and variables of substitution nodes are collected while visiting the leaves and the right variables are collected during bottom-up traversal. Furthermore, while visiting a foot node, a separating "," is inserted. The string obtained in this way is the decoration string.

1. We add a start predicate S and clauses $S(X) \rightarrow \langle \alpha, \emptyset \rangle(X)$ for all $\alpha \in I$.
2. For every $\gamma \in I \cup A$: Let L_p, R_p be the left and right symbols in σ_γ for the node at position p if this is not a substitution node. Let X_p be the symbol for the node at position p if this is a substitution node.

   We assume that $p_1, \ldots, p_k$ are the possible adjunction sites, $p_{k+1}, \ldots, p_l$ the substitution sites in γ. Then the RCG contains all clauses
   $$\langle \gamma, LPA \rangle(\sigma_\gamma) \rightarrow \langle adj, \gamma, p_1, LPA_{p_1} \rangle(L_{p_1}, R_{p_1}) \ldots \langle adj, \gamma, p_k, LPA_{p_k} \rangle(L_{p_k}, R_{p_k})$$
   $$\langle sub, \gamma, p_{k+1} \rangle(X_{p_{k+1}}) \ldots \langle sub, \gamma, p_l \rangle(X_{p_l})$$
   such that
   − If $LPA \neq \emptyset$, then $\epsilon \in \{p_1, \ldots, p_k\}$ and $LPA \subseteq LPA_\epsilon$, and
   − $\bigcup_{i=0}^{k} LPA_{p_i} = LPA \cup \Gamma(\gamma)$ where $\Gamma(\gamma)$ is either the set of arguments of γ (if γ is a head tree) or (if γ is an argument itself), the empty set.
3. For all predicates $\langle adj, \gamma, dot, LPA \rangle$ the RCG contains all clauses
   $\langle adj, \gamma, dot, LPA \rangle(L, R) \rightarrow \langle \gamma', LPA' \rangle(L, R)$ such that γ' can be adjoined at position dot in γ and

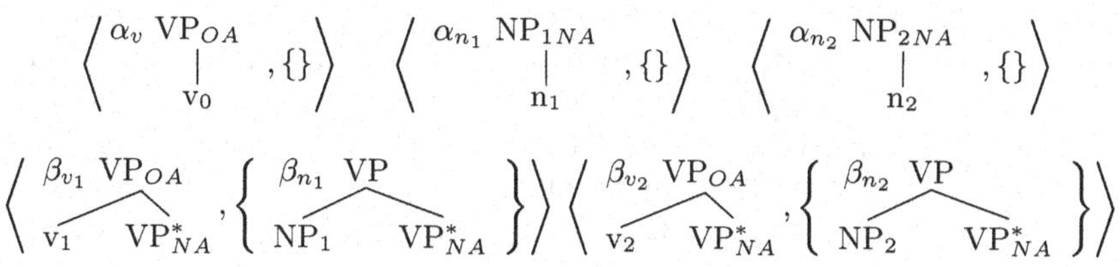

$$\left\langle \begin{array}{c} \alpha_v\ VP_{OA} \\ | \\ v_0 \end{array}, \{\} \right\rangle \quad \left\langle \begin{array}{c} \alpha_{n_1}\ NP_{1NA} \\ | \\ n_1 \end{array}, \{\} \right\rangle \quad \left\langle \begin{array}{c} \alpha_{n_2}\ NP_{2NA} \\ | \\ n_2 \end{array}, \{\} \right\rangle$$

$$\left\langle \begin{array}{c} \beta_{v_1}\ VP_{OA} \\ v_1 \quad VP^*_{NA} \end{array}, \left\{ \begin{array}{c} \beta_{n_1}\ VP \\ NP_1 \quad VP^*_{NA} \end{array} \right\} \right\rangle \Big/ \left\langle \begin{array}{c} \beta_{v_2}\ VP_{OA} \\ v_2 \quad VP^*_{NA} \end{array}, \left\{ \begin{array}{c} \beta_{n_2}\ VP \\ NP_2 \quad VP^*_{NA} \end{array} \right\} \right\rangle$$

Fig. 4. TT-MCTAG

- either $\gamma' \in LPA$ and $LPA' = LPA \setminus \{\gamma'\}$,
- or $\gamma' \notin LPA$, γ' is a head (i.e., a head tree), and $LPA' = LPA$.

4. For all predicates $\langle adj, \gamma, dot, \emptyset \rangle$ where dot in γ is no OA-node, the RCG contains a clause $\langle adj, \gamma, dot, \emptyset \rangle(\epsilon, \epsilon) \to \epsilon$.
5. For all predicates $\langle sub, \gamma, dot \rangle$ and all γ' that can be substituted into position dot in γ the RCG contains a clause $\langle sub, \gamma, dot \rangle(X) \to \langle \gamma', \emptyset \rangle(X)$.

As an example consider the TT-MCTAG from Fig. 4. For this TT-MCTAG we obtain (amongst others) the following RCG clauses:

- $\langle \alpha_v, \emptyset \rangle(L\ v_0\ R) \to \langle adj, \alpha_v, \epsilon, \emptyset \rangle(L, R)$ (only one adjunction at the root, address ϵ)
- $\langle adj, \alpha_v, \epsilon, \emptyset \rangle(L, R) \to \langle \beta_{v_1}, \emptyset \rangle(L, R) \mid \langle \beta_{v_2}, \emptyset \rangle(L, R)$ (β_{v_1} or β_{v_2} might be adjoined at ϵ in α_v, LPA (here empty) is passed)
- $\langle \beta_{v_1}, \emptyset \rangle(L\ v_1, R) \to \langle adj, \beta_{v_1}, \epsilon, \{\beta_{n_1}\} \rangle(L, R)$ (in β_{v_1}, there is only one adjunction site, address ϵ; the argument is passed to the new LPA)
- $\langle adj, \beta_{v_1}, \epsilon, \{\beta_{n_1}\} \rangle(L, R) \to$
  $\langle \beta_{n_1}, \emptyset \rangle(L, R) \mid \langle \beta_{v_1}, \{\beta_{n_1}\} \rangle(L, R) \mid \langle \beta_{v_2}, \{\beta_{n_1}\} \rangle(L, R)$ (either β_{n_1} is adjoined and removed from the LPA or another tree (β_{v_1} or β_{v_2}) is adjoined; in this case, the LPA remains)
- $\langle \beta_{v_1}, \{\beta_{n_1}\} \rangle(L\ v_1, R) \to \langle adj, \beta_{v_1}, \epsilon, \{\beta_{n_1}, \beta_{n_1}\} \rangle(L, R)$ (again, only one adjunction in β_{v_1}; the argument β_{n_1} is added to the LPA)
- $\langle \beta_{n_1}, \emptyset \rangle(L\ X, R) \to \langle adj, \beta_{n_1}, \epsilon, \emptyset \rangle(L, R)\ \langle sub, \beta_{n_1}, 1, \rangle(X)$ (adjunction to root and substitution to 1 in β_{n_1})
- $\langle adj, \beta_{n_1}, \epsilon, \emptyset \rangle(\epsilon, \epsilon) \to \epsilon$ (adjunction at root of β_{n_1} not obligatory as long as LPA is empty)
- $\langle sub, \beta_{n_1}, 1, \rangle(X) \to \langle \alpha_{n_1}, \emptyset \rangle(X)$ (substitution of α_{n_1} at address 1)
- $\langle \alpha_{n_1}, \emptyset \rangle(n_1) \to \epsilon$ (no adjunctions or substitutions at α_{n_1})

Take the input word $n_1 n_2 n_1 v_2 v_1 v_1 v_0$. The RCG derivation goes as follows[7]:
$S(n_1\ n_2\ n_1\ v_2\ v_1\ v_1\ v_0) \Rightarrow \langle \alpha_v, \emptyset \rangle(n_1\ n_2\ n_1\ v_2\ v_1\ v_1\ v_0)$
$\Rightarrow \langle adj, \alpha_v, \epsilon, \emptyset \rangle(n_1\ n_2\ n_1\ v_2\ v_1\ v_1, \epsilon)$ (adjoin at ϵ, scan v_0)
$\Rightarrow \langle \beta_{v_1}, \emptyset \rangle(n_1\ n_2\ n_1\ v_2\ v_1\ v_1, \epsilon)$ (adjoin β_{v_1})
$\Rightarrow \langle adj, \beta_{v_1}, \epsilon, \{\beta_{n_1}\} \rangle(n_1\ n_2\ n_1\ v_2\ v_1, \epsilon)$ (adj. at ϵ, scan v_1, β_{n_1} in LPA)

---

[7] In this example, we replace the ranges with the corresponding input substrings since this way the example is easier to read.

$$\Rightarrow \langle \beta_{v_1}, \{\beta_{n_1}\}\rangle(n_1\ n_2\ n_1\ v_2\ v_1, \epsilon) \qquad \text{(adjoin } \beta_{v_1})$$
$$\Rightarrow \langle adj, \beta_{v_1}, \epsilon, \{\beta_{n_1}, \beta_{n_1}\}\rangle(n_1\ n_2\ n_1\ v_2, \epsilon) \qquad \text{(adj. at } \epsilon, \text{ scan } v_1, \beta_{n_1} \text{ in LPA)}$$
$$\Rightarrow \langle \beta_{v_2}, \{\beta_{n_1}, \beta_{n_1}\}\rangle(n_1\ n_2\ n_1\ v_2, \epsilon) \qquad \text{(adjoin } \beta_{v_2})$$
$$\Rightarrow \langle adj, \beta_{v_2}, \epsilon, \{\beta_{n_2}, \beta_{n_1}, \beta_{n_1}\}\rangle(n_1\ n_2\ n_1, \epsilon) \qquad \text{(adj. at } \epsilon, \text{ scan } v_2, \beta_{n_2} \text{ in LPA)}$$
$$\Rightarrow \langle \beta_{n_1}, \{\beta_{n_2}, \beta_{n_1}\}\rangle(n_1\ n_2\ n_1, \epsilon) \qquad \text{(adjoin } \beta_{n_1} \text{ from LPA)}$$
$$\Rightarrow \langle adj, \beta_{n_1}, \epsilon, \{\beta_{n_2}, \beta_{n_1}\}\rangle(n_1\ n_2, \epsilon)\ \ \langle sub, \beta_{n_1}, 1, \rangle(n_1) \qquad \text{(adj. at } \epsilon,$$
$$\text{subst. at 1)}$$
$$\Rightarrow \langle adj, \beta_{n_1}, \epsilon, \{\beta_{n_2}, \beta_{n_1}\}\rangle(n_1\ n_2, \epsilon)\ \ \langle \alpha_{n_1}, \emptyset\rangle(n_1) \qquad \text{(subst. of } \alpha_{n_1})$$
$$\Rightarrow \langle adj, \beta_{n_1}, \epsilon, \{\beta_{n_2}, \beta_{n_1}\}\rangle(n_1\ n_2, \epsilon)\ \ \epsilon \qquad \text{(scan } n_1)$$
$$\Rightarrow \langle \beta_{n_2}, \{\beta_{n_1}\}\rangle(n_1\ n_2, \epsilon) \qquad \text{(adjoin } \beta_{n_2} \text{ from LPA)}$$
$$\Rightarrow \langle adj, \beta_{n_2}, \epsilon, \{\beta_{n_1}\}\rangle(n_1, \epsilon)\ \ \langle sub, \beta_{n_2}, 1, \rangle(n_2) \qquad \text{(adj. at } \epsilon, \text{ subst. at 1)}$$
$$\Rightarrow \langle adj, \beta_{n_2}, \epsilon, \{\beta_{n_1}\}\rangle(n_1, \epsilon)\ \ \langle \alpha_{n_2}, \emptyset\rangle(n_2) \qquad \text{(subst. of } \alpha_{n_2})$$
$$\Rightarrow \langle adj, \beta_{n_2}, \epsilon, \{\beta_{n_1}\}\rangle(n_1, \epsilon)\ \ \epsilon \qquad \text{(scan } n_2)$$
$$\Rightarrow \langle \beta_{n_1}, \emptyset\rangle(n_1, \epsilon) \qquad \text{(adjoin } \beta_{n_1} \text{ from LPA)}$$
$$\overset{*}{\Rightarrow} \langle adj, \beta_{n_1}, \epsilon, \emptyset\rangle(\epsilon, \epsilon)\ \ \langle \alpha_{n_1}, \emptyset\rangle(n_1) \qquad \text{(subst. of } \alpha_{n_1})$$
$$\overset{*}{\Rightarrow} \epsilon \qquad \text{(scan } n_1)$$

This example requires LPAs of maximal cardinality 3, i.e., a 3-TT-MCTAG. Note that with this construction, the grouping into tree sets gets lost. E.g., in our example, we do not know which of the n_1 came with which of the v_1. However, in our parser we construct the RCG only for the TT-MCTAG of a given input sentence and if the same terminal occurs more than once in the input sentence, we use different occurrences of the corresponding tree tuples. This way, we avoid using the same elementary tree twice and the grouping can be inferred from the tuple identifiers encoded in the names of the trees.

With the above construction the following can be shown:

Theorem 1. *For each k-TT-MCTAG G there is a simple RCG G' with $L(G) = L(G')$*[8].

As a corollary, we obtain that the string languages of k-TT-MCTAG are mildly context-sensitive.

To prove the theorem, we introduce TT-RCG derivation trees. These trees are obtained from an RCG derivation by turning the $\langle \gamma, LPA\rangle$ predicates into nodes and the branching predicates into edges.

Definition 11 (TT-RCG derivation tree). *Let G' be an RCG constructed from a k-TT-MCTAG as above. A tree $D_{G'}$ with node and edge labels is a TT-RCG derivation tree for G' iff*

---

[8] We suspect that the reverse does not hold. In other words, we suspect that the k-TT-MCTAG languages are properly contained in the set of languages of simple RCGs. An example of a language that is probably not in $\mathcal{L}(k\text{-TT-MCTAG})$ is the double copy language $\{www \mid w \in \{a,b\}^*\}$. The intuition is that, in order to obtain the correct dependencies, the three copies of a terminal (or their substitution slots) must be introduced in a single tree tuple. This means that two of them adjoin via node sharing. But then it is not clear how to avoid getting not only crossing but also other dependencies.

- each node in $D_{G'}$ is labeled with a predicate name $\langle \gamma, LPA \rangle$ and with a sequence of one or (if γ is an auxiliary tree) two $w \in T^*$;
- the outgoing edges of a γ-node are labeled with pairwise different adjunction and substitution sites in γ;
- if the root label is $\langle \gamma, LPA \rangle$, then there is a clause $S(X) \rightarrow \langle \gamma, LPA \rangle(X)$;
- for every node with label $\langle \gamma, LPA \rangle$ and with l daughters with node labels $\langle \gamma_i, LPA_i' \rangle$ and edge labels dot_i $(1 \leq i \leq l)$, there is a $\langle \gamma, LPA \rangle$-clause
$\langle \gamma, LPA \rangle(\sigma_\gamma) \rightarrow \langle adj, \gamma, p_1, LPA_{p_1} \rangle(L_{p_1}, R_{p_1}) \ldots$
$\ldots \langle adj, \gamma, p_k, LPA_{p_k} \rangle(L_{p_k}, R_{p_k}) \langle sub, \gamma, p_{k+1} \rangle(X_{p_{k+1}}) \ldots \langle sub, \gamma, p_l \rangle(X_{p_l})$
such that

  - for all adjunction sites p in γ, $p \notin \{dot_i \,|\, 1 \leq i \leq l\}$: $LPA_p = \emptyset$, and there is a clause $\langle adj, \gamma, p, \emptyset \rangle(\epsilon, \epsilon) \rightarrow \epsilon$
  - for all adjunction sites $p = dot_i$ in γ (for some i, $1 \leq i \leq l$): there is a clause $\langle adj, \gamma, dot_i, LPA_p \rangle(L, R) \rightarrow \langle \gamma_i, LPA_i' \rangle(L, R)$
  - all substitution sites $p = dot_i$ in γ are in $\{dot_i \,|\, 1 \leq i \leq l\}$ and there is a clause $\langle sub, \gamma, dot_i \rangle(X) \rightarrow \langle \gamma_i, \emptyset \rangle(X)$

- for every leaf with label $\langle \gamma, LPA \rangle$ and $\langle w \rangle$ (or $\langle w_1, w_2 \rangle$ resp.), there is a clause $\langle \gamma, LPA \rangle(w) \rightarrow \epsilon$ (or $\langle \gamma, LPA \rangle(w_1, w_2) \rightarrow \epsilon$ resp.).
- the sequences of strings for a mother node are computed from the daughters such that for at least one word w, the clauses leading from the mother to the daughters can be instantiated successfully, assuming an instantiation with an empty range for all variables not passed to one of the daughter predicates.

Furthermore, we call a TAG derivation tree whose nodes are equipped with the yields of the derivation trees they root (one component for initial trees, two components for auxiliary trees) and the set of arguments they dominate that actually depend on higher head trees a *decorated TAG derivation tree*.

Once these structures are defined, we can prove the correspondence between the decorated TAG derivation trees licensed in the k-TT-MCTAG G and the TT-RCG derivation trees of the RCG G'. More precisely, we show that for each decorated TAG derivation tree in G, there is an isomorphic TT-RCG derivation tree in G' and vice versa. We can show this by an induction on the height of the subtree rooted by a node. (Due to space limitations, we omit the proof here.)

4   Conclusion

This paper has investigated the relation between two grammar formalisms, TT-MCTAG and RCG. TT-MCTAG is a tree rewriting formalism that allows to adequately model the free word order in certain languages, e.g., German. RCG, on the other hand, is known to have nice formal properties: RCGs in general are polynomially parsable, simple RCGs are even mildly context-sensitive. Furthermore, parsing algorithms for simple RCGs are already available.

In this paper, we have shown how to construct for a given TT-MCTAG with a certain limitation (a so-called k-TT-MCTAG) an equivalent simple RCG. As a formal result, we obtain that the class of string languages generated by k-TT-MCTAG is contained in the class of languages generated by simple RCGs. In particular, k-TT-MCTAG are mildly context-sensitive.

As a practical result, we can use this transformation from k-TT-MCTAG to simple RCG for a 2-step k-TT-MCTAG parser that, in a first step, does the transformation and, in a second step, parses with the RCG obtained from the first step. As we have seen from the correspondence between the two derivation structures, the derivation tree of the k-TT-MCTAG can be retrieved from the RCG parse tree in a straightforward way. We have implemented this within a project that develops a TAG-based grammar for German along with a parser for this grammar[9].

References

1. Joshi, A.K., Schabes, Y.: Tree-Adjoining Grammars. In: Rozenberg, G., Salomaa, A. (eds.) Handbook of Formal Languages, pp. 69–123. Springer, Berlin (1997)
2. Boullier, P.: On TAG Parsing. In: TALN 1999, 6e conférence annuelle sur le Traitement Automatique des Langues Naturelles, Cargèse, Corse, pp. 75–84 (1999)
3. Boullier, P.: Range Concatenation Grammars. In: Proceedings of the Sixth International Workshop on Parsing Technologies (IWPT 2000), Trento, Italy, pp. 53–64 (2000)
4. Weir, D.J.: Characterizing mildly context-sensitive grammar formalisms. PhD thesis, University of Pennsylvania (1988)
5. Boullier, P.: A Proposal for a Natural Language Processing Syntactic Backbone. Technical Report 3342, INRIA (1998)
6. Joshi, A.K.: Tree adjoining grammars: How much contextsensitivity is required ro provide reasonable structural descriptions? In: Dowty, D., Karttunen, L., Zwicky, A. (eds.) Natural Language Parsing, pp. 206–250. Cambridge University Press, Cambridge (1985)
7. Boullier, P.: A Generalization of Mildly Context-Sensitive Formalisms. In: Proceedings of the Fourth International Workshop on Tree Adjoining Grammars and Related Formalisms (TAG+4), University of Pennsylvania, Philadelphia, pp. 17–20 (1998)
8. Lichte, T.: An MCTAG with Tuples for Coherent Constructions in German. In: Proceedings of the 12th Conference on Formal Grammar 2007, Dublin, Ireland (2007)
9. Kallmeyer, L.: Tree-local multicomponent tree adjoining grammars with shared nodes. Computational Linguistics 31(2), 187–225 (2005)
10. Søgaard, A., Lichte, T., Maier, W.: The complexity of linguistically motivated extensions of tree-adjoining grammar. In: Recent Advances in Natural Language Processing 2007, Borovets, Bulgaria (2007)

---

[9] See http://www.sfb441.uni-tuebingen.de/emmy/tulipa

Anti-pattern Matching Modulo

Claude Kirchner[1], Radu Kopetz[2], and Pierre-Etienne Moreau[2]

[1] INRIA Bordeaux – Sud Ouest
[2] INRIA Nancy – Grand Est

Abstract. Negation is intrinsic to human thinking and most of the time when searching for something, we base our patterns on both positive and negative conditions. In a recent work, the notion of term was extended to the one of anti-term, *i.e.* terms that may contain complement symbols.

Here we generalize the syntactic anti-pattern matching to anti-pattern matching *modulo* an arbitrary equational theory $\mathcal{E}$, and we study the specific and practically very useful case of associativity, possibly with a unity ($\mathcal{AU}$). To this end, based on the *syntacticness* of associativity, we present a rule-based associative matching algorithm, and we extend it to $\mathcal{AU}$. This algorithm is then used to solve $\mathcal{AU}$ anti-pattern matching problems. This allows us to be generic enough so that for instance, the *AllDiff* standard predicate of constraint programming becomes simply expressible in this framework. $\mathcal{AU}$ anti-patterns are implemented in the TOM language and we show some examples of their usage.

1 Introduction

Anti-patterns were introduced in [8] in order to provide a compact and expressive representation for sets of terms. Just by properly placing complement symbols in a pattern, a nice expressivity can be obtained, which can spare the user of using more complex and harder to read constructions (like disjunctions for instance).

Syntactic anti-patterns (*i.e.* when operators have no particular property) are very useful, but the anti-patterns are even more valuable when associated with equational theories, in particular with associativity, unit, and eventually with commutativity. For instance, consider the associative matching with neutral element as provided by TOM (http://tom.loria.fr) — a programming language that extends C and Java with algebraic data-types, pattern matching and strategic rewriting facilities [1]. The pattern $list(*, \neg a, *)$ denotes a list which contains at least one element different from the constant a, whereas $\neg list(*, a, *)$ denotes a list which does not contain any a ($list$ is an associative operator having empty list as its neutral element, and $*$ denotes any sublist). By using non-linearity we can express, in a single pattern, list constraints as *AllDiff* or *AllEqual*. Take for instance the pattern $list(*, x, *, x, *)$ that denotes a list with at least two equal elements (x is a variable). The complement of this, $\neg list(*, x, *, x, *)$ matches lists that have only distinct elements, *i.e. AllDiff*. In a similar way, as $list(*, x, *, \neg x, *)$ matches the lists that have at least two distinct elements, its complement $\neg list(*, x, *, \neg x, *)$ denotes any list whose elements are all equal. Without anti-patterns, these constructions would have to be expressed

C. Martín-Vide, F. Otto, and H. Fernau (Eds.): LATA 2008, LNCS 5196, pp. 275–286, 2008.

as loops, disjunctions *etc.* Of course, instead of the constant a or the variable x, we could have used any complex pattern or anti-pattern.

After presenting some general notions in Section 2, our first contribution, in Section 3, is to solve associative matching problems using a rule-based algorithm. We further adapt it to also support neutral elements. A second main contribution is to provide, in Section 4, an anti-pattern matching algorithm for an arbitrary equational theory, provided that a finitary matching algorithm is available for the given theory. We show how an equational anti-pattern matching problem can be transformed into a finite subset of equivalent equational problems. We then focus on the associative anti-patterns with neutral elements and we present a practical and efficient algorithm for solving such problems. In Section 5 we show how they are integrated in the TOM language. A survey of related work can be found in [9].

2   Terms and Anti-patterns

Terms and equality. A signature $\mathcal{F}$ is a set of function symbols, each one having a fixed arity associated to it. $\mathcal{T}(\mathcal{F}, \mathcal{X})$ is the set of *terms* built from a given finite set $\mathcal{F}$ of function symbols where constants are denoted $a, b, c, \ldots$, and a denumerable set $\mathcal{X}$ of variables denoted $x, y, z, \ldots$ A term t is said to be *linear* if no variable occurs more than once in t. The set of variables occurring in a term t is denoted by $\mathcal{V}ar(t)$. If $\mathcal{V}ar(t)$ is empty, t is called a *ground term* and $\mathcal{T}(\mathcal{F})$ is the set of ground terms.

A *substitution* σ is an assignment from $\mathcal{X}$ to $\mathcal{T}(\mathcal{F}, \mathcal{X})$, denoted $\sigma = \{x_1 \mapsto t_1, \ldots, x_k \mapsto t_k\}$ when its domain $\mathsf{Dom}(\sigma)$ is finite. Its application, written $\sigma(t)$, is defined by $\sigma(x_i) = t_i$, $\sigma(f(t_1, \ldots, t_n)) = f(\sigma(t_1), \ldots, \sigma(t_n))$ for $f \in \mathcal{F}$, and $\sigma(y) = y$ if $y \notin \mathsf{Dom}(\sigma)$. Given a term t, σ is called a *grounding substitution* for t if $\sigma(t) \in \mathcal{T}(\mathcal{F})$ (usually different from a *ground* substitution, which does not depend on t). The set of substitutions is denoted Σ. The set of grounding substitutions for a term t is denoted $\mathcal{GS}(t)$.

The ground semantics of a term $t \in \mathcal{T}(\mathcal{F}, \mathcal{X})$ is the set of all its ground instances: $[\![t]\!]_g = \{\sigma(t) \mid \sigma \in \mathcal{GS}(t)\}$. In particular, $[\![x]\!]_g = \mathcal{T}(\mathcal{F})$.

A *position* in a term is a finite sequence of natural numbers. The subterm u of a term t at position ω is denoted $t_{|\omega}$, where ω describes the path from the root of t to the root of u. $t(\omega)$ denotes the root symbol of $t_{|\omega}$. By $t[u]_\omega$ we express that the term t contains u as subterm at position ω. Positions are ordered in the classical way: $\omega_1 < \omega_2$ if ω_1 is a prefix of ω_2.

For an equational theory $\mathcal{E}$, an $\mathcal{E}$-*matching equation* (*matching equation* for short) is of the form $p \prec\!\!\!\prec_\mathcal{E} t$ where p is a term classically called a pattern and t is a term, generally considered as ground. The substitution σ is an $\mathcal{E}$-*solution of the* $\mathcal{E}$-*matching equation* $p \prec\!\!\!\prec_\mathcal{E} t$ if $\sigma(p) =_\mathcal{E} t$, and it is called an $\mathcal{E}$-*match* from p to t.

An $\mathcal{E}$-*matching system* S is a possibly existentially quantified conjunction of matching equations: $\exists \bar{x}(\wedge_i p_i \prec\!\!\!\prec_\mathcal{E} t_i)$. A substitution σ is an $\mathcal{E}$-*solution* of such a matching system if there exists a substitution ρ, with domain $\bar{x}$, such that σ is a solution of all the matching equations $\rho(p_i) \prec\!\!\!\prec_\mathcal{E} \rho(t_i)$. The set of solutions of S is denoted by $\mathcal{S}ol_\mathcal{E}(\mathsf{S})$.

An $\mathcal{E}$-*matching disjunction* D is a disjunction of $\mathcal{E}$-matching systems. Its solutions are the substitutions solution of at least one of its system constituents. Its free variables $\mathcal{F}Var(\mathsf{D})$ are defined as usual in predicate logic. We use the notation $\mathsf{D}[\mathsf{S}]$ to denote that the system S occurs in the *context* D, *i.e.* S is part of the disjunction D.

Given an equational theory $\mathcal{E}$ and two sets of terms A and B, we consider as usual that: $t \in_{\mathcal{E}} A \Leftrightarrow \exists t' \in A$ such that $t =_{\mathcal{E}} t'$; $A \subseteq_{\mathcal{E}} B \Leftrightarrow \forall t \in A$ we have $t \in_{\mathcal{E}} B$; $A =_{\mathcal{E}} B \Leftrightarrow A \subseteq_{\mathcal{E}} B$ and $B \subseteq_{\mathcal{E}} A$.

A binary operator f is called *associative* if it satisfies the equational axiom $\forall x,y,z \in \mathcal{T}(\mathcal{F},\mathcal{X}) : f(f(x,y),z) = f(x,f(y,z))$ and *commutative* if $\forall x,y \in \mathcal{T}(\mathcal{F},\mathcal{X}) : f(x,y) = f(y,x)$. A binary operator can have neutral elements — symbols of arity zero: e_f is a *left neutral* operator for f if $\forall x \in \mathcal{T}(\mathcal{F},\mathcal{X})$, $f(e_f,x) = x$; e_f is a *right neutral* operator for f if $\forall x \in \mathcal{T}(\mathcal{F},\mathcal{X})$, $f(x,e_f) = x$; e_f is a *neutral* or *unit* operator for f if it is a left and right neutral operator for f. When f is associative or associative with a unit, this is denoted $\mathcal{A}$ or $\mathcal{AU}$ respectively.

Anti-terms. An anti-term [8] is a term that may contain complement symbols, denoted by $\rceil$. The BNF of anti-terms is:

$$\mathcal{AT} ::= \mathcal{X} \mid f(\mathcal{AT},\ldots,\mathcal{AT}) \mid \rceil\mathcal{AT}, \text{ where } f \text{ respects its arity.}$$

The set of anti-terms (resp. ground anti-terms) is denoted $\mathcal{AT}(\mathcal{F},\mathcal{X})$ (resp. $\mathcal{AT}(\mathcal{F})$). Any term is an anti-term, *i.e.* $\mathcal{T}(\mathcal{F},\mathcal{X}) \subset \mathcal{AT}(\mathcal{F},\mathcal{X})$.

The free variables of an anti-term t are denoted $\mathcal{F}Var(t)$, and the non-free ones $\mathcal{NF}Var(t)$. Intuitively, a variable is free if it is not under a $\rceil$. Typically, $\mathcal{F}Var(\rceil t) = \emptyset$ and $\mathcal{F}Var(f(x,\rceil x)) = \{x\}$.

The substitutions are only active on free variables. For anti-terms, a grounding substitution is a substitution that instantiates all the free variables by ground terms. As detailed in [8], the *ground semantics* is defined as follows:

Definition 2.1. *Given an anti-term* $q \in \mathcal{AT}(\mathcal{F},\mathcal{X})$, *the* ground semantics *is defined by:* $[\![q[\rceil q']\!]_{\omega}]\!]_g = [\![q[z]_{\omega}]\!]_g \setminus [\![q[q']_{\omega}]\!]_g$, *where* z *is a fresh variable and for all* $\omega' < \omega$, $q(\omega') \neq \rceil$.

As stressed in [8], the last condition is essential as it prevents abstracting subterms in a complemented context. This would lead to counter-intuitive situations.

Example 2.1

1. $[\![h(a,\rceil b)]\!]_g = [\![h(a,z)]\!]_g \setminus [\![h(a,b)]\!]_g = \{h(a,\sigma(z)) \mid \sigma \in \mathcal{GS}(h(a,z))\} \setminus \{h(a,b)\}$
2. Non-linearity is crucial to denote for instance 'any term except those rooted by h with identical subterms':
   $[\![\rceil h(x,x)]\!]_g = [\![z]\!]_g \setminus [\![h(x,x)]\!]_g = \mathcal{T}(\mathcal{F}) \setminus \{h(\sigma(x),\sigma(x)) \mid \sigma \in \mathcal{GS}(h(x,x))\}$

The anti-terms are also called *anti-patterns*, in particular when they appear in the left-hand side of a match equation. The notions of matching equations, systems and disjunctions are extended to anti-patterns by allowing the *left-hand side* of match equations to be anti-patterns. When a match equation contains anti-patterns, we often refer to it as an *anti-pattern matching equation*. The solutions of such problems are defined later.

3 Associative Matching

To provide an equational anti-matching algorithm in the next section, we first need to make precise the matching algorithm that serves as our starting point. The rule-based presentation of an $\mathcal{AU}$ matching algorithm is also the first contribution of this paper.

In this section we focus on the particular useful case of matching modulo $\mathcal{A}$ and $\mathcal{AU}$. The reason why we chose to detail these specific theories are their tremendous usefulness in rule-based programming such as ASF+SDF [2] or MAUDE [4,5] for instance, where lists, and consequently list-matching, are omnipresent.

Since associativity and neutral element are *regular* axioms (*i.e.* equivalent terms have the same set of variables), we can apply the combination results for matching modulo the union of disjoint regular equational theories [13,15] to get a matching algorithm modulo the theory combination of an arbitrary number of $\mathcal{A}$, $\mathcal{AU}$ as well as free symbols. Therefore we study in this section matching modulo $\mathcal{A}$ or $\mathcal{AU}$ of a single binary symbol f, whose unit is denoted e_f. The only other symbols under consideration are free constants. For syntactic matching, a simple rule-based matching algorithm can be found in [3,8].

3.1 Matching Associative Patterns

By making precise this algorithm, our purpose is to provide a simple and intuitive one that can be easily proved to be correct and complete and that will be later adapted to anti-pattern matching[1]. In terms of efficiency, more appropriate solutions were developed in [4,5].

Unification modulo associativity has been extensively studied [14,10]. It is decidable, but infinitary, while $\mathcal{A}$-matching is finitary. Our algorithm $\mathcal{A}$-Matching is described in Figure 1 and is quite reminiscent from [12] although not based on a Prolog resolution strategy. It strongly relies on the *syntacticness* of the associative theory [6,7].

Proposition 3.1. *Given a matching equation* $p \ll_{\mathcal{A}} t$ *with* $p \in \mathcal{T}(\mathcal{F}, \mathcal{X})$ *and* $t \in \mathcal{T}(\mathcal{F})$, *the application of* $\mathcal{A}$-Matching *always terminates.*

If no solution is lost in the application of a transformation rule, the rule is called *preserving*. It is a *sound* rule if it does not introduce unexpected solutions.

Proposition 3.2. *The rules in* $\mathcal{A}$-Matching *are sound and preserving modulo* $\mathcal{A}$.

Proof. The rule Mutate is a direct consequence of the decomposition rules for syntactic theories presented in [7]. The rest of the rules are usual ones for which these results have been obtained for example in [3].   □

Theorem 3.1. *Given a matching equation* $p \ll_{\mathcal{A}} t$, *with* $p \in \mathcal{T}(\mathcal{F}, \mathcal{X})$ *and* $t \in \mathcal{T}(\mathcal{F})$, *the normal form w.r.t.* $\mathcal{A}$-Matching *exists and it is unique. It can only be of the following types:*

---

[1] Due to the lack of space, lengthy proofs are given in the technical report [9].

Mutate	$f(p_1, p_2) \ll_{\mathcal{A}} f(t_1, t_2)$	$\mapsto (p_1 \ll_{\mathcal{A}} t_1 \wedge p_2 \ll_{\mathcal{A}} t_2) \vee$
		$\exists x (p_2 \ll_{\mathcal{A}} f(x, t_2) \wedge f(p_1, x) \ll_{\mathcal{A}} t_1) \vee$
		$\exists x (p_1 \ll_{\mathcal{A}} f(t_1, x) \wedge f(x, p_2) \ll_{\mathcal{A}} t_2)$
SymClash$_1$	$f(p_1, p_2) \ll_{\mathcal{A}} a$	$\mapsto \perp$
SymClash$_2$	$a \ll_{\mathcal{A}} f(p_1, p_2)$	$\mapsto \perp$
ConstantClash	$a \ll_{\mathcal{A}} b$	$\mapsto \perp$ if $a \neq b$
Replacement	$z \ll_{\mathcal{A}} t \wedge S$	$\mapsto z \ll_{\mathcal{A}} t \wedge \{z \mapsto t\}S$ if $z \in \mathcal{F}Var(S)$

Utility Rules:

Delete	$p \ll_{\mathcal{A}} p$	$\mapsto \top$	PropagClash$_1$	$S \wedge \perp \mapsto \perp$
Exists$_1$	$\exists z (D[z \ll_{\mathcal{A}} t]) \mapsto D[\top]$ if $z \notin Var(D[\top])$		PropagClash$_2$	$S \vee \perp \mapsto S$
Exists$_2$	$\exists z (S_1 \vee S_2)$	$\mapsto \exists z(S_1) \vee \exists z(S_2)$	PropagSuccess$_1$	$S \wedge \top \mapsto S$
DistribAnd	$S_1 \wedge (S_2 \vee S_3)$	$\mapsto (S_1 \wedge S_2) \vee (S_1 \wedge S_3)$	PropagSuccess$_2$	$S \vee \top \mapsto \top$

Fig. 1. $\mathcal{A}$-Matching: p_i are patterns, t_i are ground terms, and S is any conjunction of matching equations. Mutate is the most interesting rule, and it is a direct consequence of the fact that associativity is a *syntactic theory*. $\wedge, \vee$ are classical boolean connectors.

1. $\top$, then p and t are identical modulo $\mathcal{A}$, i.e. $p =_{\mathcal{A}} t$;
2. $\perp$, then there is no match from p to t;
3. a disjunction of conjunctions $\bigvee_{j \in J}(\bigwedge_{i \in I} x_{i_j} \ll_{\mathcal{A}} t_{i_j})$ with $I, J \neq \emptyset$, then the substitutions $\sigma_j = \{x_{i_j} \mapsto t_{i_j}\}_{i \in I, j \in J}$ are all the matches from p to t.

Example 3.1. Applying $\mathcal{A}$-Matching for $f \in \mathcal{F}_A$, $x, y \in \mathcal{X}$, and $a, b, c, d \in \mathcal{T}(\mathcal{F})$:

$f(x, f(a, y)) \ll_{\mathcal{A}} f(f(b, f(a, c)), d)$

$\mapsto_{\text{Mutate}} (x \ll_{\mathcal{A}} f(b, f(a, c)) \wedge f(a, y) \ll_{\mathcal{A}} d) \vee$
$\exists z (f(a, y) \ll_{\mathcal{A}} f(z, d) \wedge f(x, z) \ll_{\mathcal{A}} f(b, f(a, c))) \vee$
$\exists z (x \ll_{\mathcal{A}} f(f(b, f(a, c)), z) \wedge f(z, f(a, y)) \ll_{\mathcal{A}} d)$

$\mapsto_{\text{SymClash}_1, \text{PropagClash}_2} \exists z (f(a, y) \ll_{\mathcal{A}} f(z, d) \wedge f(x, z) \ll_{\mathcal{A}} f(b, f(a, c)))$

$\mapsto_{\text{Mutate}, \text{SymClash}_1} \exists z (f(a, y) \ll_{\mathcal{A}} f(z, d) \wedge$
$((x \ll_{\mathcal{A}} b \wedge z \ll_{\mathcal{A}} f(a, c)) \vee (x \ll_{\mathcal{A}} f(b, a) \wedge z \ll_{\mathcal{A}} c)))$

$\mapsto_{\text{DistribAnd}, \text{Replacement}, \text{Mutate}, \text{SymClash}_{1,2}} \exists z (f(a, y) \ll_{\mathcal{A}} f(z, d) \wedge x \ll_{\mathcal{A}} b \wedge z \ll_{\mathcal{A}}$
$f(a, c)) \mapsto_{\text{Replacement}, \text{Exists}, \text{Mutate}, \text{SymClash}_{1,2}} x \ll_{\mathcal{A}} b \wedge y \ll_{\mathcal{A}} f(c, d).$

3.2   Matching Associative Patterns with Unit Elements

It is often the case that associative operators have a unit and we know since the early works on *e.g.* OBJ, that this is quite useful from a rule programming point of view. For example, to state *a list* L *that contains the objects* a *and* b. This can be expressed by the pattern $f(x, f(a, f(y, f(b, z))))$, where $x, y, z \in \mathcal{X}$, which will match $f(c, f(a, f(d, f(b, e))))$ but not $f(a, b)$ or $f(c, f(a, b))$. When f has for unit e_f, the previous pattern does match modulo $\mathcal{AU}$, producing the substitution $\{x \mapsto e_f, y \mapsto e_f, z \mapsto e_f\}$ for $f(a, b)$, and $\{x \mapsto c, y \mapsto e_f, z \mapsto e_f\}$ for $f(c, f(a, b))$. However, $\mathcal{A}$ is a theory with a finite equivalence class, which is not the case of $\mathcal{AU}$, and an immediate consequence is that the set of matches becomes trivially infinite. For instance, $Sol(x \ll_{\mathcal{AU}} a) = \{\{x \mapsto a\}, \{x \mapsto f(e_f, a)\}, \{x \mapsto f(e_f, f(e_f, a))\}, \ldots\}$.

In order to obtain a matching algorithm for $\mathcal{AU}$, we replace SymClash rules in $\mathcal{A}$-Matching to appropriately handle unit elements (remember that we assume, because of modularity, that we only have in $\mathcal{F}$ a single binary $\mathcal{AU}$ symbol f, and constants, including e_f):

$\mathsf{SymClash}_1^+ \quad f(p_1, p_2) \overset{\mathbf{+}}{\prec}_{\mathcal{AU}} a \longmapsto (p_1 \overset{\mathbf{+}}{\prec}_{\mathcal{AU}} e_f \wedge p_2 \overset{\mathbf{+}}{\prec}_{\mathcal{AU}} a) \vee (p_1 \overset{\mathbf{+}}{\prec}_{\mathcal{AU}} a \wedge p_2 \overset{\mathbf{+}}{\prec}_{\mathcal{AU}} e_f)$

$\mathsf{SymClash}_2^+ \quad a \overset{\mathbf{+}}{\prec}_{\mathcal{AU}} f(p_1, p_2) \longmapsto (e_f \overset{\mathbf{+}}{\prec}_{\mathcal{AU}} p_1 \wedge a \overset{\mathbf{+}}{\prec}_{\mathcal{AU}} p_2) \vee (a \overset{\mathbf{+}}{\prec}_{\mathcal{AU}} p_1 \wedge e_f \overset{\mathbf{+}}{\prec}_{\mathcal{AU}} p_2)$

In addition, we keep all other transformation rules, only changing all match symbols from $\overset{\mathbf{+}}{\prec}_{\mathcal{A}}$ to $\overset{\mathbf{+}}{\prec}_{\mathcal{AU}}$. The new system, named $\mathcal{AU}$-Matching, is clearly terminating without producing in general a minimal set of solutions. After proving its correctness, we will see what can be done in order to minimize the set of solutions.

Proposition 3.3. *The rules of $\mathcal{AU}$-Matching are sound and preserving modulo $\mathcal{AU}$.*

In order to avoid redundant solutions we further consider that all the terms are in normal form w.r.t. the rewrite system $\mathcal{U} = \{f(e_f, x) \rightarrow x, f(x, e_f) \rightarrow x\}$. Therefore, we perform a normalized rewriting [11] modulo $\mathcal{U}$. This technique ensures that before applying any rule from Figure 1, the terms are in normal forms w.r.t. $\mathcal{U}$.

4 Anti-pattern Matching Modulo

In [8], anti-patterns were studied in the case of the empty theory. In this section we generalize the matching algorithm to an arbitrary *regular* equational theory $\mathcal{E}$, that doesn't contain the symbol $\daleth$. The presented results allow the use of anti-patterns in a general context, and they constitute the main contributions of the paper.

Definition 4.1. *Given an equational theory $\mathcal{E}$ and $t \in \mathcal{T}(\mathcal{F}, \mathcal{X})$, the ground semantics of t modulo $\mathcal{E}$ is defined as: $[\![t]\!]_{g\mathcal{E}} = \{t' \mid t' \in_{\mathcal{E}} [\![t]\!]_g\}$.*

Therefore, the ground semantics of t modulo $\mathcal{E}$ is the set of all the ground terms that can be computed from the ground semantics of t by applying the axioms of $\mathcal{E}$.

Definition 4.2. *Given $q \in \mathcal{AT}(\mathcal{F}, \mathcal{X})$ and a theory $\mathcal{E}$, the ground semantics of q modulo $\mathcal{E}$ is defined recursively in the following way:*

$$[\![q[\daleth q']_\omega]\!]_{g\mathcal{E}} = \begin{cases} [\![q[z]_\omega]\!]_{g\mathcal{E}} \setminus [\![q[q']_\omega]\!]_{g\mathcal{E}}, & \text{if } \mathcal{F}\mathcal{V}ar(q[\daleth q']_\omega) = \emptyset \\[2mm] \text{otherwise } [\![\sigma(q[\daleth q']_\omega)]\!]_{g\mathcal{E}}, & \text{for all } \sigma \in \mathcal{GS}(q[\daleth q']_\omega) \end{cases}$$

where z is a fresh variable and for all $\omega' < \omega$, $q(\omega') \neq \daleth$.

When $\mathcal{E}$ is the empty theory, this definition is perfectly compatible with Definition 2.1. However, in the equational case a direct adaptation cannot be used. Consider the pattern $f(x, f(\daleth a, y))$, where f is $\mathcal{AU}$. This intuitively denotes the lists that contain at least one element different from a, like $f(b, f(a, c))$ for instance. Suppose we use Definition 2.1 to compute the ground semantics, we would get $[\![f(x, f(z, y))]\!]_{g_{\mathcal{AU}}} \setminus [\![f(x, f(a, y))]\!]_{g_{\mathcal{AU}}}$, which does not contain the term $f(b, f(a, c))$. This happens because giving different values to x, y and applying the $\mathcal{AU}$ axioms differently on the two terms, we obtain different term structures in the two sets. But this is not the intuitive semantics of anti-patterns.

Example 4.1

$$[\![\neg f(x, f(\neg a, y))]\!]_{g_{\mathcal{AU}}} = [\![z]\!]_{g_{\mathcal{AU}}} \setminus [\![f(x, f(\neg a, y))]\!]_{g_{\mathcal{AU}}} = \mathcal{T}(\mathcal{F}) \setminus \bigcup_\sigma [\![f(\sigma(x), f(\neg a, \sigma(y)))]\!]_{g_{\mathcal{AU}}}$$

$$= \mathcal{T}(\mathcal{F}) \setminus \bigcup_\sigma ([\![f(\sigma(x), f(z, \sigma(y)))]\!]_{g_{\mathcal{AU}}} \setminus [\![f(\sigma(x), f(a, \sigma(y)))]\!]_{g_{\mathcal{AU}}})$$

$$= \text{everything that is not an } f \text{ or an } f \text{ with only } a \text{ inside}$$

In the empty theory, given $q \in \mathcal{AT}(\mathcal{F}, \mathcal{X})$ and $t \in \mathcal{T}(\mathcal{F})$, the matching equation $q \prec\!\!\!\prec t$ has a solution when there exists a substitution σ such that $t \in [\![\sigma(q)]\!]_g$. This is extended to matching modulo $\mathcal{E}$ as follows:

Definition 4.3. *For all $q \in \mathcal{AT}(\mathcal{F}, \mathcal{X})$ and $t \in \mathcal{T}(\mathcal{F})$, the solutions of the anti-pattern matching equation $q \prec\!\!\!\prec_{\mathcal{E}} t$ are:*

$$Sol(q \prec\!\!\!\prec_{\mathcal{E}} t) = \{\sigma \mid t \in [\![\sigma(q)]\!]_{g_{\mathcal{E}}}, \text{ with } \sigma \in \mathcal{GS}(q)\}.$$

A general anti-pattern matching problem P is any first-order expression whose atomic formulae are anti-pattern matching equations. To define their solutions, we rely on the usual definition of validity in predicate logic:

Definition 4.4. *Given an anti-pattern matching problem P, the solutions modulo $\mathcal{E}$ are defined as: $Sol_{\mathcal{E}}(P) = \{\sigma \mid \models \sigma(P)\}$, where $\models q \prec\!\!\!\prec_{\mathcal{E}} t \Leftrightarrow \models t \in [\![q]\!]_{g_{\mathcal{E}}}$.*

Let us look at several examples of anti-pattern matching modulo in some usual equational theories:

Example 4.2. In the syntactic case we have:
- $Sol(h(\neg a, x) \prec\!\!\!\prec h(b, c)) = \{x \mapsto c\}$,
- $Sol(h(x, \neg g(x)) \prec\!\!\!\prec h(a, g(b))) = \{x \mapsto a\}$,
- $Sol(h(x, \neg g(x)) \prec\!\!\!\prec h(a, g(a))) = \emptyset$.

In the associative theory:
- $Sol(f(x, f(\neg a, y)) \prec\!\!\!\prec_A f(b, f(a, f(c, d)))) = \{x \mapsto f(b, a), y \mapsto d\}$,
- $Sol(f(x, f(\neg a, y)) \prec\!\!\!\prec_A f(a, f(a, a))) = \emptyset$.

The following patterns express that we do not want an a below an f:
- $Sol(\neg f(x, f(a, y)) \prec\!\!\!\prec_A f(b, f(a, f(c, d)))) = \emptyset$,
- $Sol(\neg f(x, f(a, y)) \prec\!\!\!\prec_A f(b, f(b, f(c, d)))) = \Sigma$.

A combination of the two previous examples, $\neg f(x, f(\neg a, y))$, would naturally correspond to "an f with only a inside":

- $Sol(\neg f(x, f(\neg a, y)) \prec\!\!\!\prec_A f(a, f(b, a))) = \emptyset$,
- $Sol(\neg f(x, f(\neg a, y)) \prec\!\!\!\prec_A f(a, f(a, a))) = \Sigma$.

Non-linearity can be also useful: $Sol(\neg f(x, x) \prec\!\!\!\prec_A f(a, f(b, f(a, b)))) = \emptyset$, but $Sol(\neg f(x, x) \prec\!\!\!\prec_A f(a, f(b, f(a, c)))) = \Sigma$. If we consider that f is also commutative, then we have the following results for matching modulo $\mathcal{AC}$: $Sol(f(x, f(\neg a, y)) \not\ll_{\mathcal{AC}} f(a, f(b, c))) = \{\{x \mapsto a, y \mapsto c\}, \{x \mapsto a, y \mapsto b\}, \{x \mapsto b, y \mapsto a\}, \{x \mapsto c, y \mapsto a\}\}$.

4.1 From Anti-pattern Matching to Equational Problems

To solve anti-pattern matching modulo, a solution is to first transform the initial matching problem into an equational one. This is performed using the following transformation rule:

$$\text{ElimAnti } q[\neg q']_\omega \not\ll_{\mathcal{E}} t \mapsto \exists z \; q[z]_\omega \not\ll_{\mathcal{E}} t \wedge \forall x \in \mathcal{FVar}(q') \; not(q[q']_\omega \not\ll_{\mathcal{E}} t)$$
$$\text{if } \forall \; \omega' < \omega, \; q(\omega') \neq \neg \text{ and } z \text{ a fresh variable}$$

An anti-pattern matching problem P not containing any $\neg$ symbol, is a first-order formula where the symbol *not* is the usual negation of predicate logic, the symbol $\not\ll_{\mathcal{E}}$ is interpreted as $=_{\mathcal{E}}$ and the symbol $\forall$ is the usual universal quantification: $\forall x \mathsf{P} \equiv not(\exists x \; not \; (\mathsf{P}))$. Therefore they are exactly $\mathcal{E}$-disunification problems.

Proposition 4.1. *The rule* ElimAnti *is sound and preserving modulo* $\mathcal{E}$.

The normal forms *w.r.t.* ElimAnti of anti-pattern matching problems are specific equational problems. Although equational problems are undecidable in general [16], even in case of $\mathcal{A}$ or $\mathcal{AU}$ theories, we will see that the specific equational problems issued from anti-pattern matching are decidable for $\mathcal{A}$ or $\mathcal{AU}$ theories.

Summarizing, if we know how to solve equational problems modulo $\mathcal{E}$, then any anti-pattern matching problem modulo $\mathcal{E}$ can be translated into equivalent equational problems using ElimAnti and further solved. These statements are formalized by the following Proposition:

Proposition 4.2. *An anti-pattern matching problem can always be translated into an equivalent equational problem in a finite number of steps.*

Solving equational problems resulting from normalization with ElimAnti can be performed with techniques like disunification for instance in the case of syntactic theory. These techniques were designed to cover more general problems. In our case, a more efficient and tailored approach can be developed. Given a finitary $\mathcal{E}$-match algorithm, a first solution would be to normalize each match equation separately, then to combine the results using replacements and some cleaning rules (as ForAll, NotOr, NotTrue, NotFalse from Figure 2). This approach can be used to effectively solve $\mathcal{A}$, $\mathcal{AU}$, and $\mathcal{AC}$ anti-pattern matching problems. We further detail the $\mathcal{AU}$ case.

4.2 A Specific Case: Matching $\mathcal{AU}$ Anti-patterns

To compute the set of solutions for an $\mathcal{AU}$ anti-pattern matching equation we develop now a specific approach.

Definition 4.5. $\mathcal{AU}$-AntiMatching: *Given an $\mathcal{AU}$ anti-pattern matching problem $q \not\ll_{\mathcal{AU}} t$, apply the rules from Figure 2, giving a higher priority to* ElimAnti.

ElimAnti	$q[\neg q']_\omega \mathrel{\not\!\!\prec}_{\mathcal{AU}} t \mapsto \exists z\, q[z]_\omega \mathrel{\not\!\!\prec}_{\mathcal{AU}} t \ \wedge\ \forall x \in \mathcal{F}Var(q')\, not(q[q']_\omega \mathrel{\not\!\!\prec}_{\mathcal{AU}} t)$
	if $\forall\, \omega' < \omega,\ q(\omega') \neq \neg$ and z a fresh variable
ForAll	$\forall \bar{y}\, not(\mathrm{D}) \mapsto not(\exists \bar{y}\ \mathrm{D})$
NotOr	$not(\mathrm{D}_1 \vee \mathrm{D}_2) \mapsto not(\mathrm{D}_1) \wedge not(\mathrm{D}_2)$
NotTrue	$not(\top) \mapsto \bot$
NotFalse	$not(\bot) \mapsto \top$

PLUS ALL THE RULES OF $\mathcal{AU}$-Matching (*Section* 3.2)

Fig. 2. $\mathcal{AU}$-AntiMatching

Note that instead of giving a higher priority to ElimAnti the algorithm can be decomposed in two steps: first normalize with ElimAnti to eliminate all $\neg$ symbols, then apply all the other rules.

We further prove that the algorithm is correct. Moreover, the normal forms of its application on an $\mathcal{AU}$ anti-pattern matching equation do not contain any $\neg$ or *not* symbols. Actually they are the same as the ones exposed in Theorem 3.1.

Proposition 4.3. *The application of* $\mathcal{AU}$*-AntiMatching is sound and preserving.*

Proof. For ElimAnti these properties were shown in the proof of Proposition 4.1. Similarly, Proposition 3.3 states the sound and preserving properties for the rules of $\mathcal{AU}$-Matching. The rest of the rules are trivial. ◻

Theorem 4.1. *The normal forms of* $\mathcal{AU}$*-AntiMatching are* $\mathcal{AU}$*-matching problems in solved form.*

$\mathcal{AU}$-AntiMatching is a general algorithm, that solves any anti-pattern matching problem. Note that it can produce 2^n matching equations, where n is the number of $\neg$ symbols in the initial problem. For instance, applying ElimAnti on $f(a, \neg b) \mathrel{\not\!\!\prec}_{\mathcal{AU}} f(a, a)$ gives $\exists z f(a, z) \mathrel{\not\!\!\prec}_{\mathcal{AU}} f(a, a) \wedge not(f(a, b) \mathrel{\not\!\!\prec}_{\mathcal{AU}} f(a, a))$. Note that all equations have the same right-hand sides $f(a, a)$, and *almost* the same left-hand sides $f(a, \_)$. Therefore, when solving the second equation for instance, we perform some matches that were already done when solving the first one. This approach is clearly not optimal, and in the following we propose a more efficient one.

4.3   A More Efficient Algorithm for $\mathcal{AU}$ Anti-patterns Matching

In this section we consider a subclass of anti-patterns, called *PureFVars*, and we present a more efficient algorithm that has the same complexity as $\mathcal{AU}$-Matching. In particular, it does no longer produce the 2^n equations introduced by $\mathcal{AU}$-AntiMatching.

Definition 4.6. *Given* $\mathcal{F},\mathcal{X}$ *we define a subclass of anti-patterns:*

$$PureFVars = \left\{ q \in \mathcal{AT}(\mathcal{F}, \mathcal{X}) \ \middle| \ \begin{array}{l} q = C[f(t_1, \ldots, t_i, \ldots, t_j, \ldots, t_n)], \\ \forall i \neq j,\ \mathcal{F}Var(t_i) \cap \mathcal{NFVar}(t_j) = \emptyset \end{array} \right\}$$

The anti-patterns in *PureFVars* are special cases of non-linearity respecting that at any position, we don't find a term that has a free variable in one of its children, and the same variable under a $\neg$ in another child. For instance, $f(x, x) \in PureFVars$, $f(\neg x, \neg x) \in PureFVars$, but $f(x, \neg x) \notin PureFVars$.

Definition 4.7. $\mathcal{AU}$-AntiMatchingEfficient: *The algorithm corresponds to $\mathcal{AU}$-AntiMatching, where the rule* ElimAnti *is replaced with the following one, and which has no longer any priority:*

$$\text{ElimAnti'} \quad \neg q \prec\!\!\!\prec_{\mathcal{AU}} t \mapsto \forall x \in \mathcal{F}Var(q) \; not(q \prec\!\!\!\prec_{\mathcal{AU}} t)$$

Note that our algorithms are finitary and based on decomposition. Therefore, when considering syntactic or regular theories the composition results for matching algorithms are still valid. Note also that $Pure\mathcal{F}Vars$ is trivially stable *w.r.t.* to this algorithm and that now the rules apply on problems that potentially contain $\neg$ symbols. For instance, we may apply the rule Mutate on $f(a, \neg b) \prec\!\!\!\prec_{\mathcal{AU}} f(a, a)$. The algorithm is still terminating, with the same arguments as in the proof of Proposition 3.1, but the proof of Proposition 3.3 is no longer valid in this new case. The correctness of the algorithm has to be established again:

Proposition 4.4. *Given* $q \prec\!\!\!\prec_{\mathcal{AU}} t$*, with* $q \in Pure\mathcal{F}Vars$*, the application of* $\mathcal{AU}$-AntiMatchingEfficient *is sound and preserving.*

This approach is much more efficient, as no duplications are being made. Let us see on a simple example: $f(x, \neg a) \prec\!\!\!\prec_{\mathcal{AU}} f(a, b) \mapsto_{\text{Mutate}}$ $(x \prec\!\!\!\prec_{\mathcal{AU}} a \wedge \neg a \prec\!\!\!\prec_{\mathcal{AU}} b) \vee D_1 \vee D_2 \mapsto_{\text{ElimAnti'}} (x \prec\!\!\!\prec_{\mathcal{AU}} a \wedge not(a \prec\!\!\!\prec_{\mathcal{AU}} b))$ $\vee D_1 \vee D_2 \mapsto_{\text{ConstantClash}} (x \prec\!\!\!\prec_{\mathcal{AU}} a \wedge not(\bot)) \vee D_1 \vee D_2 \mapsto_{\text{NotFalse,PropagSuccess}_2}$ $x \prec\!\!\!\prec_{\mathcal{AU}} a \vee D_1 \vee D_2$. We continue in a similar way for D_1, D_2 and we finally obtain the solution $\{x \mapsto a\}$.

In practice, when implementing an anti-pattern matching algorithm, one can imagine the following approach: a traversal of the term is done, and if the special non-linear case is detected (*i.e.* $\notin Pure\mathcal{F}Vars$), then $\mathcal{AU}$-AntiMatching is applied; otherwise we apply $\mathcal{AU}$-AntiMatchingEfficient. This is the method used in the TOM compiler for instance.

In this section we have given a general algorithm for solving $\mathcal{AU}$ anti-pattern matching problems, and a more efficient one for a subclass which encompasses most of the practical cases. We also conjecture that modifying the universal quantification of ElimAnti' to only quantify variables that respect the condition $\mathcal{F}Var(q_1) \cap \mathcal{NF}Var(q_2) = \emptyset$ of $Pure\mathcal{F}Vars$, would still lead to a sound and complete algorithm. For instance, when applying ElimAnti' to $f(x, \neg x)$, the variable x would not be quantified. This algorithm has been experimented and tested without showing any counter example. Proving this conjecture is part of our future work.

5   Anti-matching Modulo in Tom

Anti-patterns are successfully integrated in the TOM language for syntactic and $\mathcal{AU}$ matching. In this section we show how they can be used and we illustrate the expressiveness they add to the pattern matching capabilities of this language. It is worth mentioning that for all the theories considered, the size of the generated code is *linear* in the size of the patterns.

In order to support anti-patterns, we enriched the syntax of the TOM patterns to allow the use of operator '!' (representing '$\neg$'). For syntactic matching, here is an example of a *match* in TOM:

```
%match(s) {
  f(a(),g(b()))   -> { /* executed when f(a,g(b)) matches s   */ }
  f(!a(),g(b()))  -> { /* when f(x,g(b)) matches s with x!=a */ }
  !f(x,!g(x))     -> { /* when not 'f(x,y) matches s' or ...    */ }
  !f(x,g(y))      -> { /* action 4 */ }
}
```

Similarly to `switch/case`, an action part is executed when its corresponding pattern matches the subject s. Note that non-linear patterns are allowed. When combined with lists, anti-patterns are even more useful:

```
%match(s) {
  list(_*,a(),_*)       -> { /* executed when s contains a */ }
  list(_*,!a(),_*)      -> { /* s has one elem. diff. from a */ }
  !list(_*,a(),_*)      -> { /* s does not contain a */ }
  !list(_*,!a(),_*)     -> { /* s contains only a */ }
  list(_*,x,_*,x,_*)    -> { /* s has at least 2 equal elem. */ }
  !list(_*,x,_*,x,_*)   -> { /* s has only distinct elem. */ }
  list(_*,x,_*,!x,_*)   -> { /* s has at least 2 diff elem. */ }
  !list(_*,x,_*,!x,_*)  -> { /* when s has only equal elem. */ }
}
```

In the above patterns `list` is $\mathcal{AU}$, a `_*` stands for any sublist, `a()` is a constant and `x` is a variable that cannot be instantiated by the empty list. Note that we mainly used the constant `a()`, but any other pattern or anti-pattern could have been used instead, like in: `list(_*,f(!a(),g(b())),_*)`, or `!list(_*,f(!a(),g(b())),_*)`. There is no restriction.

The following example prints all the elements that do not appear twice or more in a list s:

```
%match(s) {
  list(_*,x,_*) && !list(_*,x,_*,x,_*) << s -> { print(x); }
}
```

For instance, if s is instantiated with the list of integers (1,2,1,3,2,1,5), the above code would output: 3 and 5. Note that the `&&` is the classical boolean connector $\wedge$ and `<<` is the $\prec$. The idea is that the first pattern selects an element from the list, and the second one verifies that it doesn't appear twice.

Without using anti-patterns, one would be forced to verify additional conditions in the action part, which would make the code more complicated and difficult to maintain (see [8], Section 6). Besides, they may improve efficiency, by verifying some conditions earlier in the matching process.

6   Conclusion

We have generalized the notion of anti-pattern matching to anti-pattern matching modulo an arbitrary regular theory $\mathcal{E}$. Because of their usefulness for rule-based programming, we chose to exemplify the anti-patterns for the $\mathcal{A}$ and $\mathcal{AU}$ theories.

What is worth noting is that the algorithms we presented are not necessarily specific to $\mathcal{AU}$, and that they can be used for other theories as well (like the empty one, $\mathcal{AC}$, *etc*), just by adapting the $\mathcal{AU}$ rules to the considered theory. This is quite interesting even for the syntactical case, as the disunification-based algorithm presented in [8] is not appropriate for an efficient implementation.

Although some of the results may at first glance seem straightforward, subtle details are not so easy to establish. The main difficulties come from matching non-linear anti-patterns, which cannot be performed using classical decomposition rules, as the semantics is not preserved.

The work in this paper opens a number of challenging directions like proving the correctness of the third algorithm presented as a conjecture. We also plan to study some theoretical properties such as the confluence, termination, and complete definition of systems that include anti-patterns. Another interesting direction is the study of unification problems in the presence of anti-patterns.

References

1. Balland, E., Brauner, P., Kopetz, R., Moreau, P.-E., Reilles, A.: Tom: Piggybacking rewriting on java. In: Baader, F. (ed.) RTA 2007. LNCS, vol. 4533, pp. 36–47. Springer, Heidelberg (2007)
2. van den Brand, M., Deursen, A., Heering, J., Jong, H., Jonge, M., Kuipers, T., Klint, P., Moonen, L., Olivier, P., Scheerder, J., Vinju, J., Visser, E., Visser, J.: The ASF+SDF Meta-Environment: a Component-Based Language Development Environment. In: Wilhelm, R. (ed.) CC 2001. LNCS, vol. 2027, pp. 365–370. Springer, Heidelberg (2001)
3. Comon, H., Kirchner, C.: Constraint solving on terms. In: Comon, H., Marché, C., Treinen, R. (eds.) CCL 1999. LNCS, vol. 2002, pp. 47–103. Springer, Heidelberg (2001)
4. Eker, S.: Associative matching for linear terms. Report CS-R9224, CWI, ISSN 0169-118X (1992)
5. Eker, S.: Associative-commutative rewriting on large terms. In: Nieuwenhuis, R. (ed.) RTA 2003. LNCS, vol. 2706, pp. 14–29. Springer, Heidelberg (2003)
6. Kirchner, C.: Computing unification algorithms, pp. 206–216 (1986)
7. Kirchner, C., Klay, F.: Syntactic theories and unification, pp. 270–277 (June 1990)
8. Kirchner, C., Kopetz, R., Moreau, P.: Anti-pattern matching. In: De Nicola, R. (ed.) ESOP 2007. LNCS, vol. 4421, pp. 110–124. Springer, Heidelberg (2007)
9. Kirchner, C., Kopetz, R., Moreau, P.: Anti-pattern matching modulo. Technical report, INRIA & LORIA Nancy (2007),
http://hal.inria.fr/inria-00129421/fr/
10. Makanin, G.S.: The problem of solvability of equations in a free semigroup. Math. USSR Sbornik 32(2), 129–198 (1977)
11. Marché, C.: Normalized rewriting: an alternative to rewriting modulo a set of equations. Journal of Symbolic Computation 21(3), 253–288 (1996)
12. Nipkow, T.: Proof transformations for equational theories, pp. 278–288 (June 1990)
13. Nipkow, T.: Combining matching algorithms: The regular case. Journal of Symbolic Computation 12(6), 633–653 (1991)
14. Plotkin, G.: Building-in equational theories. Machine Intelligence 7, 73–90 (1972)
15. Ringeissen, C.: Combining decision algorithms for matching in the union of disjoint equational theories. Information and Computation 126(2), 144–160 (1996)
16. Treinen, R.: A new method for undecidability proofs of first order theories. Journal of Symbolic Computation 14(5), 437–457 (1992)

Counting Ordered Patterns in Words Generated by Morphisms

Sergey Kitaev[1], Toufik Mansour[2], and Patrice Séébold[3]

[1] Reykjavík University, Kringlan 1, 103 Reykjavík, Iceland
sergey@ru.is
[2] Department of Mathematics, Haifa University, 31905 Haifa, Israel
toufik@math.haifa.ac.il
[3] LIRMM, Univ. Montpellier 2, CNRS, 161 rue Ada, 34392 Montpellier, France
Patrice.Seebold@lirmm.fr

Abstract. We start a general study of counting the number of occurrences of ordered patterns in words generated by morphisms. We consider certain patterns with gaps (classical patterns) and that with no gaps (consecutive patterns). Occurrences of the patterns are known, in the literature, as rises, descents, (non-)inversions, squares and p-repetitions. We give recurrence formulas in the general case, then deducing exact formulas for particular families of morphisms. Many (classical or new) examples are given illustrating the techniques and showing their interest.

Keywords: Morphisms, patterns, rises, descents, inversions, repetitions.

1 Introduction

In algebraic combinatorics, an occurrence of a pattern p in a permutation π is a subsequence of π (of the same length as that of p) whose elements are in the same relative order as those in p. For example, the permutation $\pi = 536241$ contains an occurrence of the pattern $p = 2431$. Babson and Steingrímsson introduced generalized patterns where two adjacent elements of a pattern must also be adjacent in the permutation [2].

In combinatorics on words, an occurrence of a pattern p in a word u is a factor of u having the same shape as p. For example the word $u = abaabaaabab$ contains an occurrence of the pattern $p = \alpha\alpha\beta\alpha\alpha\beta$.

Burstein [4] realized a "mixing" of these two notions by considering ordered alphabets. An occurrence of an (ordered) pattern in a word is a factor or a subsequence having the same shape, and in which the relative order of the letters is the same as that in the pattern. In [5] one computed the number of occurrences of many of ordered patterns in the Peano words. In the present paper we start a general study of counting the number of occurrences of ordered patterns in words generated by morphisms.

C. Martín-Vide, F. Otto, and H. Fernau (Eds.): LATA 2008, LNCS 5196, pp. 287–298, 2008.
© Springer-Verlag Berlin Heidelberg 2008

2 Preliminaries

2.1 Definitions and Notations

We refer to [7] for standard definitions in combinatorics on words.

Let n be a non-negative integer. The *incidence matrix* of f^n is the $k \times k$ matrix $M(f^n) = (m_{n,i,j})_{1 \le i,j \le k}$ where $m_{n,i,j}$ is the number of occurrences of the letter a_i in the word $f^n(a_j)$, i.e., $m_{n,i,j} = |f^n(a_j)|_{a_i}$.

Property 1. *For every $n \in \mathbb{N}$, $M(f)^n = M(f^n)$.*

2.2 Ordered Patterns

Let A be a totally ordered alphabet and let $\aleph$ be the ordered alphabet whose letters are the first n positive integers in the usual order (thus $\aleph = \{1, 2, \ldots, n\}$).

An *ordered pattern* is any word over $\aleph \cup \{\#\}$, $\# \notin \aleph$, without two consecutive $\#$. If a pattern contains at least one $\#$, not at the very beginning or at the very end, it is an *ordered pattern with gaps*; otherwise it is an *ordered pattern with no gaps*. Moreover, in this paper the four ordered patterns u, $\#u$, $u\#$, and $\#u\#$ are considered to be the same (but of course $u\#u$ is not the same pattern as uu). In particular, if x is a word over $\aleph$, we will write $(x\#)^\ell$ or $(\#x)^\ell$ to represent the ordered pattern $x\#x\# \cdots \#x$ containing l occurrences of the word x.

A word v over A *contains an occurrence of the ordered pattern*

$$u = u_1 \# u_2 \# \cdots \# u_n,$$

$u_i \in \aleph^+$ and $n \ge 1$, if $v = w_0 v_1 w_1 v_2 w_2 \cdots w_{n-1} v_n w_n$ and there exists a literal morphism $f : \aleph^* \to A^*$ such that $f(u_i) = v_i$, $1 \le i \le n$, and if $x, y \in \aleph$, $x < y \Rightarrow f(x) < f(y)$. Thus the word v contains an occurrence of the ordered pattern u if v contains a subsequence v' which is equal to $f(u')$ where u' is obtained from u by deleting all the occurrences of $\#$, with the additional condition that two adjacent (not separated by $\#$) letters in u must be adjacent in v. The number of different occurrences of u as an ordered pattern in v is denoted by $|v|_u$.

Example. Let $A = \{a, b, c, d, e, f\}$ with $a < b < c < d < e < f$. The word $v = eafdbc$ contains one occurrence of the ordered pattern $2\#31$, namely the subsequence efd ($|e\,afd\,b\,c|_{2\#31} = 1$). In v, the ordered pattern $2\#3\#1$ occurs in three occurrences: efd, efb, and efc ($|e\,afd\,b\,c|_{2\#3\#1} = 3$); the ordered pattern 231 does not occur in v ($|e\,afd\,b\,c|_{231} = 0$).

3 Ordered Patterns with Gaps and Morphisms

Let k be an integer ($k \ge 2$) and A the k-letter ordered alphabet $A = \{a_1 < a_2 < \cdots < a_k\}$. Let f be any morphism on A: for $1 \le i \le k$, $f(a_i) = a_{i_1} \ldots a_{i_{p_i}}$ with $p_i \ge 0$ ($p_i = 0$ if and only if $f(a_i) = \varepsilon$).

3.1 Inversions, Non-inversions, and Repetitions with Gaps of f^n

In what follows we are interested in some particular forms of ordered patterns. In accordance with permutations theory, an *inversion* (resp. *non-inversion*) is an occurrence of the ordered pattern $2\#1$ (resp. $1\#2$). *Repetitions with gaps of one letter* are occurrences of the ordered patterns $(1\#)^p$, $p \geq 1$.

Inversions and Non-inversions. Let n be a non-negative integer.

The *vector $RG(f^n)$ of non-inversions* (resp. *vector $DG(f^n)$ of inversions*) of f^n is the k vector whose i-th entry is the number of occurrences of the ordered pattern $1\#2$ (resp. $2\#1$) in the word $f^n(a_i)$, i.e.,

$$RG(f^n) = (|f^n(a_i)|_{1\#2})_{1 \leq i \leq k} \qquad DG(f^n) = (|f^n(a_i)|_{2\#1})_{1 \leq i \leq k}.$$

Our goal is to obtain recurrence formulas giving the entries of $RG(f^{n+1})$ and $DG(f^{n+1})$. Since $f^{n+1} = f^n \circ f = f \circ f^n$, we have two different ways to compute $RG(f^{n+1})$ and $DG(f^{n+1})$.

Let ℓ be an integer, $1 \leq \ell \leq k$. Either $|f^{n+1}(a_\ell)|_{1\#2}$ (resp. $|f^{n+1}(a_\ell)|_{2\#1}$) will be obtained from the word $f(a_\ell)$ and the entries of $RG(f^n)$ (resp. $DG(f^n)$) (see 1. below), or it will be computed from the values of $RG(f)$ (resp. $DG(f)$) and $f^n(a_\ell)$ (see 2. below).

1. From $f^{n+1} = f^n \circ f$:
Since $f(a_\ell) = a_{\ell_1} \ldots a_{\ell_{p_\ell}}$, the number of occurrences of the ordered pattern $1\#2$ in $f^{n+1}(a_\ell) = f^n(f(a_\ell)) = f^n(a_{\ell_1} \ldots a_{\ell_{p_\ell}})$ is obtained by adding two values:

• The number of occurrences of the ordered pattern $1\#2$ in each $f^n(a_{\ell_i})$, $1 \leq i \leq p_\ell$. Since the ℓ-th column of the incidence matrix of f indicates which letters appear in $f(a_\ell)$ (and how many times), this number is obtained by multiplying the vector $RG(f^n)$ by the ℓ-th column of $M(f)$, i.e., it is equal to $\sum_{t=1}^{k} |f^n(a_t)|_{1\#2} \cdot m_{1,t,\ell}$.

• The number of occurrences of the ordered pattern $1\#2$ in each of the $f^n(a_{\ell_i} a_{\ell_j})$, $1 \leq i < j \leq p_\ell$, where the letter corresponding to 1 is in $f^n(a_{\ell_i})$ and the letter corresponding to 2 is in $f^n(a_{\ell_j})$. In what follows we will call such an occurrence of $1\#2$ in $f^n(a_{\ell_i} a_{\ell_j})$ an *external* occurrence of the ordered pattern $1\#2$ in $f^n(a_{\ell_i} a_{\ell_j})$, and denote it $|f^n(a_{\ell_i} a_{\ell_j})|_{1\#2}^{ext}$.

The value of $|f^n(a_{\ell_i} a_{\ell_j})|_{1\#2}^{ext}$ is obtained by adding, for all the integers r, $1 \leq r \leq k - 1$, the product of the number of occurrences of the letter a_r in $f^n(a_{\ell_i})$ ($|f^n(a_{\ell_i})|_{a_r}$) by the number of occurrences of all the letters of $f^n(a_{\ell_j})$ greater than a_r ($|f^n(a_{\ell_j})|_{a_s}$, $r + 1 \leq s \leq k$). This gives

$$\sum_{r=1}^{k-1} (m_{n,r,\ell_i} \cdot \sum_{s=r+1}^{k} m_{n,s,\ell_j}).$$

The number of external occurrences of $1\#2$ in all the $f^n(a_{\ell_i} a_{\ell_j})$, $1 \leq i < j \leq p_\ell$, is thus given by

$$\sum_{1 \leq i < j \leq p_\ell} |f^n(a_{\ell_i} a_{\ell_j})|_{1\#2}^{ext} = \sum_{1 \leq i < j \leq p_\ell} (\sum_{r=1}^{k-1} (m_{n,r,\ell_i} \cdot \sum_{s=r+1}^{k} m_{n,s,\ell_j})).$$

2. From $f^{n+1} = f \circ f^n$.

Let $q_\ell = |f^n(a_\ell)| : f^{n+1}(a_\ell) = f(f^n(a_\ell)) = f(a_{\ell'_1} \ldots a_{\ell'_{q_\ell}})$. Here the number of occurrences of the ordered pattern $1\#2$ in $f^{n+1}(a_\ell)$ is obtained by adding:

- The number of occurrences of the ordered pattern $1\#2$ in each $f^n(a_{\ell'_i})$, $1 \leq i \leq q_\ell$. As above it is equal to $\sum_{t=1}^{k} |f(a_t)|_{1\#2} \cdot m_{n,t,\ell}$.
- The number of external occurrences of the ordered pattern $1\#2$ in each of the $f(a_{\ell'_i} a_{\ell'_j})$, $1 \leq i < j \leq q_\ell$. As above, this number is given by

$$\sum_{1 \leq i < j \leq q_\ell} |f(a_{\ell'_i} a_{\ell'_j})|_{1\#2}^{ext} = \sum_{1 \leq i < j \leq q_\ell} (\sum_{r=1}^{k-1} (m_{1,r,\ell'_i} \cdot \sum_{s=r+1}^{k} m_{1,s,\ell'_j})).$$

The same reasoning applies for calculating the entries of $DG(f^{n+1})$, replacing $1\#2$ by $2\#1$ and "greater" by "smaller". Thus we have the following.

Proposition 1. *For each letter $a_\ell \in A$, let p_ℓ and q_ℓ be such that $f(a_\ell) = a_{\ell_1} \ldots a_{\ell_{p_\ell}}$ and $f^n(a_\ell) = a_{\ell'_1} \ldots a_{\ell'_{q_\ell}}$. Then, for all $n \in \mathbb{N}$,*

$$|f^{n+1}(a_\ell)|_{1\#2} = \sum_{1 \leq i < j \leq q_\ell} (\sum_{r=1}^{k-1} (m_{1,r,\ell'_i} \cdot \sum_{s=r+1}^{k} m_{1,s,\ell'_j})) + \sum_{t=1}^{k} |f(a_t)|_{1\#2} \cdot m_{n,t,\ell}, \quad (1)$$

$$|f^{n+1}(a_\ell)|_{2\#1} = \sum_{1 \leq i < j \leq q_\ell} (\sum_{r=2}^{k} (m_{1,r,\ell'_i} \cdot \sum_{s=1}^{r-1} m_{1,s,\ell'_j})) + \sum_{t=1}^{k} |f(a_t)|_{2\#1} \cdot m_{n,t,\ell}. \quad (2)$$

Of course, the analysis we have done above could be realized to compute more complex ordered patterns with gaps, such as $1\#23$, $1\#2\#3$, $\cdots$ The only difficulty is to adapt the computation of external inversions and non-inversions.

Repetitions of One Letter. Let n be a non-negative integer and p a positive integer. The *vector of p-repetitions with gaps of one letter* of f^n is the k vector whose i-th entry is the number of occurrences of the ordered pattern $(1\#)^p$ in the word $f^n(a_i)$, i.e., $R_p G(f^n) = (|f^n(a_i)|_{(1\#)^p})_{1 \leq i \leq k}$. The following is obvious.

Proposition 2. *For each letter $a_\ell \in A$ and for all $n \in \mathbb{N}$,*

$$|f^n(a_\ell)|_{(1\#)^p} = \sum_{t=1}^{k} \binom{m_{n,t,\ell}}{p}. \quad (3)$$

3.2   Some Examples in the Binary Case

The Thue-Morse Morphism. The *Thue-Morse morphism* μ ([10],[9],[8]) is defined by $\mu(a_1) = a_1 a_2$, $\mu(a_2) = a_2 a_1$. It generates the famous *Thue-Morse sequence* $\mathbf{t} = \mu^\omega(a_1)$ which has been widely studied.

For every positive integers n, the incidence matrix of μ^n is

$$M(\mu^n) = \begin{bmatrix} 2^{n-1} & 2^{n-1} \\ 2^{n-1} & 2^{n-1} \end{bmatrix}.$$

Thus, from equations (1), (2), and (3) we obtain

$$|\mu^{n+1}(a_1)|_{1\#2} = |\mu^{n+1}(a_2)|_{1\#2} = 2^{2(n-1)} + |\mu^n(a_1)|_{1\#2} + |\mu^n(a_2)|_{1\#2},$$
$$|\mu^{n+1}(a_1)|_{2\#1} = |\mu^{n+1}(a_2)|_{2\#1} = 2^{2(n-1)} + |\mu^n(a_1)|_{2\#1} + |\mu^n(a_2)|_{2\#1},$$
$$|\mu^n(a_1)|_{(1\#)^p} = |\mu^n(a_2)|_{(1\#)^p} = 2 \cdot \binom{2^{n-1}}{p}.$$

Since $RG(\mu) = \begin{bmatrix} 1 & 0 \end{bmatrix}$ and $DG(\mu) = \begin{bmatrix} 0 & 1 \end{bmatrix}$, Proposition 1 gives the following well known result.

Corollary 1. *For any integer $n \geq 2$,*

$$RG(\mu^n) = DG(\mu^n) = \begin{bmatrix} 2^{2n-3} & 2^{2n-3} \end{bmatrix} \text{ and } R_pG(\mu^n) = \begin{bmatrix} 2 \cdot \binom{2^{n-1}}{p} & 2 \cdot \binom{2^{n-1}}{p} \end{bmatrix}.$$

The Fibonacci Morphism. The *Fibonacci morphism* φ is defined by $\varphi(a_1) = a_1 a_2$, $\varphi(a_2) = a_1$. It generates the well known *Fibonacci sequence* $\mathbf{f} = \varphi^\omega(a_1)$ which is the prototype of a Sturmian word (see, e.g., [7]).

Let $(F_n)_{n \geq -1}$ be the sequence of Fibonacci numbers: $F_{-1} = 0$, $F_0 = 1$, $F_n = F_{n-1} + F_{n-2}$ for $n \geq 1$. The following property of Fibonacci numbers will be useful below.

Property 2. *For every positive integer n, $F_n \cdot F_{n-2} = F_{n-1}^2 + \begin{cases} 1 & \text{if } n \text{ is even,} \\ -1 & \text{if } n \text{ is odd.} \end{cases}$*

An easy computation gives that, for every positive integer n, the incidence matrix of φ^n is $M(\varphi^n) = \begin{bmatrix} F_n & F_{n-1} \\ F_{n-1} & F_{n-2} \end{bmatrix}$.

The vector of non-inversions of φ is $RG(\varphi) = \begin{bmatrix} 1 & 0 \end{bmatrix}$. Moreover equation (1) (see Property 2) gives, for $n \geq 1$

$$|\varphi^{n+1}(a_1)|_{1\#2} = m_{n,1,1} \cdot m_{n,2,2} + |\varphi^n(a_1)|_{1\#2} + |\varphi^n(a_2)|_{1\#2}$$
$$= F_n \cdot F_{n-2} + |\varphi^n(a_1)|_{1\#2} + |\varphi^n(a_2)|_{1\#2}$$
$$= F_{n-1}^2 + |\varphi^n(a_1)|_{1\#2} + |\varphi^n(a_2)|_{1\#2} + \begin{cases} 1 & \text{if } n \text{ is even,} \\ -1 & \text{if } n \text{ is odd} \end{cases}$$

The vector of inversions of φ is $DG(\varphi) = \begin{bmatrix} 0 & 0 \end{bmatrix}$. Moreover, equation (2) gives, for $n \geq 1$

$$|\varphi^{n+1}(a_1)|_{2\#1} = m_{n,2,1} \cdot m_{n,1,2} + |\varphi^n(a_1)|_{2\#1} + |\varphi^n(a_2)|_{2\#1}$$
$$= F_{n-1}^2 + |\varphi^n(a_1)|_{2\#1} + |\varphi^n(a_2)|_{2\#1}.$$

Now, $|\varphi^{n+1}(a_2)|_{1\#2} = |\varphi^n(a_1)|_{1\#2}$ and $|\varphi^{n+1}(a_2)|_{2\#1} = |\varphi^n(a_1)|_{2\#1}$ because $\varphi(a_2) = a_1$.

From this we obtain formulas to compute, for every $n \geq 0$, $|\varphi^{n+2}(a_1)|_{1\#2}$ and $|\varphi^{n+2}(a_1)|_{2\#1}$ from the sequence of Fibonacci numbers.

Corollary 2. *For every integer $n \geq 0$,*

$$|\varphi^{n+2}(a_1)|_{2\#1} = \sum_{p=0}^{n} F_p F_{n-p}^2 \, ,$$

$$|\varphi^{n+2}(a_1)|_{1\#2} = |\varphi^{n+2}(a_1)|_{2\#1} + F_n + \begin{cases} 1 \text{ if } n \text{ is odd,} \\ -1 \text{ if } n \text{ is even.} \end{cases}$$

Regarding repetitions of one letter, $R_p G(\varphi) = \left[\binom{1}{p} + \binom{1}{p} \binom{1}{p} \right]$ and, for $n \geq 0$, the vector $R_p G(\varphi^{n+2})$ is obtained from equation (3).

Corollary 3. *For any integer $n \geq 0$, $R_p G(\varphi^{n+2}) = \left[\binom{F_{n+2}}{p} + \binom{F_{n+1}}{p} \binom{F_{n+1}}{p} + \binom{F_n}{p} \right]$.*

4   A Particular Family of Morphisms

Let k be an integer ($k \geq 2$) and A the k-letter ordered alphabet $A = \{a_1 < a_2 < \cdots < a_k\}$. In this section we are interested in morphisms f having the following particularities:

1. there exists a positive integer m such that $|f(a_1)|_{a_i} = m$, $1 \leq i \leq k$,
2. there exists a positive integer d such that $|f(a_2 \ldots a_k)|_{a_i} = d$, $1 \leq i \leq k$,
3. for every i, j, $1 \leq i, j \leq k$, $|f(a_i a_j)|_{1\#2}^{ext} = |f(a_j a_i)|_{1\#2}^{ext}$.

(Conditions 1. and 2. are particular cases of the more general situation, considered in Theorem 1 below, in which the alphabet A is partitioned in sets A_1, A_2, ..., A_n such that, for each A_i, the sum of the number of occurrences of each letter in the images of letters of A_i is the same.) In this case we are able to give direct formulas to compute $|f^{n+1}(a_1)|_{1\#2}$ and others from m, d, and n.

Proposition 3. *For every positive integer n,*

$$|f^{n+1}(a_1)|_{1\#2}$$
$$= m(d+m)^{n-1} \sum_{i=1}^{k} |f(a_i)|_{1\#2} + \frac{[m(d+m)^{n-1}-1]m(d+m)^{n-1}}{2} \sum_{j=1}^{k} |f(a_j a_j)|_{1\#2}^{ext}$$
$$+ m^2(d+m)^{2n-2} \sum_{1 \leq i < j \leq k} |f(a_i a_j)|_{1\#2}^{ext} \, ,$$

$$|f^{n+1}(a_2 \ldots a_k)|_{1\#2}$$
$$= d(d+m)^{n-1} \sum_{i=1}^{k} |f(a_i)|_{1\#2} + \frac{[d(d+m)^{n-1}-1]d(d+m)^{n-1}}{2} \sum_{j=1}^{k} |f(a_j a_j)|_{1\#2}^{ext}$$
$$+ d^2(d+m)^{2n-2} \sum_{1 \leq i < j \leq k} |f(a_i a_j)|_{1\#2}^{ext} \, .$$

Now the same reasoning can be applied for

$$|f^{n+1}(a_1)|_{2\#1} \text{ and } |f^{n+1}(a_2 \ldots a_k)|_{2\#1},$$

because of the following obvious property.

Property 3. *Let f be a morphism on A. For every non-negative integer n, and for every integers i, j, $1 \leq i, j \leq k$, $|f^n(a_i a_j)|_{1\#2}^{ext} = |f^n(a_j a_i)|_{2\#1}^{ext}$.*

Thus, using equation (2), we have the following.

Proposition 4. *For every positive integer n,*

$$|f^{n+1}(a_1)|_{2\#1}$$
$$= m(d+m)^{n-1} \sum_{i=1}^{k} |f(a_i)|_{2\#1} + \frac{[m(d+m)^{n-1}-1]m(d+m)^{n-1}}{2} \sum_{j=1}^{k} |f(a_j a_j)|_{2\#1}^{ext}$$
$$+ m^2(d+m)^{2n-2} \sum_{1 \le i < j \le k} |f(a_i a_j)|_{2\#1}^{ext},$$

$$|f^{n+1}(a_2 \ldots a_k)|_{2\#1}$$
$$= d(d+m)^{n-1} \sum_{i=1}^{k} |f(a_i)|_{2\#1} + \frac{[d(d+m)^{n-1}-1]d(d+m)^{n-1}}{2} \sum_{j=1}^{k} |f(a_j a_j)|_{2\#1}^{ext}$$
$$+ d^2(d+m)^{2n-2} \sum_{1 \le i < j \le k} |f(a_i a_j)|_{2\#1}^{ext}.$$

The previous reasoning can of course be applied if conditions 1. and 2. are verified for any partition of the alphabet (in Propositions 3 and 4 the partition is in two sets $A = \{a_1\} \cup \{a_2 \ldots a_k\}$). Then we obtain the following general result.

Theorem 1. *Let k be an integer ($k \ge 2$), and A the k-letter ordered alphabet $A = \{a_1 < a_2 < \ldots < a_k\}$. Let f be a morphism on A fulfilling the following conditions:*

- *there exist a positive integer p and a set of p positive integers $\{m_1, \ldots, m_p\}$ such that A can be partitioned into p subsets $A_1, \ldots, A_p$ with $\sum_{a \in A_\ell} |f(a)|_{a_i} = m_\ell$, $1 \le i \le k$,*
- *for every i, j, $1 \le i, j \le k$, $|f(a_i a_j)|_{1\#2}^{ext} = |f(a_j a_i)|_{1\#2}^{ext}$.*

Let $M = m_1 + \ldots + m_p$ and let $u = 1\#2$ or $u = 2\#1$. For every integer $n \ge 1$ and for each A_ℓ, $1 \le \ell \le p$,

$$\sum_{a \in A_\ell} |f^{n+1}(a)|_u = m_\ell M^{n-1} \sum_{i=1}^{k} |f(a_i)|_u + \frac{(m_\ell M^{n-1}-1)m_\ell M^{n-1}}{2} \sum_{j=1}^{k} |f(a_j a_j)|_u^{ext}$$
$$+ m_\ell^2 M^{2n-2} \sum_{1 \le i < j \le k} |f(a_i a_j)|_u^{ext}.$$

5 Examples

In this section we give a series of examples of application of Theorem 1.

5.1 The Thue-Morse Morphism

The Thue-Morse morphism (see Section 3.2) is the simplest example of a morphism fulfilling conditions 1. to 3. above. Indeed $m = d = 1$, and

$$|\mu(a_1 a_2)|_{1\#2}^{ext} = |a_1 a_2 a_2 a_1|_{1\#2}^{ext} = 1 = |a_2 a_1 a_1 a_2|_{1\#2}^{ext} = |\mu(a_2 a_1)|_{1\#2}^{ext},$$

$|\mu(a_1 a_1)|_{1\#2}^{ext} = |\mu(a_2 a_2)|_{1\#2}^{ext} = 1$. Since $|\mu(a_1)|_{1\#2} = |\mu(a_2)|_{2\#1} = 1$, and $|\mu(a_1)|_{2\#1} = |\mu(a_2)|_{1\#2} = 0$, we obtain from Propositions 3 and 4 that

$$|\mu^{n+1}(a_1)|_{1\#2} = |\mu^{n+1}(a_1)|_{2\#1} = |\mu^{n+1}(a_2)|_{1\#2} = |\mu^{n+1}(a_2)|_{2\#1} = 2^{2n-1}.$$

5.2   The Prouhet Morphisms

Let $k \geq 2$, and let A be the k-letter ordered alphabet $A = \{a_1 < \cdots < a_k\}$. The *Prouhet morphism* π_k ([9]) is defined on A by $\pi_k(a_i) = a_i a_{i+1} \ldots a_k a_1 \ldots a_{i-1}$, $1 \leq i \leq k$. As above we obtain a corollary of Theorem 1.

Corollary 4. *For every i, $1 \leq i \leq k$, and for every positive integer n,*

$$|\pi_k^{n+1}(a_i)|_{1\#2} = \frac{(k-1)k^n}{12} \left(3k^{n+1} + k - 2 \right),$$

$$|\pi_k^{n+1}(a_i)|_{2\#1} = \frac{(k-1)k^n}{12} \left(3k^{n+1} - k + 2 \right).$$

5.3   The Arshon Morphisms

Let k be any even positive integer. The morphism β_k ([1]) is defined, for every r, $1 \leq r \leq k/2$, by

$$a_{2r-1} \mapsto a_{2r-1} a_{2r} \ldots a_{k-1} a_k a_1 a_2 \ldots a_{2r-3} a_{2r-2},$$
$$a_{2r} \mapsto a_{2r-1} a_{2r-2} \ldots a_2 a_1 a_k a_{k-1} \ldots a_{2r+1} a_{2r}.$$

Corollary 5. *Let k be any even positive integer. For every i, $1 \leq i \leq k$, and for every positive integer n,*

$$|\beta_k^{n+1}(a_i)|_{1\#2} = \frac{k^{n-1}}{4} \left[k^{n+2} \cdot (k-1) + 2k \right],$$
$$|\beta_k^{n+1}(a_i)|_{2\#1} = \frac{k^{n-1}}{4} \left[k^{n+2} \cdot (k-1) - 2k \right].$$

Example. For every i, $1 \leq i \leq k$, and for every $n \geq 1$,

$$|\beta_6^{n+1}(a_i)|_{1\#2} = 6^{n-1} \cdot (45 \cdot 6^n + 3), \quad |\beta_6^{n+1}(a_i)|_{2\#1} = 6^{n-1} \cdot (45 \cdot 6^n - 3).$$

5.4   Three Other Examples

1. Let A be the four-letter ordered alphabet $A = \{a_1 < a_2 < a_3 < a_4\}$. Define the morphism f on A by $f(a_1) = a_1 a_3 a_2 a_4$, $f(a_2) = \varepsilon$, $f(a_3) = a_1 a_4$, $f(a_4) = a_2 a_3$.

The morphism f fulfills the conditions of Theorem 1. Here we choose $p = 3$, $A = A_1 \cup A_2 \cup A_3$ with $A_1 = \{a_1\}$, $A_2 = \{a_2\}$, $A_3 = \{a_3, a_4\}$, and $m_1 = m_3 = 1$, $m_2 = 0$, thus $M = 2$.

Corollary 6. *For every positive integer n,*

$$|f^{n+1}(a_1)|_{1\#2} = |f^{n+1}(a_3 a_4)|_{1\#2} = 3 \cdot 2^{n-1} \cdot (2^{n+1} + 1),$$

$$|f^{n+1}(a_1)|_{2\#1} = |f^{n+1}(a_3 a_4)|_{2\#1} = 3 \cdot 2^{n-1} \cdot (2^{n+1} - 1),$$

$$|f^{n+1}(a_2)|_{1\#2} = |f^{n+1}(a_2)|_{2\#1} = 0.$$

2. Let A be the five-letter ordered alphabet $A = \{a_1 < a_2 < a_3 < a_4 < a_5\}$. Define the morphism g on A by $g(a_1) = a_1 a_3 a_5 a_4 a_2$, $g(a_2) = a_4 a_2 a_3$, $g(a_3) = a_5 a_1$, $g(a_4) = a_1 a_5$, $g(a_5) = a_2 a_3 a_4$.

The morphism g fulfills the conditions of Theorem 1. Here we choose $p = 3$, $A = A_1 \cup A_2 \cup A_3$ with $A_1 = \{a_1\}$, $A_2 = \{a_2, a_4\}$, $A_3 = \{a_3, a_5\}$, and $m_1 = m_2 = m_3 = 1$, thus $M = 3$.

Corollary 7. *For every positive integer n,*

$$|g^{n+1}(a_1)|_{1\#2} = |g^{n+1}(a_2 a_4)|_{1\#2} = |g^{n+1}(a_3 a_5)|_{1\#2} = 3^{n-1} \cdot (5 \cdot 3^{n+1} + 2),$$
$$|g^{n+1}(a_1)|_{2\#1} = |g^{n+1}(a_2 a_4)|_{2\#1} = |g^{n+1}(a_3 a_5)|_{2\#1} = 3^{n-1} \cdot (5 \cdot 3^{n+1} - 2).$$

3. Let A be the three-letter ordered alphabet $A = \{a < b < c\}$. Define the morphism h on A by $h(a) = aba\,cab\,cac\,bab\,cba\,cbc$, $h(b) = aba\,cab\,cac\,bca\,bcb\,abc$, $h(c) = aba\,cab\,cba\,cbc\,acb\,abc$.

This morphism was proved square-free by Brandenburg in [3]. It fulfills the conditions of Theorem 1 with $p = 3$, $A = A_1 \cup A_2 \cup A_3$ with $A_1 = \{a\}$, $A_2 = \{b\}$, $A_3 = \{c\}$, and $m_1 = m_2 = m_3 = 6$, thus $M = 18$.

Corollary 8. *For every $x \in A$ and for every positive integer n,*

$$|h^{n+1}(x)|_{1\#2} = 6 \cdot 18^{n-1} \cdot (9 \cdot 18^{n+1} + 40),$$
$$|h^{n+1}(x)|_{2\#1} = 6 \cdot 18^{n-1} \cdot (9 \cdot 18^{n+1} - 40).$$

6 Ordered Patterns with No Gaps and Morphisms

6.1 Rises, Descents, and Squares of f^n

Let k be an integer ($k \geq 2$) and A the k-letter ordered alphabet $A = \{a_1 < a_2 < \cdots < a_k\}$. Let f be any morphism on A: for $1 \leq i \leq k$, $f(a_i) = a_{i_1} \ldots a_{i_{p_i}}$ with $p_i \geq 0$ ($p_i = 0$ if and only if $f(a_i) = \varepsilon$).

The *vector of rises* (resp. *vector of descents*, resp. *vector of squares of one letter*) of f^n is the k vector whose i-th entry is the number of occurrences of the ordered pattern 12 (resp. 21, resp. 11) in the word $f^n(a_i)$, i.e.,

$$R(f^n) = (|f^n(a_i)|_{12})_{1\leq i \leq k}, \quad D(f^n) = (|f^n(a_i)|_{21})_{1\leq i \leq k},$$
$$R_2(f^n) = (|f^n(a_i)|_{11})_{1\leq i \leq k}.$$

We define two sequences of k vectors, $(F(f^n))_{n\in\mathbb{N}}$ and $(L(f^n))_{n\in\mathbb{N}}$, where $F(f^n)[i]$ is the first letter of $f^n(a_i)$ and $L(f^n)[i]$ is the last letter of $f^n(a_i)$ if $f^n(a_i) \neq \varepsilon$, and $F(f^n)[i] = L(f^n)[i] = 0$ if $f^n(a_i) = \varepsilon$. Of course these two sequences take their values in a finite set: they are ultimately periodic. Thus they can be computed a priori from f.

Given a non-negative integer n, let $\aleph'$ be the subset of $\aleph$ such that, for each $i \in \aleph$, $f^n(a_i) \neq \varepsilon$ if and only if $i \in \aleph'$. We associate to the two vectors $F(f^n)$ and $L(f^n)$ an application $C_n^{12} : \aleph' \times \aleph' \to \{0, 1\}$ defined by

$$C_n^{12}(i, j) = \begin{cases} 1, & \text{if } L(f^n)[i] < F(f^n)[j] \\ 0, & \text{if } L(f^n)[i] \geq F(f^n)[j]. \end{cases}$$

Similarly we define

$$C_n^{21}(i,j) = \begin{cases} 1, & \text{if } L(f^n)[i] > F(f^n)[j] \\ 0, & \text{if } L(f^n)[i] \le F(f^n)[j], \end{cases} \quad C_n^{11}(i,j) = \begin{cases} 1, & \text{if } L(f^n)[i] = F(f^n)[j] \\ 0, & \text{if } L(f^n)[i] \ne F(f^n)[j]. \end{cases}$$

For any morphism f on A, there exists a least integer M_f ($M_f \le k$ and M_f depends only on f) such that, for every positive integer n and every $a \in A$, $f^n(a) = \varepsilon$ if and only if $f^{M_f}(a) = \varepsilon$. By convention, if f is a nonerasing morphism then $M_f = 0$. The integer M_f is known in the literature about L-systems as the *mortality exponent* of f ([6]).

Now let ℓ be an integer, $1 \le \ell \le k$. One has $f(a_\ell) = a_{\ell_1} \dots a_{\ell_{p_\ell}}$ and we denote by $\ell'_1 \dots \ell'_{p'_\ell}$ the subsequence of $\ell_1 \dots \ell_{p_\ell}$ such that $f^{n+1}(a_\ell) = f^n(a_{\ell'_1} \dots a_{\ell'_{p'_\ell}})$ for every $n \ge M_f$. This means that, for every $n \ge M_f$, a letter a_{ℓ_i} appears in $a_{\ell_1} \dots a_{\ell_{p_\ell}}$ but not in $a_{\ell'_1} \dots a_{\ell'_{p'_\ell}}$ if and only if $f^n(a_{\ell_i}) = \varepsilon$. Of course $p'_\ell \le p_\ell$, and if $M_f = 0$ then $p'_\ell = p_\ell$ for each $1 \le \ell \le k$.

Here also, as in Section 3, the number of occurrences of the ordered pattern 12 in $f^{n+1}(a_\ell) = f^n(a_{\ell_1} \dots a_{\ell_{p_\ell}}) = f^n(a_{\ell'_1} \dots a_{\ell'_{p'_\ell}})$ ($n \ge M_f$) is obtained by adding two values: (1) the number of occurrences of the ordered pattern 12 in each $f^n(a_{\ell_i})$, $1 \le i \le p_\ell$. As in the previous case, this number is equal to $\sum_{t=1}^{k} |f^n(a_t)|_{12} \cdot m_{1,t,\ell}$, and (2) the number of external occurrences of the ordered pattern 12 in $f^n(a_{\ell'_i} a_{\ell'_j})$ for each subsequence $a_{\ell'_i} a_{\ell'_j}$ of $f(a_\ell)$, $1 \le i < j \le p'_\ell$. But the only possibility for 12 to be an external occurrence in $f^n(a_{\ell'_i} a_{\ell'_j})$ is that $j = i + 1$ and the last letter of $f^n(a_{\ell'_i})$ is smaller than the first letter of $f^n(a_{\ell'_j})$. Thus, the number of occurrences of such patterns is only the number of times $L(f^n)[i] < F(f^n)[i+1]$ with $i+1 \le p'_\ell$, i.e., the number of times $C_n^{12}(\ell'_i, \ell'_{i+1}) = 1$ for $1 \le i \le p'_\ell - 1$.

We proceed similarly with the patterns 21 and 11. Consequently we have the following proposition.

Proposition 5. *For each letter $a_\ell \in A$, $f(a_\ell) = a_{\ell_1} \dots a_{\ell_{p_\ell}}$, and for all $n \ge M_f$, let $\ell'_1 \dots \ell'_{p'_\ell}$ be the subsequence of $\ell_1 \dots \ell_{p_\ell}$ such that $f^{n+1}(a_\ell) = f^n(a_{\ell'_1} \dots a_{\ell'_{p'_\ell}})$ and $f^n(a_{\ell'_i}) \ne \varepsilon$, $1 \le i \le p'_\ell$. Then*

$$|f^{n+1}(a_\ell)|_{12} = \sum_{t=1}^{k} |f^n(a_t)|_{12} \cdot m_{1,t,\ell} + \sum_{i=1}^{p'_\ell - 1} C_n^{12}(\ell'_i, \ell'_{i+1}), \tag{4}$$

$$|f^{n+1}(a_\ell)|_{21} = \sum_{t=1}^{k} |f^n(a_t)|_{21} \cdot m_{1,t,\ell} + \sum_{i=1}^{p'_\ell - 1} C_n^{21}(\ell'_i, \ell'_{i+1}), \tag{5}$$

$$|f^{n+1}(a_\ell)|_{11} = \sum_{t=1}^{k} |f^n(a_t)|_{11} \cdot m_{1,t,\ell} + \sum_{i=1}^{p'_\ell - 1} C_n^{11}(\ell'_i, \ell'_{i+1}). \tag{6}$$

6.2   Some Examples

No External Rises, No External Descents, No External Squares. Let us suppose that the morphism f is such that, for all i and j, $L(f)[i] \geq F(f)[j]$ (there are no external rises). According to equation (4), in this case, for each letter $a_\ell \in A$, $f(a_\ell) = a_{\ell_1} \ldots a_{\ell_{p_\ell}}$, and for all $n \geq M_f$,

$$|f^{n+1}(a_\ell)|_{12} = \sum_{t=1}^{k} |f^n(a_t)|_{12} \cdot m_{1,t,\ell}.$$

Moreover, if the above inequality is strict then, according to equation (5),

$$|f^{n+1}(a_\ell)|_{21} = \sum_{t=1}^{k} |f^n(a_t)|_{21} \cdot m_{1,t,\ell} + p'_\ell - 1.$$

Now if we suppose that, conversely to the previous case, the morphism f is such that, for all i and j, $L(f)[i] \leq F(f)[j]$ (there are no external descents) then we obtain the same result by switching 12 and 21 in the above formulas.

To end, if we suppose that the morphism f is such that, for all i and j, $L(f)[i] \neq F(f)[j]$ then, according to equation (6), for each letter $a_\ell \in A$, $f(a_\ell) = a_{\ell_1} \ldots a_{\ell_{p_\ell}}$, and for all $n \geq M_f$,

$$|f^{n+1}(a_\ell)|_{11} = \sum_{t=1}^{k} |f^n(a_t)|_{11} \cdot m_{1,t,\ell}.$$

The Thue-Morse Morphism. Since $R(\mu) = \begin{bmatrix} 1 & 0 \end{bmatrix}$, $D(\mu) = \begin{bmatrix} 0 & 1 \end{bmatrix}$ and $R_2(\mu) = \begin{bmatrix} 0 & 0 \end{bmatrix}$ we obtain again a well known result.

Corollary 9. *For any integer* $n \geq 0$,

$$R(\mu^{2n}) = \begin{bmatrix} \frac{4^n - 1}{3} & \frac{4^n - 1}{3} \end{bmatrix} = D(\mu^{2n}) = R_2(\mu^{2n})$$

$$R(\mu^{2n+1}) = \begin{bmatrix} \frac{2(4^n - 1)}{3} + 1 & \frac{2(4^n - 1)}{3} \end{bmatrix}$$

$$D(\mu^{2n+1}) = \begin{bmatrix} \frac{2(4^n - 1)}{3} & \frac{2(4^n - 1)}{3} + 1 \end{bmatrix}$$

$$R_2(\mu^{2n+1}) = \begin{bmatrix} \frac{2(4^n - 1)}{3} & \frac{2(4^n - 1)}{3} \end{bmatrix}.$$

The Fibonacci Morphism. Since $R(\varphi) = \begin{bmatrix} 1 & 0 \end{bmatrix}$ and $D(\varphi) = R_2(\varphi) = \begin{bmatrix} 0 & 0 \end{bmatrix}$ we have again a well known result.

Corollary 10. *For any integer* $n \geq 1$,

$$R(\varphi^n) = \begin{bmatrix} F_{n-1} & F_{n-2} \end{bmatrix}$$

$$D(\varphi^{2n}) = \begin{bmatrix} F_{2n-1} & F_{2n-2} - 1 \end{bmatrix} = R_2(\varphi^{2n+1})$$

$$R_2(\varphi^{2n}) = \begin{bmatrix} F_{2n-2} - 1 & F_{2n-3} \end{bmatrix} = D(\varphi^{2n-1}).$$

Erasing Morphisms. Let A be the four-letter ordered alphabet $A = \{a_1 < a_2 < a_3 < a_4\}$.

1. Here we consider the erasing morphism f, given in Section 5.4, defined on A by $f(a_1) = a_1a_3a_2a_4$, $f(a_2) = \varepsilon$, $f(a_3) = a_1a_4$, $f(a_4) = a_2a_3$. One has $M_f = 1$. Starting from $R(f) = \begin{bmatrix} 2 & 0 & 1 & 1 \end{bmatrix}$, we obtain the following corollary of Proposition 5.

Corollary 11. *For any integer* $n \geq 1$, $R_2(f^n) = \begin{bmatrix} 0 & 0 & 0 & 0 \end{bmatrix}$ *and*

if n *is even* $\begin{cases} R(f^n) = \begin{bmatrix} 2^n & 0 & \frac{2^{n+1}+1}{3} & \frac{2^n-1}{3} \end{bmatrix} \\ D(f^n) = \begin{bmatrix} 2^n & -1 & 0 & \frac{2^{n+1}-2}{3} & \frac{2^n-4}{3} \end{bmatrix}, \end{cases}$

if n *is odd* $\begin{cases} R(f^n) = \begin{bmatrix} 2^n & 0 & \frac{2^{n+1}-1}{3} & \frac{2^n+1}{3} \end{bmatrix} \\ D(f^n) = \begin{bmatrix} 2^n & -1 & 0 & \frac{2^{n+1}-4}{3} & \frac{2^n-2}{3} \end{bmatrix}. \end{cases}$

2. Now we consider the erasing morphism g defined on A by $g(a_1) = a_1a_2a_4a_3$, $g(a_2) = a_3$, $g(a_3) = \varepsilon$, $g(a_4) = a_1a_2a_4$. Here we have $M_f = 2$.

Corollary 12. $R(g) = \begin{bmatrix} 2 & 0 & 0 & 2 \end{bmatrix}$, $D(g) = \begin{bmatrix} 1 & 0 & 0 & 0 \end{bmatrix}$, $R_2(g) = \begin{bmatrix} 0 & 0 & 0 & 0 \end{bmatrix}$, *and, for any integer* $n \geq 2$,

$R(g^n) = \begin{bmatrix} 2^n & 0 & 0 & 2^n \end{bmatrix}$, $\qquad D(g^n) = \begin{bmatrix} 2^{n-1} + 2^{n-2} - 1 & 0 & 0 & 2^{n-1} + 2^{n-2} - 1 \end{bmatrix}$
$R_2(g^n) = \begin{bmatrix} 2^{n-2} & 0 & 0 & 2^{n-2} \end{bmatrix}$.

References

1. Arshon, S.: Démonstration de l'existence de suites asymétriques infinies. Mat. Sb. 44, 769–779 (in Russian); 777–779 (French summary) (1937)
2. Babson, E., Steingrímsson, E.: Generalized permutation patterns and a classification of the Mahonian statistics. Sém. Lothar. Comb. 44, Art. B44b (2000)
3. Brandenburg, F.-J.: Uniformly growing k-th power-free homomorphisms. Theoret. Comput. Sci. 23, 69–82 (1983)
4. Burstein, A.: Enumeration of words with forbidden patterns, Ph.D. thesis, University of Pennsylvania (1998)
5. Kitaev, S., Mansour, T., Séébold, P.: The Peano curve and counting occurrences of some patterns. J. Autom., Lang. Combin. 9(4), 439–455 (2004)
6. Levé, F., Richomme, G.: On a conjecture about finite fixed points of morphisms. Theoret. Comput. Sci. 339-1, 103–128 (2005)
7. Lothaire, M.: Algebraic Combinatorics on Words. Encyclopedia of Mathematics and its Applications, vol. 90. Cambridge University Press, Cambridge (2002)
8. Morse, M.: Recurrent geodesics on a surface of negative curvature. Trans. Amer. Math. Soc. 22, 84–100 (1921)
9. Prouhet, M.E.: Mémoire sur quelques relations entre les puissances des nombres. Comptes Rendus Acad. Sci. Paris 33, 225 (1851)
10. Thue, A.: Über die gegenseitige Lage gleicher Teile gewisser Zeichenreihen, *Vidensk.-Selsk. Skrifter. I. Mat. Nat. Kl.* 1 Kristiania. In: Nagell, T. (ed.), Universitetsforlaget, Oslo, pp. 1–67 (1912); Reprinted in *Selected Mathematical Papers of Axel Thue*, 413-478 (1977)

Literal Varieties of Languages Induced by Homomorphisms onto Nilpotent Groups

Ondřej Klíma and Libor Polák*

Department of Mathematics, Masaryk University
Janáčkovo nám 2a, 662 95 Brno, Czech Republic

Abstract. We present here new hierarchies of literal varieties of languages. Each language under consideration is a disjoint union of a certain collection of "basic" languages described here. Our classes of languages correspond to certain literal varieties of homomorphisms from free monoids onto nilpotent groups of class ≤ 2.

Keywords: Literal varieties of languages, homomorphisms onto monoids, nilpotent groups. **MSC 2000 Classification:** 68Q45 Formal languages and automata.

1 Introduction

By the classical Eilenberg's theorem, the (Boolean) varieties of recognizable languages correspond to pseudovarieties of finite monoids. The last classes appear exactly as finite members of unions of varieties of monoids. Therefore, it is natural to start our investigations with varieties of monoids.

Recognizable languages over $X_n = \{x_1, \ldots, x_n\}$ corresponding to certain varieties of groups are well-known (the notation is explained in the next section) – see [4], [14], [15], [9], [3]:

1. Boolean combinations of

$$\{\, u \in X_n^* \mid |u|_i \equiv \ell' \bmod \ell \,\}, \ i \in \{1, \ldots, n\}, \ \ell \in \mathbb{N}, \ \ell' \in \{0, \ldots, \ell - 1\}$$

for the class of all abelian groups.

2. Boolean combinations of

$$\{\, u \in X_n^* \mid |u|_i \equiv \ell' \bmod \ell \,\}, \ i \in \{1, \ldots, n\}, \ \ell' \in \{0, \ldots, \ell - 1\}$$

for the class of all abelian groups satisfying $x^\ell = 1$.

3. Boolean combinations of

$$\left\{\, u \in X_n^* \mid \binom{u}{v} \equiv r' \bmod r \,\right\}, \ v \in X_n^*, \ r \in \mathbb{N}, \ r' \in \{0, \ldots, r - 1\}$$

for the class of all nilpotent groups.

---

* Both authors were supported by the Ministry of Education of the Czech Republic under the project MSM 0021622409 and by the Grant no. 201/06/0936 of the Grant Agency of the Czech Republic.

4. Boolean combinations of

$$\{\, u \in X_n^* \mid \binom{u}{v} \equiv r' \bmod r \,\}, \ v \in X_n^*, \ |v| \le c, \ r \in \mathbb{N}, \ r' \in \{0, \dots, r-1\}$$

for the class of all nilpotent groups of class $\le c$.

Such characterizations can be refined as follows :
1'. Disjoint unions of

$$\{\, u \in X_n^* \mid |u|_1 \equiv \ell_1, \dots, |u|_n \equiv \ell_n \bmod \ell \,\}, \ \ell \in \mathbb{N}, \ \ell_1, \dots, \ell_n \in \{0, \dots, \ell-1\}$$

for the class of all abelian groups.

It is not difficult to refine the results 2, 3 and 4 in a similar way.

Recent investigations in language theory lead to the notion of a literal variety of languages (Ésik and Ito) [6] and Straubing [13]. Such classes of languages generalize the classical varieties, we postulate only the closeness with respect to inverse literal homomorphisms, not with respect to all inverse homomorphisms. Numerous examples are given in paper quoted above and in [7] and [5]. The algebraic counterpart was invented by (Ésik and Larsen)[7] and [13]. The appropriate equational logic was created by Kunc [8] and by Pin and Straubing [10].

The aim of our contribution is to find rich families of literal varieties of homomorphisms onto nilpotent groups of class ≤ 2 and to present the corresponding languages in the finer form. All the varieties of nilpotent groups of class ≤ 2 are well-known and one can find (refined) formulas for all of them. Our new classes are disjoint unions of

$$\{\, u \in X_n^* \mid |u| \equiv \ell' \bmod \ell, \ |u|_1 \equiv k_1, \dots, |u|_n \equiv k_n \bmod k,$$

$$|u|_{j,i} \equiv r_{j,i} \bmod r \text{ for all } 1 \le i < j \le n \,\} \,,$$

where $n, \ell, k, r \in \mathbb{N}$ with $r \mid k \mid \ell$ are fixed and $\ell' \in \{0, \dots, \ell-1\}$, $k_1, \dots, k_n \in \{0, \dots, k-1\}$ satisfying $k_1 + \cdots + k_n \equiv \ell' \bmod k$, $r_{j,i} \in \{0, \dots, r-1\}$ for $1 \le i < j \le n$.

The next section fixes notation. In Sections 3 and 4 we recall the basics of the classical and the literal universal algebra. Next section summarizes known results about abelian groups. The main body of our contribution is Section 6 dealing with homomorphisms onto nilpotent groups. Finally, the last section describes how to check membership to our classes of languages using minimal deterministic automata.

2   Our Languages

Let $\mathbb{N} = \{1, 2, \dots\}$, and $\mathbb{N}_0 = \mathbb{N} \cup \{0\}$ be the sets of all *positive integers*, and *non-negative integers*. The relation of *divisibility* is denoted by $|$, i.e., for $k, \ell \in \mathbb{N}_0$, $k \mid \ell$ if and only if there exists $m \in \mathbb{N}_0$ such that $mk = \ell$ and the meaning of $k \equiv \ell \bmod m$ is that $m \mid (k - \ell)$. The *greatest common divisor* of $k, \ell \in \mathbb{N}_0$ is denoted by $\gcd(k, \ell)$. For a mapping $f : B \to A$, we write $\mathrm{im} f = \{\, f(b) \mid b \in B \,\}$ and $\ker f = \{\, (b, c) \in B \times B \mid f(b) = f(c) \,\}$.

Our alphabets/sets of variables will be $X = \{x_1, x_2, \dots\}$ and, for $n \in \mathbb{N}$, $X_n = \{x_1, \dots, x_n\}$. The free semigroup (monoid) over the set Y is denoted by Y^+ and Y^*, respectively. We have $Y^* = Y^+ \cup \{1\}$ where 1 is the empty word. When using only several first variables, we write $x, y, z, \dots$ instead of $x_1, x_2, x_3, \dots$.

Let $u, v \in X^*$, $i, j \in \mathbb{N}$, $i < j$. We denote:

$|u|$ – the length of the word u,

$\binom{u}{v}$ – the number of occurrences of v in u as a subword,

in particular, we write

$|u|_i = \binom{u}{x_i}$ – the number of occurrences of x_i in u

$|u|_{j,i} = \binom{u}{x_j x_i}$ – the number of occurrences of $x_j x_i$ in u as a subword, i.e. the number of different factorizations $u = p x_j q x_i r$, $p, q, r \in X^*$.

Our basic ingredients are the following languages:
let $n, \ell, k, r \in \mathbb{N}$ with $r \mid k \mid \ell$, let $\ell' \in \{0, \dots, \ell - 1\}$, $k_1, \dots, k_n \in \{0, \dots, k - 1\}$ satisfying $k_1 + \cdots + k_n \equiv \ell' \bmod k$, $r_{j,i} \in \{0, \dots, r - 1\}$ for $1 \leq i < j \leq n$. We put

$$A(n; \ell, \ell'; k, k_1, \dots, k_n) =$$

$$= \{\, u \in X_n^* \mid |u| \equiv \ell' \bmod \ell,\ |u|_1 \equiv k_1, \dots, |u|_n \equiv k_n \bmod k \,\},$$

$$N(n; \ell, \ell'; k, k_1, \dots, k_n; r, r_{2,1}, \dots, r_{n,1}, \dots, r_{n,n-1}) =$$

$$= \{\, u \in X_n^* \mid |u| \equiv \ell' \bmod \ell,\ |u|_1 \equiv k_1, \dots, |u|_n \equiv k_n \bmod k,$$

$$|u|_{j,i} \equiv r_{j,i} \bmod r \text{ for all } 1 \leq i < j \leq n \,\}.$$

Already at this place we can mention that :

(i) Preimages of a given $A(n; \ell, \ell'; k, k_1, \dots, k_n)$ in literal homomorphisms (i.e. each letter goes to a letter) from X_m^* into X_n^* are disjoint unions of $A(m; \ell, \cdots ; k, \cdots, \dots, \cdots)$'s.

(ii) Preimages of a given $A(n; \ell, \ell'; k, k_1, \dots, k_n)$ in arbitrary homomorphisms from X_m^* into X_n^* are not of the above form since, for instance, it is not the case for $A(2; 6, 3; 2, 0, 1)$ and $f : X_2^* \to X_2^*$, $x_1 \mapsto 1$, $x_2 \mapsto x_2$.

3   Classical Universal Algebra

We recall here the basis of universal algebra of monoids. For more information see Almeida's book [1].

Let $\mathcal{M}$ denote the class of all monoids. For $\mathcal{V} \subseteq \mathcal{M}$, let $\mathrm{Fin}\, \mathcal{V}$ denote the class of all finite members from $\mathcal{V}$. A class of monoids is a *variety* if it is closed with respect to the forming of homomorphic images, submonoids and products. Similarly, a class of finite monoids is a *pseudovariety* if it is closed with respect to the forming of homomorphic images, submonoids and products of finite families.

Result 1 *(Baldwin and Berman [2]). The pseudovarieties of finite monoids are exactly the classes of the form* Fin$\mathcal{U}$ *where $\mathcal{U}$ is a union of a chain of varieties of monoids.*

Recall that an n-ary *identity* is a pair $u = v$ where $u, v \in X_n^*$, $n \in \mathbb{N}$. A monoid $M \in \mathcal{M}$ satisfies $u = v$ if ($\forall\, \alpha : X_n^* \to M$) $\alpha(u) = \alpha(v)$. In fact, the choice of n is not significant and we write $M \models u = v$ in this case. For a class $\mathcal{V} \subseteq \mathcal{M}$, we put
$$\mathsf{Id}\,\mathcal{V} = \{\, (u, v) \in X^* \times X^* \mid (\, \forall\, M \in \mathcal{V}\,)\ M \models u = v \,\} \,.$$

Let $\Pi \subseteq X^* \times X^*$ be a set of identities. We put
$$\mathsf{Mod}\,\Pi = \{\, M \in \mathcal{M} \mid (\, \forall\, \pi \in \Pi\,)\ M \models \pi \,\} \,.$$

Further, for $\pi \in X^* \times X^*$, the meaning of $\Pi \models \pi$ is
$$(\, \forall\, M \in \mathcal{M}\,)\ (\, M \models \Pi \text{ implies } M \models \pi\,) \,,$$

and Π is *closed* if ($\forall\, \pi \in X^* \times X^*$) ($\Pi \models \pi$ implies $\pi \in \Pi$).

A congruence ϱ on Y^* is *fully invariant* if for each $u, v \in Y^*$ and each endomorphism $g : Y^* \to Y^*$, $u\,\varrho\,v$ implies $g(u)\,\varrho\,g(v)$.

The following classical theorems are credited to Birkhoff:

Result 2. *The varieties of monoids are exactly the classes of the form* $\mathsf{Mod}\,\Pi$ *where $\Pi \subseteq X^* \times X^*$*

Result 3. *Let $\Pi \subseteq X^* \times X^*$. Then Π is an closed set of identities if and only if it is a fully invariant congruence.*

Result 4. *The mappings $\mathcal{V} \mapsto \mathsf{Id}\,\mathcal{V}$ and $\Pi \mapsto \mathsf{Mod}\,\Pi$ are mutually inverse bijections between the class of all varieties of monoids and the class of all fully invariant congruences on X^*.*

4   Literal Universal Algebra

Let $\mathfrak{L}$ be the category having all Y^*'s, Y is a set, as objects and
$$f \in \mathfrak{L}(Z^*, Y^*) \text{ if and only if } f(Z) \subseteq Y \,.$$

We speak about *literal* homomorphisms.

Note that we can take an arbitrary category of free monoids instead of $\mathfrak{L}$ and that classes of homomorphisms from finitely generated free monoids onto finite monoids were studied before the results which appear in [11].

Let $\mathcal{M}$ denote the class of all homomorphisms from free monoids onto monoids. Such homomorphism $\phi : A^* \twoheadrightarrow M$ is *finite* if both A and M are finite. For $\mathcal{V} \subseteq \mathcal{M}$, let **Fin**$\mathcal{V}$ denote the class of all finite members from $\mathcal{V}$ and we define:

$$\mathsf{H}\mathcal{V} = \{\, \sigma\phi : Y^* \twoheadrightarrow N \mid (\phi : Y^* \twoheadrightarrow M) \in \mathcal{V}, N \in \mathcal{M}, \sigma : M \twoheadrightarrow N \text{ surj. homom.}\},$$

$\mathbf{S}_{\mathcal{L}}\mathcal{V} = \{\phi f : Z^* \twoheadrightarrow \mathrm{im}(\phi f) \mid Z \text{ a set}, f \in \mathfrak{L}(Z^*, Y^*), (\phi : Y^* \twoheadrightarrow M) \in \mathcal{V}\}$,

$\mathbf{P}\mathcal{V} = \{(\phi_\gamma)_{\gamma \in \Gamma} : Y^* \twoheadrightarrow \mathrm{im}((\phi_\gamma)_{\gamma \in \Gamma}) \mid \Gamma \text{ a set}, (\phi_\gamma : Y^* \twoheadrightarrow M_\gamma) \in \mathcal{V} \text{ for } \gamma \in \Gamma\}$

(here $(\phi_\gamma)_{\gamma \in \Gamma} : Y^* \to \prod_{\gamma \in \Gamma} M_\gamma, u \mapsto (\phi_\gamma(u))_{\gamma \in \Gamma}$).

A more transparent definition of the product is the following: consider kernels $\ker \phi_\gamma$ of all ϕ_γ's and take the canonical homomorphism $Y^* \twoheadrightarrow Y^*/\cap_{\gamma \in \Gamma} \ker \phi_\gamma$.

A class $\mathcal{V} \subseteq \mathcal{M}$ is a $\mathfrak{L}$-*variety* (or *literal* variety) if it is closed with respect to the operators $\mathbf{H}$, $\mathbf{S}_{\mathcal{L}}$ and $\mathbf{P}$. Similarly, a class $\mathcal{X} \subseteq \mathbf{Fin}\,\mathcal{M}$ is an $\mathfrak{L}$-*pseudovariety* (or *literal* pseudovariety) of finite homomorphisms onto monoids if it is closed with respect to $\mathbf{H}$, $\mathbf{S}_{\mathcal{L}}$ and $\mathbf{P}_f$ (products of finite families).

Result 5 *([11], Theorem 3). The literal pseudovarieties of finite homomorphisms onto monoids are exactly the classes of the form* $\mathbf{Fin}\,\mathcal{U}$ *where* $\mathcal{U}$ *is a union of a chain of literal varieties of homomorphisms onto monoids.*

Let $u, v \in X_n^*$. A homomorphism $(\phi : Y^* \twoheadrightarrow M) \in \mathcal{M}$ $\mathfrak{L}$-*satisfies* (or *literally satisfies*) the identity $u = v$ if $(\forall f \in \mathfrak{L}(X_n^*, Y^*))$ $(\phi f)(u) = (\phi f)(v)$. We write $\phi \models_{\mathcal{L}} u = v$. Let $(\phi : Y^* \twoheadrightarrow M) \models u = v$ mean $M \models u = v$.

For a class $\mathcal{V} \subseteq \mathcal{M}$, we put

$$\mathbf{Id}_{\mathcal{L}}\,\mathcal{V} = \{(u, v) \in X^* \times X^* \mid (\forall \phi \in \mathcal{V})\, \phi \models_{\mathcal{L}} u = v\}.$$

Let $\Pi \subseteq X^* \times X^*$ be a set of identities. We set

$\phi \models_{\mathcal{L}} \Pi$ if $(\forall \pi \in \Pi)\, \phi \models_{\mathcal{L}} \pi$, and $\mathbf{Mod}_{\mathcal{L}}\,\Pi = \{\phi \in \mathcal{M} \mid \phi \models_{\mathcal{L}} \Pi\}$.

Further, for $\pi \in X^* \times X^*$, the meaning of $\Pi \models_{\mathcal{L}} \pi$ is

$$(\forall \phi \in \mathcal{M})\,(\phi \models_{\mathcal{L}} \Pi \text{ implies } \phi \models_{\mathcal{L}} \pi),$$

and Π is $\mathfrak{L}$-*closed* (or *literally* closed) if

$$(\forall \pi \in X^* \times X^*)\,(\Pi \models_{\mathcal{L}} \pi \text{ implies } \pi \in \Pi).$$

A congruence ϱ on Y^* is $\mathfrak{L}$-*invariant* if for each $u, v \in Y^*$ and each $g \in \mathfrak{L}(Y^*, Y^*)$, $u \varrho v$ implies $g(u) \varrho g(v)$.

The following statements are modifications of Results 2–4 :

Result 6 *([11], Theorem 1). The varieties of homomorphisms from free monoids onto monoids are exactly the classes of homomorphisms of the form* $\mathrm{Mod}_{\mathcal{L}}\Pi$ *where* $\Pi \subseteq X^* \times X^*$.

Result 7 *([12], Lemma 2.1). Let* $\Pi \subseteq X^* \times X^*$. *Then* Π *is an* $\mathfrak{L}$-*closed set of identities if and only if it is an* $\mathfrak{L}$-*invariant congruence.*

Result 8 *([12], Theorem 2.1.). The mappings* $\mathcal{V} \mapsto \mathbf{Id}_{\mathcal{L}}\,\mathcal{V}$ *and* $\Pi \mapsto \mathbf{Mod}_{\mathcal{L}}\,\Pi$ *are mutually inverse bijections between the class of all literal varieties of homomorphisms onto monoids and the class of all literally invariant congruences on* X^*.

5   Abelian Groups

The parts (i) and (ii) of the following result are well-known and the item (iii) appears in the sources quoted in the introduction.

Result 9. *(i) The varieties of monoids consisting of abelian groups are exactly*

$$\mathcal{A}(\ell) = \mathsf{Mod}\,(\,xy = yx,\ x^{\ell} = 1\,),\ \text{where}\ \ell \in \mathbb{N}\ .$$

Moreover, $\mathcal{A}(\ell) \subseteq \mathcal{A}(\ell')$ if and only if $\ell \mid \ell'$.
(ii) The corresponding fully invariant congruences are

$$\alpha(\ell) = \{\,(u,v) \in X^* \times X^* \mid (\,\forall\,i \in \mathbb{N}\,)\,|u|_i \equiv |v|_i\ \mathrm{mod}\ \ell\,\}\ .$$

(iii) For the corresponding varieties of languages $\mathscr{A}(\ell)$ we have: $(\mathscr{A}(\ell))(X_n^*)$ consists of disjoint unions of

$$\{\,u \in X_n^* \mid |u|_1 \equiv \ell_1, \ldots, |u|_n \equiv \ell_n\ \mathrm{mod}\ \ell\,\},\ \ell_1, \ldots, \ell_n \in \{0, \ldots, \ell-1\}\ .$$

The case of literal varieties of homomorphisms is solved by the following:

Result 10. *([12]) (i) The literal varieties of homomorphisms onto abelian groups are exactly*

$$\mathcal{A}(\ell, k) = \mathsf{Mod}_{\mathfrak{L}}\,(\,xy = yx,\ x^{\ell} = 1,\ x^k = y^k\,),\ \text{where}\ k, \ell \in \mathbb{N},\ k \mid \ell\ .$$

Moreover, $\mathcal{A}(\ell, k) \subseteq \mathcal{A}(\ell', k')$ if and only if $k \mid k'$ and $\ell \mid \ell'$.
(ii) The corresponding literally invariant congruences are $\alpha(\ell, k) =$

$$= \{\,(u,v) \in X^* \times X^* \mid |u| \equiv |v|\ \mathrm{mod}\ \ell\ \text{and}\ (\,\forall\,i \in \mathbb{N}\,)\,|u|_i \equiv |v|_i\ \mathrm{mod}\ k\,\}\ .$$

(iii) For the corresponding literal varieties of languages $\mathscr{A}(\ell, k)$, we have: $(\mathscr{A}(\ell, k))(X_n^*)$ consists of disjoint unions of

$$A(n; \ell, \ell'; k, k_1, \ldots, k_n)$$

where $\ell' \in \{0, \ldots, \ell-1\}$, $k_1, \ldots, k_n \in \{0, \ldots, k-1\}$, $k_1 + \cdots + k_n \equiv \ell'\ \mathrm{mod}\ k$.

6   Nilpotent Groups

Let $(G, \cdot)$ be a group. For $g, h \in G$, we define the *commutator* $[g, h]$ of g and h by $[g, h] = g^{-1}h^{-1}gh$. Further, we put $[g_1, \ldots, g_s] = [g_1, [g_2, \ldots, g_s]]$ for $s \in \mathbb{N}$, $s \geq 3$, $g_1, \ldots, g_s \in G$. A group $(G, \cdot)$ is said to be *nilpotent of the class $\leq s$ ($s \in \mathbb{N}$)* if it satisfies the identity $[x_1, \ldots, x_{s+1}] = 1$. We denote by $\mathcal{N}_s$ the class of all such groups. Let $\mathcal{N} = \bigcup_{s \in \mathbb{N}} \mathcal{N}_s$ be the class of all *nilpotent* groups and note that $\mathcal{A} = \mathcal{N}_1$ be the class of all abelian groups. Notice that each group satisfies $gh[h, g] = hg$ and that both $[gg', h] = [g, h][g', h]$ and $[g, hh'] = [g, h][g, h']$ hold in the class $\mathcal{N}_2$. We denote by $\mathcal{N}_s$ the class of all homomorphisms from $\mathcal{M}$ which are onto groups from $\mathcal{N}_s$.

We consider the case $s = 2$. By the following lemma, we can restrict our attention to the literal varieties of homomorphisms onto members of $\mathcal{N}_2$. Moreover, we can suppose that our homomorphisms are "literally periodic".

Lemma 1. *Let $\mathcal{G}$ be a variety of groups and let $\mathcal{V}$ be a literal variety of homomorphisms from free monoids onto monoids whose finite members are onto groups from $\mathcal{G}$. Then there exists a literal variety $\mathcal{W}$ of homomorphisms from free monoids onto groups from $\mathcal{G}$ and $\ell \in \mathbb{N}$ such that both $\mathcal{V}$ and $\mathcal{W}$ literally satisfy $x^\ell = 1$ and $\mathbf{Fin}\,\mathcal{W} = \mathbf{Fin}\,\mathcal{V}$.*

Proof. Let $\phi : \{a\}^* \twoheadrightarrow M$ be the *free* object in $\mathcal{V}$ over $A = \{a\}$, that is, ϕ is the product of all (we consider representatives of classes giving the same kernels) $(\phi_i : \{a\}^* \twoheadrightarrow M_i) \in \mathcal{V}$ where $i \in I$. If M is infinite, then $M = \{1, b, b^2, \dots\}$ where $b = \phi(a)$, and $b^k \neq b^\ell$ for $k \neq \ell$, $k, \ell \in \mathbb{N}_0$. Putting $b, b^2, \dots$ into a single class we get a congruence τ on M such that M/τ is finite and it is not a group, a contradiction. Consequently, M consists of pairwise different elements $1, b, b^2, \dots, b^{k+\ell-1}$ with $b^{k+\ell} = b^k$, for some $k \in \mathbb{N}_0$, $\ell \in \mathbb{N}$. Since M is a group, we have $k = 0$.

Now we show that $\mathcal{V} \models_{\mathfrak{L}} x^\ell = 1$. Indeed, for an arbitrary $(\psi : B^* \twoheadrightarrow N) \in \mathcal{V}$ consider the compositions of ψ with $\{a\}^* \to B^*$, $a \mapsto b$, $b \in B$.

Let $\mathcal{W} = \langle\, \mathbf{Fin}\mathcal{V} \,\rangle$ be a literal variety generated by $\mathbf{Fin}\,\mathcal{V}$. All members of $\mathcal{W}$ satisfy the identity $x^\ell = 1$ literally. Notice that if a monoid M with a generating set A satisfies $g^\ell = 1$ for all $g \in A$, then M is a group since $g_m^{\ell-1} \dots g_1^{\ell-1}$ is the inverse of $g_1 \dots g_m$. Moreover, each homomorphism of $\mathcal{W}$ is onto a group from $\mathcal{G}$.

Since $\mathbf{Fin}\mathcal{V} \subseteq \langle\, \mathbf{Fin}\,\mathcal{V} \,\rangle = \mathcal{W}$, we have $\mathbf{Fin}\,\mathcal{V} \subseteq \mathbf{Fin}\,\mathcal{W}$. Conversely, from $\mathcal{W} \subseteq \mathcal{V}$ it follows that $\mathbf{Fin}\,\mathcal{W} \subseteq \mathbf{Fin}\,\mathcal{V}$. ☐

For any $t \in \mathbb{N}$ and $x, y \in X$, we define $[x, y]_t = x^{t-1}y^{t-1}xy \in X^*$. If we consider $\phi : X_n^* \twoheadrightarrow G$ satisfying $x^\ell = 1$ literally then $\phi([x, y]_\ell) = [\phi(x), \phi(y)]$ for $x, y \in X_n$. From that reason $[x, y]$ will always mean $[x, y]_\ell$ for an appropriate ℓ.

A crucial role in our considerations is played by certain numerical parameters. The relationships between them are explored in the following result.

Lemma 2. *Let $(\phi : Y^* \twoheadrightarrow G) \in \mathcal{N}_2$ and let there exist $t \in \mathbb{N}$ such that $\phi \models_{\mathfrak{L}} x^t = 1$. Then there exist $m, \ell, k, r \in \mathbb{N}$, each smallest (with respect to the divisibility), satisfying*

$$\phi \models x^m = 1, \quad \phi \models_{\mathfrak{L}} x^\ell = 1, \quad \phi \models_{\mathfrak{L}} x^k = y^k, \quad \phi \models_{\mathfrak{L}} [x, y]^r = 1 \,.$$

Moreover, those parameters are related by

(a) $k \mid \ell$,
(b) $r \mid k$,
(c) If ℓ is odd or $2r \mid \ell$ then $m = \ell$; $m = 2\ell$ otherwise.

Proof. Using the Bezout's Lemma we see that, for each $\ell, \ell' \in \mathbb{N}_0$, the facts $\phi \models_{\mathfrak{L}} x^\ell = 1, x^{\ell'} = 1$ imply $\phi \models_{\mathfrak{L}} x^{\gcd(\ell, \ell')} = 1$. Therefore there exists the smallest $\ell \in \mathbb{N}_0$ (with respect to the divisibility) such that $\phi \models_{\mathfrak{L}} x^\ell = 1$. Moreover, $\phi \models_{\mathfrak{L}} x^{\ell'} = 1$ if and only if ℓ divides ℓ'.

The same is true for the literal satisfaction of $x^k = y^k$ and $[x, y]^r = 1$ and for the (usual) satisfaction of $x^m = 1$. Let $A = \phi(Y)$.

$\ell \mid m : (\, g^m = 1$ for all $g \in G \,)$ implies $(\, a^m = 1$ for all $a \in A \,)$.

$k \mid \ell$: (for all $a \in A$, $a^\ell = 1$) implies (for all $a, b \in A$, $a^\ell = b^\ell$).

$r \mid k$: Let $a, b \in A$. Then $ab^k = a \cdot a^k = a^k \cdot a = b^k \cdot a = ab^k [b, a]^k$ and thus $[b, a]^k = 1$.

By the assumptions, $\ell \in \mathbb{N}$.

The proof of (c) : For each $a, b \in A$, $p \in \mathbb{N}$, we have $(ab)^p = a^p b^p \cdot [b, a]^{\binom{p}{2}}$. If ℓ is odd or $2r \mid \ell$ then $r \mid \binom{\ell}{2}$ and thus $(ab)^\ell = 1$. For $(a_1 \ldots, a_q)^\ell$, $q \in \mathbb{N}$, use induction. Clearly $r \mid \binom{2\ell}{2}$. Conversely, if $G \models x^\ell = 1$ and ℓ is even, then, in particular, for all $a, b \in A$, we have $[b, a]^{\binom{\ell}{2}} = 1$. Thus $r \mid \binom{\ell}{2} = \frac{\ell}{2}(\ell - 1)$. Since r and $\ell - 1$ are relatively prime we have $r \mid \frac{\ell}{2}$.   □

Remark. In fact, we will not need the parameter m in our considerations. We added it to show that the relationship between identities being satisfied literally and globally is far from being trivial.

The next result helps us to understand our languages.

Lemma 3. *Let* $n, \ell, k, r \in \mathbb{N}$ *satisfy* $r \mid k \mid \ell$. *Then the classes*

$$N(n; \ell, \ell'; k, k_1, \ldots, k_n; r, r_{2,1}, \ldots, r_{n,1}, \ldots, r_{n,n-1})$$

where $\ell' \in \{0, \ldots, \ell - 1\}$, $k_1, \ldots, k_n \in \{0, \ldots, k - 1\}$ *satisfying* $k_1 + \cdots + k_n \equiv \ell' \bmod k$, $r_{j,i} \in \{0, \ldots, r - 1\}$ *for* $1 \leq i < j \leq n$ *form a partition of the set* X_n^*.

Proof. Indeed, using sequences of transpositions $px_i x_j q \to px_j x_i q$ on the word $x_i^k x_j^k$, one gets exactly the words u with the following parameters $|u|_i = |u|_j = k$, $|u|_{j,i} \in \{0, \ldots, k^2\}$. To get a word

$$u \in N(n; \ell, \ell'; k, k_1, \ldots, k_n; r, r_{2,1}, \ldots, r_{n,1}, \ldots, r_{n,n-1})$$

apply an appropriate sequence of transpositions on

$$x_1^{k_1} \ldots x_n^{k_n} \cdot x_1^k x_2^k \cdot \ldots \cdot x_1^k x_n^k \cdot \ldots \cdot x_{n-1}^k x_n^k .$$

Clearly, the union of the classes is X_n^* and they are pairwise disjoint.   □

The following result will help us to distinguish varieties with different parameters (even their finite members).

Lemma 4. *Let* $\ell, k, r \in \mathbb{N}$ *satisfy* $r \mid k \mid \ell$. *Then the formula*

$$\xi : \mathbb{Z}_r \to (\mathbb{Z}_\ell \times \mathbb{Z}_k)^{\mathbb{Z}_\ell \times \mathbb{Z}_k}, \ u \mapsto \left((a, p) \mapsto (a + up\frac{\ell}{r}, p) \right) \tag{$\diamond$}$$

correctly defines an action of the group $(\mathbb{Z}_r, +)$ *on* $(\mathbb{Z}_\ell \times \mathbb{Z}_k, +)$ *by automorphisms.*

Moreover, the formula

$$(a, p, u) \circ (b, q, v) = (a + b + uq\frac{\ell}{r}, p + q, u + v) \tag{$*$}$$

defines a group operation on the set $\mathbb{Z}_\ell \times \mathbb{Z}_k \times \mathbb{Z}_r$; in particular $(0,0,0)$ is the neutral element and $(a,p,u)^{-1} = (-a + up\frac{\ell}{r}, -p, -u)$.

This group is nilpotent of class 2. Let

$$\alpha = (1,0,0), \quad \beta = (1,1,0), \quad \gamma = (1,0,1) \ .$$

Then the set $\{\alpha, \beta, \gamma\}$ generates $(\mathbb{Z}_\ell \times \mathbb{Z}_k \times \mathbb{Z}_r, \circ)$, the number k is the smallest (with respect to the divisibility) with $\alpha^k = \beta^k = \gamma^k$ and the order of each of α, β, γ is ℓ. Finally, r is the smallest number such that $[\beta, \gamma]^r = 1$, and for each $(a,p,u), (b,q,v) \in \mathbb{Z}_\ell \times \mathbb{Z}_k \times \mathbb{Z}_r$, we have $[(a,p,u),(b,q,v)]^r = 1$.

Proof. Taking representatives $u', u'' \in \mathbb{Z}$ of $u \in \mathbb{Z}_r$, $a', a'' \in \mathbb{Z}$ of $a \in \mathbb{Z}_\ell$ and $p', p'' \in \mathbb{Z}$ of $p \in \mathbb{Z}_k$, we see that

$$a' + u'p'\frac{\ell}{r} \equiv a'' + u''p''\frac{\ell}{r} \mod \ell \ ,$$

which means that the formula $(\diamond)$ correctly defines a mapping.

Let $u \in \mathbb{Z}_r$. It is easy to see that $\xi(u)$ is a bijective homomorphisms of $(\mathbb{Z}_\ell \times \mathbb{Z}_k, +)$ onto itself.

Further, for each $u, v \in \mathbb{Z}_r$, $(a,p) \in \mathbb{Z}_\ell \times \mathbb{Z}_k$, we have $(\xi(u+v))(a,p) = \xi(u)(\xi(v)(a,p))$ and thus ξ is a homomorphism of the group $(\mathbb{Z}_r)$ into the group $\mathsf{Aut}(\mathbb{Z}_\ell \times \mathbb{Z}_k, +)$ of all automorphisms of the group $(\mathbb{Z}_\ell \times \mathbb{Z}_k, +)$.

The formula $(*)$ describes the well-known semidirect product of groups.

The commutator $[(a,p,u),(b,q,v)]$ equals to $((uq - vp)\frac{\ell}{r}, 0, 0)$ and therefore it commutes with all elements and r is as mentioned in the lemma. The rest follows from the formula

$$(a,p,u)^m = (ma + \frac{m(m-1)}{2} up\frac{\ell}{r}, mp, mu), \ m \in \mathbb{N} \ . \qquad \square$$

Lemma 5. *Let G be a finite group with a generating set A, such that $a[b,c] = [b,c]a$ for all $a, b, c \in A$. Then G is nilpotent of the class ≤ 2.*

Proof. Since G is a finite group, each element of G can be written as a product of generators, i.e. it can be viewed as a word from A^*. So, it is enough to prove that $[u,v]$ commutes with w for all $u, v, w \in A^*$.

We prove this by the induction with respect to $|u| + |v|$. It is clear for $u = 1$ or $v = 1$. If $u, v \in A$ then $[u,v]$ commutes with each generator by the assumption, hence $[u,v]$ commutes with each $w \in A^*$. So, we have proved the statement for $u, v \in A^*$ such that $|u| + |v| \leq 2$.

Now suppose that $n \in \mathbb{N}$, $n > 2$ and that the statement is true for all $u, v, w \in A^*$, $|u| + |v| < n$. Assume first, that $u = u_1 u_2$, where $u_1, u_2 \in A^+$. Then $[u_1 u_2, v] = u_2^{-1}[u_1, v]v^{-1}u_2v$. Now from the induction assumption we have that $[u_1, v]$ commutes with u_2^{-1}. Hence $[u_1 u_2, v] = [u_1, v][u_2, v]$. Again by the induction assumption both $[u_1, v]$ and $[u_2, v]$ commute with each $w \in A^*$ and we can conclude that also $[u_1 u_2, v]$ commutes with each $w \in A^*$.

The case when $u \in A$ can be treated in a similar way: we write v as a product of two shorter words. $\qquad \square$

Theorem 1. *(i) The following literal varieties of homomorphisms from free monoids onto nilpotent groups of class ≤ 2 are pairwise different:*

$$\mathcal{N}(\ell, k, r) = \mathbf{Mod}_{\mathcal{L}}(\ [x, [y, z]] = 1,\ x^\ell = 1,\ x^k = y^k,\ [x, y]^r = 1\ )$$

where $\ell, k, r \in \mathbb{N}$, $r \mid k \mid \ell$.
Moreover, $\mathcal{N}(\ell, k, r) \subseteq \mathcal{N}(\ell', k', r')$ if and only if $\ell \mid \ell'$, $k \mid k'$, $r \mid r'$.
(ii) The corresponding literally invariant congruences on X^* are of the form

$$\boldsymbol{\nu}(\ell, k, r) = \{\ (u, v) \in X^* \times X^* \mid$$

$|u| \equiv |v| \bmod \ell$, $|u|_i \equiv |v|_i \bmod k$ for $i \in \mathbb{N}$, $|u|_{j,i} \equiv |v|_{j,i} \bmod r$ for $1 \leq i < j\ \}$.

(iii) For the corresponding literal varieties of languages $\mathcal{M}(\ell, k, r)$, we have: $\mathcal{M}(\ell, k, r)(X_n^*)$ consists of disjoint unions of

$$N(n; \ell, \ell'; k, k_1, \ldots, k_n; r, r_{2,1}, \ldots, r_{n,1}, \ldots, r_{n,n-1})$$

where $\ell' \in \{0, \ldots, \ell - 1\}$, $k_1, \ldots, k_n \in \{0, \ldots, k - 1\}$, $k_1 + \cdots + k_n \equiv \ell' \bmod k$,

$$r_{j,i} \in \{0, \ldots, r - 1\} \text{ for } 1 \leq i < j \leq n\ .$$

Proof. (i) It follows from Lemmas 2, 4 and 5.

(ii) Claim 1. $\boldsymbol{\nu}(\ell, k, r)$ is a literally invariant congruence on X^*. Clearly, the mentioned relation is an equivalence on the set X^*. Now notice that for arbitrary $u \in X^*$, $i, j \in \mathbb{N}$, $i \neq j$, $\xi : X \to X$, we have:

(a) $|x_i u| = |u| + 1$,
(b) $|x_i u|_i = |u|_i + 1$, $|x_i u|_j = |u|_j$,
(c) $|x_i u|_{i,j} = |u|_{i,j} + |u|_j$,
(d) $|\xi(u)| = |u|$,
(e) $|\xi(u)|_i = \sum_{s=1,\ldots,p} |u|_{i_s}$ where $\xi^{-1}(x_i) = \{x_{i_1}, \ldots, x_{i_p}\}$, $i_1, \ldots, i_p$ pairwise different,
(f) $|\xi(u)|_{i,j} = \sum_{s=1,\ldots,p,\ t=1,\ldots,q} |u|_{i_s,j_t}$ with $i_1, \ldots, i_p$ as above and $\xi^{-1}(x_j) = \{x_{j_1}, \ldots, x_{j_q}\}$, $j_1, \ldots, j_q$ pairwise different.

Let $(u, v) \in \boldsymbol{\nu}(\ell, k, r)$. Then $(x_i u, x_i v) \in \boldsymbol{\nu}(\ell, k, r)$ by (a) – (c) and $(u x_i, v x_i) \in \boldsymbol{\nu}(\ell, k, r)$ by their duals. Finally, $(\xi(u), \xi(v)) \in \boldsymbol{\nu}(\ell, k, r)$ due to (d) – (f).

Claim 2. $\boldsymbol{\nu}(\ell, k, r)$ is generated as literally invariant congruence by the set

$$\{\ ([x, [y, z]], 1),\ (x^\ell, 1),\ (x^k, y^k),\ ([x, y]^r, 1) \mid x, y, z \in X\ \}\ .$$

Indeed, let $(u, v) \in \boldsymbol{\nu}(\ell, k, r)$ and let all the variables of u and v be among $x_1, \ldots, x_n$. The identity $u = u'$ where

$$u' = x_1^{|u|_1} \ldots x_n^{|u|_n} [x_2, x_1]^{|u|_{2,1}} \ldots [x_n, x_1]^{|u|_{n,1}} \ldots [x_n, x_{n-1}]^{|u|_{n,n-1}}$$

is valid in each nilpotent group of class ≤ 2. Similarly for $v = v'$ where

$$v' = x_1^{|v|_1} \ldots x_n^{|v|_n} [x_2, x_1]^{|v|_{2,1}} \ldots [x_n, x_1]^{|v|_{n,1}} \ldots [x_n, x_{n-1}]^{|v|_{n,n-1}}\ .$$

Since $[x_j, x_i] = x_j^{\ell-1} x_i^{\ell-1} x_j x_i$, we have

$$|u'| \equiv |u| \bmod \ell, \ |u'|_i \equiv |u|_i \bmod \ell \text{ for each } i \in \mathbb{N} \ ,$$

$$|u'|_{i,j} \equiv |u|_{i,j} \bmod r \text{ for each } i, j \in \mathbb{N}, \ i \neq j \ .$$

and similarly for v.

Now we will rewrite u' to v'. Since $[x, y]^r = 1$ literally we are done with commutators. Then we literally use $x^k = y^k$

to rewrite $x_1^{|u|_1}$ to $x_1^{|v|_1} x_2^{|u|_1-|v|_1}$ to get $x_1^{|v|_1} x_2^{|u|_1-|v|_1+|u|_2} x_3^{|u|_3} \dots x_n^{|u|_n}$,

then to rewrite $x_2^{|u|_1-|v|_1+|u|_2}$ to $x_2^{|v|_2} x_3^{|u|_1-|v|_1+|u|_2-|v|_2}$

to get $x_1^{|v|_1} x_2^{|v|_2} x_3^{|u|_1-|v|_1+|u|_2-|v|_2+|u|_3} x_4^{|u|_4} \dots x_n^{|u|_n}$ and so on .

Finally, we use $x^\ell = 1$ literally to rewrite $x_n^{|u|_1-|v|_1+\dots+|u|_{n-1}-|v|_{n-1}+|u|_n}$ to $x_n^{|v|_n}$. We succeed since $|u| \equiv |v| \bmod \ell$ literally.

(iii) : The restriction of the relation $\nu(\ell, k, r)$ onto $X_n^* \times X_n^*$ has exactly the classes $N(n; \ell, \ell'; k, k_1, \dots, k_n; r, r_{2,1}, \dots, r_{n,1}, \dots, r_{n,n-1})$. The result follows from [12] Theorem 5.1.2. $\square$

Remarks. 1. Putting $r = 1$ we get all the results concerning homomorphisms onto abelian groups.
2. The case $k = \ell = m$ corresponds to the classical varieties of nilpotent groups of the class ≤ 2.

7   Automata

Let $\mathcal{A} = (Q, A, \cdot, i, T)$ be a complete deterministic automaton, i.e. Q is a non-empty finite set of states, A is an alphabet, $\cdot : Q \times A \to Q$ is the action by letters, $i \in Q$ is the initial state, and $T \subseteq Q$ is the set of all terminal states.

For any $a \in A$, we have the action $\alpha_a : Q \to Q$, $q \mapsto q \cdot a$. We speak about an action of the *first type*. If each action of the first type is a permutation of the set Q then we can define actions of the second type in the following way. For given $a, b \in A$ we define the action $\beta_{a,b}$ as a mapping $\beta_{a,b} : Q \to Q$ by the rule

$$p \cdot \beta_{a,b} = q \quad \text{if and only if} \quad (\exists r \in Q)(r \cdot ba = p \text{ and } r \cdot ab = q) \ .$$

Because actions by a and b are permutations, the state r in the defining property is uniquely determined. Hence the previous definition is correct and moreover, the action $\beta_{a,b}$ is also a permutation of the set Q.

We say that an automaton $\mathcal{A}$ is *2-nilpotent* if and only if it is a complete deterministic automaton in which the actions of the first type are permutations and commute with the actions of the second type.

For a 2-nilpotent automaton $\mathcal{A}$, we define $\ell_{\mathcal{A}}$ as the least common multiple of the lengths of all cycles in all actions of the first type; $k_{\mathcal{A}}$ as the smallest number such that the $k_{\mathcal{A}}$-th iteration of each action of the first type gives the same mapping from Q to Q; $r_{\mathcal{A}}$ as the least common multiple of the lengths of all cycles in all actions of the second type.

Proposition 1. *Let L be a language over X_n with minimal automaton $\mathcal{A}(L)$. Then $L \in \mathcal{M}(\ell, k, r)(X_n^*)$ if and only if $\mathcal{A}(L)$ is 2-nilpotent and $\ell_{\mathcal{A}_L}|\ell$, $k_{\mathcal{A}_L}|k$ and $r_{\mathcal{A}_L}|r$.*

Proof. It is a consequence of Theorem 1 and the well-known fact that transformation monoid of a minimal automaton $\mathcal{A}(L)$ is isomorphic to the syntactic monoid of L.   □

References

1. Almeida, J.: Finite Semigroups and Universal Algebra. World Scientific, Singapore (1994)
2. Baldwin, J., Berman, J.: Varieties and finite closure conditions. Colloq. Math. 35, 15–20 (1976)
3. Carton, O., Pin, J.-E., Soler-Escrivà, X.: Languages recognized by finite supersoluble groups (submitted)
4. Eilenberg, S.: Automata, Languages and Machines, vol. A,B. Academic Press, New York (1974-1976)
5. Ésik, Z.: Extended temporal logic on finite words and wreath product of monoids with distinguished generators. In: Ito, M., Toyama, M. (eds.) DLT 2002. LNCS, vol. 2450, pp. 43–58. Springer, Heidelberg (2003)
6. Ésik, Z., Ito, M.: Temporal logic with cyclic counting and the degree of aperiodicity of finite automata. Acta Cybernetica 16, 1–28 (2003); a preprint BRICS 2001
7. Ésik, Z., Larsen, K.G.: Regular languages defined by Lindström quantifiers. Theoretical Informatics and Applications 37, 197–242 (2003); preprint BRICS 2002
8. Kunc, M.: Equational description of pseudovarieties of homomorphisms. Theoretical Informatics and Applications 37, 243–254 (2003)
9. Pin, J.-E.: Syntactic semigroups. In: Rozenberg, G., Salomaa, A. (eds.) Handbook of Formal Languages, ch. 10. Springer, Heidelberg (1997)
10. Pin, J.-E., Straubing, H.: Some results on C-varieties. Theoretical Informatics and Applications 39, 239–262 (2005)
11. Polák, L.: On varieties, generalized varieties and pseudovarieties of homomorphisms. In: Contributions to General Algebra, vol. 16, pp. 173–187. Verlag Johannes Heyn, Klagenfurt (2005)
12. Polák, L.: On varieties and pseudovarieties of homomorphisms onto abelian groups. In: Proc. International Conference on Semigroups and Languages, Lisboa 2005, pp. 255–264. World Scientific Publishing, Singapore (2007)
13. Straubing, H.: On logical descriptions of regular languages. In: Rajsbaum, S. (ed.) LATIN 2002. LNCS, vol. 2286, pp. 528–538. Springer, Heidelberg (2002)
14. Thérien, D.: Languages of Nilpotent and Solvable Groups. In: Proceedings of the 6th ICALP. LNCS, vol. 71, pp. 616–632 (1979)
15. Thérien, D.: Subwords counting and nilpotent groups. In: Cummings, L. (ed.) Combinatorics on Words, Progress and Perspectives, pp. 293–306. Academic Press, London (1983)

Characterization of Star-Connected Languages Using Finite Automata

Barbara Klunder

Faculty of Mathematics and Computer Science, Nicolaus Copernicus University
Toruń, Poland
klunder@mat.uni.torun.pl

Abstract. In this paper we characterize star-connected languages using finite automata: any language is star-connected if and only if it is accepted by a finite automaton with cycles which are proper compositions of connected cycles. Star-connected (flat) languages play an important role in the theory of recognizable languages of monoids with partial commutations. In addition we introduce a flat counterpart of concurrent star operation used in the theory of recognizable languages.

The theory of traces (i.e. monoids with partial commutations) has two independent origins: combinatorial problems and the theory of concurrent systems. Since the fundamental Mazurkiewicz's paper [6], trace languages are regarded as a powerful means for description of behaviours of concurrent systems. The formal language theory over traces, limited to recognizable and rational trace languages, is the subject of [8]. It is known that rational expressions with the classical meaning are useless for expressing recognizable trace languages. For example classical iteration $T^\star$ of a recognizable trace language T need not be recognizable.

We assume the reader knows basic notions of the theory of free monoids: finite automata, regular expressions (named, from now on, *rational expressions*) and regular languages. A language $L \subseteq A^\star$ is *recognizable* if it is recognizable by a finite automaton, and is *rational* if it is defined by a rational expression.

The classes **Rec** of recognizable and **Rat** of rational subsets are different in all trace monoids which are not free. In spite of this, rational expressions over trace monoids are proved to be quite useful for description of recognizable subsets of arbitrary trace monoids. Namely, in any trace monoid, the class **Rec** can be characterized with some special kind of rational expressions, called the star-connected rational expressions. By virtue of that fact, the role of rational expressions in trace monoids appears not less important than in free monoids.

1 Preliminaries

We assume the reader knows basics of formal language theory. Nevertheless, we recall briefly some of them, in order to fix the notation, used in the paper. The set of non-negative integers is denoted by $\mathbb{N}$.

C. Martín-Vide, F. Otto, and H. Fernau (Eds.): LATA 2008, LNCS 5196, pp. 311–320, 2008.
© Springer-Verlag Berlin Heidelberg 2008

1.1   Flat Languages and Trace Languages

An *alphabet* A is a finite set of *letters*. Strings of letters of A are *words* over A; the set of all finite words over A is denoted by $A^\star$, the empty word is denoted by ε. The set $A^\star$ with the operation of concatenation is the *free monoid*. The *length* of $w \in A^\star$ is denoted by $|w|$. The set of letters occurring in a word $w \in A^\star$ is denoted by $\mathrm{Alph}(w)$. Subsets of $A^\star$ are called *languages*. In order to avoid possible confusions with trace languages, defined below, subsets of free monoids will be called *flat languages*. Basic operations on languages are: set-theoretical operations of *union*, *intersection* and *difference*, and algebraic operations of *concatenation*: $XY = \{xy | x \in X \,\&\, y \in Y\}$, *power*: $X^0 = \{\varepsilon\}$, $X^{n+1} = XX^n$, and *iteration* (or the *star operation*): $X^\star = \bigcup\{X^n | n \geq 0\}$ (as usual $X^+ = \bigcup\{X^n | n \geq 1\}$).

Let $I \subseteq A \times A$ be a symmetric and irreflexive relation on A. Such a relation is named the *independency relation*; it expresses possibilities of concurrent executions of atomic actions of systems. The complement $D = A \times A \setminus I$ of I is named the *dependency relation*. One can extend the relations I and D onto the whole $A^\star$:

$$u I v \text{ iff } \forall a \in \mathrm{Alph}(u),\ \forall b \in \mathrm{Alph}(v)\colon a I b$$

$$u D v \text{ iff } \exists a \in \mathrm{Alph}(u),\ \exists b \in \mathrm{Alph}(v)\colon a D b$$

The pair (A, I) or (A, D) is said to be a *concurrent alphabet*. Given a concurrent alphabet (A, I), the *trace monoid* $A^\star/I$ is the quotient of the free monoid $A^\star$ by the least congruence on $A^\star$ containing the relation $\{ab = ba | aIb\}$. Members of $A^\star/I$ are called *traces*, and sets of traces (i.e. subsets of $A^\star/I$) are called *trace languages*. Operations on trace languages - *union, intersection, difference, concatenation, power* and *iteration* - are defined exactly in the same way as the respective operations on flat languages.

Let (A, I) be a concurrent alphabet. Any word $w \in A^\star$ induces a trace $[w] \in A^\star/I$ - the congruence class of w. Any flat language $L \subseteq A^\star$ induces a trace language $[L] = \{[w] | w \in L\}$ - the set of all traces induced by members of L. Any class $\mathcal{R} \subseteq 2^{A^\star}$ of flat languages induces a class of trace languages $[\mathcal{R}] \subseteq 2^{A^\star/I}$ - the set of all trace languages induced by flat languages of $\mathcal{R}$.

Let $T \subseteq A^\star/I$ be a trace language. The *flattening* of T is a flat language $\bigcup T = \{w \in A^\star | [w] \in T\}$ - the union of traces in T, viewed as subsets of $A^\star$. A flat language L is said to be *closed* (w.r.t. I) iff $L = \bigcup[L]$.

1.2   Rational and Recognizable Languages

A language $L \subseteq A^\star$ is *recognizable* if it is recognizable by a finite automaton, and is *rational* if it is defined by a rational expression. The well known Kleene Theorem says that both the notions are equivalent in finitely generated free monoids. For this reason, the classes of recognizable and rational subsets of $A^\star$ will be uniformly denoted by $\mathbf{Reg}_A$, and its members be called *regular languages*.

A trace language $T \subseteq A^\star/I$ is *rational* iff $T = [L]$ for some $L \in \mathbf{Reg}_A$, and is *recognizable* iff $\bigcup T = \{w \in A^\star | [w] \in T\} \in \mathbf{Reg}_A$. The classes of *rational* and *recognizable* trace languages in $A^\star/I$ will be denoted, respectively, by $\mathbf{Rat}_{A,I}$ and $\mathbf{Rec}_{A,I}$. Indices A and I will be omitted, if it will not lead to a confusion. It is easily seen that $\mathbf{Rec} \subseteq \mathbf{Rat}$ and the inclusion is proper, whenever $I \neq \emptyset$.

1.3 Connected Words, Traces and Languages

Definition 1 (Connected words, traces and languages). *Let (A, I) be a concurrent alphabet; let $D = A \times A \setminus I$ be the dependency relation.*

- *A word $w \in A^\star$ is connected (w.r.t. D) iff the graph $D|_{Alph(w)}$ is connected;*
- *A flat language $L \subseteq A^\star$ is connected iff all its members are connected;*
- *A trace $[w] \in A^\star/I$ is connected iff the word w is connected;*
- *A trace language $T \subseteq A^\star/I$ is connected iff all its members are connected.*

Connected trace languages play an important role in trace theory. We know that the star may destroy the recognizability. It is not the case, if the iterated language is connected. The following result is due to E. Ochmański [7] and, in a more general framework of semicommutations, to Clerbout/Latteux [1].

Proposition 1. *Let $A^\star/I$ be a trace monoid.*
If $T \subseteq A^\star/I$ is recognizable and connected, then $T^\star$ is recognizable.

Example 1. Let $(X, I) = (\{a, b\}, \{(a, b), (b, a)\})$, the trace $[ab]$ is not connected and has a decomposition $[ab] = [a][b] = [b][a]$, of course $[a], [b]$ are connected traces. The trace language $T = \{[ab]\}$ is an example of a recognizable language such that $T^\star$ is not recognizable (see [7]).

Definition 2 (Concurrent star). *Let $M = A^\star/I$ be a trace monoid and let α, γ be nonempty traces in M. The trace γ is a component of α iff γ is connected and $\alpha = \beta\gamma$ for some $\beta \in M$, such that $Alph(\beta) \times Alph(\gamma) \subseteq I$. The decomposition of a trace $\alpha \neq [\varepsilon]$ is the set $/\alpha/$ of all components of α, The decomposition of $[\varepsilon]$ is defined as $/[\varepsilon]/ = \{[\varepsilon]\}$. The decomposition of a trace language $T \subseteq M$ is the trace language $/T/ = \bigcup\{/\alpha/ \; ; \; \alpha \in T\}$. The concurrent star of T is defined as $T^\otimes = /T/^\star$.*

The decomposition operation of trace languages has the following interesting property (see [7]).

Lemma 1. *Let $A^\star/I$ be a trace monoid. If $T \subseteq A^\star/I$ is recognizable, then $/T/$ is recognizable.*

The following corollary holds

Corollary 1. *Let $A^\star/I$ be a trace monoid. If $T \subseteq A^\star/I$ is recognizable, then $T^\otimes$ is recognizable.*

2   Star-Connected Expressions and Languages

Let (A, I) be the fixed concurrent alphabet.

Definition 3 (Rational expressions and languages)

- *A rational expression (over A) is any string over the extended alphabet $A \cup \{\emptyset, \varepsilon, \cup, \cdot, \star, (,)\}$, built with the following rules: Any atomic expression (i.e., $\emptyset$, ε and a, for $a \in A$) is a rational expression. If R and S are rational expressions, then $(R \cup S)$, (RS) and $R^\star$ are rational expressions.*
- *Assuming the natural semantics for symbols $\cup$, $\cdot$ and $\star$ (union, concatenation and iteration, respectively), any rational expression R defines, in a unique way, a flat language $L(R)$.*
- *The trace language defined by rational expression R is the trace language $[L(R)]$ induced by the flat language $L(R)$.*
- *A rational expression R is said to be connected iff the flat language $L(R)$ is connected. The last condition is equivalent to "the trace language $[L(R)]$ is connected", because, for any $L \subseteq A^\star$, the flat language L is connected iff the trace language $[L]$ is connected.*

Definition 4 (Star-connected expressions and languages)

- *A rational expression (over A) is star-connected iff it has a connected expression under any of its stars. Formally: Any atomic expression ($\emptyset, \varepsilon$, and a, for $a \in A$) is star-connected. If R and S are star-connected, then $(R \cup S)$ and (RS) are star-connected as well. If, moreover, R is connected, then $R^\star$ is also star-connected. The class of star-connected (w.r.t. I) rational expressions (over A) is the least class of rational expressions built with the above rules.*
- *A flat language $L \subseteq A^\star$ is star-connected iff there is a star-connected expression R such that $L = L(R)$. The class of all star-connected (w.r.t. I) languages (over A) will be denoted by $\mathbf{StarCon}_I(A)$.*
- *A trace language $T \subseteq A^\star/I$ is star-connected iff there is a star-connected expression R such that $T = [L(R)]$. Or equivalently: iff there is a star-connected flat language $L \subseteq A^\star$ such that $T = [L]$.*

Star-connected flat languages induce the whole class of recognizable trace languages. For the proof see [7,8].

Theorem 1 $\mathbf{Rec}_{A,I} = [\mathbf{StarCon}_I(A)]$, *i.e.* $T \in \mathbf{Rec}_{A,I}$ iff $(\exists L \in \mathbf{StarCon}_I (A))$ $T = [L]$.

Example 2 Any star-connected language has infinitely many rational expressions, defining it. For instance:

$$(a \cup b)^\star = (a^\star b^\star)^\star$$
$$(a \cup b)^\star b = (a^\star \cup bb^\star a)^\star bb^\star$$

Notice that, if aIb, then the left-hand-side expressions are star-connected, whereas the right-hand-side ones are not.

Example 3 Let $X = \{x_1, \ldots, x_n\}$ be any subset of A. Let S_n denote the set of all permutations of $\{1, \ldots, n\}$. Then the expression

$$R_X = \bigcup_{\sigma \in S_n} X^* x_{\sigma(1)} X^* x_{\sigma(2)} \dots X^* x_{\sigma(n)} X^*$$

is star-connected and defines the language $L_X = \{w \in A^* | Alph(w) = X\}$.

Example 4 Let $\mathcal{C}(A, I) = \{X \subseteq A \mid (X, D|_X)$ is connected$\}$. Then

$$C_{(A,I)} = \bigcup_{X \in \mathcal{C}(A,I)} R_X$$

is star-connected expression which defines the language $Con(A, I)$ of all connected words over the alphabet (A, I).

In [4] we proved the following lemma.

Lemma 2 (About intersections) *Let L be star-connected flat language and let $X \subseteq A$. Then the flat language $L \cap L_X$ is star-connected. If L is defined by a star-connected expression R then we can effectively find the star-connected expression R_X defining $L \cap L_X$.*

From Example 4 we obtain immediately the following corollary.

Corollary 2 *Let L be any star-connected flat language. Let X be any subset of A. Then the flat languages $Con(X, I) = X^* \cap Con(A, I), L \cap Con(X, I)$ are star-connected.*

Using the last corollary we shall characterize automata accepting star-connected languages.

3   Automata Accepting Star-Connected Languages

A *finite automaton* $\mathbf{A} =< A, Q, \delta, q, F >$ *(with ε-transitions)* over A consists of a finite set Q of states, a finite set $\delta \subseteq Q \times A \times Q$ ($\delta \subseteq Q \times (A \cup \{\varepsilon\}) \times Q$) of transitions, an initial state $q \in Q$ and a set $F \subseteq Q$ of final states. A *path* of $\mathbf{A}$ is a sequence $s_0 a_1 s_1 a_2 s_2 \dots a_n s_n$ such that $n > 0$ and $(s_i, a_{i+1}, s_{i+1}) \in \delta$ for all $i = 0, \dots n - 1$. The word $w = a_1 a_2 \dots a_n$ is called a *label* of the path. $\mathbf{A}$ defines the language $L(\mathbf{A} =< A, Q, \delta, q, F >)$ of labels of all paths starting in q and ending in some state $f \in F$. When F is a singleton ($F = \{f\}$) we identify it with the element f.

Let $I \subseteq A \times A$ be an independency on A.

Definition 5 (Composition of cycles) *Let $\mathbf{A} =< A, Q, \delta, q, F >$ be an automaton. A cycle of $\mathbf{A}$ is a path $s_0 a_1 s_1 a_2 s_2 \dots a_n s_n$ such that $n > 0$ and $s_0 = s_n$; a cycle is simple if $s_i \neq s_j$ whenever $i \neq j$ (for $i, j = 0, \dots, n - 1$). A cycle is connected if its label w is a connected word. If for some $k \geq 0$ and $i_0 = 0 < i_1 < \dots < i_k < i_{k+1} = n$ equations $s_{i_j} = s_0$ hold the cycle is a composition in s_0 of cycles $s_{i_j} a_{i_j+1} \dots s_{i_{j+1}}$ for $0 \leq j \leq k$. If the sequence $i_0 = 0 < i_1 < \dots < i_k < i_{k+1} = n$ contains all occurrences of s_0 then the composition is proper.*

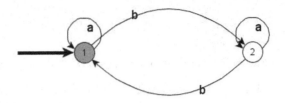

Fig. 1. Automaton with connected simple cycles. Let aIb. Then every simple cycle of **A** is connected, but the cycle with a label $ababa$ starting in initial state 1 is not isomorphic to any cycle which is a proper composition of connected cycles. We can show that the language $L(\mathbf{A})$ of all words with even occurrence of b is not star-connected.

For every $0 < i < n$ the cycle $s_i a_{i+1} \ldots a_n s_n a_1 s_1 \ldots a_i s_i$ is *isomorphic* to a cycle $s_0 a_1 s_1 a_2 s_2 \ldots a_n s_n$ of a given automaton **A**.

It is easy to see that for any $R \subseteq A^\star$, R is star-connected iff $R \setminus \{\varepsilon\}$ is star-connected.

Let us denote $\mathbf{StarCon}_I(A^+) = \{R \subseteq A^\star | R$ is star-connected and $\varepsilon \notin R\}$. Then $\mathbf{StarCon}_I(A^+)$ is the least family of A-languages containing all singletons, except $\{\varepsilon\}$, and closed under union, concatenation and $^+$-operation applied to connected languages.

Lemma 3 *For every star-connected language $R \in \mathbf{StarCon}_I(A^+)$ there exists an automaton $\mathbf{A} =< A, Q, \delta, q, f >$ with two different states $q, f \in Q$, initial and final, respectively, such that*

1. *$R = L(\mathbf{A})$, i.e. R is accepted by $\mathbf{A}$;*
2. *every simple cycle of $\mathbf{A}$ is connected;*
3. *for every cycle $s_0 a_1 s_1 a_2 s_2 \ldots a_n s_n$ there exist $0 \leq i \leq n$ such that the isomorphic cycle $s_i a_{i+1} \ldots a_n s_n a_1 s_1 \ldots a_i s_i$ is a proper composition in s_i of connected cycles;*
4. *for every cycle $s_0 a_1 s_1 a_2 s_2 \ldots a_n s_n$ such that $s_i = f$ for some $0 \leq i \leq n$ the isomorphic cycle $s_i a_{i+1} \ldots a_n s_n a_1 s_1 \ldots a_i s_i$ is a proper composition in s_i of connected cycles.*

Proof The construction of $\mathbf{A}$ is inductive. It adopts the classical inductive construction (see [3], for instance) to obtain an automaton without ε-transitions. Because the same argument was applied in [5] we only consider the case of $^+$ operation.

Let $\mathbf{A} =< A, Q, \delta, q, f >$ accept a connected language $S \in \mathbf{StarCon}_I(A^+)$, and assume that $\mathbf{A}$ satisfies all conditions of the lemma. Let $a \in A$, $S^a = \{s \in Q | (q, a, s) \in \delta\}$ and

$$\delta^+ = \delta \cup \delta_+ \text{ where } \delta_+ = \bigcup_{a \in A} \{(f, a, s) | s \in S^a\}$$

Then $\mathbf{A}^+ =< A, Q, \delta^+, q, f >$ accepts S^+. Every simple cycle $fasa_1 \ldots f$ with a transition (f, a, s) is either a cycle of $\mathbf{A}$ or its label aw is accepted by $\mathbf{A}$ so every simple cycle of $\mathbf{A}^+$ is connected. Let $\pi : s_0 a_1 s_1 a_2 s_2 \ldots a_n s_0$ be a cycle of $\mathbf{A}^+$.

If π does not contain transitions from $\delta^+ \setminus \delta$, then it obviously satisfies the last condition. So we can assume that the set $F = \{i \le n | s_i = f \wedge (f, a_{i+1}, s_{i+1}) \in \delta^+ \setminus \delta\}$ is nonempty and $0 \in F$ i.e. $s_0 = f$. Let $F_1 = \{i_1, \ldots, i_k\}$, $i_j < i_{j+1}$ for $0 < j < k$, be the set of all occurrences of f in π. We claim that π is a composition in f of connected cycles. Indeed, for every $j < k$ such that $s_{i_j} \in F$ the label of a cycle $s_{i_j} a_{i_j+1} \ldots a_{i_{j+1}} s_{i_{j+1}}$ is accepted by $\mathbf{A}$. If $s_{i_j} \in F_1 \setminus F$ then $s_{i_j} a_{i_j+1} \ldots a_{i_{j+1}} s_{i_{j+1}}$ is a cycle of $\mathbf{A}$ such that f occurs in it only on positions i_j and i_{j+1}. Thus this cycle is connected by the induction hypothesis (4).

In this way we obtain stronger version of Lemma 3.1 [4]. This result implies the following modification of the classical pumping lemma for a star-connected language L (see [4]):

There exist $m > 0$ such that for all $z \in L$ if $|z| \ge m$ then for some $u, v_1, v_2, v_3, w \in A^*$: $z = uv_1v_2v_3w$, words v_2, v_3v_1 are nonempty and connected, $uv_1(v_2 \cup v_3v_1)^* v_3 w \subseteq L$.

In this assertion we can require that $|v_1v_2v_3| \le m$. Using this property of star-connected languages it is easy to prove that the language $L = (a \cup ba^*b)^*$ is not star-connected. For $m > 0$ the word $(ab)^{2m}$ does not satisfy this condition.

Definition 6 *Let X be any subset of A and $\pi_X : A \mapsto X^*$ be defined by the equation:*

$$\pi_X(a) = \begin{cases} a & \text{if } a \in X \\ \varepsilon & \text{otherwise.} \end{cases}$$

Then the unique extension of π_X to A^* we call the projection (on X) and denote in the same way.

Now we can formulate the fundamental result of this section.

Proposition 2 *Let (A, I) be any concurrent alphabet. Let $X \subseteq A$ be such that X and $A \setminus X$ are independent: $X \times (A \setminus X) \subseteq I$. Let $\mathbf{A} = <A, Q, \delta, q, F>$ be any automaton satisfying conditions 2 and 3 of Lemma 3. Then the language $\pi_X(L(\mathbf{A}))$ is star-connected.*

Proof Let us define for $S \subseteq Q$ and $i, j \in Q$ the set L_{ij}^S of all projections of labels of paths of $\mathbf{A}$ starting in i, ending in j and going through the states of S. We shall prove that all those languages are star-connected by the induction on cardinality of S. Clearly

$$L_{ij}^{\emptyset} = \begin{cases} \{\pi_X(a) | a \in A \wedge (i, a, j) \in \delta\} & \text{if } i \ne j \\ \{\pi_X(a) | a \in A \wedge (i, a, j) \in \delta\} \cup \{\varepsilon\} & \text{otherwise.} \end{cases}$$

and if $S \ne \emptyset$

$$L_{ij}^S = \bigcup_{k \in S} L_{ik}^{S \setminus \{k\}} (L_{kk}^{S \setminus \{k\}})^* L_{kj}^{S \setminus \{k\}} \cup \bigcup_{k \in S} L_{ij}^{S \setminus \{k\}}$$

But in the induction step we have to prove the languages $(L_{kk}^{S\setminus\{k\}})^\star$ are star-connected. We claim that

$$(L_{kk}^{S\setminus\{k\}})^\star = \bigcup_{j\in S} L_{kj}^{S\setminus\{j\}}(L_{jj}^{S\setminus\{j\}} \cap Con(X,I))^\star L_{jk}^{S\setminus\{j\}}$$

Indeed, $(L_{kk}^{S\setminus\{k\}})^\star$ is in fact the set of projections of all labels of cycles starting in k and visiting only states of S. Every such cycle is isomorphic to a cycle which is a proper composition in some state j of connected cycles, by Lemma 3 (3). If $X \times (A \setminus X) \subseteq I$ then for every connected word w (being a label of a cycle) either $Alph(w) \subseteq X$ or $Alph(w) \subseteq A \setminus X$. Finally, by the induction hypothesis and Corollary 2 the intersection of $L_{jj}^{S\setminus\{j\}}$ with the language $Con(X,I)$ of all connected words over the alphabet X is star-connected.

As a consequence of Lemma 3 and Proposition 2 we obtain the main theorem.

Theorem 2 (about characterization) *Let (A,I) be any concurrent alphabet. Then the language $L \subseteq A^\star$ is star-connected iff L is accepted by some automaton satisfying conditions 2 and 3 of Lemma 3.*

4   Concurrent Star Operation

Let $w \in A^\star$ and X,Y be two independent subsets of $Alph(w)$ ($X \times Y \subseteq I$) such that $X \cup Y = Alph(w)$. It is easy to see that if $\pi_X(w) = u$ is connected then $[u]$ is a component of $[w]$. On the other hand every component of the trace $[w]$ can be obtained in this way. But Proposition 2 allows us to introduce flat counterpart of decomposition and concurrent star operation (see Definition 2).

Lemma 4 *Let L be star-connected flat language. Then the following languages are star-connected:*

1. *the decomposition of L:*

$$/L/ = \bigcup_{\{X\in\mathcal{C}(A,I)|X\times(A\setminus X)\subseteq I\}} \pi_X(L) \cap L_X$$

2. *the concurrent iteration of L: $L^\otimes = /L/^\star$.*

If L is defined by a star-connected expression R then we can effectively find star-connected expressions R_d, R_c such that $L(R_d) = /L/$ and $L(R_c) = L^\otimes$.

The poof of the above lemma follows easily from the previous considerations. Because we characterized star-connected languages in terms of transitions of some automaton accepting it we can expect that the set $\mathbf{StarCon}_I(A)$ is closed under some classical operations too.

Corollary 3 *Let $L \in \mathbf{StarCon}_I(A)$ and $M \subseteq A^\star$. Then the language $L/M = \{w|\exists_{v \in M} wv \in L\}$ is star-connected.*

The proof is easy if we realize that it is enough to change the set of accepting states of automaton accepting L (and satisfying assertions of Lemma 3).

On the other hand it is natural to ask if the property characterizing automata accepting star-connected languages is preserved in the process of determinization or minimalization. The positive answer to this question implies the decidability of the following problem:

Is a regular language L star-connected?

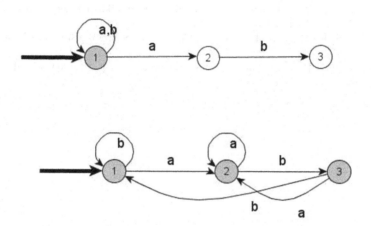

Fig. 2. Deterministic automaton without connected simple cycles. Let aIb. The second automaton (without connected cycles) is the result of determinization of the first one.

The automaton given on the next figure shows that the property characterizing automata accepting star-connected languages is not preserved in the process of minimalization.

The status of the problem "Is a regular language L star-connected?" is unknown.

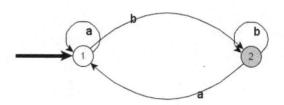

Fig. 3. Minimal deterministic automaton without connected simple cycles. Let aIb. Then the simple cycle $1b2a1$ of $\mathbf{A}$ is not connected, but the language accepted by $\mathbf{A}$ of all words ending with b is defined by the following star-connected expression: $(a \cup b)^* b$.

References

1. Clerbout, M., Latteux, M.: Semi-commutations. Information & Computation 73, 59–74 (1987)
2. Diekert, V., Rozenberg, G. (eds.): The Book of Traces. World Scientific, Singapore (1995)
3. Hopcroft, J.E., Motwani, R., Ullman, J.D.: Introduction to Automata Theory, Languages and Computation. Addison-Wesley, Reading (2001)
4. Klunder, B.: Star-Connected Flat Languages and Automata. Fundamenta Informaticae 72(1-3), 235–243 (2006)
5. Klunder, B., Ochmański, E., Stawikowska, K.: On Star-Connected Flat Languages. Fundamenta Informaticae 67(1-3), 93–105 (2005)
6. Mazurkiewicz, A.: Concurrent Program Schemes and Their Interpretations. Report DAIMI-PB-78, Aarhus University (1977)
7. Ochmański, E.: Regular Behaviour of Concurrent Systems. Bulletin of EATCS 27, 56–67 (1985)
8. Ochmański, E.: Recognizable Trace Languages. In: [2], pp. 167–204. World Scientific, Singapore (1995)

Match-Bounds with Dependency Pairs for Proving Termination of Rewrite Systems

Martin Korp and Aart Middeldorp

Institute of Computer Science
University of Innsbruck
Austria

Abstract. The match-bound technique is a recent and elegant method to prove the termination of rewrite systems using automata techniques. To increase the applicability of the method we incorporate it into the dependency pair framework. The key to this is the introduction of two new enrichments which take the special properties of dependency pair problems into account.

1 Introduction

The use of word automata for proving the termination of string rewrite systems was proposed by Geser, Hofbauer, and Waldmann [5]. In [8] tree automata were used to cover left-linear term rewrite systems. We extended the latter work to arbitrary term rewrite systems in [17] by considering so-called quasideterministic tree automata. Variations and improvements of the basic method for string rewrite systems are discussed in [6,7]. The fact that the method has been implemented in several different termination provers ([10,16,21,22,23]) is a clear witness of the success of the automata based approach.

In this paper we integrate the method into the dependency pair framework [11,20]. To guarantee a successful integration we need to modularise the method in order to be able to simplify dependency pair problems. We achieve this by introducing two new enrichments which exploit the special properties of dependency pair problems.

The remainder of the paper is organised as follows. In the next section we recall basic definitions concerning the automata theory approach for proving termination and the dependency pair framework. In Section 3 we introduce the concept of e-DP-bounds which is based on two new enrichments that allow us to simplify dependency pair problems. In Section 4 we consider right-hand sides of forward closures to reduce the set of terms which have to be considered. To simplify the discussion, we restrict ourselves to left-linear rewrite systems. The extension to non-left-linear term rewrite systems is sketched in Section 5. Experimental data is presented in Section 6. Missing proofs can be found in the full version, which is available from `http://cl-informatik.uibk.ac.at/~mkorp/`

C. Martín-Vide, F. Otto, and H. Fernau (Eds.): LATA 2008, LNCS 5196, pp. 321–332, 2008.
© Springer-Verlag Berlin Heidelberg 2008

2 Preliminaries

We assume familiarity with term rewriting [2] and tree automata [3]. General knowledge of the match-bound technique [5,8] and dependency pairs [1,11,13,15] will be helpful.

Match-Bounds

Let $\mathcal{F}$ be a signature, $\mathcal{R}$ a term rewrite system (TRS for short) over $\mathcal{F}$, and $L \subseteq \mathcal{T}(\mathcal{F})$ a set of ground terms. The set $\{t \in \mathcal{T}(\mathcal{F}) \mid s \to_{\mathcal{R}}^* t \text{ for some } s \in L\}$ of descendants of L is denoted by $\to_{\mathcal{R}}^*(L)$. Given a set $N \subseteq \mathbb{N}$ of natural numbers, the signature $\mathcal{F} \times N$ is denoted by $\mathcal{F}_N$. Here function symbols (f, n) with $f \in \mathcal{F}$ and $n \in N$ have the same arity as f and are written as f_n. The mappings $\mathrm{lift}_c \colon \mathcal{F} \to \mathcal{F}_{\mathbb{N}}$, $\mathrm{base} \colon \mathcal{F}_{\mathbb{N}} \to \mathcal{F}$, and $\mathrm{height} \colon \mathcal{F}_{\mathbb{N}} \to \mathbb{N}$ are defined as $\mathrm{lift}_c(f) = f_c$, $\mathrm{base}(f_i) = f$, and $\mathrm{height}(f_i) = i$ for all $f \in \mathcal{F}$ and $c, i \in \mathbb{N}$. They are extended to terms and to sets of terms in the obvious way. Let $\mathcal{R}$ be a TRS over the signature $\mathcal{F}$ and e a function that maps every rewrite rule $l \to r \in \mathcal{R}$ to a nonempty subset of $\mathcal{F}Pos(l)$, where $\mathcal{F}Pos(l) = \{p \in Pos(l) \mid l(p) \in \mathcal{F}\}$. The TRS $e(\mathcal{R})$ over the signature $\mathcal{F}_{\mathbb{N}}$ consists of all rewrite rules $l' \to \mathrm{lift}_c(r)$ for which there exists a rule $l \to r \in \mathcal{R}$ such that $\mathrm{base}(l') = l$ and $c = 1 + \min\{\mathrm{height}(l'(p)) \mid p \in e(l, r)\}$. Let $c \in \mathbb{N}$. The restriction of $e(\mathcal{R})$ to the signature $\mathcal{F}_{\{0,\ldots,c\}}$ is denoted by $e_c(\mathcal{R})$. We consider three concrete functions e in the following: $\mathrm{top}(l \to r) = \{\epsilon\}$, $\mathrm{roof}(l \to r) = \{p \in \mathcal{F}Pos(l) \mid Var(r) \subseteq Var(l|_p)\}$, and $\mathrm{match}(l \to r) = \mathcal{F}Pos(l)$. Let $e \in \{\mathrm{top}, \mathrm{roof}, \mathrm{match}\}$ and L a set of terms. The TRS $\mathcal{R}$ is called e-bounded for L if there exists a $c \in \mathbb{N}$ such that the maximum height of function symbols occurring in terms in $\to_{e(\mathcal{R})}^*(\mathrm{lift}_0(L))$ is at most c. If we want to precise the bound c, we say that $\mathcal{R}$ is e-bounded for L by c. If we do not specify the set of terms L then it is assumed that $L = \mathcal{T}(\mathcal{F})$.

Theorem 1 (Geser et al. [8]). *If a left-linear TRS $\mathcal{R}$ is top-bounded, roof-bounded, or both right-linear and match-bounded for a language L then $\mathcal{R}$ is terminating on L.* $\qquad\square$

In order to prove that a TRS $\mathcal{R}$ is e-bounded for some language L and some $e \in \{\mathrm{top}, \mathrm{roof}, \mathrm{match}\}$, the idea is to construct a tree automaton that is compatible with $e(\mathcal{R})$ and $\mathrm{lift}_0(L)$. A tree automaton $\mathcal{A} = (\mathcal{F}, Q, Q_f, \Delta)$ is said to be *compatible* with some TRS $\mathcal{R}$ and some language L if $L \subseteq \mathcal{L}(\mathcal{A})$ and for each rewrite rule $l \to r \in \mathcal{R}$ and state substitution $\sigma \colon Var(l) \to Q$ such that $l\sigma \to_{\Delta}^* q$ it holds that $r\sigma \to_{\Delta}^* q$.

Dependency Pairs

The dependency pair method [1] is a powerful approach for proving termination of TRSs. The dependency pair framework [12,20] is a modular reformulation and improvement of this approach. We present a simplified version which is sufficient for our purposes.

Let $\mathcal{R}$ be a TRS over a signature $\mathcal{F}$. The signature $\mathcal{F}$ is extended with symbols $f^\sharp$ for every symbol $f \in \{\mathrm{root}(l) \mid l \to r \in \mathcal{R}\}$, where $f^\sharp$ has the same arity as f, resulting in the signature $\mathcal{F}^\sharp$. If $t \in \mathcal{T}(\mathcal{F}, \mathcal{V})$ with $\mathrm{root}(t)$ defined then $t^\sharp$ denotes the term that is obtained from t by replacing its root symbol with $\mathrm{root}(t)^\sharp$. If $l \to r \in \mathcal{R}$ and t is a subterm of r with a defined root symbol that is not a proper subterm of l then the rule $l^\sharp \to t^\sharp$ is a *dependency pair* of $\mathcal{R}$. The set of dependency pairs of $\mathcal{R}$ is denoted by $\mathrm{DP}(\mathcal{R})$. A *DP problem* is a pair of TRSs $(\mathcal{P}, \mathcal{R})$ such that symbols in $\{\mathrm{root}(l), \mathrm{root}(r) \mid l \to r \in \mathcal{P}\}$ do neither occur in $\mathcal{R}$ nor in proper subterms of the left and right-hand sides of rules in $\mathcal{P}$. The problem is said to be *finite* if there is no infinite sequence $s_1 \xrightarrow{\epsilon}_{\mathcal{P}} t_1 \xrightarrow{*}_{\mathcal{R}} s_2 \xrightarrow{\epsilon}_{\mathcal{P}} t_2 \xrightarrow{*}_{\mathcal{R}} \cdots$ such that all terms $t_1, t_2, \ldots$ are terminating with respect to $\mathcal{R}$. Such an infinite sequence is said to be *minimal*. Here the ϵ in $\xrightarrow{\epsilon}_{\mathcal{P}}$ denotes that the application of the rule in $\mathcal{P}$ takes place at the root position. The main result underlying the dependency pair approach states that a TRS $\mathcal{R}$ is terminating if and only if the DP problem $(\mathrm{DP}(\mathcal{R}), \mathcal{R})$ is finite.

In order to prove finiteness of a DP problem a number of so-called *DP processors* have been developed. DP processors are functions that take a DP problem as input and return a set of DP problems as output. In order to be employed to prove termination they need to be *sound*, that is, if all DP problems in a set returned by a DP processor are finite then the initial DP problem is finite. In addition, to ensure that a DP processor can be used to prove non-termination it must be *complete* which means that if one of the DP problems returned by the DP processor is not finite then the original DP problem is not finite.

3   DP-Bounds

To prove finiteness of a DP problem $(\mathcal{P}, \mathcal{R})$ it must be shown that it admits no minimal rewrite sequence. This is done by removing step by step those rewrite rules in $\mathcal{P}$ which cannot be used infinitely often in any minimal rewrite sequence. In each step a different DP processor can be applied. As soon as $\mathcal{P}$ is empty, we can conclude that the DP problem $(\mathcal{P}, \mathcal{R})$ is finite.

It is easy to incorporate the match-bound technique into the DP framework by defining a processor that checks for e-boundedness of $\mathcal{P} \cup \mathcal{R}$.

Theorem 2. *The DP processor*

$$(\mathcal{P}, \mathcal{R}) \mapsto \begin{cases} \varnothing & \text{if } \mathcal{P} \cup \mathcal{R} \text{ is left-linear and either top-bounded,} \\ & \text{roof-bounded, or both linear and match-bounded} \\ & \text{for } \mathcal{T}(\mathcal{F}) \\ \{(\mathcal{P}, \mathcal{R})\} & \text{otherwise} \end{cases}$$

where $\mathcal{F}$ is the signature of $\mathcal{P} \cup \mathcal{R}$, is sound and complete.   □

This DP processor either succeeds by proving that the combined TRS $\mathcal{P} \cup \mathcal{R}$ is e-bounded or, when the e-boundedness of $\mathcal{P} \cup \mathcal{R}$ cannot be proved, it returns the initial DP problem. Since the construction of a compatible tree automaton does not terminate for TRSs that are not e-bounded, the latter situation typically does not happen. Hence the DP processor of Theorem 2 is applicable only at the leaves of the DP search tree, which means that it can be used only as a last option in a termination proving strategy. So it cannot cooperate with other DP processors.

Below we address this problem by adapting the match-bound technique in such a way that it can remove single rules of $\mathcal{P}$. We introduce two new enrichments top-DP$(\mathcal{P}, s \to t, \mathcal{R})$ and match-DP$(\mathcal{P}, s \to t, \mathcal{R})$ to achieve this. The basic idea behind these TRSs is that every height increasing infinite sequence descends from an infinite sequence of $(\mathcal{P}, \mathcal{R})$ in which the rule $s \to t$, which is to be removed from $\mathcal{P}$, is used infinitely often.

To simplify the presentation we consider only left-linear TRSs. The extension to non-left-linear TRSs is briefly discussed in Section 5.

Definition 3. *Let $\mathcal{S}$ be a TRS over a signature $\mathcal{F}$. The TRS e-DP$(\mathcal{S})$ over the signature $\mathcal{F}_\mathbb{N}$ consists of all rules $l' \to \mathrm{lift}_c(r)$ such that base$(l') \to r \in \mathcal{S}$ and*

$$c = \min\left(\{\mathrm{height}(l'(\epsilon))\} \cup \{1 + \mathrm{height}(l'(p)) \mid p \in e(\mathrm{base}(l'), r)\}\right)$$

Given a DP problem $(\mathcal{P}, \mathcal{R})$ and a rule $s \to t \in \mathcal{P}$, the TRS e-DP$(\mathcal{P}, s \to t, \mathcal{R})$ is defined as the union of e-DP$((\mathcal{P} \setminus \{s \to t\}) \cup \mathcal{R})$ and $e(s \to t)$. The restriction of e-DP$(\mathcal{P}, s \to t, \mathcal{R})$ to the signature $\mathcal{F}_{\{0,\ldots,c\}}$ is denoted by e-DP$_c(\mathcal{P}, s \to t, \mathcal{R})$.

Example 4. Consider the DP problem $(\mathcal{P}, \mathcal{R})$ with $\mathcal{R}$ consisting of the rewrite rules $\mathsf{f}(\mathsf{g}(x), y) \to \mathsf{g}(\mathsf{h}(x, y))$ and $\mathsf{h}(x, y) \to \mathsf{f}(x, \mathsf{g}(y))$, and $\mathcal{P} = \mathrm{DP}(\mathcal{R})$ consisting of $\mathsf{F}(\mathsf{g}(x), y) \to \mathsf{H}(x, y)$ and $\mathsf{H}(x, y) \to \mathsf{F}(x, \mathsf{g}(y))$. Let $s \to t$ be the first of the two dependency pairs. Then match-DP$(\mathcal{R})$ contains the rules

$$\mathsf{f}_0(\mathsf{g}_0(x), y) \to \mathsf{g}_0(\mathsf{h}_0(x, y)) \qquad\qquad \mathsf{h}_0(x, y) \to \mathsf{f}_0(x, \mathsf{g}_0(y))$$
$$\mathsf{f}_0(\mathsf{g}_1(x), y) \to \mathsf{g}_0(\mathsf{h}_0(x, y)) \qquad\qquad \mathsf{h}_1(x, y) \to \mathsf{f}_1(x, \mathsf{g}_1(y))$$
$$\mathsf{f}_2(\mathsf{g}_0(x), y) \to \mathsf{g}_1(\mathsf{h}_1(x, y)) \qquad\qquad \cdots$$

match-DP$(\mathcal{P} \setminus \{s \to t\})$ contains

$$\mathsf{H}_0(x, y) \to \mathsf{F}_0(x, \mathsf{g}_0(y)) \qquad\qquad \mathsf{H}_1(x, y) \to \mathsf{F}_1(x, \mathsf{g}_1(y))$$
$$\mathsf{H}_2(x, y) \to \mathsf{F}_2(x, \mathsf{g}_2(y)) \qquad\qquad \cdots$$

and match$(s \to t)$ contains

$$\mathsf{F}_0(\mathsf{g}_0(x), y) \to \mathsf{H}_1(x, y) \qquad\qquad \mathsf{F}_1(\mathsf{g}_0(x), y) \to \mathsf{H}_1(x, y)$$
$$\mathsf{F}_0(\mathsf{g}_1(x), y) \to \mathsf{H}_1(x, y) \qquad\qquad \cdots$$

The union of these three infinite TRSs constitutes match-DP$(\mathcal{P}, s \to t, \mathcal{R})$. If we replace match$(s \to t)$ by match-DP$(\{s \to t\})$, which consists of the rules

$$\mathsf{F}_0(\mathsf{g}_0(x), y) \to \mathsf{H}_0(x, y) \qquad\qquad \mathsf{F}_1(\mathsf{g}_0(x), y) \to \mathsf{H}_1(x, y)$$
$$\mathsf{F}_0(\mathsf{g}_1(x), y) \to \mathsf{H}_0(x, y) \qquad\qquad \cdots$$

we obtain the TRS match-DP$(\mathcal{P} \cup \mathcal{R})$.

The idea now is to use the enrichment $e\text{-DP}(\mathcal{P}, s \to t, \mathcal{R})$ to simplify the DP problem $(\mathcal{P}, \mathcal{R})$ into $(\mathcal{P} \setminus \{s \to t\}, \mathcal{R})$. For that we need the property defined below.

Definition 5. *Let $(\mathcal{P}, \mathcal{R})$ be a DP problem and let $s \to t \in \mathcal{P}$. We call $(\mathcal{P}, \mathcal{R})$ e-DP-bounded for $s \to t$ and a set of terms L if there exists a $c \in \mathbb{N}$ such that the height of function symbols occurring in terms in $\to^*_{e\text{-DP}(\mathcal{P}, s \to t, \mathcal{R})}(\mathrm{lift}_0(L))$ is at most c.*

Moreover, we need to ensure that every restriction of $e\text{-DP}(\mathcal{P}, s \to t, \mathcal{R})$ to a finite signature does not admit minimal rewrite sequences with infinitely many $\xrightarrow{\epsilon}_{e(s \to t)}$ rewrite steps. For $e = \text{top}$ this is shown below. Note that if we would use $e\text{-DP}(\mathcal{P} \cup \mathcal{R})$ instead of $e\text{-DP}(\mathcal{P}, s \to t, \mathcal{R})$ then this property would not hold because every rewrite sequence in $\mathcal{P} \cup \mathcal{R}$ can be simulated by an $e\text{-DP}_0(\mathcal{P} \cup \mathcal{R})$-sequence.

Lemma 6. *Let $(\mathcal{P}, \mathcal{R})$ be a DP problem, let $s \to t \in \mathcal{P}$, and let $c \geqslant 0$. The TRS $\text{top-DP}_c(\mathcal{P}, s \to t, \mathcal{R})$ does not admit rewrite sequences with infinitely many $\xrightarrow{\epsilon}_{\text{top}(s \to t)}$ rewrite steps.*

Proof. Assume to the contrary that there is such an infinite rewrite sequence

$$s_1 \xrightarrow{\epsilon}_{\text{top}(s \to t)} t_1 \to^*_{\text{top-DP}((\mathcal{P} \setminus \{s \to t\}) \cup \mathcal{R})} s_2 \xrightarrow{\epsilon}_{\text{top}(s \to t)} t_2 \to^*_{\text{top-DP}((\mathcal{P} \setminus \{s \to t\}) \cup \mathcal{R})} \cdots$$

Because the root symbols in $\mathcal{P}$ do not appear anywhere else in $\mathcal{P}$ or $\mathcal{R}$, we know that only rewrite rules from $\text{top-DP}(\mathcal{P} \setminus \{s \to t\})$ and $\text{top}(s \to t)$ are applied at root positions. Every rewrite rule $l \to r$ in $\text{top-DP}(\mathcal{P} \setminus \{s \to t\})$ has the property that $\text{height}(l(\epsilon)) = \text{height}(r(\epsilon))$. Hence $\text{height}(t_i(\epsilon)) = \text{height}(s_{i+1}(\epsilon))$ for all $i \geqslant 1$. By definition, for every $l \to r \in \text{top}(s \to t)$ we have $\text{height}(r(\epsilon)) = \text{height}(l(\epsilon)) + 1$ and thus $\text{height}(t_i(\epsilon)) = \text{height}(s_i(\epsilon)) + 1$ for all $i \geqslant 1$. It follows that $\text{height}(t_{c+1}(\epsilon)) \geqslant c + 1$, contradicting the assumption. $\square$

Theorem 7. *Let $(\mathcal{P}, \mathcal{R})$ be a DP problem and let $s \to t \in \mathcal{P}$ such that $(\mathcal{P}, \mathcal{R})$ is top-DP-bounded for $s \to t$ and a set of terms L. If $\mathcal{P} \cup \mathcal{R}$ is left-linear then $(\mathcal{P}, \mathcal{R})$ is finite for L if and only if $(\mathcal{P} \setminus \{s \to t\}, \mathcal{R})$ is finite for L.*

Proof. The only-if direction is trivial. For the if direction, suppose that the DP problem $(\mathcal{P} \setminus \{s \to t\}, \mathcal{R})$ is finite for L. If $(\mathcal{P}, \mathcal{R})$ is not finite for L then there exists a minimal rewrite sequence

$$s_1 \xrightarrow{\epsilon}_{s \to t} t_1 \to^*_{(\mathcal{P} \setminus \{s \to t\}) \cup \mathcal{R}} s_2 \xrightarrow{\epsilon}_{s \to t} t_2 \to^*_{(\mathcal{P} \setminus \{s \to t\}) \cup \mathcal{R}} s_3 \xrightarrow{\epsilon}_{s \to t} \cdots$$

with $s_1 \in L$. Due to left-linearity, this sequence can be lifted to an infinite $\text{top-DP}(\mathcal{P}, s \to t, \mathcal{R})$ rewrite sequence starting from $\mathrm{lift}_0(s_1)$. Since the original sequence contains infinitely many $\xrightarrow{\epsilon}_{s \to t}$ rewrite steps the lifted sequence contains infinitely many $\xrightarrow{\epsilon}_{\text{top}(s \to t)}$ rewrite steps. Moreover, because $(\mathcal{P}, \mathcal{R})$ is top-DP-bounded for L, there is a $c \geqslant 0$ such that the height of every function symbol occurring in a term in the lifted sequence is at most c. Hence the employed rules must come from $\text{top-DP}_c(\mathcal{P}, s \to t, \mathcal{R})$ and therefore $\text{top-DP}_c(\mathcal{P}, s \to t, \mathcal{R})$ contains an infinite rewrite sequence consisting of infinitely many $\xrightarrow{\epsilon}_{\text{top}(s \to t)}$ rewrite steps. This however is excluded by Lemma 6. $\square$

If we restrict Lemma 6 to *minimal* rewrite sequences, it also holds for $e =$ match provided $\mathcal{P}$ and $\mathcal{R}$ are non-duplicating. The proof is considerably more complicated and omitted for reasons of space.

Lemma 8. *Let $(\mathcal{P}, \mathcal{R})$ be a DP problem, let $s \rightarrow t \in \mathcal{P}$, and let $c \geqslant 0$. If $\mathcal{P} \cup \mathcal{R}$ is non-duplicating then the TRS $\text{match-DP}_c(\mathcal{P}, s \rightarrow t, \mathcal{R})$ does not admit minimal rewrite sequences with infinitely many $\xrightarrow{\epsilon}_{\text{match}(s \rightarrow t)}$ rewrite steps.* □

Theorem 9. *Let $(\mathcal{P}, \mathcal{R})$ be a DP problem and let $s \rightarrow t \in \mathcal{P}$ such that $(\mathcal{P}, \mathcal{R})$ is match-DP-bounded for $s \rightarrow t$ and a set of terms L. If $\mathcal{P} \cup \mathcal{R}$ is linear then $(\mathcal{P}, \mathcal{R})$ is finite for L if and only if $(\mathcal{P} \setminus \{s \rightarrow t\}, \mathcal{R})$ is finite for L.*

Proof. Similarly to the proof of Theorem 7, using Lemma 8 instead of Lemma 6. Note that in the presence of left-linearity, the non-duplicating requirement in Lemma 8 is equivalent to linearity. □

We conjecture that Lemma 6 also holds for $e = $ roof. A positive solution is important as roof-bounds are strictly more powerful than top-bounds [8,17].

Theorem 10. *The DP processor*

$$(\mathcal{P}, \mathcal{R}) \mapsto \begin{cases} \{(\mathcal{P} \setminus \{s \rightarrow t\}, \mathcal{R})\} & \textit{if } (\mathcal{P}, \mathcal{R}) \textit{ is left-linear and top-DP-bounded} \\ & \textit{or linear and match-DP-bounded for } s \rightarrow t \\ & \textit{and } \mathcal{T}(\mathcal{F}) \\ \{(\mathcal{P}, \mathcal{R})\} & \textit{otherwise} \end{cases}$$

where $\mathcal{F}$ is the signature of $\mathcal{P} \cup \mathcal{R}$, is sound and complete.

Proof. Immediate consequence of Theorems 7 and 9. □

Example 11. We show that the DP problem $(\mathcal{P}, \mathcal{R})$ of Example 4 over the signature $\mathcal{F} = \{\mathsf{a}, \mathsf{f}, \mathsf{g}, \mathsf{h}, \mathsf{F}, \mathsf{H}\}$ is match-DP-bounded for $\mathsf{F}(\mathsf{g}(x), y) \rightarrow \mathsf{H}(x, y)$ by constructing a compatible tree automaton. As starting point we consider the initial tree automaton

$$\mathsf{a}_0 \rightarrow 1 \qquad \mathsf{f}_0(1,1) \rightarrow 1 \qquad \mathsf{g}_0(1) \rightarrow 1$$
$$\mathsf{h}_0(1,1) \rightarrow 1 \qquad \mathsf{F}_0(1,1) \rightarrow 2 \qquad \mathsf{H}_0(1,1) \rightarrow 2$$

which accepts the set of all ground terms that have F_0 or H_0 as root symbol and a_0, f_0, g_0, and h_0 below the root. Since $\mathsf{F}_0(\mathsf{g}_0(x), y) \rightarrow_{\text{match}(s \rightarrow t)} \mathsf{H}_1(x, y)$ and $\mathsf{F}_0(\mathsf{g}_0(1), 1) \rightarrow^* 2$, we add the transition $\mathsf{H}_1(1, 1) \rightarrow 2$. Next we consider $\mathsf{H}_1(x, y) \rightarrow_{\text{match-DP}(\mathcal{P} \setminus \{s \rightarrow t\})} \mathsf{F}_1(x, \mathsf{g}_1(y))$ with $\mathsf{H}_1(1, 1) \rightarrow 2$. By adding the transitions $\mathsf{F}_1(1, 3) \rightarrow 2$ and $\mathsf{g}_1(1) \rightarrow 3$ this compatibility violation is solved. After that the rewrite rule $\mathsf{F}_1(\mathsf{g}_0(x), y) \rightarrow_{\text{match}(s \rightarrow t)} \mathsf{H}_1(x, y)$ and the derivation $\mathsf{F}_1(\mathsf{g}_0(1), 3) \rightarrow^* 2$ give rise to the transition $\mathsf{H}_1(1, 3) \rightarrow 2$. Finally we have $\mathsf{H}_1(x, y) \rightarrow_{\text{match-DP}(\mathcal{P} \setminus \{s \rightarrow t\})} \mathsf{F}_1(x, \mathsf{g}_1(y))$ and $\mathsf{H}_1(1, 3) \rightarrow 2$. In order to ensure $\mathsf{F}_1(1, \mathsf{g}_1(3)) \rightarrow^* 2$ we reuse the transition $\mathsf{F}_1(1, 3) \rightarrow 2$ and add the new transition $\mathsf{g}_1(3) \rightarrow 3$. After that step, the obtained tree automaton is compatible with match-DP($\mathcal{P}, s \to t, \mathcal{R}$). Hence the DP problem $(\mathcal{P}, \mathcal{R})$ is match-DP-bounded for $\mathsf{F}(\mathsf{g}(x), y) \to \mathsf{H}(x, y)$ by 1. Applying the DP processor of Theorem 10 yields the new DP problem $(\{\mathsf{H}(x, y) \to \mathsf{F}(x, \mathsf{g}(y))\}, \mathcal{R})$, which is easily (and automatically by numerous DP processors) shown to be finite. We note that the DP processor of Theorem 2 fails on $(\mathcal{P}, \mathcal{R})$.

To ensure that the TRS e-DP$(\mathcal{P}, s \to t, \mathcal{R})$ can assist to prove finiteness of the DP problem $(\mathcal{P}, \mathcal{R})$, it is crucial that every minimal rewrite sequence in $(\mathcal{P}, \mathcal{R})$ with infinitely many $\xrightarrow{\epsilon}_{s \to t}$ rewrite steps can be simulated by an infinite height increasing sequence in e-DP$(\mathcal{P}, s \to t, \mathcal{R})$. To this end it is important that rewrite rules in e-DP$((\mathcal{P} \setminus \{s \to t\}) \cup \mathcal{R})$ do not propagate the minimal height of the contracted redex unless the height of the root symbol of the redex is minimal. This is the reason for the slightly complicated definition of c in Definition 3. The following example shows what goes wrong if we would simplify the definition.

Example 12. Consider the DP problem $(\mathcal{P}, \mathcal{R})$ with $\mathcal{R}$ consisting of the rewrite rules $\mathsf{f}(x) \to \mathsf{g}(x)$ and $\mathsf{g}(\mathsf{a}(x)) \to \mathsf{f}(\mathsf{a}(x))$ and $\mathcal{P} = \mathrm{DP}(\mathcal{R})$ consisting of $\mathsf{F}(x) \to \mathsf{G}(x)$ and $\mathsf{G}(\mathsf{a}(x)) \to \mathsf{F}(\mathsf{a}(x))$. The DP problem $(\mathcal{P}, \mathcal{R})$ is not finite because the term $\mathsf{G}(\mathsf{a}(x))$ admits a minimal rewrite sequence. If we change the definition of c in Definition 3 to

$$c = \min \{\mathrm{height}(l'(p)) \mid p \in e(\mathrm{base}(l'), r)\}$$

then for $s \to t = \mathsf{F}(x) \to \mathsf{G}(y)$ we have

$$\mathsf{F}_0(\mathsf{a}_0(x)) \to_{\mathrm{match}(s \to t)} \mathsf{G}_1(\mathsf{a}_0(x)) \to_{\mathrm{match\text{-}DP}(\mathcal{P} \setminus \{s \to t\})} \mathsf{F}_0(\mathsf{a}_0(x))$$

and it would follow that $(\mathcal{P}, \mathcal{R})$ is match-DP-bounded for $\mathsf{F}(x) \to \mathsf{G}(x)$. However, removing this rule from $\mathcal{P}$ would leave a finite DP problem and hence we would falsely conclude termination of the original TRS $\mathcal{R}$.

4   Forward Closures

When proving the termination of a TRS $\mathcal{R}$ that is non-overlapping [9] or right-linear [4] it is sufficient to restrict attention to the set $\mathrm{RFC}_{\mathrm{rhs}(\mathcal{R})}(\mathcal{R})$ of right-hand sides of forward closures. This set is defined as the closure of the right-hand sides of the rules in $\mathcal{R}$ under variable renaming and narrowing. More formally, $\mathrm{RFC}_L(\mathcal{R})$ is the least extension of L such that

- $t[r]_p \sigma \in \mathrm{RFC}_L(\mathcal{R})$ whenever $t \in \mathrm{RFC}_L(\mathcal{R})$ and there exist a position $p \in \mathcal{FP}\mathrm{os}(t)$ and a fresh variant $l \to r$ of a rewrite rule in $\mathcal{R}$ with σ a most general unifier of $t|_p$ and l,
- $t\sigma \in \mathrm{RFC}_L(\mathcal{R})$ whenever $t \in \mathrm{RFC}_L(\mathcal{R})$ and σ is a variable renaming.

Dershowitz [4] obtained the following result.

Theorem 13. *A right-linear TRS $\mathcal{R}$ is terminating if and only if $\mathcal{R}$ is terminating on* $\mathrm{RFC}_{\mathrm{rhs}(\mathcal{R})}(\mathcal{R})$. $\qquad\square$

In our setting we can benefit from the properties of DP problems.

Lemma 14. *Let $(\mathcal{P}, \mathcal{R})$ be a DP problem. If $\mathcal{P}$ and $\mathcal{R}$ are right-linear then $(\mathcal{P}, \mathcal{R})$ is finite if and only if it is finite on $\mathrm{RFC}_{\mathrm{rhs}(\mathcal{P})}(\mathcal{P} \cup \mathcal{R})$.*

Proof. Easy consequence of Theorem 13 and the definition of DP problems. □

Lemma 14 can be used in connection with the DP processor of Theorem 2. For the DP processor of Theorem 10 we can do better.

Lemma 15. *Let $(\mathcal{P}, \mathcal{R})$ be a DP problem and $s \to t \in \mathcal{P}$. If $\mathcal{P}$ and $\mathcal{R}$ are right-linear then $(\mathcal{P}, \mathcal{R})$ admits a minimal rewrite sequence with infinitely many $\xrightarrow{\epsilon}_{s \to t}$ rewrite steps if and only if it admits such a sequence starting from a term in $\mathrm{RFC}_{\{t\}}(\mathcal{P} \cup \mathcal{R})$.* □

In general $\mathrm{RFC}_L(\mathcal{P} \cup \mathcal{R})$ is not computable. We can however over-approximate $\mathrm{RFC}_L(\mathcal{P} \cup \mathcal{R})$ by using tree automata as described in [8] and [17].

5   Raise-DP-Bounds

The reason why e-DP-bounds can be used only for DP problems $(\mathcal{P}, \mathcal{R})$ consisting of left-linear TRSs $\mathcal{P}$ and $\mathcal{R}$ is that without left-linearity, rewrite sequences in $(\mathcal{P}, \mathcal{R})$ cannot be lifted to sequences in e-DP$(\mathcal{P}, s \to t, \mathcal{R})$, cf. the proof of Theorem 7. As described in [17] one can solve that problem by using raise rules.

Definition 16. *Let $\mathcal{F}$ be a signature. The TRS $\mathrm{raise}(\mathcal{F})$ over the signature $\mathcal{F}_{\mathbb{N}}$ consists of all rules $f_i(x_1, \ldots, x_n) \to f_{i+1}(x_1, \ldots, x_n)$ with f an n-ary function symbol in $\mathcal{F}$, $i \in \mathbb{N}$, and $x_1, \ldots, x_n$ pairwise different variables. For terms $s, t \in \mathcal{T}(\mathcal{F}_{\mathbb{N}}, \mathcal{V})$ we write $s \leqslant t$ if $s \to^*_{\mathrm{raise}(\mathcal{F})} t$ and $s \uparrow t$ for the least term u with $s \leqslant u$ and $t \leqslant u$. Furthermore, this notion is extended to $\uparrow S$ for finite nonempty sets $S \subset \mathcal{T}(\mathcal{F}_{\mathbb{N}}, \mathcal{V})$ in the obvious way.*

The raise rules are used below to modify the rewrite relation associated to e-DP$(\mathcal{P}, s \to t, \mathcal{R})$ in such a way that non-left-linear rules are handled properly.

Definition 17. *Let $(\mathcal{P}, \mathcal{R})$ be a DP problem over a signature $\mathcal{F}$. We define the relation $\xrightarrow{\geqslant}_{e\text{-DP}(\mathcal{P}, s \to t, \mathcal{R})}$ on $\mathcal{T}(\mathcal{F}_{\mathbb{N}}, \mathcal{V})$ as follows: $s \xrightarrow{\geqslant}_{e\text{-DP}(\mathcal{P}, s \to t, \mathcal{R})} t$ if and only if there exist a rewrite rule $l \to r \in e\text{-DP}(\mathcal{P}, s \to t, \mathcal{R})$, a position $p \in \mathcal{P}os(s)$, a context C, and terms $s_1, \ldots, s_n$ such that $l = C[x_1, \ldots, x_n]$ with all variables displayed, $s|_p = C[s_1, \ldots, s_n]$, $\mathrm{base}(s_i) = \mathrm{base}(s_j)$ whenever $x_i = x_j$, and $t = s[r\theta]_p$. Here the substitution θ is defined as follows: $\theta(x) = \uparrow \{s_i \mid x_i = x\}$ if $x \in \{x_1, \ldots, x_n\}$ and $\theta(x) = x$ otherwise.*

Definition 18. *Let $(\mathcal{P}, \mathcal{R})$ be a DP problem and let $s \to t \in \mathcal{P}$. We say that $(\mathcal{P}, \mathcal{R})$ is e-raise-DP-bounded for $s \to t$ and a set of terms L if there exists a $c \in \mathbb{N}$ such that the height of function symbols occurring in terms in $\xrightarrow{\geqslant}{}^*_{e\text{-DP}(\mathcal{P}, s \to t, \mathcal{R})}(\mathrm{lift}_0(L))$ is at most c.*

For left-linear TRSs e-raise-DP-boundedness coincides with e-DP-boundedness. The following result is a straightforward adaption of Theorem 10.

Theorem 19. *The DP processor*

$$(\mathcal{P}, \mathcal{R}) \mapsto \begin{cases} \{(\mathcal{P} \setminus \{s \to t\}, \mathcal{R})\} & \text{if } (\mathcal{P}, \mathcal{R}) \text{ is top-raise-DP-bounded or non-} \\ & \text{duplicating and match-raise-DP-bounded} \\ & \text{for } s \to t \text{ and } \mathcal{T}(\mathcal{F}) \\ \{(\mathcal{P}, \mathcal{R})\} & \text{otherwise} \end{cases}$$

where $\mathcal{F}$ is the signature of $\mathcal{P} \cup \mathcal{R}$, is sound and complete. □

In [17] we showed that deterministic tree automata—a common approach to handle non-linearity with automata techniques (cf. [3,18,19])—are unsuitable. The problem with deterministic automata is that during the construction of a compatible tree automaton $\mathcal{A}$, it can happen that $\mathcal{A}$ becomes non-deterministic. Making $\mathcal{A}$ deterministic could lead to the removal of transitions that were added in earlier stages to ensure compatibility. However as soon as we add those transitions again, they are removed since they cause $\mathcal{A}$ to be non-deterministic. In [17] we introduced *quasi-deterministic* tree automata to solve this problem. This carries over to the present setting without any problems.

6   Experiments

The techniques described in the preceding sections are implemented in T$_T$T$_2$ [21]. T$_T$T$_2$ is written in OCaml[1] and consists of about 25000 lines of code. About 20% is used to implement the match-bound technique.

An important criterion for the success of e(-raise)-DP-bounds is the choice of the rewrite rule from $\mathcal{P}$ that should be removed from the DP problem $(\mathcal{P}, \mathcal{R})$ under consideration. To find a suitable rule, T$_T$T$_2$ simply starts the construction of a (quasi-)compatible tree automaton for each $s \to t \in \mathcal{P}$ in parallel. As soon as one of the processes terminates the procedure stops and returns the corresponding rule.

Below we report on the experiments we performed with T$_T$T$_2$ on the 1321 TRSs in version 4.0 of the Termination Problem Data Base that fulfill the variable condition, i.e., $\mathcal{V}ar(r) \subseteq \mathcal{V}ar(l)$ for each rewrite rule $l \to r \in \mathcal{R}$.[2] All tests were performed on a workstation equipped with an Intel® Pentium™ M processor running at a CPU rate of 2 GHz and 1 GB of system memory. Our results are summarized in Table 1.

We list the number of successful termination attempts, the average system time needed to prove termination (measured in milliseconds), and the number of timeouts. For all experiments we used a 60 seconds time limit. Besides the recursive SCC algorithm [14] and the improved estimated dependency graph processor [12], we use the following four DP processors:

---

[1] http://caml.inria.fr/
[2] http://www.lri.fr/~marche/tpdb

Table 1. Summary

	sp	no RFC no ur spb	no RFC no ur spd	no RFC ur spb	no RFC ur spd	RFC no ur spb	RFC no ur spd	RFC ur spb	RFC ur spd
# successes	497	558	587	584	612	574	588	605	615
average time	150	95	228	96	246	114	189	130	197
# timeouts	12	763	734	737	709	747	733	716	706

s the subterm criterion of [15],
p polynomial orderings with 0/1 coefficients [13],
b the DP processor of Theorem 2 extended to non-left-linear TRSs,
d the DP processor of Theorem 10 for left-linear TRSs and the one of Theorem 19
   for non-left-linear TRSs.

For the latter two, if the DP problem is non-duplicating we take $e = $ match. For duplicating problems we take $e = $ roof for b and $e = $ top for d.

A widely used approach to increase the power of DP processors is to consider only those rewrite rules of $\mathcal{R}$ which are usable [13,15]. Since in general usable rules (ur in Table 1) do not preserve the minimality of rewrite sequences for duplicating TRSs [13], it must be guaranteed that the DP processors of Theorem 2, 10 and 19 do not rely on the minimality of infinite rewrite sequences. For the DP processor of Theorem 2 this is obviously the case, since e(-raise)-bounds take all infinite rewrite sequences into account. For the DP processor of Theorem 10 with $e = $ top this follows from Lemma 6. For $e = $ match there is also no problem since $e = $ match can only be used for non-duplicating systems and it is known that usable rules can be used without restrictions for non-duplication systems ([11, Example 29] and [15, Theorem 23]).

The advantage of the DP processors of Theorems 10 and 19 over the naive one of Theorem 2 is clear, although the difference decreases when usable rules and RFC are in effect. There are two TRSs that could not be proved terminating by any tool participating in last year's termination competition[3] but which can now be handled by T$_\mathsf{T}$T$_2$ due to the results of this paper: secret05-teparla3 and secret06-matchbox-gen-25.

Example 20. The TRS secret06-matchbox-gen-25 ($\mathcal{R}$ in the following) consists of the following rewrite rules:

$$\mathsf{c}(\mathsf{c}(z, x, \mathsf{a}), \mathsf{a}, y) \rightarrow \mathsf{f}(\mathsf{f}(\mathsf{c}(y, \mathsf{a}, \mathsf{f}(\mathsf{c}(z, y, x)))))$$
$$\mathsf{f}(\mathsf{f}(\mathsf{c}(\mathsf{a}, y, z))) \rightarrow \mathsf{b}(y, \mathsf{b}(z, z))$$
$$\mathsf{b}(\mathsf{a}, f(\mathsf{b}(\mathsf{b}(z, y), \mathsf{a}))) \rightarrow z$$

The dependency graph contains one strongly connected component, consisting of the dependency pairs

$$1: \mathsf{C}(\mathsf{c}(z, x, \mathsf{a}), \mathsf{a}, y) \rightarrow \mathsf{C}(y, \mathsf{a}, \mathsf{f}(\mathsf{c}(z, y, x)))$$
$$2: \mathsf{C}(\mathsf{c}(z, x, \mathsf{a}), \mathsf{a}, y) \rightarrow \mathsf{C}(z, y, x)$$

---

[3] http://www.lri.fr/~marche/termination-competition/2007

Hence termination of $\mathcal{R}$ is reduced to finiteness of the DP problem $(\{1, 2\}, \mathcal{R})$. This problem is top-DP-bounded for rule 1; the compatible tree automaton computed by $\mathsf{T_T T_2}$ consists of the following transitions:

$\mathsf{a}_0 \to 1$	$\mathsf{c}_0(2, 2, 2) \to 4$	$\mathsf{C}_0(1, 5, 1) \to 3$	$\mathsf{f}_1(13) \to 1, 10, 14$
$\mathsf{a}_1 \to 6$	$\mathsf{c}_1(1, 1, 1) \to 10$	$\mathsf{C}_0(2, 1, 5) \to 3$	$\mathsf{f}_1(17) \to 4$
$\mathsf{b}_0(1, 1) \to 1$	$\mathsf{c}_1(1, 2, 1) \to 14$	$\mathsf{C}_1(5, 6, 8) \to 3$	$\mathsf{f}_1(20) \to 21$
$\mathsf{b}_1(1, 1) \to 9$	$\mathsf{c}_1(1, 5, 1) \to 7$	$\mathsf{f}_0(1) \to 1$	$\mathsf{f}_1(22) \to 23$
$\mathsf{b}_1(1, 9) \to 1$	$\mathsf{c}_1(1, 6, 11) \to 12$	$\mathsf{f}_0(4) \to 5$	$\mathsf{f}_1(23) \to 12$
$\mathsf{b}_1(6, 18) \to 1, 10, 14$	$\mathsf{c}_1(1, 11, 1) \to 20$	$\mathsf{f}_1(7) \to 8$	$\mathsf{f}_1(24) \to 25$
$\mathsf{b}_1(6, 19) \to 4$	$\mathsf{c}_1(1, 15, 1) \to 24$	$\mathsf{f}_1(10) \to 11$	$\mathsf{f}_1(26) \to 27$
$\mathsf{b}_1(11, 11) \to 18$	$\mathsf{c}_1(2, 6, 15) \to 16$	$\mathsf{f}_1(12) \to 13$	$\mathsf{f}_1(27) \to 16$
$\mathsf{b}_1(15, 15) \to 19$	$\mathsf{c}_1(11, 6, 21) \to 22$	$\mathsf{f}_1(14) \to 15$	$1 \to 2, 9$
$\mathsf{c}_0(1, 1, 1) \to 1$	$\mathsf{c}_1(15, 6, 25) \to 26$	$\mathsf{f}_1(16) \to 17$	$6 \to 1$

Hence the DP processor of Theorem 10 is applicable. This results in the new DP problem $(\{2\}, \mathcal{R})$, which is proved finite by the subterm criterion with the simple projection $\pi(C) = 1$.

7   Conclusion

In this paper we showed how the match-bound technique can be incorporated into the dependency pair framework. We introduced two new enrichments which take care of the special properties of DP problems. We also showed how to strengthen the method by taking right-hand sides of forward closures into account. Experimental results demonstrated the usefulness of our approach.

An important open question is whether we can use the roof enrichment in this setting. To ensure soundness of roof(-raise)-DP-bounds, it has to be proved that no restriction of roof-$\mathrm{DP}(\mathcal{P}, s \to t, \mathcal{R})$ to a finite signature admits a minimal rewrite sequence with infinitely many $\xrightarrow{\epsilon}_{\mathrm{roof}(s \to t)}$ ($\xrightarrow{\geq}_{\mathrm{roof}(s \to t)}$) rewrite steps. We conjecture that this claim holds for arbitrary $\mathcal{P}$ and $\mathcal{R}$. We anticipate that a positive solution would make additional termination proofs possible.

References

1. Arts, T., Giesl, J.: Termination of term rewriting using dependency pairs. TCS 236, 133–178 (2000)
2. Baader, F., Nipkow, T.: Term Rewriting and All That. Cambridge University Press, Cambridge (1998)
3. Comon, H., Dauchet, M., Gilleron, R., Jacquemard, F., Lugiez, D., Tison, S., Tommasi, M.: Tree automata techniques and applications (2002), www.grappa.univ-lille3.fr/tata
4. Dershowitz, N.: Termination of linear rewriting systems (preliminary version). In: Proc. 8th ICALP, vol. 115, pp. 448–458 (1981)

5. Geser, A., Hofbauer, D., Waldmann, J.: Match-bounded string rewriting systems. AAECC 15(3-4), 149–171 (2004)
6. Geser, A., Hofbauer, D., Waldmann, J.: Termination proofs for string rewriting systems via inverse match-bounds. JAR 34(4), 365–385 (2005)
7. Geser, A., Hofbauer, D., Waldmann, J., Zantema, H.: Finding finite automata that certify termination of string rewriting systems. International Journal of Foundations of Computer Science 16(3), 471–486 (2005)
8. Geser, A., Hofbauer, D., Waldmann, J., Zantema, H.: On tree automata that certify termination of left-linear term rewriting systems. I&C 205(4), 512–534 (2007)
9. Geupel, O.: Overlap closures and termination of term rewriting systems. Report MIP-8922, Universität Passau (1989)
10. Giesl, J., Schneider-Kamp, P., Thiemann, R.: AProVE 1.2: Automatic termination proofs in the dependency pair framework. In: Furbach, U., Shankar, N. (eds.) IJCAR 2006. LNCS (LNAI), vol. 4130, pp. 281–286. Springer, Heidelberg (2006)
11. Giesl, J., Thiemann, R., Schneider-Kamp, P.: The dependency pair framework: Combining techniques for automated termination proofs. In: Proc. 11th LPAR 2004. LNCS (LNAI), vol. 3425, pp. 301–331. Springer, Heidelberg (2004)
12. Giesl, J., Thiemann, R., Schneider-Kamp, P.: Proving and disproving termination of higher-order functions. In: Gramlich, B. (ed.) FroCos 2005. LNCS (LNAI), vol. 3717, pp. 216–231. Springer, Heidelberg (2005)
13. Giesl, J., Thiemann, R., Schneider-Kamp, P., Falke, S.: Mechanizing and improving dependency pairs. JAR 37(3), 155–203 (2006)
14. Hirokawa, N., Middeldorp, A.: Automating the dependency pair method. I&C 199(1,2), 172–199 (2005)
15. Hirokawa, N., Middeldorp, A.: Tyrolean termination tool: Techniques and features. I&C 205(4), 474–511 (2007)
16. Jambox, http://joerg.endrullis.de/
17. Korp, M., Middeldorp, A.: Proving termination of rewrite systems using bounds. In: Baader, F. (ed.) RTA 2007. LNCS, vol. 4533, pp. 273–287. Springer, Heidelberg (2007)
18. Middeldorp, A.: Approximating dependency graphs using tree automata techniques. In: Goré, R.P., Leitsch, A., Nipkow, T. (eds.) IJCAR 2001. LNCS (LNAI), vol. 2083, pp. 593–610. Springer, Heidelberg (2001)
19. Nagaya, T., Toyama, Y.: Decidability for left-linear growing term rewriting systems. I&C 178(2), 499–514 (2002)
20. Thiemann, R.: The DP Framework for Proving Termination of Term Rewriting. PhD thesis, RWTH Aachen, Available as technical report AIB-2007-17 (2007)
21. Tyrolean Termination Tool 2, http://colo6-c703.uibk.ac.at/ttt2
22. Waldmann, J.: Matchbox: A tool for match-bounded string rewriting. In: van Oostrom, V. (ed.) RTA 2004. LNCS, vol. 3091, pp. 85–94. Springer, Heidelberg (2004)
23. Zantema, H.: Termination of rewriting proved automatically. JAR 34(2), 105–139 (2005)

Further Results on Insertion-Deletion Systems with One-Sided Contexts

Alexander Krassovitskiy[1], Yurii Rogozhin[1,2], and Serghey Verlan[2,3]

[1] Rovira i Virgili University,
Research Group on Mathematical Linguistics,
Pl. Imperial Tàrraco 1, 43005 Tarragona, Spain
alexander.krassovitskiy@estudiants.urv.cat
[2] Institute of Mathematics and Computer Science
Academy of Sciences of Moldova
5, str. Academiei, MD-2028, Chişinău, Moldova
rogozhin@math.md
[3] LACL, Département Informatique, Université Paris Est,
61, av. Général de Gaulle, 94010 Créteil, France
verlan@univ-paris12.fr

Abstract. In this article we continue the investigation of insertion-deletion systems having a context only on one side of insertion or deletion rules. We show a counterpart of the results obtained in (Matveevici et al., 2007) by considering corresponding systems and exchanging deletion and insertion parameters. We prove three computational completeness results and one non-completeness result for these systems. We also solve the remaining open problem concerning the generative power of insertion-deletion systems having both contexts by proving the computational completeness of systems having a context-free insertion of two symbols and a contextual deletion of one symbol.

Keywords: Insertion-deletion systems, universality, computational non-completeness.

1 Introduction

The operations of insertion and deletion are fundamental in formal language theory, and generative mechanisms based on them were considered (with linguistic motivation) for some time, see [6] and [2]. Related formal language investigations can be found in several places; we mention only [3], [5], [8], [10]. In the last years, the study of these operations has received a new motivation from molecular computing see [1], [4], [11], [13].

In general form, an insertion operation means adding a substring to a given string in a specified (left and right) context, while a deletion operation means removing a substring of a given string from a specified (left and right) context. A finite set of insertion-deletion rules, together with a set of axioms provide a language generating device (an InsDel system): starting from the set of initial

C. Martín-Vide, F. Otto, and H. Fernau (Eds.): LATA 2008, LNCS 5196, pp. 333–344, 2008.

strings and iterating insertion-deletion operations as defined by the given rules we get a language. The number of axioms, the length of the inserted or deleted strings, as well as the length of the contexts where these operations take place are natural descriptional complexity measures in this framework. As expected, insertion and deletion operations with context dependence are very powerful, leading to characterizations of recursively enumerable languages. Most of the papers mentioned above contain such results, in many cases improving the complexity of insertion-deletion systems previously available in the literature. However, the power of the above operations is not necessarily related to the used context: the paper [7] contains an unexpected result: context-free insertion-deletion systems with one axiom are already universal, they can generate any recursively enumerable language. Moreover, this result can be obtained by inserting and deleting strings of a rather small length, at most three.

The further study of context-free insertion-deletion systems led in [14] to the complete description of this class. In particular, it was shown that if inserted or deleted strings are at most of length two, then a specific subclass of the family of context-free languages is obtained. This result showed that the traditional complexity measures for insertion-deletion systems, in particular the total weight based on the size of contexts, need a revision, because both systems from [13] and [14] have same total weight, but different computational power. In the same article, new complexity measures taking in account the sizes of both left and right context were proposed.

The article [9] investigates insertion-deletion systems which use a context only on one side of insertion or deletion rules. Such systems are very similar to context-free insertion-deletion systems where insertions or deletions are uncontrollable and may happen an arbitrary number of times at any place. It shows three computational completeness results and one non-completeness result based on different combination of parameters. Moreover, this article uses a new technique to prove the computational completeness results by simulating previously known insertion-deletion systems.

In this article we continue the investigation of insertion-deletion systems with one-sided contexts. We prove the counterpart of the results from [9] by considering corresponding systems and exchanging deletion and insertion parameters. Surprisingly, the obtained classes have the same computational completeness properties as their counterparts. Hence, we prove three completeness and one non-completeness result. A table with a summary of these results might be found in Section 4.

We also show a computational completeness for insertion-deletion systems which insert two symbols in a context-free manner, but delete one symbol in one-symbol left and right context. This result closes the last open problem concerning the generative power of insertion-deletion systems having both contexts.

The article refines the borderline between universality and non-universality (and even decidability) for insertion-deletion systems and leaves a number of open problems related to other combinations of parameters.

2 Prerequisites

All formal language notions and notations we use here are elementary and standard. The reader can consult any of the many monographs in this area – for instance, [12] – for the unexplained details.

We denote by $|w|$ the length of word w and by $card(A)$ the cardinality of the set A.

An *InsDel system* is a construct $ID = (V, T, A, I, D)$, where V is an alphabet, $T \subseteq V$, A is a finite language over V, and I, D are finite sets of triples of the form (u, α, v), $\alpha \neq \varepsilon$ of strings over V, where ε denotes the empty string. The elements of T are *terminal* symbols (in contrast, those of $V - T$ are called nonterminals), those of A are *axioms*, the triples in I are *insertion rules*, and those from D are *deletion rules*. An insertion rule $(u, \alpha, v) \in I$ indicates that the string α can be inserted in between u and v, while a deletion rule $(u, \alpha, v) \in D$ indicates that α can be removed from the context (u, v). Stated otherwise, $(u, \alpha, v) \in I$ corresponds to the rewriting rule $uv \rightarrow u\alpha v$, and $(u, \alpha, v) \in D$ corresponds to the rewriting rule $u\alpha v \rightarrow uv$. We denote by $\Longrightarrow_{ins}$ the relation defined by an insertion rule (formally, $x \Longrightarrow_{ins} y$ iff $x = x_1 uv x_2, y = x_1 u\alpha v x_2$, for some $(u, \alpha, v) \in I$ and $x_1, x_2 \in V^*$) and by $\Longrightarrow_{del}$ the relation defined by a deletion rule (formally, $x \Longrightarrow_{del} y$ iff $x = x_1 u\alpha v x_2, y = x_1 uv x_2$, for some $(u, \alpha, v) \in D$ and $x_1, x_2 \in V^*$). We refer by $\Longrightarrow$ to any of the relations $\Longrightarrow_{ins}, \Longrightarrow_{del}$, and denote by $\Longrightarrow^*$ the reflexive and transitive closure of $\Longrightarrow$ (as usual, $\Longrightarrow^+$ is its transitive closure).

The language generated by ID is defined by $L(ID) = \{w \in T^* \mid x \Longrightarrow^* w,$ for all $x \in A\}$.

The complexity of an InsDel system $ID = (V, T, A, I, D)$ is traditionally described by the vector $(n, m; p, q)$ called *weight*, where

$$n = \max\{|\alpha| \mid (u, \alpha, v) \in I\},$$
$$m = \max\{|u| \mid (u, \alpha, v) \in I \text{ or } (v, \alpha, u) \in I\},$$
$$p = \max\{|\alpha| \mid (u, \alpha, v) \in D\},$$
$$q = \max\{|u| \mid (u, \alpha, v) \in D \text{ or } (v, \alpha, u) \in D\},$$

The *total weight* of ID is the sum $\gamma = m + n + p + q$.

However, it was shown in [14] that this complexity measure is not accurate and it cannot distinguish between universality and non-universality cases (there are families having same total weight but not the same computational power). In the same article it was proposed to use the length of each context instead of the maximum. More exactly,

$$n = \max\{|\alpha| \mid (u, \alpha, v) \in I\},$$
$$m = \max\{|u| \mid (u, \alpha, v) \in I\},$$
$$m' = \max\{|v| \mid (u, \alpha, v) \in I\},$$
$$p = \max\{|\alpha| \mid (u, \alpha, v) \in D\},$$
$$q = \max\{|u| \mid (u, \alpha, v) \in D\},$$
$$q' = \max\{|v| \mid (u, \alpha, v) \in D\}.$$

Hence the complexity of an insertion-deletion system will be described by the vector $(n, m, m'; p, q, q')$ that we call *size*. We also denote by $INS_n^{m,m'} DEL_p^{q,q'}$ corresponding families of insertion-deletion systems. Moreover, we define the total weight of the system as the sum of all numbers above: $\psi = n + m + m' + p + q + q'$. Since it is known from [14] that systems using a context-free insertion or deletion of one symbol are not powerful, we additionally require $n + m + m' \geq 2$ and $p + q + q' \geq 2$.

If some of the parameters n, m, m', p, q, q' is not specified, then we write instead the symbol $*$. In particular, $INS_*^{0,0} DEL_*^{0,0}$ denotes the family of languages generated by *context-free InsDel systems*. If one of numbers from the couples m, m' and/or q, q' is equal to zero (while the other is not), then we say that corresponding families have a one-sided context.

InsDel systems of a "sufficiently large" weight can characterize RE, the family of recursively enumerable languages. A collection of these results may be found in Section 4.

3   Main Results

In this section we present the main results of the paper. We start with the following lemma:

Lemma 1. *For any insertion-deletion system $ID = (V, T, A, I, D)$ having the size $(n, 0, 0; p, 0, 0)$ it is possible to construct an insertion-deletion system $ID_2 = (V \cup \{X, Y\}, T, A_2, I, D \cup D')$ of size $(n, 0, 0; p, 0, 0)$ such that $L(ID_2) = L(ID)$ and all insertions and deletions in ID_2 are made inside the site $X \ldots Y$ and $D' = \{(\varepsilon, X, \varepsilon), (\varepsilon, Y, \varepsilon)\}$.*

Proof. Consider $A_2 = \{XwY \mid w \in A\}$. To prove the statement it is enough to observe that for any terminal derivation in ID it is possible to do the same derivation inside the site $X \ldots Y$ and after that erase these surrounding symbols (X and Y are like parentheses that surround the derivation site).

We also recall the following lemma proved in [9]:

Lemma 2. *For any insertion-deletion system $ID = (V, T, A, I, D)$ having the size $(n, m, m'; p, q, q')$ it is possible to construct an insertion-deletion system $ID_2 = (V \cup \{X, Y\}, T, A_2, I_2, D_2 \cup D'_2)$ having same size such that $L(ID_2) = L(ID)$. Moreover, all rules from I_2 have the form (u, α, v), where $|u| = m$, $|v| = m'$, all rules from D_2 have the form (u', α, v'), where $|u'| = q$, $|v'| = q'$ and $D'_2 = \{(\varepsilon, X, \varepsilon), (\varepsilon, Y, \varepsilon)\}$.*

Now we prove the following theorem which is a counterpart of the result from [11] where a system of size $(1, 1, 1; 2, 0, 0)$ is presented.

Theorem 3. $INS_2^{0,0}DEL_1^{1,1} = RE$.

Proof. The proof of the theorem is based on a simulation of insertion-deletion systems of size $(2,0,0;3,0,0)$. It is known that these systems generate any recursively enumerable language [7]. Consider $ID = (V,T,A,I,D)$ to be such a system. Now we construct a system $ID_2 = (V_2,T,A,I_2,D_2)$ of size $(2,0,0;1,1,1)$ that will generate same language as ID.

It is clear that in order to show the inclusion $L(ID) \subseteq L(ID_2)$ it is sufficient to show how a deletion rule $(\varepsilon, abc, \varepsilon) \in D$, with $a,b,c \in V$, may be simulated by using rules of system ID_2, *i.e.*, insertion rules of type $(\varepsilon, xy, \varepsilon)$ and deletion rules of type (a', y', b'), with $a', b' \in V_2 \cup \{\varepsilon\}$, $x, y, y' \in V_2$.

We may suppose that for any deletion rule $(\varepsilon, abc, \varepsilon)$ of ID following conditions hold:

a. $a \neq b \neq c$.

b. Insertions are made inside the site $X \ldots Y$, where X and Y are from lemma 1.

Indeed, if condition (a) does not hold, *i.e.*, we have a rule $(\varepsilon, aac, \varepsilon)$, then we replace this rule by an insertion rule $(\varepsilon, AA', \varepsilon)$ and two deletion rules $(\varepsilon, aA, \varepsilon)$ and $(\varepsilon, A'ac, \varepsilon)$. If a deletion rule $(\varepsilon, aaa, \varepsilon)$ is present, then it can be replaced by two insertion rules $(\varepsilon, AA', \varepsilon)$, $(\varepsilon, BB', \varepsilon)$ and three deletion rules $(\varepsilon, aA, \varepsilon)$, $(\varepsilon, A'aB, \varepsilon)$ and $(\varepsilon, B'a, \varepsilon)$.

Consider $V_2 = V \cup \{L_i, L_i', R_i, R_i', K_i, K_i' \mid 1 \leq i \leq card(D)\}$.

Let us label all rules from D by integer numbers. Consider now a rule $i : (\varepsilon, abc, \varepsilon) \in D$, where $1 \leq i \leq card(D)$ is the label of the rule. We introduce following insertion rules in I_2:

$$(\varepsilon, L_i L_i', \varepsilon) \tag{1}$$
$$(\varepsilon, R_i' R_i, \varepsilon) \tag{2}$$
$$(\varepsilon, K_i K_i', \varepsilon) \tag{3}$$

and following deletion rules in D_2 ($l, m \in V$):

$$(L_i, L_i', a) \tag{4}$$
$$(L_i, a, b) \tag{5}$$
$$(c, R_i', R_i) \tag{6}$$
$$(b, c, R_i) \tag{7}$$
$$(L_i, b, R_i) \tag{8}$$
$$(K_i, K_i', L_i) \tag{9}$$
$$(K_i, L_i, R_i) \tag{10}$$
$$(K_i, R_i, m) \tag{11}$$
$$(l, K_i, m) \tag{12}$$

We say that these rules are i-related.

The rule $i : (\varepsilon, abc, \varepsilon) \in D$ is simulated as follows. We first perform two insertions:

$$w_1 abc w_2 \Longrightarrow^1 w_1 L_i L_i' abc w_2 \Longrightarrow^2 w_1 L_i L_i' abc R_i' R_i w_2$$

And after that some deletions

$$w_1 L_i L_i' abc R_i' R_i w_2 \Longrightarrow^4 w_1 L_i abc R_i' R_i w_2 \Longrightarrow^6 w_1 L_i abc R_i w_2 \Longrightarrow^5$$
$$w_1 L_i bc R_i w_2 \Longrightarrow^7 w_1 L_i b R_i w_2 \Longrightarrow^8 w_1 L_i R_i w_2$$

Now we delete symbols $L_i R_i$ using same technique as above with the help of $K_i K_i'$

$$w_1 L_i R_i w_2 \Longrightarrow^3 w_1 K_i K_i' L_i R_i w_2 \Longrightarrow^9 w_1 K_i L_i R_i w_2 \Longrightarrow^{10} w_1 K_i R_i w_2 \Longrightarrow^{11}$$
$$w_1 K_i w_2 \Longrightarrow^{12} w_1 w_2$$

Hence, $L(ID) \subseteq L(ID_2)$. Now in order to prove the converse inclusion, we observe that we perform insertions of non-terminal symbols from V_2. After performing any of these insertions, the whole sequence of insertion and deletion rules above must be performed, otherwise some non-terminal symbols are left and cannot be deleted any more. Moreover, the above sequence permits to eliminate three symbols abc in a string. Indeed, the symbol L_i deletes a if and only if it was inserted one time at the left of a. Similarly, R_i deletes c if it was inserted one time at the right of c. Now, R_i and L_i are eliminated if and only if they meet, this means that they delete b (and of course a and c). In order to delete L_i, $K_i K_i'$ must be inserted before it. Symbol K_i deletes symbols K_i', L_i and R_i, and only after that it is eliminated.

Now we prove the following theorem which is a counterpart of the result from [9] where a system of size $(1,1,2;1,1,0)$ is presented.

Theorem 4. $INS_1^{1,0} DEL_1^{1,2} = RE$.

Proof. The proof of the theorem is based on a simulation of insertion-deletion systems of size $(1,1,1;1,1,1)$. It is known that these systems generate any recursively enumerable language [13]. Consider $ID = (V, T, A, I, D)$ to be such a system. Now we construct a system $ID_2 = (V_2, T, A, I_2, D_2)$ of size $(1,1,0;1,1,2)$ that will generate the same language as ID.

Using Lemma 2 it is clear that in order to show the inclusion $L(ID) \subseteq L(ID_2)$ it is sufficient to show how an insertion rule $(a, x, b) \in I$, with $a, b, x \in V$, may be simulated by using rules of system ID_2, i.e., insertion rules of type (a', x', ε) and deletion rules of type $(a', y', b'c')$, with $a', b', c' \in V_2 \cup \{\varepsilon\}$, $x', y' \in V_2$.

We may suppose that for any rule $(a, x, b) \in I$ it holds $x \neq b$. Indeed, if this is not the case then this rule may be replaced by two insertion rules (a, B, b), (a, b, B) and one deletion rule (b, B, b).

Consider $V_2 = V \cup \{A_i \mid 1 \leq i \leq card(I)\}$.

Let us label all rules from I by integer numbers. Consider now a rule $i : (a, x, b) \in I$, where $1 \leq i \leq card(I)$ is the label of the rule. We introduce insertion rules (a, A_i, ε), (A_i, x, ε) in I_2 and the deletion rule (a, A_i, xb) in D_2. We say that these rules are i-related. The rule $i : (a, x, b) \in I$ is simulated as follows. We first perform insertions of A_i and x:

$$w_1 ab w_2 \Longrightarrow^+ w_1 a(A_i)^+ b w_2 \Longrightarrow^+ w_1 a(A_i(x)^+)^+ b w_2$$

And after that one deletion (it is applicable to the string $w_1aA_ixbw_2$)

$$w_1aA_ixbw_2 \Longrightarrow w_1axbw_2$$

Hence, $L(ID) \subseteq L(ID_2)$. Now in order to prove the converse inclusion, we observe that we perform insertion of non-terminal symbol A_i from V_2. After performing this insertion, the only way to get rid of this symbol is to erase it with the introduced deletion rule. But this means that x is inserted between a and b. To conclude the proof we remark that if more than one A_i or x are inserted, then it is impossible to eliminate the corresponding symbol A_i.

Now we prove the following theorem which is a counterpart of the result from [9] where a system of size $(2, 0, 2; 1, 1, 0)$ is presented.

Theorem 5. $INS_1^{1,0}DEL_2^{0,2} = RE$.

Proof. The proof of the theorem is based on a simulation of insertion-deletion systems of size $(1, 1, 0; 1, 1, 2)$ from Theorem 4. Let $ID = (V, T, A, I, D)$ be such a system. Now we construct a system $ID_2 = (V_2, T, A, I_2, D_2)$ of size $(1, 1, 0; 2, 0, 2)$ that will generate the same language as ID.

From Lemma 2 it is clear that in order to show the inclusion $L(ID) \subseteq L(ID_2)$ it is sufficient to show how a deletion rule $(a, x, bc) \in D$, with $a, b, c, x \in V$, may be simulated by using rules of system ID_2, i.e., insertion rules of type (a, x, ε) and deletion rules of type $(\varepsilon, x'y', b'c')$, with $b', c' \in V_2 \cup \{\varepsilon\}$, $x', y' \in V_2$.

We may suppose that for any rule $(a, x, bc) \in D$ it does not hold $x = b = c$. Indeed, if this is not the case then this rule may be replaced by an insertion rule (x, D_x, ε) and two deletion rules (a, x, D_xx), (a, D_x, xx).

Consider $V_2 = V \cup \{A_i \mid 1 \leq i \leq card(D)\}$.

Let us label all rules from D by integer numbers. Consider now a rule $i : (a, x, bc) \in D$, where $1 \leq i \leq card(D)$ is the label of the rule. We introduce insertion rule (a, A_i, ε) in I_2 and the deletion rule (ε, A_ix, bc) in D_2. The rule $i : (a, x, bc) \in D$ is simulated as follows. We first perform insertions of A_i:

$$w_1axbcw_2 \Longrightarrow^+ w_1a(A_i)^+xbcw_2$$

And after that one deletion (it is applicable to the string $w_1aA_ixbcw_2$)

$$w_1aA_ixbcw_2 \Longrightarrow w_1abcw_2$$

Hence, $L(ID) \subseteq L(ID_2)$. Now in order to prove the converse inclusion, we observe that we perform insertion of non-terminal symbol A_i from V_2. After performing this insertion, the only way to get rid of this symbol is to erase it with the introduced deletion rule. But this means that x is deleted between a and b. To conclude the proof we remark that if more than one A_i is inserted, then it is impossible to eliminate the corresponding symbol A_i.

Now we prove the following theorem which is a counterpart of the result from [9] where a system of size $(2, 0, 1; 2, 0, 0)$ is presented.

Theorem 6. $INS_2^{0,0}DEL_2^{0,1} = RE$.

Proof. The proof of the theorem is based on a simulation of insertion-deletion systems of size (2,0,0;3,0,0) from [7]. Let $ID = (V, T, A, I, D)$ be such a system. Now we construct a system $ID_2 = (V_2, T, A, I_2, D_2)$ of size $(2, 0, 0; 2, 0, 1)$ that will generate the same language as ID.

It is clear that in order to show the inclusion $L(ID) \subseteq L(ID_2)$ it is sufficient to show how a deletion rule $(\varepsilon, abc, \varepsilon) \in D$, with $a, b, x \in V$, may be simulated by using rules of system ID_2, i.e., insertion rules of type $(\varepsilon, xy, \varepsilon)$ and deletion rules of type $(\varepsilon, x'y', b')$ or (ε, x', b'), with $b' \in V_2 \cup \{\varepsilon\}$, $x', y' \in V_2$. From Lemma 1 we can assume that all insertions and deletions are done inside the site $X...Y$.

Consider $V_2 = V \cup \{A_i^{(j)}, B_i^{(j)}, | j \in \{1, 2, 3, 4\}, 1 \le i \le card(D)\}$.

Let us label all rules from D by integer numbers. Consider now a rule i : $(\varepsilon, abc, \varepsilon) \in D$, where $1 \le i \le card(D)$ is the label of the rule. We introduce following insertion rules in I_2:

$$(\varepsilon, A_i^{(1)} B_i^{(1)}, \varepsilon) \tag{13}$$

$$(\varepsilon, A_i^{(2)} B_i^{(2)}, \varepsilon) \tag{14}$$

$$(\varepsilon, A_i^{(3)} B_i^{(3)}, \varepsilon) \tag{15}$$

$$(\varepsilon, A_i^{(4)} B_i^{(4)}, \varepsilon) \tag{16}$$

and following deletion rules in D_2:

$$(\varepsilon, aA_i^{(1)}, B_i^{(1)}) \tag{17}$$

$$(\varepsilon, bA_i^{(2)}, B_i^{(2)}) \tag{18}$$

$$(\varepsilon, cA_i^{(3)}, B_i^{(3)}) \tag{19}$$

$$(\varepsilon, B_i^{(2)} B_i^{(3)}, A_i^{(4)}) \tag{20}$$

$$(\varepsilon, B_i^{(1)} A_i^{(4)}, B_i^{(4)}) \tag{21}$$

$$(\varepsilon, B_i^{(4)}, \varepsilon) \tag{22}$$

The rule i : $(\varepsilon, abc, \varepsilon) \in D$ is simulated as follows. At first we perform insertions of $A_i^{(j)} B_i^{(j)}$, $j \in \{1, 2, 3, 4\}$ using rules (13) – (16):

$$w_1 abc w_2 \Longrightarrow^+ w_1 a A_i^{(1)} B_i^{(1)} b A_i^{(2)} B_i^{(2)} c A_i^{(3)} B_i^{(3)} A_i^{(4)} B_i^{(4)} w_2$$

After that deletion rules (17) – (19) are be applied:

$$w_1 a A_i^{(1)} B_i^{(1)} b A_i^{(2)} B_i^{(2)} c A_i^{(3)} B_i^{(3)} A_i^{(4)} B_i^{(4)} w_2 \Longrightarrow^+ w_1 B_i^{(1)} B_i^{(2)} B_i^{(3)} A_i^{(4)} B_i^{(4)} w_2$$

Now the remaining introduced symbols are removed:

$$w_1 B_i^{(1)} B_i^{(2)} B_i^{(3)} A_i^{(4)} B_i^{(4)} w_2 \Longrightarrow^{20} w_1 B_i^{(1)} A_i^{(4)} B_i^{(4)} w_2 \Longrightarrow^{21}$$
$$\Longrightarrow^{21} w_1 B_i^{(4)} w_2 \Longrightarrow^{22} w_1 w_2$$

Thus, we obtain string $w_1 w_2$, so we model rule $i : (\varepsilon, abc, \varepsilon) \in D$ correctly.

Hence, $L(ID) \subseteq L(ID_2)$. Now in order to prove the converse inclusion, we observe that we perform insertion of non-terminal symbol from V_2. After performing this insertion, the deletion rule above must be performed, otherwise some non-terminal symbol are left and cannot be deleted any more.

Now we consider the class $INS_1^{1,0} DEL_1^{1,1}$. We show that this class is not complete. Firstly we prove the following lemma which shows that the deletion of terminal symbols may be excluded.

Lemma 7. *For any insertion-deletion system $ID = (V, T, A, I, D)$ having the size $(1, 1, 0; 1, 1, 1)$ there is a system $ID' = (V \cup V', T, A \cup A', I \cup I', D')$ such that $L(ID') = L(ID)$. Moreover, for any rule $(a, b, c) \in D'$ it holds $b \notin T$.*

Proof. Indeed, we can transform system $ID = (V, T, A, I, D)$ to an equivalent system $ID' = (V \cup V', T, A \cup A', I \cup I', D')$ as follows:

Any rule $(a, b, c) \in D$, $a, c \in V$, $b \in V \setminus T$ will also be part of D'. Now consider a rule $(A, t, C) \in D$, $A, C \in V$, $t \in T$. Then we add the rule (A, N_t, C) to D', where $N_t \in V'$ is a new nonterminal. Moreover, we add also following rules to I' and strings to A':

- If $w_1 t w_2 \in A$, then we add $w_1 N_t w_2$ to A', where $w_1, w_2 \in V^*$,
- if $(t, A, \varepsilon) \in I$, then we add (N_t, A, ε) to I',
- if $(A, t, \varepsilon) \in I$, then we add (A, N_t, ε) to I',
- if $(A, C, t) \in D$, then we add (A, C, N_t) to D', and
- if $(t, A, C) \in D$, then we add (N_t, A, C) to D'.

It is clear that $L(ID') = L(ID)$ because there is no difference between erasing t or N_t.

The following result shows that the class $INS_1^{1,0} DEL_1^{1,1}$ is not computationally complete.

Theorem 8. $REG \setminus INS_1^{1,0} DEL_1^{1,1} \neq \emptyset$.

Proof. Consider the regular language $L = (ba)^+$. We claim that there is no insertion-deletion system Γ of size $(1,1,0;1,1,1)$ such that $L(\Gamma) = L$.

We shall prove the above statement by contradiction. Suppose there is such system $\Gamma = (V, \{a, b\}, A, I, D)$ and $L(\Gamma) = L$. From lemma 7 we can suppose that Γ does not delete terminal symbols.

Table 1. Known results on insertion-deletion systems

Nb.	γ	$(n, m; p, q)$	family	references	ψ	$(n, m, m'; p, q, q')$
1	6	$(3, 0; 3, 0)$	RE	[7]	6	$(3, 0, 0; 3, 0, 0)$
2	5	$(1, 2; 1, 1)$	RE	[4,11]	8	$(1, 2, 2; 1, 1, 1)$
3	5	$(1, 2; 2, 0)$	RE	[4,11]	7	$(1, 2, 2; 2, 0, 0)$
4	5	$(2, 1; 2, 0)$	RE	[4,11]	6	$(2, 1, 1; 2, 0, 0)$
5	5	$(1, 1; 1, 2)$	RE	[13]	8	$(1, 1, 1; 1, 2, 2)$
6	5	$(2, 1; 1, 1)$	RE	[13]	7	$(2, 1, 1; 1, 1, 1)$
7	5	$(2, 0; 3, 0)$	RE	[7]	5	$(2, 0, 0; 3, 0, 0)$
8	5	$(3, 0; 2, 0)$	RE	[7]	5	$(3, 0, 0; 2, 0, 0)$
9	4	$(1, 1; 2, 0)$	RE	[11]	5	$(1, 1, 1; 2, 0, 0)$
10	4	$(1, 1; 1, 1)$	RE	[13]	6	$(1, 1, 1; 1, 1, 1)$
11	4	$(2, 0; 2, 0)$	$\subsetneq CF$	[14]	4	$(2, 0, 0; 2, 0, 0)$
12	$m+1$	$(m, 0; 1, 0)$	$\subsetneq CF$	[14]	$-$	$(m, 0, 0; 1, 0, 0)$
13	$p+1$	$(1, 0; p, 0)$	$\subsetneq REG$	[14]	$-$	$(1, 0, 0; p, 0, 0)$
14	4	$(2, 0; 1, 1)$	RE	Theorem 3	5	$(2, 0, 0; 1, 1, 1)$

Consider a terminal derivation in Γ:

$w \Longrightarrow^+ w_f$, where $w \in A$ and $w_f \in (ba)^+$. Now consider an arbitrary ba block of w_f ($w_f = \alpha ba\beta$, $\alpha, \beta \in (ba)^*$) and take its letter a. Since there are no deletion rules in Γ this letter is either inserted by an insertion rule or it was a part of an axiom. We may omit the latter case by taking a derivation that produces a string that is long enough. We may also omit the case when this letter a was inserted by a rule $(\varepsilon, a, \varepsilon) \in I$, because in this case a may be inserted at any place in the final string, in particular a string $\alpha baa\beta$ might be obtained. Now suppose that this letter was inserted using a rule $(z, a, \varepsilon) \in I$, $z \in V$:

$$w \Longrightarrow^* w_1 z w_2 \Longrightarrow w_1 z a w_2 \Longrightarrow^* \alpha ba\beta = w_f. \tag{23}$$

This means that:

$$\begin{aligned} w_1 z &\Longrightarrow^* \alpha b \\ w_2 &\Longrightarrow^* \beta \end{aligned} \tag{24}$$

Now we remark that symbol a might be inserted twice:

$$w \Longrightarrow^* w_1 z w_2 \Longrightarrow w_1 z a w_2 \Longrightarrow w_1 z a a w_2. \tag{25}$$

From (25) and (24) we obtain:

$$w \Longrightarrow^* w_1 z a a w_2 \Longrightarrow^* \alpha baa\beta$$

which is a contradiction.

4   Complexity Measures

We collect known results on insertion-deletion systems in the two tables below. We indicate both traditional measures and measures proposed in [14] (see Section 2 for definitions). The first table contains the systems with both contexts and the second table concentrates on systems with one-sided contexts. In table 2 we do not present the symmetrical variants which have same generation capabilities.

Table 2. Known results on insertion-deletion systems with one-sided contexts

Nb.	γ	$(n, m; p, q)$	family	references	ψ	$(n, m, m'; p, q, q')$
15	5	$(1, 2; 1, 1)$	RE	[9]	6	$(1, 1, 2; 1, 1, 0)$
16	6	$(2, 2; 1, 1)$	RE	[9]	6	$(2, 0, 2; 1, 1, 0)$
17	5	$(2, 1; 2, 0)$	RE	[9]	5	$(2, 0, 1; 2, 0, 0)$
18	4	$(1, 1; 1, 1)$	$\subsetneq RE$	[9]	5	$(1, 1, 1; 1, 1, 0)$
19	5	$(1, 1; 1, 2)$	RE	Theorem 4	6	$(1, 1, 0; 1, 1, 2)$
20	6	$(1, 1; 2, 2)$	RE	Theorem 5	6	$(1, 1, 0; 2, 0, 2)$
21	5	$(2, 0; 2, 1)$	RE	Theorem 6	5	$(2, 0, 0; 2, 0, 1)$
22	4	$(1, 1; 1, 1)$	$\subsetneq RE$	Theorem 8	5	$(1, 1, 0; 1, 1, 1)$

5   Conclusions

In this article we have investigated insertion-deletion systems having a one-sided context, in particular, systems with minimal insertion. We showed that systems of size $(1, 1, 0; 1, 1, 2)$ and $(1, 1, 0; 2, 0, 2)$ generate all recursively enumerable languages, while systems of size $(1, 1, 0; 1, 1, 1)$ are not computationally complete and that they cannot generate the language $(ba)^{+}$. We also have considered systems with a minimal context-free insertion and we showed that systems of size $(2, 0, 0; 2, 0, 1)$ also generate all recursively enumerable languages.

We remark that these results are a counterpart of the results from [9]. Moreover, corresponding systems have the same computational completeness properties. These results leave an open question about generative power for 10 classes of insertion-deletion systems having the total weight equal to 5.

We also solved the last open problem concerning the generative power of insertion-deletion systems having both contexts by showing that systems of size $(2, 0, 0; 1, 1, 1)$ are computationally complete.

We note that the proof of the above results is based on a simulation of classes of insertion-deletion systems (previously simulations of Chomsky grammars were used) and the corresponding approach was firstly used in [9].

Acknowledgments

The first author acknowledges the grant of Ramon y Cajal from University Rovira i Virgili 2005/08. The second author acknowledges the support of European Commission, project MolCIP, MIF1-CT-2006-021666. The second and the third author acknowledge the Science and Technology Center in Ukraine, project 4032.

References

1. Daley, M., Kari, L., Gloor, G., Siromoney, R.: Circular contextual insertions/deletions with applications to biomolecular computation. In: Proc. of 6th Int. Symp. on String Processing and Information Retrieval, SPIRE1999, Cancun, Mexico, pp. 47–54 (1999)
2. Galiukschov, B.S.: Semicontextual grammars. In: Matematika Logica i Matematika Linguistika, Tallin University, pp. 38–50 (1981) (in Russian)

3. Kari, L.: On insertion and deletion in formal languages, PhD Thesis, University of Turku (1991)
4. Kari, L., Păun, G., Thierrin, G., Yu, S.: At the crossroads of DNA computing and formal languages: characterizing RE using insertion-deletion systems. In: Proc. of 3rd DIMACS Workshop on DNA Based Computing, Philadelphia, pp. 318–333 (1997)
5. Kari, L., Thierrin, G.: Contextual insertion/deletion and computability. Information and Computation 131(1), 47–61 (1996)
6. Marcus, S.: Contextual grammars. Rev. Roum. Math. Pures Appl. 14, 1525–1534 (1969)
7. Margenstern, M., Păun, G., Rogozhin, Y., Verlan, S.: Context-free insertion-deletion systems. Theoretical Computer Science 330, 339–348 (2005)
8. Martin-Vide, C., Păun, G., Salomaa, A.: Characterizations of recursively enumerable languages by means of insertion grammars. Theoretical Computer Science 205(1-2), 195–205 (1998)
9. Matveevici, A., Rogozhin, Y., Verlan, S.: Insertion-Deletion Systems with One-Sided Contexts. In: Durand-Lose, J., Margenstern, M. (eds.) MCU 2007. LNCS, vol. 4664, pp. 205–217. Springer, Heidelberg (2007)
10. Păun, G.: Marcus contextual grammars. Kluwer Academic Publishers, Dordrecht (1997)
11. Păun, G., Rozenberg, G., Salomaa, A.: DNA Computing. New Computing Paradigms. Springer, Berlin (1998)
12. Rozenberg, G., Salomaa, A. (eds.): Handbook of Formal Languages. Springer, Berlin (1997)
13. Takahara, A., Yokomori, T.: On the computational power of insertion-deletion systems. In: Proc. of 8th International Workshop on DNA-Based Computers, DNA8, Sapporo, Japan, June 10–13, 2002. LNCS, vol. 2568, pp. 269–280 (2003) (revised papers)
14. Verlan, S.: On minimal context-free insertion-deletion systems. In: Mereghetti, C., Palano, B., Pighizzini, G., Wotschke, D. (eds.) Seventh International Workshop on Descriptional Complexity of Formal Systems, Como, Italy, June 30 - July 2, 2005, pp. 285–292 (2005); Technical repport no. 06-05, University of Milan. Journal of Automata Languages and Combinatorics (in publication)

On Regularity-Preservation
by String-Rewriting Systems

Peter Leupold*

Department of Mathematics, Faculty of Science
Kyoto Sangyo University
Kyoto 603-8555, Japan
leupold@cc.kyoto-su.ac.jp

Abstract. When using string-rewriting systems in the context of formal languages, one of the most common questions is whether they preserve regularity. A class of string-rewriting systems that has received attention lately are idempotency relations. They were mainly used to generate languages starting from a single word.

Here we apply these relations to entire languages and investigate whether they preserve regularity. For this, it turns out to be convenient to define two more general classes of string-rewriting systems, the k-period expanding and the k-period reducing ones. We show that both preserve regularity. This implies regularity preservation for many classes of idempotency relations.

1 Introduction

The root of our investigations lies in the operation called duplication and introduced by Dassow et al. [3], who rediscovered a result shown earlier by Bovet and Varricchio [2] for so-called copy systems introduced by Ehrenfeucht and Rozenberg [4]. Mainly, a string-rewriting system that duplicates factors via rules $u \to u^2$ is applied iteratively to a word; then the question is whether the resulting language is regular or context-free. Later, this operation was also applied to entire languages rather than single words [7]. On the other hand, the duplication of words was also generalized to so-called idempotency languages [8]. These are generated by rules $u^m \to u^n$ for any fixed m and n rather than only by rules $u^1 \to u^2$.

Another line of research has dealt with classes of string-rewriting like monadic or prefix rewriting systems. It was investigated whether the result is regular/context-free if they are iteratively applied to regular/context-free languages. An example is the work of Hofbauer and Waldmann on deleting string-rewriting systems [5] which provides many references to earlier work. Also the book by Book and Otto contains a few results in this direction [1].

---

* This work was done, while the author was funded as a post-doctoral fellow by the Japanese Society for the Promotion of Science under number P07810.

C. Martín-Vide, F. Otto, and H. Fernau (Eds.): LATA 2008, LNCS 5196, pp. 345–356, 2008.

So far, investigations on idempotency relations have been focused on whether they produce regular or context-free languages when applied to singleton languages. In the context of the work on regularity preservation of string-rewriting systems it seems even more interesting to look at their behaviour when applied to entire languages. This is the object of this article, which therefore in some sense brings the two lines of research described above together. We consider only the length-bounded variants of idempotency relations; only here the underlying rewriting-systems are finite and therefore they are more tractable in this context.

In the section about uniformly length-bounded systems we actually treat a more general class of systems. Namely, we abandon the restriction that all rules must be of the form $u^m \rightarrow u^n$ for fixed m and n. First we define k-period-expanding string-rewriting systems, where only $m \leq n$ is required. Then we also consider the somewhat inverse class of k-period-reducing string-rewriting systems, which are characterized by the condition $m \geq n$. Both of these classes are shown to preserve regularity. Finally, we show that also finite unions of k-expanding and k-reducing systems, so-called k-periodic systems preserve regularity. It is a direct consequence of these results that all relations $=^k\bowtie_m^n$ preserve regularity except for $k \geq 2$, $m = 0$, and $n \geq 2$. For the latter cases it is already known that they generate non-regular languages from single words [8].

There are less results about idempotency relations with just an upper bound on the left side of rules. Mainly a result on k-period reducing systems can be adapted and yields that length-decreasing bounded idempotency relations preserve regularity. The proof also shows that their inverses preserve context-freeness. We also solve a problem left open in earlier work [7]: duplications with length three preserve regularity.

2   String-Rewriting Systems

Terms and notation from general formal language theory, logic and set theory are assumed to be known by the reader. Let $w[i]$ denote the i-th letter of a word w for $1 \leq i \leq |w|$, where $|w|$ is w's length. By $w[i \ldots j]$ we denote the factor of a word w, which begins in position i and ends in j. A word w has a positive integer k as a *period*, iff for all i, j such that $i \equiv j (\mathrm{mod}\ k)$ we have $w[i] = w[j]$, if both $w[i]$ and $w[j]$ are defined. $u \subset_{\mathsf{pref}} v$ means that u is a prefix of v, $\subset_{\mathsf{suff}}$ is our symbol for suffix. Two words u and v are *conjugates* iff there exists a factorization $u = rs$ such that $v = sr$. If not specified otherwise, the alphabet we use will be denoted by Σ.

In our notation on string-rewriting systems we mostly follow Book and Otto [1] and define a *string-rewriting system (SRS)* R on Σ to be a subset of $\Sigma^* \times \Sigma^*$. Its single-step reduction relation is defined as $u \rightarrow_R v$ iff there exists $(\ell, r) \in R$ such that for some u_1, u_2 we have $u = u_1 \ell u_2$ and $v = u_1 r u_2$. We also write simpler just $\rightarrow$, if it is clear which is the underlying rewriting system. By $\xrightarrow{*}$ we denote the relation's reflexive and transitive closure, which is called the *reduction relation* or *rewrite relation*. The inverse of a single-step reduction relation $\rightarrow$ is $\rightarrow^{-1} := \{(r, \ell) : (\ell, r) \in \rightarrow\}$. Note that we also use the notation $u \rightarrow v$ for rewrite rules, mainly when speaking about rules in a natural language sentence to make it graphically clear that we are speaking about a rewrite rule and not some other ordered pair. All the SRSs in this article will be finite.

An SRS is said to be *confluent*, iff for all $w, w_1, w_2 \in \Sigma^*$ always $w_1 \overset{*}{\leftarrow} w \overset{*}{\rightarrow} w_2$ implies the existence of some w' such that $w_1 \overset{*}{\rightarrow} w' \overset{*}{\leftarrow} w_2$. Here we use $w_1 \leftarrow w$ as a sometimes convenient way of writing $w \rightarrow w_1$.

By imposing restrictions on the format of the rewriting rules, many special classes of rewriting systems can be defined. Following Hofbauer and Waldmann [5], we will call a rule (ℓ, r) *context-free* (*inverse context-free*), if $|\ell| \leq 1$ ($|r| \leq 1$). A system is *monadic*, if it is inverse context-free and for all its rewrite rules (ℓ, r) we have $|\ell| > |r|$. Finally, we define *deleting* SRSs again following Hofbauer and Waldmann [5]. For these, we need a precedence, i.e. a irreflexive partial ordering $<$ on the alphabet. This is extended to words by defining that $u < v$ holds iff u and v do not use the same set of letters, and for every letter x which occurs in u there exists a letter y which occurs in v such that $x < y$. Now a SRS over the alphabet Σ is called $<$-*deleting*, iff it is a subset of $<^{-1}$; this means every right side of a rule is smaller than the corresponding left side with respect to $<$. More general, a SRS is called *deleting* iff it is deleting for some precedence. Hofbauer and Waldmann have shown that all deleting SRSs preserve regularity.

The *bounded idempotency* relations, which are one of the origins of the work here were first defined in [8]. For fixed parameters m, n, and k they are the rewrite relations

$$u^{\leq k}\bowtie_m^n v :\Leftrightarrow \exists z[z \in \Sigma^+ \wedge u = u_1 z^m u_2 \wedge v = u_1 z^n u_2 \wedge |z| \leq k]$$

and the corresponding SRSs are $\{(z^m, z^n) : |z| \leq k\}$. We will denote it by the same symbol $\leq k\bowtie_m^n$; no confusion should arise. A restricted version are the *uniformly bounded idempotency* relations

$$u^{=k}\bowtie_m^n v :\Leftrightarrow \exists z[z \in \Sigma^+ \wedge u = u_1 z^m u_2 \wedge v = u_1 z^n u_2 \wedge |z| = k]$$

and the corresponding SRSs are $\{(z^m, z^n) : |z| = k\}$. We denote the languages generated by these relations from a word w by $w^{\leq k\bowtie_m^n} := \{u : w({\leq k}\bowtie_m^n)^* u\}$ and $w^{=k\bowtie_m^n} := \{u : w({=k}\bowtie_m^n)^* u\}$.

For a string-rewriting system R and a language L we denote the set of all descendants of words from L modulo R by $R^*(L)$ following Hofbauer and Waldmann [5]. In the case of idempotency relations, however, we will still use the established notation $L^{\bowtie_m^n}$ meaning the same as $(\bowtie_m^n)^*(L)$. A class of languages $\mathcal{C}$ is said to be closed under (rewriting by) a class $\mathcal{S}$ of SRSs, iff the following holds: $\forall L, R[L \in \mathcal{C} \wedge R \in \mathcal{S} \Rightarrow R^*(L) \in \mathcal{C}]$.

3   Uniformly Length-Bounded Systems

Idempotency relations without restrictions on their rules' lengths often generate very complicated structures. The relations with length bound are in general much more accessible, especially the ones with uniform length bound. The main reasons for this are that on the one hand they are closely related to periodicity and thus tools from that field can be used; on the other hand, the underlying SRSs are finite and thus more tractable. Here we will not only use periodicity as a tool, but we will define a new class of SRSs based on periodicity in their rules. These will include almost the entire class of uniformly length bounded idempotency relations. Thus regularity preservation of the latter will be implied by our results.

But first we recall a single non-closure result that follows directly from prior work on the idempotency closure of words [8].

Proposition 1. *String-rewriting systems* $=^k\bowtie_m^n$ *do not preserve regularity for* $k \geq 2$, $m = 0$, *and* $n \geq 2$.

In the course of this section, we will see that these are actually the only combinations of parameters, for which regularity is not preserved. Now we define a class of SRSs all of whose rules increase the length of factors with period k.

Definition 2. An SRS is called k-*period-expanding*, if for all of its rules (ℓ, r)

(i) ℓ is non-empty,
(ii) $\ell, r \in w^*$ for a word w of length k, and
(iii) $\ell \subset_{\text{pref}} r$.

Thus the left sides of all rules of a k-period-expanding SRS have period k and the corresponding right sides add repetitions of that period — therefore the name. Now we establish an interesting property of this type of SRS.

Lemma 3. k-*period-expanding SRS are confluent.*

Proof. It is known that the diamond property implies confluence [1]. Therefore it suffices to show for k-period-expanding SRS that for every pair of derivation steps $w_1 \leftarrow u \rightarrow w_2$ there exists a word v such that $w_1 \rightarrow v \leftarrow w_2$. So let two words w_1 and w_2 be direct successors of another word u via such a k-period-expanding SRS R.

If the factors in u, where the rules are applied, do not overlap, then obviously in both cases the respectively other rule can be applied afterwards and one arrives at a common descendant v. So let two application sites r^m and s^i in u for rules $r^m \rightarrow r^n$ and $s^i \rightarrow s^j$ overlap. Without loss of generality, let r^m occur first from the left. If s^i is completely inside of r^m, then s and r are conjugates as both have length k. The result of applying the rules in either order will be r^{n+j-i}.

If s^i is not completely inside of r^m, then let us call u' the factor from the start of r^m till the end of s^i such that $u = u_1 u' u_2$ for some $u_1, u_2 \in \Sigma^*$. Now we can interpret the application of $r^m \rightarrow r^n$ as the insertion of r^{n-m} just in front of u'; equally $s^i \rightarrow s^j$ amounts to the insertion of s^{j-i} just after u'. Since application of these rules leaves u' unchanged, the two derivations

$$u_1 u' u_2 \rightarrow u_1 r^{n-m} u' u_2 \rightarrow u_1 r^{n-m} u' s^{j-i} u_2$$

and

$$u_1 u' u_2 \rightarrow u_1 u' s^{j-i} u_2 \rightarrow u_1 r^{n-m} u' s^{j-i} u_2$$

are possible, and the fact that they result in the same word with only two steps each concludes our proof. $\qquad\square$

The proof shows even more than the lemma states: all rules can be applied from left to right, that is in an order such that the prefix left of an application site will never be altered by another rule. Thus in some sense the different rule applications are independent from each other. This will help us in showing that they preserve regularity.

Proposition 4. *k-period-expanding SRSs preserve regularity.*

Proof. Let R be a k-period-expanding SRS. Let the longest left side of a rule from R have length km. We will insert additional symbols from the alphabet $\Gamma := \{[w^i] : |w| = k \wedge i \le m\} \cup \{\nabla\}$ into the words of a given language L. The $[w^i]$ will mark positions that are preceded by a factor w^i in the original word. ∇ is an auxiliary symbol, which is used to construct a deleting SRS S that essentially simulates R. Since it is deleting it preserves regularity and thus $R^*(L)$ is regular if L is.

First we describe informally the gsm mapping g, which introduces the symbols of Γ. Reading an input word from left to right, the gsm needs to remember at any given point the last km letters of the input. If they have a suffix w^i, which is the left side of a rule from R, then the letter $[w^i]$ must be output; notice that there can be several such letters to output. After each $[w^i]$ an arbitrary number of ∇ is written. Then the gsm advances and writes also the letter from Σ, which it reads, on the output.

Now we define the SRS that will work on the words produced by g. It simulates the rules from R by inserting the newly produced symbols to the left of the corresponding symbols from Γ, and deleting, in some sense consuming one ∇ in every step.

$$S := \{([w^i]\nabla, w^j[w^i]) : (w^i, w^{i+j}) \in R\}$$

This is a deleting SRS as for a precedence where ∇ is greater than all the other symbols, since all the rules delete ∇. Finally, to obtain $R^*(L)$ we need to delete all the symbols from Γ. This is done by the morphism

$$\delta := \begin{cases} x & \text{if } x \in \Sigma \\ \lambda & \text{if } x \in \Gamma. \end{cases}$$

Now we try to prove the inclusion $R^*(L) \subset \delta((S)^*(g(L)))$. Obviously $L = \delta(g(L))$. Further it should be clear that the rules from S can simulate the rules from R in the sense that if for some $w \in \Sigma^*$ we have $w \rightarrow_R w'$, then there is also $g(w) \rightarrow_S w''$ such that $w' = \delta(w'')$. So the first crucial fact here is that also further applications of rules to w' can be simulated starting from w''; this is not obvious, because $g(w') \ne w''$. The difference is that the second word contains less symbols from Γ since the rules from S do not create these. Thus these are missing in the newly created factor. This factor and a preceding factor (to which the rule was applied) have period k.

The one problem here is if some periodic factor in the original word is not long enough to be the application side for a rule from R, but through application of shorter rules this one can become applicable. Then the corresponding symbol from Γ is not there. See the following Example 5 for an illustration. In these cases an iteration of the process is necessary. In every iteration, k-periodic factors that allow rule applications are expanded as far as possible, in the next iteration longer rules will be applicable, too. Therefore the maximum number of iterations necessary is the number of different rules in R. To see this, observe that rule applications to a k-periodic factor can be ordered in such a way that first all applications of the rule with the shortest left side are done, then applications of the rule with the second shortest left-hand side etc. The first of these blocks of applications of the same rule will be possible in the first iteration, the second one in the second iteration and so forth. Thus we have

$$R^*(L) \subset \underbrace{\delta((S)^*(g(\ldots \delta((S)^*(g(\, L)))\ldots)))}_{|R| \text{ times}}.$$

The inverse inclusion does not need further arguments. It is clear that rewriting the symbols of Γ does not produce anything that is outside of $R^*(L)$ after the application of δ.

In conclusion, we have shown the equality

$$R^*(L) = \underbrace{\delta((S)^*(g(\ldots \delta((S)^*(g(\, L)))\ldots)))}_{|R| \text{ times}},$$

and since all the finitely many operations on the right hand side preserve regularity this proves the proposition.   □

We now illustrate with an example, why so many iterations of the procedure can be necessary to fully simulate the original SRS.

Example 5. We consider the SRS $R = \{(a, a^6), (a^8, a^{15}), (a^{17}, a^{21})\}$ and the regular language $L = \{a\}$. Applying the construction from the proof of Proposition 4, in one iteration only a symbol for simulating the first rule is inserted, the resulting language $\delta(S^*(g(L)))$ is $\{a^{5i+1} : i \geq 0\}$. In a second iteration, symbols for the other two rules are inserted, too. However, in the word a^{11} no symbol for the second rule is inserted, because the word is too short. Analysis of all possible derivations shows that therefore $a^{22} \notin \delta(S^*(g(\delta(S^*(g(L))))))$ although via R the derivation $a \to a^6 \to a^{11} \to a^{18} \to a^{22}$ is possible. Thus for this three-rule system three iterations of the procedure are necessary.

Since a large class of uniformly bounded idempotency relations falls in the class of k-period-expanding SRSs, we obtain an immediate corollary.

Corollary 6. *String-rewriting systems* $=^k\bowtie_m^n$ *preserve regularity for* $k \geq 0$, $m > 0$, *and* $n \geq m$.

Looking at the SRS S from the proof, we also see that the left sides of all rules consist of one letter of the form $[w^i]$ and one ∇. If we simply delete all the ∇ from the proof, the language generated is still the same, only S is not deleting any more. Instead, now S is context-free and this observation provides us with another closure property.

Corollary 7. *String-rewriting systems* $=^k\bowtie_m^n$ *preserve context-freeness for* $k \geq 0$, $m > 0$, *and* $n \geq m$.

Let us look a moment at the reason for the cases $m = 0$ not to be included here. The proof of Proposition 4 does not work, because rules with empty left side are applicable anywhere. Thus after every rule application another iteration of the process would be necessary, and there is no bound on this number. We now define a class of SRSs somewhat inverse to the k-period expanding ones, namely ones that reduce the length of periodic factors. Note that here right sides of length 0, i.e. deletions, are not excluded.

Definition 8. An SRS is called k-*period-reducing*, if for all of its rules (ℓ, r)

(i) $\ell, r \in w^*$ for a word w of length k and
(ii) $r \subset_{\mathrm{pref}} \ell$.

Also here, we can show that all systems of this class preserve regularity.

Proposition 9. k-*period-reducing SRSs preserve regularity.*

Proof. For a given regular language L and a k-period-reducing SRSs R, we will define a context-free SRS T such that T^{-1} simulates R. Since the inverse context-free SRS T^{-1} we construct is monadic, and since monadic SRSs preserve regularity, our claim follows.

First, we transform words from Σ^+ into a redundant representation, where every letter contains also the information about the $mk-1$ following ones, where mk is the length the longest right side of a rule in R. This way, rewrite rules from R can be simulated by ones with a right side of length only one, i.e. by inverse context-free ones.

First off we define the mapping $\phi : \Sigma^+ \mapsto ((\Sigma \cup \{\square\})^{mk})^+$ as follows. We delimit with $(\ldots)$ letters from $(\Sigma \cup \{\square\})^{mk}$ and with $[\ldots]$ factors of a word as usual. The image of a word u is

$$u \mapsto (u[1 \ldots mk])\,(u[2 \ldots mk+1]) \cdots (u[|u| - mk + 1 \ldots |u|]) \cdot$$
$$(u[|u| - mk + 2 \ldots |u|]\square) \cdots (u[|u|]\square^{mk-1}).$$

Thus every letter contains also the information about the mk following ones from the original word u. At the end of the word letters are filled up with the space symbol $\square$. ϕ is a gsm mapping and as such preserves regularity.

This encoding can be reversed by a letter-to-letter morphism h defined as $h(x) := x[1]$ if $x[1] \in \Sigma$, for the other case we select for the sake of completeness some arbitrary letter a and set $h(x) := a$ if $x[1] = \square$; the latter case will never occur in our context. It is clear that $h(\phi(u)) = u$ for words from Σ^*. Both mappings are extended to languages in the canonical way such that $\phi(L) := \{\phi(u) : u \in L\}$ and $h(L) := \{h(u) : u \in L\}$.

Now we define the string-rewriting system T over the alphabet $(\Sigma \cup \{\Box\})^{mk}$ as follows:

$$T := \{(u^i vv'), (\phi(u^j v)[1 \ldots |u^{j-i}|]) : (u^j, u^i) \in R \wedge |u^i vv'| = mk \wedge$$
$$u^i v \in \Sigma^+ \wedge v' \in \{\Box\}^*\}.$$

A letter $(u^i vv')$ is replaced by the image of $u^j v$ under ϕ minus the suffix of letters that are already there in the image of $u^i v$. In this way, application of rules from T keeps this space symbol only in the last letters of our words. If we have $w \xrightarrow{*}_R w'$, then clearly also $\phi(w') \xrightarrow{*}_T \phi(w)$ and thus $\phi(w) \xrightarrow{*}_{T^{-1}} \phi(w')$. This shows that $R^*(w) \subset h((T^{-1})^* \phi(w))$. For the inverse inclusion, let us take a look at what rules from T^{-1} do. For any such rule (ℓ, r) we have $h(\ell) = u^{j-i}u[1]$ and $h(r) = u[1]$ for some rule $u^j \to u^i$ from R. Thus exactly the same subword is deleted. Further examination of the rules and their contexts show that also $w \xrightarrow{(u^j, u^i)} w'$ iff $\phi(w) \xrightarrow{(\ell, r)} \phi(w')$. Thus only images under ϕ of words in $R^*(w)$ can be reached. Example 10 following this proof will further illustrate this.

As all the rules of T have left sides of length one and right sides of length greater than one, their inverses are all monadic, i.e. the system T^{-1} is monadic. Monadic string-rewriting systems are known to preserve regularity, see for example the textbook by Book and Otto [1].

Summarizing, we can obtain $R^*(L)$ by a series of regularity-preserving operations in the following way:

$$R^*(L) = h((T^{-1})^*(\phi(L))).$$ $\Box$

Example 10. Let R be a 1-period reducing SRS which contains a rule $a^3 \to a^2$ and whose longest left-hand side of a rule is of length 4. Then the reduction $ba^3bc \to_R ba^2bc$ is possible. The SRS T constructed in the proof of Proposition 9 has a rule $[a^2bc] \to [a^3b][a^2bc]$. The inverse is applicable to the word $\phi(ba^3bc) = [ba^3][a^3b][a^2bc][abc\Box][bc\Box\Box][c\Box\Box\Box]$ where it deletes the letter $[a^3b]$. The result is exactly $\phi(ba^2bc)$ and in this way the original rule is simulated. Here it is clearly visible how we can know from only looking at the letter $[a^2bc]$ that also the following two start with an a and thus in the original word a rule with the left-hand side a^3 is applicable.

Once more, the consequences for idempotency systems are immediate.

Corollary 11. *String-rewriting systems* $^{=k}\bowtie_m^n$ *preserve regularity for* $k \geq 0$, $m \geq 0$, *and* $m > n$.

Further, we have stated already in the proof that T is context-free, and thus it preserves context-freeness.

Corollary 12. *String-rewriting systems* $^{=k}\bowtie_m^n$ *preserve context-freeness for* $k \geq 0$, $m \geq 0$, *and* $m < n$.

Proof. Let T be the string-rewriting system and h and ϕ be the mappings from the proof of Proposition 9 constructed for $^{=k}\bowtie_m^n$. From the argumentation there we can see that

$$L^{\,=k}\bowtie_m^n = h(T^*(\phi(L))).$$

Since T is context-free and since this class of string-rewriting system preserves context-freeness, also $L^{\,=k}\bowtie_m^n$ is context-free for the given combinations of parameters.   □

Now we can define a more general class of SRSs that can expand as well as reduce factors of period k and we can show that also these preserve regularity and context-freeness.

Definition 13. Any union of finitely many k-period-expanding and k-period reducing SRSs is called a *k-periodic SRS*.

Proposition 14. *k-periodic SRSs preserve regularity.*

Proof. We can combine the proofs for k-period-expanding and k-period-reducing systems. First, observe that all factors of a word that have period k can be considered independently in the following sense: application of a rule to one of them does not affect applicability of rules in other such factors. That is, in doing a reduction via R we can look at the first such block of a word, apply all the rules that are to be applied there, then go to the second block etc.

Further, observe that such a block has the form $v^i v'$ for a word v of length k and a word v' shorter than k. Application of any rule in this factor will only change the exponent i. Thus we can look at k-period-expanding rules as additions to the exponent, at k-period-reducing rules as subtractions.

Over integers, a series of additions and subtractions can be done in any order, the result is always the same. Since $v^+ v'$ can only represent non-negative integers, in our case we just have to be sure that the intermediary results are always non-negative; by doing first all the additions, then the subtractions, this is ensured. This shows us that we can reorder the application of rules in such a way that first all of the k-period-expanding ones are applied, then the k-period-reducing ones. Only the original order of the length-increasing rules must be preserved, because some long left side might be created only by earlier application of shorter ones.

We can partition a k-periodic SRS R into two systems $R_\nearrow$ and $R_\searrow$, the first of which contains all the k-period-expanding rules while the second one contains all the k-period-reducing rules. Now we construct for $R_\nearrow$ the SRS S from the proof of Proposition 4 and the corresponding mappings; for $R_\searrow$ we construct the SRS T from the proof of Proposition 9 and the corresponding mappings. The considerations above then show us that

$$R^*(L) = h((T^{-1})^*(\phi(\underbrace{\delta((S)^*(g(\ldots \delta((S)^*(g(}_{|R_\nearrow|\text{ times}} L)))\ldots))))))$$

which proves the proposition's claim.   □

In most cases, k-periodic SRSs preserve also context-freeness. Only rules of the form $\lambda \to u$ must be excluded as can be seen from the various results in this section.

4   Length-Bounded Relations

From prior work on the idempotency closure of single words we can deduce several results. Namely all the cases, where already the closure of a single word is not regular cannot preserve regularity in general. Thus we have the following.

Proposition 15. *String-rewriting systems* $^{\leq k}\bowtie_m^n$ *do not preserve regularity for the following combinations of parameters:*

- $k \geq 1$, $m = 0$, $n \geq 2$,
- $k \geq 4$, $m = 1$, $n \geq 2$.

The second clause leaves the question of regularity preservation open for $k \leq 3$ and this is the case we will treat now for $n = 2$.

Proposition 16. *String-rewriting systems* $^{\leq k}\bowtie_1^2$ *preserve regularity for* $k \leq 3$.

Proof. Since the left sides of all rules have at most three different letters, we can restrict our attention to an alphabet of three letters here. We will decompose the SRS $^{\leq 3}\bowtie_1^2$ into three regularity preserving ones. For this, notice that a rule (a^2, a^4) can easily be simulated by two applications of the rule (a, a^2). So with left sides of length two, only rules $(ab, (ab)^2)$ with $a \neq b$ are necessary to obtain the entire language. Similar considerations show that with left sides of length three only rules $(abc, (abc)^2)$ for $a \neq b \neq c$ are necessary. For example, the result of $(aba, (aba)^2)$ can be obtained by first applying $(ab, (ab)^2)$ and then (a, a^2). Let us denote the sets of rules of the latter two types by R_{ab} and R_{abc} respectively.

Let us define the *letter sequence* $\mathsf{seq}(u)$ of a word u as follows: any word u can be uniquely factorized as $u = x_1^{i_1} x_2^{i_2} \cdots x_\ell^{i_\ell}$ for some integers $\ell \geq 0$ and $i_1, i_2, \ldots, i_\ell \geq 1$ and for letters $x_1, x_2, \ldots, x_\ell$ such that always $x_j \neq x_{j+1}$; then $\mathsf{seq}(u) := x_1 x_2 \cdots x_\ell$. Intuitively speaking, every block of several adjacent occurrences of the same letter is reduced to just one occurrence.

Now we can identify two classes of rules: Those from $^{=1}\bowtie_1^2$ only change the length of a consecutive block of occurrences of the same letter, but leave the letter sequence unchanged. The other rules from R_{ab} and R_{abc} do not change such lengths but change the letter sequence. Changing the letter sequence does not affect the possibilities of later expanding a block of letters, therefore the latter action can be postponed and we see that $L^{\leq 3}\bowtie_1^2 = ((R_{ab} \cup R_{abc})^*(L))^{=1}\bowtie_1^2$.

For the SRS $R_{ab} \cup R_{abc}$ it is obvious that the diamond property holds. Therefore we can interchange the sequence in which rules are applied. This way we can order them to first apply all the rules from R_{abc}, and thus we see that

$$L^{\leq 3}\bowtie_1^2 = ((L^{=3}\bowtie_1^2)^{=2}\bowtie_1^2)^{=1}\bowtie_1^2.$$

The three systems applied on the right hand side preserve regularity by Proposition 4, and thus also their composition does.

Now it is clear that also $L^{\leq 2}\bowtie_1^2 = (L^{=2}\bowtie_1^2)^{=1}\bowtie_1^2$. □

This also solves the problem whether $\leq^3\bowtie_1^2$ preserves regularity, which was left open in earlier work [7]. Through personal communication with Masami Ito we know that also he has a proof for this fact, which is, however, based on a automata-theoretic construction.

Proposition 17. *String-rewriting systems $\leq^k\bowtie_m^n$ preserve regularity for $m \geq n$.*

Proof. This proof goes along lines completely analogous to the one of Proposition 9. Therefore we omit here the definition of the mappings ϕ and h and only define the modified, context-free SRS T derived from a given SRS $\leq^k\bowtie_m^n$.

$$T := \{[u^i], (\phi(u^j)[1 \ldots |u^{j-i}|]) : (u^j, u^i) \in \leq^k\bowtie_m^n \}.$$

The differences to the system T from proof the of Proposition 9 are that the length of u is not fixed, while the parameters m and n are.

Again, we can obtain $L^{\leq^k\bowtie_m^n}$ by a series of regularity-preserving operations in the following way:

$$L^{\leq^k\bowtie_m^n} = h((T^{-1})^*(\phi(L))). \qquad \square$$

Again, the consequence corresponding to Corollary 12 is true.

Corollary 18. *String-rewriting systems $\leq^k\bowtie_m^n$ preserve context-freeness for $m \leq n$.*

5   Outlook

When looking at the languages generated by idempotency relations without length bounds from single words, we see that almost all cases generate non-regular languages, Thus the questions treated here are not of great interest in that context. However, when the size of the alphabet is limited to two or even one letter, the picture changes [7,9]. The reason for this is mainly that often –as in the case of duplication– there exists a finite SRS, which is equivalent to the infinite one. For these cases also the closure of regular languages under the respective SRSs is an interesting question. The cases of $\bowtie_0^1$ and $\bowtie_1^0$, however, are regular over any alphabet size as shown in work summarized by Ito [6].

References

1. Book, R., Otto, F.: String-Rewriting Systems. Springer, Berlin (1993)
2. Bovet, D.P., Varricchio, S.: On the Regularity of Languages on a Binary Alphabet Generated by Copying Systems. Information Processing Letters 44, 119–123 (1992)
3. Dassow, J., Mitranaand, V., Păun, G.: On the Regularity of Duplication Closure. Bull. EATCS 69, 133–136 (1999)
4. Ehrenfeucht, A., Rozenberg, G.: On Regularity of Languages Generated by Copying Systems. Discrete Applied Mathematics 8, 313–317 (1984)

5. Hofbauer, D., Waldmann, J.: Deleting String-Rewriting Systems preserve regularity. In: Theoretical Computer Science, vol. 327, pp. 301–317 (2004)
6. Ito, M.: Algebraic Theory of Automata and Languages. World Scientific, New Jersey (2004)
7. Ito, M., Leupold, P., Shikishima-Tsuji, K.: Closure of Language Classes under Bounded Duplication. In: H. Ibarra, O., Dang, Z. (eds.) DLT 2006. LNCS, vol. 4036, pp. 238–247. Springer, Heidelberg (2006)
8. Leupold, P.: Languages Generated by Iterated Idempotencies. Theoretical Computer Science 370(1-3), 170–185 (2007)
9. Leupold, P.: General Idempotency Languages over Small Alphabets. Journal of Automata, Languages and Combinatorics (accepted for publication)

Minimizing Deterministic Weighted Tree Automata

Andreas Maletti*

International Computer Science Institute
1947 Center Street, Suite 600
Berkeley, CA 94704, USA
maletti@icsi.berkeley.edu

Abstract. The problem of efficiently minimizing deterministic weighted tree automata (wta) is investigated. Such automata have found promising applications as language models in Natural Language Processing. A polynomial-time algorithm is presented that given a deterministic wta over a commutative semifield, of which all operations including the computation of the inverses are polynomial, constructs an equivalent minimal (with respect to the number of states) deterministic and total wta. If the semifield operations can be performed in constant time, then the algorithm runs in time $O(rmn^4)$ where r is the maximal rank of the input symbols, m is the number of transitions, and n is the number of states of the input wta.

1 Introduction

Weighted tree automata (wta) [9,8,20,6] are a joint generalization of weighted string automata [21] and tree automata [14,15]. Weighted string automata have successfully been applied as language models in Natural Language Processing largely due to their ability to easily incorporate n-gram models. Several toolkits (e.g., CARMEL [16], FIRE STATION [13], and OPENFST [1]) enable language engineers to rapidly prototype and develop language models because of the standardized implementation model and the consolidated algorithms made available by the toolkits.

In recent years, the trend toward more syntactical approaches in Natural Language Processing [19] sparked renewed interest in tree-based devices. The weighted tree automaton is the natural tree-based analogue of the weighted string automaton. First experiments with toolkits (e.g., TIBURON [24]) based on tree-based devices show that the situation is not as consolidated here. In particular, many basic algorithms are missing in the weighted setting.

In general, a wta processes a given input tree stepwise using a locally specified transition behavior. During this process transition weights are combined using the

---

* Author on leave from *Technische Universität Dresden, Faculty of Computer Science, 01062 Dresden, Germany* with the help of financial support by a DAAD (German Academic Exchange Service) grant.

C. Martín-Vide, F. Otto, and H. Fernau (Eds.): LATA 2008, LNCS 5196, pp. 357–372, 2008.

operations (addition and multiplication) of a semiring to form the weight associated with the input tree. Altogether, the wta thus recognizes (or computes) a mapping $\varphi\colon T_\Sigma \to A$ where T_Σ is the set of all input trees and A is the carrier set of the semiring. Such a mapping is also called a tree series, and if it can be computed by a wta, then it is recognizable. The deterministically recognizable tree series are exactly those recognizable tree series that can be computed by deterministic wta. Recognizable and deterministically recognizable tree series have been thoroughly investigated (see [20,5] and references provided therein). In fact, [6] and [23] show which recognizable tree series are also deterministically recognizable.

In this contribution, we consider deterministically recognizable tree series. To the author's knowledge, we propose the first polynomial-time minimization algorithm for deterministic wta over semifields. A MYHILL-NERODE theorem for tree series recognized by such automata is known [3]. However, it only asserts the existence of a unique, up to slight changes of representation, minimal (with respect to the number of states) deterministic wta recognizing a given tree series. The construction of such a wta, which is given in [3], is not effective, but with the help of the pumping lemma of [4] an exponential-time algorithm, which given a deterministic wta constructs an equivalent minimal deterministic and total wta, could easily be derived. For (not necessarily deterministic) wta over fields the situation is similar. In [9,7] the existence of a unique, up to slight changes of representation, minimal wta is proved. Moreover, [7] shows that minimization is effective by providing the analogue to the pumping argument already mentioned above in this more general setting. However, the trivially obtained algorithm is exponential.

ANGLUIN [2] learning algorithms exist for both general [17] and deterministic [11,22] wta. In principle, those polynomial-time learning algorithms could also be used for minimization since they produce minimal wta recognizing the taught tree series. However, this also requires us to implement the oracle, which answers coefficient and equivalence queries. Although equivalence is decidable in polynomial time in both cases [26,4], a simple implementation would return counterexamples of exponential size, which would yield an exponential-time minimization algorithm. Clearly, this can be avoided by the method presented in this contribution.

Finally, let us mention the minimization procedures [25,12] for deterministic weighted string automata. They rely on a weight normal-form obtained by a procedure called *pushing*. After this normal form is obtained, the weight of a transition is treated as an input symbol and the automaton is minimized as if it were unweighted. We do not follow this elegant approach here because we might have to explore several distributions of the weight to the input states of a transition (in a tree automaton a transition can have any number of input states whereas in a string automaton it has exactly one) during pushing. It remains open whether there is an efficient heuristic that prescribes how to distribute the weight such that we obtain a minimal deterministic wta recognizing the given series after the unweighted minimization.

Here we give a direct minimization construction, which uses partition refinement as in the unweighted case [10]. To this end, we first define the MYHILL-NERODE relation on states of the deterministic input wta. This definition, as well as the MYHILL-NERODE relation on tree series [3], will include a scaling factor and Algorithm 2 will determine those scaling factors. In the refinement process (see Definition 13) we check for the congruence property (as in the unweighted case) and the consistency of the weight placement on the transitions. Overall, our algorithm runs in time $O(rmn^4)$ where r is the maximal rank of the input symbols, m is the number of transitions, and n is the number of states of the input wta.

2 Preliminaries

The set of nonnegative integers is $\mathbb{N}$. Given $l, u \in \mathbb{N}$ we denote $\{i \in \mathbb{N} \mid l \le i \le u\}$ simply by $[l, u]$. Let $n \in \mathbb{N}$ and Q a set. We write Q^n for the n-fold CARTESIAN product of Q. The empty tuple $() \in Q^0$ is sometimes displayed as ε. We reserve the use of a special symbol $\square \notin Q$. The set of n-ary contexts over Q, denoted by $C_n(Q)$, is $\bigcup_{i+j+1=n} Q^i \times \{\square\} \times Q^j$. Given $C \in C_n(Q)$ and $q \in Q$ we write $C[q]$ to denote the tuple of Q^n obtained from C by replacing $\square$ by q.

An equivalence relation $\equiv$ on Q is a reflexive, symmetric, and transitive subset of Q^2. Let $\equiv$ and $\equiv'$ be equivalence relations on Q. Then $\equiv$ is a refinement of $\equiv'$ if $\equiv \subseteq \equiv'$. The equivalence class of $q \in Q$ is $[q]_\equiv = \{q' \in Q \mid q' \equiv q\}$. Whenever $\equiv$ is obvious from the context, we simply omit it. The system $(Q/\equiv) = \{[q] \mid q \in Q\}$ actually forms a partition of Q; i.e., a system Π of subsets (also called blocks) of Q such that $\bigcup_{P \in \Pi} P = Q$ and $P \cap P' = \emptyset$ for every $P, P' \in \Pi$ with $P \ne P'$. A mapping $r \colon (Q/\equiv) \to Q$ is a representative mapping if $r(P) \in P$ for every $P \in (Q/\equiv)$. The number of blocks of $(Q/\equiv)$ is denoted by index$(\equiv)$. Let Π be any partition on Q and $F \subseteq Q$. The equivalence relation $\equiv_\Pi$ on Q is defined for every $p, q \in Q$ by $p \equiv_\Pi q$ if and only if $\{p, q\} \subseteq P$ for some block $P \in \Pi$. We say that Π saturates F if $\equiv_\Pi$ is a refinement of $\equiv_{\{F, Q \setminus F\}}$; i.e., $\bigcup_{P \in \Pi'} P = F$ for some $\Pi' \subseteq \Pi$.

An alphabet is a finite and nonempty set of symbols. A ranked alphabet (Σ, rk) is an alphabet Σ and a mapping $\text{rk} \colon \Sigma \to \mathbb{N}$. Whenever rk is clear from the context, we simply drop it. The subset of n-ary symbols of Σ is $\Sigma_n = \{\sigma \in \Sigma \mid \text{rk}(\sigma) = n\}$. The set $T_\Sigma(Q)$ of Σ-trees indexed by Q is inductively defined to be the smallest set such that $Q \subseteq T_\Sigma(Q)$ and $\sigma(t_1, \ldots, t_n) \in T_\Sigma(Q)$ for every $\sigma \in \Sigma_n$ and $t_1, \ldots, t_n \in T_\Sigma(Q)$. We write T_Σ for $T_\Sigma(\emptyset)$. The mapping $\text{var} \colon T_\Sigma(Q) \to \mathcal{P}(Q)$, where $\mathcal{P}(Q)$ is the power set of Q, is inductively defined by $\text{var}(q) = \{q\}$ for every $q \in Q$ and $\text{var}(\sigma(t_1, \ldots, t_n)) = \bigcup_{i=1}^n \text{var}(t_i)$ for every $\sigma \in \Sigma_n$ and $t_1, \ldots, t_n \in T_\Sigma(Q)$. For every $P \subseteq Q$, we use $\text{var}_P(t)$ as a shorthand for $\text{var}(t) \cap P$. Moreover, we use $|t|_q$ to denote the number of occurrences of $q \in Q$ in $t \in T_\Sigma(Q)$. Finally, we define the height and size of a tree with the help of the mappings $\text{ht}, \text{size} \colon T_\Sigma(Q) \to \mathbb{N}$ inductively for every $q \in Q$ by $\text{ht}(q) = \text{size}(q) = 1$ and $\text{ht}(\sigma(t_1, \ldots, t_n)) = 1 + \max\{\text{ht}(t_i) \mid i \in [1, n]\}$ and $\text{size}(\sigma(t_1, \ldots, t_n)) = 1 + \sum_{i=1}^n \text{size}(t_i)$ for every $\sigma \in \Sigma_n$ and $t_1, \ldots, t_n \in T_\Sigma(Q)$. Note that $\max \emptyset = 0$.

The set $C_\Sigma(Q)$ of Σ-contexts indexed by Q is defined as the smallest set such that $\square \in C_\Sigma(Q)$ and $\sigma(t_1, \ldots, t_{i-1}, C, t_{i+1}, \ldots, t_n) \in C_\Sigma(Q)$ for every $\sigma \in \Sigma_n$ with $n \geq 1$, index $i \in [1, n]$, $t_1, \ldots, t_n \in T_\Sigma(Q)$, and $C \in C_\Sigma(Q)$. We write C_Σ for $C_\Sigma(\emptyset)$. Note that $C_\Sigma(Q) \subseteq T_\Sigma(Q \cup \{\square\})$. Next we recall substitution. Let V be an alphabet (possibly containing $\square$), $v_1, \ldots, v_n \in V$ be pairwise distinct, and $t_1, \ldots, t_n \in T_\Sigma(V)$. Then we denote by $t[v_i \leftarrow t_i \mid 1 \leq i \leq n]$ the tree obtained from t by replacing every occurrence of q_i by t_i for every $i \in [1, n]$. We abbreviate $C[\square \leftarrow t]$ simply by $C[t]$ for every $C \in C_\Sigma(Q)$ and $t \in T_\Sigma(V)$.

A (commutative) semiring is a tuple $\mathcal{A} = (A, +, \cdot, 0, 1)$ such that $(A, +, 0)$ and $(A, \cdot, 1)$ are commutative monoids; $a \cdot 0 = 0 = 0 \cdot a$ for every $a \in A$; and $\cdot$ distributes over $+$ from both sides. The semiring $\mathcal{A}$ is a semifield if for every $a \in A \setminus \{0\}$ there exists $a^{-1} \in A$ such that $a \cdot a^{-1} = 1$. A tree series is a mapping $\varphi \colon T \to A$ where $T \subseteq T_\Sigma(Q)$. The set of all such tree series is denoted by $A\langle\!\langle T \rangle\!\rangle$. For every $\varphi \in A\langle\!\langle T \rangle\!\rangle$ and $t \in T$, the coefficient $\varphi(t)$ is usually denoted by (φ, t).

A *weighted tree automaton* [9,8,20,6] (for short: wta) is a tuple $M = (Q, \Sigma, \mathcal{A}, \mu, \nu)$ such that (i) Q is an alphabet of states; (ii) Σ is a ranked alphabet; (iii) $\mathcal{A} = (A, +, \cdot, 0, 1)$ is a (commutative) semiring; (iv) $\mu = (\mu_n)_{n \geq 0}$ with $\mu_n \colon \Sigma_n \to A^{Q^n \times Q}$; and (v) $\nu \in A^Q$ is a final weight vector. The semantics of M is the tree series $\varphi_M \in A\langle\!\langle T_\Sigma \rangle\!\rangle$ given by $(\varphi_M, t) = \sum_{q \in Q} h_\mu(t)_q \cdot \nu_q$ (or simply the scalar product $h_\mu(t) \cdot \nu$) where $h_\mu \colon T_\Sigma \to A^Q$ is inductively defined by

$$h_\mu(\sigma(t_1, \ldots, t_n))_q = \sum_{q_1, \ldots, q_n \in Q} \mu_n(\sigma)_{(q_1, \ldots, q_n), q} \cdot \prod_{i=1}^{n} h_\mu(t_i)_{q_i}$$

for every $\sigma \in \Sigma_n$, $q \in Q$, and $t_1, \ldots, t_n \in T_\Sigma$. The wta M is said to recognize φ_M and two wta are *equivalent* if they recognize the same tree series.

The wta M is *deterministic and total* [6] if for every $\sigma \in \Sigma_n$ and $w \in Q^n$ there exists exactly one $q \in Q$ such that $\mu_n(\sigma)_{w,q} \neq 0$. Since we will exclusively deal with deterministic and total wta over semifields from now on, we will use the following representation: $M = (Q, \Sigma, \mathcal{A}, \delta, c, \nu)$ where $\delta \subseteq \bigcup_{n \geq 0} Q^n \times \Sigma_n \times Q$ is finite and $c \colon \delta \to A \setminus \{0\}$. In particular, $(w, \sigma, q) \in \delta$ if and only if $\mu_n(\sigma)_{w,q} \neq 0$, and for every $\tau = (w, \sigma, q) \in \delta$ we have $c(\tau) = \mu_n(\sigma)_{w,q}$. The determinism and totality restriction ensures that δ can be represented as $(\delta_\sigma)_{\sigma \in \Sigma}$ with $\delta_\sigma \colon Q^n \to Q$. We extend δ to a mapping $\delta \colon T_\Sigma(Q) \to Q$ as follows: $\delta(q) = q$ for every $q \in Q$ and $\delta(\sigma(t_1, \ldots, t_n)) = \delta_\sigma(\delta(t_1), \ldots, \delta(t_n))$ for every $\sigma \in \Sigma_n$ and $t_1, \ldots, t_n \in T_\Sigma(Q)$. A state $q \in Q$ is useful if there exists $t \in T_\Sigma$ such that $\delta(t) = q$. The deterministic and total wta M is said to have no useless states if all states of Q are useful.

Similarly, c can be represented as $(c_\sigma)_{\sigma \in \Sigma}$ with $c_\sigma \colon Q^n \to A \setminus \{0\}$. Due to the semifield restriction, this can be extended to a mapping $c \colon T_\Sigma(Q) \to A \setminus \{0\}$ by $c(q) = 1$ for every $q \in Q$ and $c(\sigma(t_1, \ldots, t_n)) = c_\sigma(\delta(t_1), \ldots, \delta(t_n)) \cdot \prod_{i=1}^{n} c(t_i)$ for every $\sigma \in \Sigma_n$ and $t_1, \ldots, t_n \in T_\Sigma(Q)$. It is then easy to show that $(\varphi_M, t) = c(t) \cdot \nu_{\delta(t)}$ for every $t \in T_\Sigma$. In fact, we extend φ_M to a tree series of $A\langle\!\langle T_\Sigma(Q) \rangle\!\rangle$ by defining $(\varphi_M, t) = c(t) \cdot \nu_{\delta(t)}$ for every $t \in T_\Sigma(Q)$. The following property, which will be used without explicit mention in the sequel, follows immediately.

Proposition 1 (cf. [3, Theorem 1]). *We have $(\varphi_M, t) = 0$ if and only if $\nu_{\delta(t)} = 0$ for every $t \in T_\Sigma(Q)$. Moreover,*

$$c(t[q_i \leftarrow t_i \mid 1 \le i \le n]) = c(t) \cdot \prod_{i=1}^{n} c(t_i)^{|t|_{q_i}}$$

for all pairwise distinct $q_1, \ldots, q_n \in Q$ and $t_1, \ldots, t_n \in T_\Sigma(Q)$ such that $\delta(t_i) = q_i$ for every $i \in [1, n]$.

Finally, let us recall the MYHILL-NERODE congruence relation [3] for tree series. To this end, we first recall Σ-algebras and congruences. A Σ-algebra (S, f) consists of a carrier set S and $f = (f_\sigma)_{\sigma \in \Sigma}$ such that $f_\sigma \colon S^n \to S$ for every $\sigma \in \Sigma_n$. The term Σ-algebra is given by $(T_\Sigma, \overline{\Sigma})$ where $\overline{\Sigma} = (\overline{\sigma})_{\sigma \in \Sigma}$ with $\overline{\sigma}(t_1, \ldots, t_n) = \sigma(t_1, \ldots, t_n)$ for every $\sigma \in \Sigma_n$ and $t_1, \ldots, t_n \in T_\Sigma$. In the sequel, we will drop the overlining. Note that (Q, δ) is a Σ-algebra. Let $\equiv$ be an equivalence relation on S. Then $\equiv$ is a congruence of (S, f) if for every $\sigma \in \Sigma_n$ and $s_1, \ldots, s_n, t_1, \ldots, t_n \in S$ such that $s_i \equiv t_i$ for every $i \in [1, n]$ we also have $f_\sigma(s_1, \ldots, s_n) \equiv f_\sigma(t_1, \ldots, t_n)$.

Let $\varphi \in A\langle\langle T_\Sigma \rangle\rangle$. The MYHILL-NERODE [3] relation $\equiv_\varphi \subseteq T_\Sigma \times T_\Sigma$ is defined for every $t, u \in T_\Sigma$ by $t \equiv_\varphi u$ if and only if there exists $a \in A \setminus \{0\}$ such that $(\varphi, C[t]) = a \cdot (\varphi, C[u])$ for every $C \in C_\Sigma$. We note that $\equiv_\varphi$ is a congruence of (T_Σ, Σ) [3, Lemma 5].

3   Myhill-Nerode Relation

In this section, we recall the theoretical foundations for the minimization procedure and introduce the MYHILL-NERODE relation on states of a deterministic and total wta. We keep it short because most of the material is only slightly adapted. Readers who are familiar with the MYHILL-NERODE congruence $\equiv_\varphi$ for a tree series φ may decide to read only Definition 2 and proceed to the next section. For the rest of the paper, let $M = (Q, \Sigma, \mathcal{A}, \delta, c, \nu)$ be a deterministic and total wta without useless states and $\mathcal{A} = (A, +, \cdot, 0, 1)$ a semifield, of which multiplication and calculation of inverses can be performed in constant time. Note that, depending on the actual semifield used, this might be an unrealistic assumption, but it simplifies the complexity analysis and is typically true for the fixed-precision arithmetic implemented on stock hardware. Finally, let $\varphi = \varphi_M$.

Definition 2 (cf. [3, page 8]). *The MYHILL-NERODE relation $\equiv \subseteq Q \times Q$ is defined for every $p, q \in Q$ by $p \equiv q$ if and only if there exists $a \in A \setminus \{0\}$ such that $(\varphi, C[p]) = a \cdot (\varphi, C[q])$ for every $C \in C_\Sigma$. We denote such a scaling factor a by $a_{p,q}$ for every $p, q \in Q$ such that $p \equiv q$.*

Note the similarity with the definition of the MYHILL-NERODE relation $\equiv_\varphi$ [3]. This similarity allows us to retain some of the useful properties of $\equiv_\varphi$. In the remainder of this section, we show some of those properties. We start with the fact that $\equiv$ is a congruence relation on (Q, δ).

Proposition 3 (cf. [3, Lemma 5]). *The relation $\equiv$ is a congruence relation on (Q, δ).*

The next lemma introduces a more restricted variant of $\equiv$ and relates it to $\equiv$. The more restricted variant will be very useful in the sequel and allows us to avoid an exponential blow-up.

Lemma 4. *The relation $\equiv' \subseteq Q \times Q$, which is given for every $p, q \in Q$ by $p \equiv' q$ if and only if there exists $a \in A \setminus \{0\}$ such that $(\varphi, C[p]) = a \cdot (\varphi, C[q])$ for every $C \in C_\Sigma(Q)$, coincides with $\equiv$.*

In fact, for every $p, q \in Q$ such that $p \equiv q$ the scaling factor $a_{p,q}$ also verifies $p \equiv' q$. This leads to the second main property (the first being the congruence property) that we will later use for refinement (see Proposition 12). The final lemma of this section establishes the relation of $\equiv$ to $\equiv_\varphi$. Note that $\mathrm{index}(\equiv_\varphi)$ coincides with the number of states of a minimal deterministic and total wta recognizing φ [3, Theorem 3].

Lemma 5. $\mathrm{index}(\equiv) = \mathrm{index}(\equiv_\varphi)$.

Proof. Let $t, u \in T_\Sigma$ such that $t \equiv_\varphi u$. There exists $a \in A \setminus \{0\}$ such that $(\varphi, C[t]) = a \cdot (\varphi, C[u])$ for every $C \in C_\Sigma$. We reason as follows:

$$c(t) \cdot (\varphi, C[\delta(t)]) = (\varphi, C[t]) = a \cdot (\varphi, C[u]) = a \cdot c(u) \cdot (\varphi, C[\delta(u)]) \ .$$

Since $a \cdot c(t)^{-1} \cdot c(u)$ does not depend on C, we obtain $\delta(t) \equiv \delta(u)$. Since M has no useless states, $\equiv$ thus has at most as many equivalence classes as $\equiv_\varphi$. For the converse, let $p, q \in Q$ such that $p \equiv q$. Moreover, let $t, u \in T_\Sigma$ be such that $\delta(t) = p$ and $\delta(u) = q$. Then analogous to the above we can prove that $t \equiv_\varphi u$. Hence, $\mathrm{index}(\equiv)$ and $\mathrm{index}(\equiv_\varphi)$ coincide. $\qquad\square$

4   Minimization Algorithm

In this section, we will develop our minimization algorithm for deterministic wta. Throughout, let $F = \{q \in Q \mid \nu_q \neq 0\}$. Note that any deterministic wta M' can be converted in linear time (in the number of transitions) into an equivalent deterministic and total wta without useless states. In contrast to the classical minimization algorithm for deterministic unweighted tree automata, we need to determine the scaling factor $a_{p,q}$ (see Definition 2) for each pair (p, q) of equivalent states. We will use the concept of a *sign of life* to help us determine it.

Definition 6. *A state $q \in Q$ is* live *if $\delta(C[q]) \in F$ for some context $C \in C_\Sigma(Q)$. Such a context is called a* sign of life *of q. If no sign of life of q exists, then q is* dead.

Roughly speaking, a state q is live if some final state can be reached from it. A sign of life of q shows one such path. Note that the sign-of-life context may contain states. Our first task is to determine signs of life for all states. This also identifies live and dead states. Let n be the number of states of M, m the number of transitions of M, and r the maximal rank of the symbols in Σ. To simplify the complexity statements, suppose that $r \geq 1$. Note that consequently $m \geq n$.

Algorithm 1. COMPUTESOL(M): Compute signs of life and initial partition.

Require: deterministic and total wta $M = (Q, \Sigma, \mathcal{A}, \delta, c, \nu)$
    $D \leftarrow Q \setminus F$ // unexplored states
2:   $\mathrm{sol} \leftarrow \{(q, \square) \mid q \in F\}$ // final states have trivial sign of life
    $T \leftarrow \{(w, \sigma, q) \in \delta \mid q \in F\}$ // add all transitions leading to a final state to FIFO queue
4: **while** $T \neq \emptyset$ **do**
    let $\tau = ((q_1, \ldots, q_k), \sigma, q) \in T$ // get first element in FIFO queue T
6:   $I \leftarrow \{i \in [1, k] \mid q_i \in D, \forall j \in [1, i-1]: q_j \neq q_i\}$ // select indexes of unexplored states
    $\mathrm{sol} \leftarrow \mathrm{sol} \cup \{(q_i, \mathrm{sol}(q)[\sigma(q_1, \ldots, q_{i-1}, \square, q_{i+1}, \ldots, q_k)]) \mid i \in I\}$ // add signs of life
8:   $P \leftarrow \{q_i \mid i \in I\}$ // new live states
    $D \leftarrow D \setminus P$ // remove new live states from unexplored states
10:   $T \leftarrow (T \setminus \{\tau\}) \cup \{(w, \gamma, p) \in \delta \mid p \in P\}$ // add all transitions leading to new live states
    end while
12: $\Pi \leftarrow \{\{q \in Q \setminus D \mid \mathrm{ht}(\mathrm{sol}(q)) = i\} \mid i \geq 1\} \cup \{D\}$ // group states by height of sign of life
    return (Π, sol, D)

Algorithm 1 returns an initial partition Π, signs of life sol, and the set D of dead states. Let us make two remarks. First, it is essential that T is handled as a FIFO queue with additions at the end and removal at the beginning. This guarantees that the height of the constructed signs of life is minimal. Second, $\emptyset$ might be an element of Π. To avoid complicated case distinctions, we permit this slightly nonstandard behavior, which does not affect the correctness of our algorithms.

Lemma 7. *Let* (Π, sol, D) *be the result of running Algorithm 1 on* M. *Then* sol(q) *is a sign of life of* q *of size at most* rn *for every state* q *of* $Q \setminus D$, D *is the set of all dead states,* Π *saturates* F, *and* $\equiv$ *is a refinement of* $\equiv_\Pi$. *Moreover, Algorithm 1 can be implemented to run in time* $O(rm)$.

Proof. Lines 1–3 run in time $O(m)$ because $m \geq n$. Clearly, each transition can be added at most once to T, so lines 4–11 can be executed at most m times. Lines 6–8 can be executed in time $O(r)$; note that this requires a list representation of the signs of life (i.e., a sign of life is a list of pairs consisting of a transition and an integer indicating the position of $\square$) to avoid the creation and/or copying of transitions. If we suppose that access to the list of transitions leading to a certain state is constant (which can be achieved by a $O(m)$ preprocessing step sorting the transitions in n buckets), then line 10 can be executed in $O(r)$ time. Since $r \geq 1$, we obtain a running time of $O(rm)$. Finally, we note that the partition constructed in line 12 could have been constructed during the loop at no additional expense; we presented it this way for clarity.

Next, we prove that sol(q) is indeed a sign of life of q for every $q \in Q \setminus D$. The trivial contexts added for each final state in line 2 are obviously signs of life.

It remains to show that C, the second component in the pair of line 7, is a sign of life of q_i. By induction hypothesis, we may assume that $\mathrm{sol}(q)$ is a sign of life of q; i.e., $\delta(\mathrm{sol}(q)[q]) \in F$. Since M is deterministic and $((q_1, \ldots, q_k), \sigma, q) \in \delta$, we obtain $\delta(C[q_i]) = \delta(\mathrm{sol}(q)[q]) \in F$ and thus C is a sign of life of q_i. It is obvious that Π saturates F. We leave the proof of the fact that D is indeed the set of all dead states to the reader.

Clearly, final states are assigned a sign of life of height 1 and size $1 \leq r$. Further signs of life are always constructed as $C = \mathrm{sol}(q)[\sigma(q_1, \ldots, q_{i-1}, \Box, q_{i+1}, \ldots, q_k)]$ (see line 7). Thus, the height (respectively, the size) of the new sign of life C is 1 (respectively, at most r) greater than that of the sign of life $\mathrm{sol}(q)$. Consequently, height and size of every sign of life $\mathrm{sol}(q)$ with $q \in Q \setminus D$ are at most n and rn, respectively. Finally, we have to show that $\equiv$ is a refinement of $\equiv_\Pi$. To this end, we have to show that $p \equiv q$ implies that $p \equiv_\Pi q$ for every $p, q \in Q$. Let $p, q \in Q$ be such that $p \equiv q$. Clearly, p and q share all signs of life (see Lemma 4). Consequently, if p is dead, then also q must be dead, and in that case, $p \equiv_\Pi q$. Otherwise, p and q are live. We already remarked that Algorithm 1 computes signs of life that are minimal in height; the proof of that statement is left as an exercise. The height-minimal sign of life $\mathrm{sol}(p)$ must be a sign of life of q as well, and consequently, $\mathrm{ht}(\mathrm{sol}(p)) = \mathrm{ht}(\mathrm{sol}(q))$, which yields $p \equiv_\Pi q$.   □

We allow contexts of $C_\Sigma(Q)$ instead of only contexts of C_Σ as signs of life in order to obtain the linear size complexity given in Lemma 7. The more common approach to use contexts of C_Σ would yield signs of life, whose size might be exponential in n. Since we will run M on signs of life, this would have led to an exponential time complexity.

The principal approach of the minimization algorithm is partition refinement as, for example, in the classical minimization algorithm for minimizing unweighted deterministic tree automata [10]. We successively refine the initial partition returned by Algorithm 1 until $\equiv$ is reached. Before we turn to more detail, let us introduce the main data structure.

Definition 8. *Let Π be a partition of Q that saturates F, $L \subseteq Q$ be the set of live states, $\mathrm{sol}: L \to C_\Sigma(Q)$ be such that $\mathrm{sol}(q)$ is a sign of life of q for every $q \in L$, $f: L \to A \setminus \{0\}$, and $r: (\Pi \setminus \{\emptyset, Q \setminus L\}) \to Q$ a representative mapping. Then $(\Pi, \mathrm{sol}, f, r)$ is a* stage *if*

(i) $\equiv$ is a refinement of $\equiv_\Pi$;
(ii) $\mathrm{sol}(q) = \Box$ for every $q \in F$; and
(iii) for every $q \in L$ we have $(\varphi, \mathrm{sol}(p)[q]) = f(q) \cdot (\varphi, \mathrm{sol}(p)[p])$ where $p = r([q])$.

The stage is stable *if additionally*

(iv) $\equiv_\Pi$ is a congruence of (Q, δ); and
(v) for every $q \in L$, $\sigma \in \Sigma_n$, and $C \in C_n(Q)$ such that $\delta_\sigma(C[q]) \in L$

$$f(q)^{-1} \cdot c_\sigma(C[q]) \cdot f(\delta_\sigma(C[q])) = c_\sigma(C[p]) \cdot f(\delta_\sigma(C[p]))$$

where $p = r([q])$.

In a stage, we have a partition, signs of life, and two new components. The mapping r assigns to each nonempty block (apart from the block of dead states) of the partition a representative and the mapping f assigns to each live state the scaling factor to the representative of its block [see Condition (iii)]. A stable stage additionally requires $\equiv_\Pi$ to be a congruence of (Q, δ) and Condition (v), which is of paramount importance in the implementation (as a wta) of a stable stage. Let us show how to derive a deterministic and total wta recognizing φ from a stable stage.

Definition 9 (cf. [3, Definition 4]). Let $S = (\Pi, \text{sol}, f, r)$ be a stable stage and D the set of dead states. The wta $M_S = (\Pi \setminus \{\emptyset\}, \Sigma, \mathcal{A}, \delta', c', \nu')$ is constructed as follows: for every $\sigma \in \Sigma_k$ and $q_1, \ldots, q_k \in Q$ let

- $\delta'_\sigma([q_1], \ldots, [q_k]) = [\delta_\sigma(q_1, \ldots, q_k)]$;
- $c'_\sigma([q_1], \ldots, [q_k]) = 1$ if $\delta_\sigma(q_1, \ldots, q_k) \in D$ and otherwise

$$c'_\sigma([q_1], \ldots, [q_k]) = \prod_{i=1}^{k} f(q_i)^{-1} \cdot c_\sigma(q_1, \ldots, q_k) \cdot f(\delta_\sigma(q_1, \ldots, q_k))$$

- $\nu'_B = \nu_{r(B)}$ for every $B \in \Pi \setminus \{\emptyset, D\}$ and if $D \in \Pi \setminus \{\emptyset\}$, then $\nu'_D = 0$.

The construction of M_S can be implemented to run in time $O(rm)$. However, some remarks are required here. First, δ' is well-defined because $\equiv_\Pi$ is a congruence on (Q, δ). Second, let us consider the definition of c'. Suppose that $p_1, \ldots, p_k, q_1, \ldots, q_k \in Q$ such that $p_i \equiv_\Pi q_i$ for every $i \in [1, k]$. By the congruence property and Condition (i) of Definition 8, the case distinction is well-defined. We show that

$$\prod_{i=1}^{k} f(p_i)^{-1} \cdot c_\sigma(p_1, \ldots, p_k) \cdot f(\delta_\sigma(p_1, \ldots, p_k))$$

$$= \prod_{i=2}^{k} f(p_i)^{-1} \cdot c_\sigma(r([p_1]), p_2, \ldots, p_k) \cdot f(\delta_\sigma(r([p_1]), p_2, \ldots, p_k))$$

$$= \ldots$$

$$= c_\sigma(r([p_1]), \ldots, r([p_k])) \cdot f(\delta_\sigma(r([p_1]), \ldots, r([p_k])))$$

$$= c_\sigma(r([q_1]), \ldots, r([q_k])) \cdot f(\delta_\sigma(r([q_1]), \ldots, r([q_k])))$$

$$= \ldots$$

$$= \prod_{i=2}^{k} f(q_i)^{-1} \cdot c_\sigma(r([q_1]), q_2, \ldots, q_k) \cdot f(\delta_\sigma(r([q_1]), q_2, \ldots, q_k))$$

$$= \prod_{i=1}^{k} f(q_i)^{-1} \cdot c_\sigma(q_1, \ldots, q_k) \cdot f(\delta_\sigma(q_1, \ldots, q_k))$$

by repeated application of Condition (v) of Definition 8, which proves the well-definedness of c'. It is obvious that M_S has index($\equiv_\Pi$) many states. We should finally show that M_S recognizes φ.

Theorem 10. *Let* $S = (\Pi, \text{sol}, \text{f}, \text{r})$ *be a stable stage and* $M_S = (Q', \Sigma, \mathcal{A}, \delta', c', \nu')$. *Then* M_S *is a minimal deterministic and total wta recognizing* φ.

Proof. Let us first show that M_S recognizes φ. From the classical minimization construction it should be clear that $\delta(t) \in \delta'(t)$ for every $t \in T_\Sigma$. We prove the statement $c(t) = c'(t) \cdot f(\delta(t))^{-1}$ by induction on t. Let $t = \sigma(t_1, \ldots, t_k)$ for some $\sigma \in \Sigma_k$ and $t_1, \ldots, t_k \in T_\Sigma$. By definition

$$c(t) = c_\sigma(\delta(t_1), \ldots, \delta(t_k)) \cdot \prod_{i=1}^{k} c(t_i) = c_\sigma(\delta(t_1), \ldots, \delta(t_k)) \cdot \prod_{i=1}^{k} \Big(c'(t_i) \cdot f(\delta(t_i))^{-1} \Big)$$

where the last equality is by induction hypothesis. We finish the proof of the auxiliary statement using the definition of c' and $\delta(t_i) \in \delta'(t_i)$

$$c(t) = c'_\sigma(\delta'(t_1), \ldots, \delta'(t_k)) \cdot \prod_{i=1}^{k} c'(t_i) \cdot f(\delta_\sigma(\delta(t_1), \ldots, \delta(t_k)))^{-1} = c'(t) \cdot f(\delta(t))^{-1} \ .$$

We now continue for every $t \in T_\Sigma$ with a case distinction. Let $q = \delta(t)$. If $q \notin F$, then $(\varphi, t) = 0$ and $(\varphi_{M_S}, t) = 0$ because $\nu'_{[q]} = 0$ (recall that Π saturates F). If $q \in F$, then

$$(\varphi_{M_S}, t) = c'(t) \cdot \nu'_{[q]} = c(t) \cdot f(q) \cdot \nu_{r([q])} = c(t) \cdot \nu_q = (\varphi, t)$$

by Conditions (ii) and (iii) of Definition 8 since $\text{sol}(\text{r}([q])) = \square$. Thus, we proved that M_S recognizes φ. By Lemma 5 and [3, Theorem 3], index($\equiv$) is the number of states of a minimal deterministic and total wta recognizing φ. The wta M_S has index($\equiv_\Pi$) states and by Condition (i) of Definition 8, $\equiv$ is a refinement of $\equiv_\Pi$, hence index($\equiv_\Pi$) $\leq$ index($\equiv$). Consequently, $\equiv_\Pi$ coincides with $\equiv$ and M_S is a minimal deterministic and total wta recognizing φ. $\square$

Note that the above theorem also shows that $(\Pi, \text{sol}, \text{f}, \text{r})$ can only be a stable stage if $\equiv_\Pi$ coincides with $\equiv$. Since we already have a suitable initial partition that saturates F along with suitable signs of life, our next step is to determine the scaling factors (see Definition 2). To this end, we employ Algorithm 2, which is given a partition and computes the scaling factor for each element relative to a chosen representative of its block. If it cannot compute such a scaling factor (which only happens when the sign of life of the representative is not a sign of life of the considered state), then it splits the state from its current block. The algorithm completely ignores the block of dead states. This can be done because the block of dead states will never be split (since $p \equiv q$ whenever both p and q are dead). The idea of our minimization algorithm is to refine the initial partition returned by Algorithm 1 until we reach $\equiv$. Algorithm 2 completes a suitable partition to a stage.

Lemma 11. *Given signs of life of Algorithm 1 and a partition* Π *such that* $\equiv$ *is a refinement of* $\equiv_\Pi$, *Algorithm 2 can be implemented to run in time* $O(rn^3)$ *and returns a stage* $(\Pi', \text{sol}, \text{f}, \text{r})$ *such that* $\equiv_{\Pi'}$ *is a refinement of* $\equiv_\Pi$.

Algorithm 2. COMPLETE(M, Π, sol, D): Compute scaling factors.

Require: deterministic and total wta $M = (Q, \Sigma, \mathcal{A}, \delta, c, \nu)$, set D of dead states, sign of life sol(q) for every $q \in Q \setminus D$, and a partition Π such that $\equiv$ is a refinement of $\equiv_\Pi$

$\ f \leftarrow \emptyset$ // map of scaling factors
2. $\ r \leftarrow \emptyset$ // map of representatives
$\ \Pi \leftarrow \Pi \setminus \{D\}$ // remove block of dead states
4. $\ \Pi' \leftarrow \{D\}$ // output partition containing block of dead states
 **while** $\Pi \neq \emptyset$ and $\Pi \neq \{\emptyset\}$ **do**
6. $\quad$ let $P \in \Pi$ and $p \in P$ // select new block and representative
$\quad P' \leftarrow \{q \in P \mid \delta(\text{sol}(p)[q]) \in F\}$ // collect states that share sign of life sol(p)
8. $\quad f \leftarrow f \cup \{(q, c(\text{sol}(p)[q]) \cdot c(\text{sol}(p)[p])^{-1}) \mid q \in P'\}$ // add scaling factors
$\quad r \leftarrow r \cup \{(P', p)\}$ // add representative for P'
10. $\quad \Pi' \leftarrow \Pi' \cup \{P'\}$ // block P' is processed
$\quad \Pi \leftarrow (\Pi \setminus \{P\}) \cup \{P \setminus P'\}$ // remove old block and add new block $P \setminus P'$
12. **end while**
 **return** (Π', sol, f, r)

Proof. We defer correctness for the moment. The loop in lines 5–12 can be executed at most n times as each time at least one state is processed. Evaluating the sign of life takes at most $O(rn)$ since the size of any sign of life is at most rn. Thus, lines 7–8 execute in $O(rn^2)$ and the whole algorithm runs in time $O(rn^3)$. Now, let us consider correctness. It should be clear that Condition (ii) holds and Condition (iii) holds because it is enforced in line 8. It remains to check Condition (i). Clearly, $\equiv_{\Pi'}$ is a refinement of $\equiv_\Pi$ and by assumption $\equiv$ is a refinement of $\equiv_\Pi$. Let $p, q \in Q$ such that $p \equiv q$. Consequently, $p \equiv_\Pi q$. Let $p' \in Q$ be such that $p \equiv_\Pi p'$ (cf. the selection in line 6). Then $\delta(\text{sol}(p')[p]) \in F$ if and only if $\delta(\text{sol}(p')[q]) \in F$ because $(\varphi, \text{sol}(p')[p]) \neq 0$ if and only if $(\varphi, \text{sol}(p')[q]) \neq 0$ by $p \equiv q$. This yields that independently of the selection of the representative in line 6, p and q cannot be split in line 7, and hence $p \equiv_{\Pi'} q$. $\qquad\square$

Note that Π' saturates F whenever Π does because $\equiv_{\Pi'}$ is a refinement of $\equiv_\Pi$. As in the unweighted case [10], we use a partition refinement algorithm to compute the MYHILL-NERODE relation. To this end, we start with the initial partition computed by Algorithm 1 and complete it to a stage with the help of Algorithm 2. Then we refine according to Conditions (iv) and (v) of Definition 8. For this to work, variants of these conditions should also be fulfilled by $\equiv$. We already proved in Proposition 3 that $\equiv$ is a congruence of (Q, δ) as in the classical unweighted case. Second, the weights of transitions from equivalent states leading to live states must obey a certain compatibility requirement, which we show in the next proposition.

Proposition 12. *For every* $\sigma \in \Sigma_k$, $C \in C_k(Q)$, *and* $p, q \in Q$ *such that* $p \equiv q$ *and* $\delta_\sigma(C[p])$ *is live, we have* $a_{p,q}^{-1} \cdot c_\sigma(C[p]) \cdot a_{p',q'} = c_\sigma(C[q])$ *where* $p' = \delta_\sigma(C[p])$ *and* $q' = \delta_\sigma(C[q])$.

Proof. Since p' is live, there exists a context $C' \in C_\Sigma(Q)$ such that $\delta(C'[p']) \in F$. Consider the context $C'' = C'[\sigma(C)]$. Since $p \equiv q$, and thus $p \equiv' q$ by Lemma 4, it follows that $(\varphi, C''[p]) = a_{p,q} \cdot (\varphi, C''[q])$. In addition, $p' \equiv q'$ because $\equiv$ is a congruence. Now we compute as follows:

$$c_\sigma(C[p]) \cdot (\varphi, C'[p']) = (\varphi, C''[p]) = a_{p,q} \cdot (\varphi, C''[q])$$
$$= a_{p,q} \cdot c_\sigma(C[q]) \cdot (\varphi, C'[q']) = a_{p,q} \cdot c_\sigma(C[q]) \cdot a_{p',q'}^{-1} \cdot (\varphi, C'[p']) \ .$$

Since $(\varphi, C'[p']) \neq 0$ because $\delta(C'[p']) \in F$, we obtain the statement by cancelling $(\varphi, C'[p'])$. □

In the classical unweighted case, only the congruence property is used to refine. The additional constraint basically restricts the weights on the transitions whereas the congruence property only restricts the presence/absence of transitions. Altogether, the previous proposition suggests the following refinement step.

Definition 13. *Let $(\Pi, \mathrm{sol}, \mathrm{f}, \mathrm{r})$ be a stage and D be the set of dead states. Then the refinement $\mathrm{REFINE}(M, \Pi, \mathrm{sol}, \mathrm{f}, \mathrm{r}, D)$ is defined to be the partition Π' such that for every $p, q \in Q$ we have $p \equiv_{\Pi'} q$ if and only if $p \equiv_\Pi q$ and for every $\sigma \in \Sigma_k$ and $C \in C_k(Q)$*

(i) $\delta_\sigma(C[p]) \equiv_\Pi \delta_\sigma(C[q])$; and
(ii) $f(p)^{-1} \cdot c_\sigma(C[p]) \cdot f(\delta_\sigma(C[p])) = f(q)^{-1} \cdot c_\sigma(C[q]) \cdot f(\delta_\sigma(C[q]))$, if $\delta_\sigma(C[p])$ is live.

The following lemma shows that REFINE refines in the desired manner. In particular, whenever $\equiv$ is a refinement of $\equiv_\Pi$, then $\equiv$ is also a refinement of $\equiv_{\Pi'}$. Thus, we only refine to the level of $\equiv$ and never beyond. This simple property follows in a straightforward manner from Definition 13 and Proposition 12.

Lemma 14. *Let $(\Pi, \mathrm{sol}, \mathrm{f}, \mathrm{r})$ be a stage and D the set of dead states. $\mathrm{REFINE}(M, \Pi, \mathrm{sol}, \mathrm{f}, \mathrm{r}, D)$ can be implemented to run in time $O(rmn^2)$. Moreover, the resulting partition Π' is such that $\equiv_{\Pi'}$ is a refinement of $\equiv_\Pi$, and if $\equiv$ is a refinement of $\equiv_\Pi$, then $\equiv$ is a refinement of $\equiv_{\Pi'}$.*

Again note that whenever Π saturates F, then also Π' saturates F simply because $\equiv_{\Pi'}$ is a refinement of $\equiv_\Pi$. We are now ready to state the main minimization algorithm.

Algorithm 3. Minimization of deterministic wta

Require: deterministic and total wta $M = (Q, \Sigma, \mathcal{A}, \delta, c, \nu)$ without useless states
   $(\Pi', \mathrm{sol}, D) \leftarrow \mathrm{COMPUTESOL}(M)$ // see Algorithm 1; complexity: $O(rm)$
2. **repeat**
   $(\Pi, \mathrm{sol}, \mathrm{f}, \mathrm{r}) \leftarrow \mathrm{COMPLETE}(M, \Pi', \mathrm{sol}, D)$ // see Algorithm 2; complexity: $O(rn^3)$
4. $\Pi' \leftarrow \mathrm{REFINE}(M, \Pi, \mathrm{sol}, \mathrm{f}, \mathrm{r}, D)$ // see Definition 13; complexity: $O(rmn^2)$
   until $\Pi' = \Pi$
6. **return** $M_{(\Pi, \mathrm{sol}, \mathrm{f}, \mathrm{r})}$ // see Definition 9; complexity $O(rm)$

Theorem 15. *A minimal deterministic and total wta M' recognizing φ can be obtained in time $O(rmn^4)$.*

Proof. The loop in lines 2–5 in Algorithm 3 can be entered at most n times, which immediately yields the required time bound using Lemmata 7, 11, and 14. The initial partition Π' in line 1 saturates F and so does every subsequent partition. Moreover, by Lemmata 11 and 14, $\equiv$ is a refinement of every subsequent $\equiv_\Pi$ and $\equiv_{\Pi'}$ because $\equiv$ is a refinement of $\equiv_{\Pi'}$ in line 1 by Lemma 7. Consequently, every Π is a stage by Lemma 11, and if $\Pi = \Pi'$ in line 5, then $(\Pi, \text{sol}, \text{f}, \text{r})$ is a stable stage. By Theorem 10, the wta returned in line 6 is a minimal deterministic and total wta recognizing φ. $\qquad\square$

5 A Small Example

Let us discuss the example of [22] (with one minor modification), which presents a simplistic wta for simple English sentences. It penalizes long sentences by decreasing their score. The score will be a real number and we will use $(\mathbb{R}, +, \cdot, 0, 1)$ as the underlying field. Our ranked alphabet is

$$\Sigma = \{\sigma, \text{Alice}, \text{Bob}, \text{loves}, \text{hates}, \text{ugly}, \text{nice}, \text{mean}\}$$

of which σ is binary and all other symbols have rank 0. We abbreviate the multi-letter symbols by their first letter (e.g., Alice by just A). As states we have $Q = \{\text{NN}, \text{VB}, \text{ADJ}, \text{VP}, \text{NP}, \text{S}, \bot\}$ of which only S is final (with $\nu_S = 1$). Transitions and transition weights are given as follows:

$$\delta_\sigma(\text{NN}, \text{VP}) = \text{S} \quad \delta_\sigma(\text{NP}, \text{VP}) = \text{S} \quad \delta_\sigma(\text{VB}, \text{NN}) = \text{VP} \quad \delta_\sigma(\text{VB}, \text{NP}) = \text{VP}$$
$$c_\sigma(\text{NN}, \text{VP}) = 0.5 \quad c_\sigma(\text{NP}, \text{VP}) = 0.5 \quad c_\sigma(\text{VB}, \text{NN}) = 0.5 \quad c_\sigma(\text{VB}, \text{NP}) = 0.5$$
$$\delta_\sigma(\text{ADJ}, \text{NN}) = \text{NP} \quad \delta_\sigma(\text{ADJ}, \text{NP}) = \text{NP}$$
$$c_\sigma(\text{ADJ}, \text{NN}) = 0.5 \quad c_\sigma(\text{ADJ}, \text{NP}) = 0.5$$

and

$$\delta_A() = \text{NN} \quad \delta_B() = \text{NN} \quad \delta_l() = \text{VB} \quad \delta_h() = \text{VB} \quad \delta_u() = \text{ADJ} \quad \delta_n() = \text{ADJ} \quad \delta_m() = \text{ADJ}$$
$$c_A() = 0.5 \quad c_B() = 0.5 \quad c_l() = 0.5 \quad c_h() = 0.5 \quad c_u() = 0.33 \quad c_n() = 0.33 \quad c_m() = 0.33 \ .$$

For all remaining combinations (x, y) we set $\delta_\sigma(x, y) = \bot$ and $c_\sigma(x, y) = 1$. Now, we completely specified our deterministic and total input wta M, which has no useless states.

Next, we compute signs of life according to Algorithm 1. It may return $(\Pi', \text{sol}, \{\bot\})$ where $\Pi' = \{\{S\}, \{\text{VP}, \text{NP}, \text{NN}\}, \{\text{ADJ}, \text{VB}\}, \{\bot\}\}$ and the signs of life are

$$\text{sol}(S) = \Box \qquad \text{sol}(\text{NP}) = \sigma(\Box, \text{VP}) \qquad \text{sol}(\text{ADJ}) = \sigma(\sigma(\Box, \text{NN}), \text{VP})$$
$$\text{sol}(\text{VP}) = \sigma(\text{NN}, \Box) \qquad \text{sol}(\text{NN}) = \sigma(\Box, \text{VP}) \qquad \text{sol}(\text{VB}) = \sigma(\text{NN}, \sigma(\Box, \text{NN})) \ .$$

Next, we call COMPLETE($M, \Pi', \mathrm{sol}, \{\bot\}$), which may return the stage $(\Pi, \mathrm{sol}, \mathrm{f}, \mathrm{r})$ where

- $\Pi = \{\{S\}, \{VP\}, \{NP, NN\}, \{ADJ\}, \{VB\}, \{\bot\}\}$;
- $f(x) = 1$ for all live states x; and
- $r(\{NP, NN\}) = NP$ and $r(\{x\}) = x$ for all other live states x.

Finally, we refine this partition, but NP and NN will not be split. Thus, we construct the deterministic and total wta $M_{(\Pi, \mathrm{sol}, \mathrm{f}, \mathrm{r})} = (\Pi, \Sigma, \mathbb{R}, \delta', c', \nu')$ with the final state $\{S\}$ (with $\nu'_{\{S\}} = 1$). Transitions and transition weights are given as follows (we drop the parentheses from the singleton sets):

$$\delta'_A() = \{NN, NP\} \quad \delta_B() = \{NN, NP\} \quad \delta'_l() = VB \quad \delta_h() = VB \quad \delta'_u() = ADJ$$
$$c'_A() = 0.5 \quad\quad c'_B() = 0.5 \quad\quad c'_l() = 0.5 \quad c'_h() = 0.5 \quad c'_u() = 0.33$$

$$\delta'_n() = ADJ \quad\quad \delta'_m() = ADJ$$
$$c'_n() = 0.33 \quad\quad c'_m() = 0.33$$

and

$$\delta'_\sigma(\{NN, NP\}, VP) = S \quad \delta'_\sigma(VB, \{NN, NP\}) = VP \quad \delta'_\sigma(ADJ, \{NN, NP\}) = \{NN, NP\}$$
$$c'_\sigma(\{NN, NP\}, VP) = 0.5 \quad c'_\sigma(VB, \{NN, NP\}) = 0.5 \quad c'_\sigma(ADJ, \{NN, NP\}) = 0.5 \ .$$

For all remaining combinations (x, y) we have $\delta'_\sigma(x, y) = \{\bot\}$ and $c'_\sigma(x, y) = 1$. Note that a different minimal deterministic wta was obtained in [22]; note that this different wta cannot be obtained by our algorithm (since all transitions not involving NP and NN are essentially kept).

6   Conclusion and Open Problems

We presented the first polynomial-time minimization algorithm for deterministic weighted tree automata over semifields. If we suppose that the semifield operations can be performed in constant time, then our algorithm runs in time $O(rmn^4)$. In fact, our algorithm works equally well for wta with final states (i.e., $\nu_q \in \{0, 1\}$ for every $q \in Q$) because it then returns a minimal equivalent wta with final states. This contrasts the situation encountered with the pushing strategy of [25,12], which needs final weights in general.

Finally, let us mention some open problems. Can a HOPCROFT-like strategy [18] improve the presented algorithm? A more detailed complexity analysis should be conducted to obtain a tighter bound on the time complexity of the algorithm. Can minimization be performed in a similar manner as presented in [25,12] for deterministic weighted string automata? This might lead to an algorithm that outperforms our algorithm. Finally, the theoretical foundations for minimization of (even nondeterministic) weighted tree automata over fields have been laid in [9,7], but to the author's knowledge a polynomial-time minimization algorithm is still missing.

References

1. Allauzen, C., Riley, M., Schalkwyk, J., Skut, W., Mohri, M.: OpenFst — a general and efficient weighted finite-state transducer library. In: Holub, J., Žďárek, J. (eds.) CIAA 2007. LNCS, vol. 4783, pp. 11–23. Springer, Heidelberg (2007)
2. Angluin, D.: Learning regular sets from queries and counterexamples. Inform. and Comput. 75(2), 87–106 (1987)
3. Borchardt, B.: The Myhill-Nerode theorem for recognizable tree series. In: Ésik, Z., Fülöp, Z. (eds.) DLT 2003. LNCS, vol. 2710, pp. 146–158. Springer, Heidelberg (2003)
4. Borchardt, B.: A pumping lemma and decidability problems for recognizable tree series. Acta Cybernet 16(4), 509–544 (2004)
5. Borchardt, B.: The Theory of Recognizable Tree Series. PhD thesis, Technische Universität Dresden (2005)
6. Borchardt, B., Vogler, H.: Determinization of finite state weighted tree automata. J. Autom. Lang. Combin. 8(3), 417–463 (2003)
7. Bozapalidis, S.: Effective construction of the syntactic algebra of a recognizable series on trees. Acta Inform. 28(4), 351–363 (1991)
8. Bozapalidis, S.: Equational elements in additive algebras. Theory Comput. Systems 32(1), 1–33 (1999)
9. Bozapalidis, S., Louscou-Bozapalidou, O.: The rank of a formal tree power series. Theoret. Comput. Sci. 27(1–2), 211–215 (1983)
10. Comon-Lundh, H., Dauchet, M., Gilleron, R., Jacquemard, F., Lugiez, D., Tison, S., Tommasi, M.: Tree automata—techniques and applications (2007), http://tata.gforge.inria.fr/
11. Drewes, F., Vogler, H.: Learning deterministically recognizable tree series. J. Autom. Lang. Combin. 12(3), 332–354 (2007)
12. Eisner, J.: Simpler and more general minimization for weighted finite-state automata. In: Human Language Technology Conf. of the North American Chapter of the Association for Computational Linguistics, pp. 64–71 (2003)
13. Frishert, M., Cleophas, L.G., Watson, B.W.: Fire station: An environment for manipulating finite automata and regular expression views. In: Domaratzki, M., Okhotin, A., Salomaa, K., Yu, S. (eds.) CIAA 2004. LNCS, vol. 3317, pp. 125–133. Springer, Heidelberg (2005)
14. Gécseg, F., Steinby, M.: Tree Automata. Akadémiai Kiadó, Budapest (1984)
15. Gécseg, F., Steinby, M.: Tree languages. In: Handbook of Formal Languages, ch.1, vol. 3, pp. 1–68. Springer, Heidelberg (1997)
16. Graehl, J.: Carmel finite-state toolkit. ISI/USC (1997), http://www.isi.edu/licensed-sw/carmel
17. Habrard, A., Oncina, J.: Learning multiplicity tree automata. In: Sakakibara, Y., Kobayashi, S., Sato, K., Nishino, T., Tomita, E. (eds.) ICGI 2006. LNCS (LNAI), vol. 4201, pp. 268–280. Springer, Heidelberg (2006)
18. Hopcroft, J.E.: An $n \log n$ algorithm for minimizing states in a finite automaton. In: Theory of Machines and Computations, pp. 189–196. Academic Press, London (1971)
19. Knight, K., Graehl, J.: An overview of probabilistic tree transducers for natural language processing. In: Gelbukh, A. (ed.) CICLing 2005. LNCS, vol. 3406, pp. 1–24. Springer, Heidelberg (2005)
20. Kuich, W.: Formal power series over trees. In: Proc. 3rd Int. Conf. Developments in Language Theory, pp. 61–101. Aristotle University of Thessaloniki (1998)

21. Kuich, W., Salomaa, A.: Semirings, Automata, Languages. Monographs in Theoretical Computer Science. An EATCS Series, vol. 5. Springer, Heidelberg (1986)
22. Maletti, A.: Learning deterministically recognizable tree series — revisited. In: Bozapalidis, S., Rahonis, G. (eds.) CAI 2007. LNCS, vol. 4728, pp. 218–235. Springer, Heidelberg (2007)
23. May, J., Knight, K.: A better n-best list: Practical determinization of weighted finite tree automata. In: Proc. North American Chapter of the Association for Computational Linguistics, pp. 351–358 (2006)
24. May, J., Knight, K.: Tiburon: A weighted tree automata toolkit. In: H. Ibarra, O., Yen, H.-C. (eds.) CIAA 2006. LNCS, vol. 4094, pp. 102–113. Springer, Heidelberg (2006)
25. Mohri, M.: Minimization algorithms for sequential transducers. Theoret. Comput. Sci. 234(1–2), 177–201 (2000)
26. Seidl, H.: Deciding equivalence of finite tree automata. SIAM J. Comput. 19(3), 424–437 (1990)

Lower Bounds for Generalized Quantum Finite Automata

Mark Mercer

Département d'Informatique,
Université de Sherbrooke, QC, Canada

Abstract. We obtain several lower bounds on the language recognition power of Nayak's *generalized quantum finite automata (GQFA)* [12]. Techniques for proving lower bounds on Kondacs and Watrous' *one-way quantum finite automata (KWQFA)* were introduced by Ambainis and Freivalds [2], and were expanded in a series of papers. We show that many of these techniques can be adapted to prove lower bounds for GQFAs. Our results imply that the class of languages recognized by GQFAs is not closed under union. Furthermore, we show that there are languages which can be recognized by GQFAs with probability $p > 1/2$, but not with $p > 2/3$.

Quantum finite automata (QFA) are online, space-bounded models of quantum computation. Similar to *randomized finite automata* [16] where the state is a random variable over a finite set, the state of a QFA is a quantum superposition of finite dimension. The machine processes strings $w \in \Sigma^*$ by applying a sequence of state transformations specified by the sequence of letters in w, and the output of the machine is determined by a measurement of the machine state. A central problem is to characterize the language recognition power of QFAs.

Most quantized versions of classical computation devices (such as quantum circuits [17]) are at least as powerful as their classical counterparts. It is not clear that this should be the case for quantum finite automata. Typically, the execution of classical computation on a quantum device is performed by converting classical computation into reversible computation using standard techniques such as in [5]. The most general definitions of QFAs [8] are equal in language recognition power to deterministic finite automata. However, such definitions require a nonconstant sized (but not directly accessible) memory for bookkeeping. This in some sense violates the spirit of the definition of a finite machine.

For this reason, most QFA research has been focused on the case where the transformations are limited to various combinations of unitary transformations and projective measurements on a finite dimensional state. In this case, the class of languages recognized by these QFAs is a strict subset of the regular languages. It is important to note that, despite this limit on language recognition power, there is a sense in which QFAs can be more powerful than their deterministic counterparts. In particular, there are languages which can be recognized by QFAs using exponentially fewer states than the smallest deterministic or randomized finite automaton [2,6].

C. Martín-Vide, F. Otto, and H. Fernau (Eds.): LATA 2008, LNCS 5196, pp. 373–384, 2008.
© Springer-Verlag Berlin Heidelberg 2008

The simplest type of QFA is the *measure-once QFA (MOQFA)* model of Moore and Crutchfield [11]. These QFAs are limited to recognizing those languages whose minimal automaton is such that each letter induces a permutation on the states. Two types of generalizations of the MOQFA model have been considered. In the first type, the machine is allowed to halt before reading the entire input word. This corresponds to *Kondacs and Watrous' one-way QFAs (KWQFAs)* [10]. The second type allows state transformations to include the application of quantum measurements, which generates some classical randomness in the system. This corresponds to Ambainis et. al's *Latvian QFAs (LQFAs)* [1].

Nayak [12] investigated a model called *generalized QFAs (GQFAs)*, which generalize both KWQFAs and LQFAs. This paper introduced new entropy-based techniques which were used to show that GQFAs cannot recognize the language $\Sigma^* a$. These techniques have since been used to obtain lower bounds on quantum random access codes [12] and quantum communication complexity [13]. However, no further lower bounds have been shown for GQFAs.

In a series of papers [2,7,4,3], a number lower bounds on the power of KWQFA were shown. These results identify limits on the computational advantage of KWQFAs over MOQFAs. The main tool used in these results was a technical lemma which is used to decompose the state space of a KWQFA into two subspaces (called the *ergodic* and *transient* subspaces) in which the state transitions have specific behaviors. In this paper, we show that this lemma, and many of the same results, can be adapted to the case of GQFA. The framework of our proof follows the basic outline of [2], however we must overcome a number of technical hurdles which arise from allowing classical randomness in the state.

Following [4], we can use the lemma to show that a certain property of the minimal automaton for L implies that L is not recognizable by a GQFA. We use this result to show that the class of languages recognized by this model is not closed under union. Furthermore, we show the existence of languages which can be recognized by GQFA with probability $p = 2/3$ but not $p > 2/3$. These results highlight the key similarities and differences between KWQFA and GQFA.

The paper is organized as follows. In Section 1 we give definitions and basic properties of GQFA and we review the necessary background. In Section 2 we prove the main technical lemma and in Section 3 we apply this lemma to prove the remaining results. We conclude with a brief discussion of open problems and future work.

1   Introduction

Let us review some concepts from quantum mechanics. See e.g. [14] for more details on the mathematics of quantum computation. We use the notation $|\psi\rangle$ to denote vectors in $\mathbb{C}^n$, and we denote by $\langle\psi|$ the dual of $|\psi\rangle$.

Let Q be a finite set with $|Q| = n$, and let $\{|q\rangle\}_{q \in Q}$ be an orthonormal basis for $\mathbb{C}^n$. Then a *superposition* over Q is a vector $|\psi\rangle = \sum_q \alpha_q |q\rangle$ which satisfies $\langle\psi|\psi\rangle = \sum_q |\alpha_q|^2 = 1$. We say α_q is the *amplitude* with which $|\psi\rangle$ is in state q. The state space of a QFA will be a superposition over a finite set Q.

We consider two types of operations on superpositions. First, a *unitary transformation* U is a linear operator on $\mathbb{C}^n$ such that the conjugate transpose $U^\dagger$ of U satisfies $U^\dagger U = UU^\dagger = I$. Unitary operators are exactly those which preserve the inner product, thus unitary matrices map superpositions to superpositions. The second type of operation is projective measurements. Such measurements are specified by a set $\mathcal{M} = \{P_i\}$ of orthonormal projectors on $\mathbb{C}^n$ satisfying $\sum_i P_i = I$. The *outcome* of is the measurement $\mathcal{M}$ on state $|\psi\rangle$ is the random variable which takes the value i with probability $\||P_i|\psi\rangle\|^2$. If the outcome of the measurement is i, the state is transformed to $|\psi'\rangle = P_i|\psi\rangle/\||P_i|\psi\rangle\|$. Note that measurement induces a probabilistic transformation on the state. Measurements describe the interface by which we obtain observations from a quantum system, but they also model *decoherence*, the process by which a quantum system becomes a probabilistic system through interaction with the environment (c.f. Chapter 8 of [14]).

A *generalized QFA* (GQFA) [12] is given by a tuple of the form:

$$M = (\Sigma, Q, q_0, \{U_a\}_{a\in\Gamma}, \{\mathcal{M}_a\}_{a\in\Gamma}, Q_{acc}, Q_{rej}).$$

The set Σ is the input alphabet. The working alphabet will be $\Gamma = \Sigma \cup \{\text{¢}, \$\}$. The set Q is finite set of state indices with $q_0 \in Q$, $Q_{acc}, Q_{rej} \subseteq Q$. On input $w \in \Sigma^*$, M will process the letters of the string $\text{¢}w\$$ from left to right. The ¢ and $ characters are present to allow for pre- and post- processing of the state. The sets $\{U_a\}_{a\in\Gamma}$ and $\{\mathcal{M}_a\}_{a\in\Gamma}$ are collections of unitary transformations and projective measurements.

The state of the machine is expressed as a superposition over Q, and the initial state is $|q_0\rangle$. When a letter $a \in \Gamma$ is read, a state transformation is made in the manner we describe below. After each letter is read, the machine may decide to halt and accept the input, to halt and reject the input, or to continue processing the string. The set Q is partitioned into three parts: an *accepting* set (Q_{acc}), a *rejecting* set (Q_{rej}) and a *nonhalting* set ($Q_{non} = Q - Q_{acc} \cup Q_{rej}$). We define $P_{acc} = \sum_{q\in Q_{acc}} |q\rangle\langle q|$ and we likewise define P_{rej} and P_{non}. Finally, we define $\mathcal{M}_H = \{P_{acc}, P_{rej}, P_{non}\}$.

Suppose that after reading some input prefix the machine is in state $|\psi\rangle$. To process $a \in \Gamma$, we first apply the unitary U_a, then the measurement $\mathcal{M}_a$ (recall that this is a probabilistic transformation), then the measurement $\mathcal{M}_H$. If the outcome of the measurement $\mathcal{M}_H$ is *acc* or *rej*, then the machine halts and accepts or rejects accordingly. Otherwise, the outcome of the $\mathcal{M}_H$ was *non* and the machine reads the next symbol in the string[1].

The GQFA defined above will behave stochastically. We will be interested in what languages can be recognized by this machine with bounded error. For $p > \frac{1}{2}$ we say that language $L \subseteq \Sigma^*$ is *recognized by M with probability p* if all

---

[1] The original definition allowed ℓ alternations of unitary operators and measurements per letter. However, such alternations can be simulated by a single transformation and measurement (Claim 1 of [1]) and so this change does not limit the class of transformations allowed by GQFAs.

words are correctly distinguished with probability at least p. We say that L is *recognized with bounded error* if there is a $p > \frac{1}{2}$ such that L is recognized with probability p.

Here are some basic facts about GQFAs. For all p, the class of languages recognized by GQFA with probability p is closed under complement, inverse morphisms, and word quotient. We also make note of the relationship between GQFAs and other QFA definitions. Firstly, in the case that each $\mathcal{M}_a$ is equal to the trivial measurement $\{I\}$ (i.e. so that $\mathcal{M}_H$ is the only measurement applied to the state), we obtain KWQFAs as a special case. Second, in the case that we are promised that the machine does not halt until the entire input is read, then we have the special case of Ambainis et al's LQFAs. If both of these conditions hold, we obtain MOQFAs.

In this paper we will see that many of the lower bounds for KWQFAs apply also to GQFAs. It should be noted, however, that GQFA are strictly more powerful than KWQFA. In [1] it was shown that any language L whose transition monoid is a *block group* [15] can be recognized by an LQFA with probability $1 - \varepsilon$ for any $\varepsilon > 0$. This language class corresponds exactly to the boolean closure of languages of the form $L_0 a_1 L_1 \ldots a_k L_k$, where the a_i's are letters and the L_i's are languages recognized by permutation automata. On the other hand, KWQFA cannot recognize $\Sigma^* a \Sigma^* b \Sigma^*$ with probability more than 7/9 [2]. It was moreover shown in [1] that LQFA cannot recognize the languages $a\Sigma^*$ or $\Sigma^* a$. We will need these properties in order to prove our results.

Furthermore it is known that KWQFA, and hence GQFA, can recognize languages which cannot be recognized by LQFA. For example KWQFA can simulate a certain type of reversible automaton where $\delta(q_1, x) = \delta(q_2, x) = q_2$ is permitted only when q_2 is a sink. These machines, and the class of languages which they recognize, were considered in [9]. Machines of this type can recognize $a\Sigma^*$, so KWQFA can recognize languages which cannot be recognized by LQFA.

Finally, a few notes about density matrices. Recall that the state of a GQFA after reading some input prefix is a random variable. In other words, the state is taken from a probability distribution $\mathcal{E} = \{(p_j, |\psi_j\rangle)\}$ of superpositions, where $|\psi_j\rangle$ occurs with probability p_j. Such systems are called *mixed states*. The measurement statistics which can be obtained from transforming and measuring a mixed state can be described succinctly in terms of *density matrices*. In our case it will be sufficient to identify a mixed state with its density matrix.

The density matrix corresponding to $\mathcal{E}$ is $\rho = \sum_j p_j |\psi_j\rangle\langle\psi_j|$. Density matrices are positive operators so their eigenvalues are nonnegative real. For a operator M we denote by $Tr(M)$ the *trace*, or the sum of the eigenvalues, of M. In the case of density matrices we have $Tr(\rho) = 1$. Unitary operators U transform density matrices according to the rule $\rho \mapsto U^\dagger \rho U$. A measurement $\mathcal{M} = \{P_i\}$ will transform the states by the rule $\rho \mapsto \sum_i P_i \rho P_i$ in the case that the outcome is unknown, or by $\rho \mapsto P_i \rho P_i / Tr(P_i \rho)$ if the outcome is known to be i.

Density matrices are examples of normal matrices. The spectral decomposition theorem states that every normal matrix can be decomposed as $\rho = \sum_i \lambda_i |\phi_i\rangle\langle\phi_i|$, where $\{|\phi_i\rangle\}$ is a set of orthonormal eigenvectors of ρ and λ_i

is the eigenvalue corresponding to $|\phi_i\rangle$. We say that the *support* of ρ, or $supp(\rho)$, is the space spanned by the nonzero eigenvectors of ρ.

2 Technical Results

Fix a GQFA M. We will be using density matrices weighted by a factor $p \in [0,1]$ to describe the state of M on reading some prefix $\mathetax{c}w$. Let A_a be the mapping $\rho \mapsto \sum_i P_{a,i} U_a \rho U_a^\dagger P_{a,i}$, and let $A'_a = P_{non}(A_a\rho)P_{non}$. Furthermore for $w = w_1 \ldots w_n \in \Sigma^*$, we define $A'_w = A'_{w_n} \cdots A'_{w_1}$. Then $A'_w\rho$ is a scaled density matrix such that $Tr(A'_w\rho) = pTr(\rho)$, where p is the probability of not halting in the process of reading w while in state ρ. Let $\rho_w = A'_{\matheta{c}w}|q_0\rangle\langle q_0|$. Then $Tr(\rho_w)$ is the probability of not halting while processing $\mathetax{c}w$, and the density matrix describing the machine state in the case that it has not halted is $\rho_w/Tr(\rho_w)$.

We first state a technical lemma which gives an important characterization of the behaviour of a GQFA machine. It is the counterpart to Lemma 1 of [2]. This, along with its extension (Lemma 2), will be instrumental in proving the later results.

Lemma 1. *For every $w \in \Sigma^*$ there exists a pair E_1, E_2 of orthonormal subspaces of $\mathbb{C}^n$ such that $\mathbb{C}^n = E_1 \oplus E_2$ and for all weighted density matrices ρ over $\mathbb{C}^n$ we have:*

1. *If $supp(\rho) \subseteq E_1$, then $supp(A'_w\rho) \subseteq E_1$ and $Tr(A'_w\rho) = Tr(\rho)$.*
2. *If $supp(\rho) \subseteq E_2$, then $supp(A'_w\rho) \subseteq E_2$ and $\lim_{k\to\infty}Tr((A'_w)^k\rho) = 0$.*

The E_1 and E_2 parts of the state are called the *ergodic* and *transient* parts. Suppose M is in state ρ, and suppose that ρ satisfies $supp(\rho) \subseteq E_1$. Then $Tr(A'_w\rho) = Tr(\rho)$ would imply that M did not halt in the process of reading w. Thus, M is behaving exactly as an LQFA. Suppose now that M is in state ρ, then the fact $\lim_{k\to\infty}Tr((A'_w)^k\rho) = 0$ implies that the probability that M does not halt after reading w^k tends to 0 as $k \to \infty$. In general $supp(\rho)$ will be partially in E_1 and partially in E_2.

Proof: The proof proceeds as in [2]. We first show how to do this for the case that $|w| = 1$, and then we sketch how to extend it to arbitrary length words. Let $w = a$. We first construct the subspace E_1 of $\mathbb{C}^n$. E_2 will be the orthogonal complement of E_1. Let

$$E_1^1 = span(\{|\psi\rangle : Tr(A'_a|\psi\rangle\langle\psi|) = Tr(|\psi\rangle\langle\psi|)\}).$$

Equivalently, $E_1^1 = span\{|\psi\rangle : supp(A_a(|\psi\rangle\langle\psi|)) \subseteq S_{non}\}$ where S_{non} is the nonhalting subspace. We claim that $supp(\rho) \in E_1^1$ implies that $supp(A_a(\rho)) \in S_{non}$. By linearity it is sufficient to show this for $\rho = |\psi\rangle\langle\psi|$. Essentially, we need to show that the condition of $|\psi\rangle$ satisfying $Tr(A'|\psi\rangle\langle\psi|) = Tr(|\psi\rangle\langle\psi|)$ is closed under linear combinations. Suppose that $|\psi\rangle = \sum_j \alpha_j|\psi_j\rangle$, with $|\psi_j\rangle$ satisfying $supp(A_a(|\psi_j\rangle\langle\psi_j|)) \in S_{non}$ and $\sum_j |\alpha_j|^2 = 1$. Then:

$$\|\sum_i P_{halt}P_{a,i}U_a(\sum_j \alpha_j|\psi_j\rangle)\|^2 \leq \sum_{i,j}\|\alpha_j P_{halt}P_{a,i}U_a|\psi_j\rangle\|^2 = 0,$$

and thus $supp(A_a|\psi\rangle\langle\psi|) \in S_{non}$. Thus, for mixed states ρ we have $supp(A_a\rho) \in S_{non}$ if and only if $supp(\rho) \in E_1^1$. For general $i > 2$, let:

$$E_1^i = span(\{|\psi\rangle : supp(A_a|\psi\rangle\langle\psi|) \in E_1^{i-1} \wedge Tr(A_a'|\psi\rangle\langle\psi|) = Tr(|\psi\rangle\langle\psi|)\}).$$

As before, for weighted density matrices ρ, we can interchange the condition $Tr(A_a'\rho) = Tr(\rho)$ for $supp(A_a\rho) \subseteq S_{non}$.

Observe that $E_1^i \subseteq E_1^{i+1}$ for all i. Since the dimension of each of these spaces is finite, there must be an i_0 such that $E_1^{i_0} = E_1^{i_0+j}$ for all $j > 0$. We define $E_1 = E_1^{i_0}$, and set E_2 to be the orthogonal complement of E_1.

It is clear that the first condition of the lemma is true for mixed states with support in E_1. For the second part, it will be sufficient to show the following proposition, which implies that the probability with which the machine will halt while reading a^j is bounded by a constant.

Proposition 1. *Let* $j \in \{1, \ldots, i_0\}$. *There is a constant* $\delta_j > 0$ *such that for any* $|\psi\rangle \in E_2^j$ *there is an* $l \in \{0, \ldots, j-1\}$ *such that* $Tr(P_{halt}A_a(A_a')^l(|\psi\rangle\langle\psi|)) \geq \delta_j$.

Proof: We proceed by induction on j. Let $\mathcal{H} = \bigoplus_{k=1}^{m_a} \mathbb{C}^n$. Let $P_k : E_2^1 \to \mathcal{H}$ be the projector into the kth component of $\mathcal{H}$, and let $T_1 : E_2^1 \to \mathcal{H}$ be the function $T_1|\psi\rangle = \sum_k P_k P_{halt} P_{a,k} A_a|\psi\rangle$. Observe that $\|T_1|\psi\rangle\|^2$ is the probability of halting when a is read while the machine is in state $|\psi\rangle\langle\psi|$. By the previous discussion, $Tr(A_a'|\psi\rangle\langle\psi|) = 1 - \|T_1|\psi\rangle\|^2$. Define $\|T_1\| = min_{\||\psi\rangle\|=1}\|T_1|\psi\rangle\|$. Note that the minimum exists since the set of unit vectors in $\mathbb{C}^n$ is a compact space. Also, let $\delta_1 = \|T_1\|^2$. Then $\delta_1 > 0$, otherwise there would be a vector $|\psi\rangle \in E_2^1$ such that $supp(A_a|\psi\rangle\langle\psi|) \in S_{non}$, a contradiction.

Now assume that δ_{j-1} has been found. We need to show that, for $|\psi\rangle \in E_2^j$, either a constant sized portion of $|\psi\rangle$ is sent into the halting subspace, or it is mapped to a vector on which we can apply the inductive assumption. We construct two functions $T_{j,halt}, T_{j,non} : E_2^j \to \mathcal{H}$ defined by:

$$T_{j,halt}|\psi\rangle = \sum_{k=1}^{m_a} P_k P_{halt} P_{a,k} A_a|\psi\rangle,$$

$$T_{j,non}|\psi\rangle = \sum_{k=1}^{m_a} P_k P_{E_2^{j-1}} P_{non} P_{a,k} A_a|\psi\rangle.$$

Then the quantity $\|T_{j,halt}|\psi\rangle\|^2$ is the probability of halting while reading a, and $\|T_{j,non}|\psi\rangle\|^2 = Tr(P_{E_2^{j-1}}A_a'|\psi\rangle\langle\psi|)$. Note that for all vectors $|\psi\rangle \in E_2^j$ we must have either $\|T_{j,halt}|\psi\rangle\| \neq 0$ or $\|T_{j,non}|\psi\rangle\| \neq 0$, otherwise $|\psi\rangle$ is in E_1^j, a contradiction. This implies that $\|T_{j,non} \oplus T_{j,halt}\| > 0$. Note also that $\|T_{j,non} \oplus T_{j,halt}\| \leq 1$.

Define $\delta_j = \delta_{j-1}\frac{\|T_{j,non}\oplus T_{j,halt}\|^2}{2m_a}$. Take any unit vector $|\psi\rangle \in E_2^j$. Then $\|(T_{j,non}\oplus T_{j,halt})|\psi\rangle\| \geq \|T_{j,non} \oplus T_{j,halt}\|$. Recall that the range of $T_{j,non} \oplus T_{j,halt}$ is $\bigoplus_{k=1}^{m_a} \mathbb{C}^n \oplus \bigoplus_{k=1}^{m_a} \mathbb{C}^n$. In one of these subspaces, $(T_{j,non} \oplus T_{j,halt})|\psi\rangle$ has size at least $\frac{1}{\sqrt{2\cdot m_a}}$. If it is in one of the last m_a subspaces, corresponding to $T_{j,halt}$ part, then there is nothing further to prove. Otherwise, assume that this component is in one of the subspaces corresponding to the $T_{j,non}$ part. In particular, there is a k such that $|\phi\rangle = P_{non}P_{a,k}A_a|\psi\rangle$ satisfies:

$$\|P_{E_2^{j-1}}|\phi\rangle\|^2 \geq \frac{1}{2 \cdot m_a}.$$

We can split $|\phi\rangle$ into $|\phi_1\rangle + |\phi_2\rangle$, with $|\phi_i\rangle \in E_i^{j-1}$. By the inductive hypothesis, there is an $l < j - 1$ such that $Tr(P_{halt}A_a(A_a')^l(|\phi_2\rangle\langle\phi_2|)) \geq \delta_{j-1}Tr(|\phi_2\rangle\langle\phi_2|)$. Furthermore, the first condition of the lemma implies that for every choice of $(k_1, \ldots, k_l) \in [m^a]^l$,

$$P_{halt}P_{a,k_l}U_a P_{a,k_{l-1}}U_a \cdots P_{a,k_1}U_a|\phi_1\rangle = \mathbf{0}.$$

This implies $Tr(P_{halt}A_a(A_a')^l(|\phi_1\rangle\langle\phi_1|)) = 0$ and $Tr(P_{halt}A_a(A_a')^l(|\phi_2\rangle\langle\phi_1|)) = Tr(P_{halt}A_a(A_a')^l(|\phi_2\rangle\langle\phi_1|)) = 0$. Together, we obtain:

$$
\begin{aligned}
&Tr(P_{halt}A_a(A_a')^l|\phi\rangle\langle\phi|) \\
&= Tr(P_{halt}(A_a')^l(|\phi_1\rangle\langle\phi_1| + |\phi_1\rangle\langle\phi_2| + |\phi_2\rangle\langle\phi_1| + |\phi_2\rangle\langle\phi_2|)) \\
&= Tr(P_{halt}A_a(A_a')^l(|\phi_1\rangle\langle\phi_1|)) + Tr(P_{halt}A_a(A_a')^l(|\phi_1\rangle\langle\phi_2|)) \\
&\quad + Tr(P_{halt}A_a(A_a')^l(|\phi_2\rangle\langle\phi_1|)) + Tr(P_{halt}A_a(A_a')^l(|\phi_2\rangle\langle\phi_2|)) \\
&= Tr(P_{halt}A_a(A_a')^l(|\phi_2\rangle\langle\phi_2|)) \geq \delta_{j-1}\frac{\|T_{j,non} \oplus T_{j,halt}\|^2}{2m_a}.
\end{aligned}
$$

This concludes the proof of the proposition. $\qquad\square$

Proposition 2. *Let U_a be the unitary transformation that is applied when a is read. Then $U_a = U_a^1 \oplus U_a^2$, where U_a^i acts unitarily on subspace E_i.*

Proof: By the unitarity of U_a, it is sufficient to show that $|\psi\rangle \in E_1$ implies $U_a|\psi\rangle \in E_1$. By definition of E_1, $|\psi\rangle \in E_1$ implies that all of the vectors $P_{a,i}U_a|\psi\rangle$ are in E_1. But $U_a|\psi\rangle = \sum_i P_{a,i}U_a|\psi\rangle$, and thus $U_a|\psi\rangle \in E_1$ since E_1 is a subspace. $\qquad\square$

We are now ready to prove the second part of the lemma. We first show that $|\psi\rangle \in E_2$ implies $supp(A_a|\psi\rangle\langle\psi|) \subseteq E_2$. Let $|\psi'\rangle = U_a|\psi\rangle$. Then $A_a|\psi\rangle\langle\psi| = \sum_i |\psi_i\rangle\langle\psi_i|$, where $|\psi_i\rangle = P_{a,i}U_a|\psi\rangle$. Split $|\psi_i\rangle$ into vectors $|\psi_{i,1}\rangle + |\psi_{i,2}\rangle$, with $|\psi_{i,1}\rangle \in E_1$ and $|\psi_{i,2}\rangle \in E_2$. We claim that either $|\psi_{i,1}\rangle$ or $|\psi_{i,2}\rangle$ are trivial vectors. Suppose $\||\psi_{i,1}\rangle\| \neq 0$, and consider the intersection of the image of $P_{a,i}$ in the space spanned by $|\psi_{i,1}\rangle$ and $|\psi_{i,2}\rangle$. Now $|\psi_{i,1}\rangle$ implies that $U_a^{-1}|\psi_{i,1}\rangle \in E_1$ and thus $P_{a,i}|\psi_{i,1}\rangle \in E_1$, which implies $|\psi_i\rangle \in E_1$.

Now since each $|\psi_i\rangle$ satisfies $|\psi_i\rangle \in E_1$ or $|\psi_i\rangle \in E_2$, then we are done since the fact that the $|\psi_i\rangle$'s are orthonormal and sum to $U_a|\psi\rangle \in E_2$ implies that $|\psi_i\rangle \in E_2$ for all i. Thus, $|\psi\rangle \in E_2$ implies $span(A_a|\psi\rangle\langle\psi|) \subseteq E_2$.

Now supposing $supp(\rho) \in E_2$, we can repeatedly apply Proposition 1 to show that $Tr((A_a')^k(\rho)) \to 0$ as $k \to \infty$. To apply the claim to a general mixed state, we first use the spectral decomposition to show that the mixed state is equivalent to an ensemble of at most n pure states.

To construct E_1 and E_2 for $w = w_1 \ldots w_n$, we define $E_1^0 = S_{non}$ and E_1^k to be the set of all vectors $|\psi\rangle$ such that $Tr(A'_{w_k \mod n+1}|\psi\rangle\langle\psi|) = 1$ and $supp$ $(A'_{w_k \mod n+1}|\psi\rangle\langle\psi|) \in E_1^{k-1}$, and we follow the proof as above. The proof of the first part of the theorem and of the claim will generalize since the proof does not make use of the fact that the transformation and measurement defining E_1^j is the same as that of E_1^{j+1}. Proposition 2 will apply to w_i for all i. $\qquad \square$

Lemma 2. *Let M be an n-state GQFA over alphabet Σ, and let $x, y \in \Sigma^*$. Then there exists a pair E_1, E_2 of orthonormal subspaces of $\mathbb{C}^n$ such that $\mathbb{C}^n = E_1 \oplus E_2$ and for all weighted density matrices ρ over $\mathbb{C}^n$ we have:*

1. *If $supp(\rho) \subseteq E_1$, then for all $w \in (x \cup y)^*$, $supp(A'_w\rho) \subseteq E_1$, and $Tr(A'_w\rho) = Tr(\rho)$.*
2. *If $supp(\rho) \subseteq E_2$, then $supp(A'_w\rho) \subseteq E_2$ and for all $\varepsilon > 0$ there exists a word $w \in (x \cup y)^*$ such that $Tr(A'_w\rho) \leq \varepsilon$.*

Proof: This is the counterpart of Lemma 2.3 of [4]. Let E_1^w be the subspace constructed as in Lemma 1. Define $E_1 = \cap_{w \in (x \cup y)^*} E_1^w$, and let E_2 be the orthogonal complement of E_1.

Suppose $supp(\rho) \subseteq E_2$. If there is a $w \in (x \cup y)^*$ such that $supp(\rho) \subseteq E_2^w$, we can directly apply the argument from the previous lemma to show that $Tr((A'_w)^j\rho) \to 0$ as $j \to \infty$. However such a w may not exist so a stronger argument is necessary. As the application of an A'_w transformation can only decrease the trace of ρ, for any ε there exists a $t \in (x \cup y)^*$ such that for all $w \in (x \cup y)^*$, $Tr(A'_t\rho) - Tr(A'_{tw}) \leq \varepsilon$. For all i let t_i be a such a string for $\varepsilon = \frac{1}{2^i}$. Consider the sequence $\rho_1, \rho_2, \ldots$ defined by $\rho_i = A'_{t_i}\rho$. The set of weighted density matrices form a compact, closed space with respect to the trace metric, and so this sequence of must have a limit point ρ.

We claim that $Tr(\rho) = 0$. Suppose not. The support of ρ is in E_2, so there must be some word $w \in (x \cup y)^*$ such that $Tr(A'_w\rho) < Tr(\rho)$. This contradicts the assumption that ρ is a limit point. $\qquad \square$

Finally we note a very simple fact that will allow us to extend impossibility results for LQFA to GQFA:

Fact 1. *Let M be a GQFA. Let E_1 be the subspace defined as in Lemma 2, and suppose that the state of the machine ρ on reading the ¢ character satisfies $supp(\rho) \in E_1$. Then there is an LQFA M' such that, for all $w \in (x \cup y)^*$ the state of M on reading w is isomorphic to the state of M' on reading w.*

3   Applications

We now apply the results of the previous section to prove several fundamental properties of GQFAs. The first result is a formal condition for recognizability by GQFAs:

Theorem 1. *Let M_L be the minimal automaton for $L \subseteq \Sigma^*$ and let F be the accepting set. If there exists words $x, y, z_1, z_2 \subseteq \Sigma^*$ and states q_0, q_1, q_2 such that $\delta(q_0, x) = q_1$, $\delta(q_0, y) = q_2$, $\delta(q_1, x) = \delta(q_1, y) = q_1$, $\delta(q_2, x) = \delta(q_2, y) = q_2$, $\delta(q_1, z_1) \in F$, $\delta(q_2, z_1) \notin F$, $\delta(q_1, z_2) \notin F$, $\delta(q_2, z_2) \in F$, then L cannot be recognized by GQFA with probability $p > \frac{1}{2}$.*

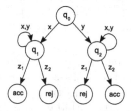

Fig. 1. The forbidden construction of Theorem 1

Proof: Suppose that L satisfies the conditions of the theorem, and suppose that M recognizes L with probability $p > \frac{1}{2}$. By closure under left quotient, we can assume that the state q_0 in the forbidden construction is also the initial state of the minimal automaton for L.

Let $\rho_w = A'_{\mathbb{C}w} |q_0\rangle \langle q_0|$. The basic outline of the proof is that we will use Lemma 2 to find two words $w_1 \in x(x \cup y)^*$, $w_2 \in y(x \cup y)^*$ such that ρ_{w_1} and ρ_{w_2} have similar output behavior. We then analyze the acceptance probabilities of the words $w_1 z_1$, $w_1 z_2$, $w_2 z_1$, and $w_2 z_2$ to arrive at a contradiction.

Let E_1 and E_2 be subspaces which meet the conditions of Lemma 2 with respect to x and y. Note that if the support of ρ is in E_1, M will not halt while reading $w \in (x \cup y)^*$, and in this case M can be simulated by an LQFA. Let P_{E_i} be the projection onto subspace E_i. We claim that for all $\varepsilon > 0$ there exists $u, v \in (x \cup y)^*$ such that $\|Tr(P_{E_1}\rho_{xu} - P_{E_1}\rho_{yv})\|_t \le \varepsilon$. Suppose to the contrary that there exists $\varepsilon > 0$ such that $\|Tr(P_{E_1}\rho_{xu} - P_{E_1}\rho_{yv})\|_t > \varepsilon$ for all u, v. Then there exists an LQFA which can recognize the language $x(x \cup y)^*$ with bounded error, contradicting the fact that LQFA is closed under inverse morphisms and cannot recognize $a\Sigma^*$ [1]. Let $\delta = p - \frac{1}{2}$ and let $\varepsilon = \frac{\delta}{4}$.

By Lemma 2, for all ε' we can find $u' \in (x \cup y)^*$ such that $Tr(P_{E_2}\rho_{xuu'}) < \varepsilon'$. Furthermore we can find $v' \in (x \cup y)^*$ such that $Tr(P_{E_2}\rho_{xuu'v'}) < \varepsilon'$ and $Tr(P_{E_2}\rho_{yvu'v'}) < \varepsilon'$. Let $w_1 = xuu'v'$ and $w_2 = yvu'v'$, and let $\varepsilon' = \frac{\delta}{4}$.

Let $p_{i,acc}$ ($p_{i,rej}$) be the probability with which M accepts (rejects) while reading w_i. Furthermore let $q_{ij,acc}$ (resp $q_{ij,rej}$) be the probability that M accepts if the state of the machine is ρ_{w_i} and the string $z_j\$$ is read. Since $\|\rho_{w_1} - \rho_{w_2}\|_t \le \|\rho_{xu} - \rho_{yv}\|_t = \frac{\delta}{2} \le \varepsilon$, $q_{1j,acc}$ (and likewise $q_{1j,rej}$) can be different from $q_{2j,acc}$ by a factor of at most $\frac{\delta}{2}$. As a consequence, one of the words $w_1 z_1$, $w_1 z_2$, $w_2 z_1$, or $w_2 z_2$ must not be classified correctly. Suppose e.g. that $w_1 z_1$, $w_1 z_2$, and $w_2 z_1$ are classified correctly. Since $q_{11,rej}$ differs from $q_{21,rej}$ by a factor of at most $\frac{\delta}{2}$, the fact that $w_1 z_1$ is accepted and $w_2 z_1$ is rejected implies that $p_{2,rej} > p_{1,rej} + \delta$. since $q_{12,rej}$ differs from $q_{22,rej}$ by at most a factor of $\frac{\delta}{2}$, will be rejected with probability greater than $1 - p$, a contradiction. The other cases are similar.   □

We now apply Theorem 1 to prove nonclosure under union.

Theorem 2. *The class of languages recognized by GQFA with bounded error is not closed under union.*

Proof: Let A, B_0, B_1 be languages over $\Sigma = \{a, b\}$ defined as follows. Let $A = \{w : |w|_a \mod 2 = 0\}$, $B_0 = (aa)^*b\Sigma^*$, and $B_1 = a(aa)^*b\Sigma^*$. Finally, let $L_1 = (\overline{A} \cap a^*) \cup (A \cap B_1)$, and let $L_2 = (A \cap a^*) \cup (\overline{A} \cap B_0)$. The union $L_1 \cup L_2$ consists of the strings containing either no b's or an odd number of a's after the first b.

In Theorem 3.2 of [4], the languages L_1 and L_2 were shown to be recognizable by KWQFAs with probability of correctness $2/3$, thus they can also be recognized by GQFA with this probability of correctness. On the other hand, the minimal automaton of $L_1 \cup L_2$ contains the forbidden construction of Theorem 1.   □

In [2] it was shown that there exists languages L and constants $p > \frac{1}{2}$ such that L can be recognized by KWQFA with bounded probability, but not with probability p. Furthermore, it was demonstrated that certain properties of the minimal automaton for L would imply that L is not recognized with probability p. We will show that a similar situation holds for GQFAs.

Theorem 3. *If the minimal DFA M_L for L contains states q_0, q_1, q_2, such that for some words x, y, z_1, z_2 we have $\delta(q_0, x) = \delta(q_1, x) = \delta(q_1, y) = q_1$, $\delta(q_0, y) = \delta(q_2, y) = \delta(q_2, x) = q_2$, $\delta(q_2, z_2) \in F$, $\delta(q_2, z_1) \notin F$, then L cannot be recognized by GQFA with probability $p > \frac{2}{3}$.*

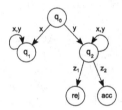

Fig. 2. The forbidden construction of Theorem 3

Proof: Suppose that the GQFA M recognizes L with probability $p > 2/3$. Since $q_2 \neq q_3$ and by closure under complement, there exists a word z_3 such that $xz_3 \in L$ and $yz_3 \notin L$. We can also assume by closure under left quotient that q_1 is the initial state. As in Lemma 2, split $\mathbb{C}^n$ into subspaces E_1 and E_2 with respect to x and y.

For all ε, we can find $w_1 \in x(x \cup y)^*$ and $w_2 \in y(x \cup y)^*$ such that $\|\rho_{w_1} - \rho_{w_2}\|_t \leq \varepsilon$, $Tr(P_{E_2}\rho_w) < \varepsilon$, $Tr(P_{E_2}\rho_w) < \varepsilon$. let p_i be the probability that M rejects while reading w_i, and let p_{i3} be the probability of rejecting when M is in state q_i and reads z_3. By setting ε, the difference between p_{13} and p_{23} can be made arbitrarily small, so that $p_1 + p_{13} \leq (1 - p) < 1/3$ and $p_2 + p_{23} \geq p > 2/3$ imply that $p_2 - p_1 > 1/3$. Thus M rejects while reading w_2 with probability greater than $1/3$, contradicting the assumption that $w_2 z_2$ is accepted with probability greater than $2/3$.   □

Corollary 1. *There is a language L which can be recognized by GQFAs with probability $p = 2/3$, but not with $p > 2/3$.*

To see this, note that the constructions for L_1 and L_2 in [4] achieve the optimal probability of correctness.

4 Discussion

We have shown that several of the known lower proofs for KWQFA can be adapted to the case of GQFA. In particular, we have shown that the class of languages recognized by GQFA is not closed under union, and there exists languages which can be recognized by GQFA with probability $p = 2/3$ but not $p > 2/3$. Both KWQFA and GQFA are permitted to halt before the end, and the lack of robustness in these models seems to arise from this feature. By comparison, the classes of languages recognized by MOQFA and LQFA respectively are closed under union, and any language recognized with probability $p > 1/2$ by these machines can be recognized with probability $1 - \varepsilon$ for any $\varepsilon > 0$.

We note here that not all of the KWQFA lower bound results hold for GQFA. For example, it was shown that $a^* b^*$ can be recognized by KWQFA with probability $p \approx 0.68$ but not $p > 7/9$, while this language can be recognized by GQFA with probability $1 - \varepsilon$ for any $\varepsilon > 0$. Several other KWQFA lower bounds were shown in [4,3], and we can clarify the relationship between the two models by identifying which of these results extend to GQFAs. It is still not known whether the class of languages recognized with bounded error by GQFA is strictly larger than the class recognized by KWQFA. We conjecture that the language class is indeed larger and that a proof would involve the fact that the probability with which KWQFAs can recognize $\Sigma^* a_1 \Sigma^* \ldots a_k \Sigma^*$ tends to $1/2$ as $k \to \infty$.

References

1. Ambainis, A., Beaudry, M., Golovkins, M., Kikusts, A., Mercer, M., Thérien, D.: Algebraic results on quantum automata. Theory of Computing Systems 38, 165–188 (2006)
2. Ambainis, A., Freivalds, R.: 1-way quantum finite automata: strengths, weaknesses and generalizations. In: 39th Annual Symposium on Foundations of Computer Science, pp. 332–341 (1998)
3. Ambainis, A., Ķikusts, A.: Exact results for accepting probabilities of quantum automata. Theoretical Computer Science 295(1–3), 3–25 (2003)
4. Ambainis, A., Ķikusts, A., Valdats, M.: On the class of languages recognizable by 1-way quantum finite automata. In: Ferreira, A., Reichel, H. (eds.) STACS 2001. LNCS, vol. 2010, pp. 75–86. Springer, Heidelberg (2001)
5. Bennett, C.H.: Logical reversibility of computation. IBM Journal of Research and development 6, 525–532 (1973)

6. Bertoni, A., Mereghetti, C., Palano, B.: Quantum computing: 1-way quantum automata. In: Ésik, Z., Fülöp, Z. (eds.) DLT 2003. LNCS, vol. 2710, pp. 1–20. Springer, Heidelberg (2003)

7. Brodsky, A., Pippenger, N.: Characterizations of 1-way quantum finite automata. SIAM Journal on Computing 31(5), 1456–1478 (2002)

8. Ciamarra, M.P.: Quantum reversibility and a new model of quantum automaton. Fundamentals of Computation Theory 13, 376–379 (2001)

9. Golovkins, M., Pin, J.-É.: Varieties generated by certain models of reversible finite automata. In: Chen, D.Z., Lee, D.T. (eds.) COCOON 2006. LNCS, vol. 4112, pp. 83–93. Springer, Heidelberg (2006)

10. Kondacs, A., Watrous, J.: On the power of quantum finite state automata. In: 38th Annual Symposium on Foundations of Computer Science, pp. 66–75. IEEE Computer Society Press, Los Alamitos (1997)

11. Moore, C., Crutchfield, J.: Quantum automata and quantum grammars. Theoretical Computer Science 237(1-2), 275–306 (2000)

12. Nayak, A.: Optimal lower bounds for quantum automata and random access codes. In: 40th Annual Symposium on Foundations of Computer Science, pp. 369–377 (1999)

13. Nayak, A., Salzman, J.: On communication over an entanglement-assisted quantum channel. In: Proceedings of the Thirty-Fourth Annual ACM Symposium on the Theory of Computing, pp. 698–704 (2002)

14. Nielsen, M., Chuang, I.: Quantum Computation and Quantum Information. Cambridge University Press, Cambridge (2000)

15. Pin, J.-É.: k BG=PG, a success story. In: Fountain, J. (ed.) NATO Advanced Study Institute Semigroups, Formal Languages, and Groups, pp. 33–47. Kluwer Academic Publishers, Dordrecht (1995)

16. Rabin, M.: Probabilistic automata. Information and Control 6(3), 230–245 (1963)

17. Yao, A.C.-C.: Quantum circuit complexity. In: Proceedings of the 36th annual Symposium on Foundations of Computer Science, pp. 352–361 (1993)

How Many Figure Sets Are Codes?

Małgorzata Moczurad and Włodzimierz Moczurad

Institute of Computer Science, Jagiellonian University,
Nawojki 11, 30-072 Kraków, Poland
{mmoczurad,wkm}@ii.uj.edu.pl

Abstract. Defect theorem, which provides a kind of dimension property for words, does not hold for two-dimensional figures (labelled polyominoes), except for some small sets. We thus turn to the analysis of asymptotic density of figure codes. Interestingly, it can often be proved to be 1, even in those cases where the defect theorem fails. Hence it reveals another weak dimension property which does hold for figures, i.e., non-codes are rare.

We show that the asymptotic densities of codes among the following sets are all equal to 1: (ordinary) words, square figures and small sets of dominoes, where small refers to cardinality ≤ 3. The latter is a borderline case for the defect theorem and additionally exhibits interesting properties at different alphabet sizes.

Keywords: Polyominoes, codes, asymptotic density.

1 Introduction

Counting polyominoes or figures of a given size is hard; e.g. no exact formula or generating function is known for the sequence (p_n) describing the number of polyominoes of size n; cf. [14]. On the other hand, the problem is trivial for words. Counting polyomino or figure codes is at least as hard as counting polyominoes. Codicity is a "semantic" property that does not admit a simple "syntactic" characterization. Hence powerful combinatoric tools, like e.g. generating functions, come to no rescue. The problem whether a given set of figures, even polyominoes, is a code is undecidable in general, cf. [1]. Again, the problem is easy for sets of words, cf. [2].

The defect theorem is one of the fundamental results of combinatorics on words, providing a kind of dimension property of words (cf. Lothaire [7,8]). Various authors have studied its variants and extensions to other structures, including trees and two-dimensional figures, e.g. [4,5,6,9,10]. See Harju and Karhumäki [3] for a comprehensive treatment of the field. In its classical version, the defect theorem states that if $X \subseteq A^*$ is a finite non-code, i.e., there exists a word in X^* with two different X-factorizations, then there exists a code $Y \subseteq A^*$ such that $X \subseteq Y^*$ and $|Y| < |X|$. More precisely, Y can be taken to be the free hull of X (the smallest free submonoid of A^* containing X). Whilst it has been shown that the defect property can be extended to trees, the property is not satisfied in many simple cases of plane figures.

C. Martín-Vide, F. Otto, and H. Fernau (Eds.): LATA 2008, LNCS 5196, pp. 385–396, 2008.
© Springer-Verlag Berlin Heidelberg 2008

We will be computing asymptotic density in the following sense, cf. [12,15]: Given a set of objects, choose those of size n and count those that have a desired property. We are interested in the proportion of objects with the desired property (codicity in our case) as n tends to infinity. The measure of size will be defined separately for different classes of figures.

The main result of the paper is that for some classes of figures this proportion tends to 1 as the size of the figures tends to infinity. This result may be interpreted probabilistically as follows: the probability that a randomly chosen set of figures is a code approaches 1 when the figures are large. As a by-product, we exhibit classes of codes that have density 1 among all codes. We also demonstrate a non-obvious link between the alphabet size and the form of domino figure codes.

2   Definitions and Notations

Let A be a finite alphabet. We use the usual notation of A^* to denote the free monoid over A, and X^* to denote the submonoid generated by $X \subseteq A^*$. A set of words $X \subseteq A^*$ is a code, if X^* is free over X, i.e., every word in X^* has a unique factorization over X.

A *figure* (or a *brick*) is a partial mapping $x : \mathbb{Z}^2 \to A$, where the domain of x is a polyomino, i.e., a finite and connected union of lattice points (or unit squares) in $\mathbb{R}^2$. In other words, it is a polyomino with the cells labelled with the symbols of A. If $|A| = 1$, there is an obvious natural correspondence between figures and polyominoes. The set of all figures over A is denoted by $A^{\bowtie}$.

Given a set of figures $X \subseteq A^{\bowtie}$, the set of all figures tilable with (translated copies of) the elements of X is denoted by $X^{\bowtie}$. Note that we do not allow rotations of figures. $X \subseteq A^{\bowtie}$ is a *figure code*, if every element of $X^{\bowtie}$ admits exactly one tiling with the elements of X.

The terms *rectangles, squares* and *dominoes* will refer to figures with respective domains. In particular, the domain of a domino is a $1 \times n$ or $n \times 1$ rectangle with $n \geq 1$ (a *vertical domino* and a *horizontal domino,* respectively). Note that since we actually consider all figures up to a translation, horizontal dominoes can be identified with words.

To formalize the counting principle, let $\mathcal{F}$ be the set of objects and let $\mathcal{A} \subseteq \mathcal{F}$ be the set of objects having the desired property. The *asymptotic density* (or *asymptotic probability*) $\mu_{\mathcal{F}}(\mathcal{A})$ of $\mathcal{A}$ in $\mathcal{F}$ is defined as

$$\mu_{\mathcal{F}}(\mathcal{A}) = \lim_{n \to \infty} \frac{|\{X \in \mathcal{A} : \|X\| = n\}|}{|\{X \in \mathcal{F} : \|X\| = n\}|},$$

where $\|X\|$ denotes the size of X. The existence and value of $\mu_{\mathcal{F}}(\mathcal{A})$ depend on the choice of $\| \cdot \|$. If $\mu_{\mathcal{F}}(\mathcal{A})$ exists, it is within $[0, 1]$. Note that μ is not a probability in the classical sense since the enumerable additivity axiom does not hold.

For the sake of clarity, in the sequel we assume a two-element alphabet $A = \{a, b\}$. All results can be easily generalized to any $|A| \geq 2$. Dominoes are the only exception where we deal separately with two alphabet sizes, $|A| = 2$ and $|A| = 3$. In all cases the results become trivial when $|A| = 1$.

3   Counting Square Figure Codes

We use the following size measure for sets of figures. Let $\mathcal{S}_k \subset \mathcal{P}(A^{\bowtie})$ denote the family of all sets containing k squares; $\mathcal{S}_k^C \subseteq \mathcal{S}_k$ will denote the family of codes containing k squares. Define the size of a set $X \in \mathcal{S}_k$ as the size of the largest square in X, i.e., $\|X\| = \max\{\operatorname{len} x : x \in X\}$, where $\operatorname{len} x$ denotes the edge length of x.

We define the numbers $S_{k,=n}$, $S_{k,<n}$ and $S_{k,\leq n}$ as follows:

- $S_{k,=n}$ is the number of sets containing k squares, each of them of size $n \times n$
- $S_{k,<n}$ is the the number of sets containing k squares, each of them strictly smaller than $n \times n$
- $S_{k,\leq n} = |\{X \in \mathcal{S}_k : \|X\| = n\}|$ is the number of sets containing k squares, with the biggest one of size $n \times n$.

Similarly, $S_{k,=n}^C$, $S_{k,<n}^C$ and $S_{k,\leq n}^C$ will denote the number of codes containing k squares of respective sizes. All numbers defined above are assumed to be 1 when $k = 0$.

Obviously $S_{1,=n} = 2^{n^2}$ and for arbitrary $n \geq 1$, $k \geq 0$ we have $S_{k,=n} = \binom{S_{1,=n}}{k}$.

We are interested in finding the asymptotic density of k-element codes, i.e., the limit

$$\mu_{\mathcal{S}_k}(\mathcal{S}_k^C) = \lim_{n \to \infty} \frac{S_{k,\leq n}^C}{S_{k,\leq n}} = \lim_{n \to \infty} \frac{|\{X \in \mathcal{S}_k^C : \|X\| = n\}|}{|\{X \in \mathcal{S}_k : \|X\| = n\}|}.$$

The following propositions, stating basic combinatorial properties of the $S_{k,\ldots n}$ numbers, are easily proved. Proofs can be found in [11].

Proposition 1. *For any* $k, n \geq 1$

$$S_{k,<n} = \sum_{i=0}^{n-1} S_{k,\leq i}$$

$$S_{k,\leq n} = \sum_{i=1}^{k} S_{i,=n} \cdot S_{k-i,<n}$$

These numbers can also be expressed non-recursively as:

Proposition 2. *For any* $k, n \geq 1$

$$S_{k,<n} = \binom{\sum_{i=0}^{n-1} S_{1,=i}}{k}$$

$$S_{k,\leq n} = \binom{\sum_{i=0}^{n} S_{1,=i}}{k} - \binom{\sum_{i=0}^{n-1} S_{1,=i}}{k}$$

Proposition 3

$$\forall k \geq 1 \; \exists c = c(k): \; S_{k,<n} < \frac{c}{2^n} \cdot 2^{kn^2}$$

$$\forall k \geq 1 \; \exists c = c(k): \; S_{k,\leq n} < \left(\frac{1}{k!} + \frac{c}{2^n}\right) 2^{kn^2}$$

Recall that $S_{k,\leq n}^C$ is the number of codes containing k figures, with the biggest one of size $n \times n$. Any set containing k squares of fixed size is a code, hence $S_{k,\leq n}^C > S_{k,=n}$. We can now compute the approximate proportion:

$$\frac{S_{k,\leq n}^C}{S_{k,\leq n}} > \frac{S_{k,=n}}{S_{k,\leq n}}$$

$$> \frac{\frac{1}{k!} S_{1,=n}(S_{1,=n} - 1)...(S_{1,=n} - (k-1))}{\frac{1}{k!} 2^{kn^2} + \frac{c(k)}{2^n} 2^{kn^2}}$$

$$= \frac{2^{n^2}(2^{n^2} - 1)...(2^{n^2} - (k-1))}{2^{kn^2} + \frac{c(k) \cdot k!}{2^n} 2^{kn^2}}$$

$$= \frac{(1 - \frac{1}{2^{n^2}})...(1 - \frac{k-1}{2^{n^2}})}{1 + \frac{c(k) \cdot k!}{2^n}}$$

Since $S_{k,\leq n}^C / S_{k,\leq n}$ is bounded by 1, the limit is

$$\mu_{S_k}(S_k^C) = \lim_{n \to \infty} \frac{(1 - \frac{1}{2^{n^2}})...(1 - \frac{k-1}{2^{n^2}})}{1 + \frac{c(k) \cdot k!}{2^n}}$$

$$= 1$$

We thus have:

Theorem 1. *For any fixed k, the density of codes among sets containing k squares, $\mu_{S_k}(S_k^C)$, is equal to 1.*

Note that the codes with squares of fixed size are enough to make the density of codes equal to 1, implying that the density of sets of fixed-size squares among all codes composed of squares is 1. This is not true in the one-dimensional case of words; the density of fixed-length word codes is strictly less than 1, see Remark 1.

4   Counting Word Codes

We now consider ordinary word codes. Although they differ from figure codes in that codicity testing is decidable, their probabilistic behaviour is similar in that the density of codes is equal to 1.

By $\mathcal{W}_k \subset \mathcal{P}(A^*)$ we denote the family of all sets containing k words; $\mathcal{W}_k^C \subseteq \mathcal{W}_k$ is the family of codes containing k words. The size $\|X\|$ of a set $X \in \mathcal{W}_k$ is the length of the longest word in X.

Similarly to the figure case, $W_{k,=n}$, $W_{k,<n}$ and $W_{k,\leq n}$ denote the number of sets containing k words with respective lengths (all of length n, all shorter than n, the longest one of length n) and $W^C_{k,=n}$, $W^C_{k,<n}$ and $W^C_{k,\leq n}$ denote the number of codes containing k words with respective lengths.

Once again, we are interested in finding the asymptotic density of k-element codes, i.e., the limit

$$\mu_{\mathcal{W}_k}(\mathcal{W}^C_k) = \lim_{n\to\infty} \frac{W^C_{k,\leq n}}{W_{k,\leq n}} = \lim_{n\to\infty} \frac{|\{X \in \mathcal{W}^C_k : \|X\| = n\}|}{|\{X \in \mathcal{W}_k : \|X\| = n\}|}.$$

Clearly $W_{1,=n} = 2^n$ and basic combinatorial properties follow those of the $S_{k,\ldots n}$ numbers.

Proposition 4. *For any* $k, n \geq 1$

$$W_{k,<n} = \sum_{i=0}^{n-1} W_{k,\leq i}$$

$$W_{k,\leq n} = \sum_{i=1}^{k} W_{i,=n} \cdot W_{k-i,<n}$$

Proposition 5. *For any* $k, n \geq 1$

$$W_{k,<n} = \binom{\sum_{i=0}^{n-1} W_{1,=i}}{k}$$

$$W_{k,\leq n} = \binom{\sum_{i=0}^{n} W_{1,=i}}{k} - \binom{\sum_{i=0}^{n-1} W_{1,=i}}{k}$$

Proposition 6

$$\lim_{n\to\infty} \frac{W_{k,\leq n}}{2^{kn}} = \frac{2^k - 1}{k!}$$

Remark 1 As noted earlier, the density of fixed-length word codes is strictly less than 1, hence different estimates have to be used. Note that for $k, n \geq 1$

$$\frac{W^C_{k,=n}}{W_{k,\leq n}} = \frac{\binom{2^n}{k}}{\binom{2^{n+1}-1}{k} - \binom{2^n-1}{k}}$$

$$= \frac{(1 - \frac{k-1}{2^n})\ldots(1 - \frac{1}{2^n})}{(2 - \frac{k}{2^n})\ldots(2 - \frac{1}{2^n}) - (1 - \frac{k}{2^n})\ldots(1 - \frac{1}{2^n})}$$

$$\xrightarrow{n\to\infty} \frac{1}{2^k - 1}$$

$$\xrightarrow{k\to\infty} 0$$

We now estimate $W^C_{k,\leq n}$, the number of codes containing k words, with the longest one of length n.

Lemma 1

$$W_{k,\leq n} - W^C_{k,\leq n} < kn \cdot W_{k-1,\leq n}$$

Proof The number on the left-hand side is the number of all non-codes (with given cardinality and size). This is less than e.g. the number of all non-prefix sets. Every non-prefix set can be formed by taking one of the $W_{k-1,\leq n}$-type sets and adding a word that is a prefix of one of words already taken. There are no more than $(k-1)(n-1)$ ways of choosing the prefix, thus the number of non-prefix sets is strictly less than $kn \cdot W_{k-1,\leq n}$.   □

Lemma 2

$$\lim_{n \to \infty} \frac{W_{k,\leq n} - W^C_{k,\leq n}}{2^{kn}} = 0$$

The approximate proportion of codes among words can now be computed as

$$\frac{W^C_{k,\leq n}}{W_{k,\leq n}} = \frac{W_{k,\leq n} - (W_{k,\leq n} - W^C_{k,\leq n})}{W_{k,\leq n}}$$

$$= 1 - \frac{W_{k,\leq n} - W^C_{k,\leq n}}{2^{kn}} \cdot \frac{2^{kn}}{W_{k,\leq n}}$$

Hence

$$\lim_{n \to \infty} \frac{W^C_{k,\leq n}}{W_{k,\leq n}} = 1 - 0 \cdot \frac{k!}{2^k - 1}$$

$$= 1$$

We have now proved

Theorem 2. *For any fixed k, the density of codes among sets containing k words, $\mu_{\mathcal{W}_k}(\mathcal{W}^C_k)$, is equal to 1.*

Note that the estimate used to prove Theorem 2 is based on prefix codes alone. Consequently, the density of prefix codes among all word codes is equal to 1.

5   Counting Domino Figure Codes

Counting domino codes is harder than counting e.g. the squares, since there seems to be no "easy" subclass like fixed-size squares. First, we will count two-element (non-)codes, then we will consider three-domino codes with alphabet of size 2 or 3. The latter case uses a corollary of the defect theorem which provides additional bound for the number of non-codes.

By $\mathcal{D}_k \subset \mathcal{P}(A^{\bowtie})$ we denote the family of all sets containing k dominoes; $\mathcal{D}^C_k \subseteq \mathcal{D}_k$ is the family of codes containing k dominoes. The length of a $1 \times n$ or $n \times 1$ domino is defined to be n and the size $\|X\|$ of a set $X \in \mathcal{D}_k$ is the length of the longest domino in X.

Following the already established convention, $D_{k,=n}$, $D_{k,<n}$ and $D_{k,\leq n}$ denote numbers of sets containing k dominoes with respective lengths (all of length n, all

shorter than n, the longest one of length n) and $D_{k,=n}^C$, $D_{k,<n}^C$ and $D_{k,\leq n}^C$ denote the number of codes containing k dominoes with respective lengths. Additionally, NC superscript will be used to denote the respective numbers of non-codes.

We are obviously interested in finding the asymptotic density of k-element codes, i.e., the limit

$$\mu_{\mathcal{D}_k}(\mathcal{D}_k^C) = \lim_{n\to\infty} \frac{D_{k,\leq n}^C}{D_{k,\leq n}} = \lim_{n\to\infty} \frac{|\{X \in \mathcal{D}_k^C : \|X\| = n\}|}{|\{X \in \mathcal{D}_k : \|X\| = n\}|}.$$

In the present paper we deal with the case of $k = 2$ (simple) and $k = 3$. Note that, as opposed to words and squares, there is no simple subclass of domino codes of density 1. Hence, more specific enumerations have to be considered.

The following properties of the $D_{k,\ldots n}$ numbers are easily proved.

Proposition 7

$$D_{1,=n} = \begin{cases} 1, & n = 0 \\ 2, & n = 1 \\ 2 \cdot 2^n, & n \geq 2 \end{cases}$$

Proposition 8. *For any* $k, n \geq 1$

$$D_{k,<n} = \sum_{i=0}^{n-1} D_{k,\leq i}$$

$$D_{k,\leq n} = \sum_{i=1}^{k} D_{i,=n} \cdot D_{k-i,<n}$$

Proposition 9. *For any* $k, n \geq 1$

$$D_{k,<n} = \binom{\sum_{i=0}^{n-1} D_{1,=i}}{k}$$

$$D_{k,\leq n} = \binom{\sum_{i=0}^{n} D_{1,=i}}{k} - \binom{\sum_{i=0}^{n-1} D_{1,=i}}{k}$$

By the above Propositions we immediately get

Corollary 1

$$D_{k,<n} = \binom{\sum_{i=0}^{n-1} D_{1,=i}}{k}$$

$$= \binom{2^n - 5}{k}$$

$$D_{k,\leq n} = D_{k,<n+1} - D_{k,<n}$$

$$= \binom{2^{n+1} - 5}{k} - \binom{2^n - 5}{k}$$

$$= \Theta(2^{kn})$$

5.1   Figures Defect

Below we quote two specific formulations of the defect theorem for figures. Proofs can be found in [13]. Note that the defect theorem does not hold for larger sets of figures; this is summarized in the following table:

Figures/set size	2	3	≥ 4
Squares	+	−	−
Dominoes	+	+	−
Rectangles	+	?	−
Unrestricted	?	−	−

Theorem 3. *Let $X = \{k, l\} \subseteq A^{\bowtie}$ be a non-code containing two rectangles. Then there exists a common rectangular tiler for k, l, i.e., a rectangle $t \in A^{\bowtie}$ such that $k, l \in \{t\}^{\bowtie}$.*

Corollary 2. *Let $X = \{k, l\} \in \mathcal{D}_2$ be a non-code containing two dominoes. Then (i) k and l have the same orientation, i.e., both are horizontal or both are vertical, or (ii) k and l use just one label.*

Note that in the former case the dominoes can be identified with words.

Theorem 4. *Let $X \in \mathcal{D}_3 \subseteq \mathcal{P}(A^{\bowtie})$ be a non-code containing three dominoes. Then there exists a code $Y \in \mathcal{D}_1 \cup \mathcal{D}_2$ such that Y tiles the dominoes of X.*

The following proposition is a consequence of a detailed case analysis that appears in the proof of the above theorem.

Proposition 10. *Let $X \in \mathcal{D}_3 \subseteq \mathcal{P}(A^{\bowtie})$ be a non-code containing one vertical and two horizontal dominoes. If X uses at least three labels, the vertical domino uses just one label.*

5.2   Codes in $\mathcal{D}_2$

Because of Corollary 2, the number of two-domino non-codes can be computed as

$$2W_{2,\leq n}^{NC} + m(n),$$

where $m(n)$ is the number of "monochromatic" sets with one horizontal and one vertical domino. The factor 2 appears because every non-code of the $W_{2,\leq n}^{NC}$ type can be mapped to two $D_{2,\leq n}^{NC}$ non-codes, one with horizontal dominoes and one with vertical ones.

Now $m(n)$ is simply $2(1 + 2(n-2)) = O(n)$, where (i) the outermost 2 corresponds to the choice of one of two possible labels, (ii) the term 1 describes the set with both dominoes of length n, (iii) $n-2$ corresponds to possible choices for the shorter domino (length in $2...n-1$; note that the case of shorter domino of length 1 is covered by the W^{NC} term), (iv) there are two possible orientations (horizontal longer or shorter), hence the inner factor 2.

Lemma 1, combined with the above, gives us

$$D_{2,\leq n}^{NC} = 2W_{2,\leq n}^{NC} + m(n)$$
$$< 2 \cdot 2nW_{1,\leq n} + m(n)$$
$$< n2^{n+3} + m(n)$$
$$= O(n2^n)$$

Using Corollary 1, we thus obtain

Proposition 11

$$\lim_{n\to\infty} \frac{D_{2,\leq n}^{NC}}{D_{2,\leq n}} = \frac{O(n2^n)}{\Theta(2^{2n})}$$
$$= 0$$

Hence we have proved

Theorem 5. *The density of codes among sets containing two dominoes, $\mu_{\mathcal{D}_2}(\mathcal{D}_2^C)$, is equal to 1.*

5.3 Codes in $\mathcal{D}_3$

We start with the observation that sets in $\mathcal{D}_3$ fall into two categories: (1) sets containing three dominoes of the same orientation, i.e., three horizontal or three vertical ones, (2) sets containing one vertical and two horizontal dominoes, or vice-versa. Case (1) reduces to counting word codes; case (2) requires a detailed analysis.

Lemma 3. *Let $X = \{v, h_1, h_2\} \in \mathcal{D}_3$ contain a vertical domino v and two horizontal dominoes, h_1 and h_2. Let a be the label of the topmost cell of v, i.e., $v = (a...)^T$. If X is not a code, then $X' = \{a, h_1, h_2\} \subseteq A^*$ is not a code (in the word sense).*

Proof. Since X is not a code, there exists a minimal (in the sense of domain inclusion) figure F with two different X-factorizations. We consider rows of the figure defined as maximal horizontal words contained within its domain. Notice that the X-factorizations define two different X'-factorizations of the topmost row of F. Hence X' is not a code. □

We now estimate from above the number of $\{a, h_1, h_2\}$-type non-codes. Without loss of generality, we assume that len $h_1 = n$ and len $h_2 = l \leq n$. Consider the following cases: (2.1) one of $\{a, h_i\}(i = 1, 2)$ is a non-code, (2.2) $\{h_1, h_2\}$ is a non-code, (2.3) all 2-subsets are codes.

Proposition 12. *The number of $\{a, h_1, h_2\}$-type non-codes specified above is $O(p(n)2^n)$, where $p(n)$ is polynomial in n.*

Proof. Case (2.1): Since $\{a, h_i\}$ is not a code, $h_i \in a^*$ and the remaining choices for h_j, $j \neq i$, are $O(n2^n)$.

Case (2.2): By the defect theorem there exists a common factor of h_1 and h_2, i.e., $h_1 = w^{k_1}$ and $h_2 = w^{k_2}$. Choices for w are $O(2^n)$, choices for k_1 and k_2 are $O(n^2)$, giving $O(n^2 2^n)$ in total.

Case (2.3): Obviously, there exists a word with two different factorizations over $\{a, h_1, h_2\}$. Assume that h_1 is non-overlapping. Thus, it is wholly covered by a and h_2, with parts of h_2 possibly extending beyond h_1. The number of choices can now be estimated as

$$2^l \sum_{p_l, p_r} c(n - p_l - p_r),$$

where p_l and p_r signify lengths of those parts of h_2 that cover the left and right end of h_1, respectively (thus p_l and p_r run over $0...n$ with $p_l + p_r \leq n$); $c(m)$ is the number of choices for covering a subword of length m with a and h_2. The factor 2^l is the number of choices for h_2.

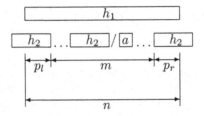

Substituting m for $n - p_l - p_r$ we lose track of the position of the h_2/h_1 overlaps; thus we introduce a factor of $n + 1$ and get

$$2^l \sum_{p_l, p_r} c(n - p_l - p_r) \leq 2^l \sum_{m=0}^{m=n} (n+1) c(m)$$

$$\leq 2^l \sum_{m=0}^{m=n} (n+1) c(n)$$

$$= 2^l (n+1)^2 c(n)$$

$$= 2^l (n+1)^2 (1 + c'(n))$$

$$\leq 2^l (n+1)^2 (1 + 2^{n-l+2} - 1)$$

$$= (n+1)^2 2^{n+2}$$

where $c'(n)$ denotes the number of choices for covering a word of length n with a and h_2, with at least one occurrence of h_2.

A similar estimation can be used if h_1 does overlap itself. Note that cases (2.1) and (2.2) are not disjoint, but their intersection is trivial.

Finally, note that the sum of numbers in (2.1), (2.2) and (2.3) is $O(p(n)2^n)$. $\square$

Proposition 13

$$\lim_{n\to\infty} \frac{D^{NC}_{3,\leq n}}{D_{3,\leq n}} = \frac{O(p(n)2^{2n})}{\Theta(2^{3n})}$$
$$= 0$$

Proof. By Lemma 1, the number of non-codes in case (1) is $2W^{NC}_{3,\leq n} = O(p(n)2^{2n})$. By Lemma 3, the number of non-codes in case (2) is bounded by the number of $\{a, h_1, h_2\}$-type non-codes multiplied by 2^n, the number of choices for the vertical domino. By Proposition 12, this is again $O(p(n)2^{2n})$. Using Corollary 1 we get the required limit. $\square$

Finally we have arrived at

Theorem 6. *The density of codes among sets containing three dominoes, $\mu_{\mathcal{D}_3}(\mathcal{D}^C_3)$, is equal to 1.*

5.4 Codes in $\mathcal{D}_3$ with $|A| \geq 3$

This time we assume an alphabet of at least three letters. The estimates for $|A| = 2$ can be used with base 3 replacing 2 throughout. However, it is interesting to observe that we can make a better estimate for non-codes resulting from case (2.3) of the previous section.

By Proposition 10, each $\{a, h_1, h_2\}$-type non-code gives rise to just one non-code in $\mathcal{D}_3$. Thus their number is $O(p(n)3^n)$.

6 Conclusions

We obviously conjecture that the asymptotic density of domino codes is equal to 1 for all cardinalities of the sets. In fact, we conjecture this is the case for all non-trivial classes of figures, including the class of all figures.

References

1. Beauquier, D., Nivat, M.: A codicity undecidable problem in the plane. Theoret. Comp. Sci. 303, 417–430 (2003)
2. Berstel, J., Perrin, D.: Theory of Codes. Academic Press, London (1985)
3. Harju, T., Karhumäki, J.: Many aspects of the defect effect. Theoret. Comp. Sci. 324, 35–54 (2004)
4. Karhumäki, J., Mantaci, S.: Defect Theorems for Trees. Fundam. Inform. 38, 119–133 (1999)
5. Karhumäki, J., Maňuch, J.: Multiple factorizations of words and defect effect. Theoret. Comp. Sci. 273, 81–97 (2002)
6. Karhumäki, J., Maňuch, J., Plandowski, W.: A defect theorem for bi-infinite words. Theoret. Comp. Sci. 292, 237–243 (2003)

7. Lothaire, M.: Combinatorics on Words. Cambridge University Press, Cambridge (1997)
8. Lothaire, M.: Algebraic Combinatorics on Words. Cambridge University Press, Cambridge (2002)
9. Mantaci, S., Restivo, A.: Codes and equations on trees. Theoret. Comp. Sci. 255, 483–509 (2001)
10. Mañuch, J.: Defect Effect of Bi-infinite Words in the Two-element Case. Discrete Mathematics & Theoretical Computer Science 4, 273–290 (2001)
11. Moczurad, M., Moczurad, W.: Asymptotic density of brick and word codes. Ars Combinatoria 83, 169–177 (2007)
12. Moczurad, M., Tyszkiewicz, J., Zaionc, M.: Statistical properties of simple types. Math. Struct. in Comp. Science 10, 575–594 (2000)
13. Moczurad, W.: Defect theorem in the plane. Theoret. Informatics Appl. 41, 403–409 (2007)
14. Wilf, H.: Generatingfunctionology. Academic Press, London (1994)
15. Yeats, K.: Asymptotic Density in Combined Number Systems. New York J. Math. 8, 63–83 (2002)

On Alternating Phrase-Structure Grammars[*]

Etsuro Moriya[1] and Friedrich Otto[2]

[1] Advanced Research Institute for Science and Engineering and
Department of Mathematics, School of Education,
Waseda University, Shinjuku-ku, Tokyo, 169-8050, Japan
moriya@waseda.jp
[2] Fachbereich Elektrotechnik/Informatik, Universität Kassel
34109 Kassel, Germany
otto@theory.informatik.uni-kassel.de

Abstract. We study several extensions of the notion of alternation from context-free grammars to context-sensitive and arbitrary phrase-structure grammars. Thereby new grammatical characterizations are obtained for the class of languages that are accepted by alternating push-down automata.

1 Introduction

Alternation is a powerful concept that was first introduced by Chandra and Stockmeyer [2,3] for general Turing machines and then by Ladner, Lipton, and Stockmeyer [9,10] for pushdown automata. Thereafter this notion has been studied for a variety of other devices. In particular, in [13] one of the authors introduced the concept of *alternating context-free grammars* (ACFG for short) by distinguishing between *existential* and *universal variables* (nonterminals) with the aim of deriving a grammatical characterization for the class of languages that are accepted by alternating pushdown automata (APDA for short).

As no such characterization was obtained in [13], further studies of the notion of alternation for context-free grammars and pushdown automata followed (see, e.g., [4,14,15,16]). Also Okhotin's *conjunctive grammars* [18] can be interpreted as a variant of ACFGs in which the effect of universal steps is localized. In [7] the class of languages that are accepted by APDAs was finally characterized through *linear-erasing* ACFGs. Further, inspired by the notion of *context-free grammar with states* of Kasai [8], the *state-alternating context-free grammar* (sACFG for short) was introduced in [14] by distinguishing between *existential* and *universal states*. Thus, while in an ACFG the variable on the lefthand side of a production determines whether this production is to be used in an existential or a universal fashion, it is the states that make this distinction in an sACFG. For each ACFG G, an sACFG G' can be constructed such that G and G' generate the same language, but it is still open whether or not the converse is true. At

---

[*] Major parts of this work were done while Etsuro Moriya was visiting at the Fachbereich Elektrotechnik/Informatik, Universität Kassel.

C. Martín-Vide, F. Otto, and H. Fernau (Eds.): LATA 2008, LNCS 5196, pp. 397–408, 2008.
© Springer-Verlag Berlin Heidelberg 2008

least for linear context-free grammars, and therewith in particular for right-linear (that is, regular) grammars, it has been shown that the two notions of alternation have the same expressive power. Actually, both types of alternating right-linear grammars just generate the regular languages. Further, it turned out that sACFGs working in leftmost derivation mode generate exactly those languages that are accepted by APDAs [14]. In this way another grammatical characterization for this class of languages was obtained.

In [16] the authors studied a different way of defining the notion of alternation for pushdown automata. Instead of distinguishing between existential and universal states as in [9,10], here the pushdown symbols are used for this purpose. The *stateless* variant of this so-called *stack-alternating pushdown automaton* accepts exactly those languages that are generated by ACFGs in leftmost derivation mode [16]. However, in general *stack-alternating pushdown automata* are equivalent in expressive power to the original variant of the APDA. It is known that the class of languages these automata accept coincides with the deterministic time complexity class $\mathsf{ETIME} = \bigcup_{c>0} \mathsf{DTIME}(c^n)$ as well as with the alternating space complexity class $\mathsf{ALINSPACE}$, that is, the class of languages that are accepted by *alternating linear bounded automata* (ALBA) [3,10]. As in the classical (non-alternating) setting pushdown automata correspond to context-free grammars and linear bounded automata correspond to context-sensitive grammars, the above results raise the question about the expressive power of alternating context-sensitive grammars.

In this paper we carry the notion of alternation over to general phrase structure and context-sensitive grammars. In fact, we consider both types of alternation for grammars mentioned above. By distinguishing between existential and universal variables we obtain the *alternating phrase-structure grammars* (APSG) and the *alternating context-sensitive grammars* (ACSG). By considering grammars with states, for which we distingush between existential and universal states, we obtain the *state-alternating phrase-structure grammars* (sAPSG) and the *state-alternating context-sensitive grammars* (sACSG). For state-alternating grammars it is rather straightforward to define the notion of derivation. However, for the other type of alternating grammars there are various different ways for defining the corresponding derivation relation. We will consider two such definitions, and we will prove that they are in fact equivalent in a weak sense, that is, for a fixed alternating grammar the two definitions yield different languages, but to each alternating grammar working with the one notion of derivation, there is another grammar of the same type that is working with the other notion of derivation, and that generates the same language. In addition, we will consider two modes of derivation: *leftmost* derivations and *unrestricted* derivations.

Actually, it will turn out that for phrase-structure and for context-sensitive grammars, the state-alternating variant is equivalent to the alternating variant. This equivalence is valid for both leftmost derivations and unrestricted derivations. With respect to unrestricted derivations APSGs just give another characterization for the class RE of recursively enumerable languages. However, with respect to leftmost derivations, they have the same generative power as sACFGs. This can be interpreted as the counterpart to the corresponding result for non-alternating grammars, which states that in leftmost mode general phrase-structure grammars can only generate context-free languages [12]. Our second main result states that with respect to unrestricted derivations ACSGs generate exactly those languages that are accepted by alternating linear bounded automata. As ALBAs and APDAs accept the same languages, we see that APSGs (working in leftmost mode) and ACSGs (working in unrestricted mode) give new grammatical characterizations for the class of languages that are accepted by APDAs. Finally, when working in leftmost mode, ACSGs generate a subclass of this class of languages. It remains open, however, whether this is a proper subclass. These facts should be compared to the fact that no inclusion relation is known between the class of languages generated by sACFGs (or ACFGs) in leftmost mode and the class of languages generated by sACFGs (or ACFGs) in unrestricted mode.

This paper is structured as follows. In Section 2 the basic definitions of alternating and state-alternating grammars are given, and two different ways of defining the notion of *derivation* for alternating grammars are considered. In Section 3 we derive the announced results on the relationships between alternating grammars and state-alternating grammars. The main result of this section states that, with respect to leftmost derivations, APSGs (and therewith sAPSGs) have the same expressive power as sACFGs. Finally, Section 4 is devoted to the study of the relationship between ACSGs on the one hand and alternating linear bounded automata on the other hand. The paper closes with Section 5, where some open problems are presented.

2   Two Types of Alternating Grammars

An *alternating phrase-structure grammar* is a grammar $G = (V, U, \Sigma, P, S)$, where V is a set of variables (or nonterminals), $U \subseteq V$ is a set of *universal* variables, while the variables in $V \smallsetminus U$ are called *existential*, Σ is a set of terminals, S is the start symbol, and P is a set of productions, where $(\ell, r) \in P$ implies that $\ell, r \in (V \cup \Sigma)^*$, and ℓ contains at least one variable. If $|\ell| \leq |r|$ holds for all productions $(\ell, r) \in P$, then G is called an *alternating context-sensitive grammar*, and if $\ell \in V$ holds for all productions $(\ell, r) \in P$, then G is called an *alternating context-free grammar*. By APSG (ACSG, ACFG) we denote the class of all alternating phrase-structure (context-sensitive, context-free) grammars.

It remains to specify the way in which derivations are performed by an alternating grammar G. In particular, we must determine a way to distinguish between existential and universal derivation steps. There are various options.

First of all we can use a specific nonterminal occurring in a sentential form α to determine whether α itself is existential or universal. For example, we could use the leftmost variable occurring in α for that, that is, if $\alpha = xA\beta$, where $x \in \Sigma^*$, $A \in V$, and $\beta \in (V \cup \Sigma)^*$, then we call α an *existential sentential form* if $A \in V \smallsetminus U$, and we call α a *universal sentential form* if $A \in U$. To apply a derivation step to α, we nondeterministically choose a substring ℓ of α that occurs as the left-hand side of one or more rules of P. Now if α is existential, then one of these rules is chosen, and $\alpha = \gamma\ell\delta$ is rewritten into $\gamma r\delta$, where $(\ell, r) \in P$ is the rule chosen. If α is universal, then let $(\ell, r_1), \ldots, (\ell, r_m)$ be those rules of P with left-hand side ℓ. Now all these productions are applied simultaneously, thus giving a finite number of successor sentential forms $\gamma r_1\delta, \ldots, \gamma r_m\delta$. In this way a derivation is not a linear chain, but it has the form of a tree. A terminal word w can be derived from G, if there exists a finite derivation tree in the above sense such that the root is labelled with the start symbol S and all leaves are labelled with w. Observe that in this way the rules themselves are neither existential nor universal, but that it purely depends on the type of the leftmost variable in the actual sentential form whether the next derivation step is existential or universal. Below we will use the notation $\Rightarrow_G^c$ to denote this derivation relation. By $L^c(G)$ we denote the language that is generated by G using this relation.

Alternatively, we can use a distinguished occurrence of a variable in the left-hand side of a rule to declare that rule as being existential or universal. Of course, this must be done in a consistent way, that is, for all rules with the same left-hand side, the same variable occurrence must be chosen. Then for $\alpha = \gamma\ell\delta$, if ℓ is existential, then α is rewritten into $\gamma r\delta$, where $(\ell, r) \in P$ is one of the rules with left-hand side ℓ, and if ℓ is universal, then α is rewritten simultaneously into $\gamma r_1\delta, \ldots, \gamma r_m\delta$, where $(\ell, r_1), \ldots, (\ell, r_m)$ are all the rules in P with left-hand side ℓ. Here we use the following convention: ℓ is universal (or existential, resp.) if the leftmost variable occurring in ℓ is universal (or existential, resp.). We will use the notation $\Rightarrow_G$ to denote this derivation relation. By $L(G)$ we denote the language that is generated by G using this derivation relation.

The following example demonstrates that the derivation relations $\Rightarrow_G^c$ and $\Rightarrow_G$ will in general yield different languages.

Example 1. Let $G = (\{S, A, B\}, \{B\}, \{a, b, c\}, P, S)$ with $P = \{S \rightarrow AB, A \rightarrow a,$ $A \rightarrow ab, B \rightarrow c, B \rightarrow bc\}$. Then with respect to $\Rightarrow_G^c$, G generates the language $L^c(G) = \{ac, abc, abbc\}$, while with respect to $\Rightarrow_G$, we only obtain the language $L(G) = \{abc\}$. The reason is the fact that with respect to $\Rightarrow_G^c$, the rules with left-hand side B can be applied in existential fashion as long as the variable A is still present in the actual sentential form.

However we have the following results [17].

Proposition 1. *For each alternating phrase-structure grammar G, there exists an alternating phrase-structure grammar G' such that $L(G') = L^c(G)$.*

Proposition 2. *For each alternating phrase-structure grammar G, there exists an alternating phrase-structure grammar G' such that $L^c(G') = L(G)$.*

These results also hold for the special case of alternating context-sensitive grammars. Thus, we see that for context-sensitive as well as for general phrase-structure grammars, both definitions of alternation yield the same expressive power. Therefore we restrict our attention in the rest of this paper to alternating grammars for which the leftmost variable occurring in the lefthand side of a production determines whether the production itself is existential or universal.

In addition to the unrestricted derivation mode, we are also interested in the so-called *leftmost* derivation mode. A derivation step $\alpha = \gamma\ell\delta \Rightarrow_G \beta$, respectively $\alpha = \gamma\ell\delta \Rightarrow_G (\gamma r_1\delta, \ldots, \gamma r_m\delta)$, is called *leftmost* if $\gamma \in \Sigma^*$, that is, this step involves the leftmost variable occurrence in α. By $L_{\mathsf{lm}}(G)$ we denote the language consisting of all terminal words that G generates by leftmost derivations. It is obvious that with respect to leftmost derivations the above two definitions of the derivation process of an alternating grammar coincide, if in both definitions the leftmost variable occurrence is chosen.

In [12] it is shown that the language $L_{\mathsf{lm}}(G)$ is context-free if $G = (V, \Sigma, S, P)$ is a phrase-structure grammar such that each rule $(\ell \to r) \in P$ has the structure

$$\ell = x_0 A_1 x_1 \cdots x_{n-1} A_n x_n \to x_0 \beta_1 x_1 \cdots x_{n-1} \beta_n x_n = r$$

for some $n \geq 1$, where $x_0, x_i \in \Sigma^*$, $A_i \in V$, and $\beta_i \in (V \cup \Sigma)^*$ for all $1 \leq i \leq n$ (see, e.g., [11], p. 198). To obtain a corresponding result, we restrict our attention to alternating phrase-structure grammars $G = (V, U, \Sigma, P, S)$ that satisfy the following condition: each rule $(\ell \to r) \in P$ has the form $\ell = xA\alpha \to x\beta = r$, where $x \in \Sigma^*$, $A \in V$, and $\alpha, \beta \in (V \cup \Sigma)^*$. It can be shown that this restriction does not influence the expressive power of alternating phrase-structure grammars as far as the unrestricted derivation mode is concerned. Obviously, this restriction contains the above restriction as a special case, and it is satisfied by all grammars for which the lefthand side of each production begins with a nonterminal.

We denote the class of languages generated by grammars of type X in leftmost derivation mode by $\mathcal{L}_{\mathsf{lm}}(\mathsf{X})$, while $\mathcal{L}(\mathsf{X})$ is used to denote the class of languages generated by these grammars in unrestricted derivation mode.

In [14] also the *state-alternating context-free grammar* (sACFG) was introduced. Analogously, we define the *state-alternating phrase-structure grammar* as an 8-tuple $G = (Q, U, V, \Sigma, P, S, q_0, F)$, where Q is a finite set of states, $U \subseteq Q$ is a set of *universal states*, while the states in $Q \setminus U$ are called *existential states*, V is a finite set of variables, Σ is a set of terminals, $S \in V$ is the start symbol, $q_0 \in Q$ is the initial state, and $F \subseteq Q$ is a set of final states. Finally, P is a finite set of productions of the form $(p, \ell) \to (q, r)$, where $p, q \in Q$, $\ell \in (V \cup \Sigma)^* \cdot V \cdot (V \cup \Sigma)^*$, and $r \in (V \cup \Sigma)^*$. The *derivation relation* $\Rightarrow_G^*$ is defined on the set $Q \times (V \cup \Sigma)^*$ of *extended sentential forms*. Let $p \in Q$ and $\alpha \in (V \cup \Sigma)^*$. If p is an existential state, that is, $p \in Q \setminus U$, then $(p, \alpha) \Rightarrow_G (q, \alpha_1 r \alpha_2)$, if $\alpha = \alpha_1 \ell \alpha_2$, and there exists a production of the form $(p, \ell) \to (q, r)$. If p is a universal state, α has the factorization $\alpha = \alpha_1 \ell \alpha_2$, and $(p, \ell) \to (q_i, r_i)$ $(1 \leq i \leq k)$ are all the productions with lefthand side (p, ℓ), then $(p, \alpha) \Rightarrow_G ((q_1, \alpha_1 r_1 \alpha_2), \ldots, (q_k, \alpha_1 r_k \alpha_2))$, that is, all these productions are applied in parallel to the chosen occurrence of the substring ℓ, and following this step all these sentential forms are rewritten further, independently of each other. In this way a derivation tree is obtained.

The language $L(G)$ that is generated by G consists of all words $w \in \Sigma^*$ for which there exists a derivation tree such that the root is labelled with (q_0, S) and all leaves are labelled with pairs of the form (p, w) with $p \in F$. Note that the labels of different leaves may differ in their first components.

If $|\ell| \leq |r|$ holds for all productions $(p, \ell) \to (q, r)$ of P, then G is called a *state-alternating context-sensitive grammar*, and if $\ell \in V$ for all productions

$(p, \ell) \rightarrow (q, r)$ of P, then G is a *state-alternating context-free grammar*. By sACFG, sACSG, and sAPSG we denote the classes of state-alternating context-free, context-sensitive, and general phrase-structure grammars, respectively. As before we are interested in the expressive power of these grammars with respect to the leftmost and the unrestricted derivation modes. It is known that the class of languages $\mathcal{L}_{lm}(\mathsf{sACFG})$ coincides with the class of languages that are accepted by alternating pushdown automata ([14] Theorem 6.4).

3   Alternation Versus State-Alternation

First we consider the generative power of alternating grammars with respect to the leftmost derivation mode. Recall that we require that each production of an alternating grammar is of the form $(xA\alpha \rightarrow x\beta)$, where $x \in \Sigma^*$, $A \in V$, and $\alpha, \beta \in (V \cup \Sigma)^*$. For state-alternating grammars, we require analogously that each production is of the form $((p, xA\alpha) \rightarrow (q, x\beta))$, where p and q are states.

Lemma 1. $\mathcal{L}_{lm}(\mathsf{ACSG}) \subseteq \mathcal{L}_{lm}(\mathsf{sACSG})$.

Proof. Let $G = (V, U, \Sigma, P, S)$ be an ACSG that satisfies the above condition. We construct an sACSG $G' = (Q, Q^\forall, V, \Sigma, P', S, [?], \{[?]\})$ satisfying $L_{lm}(G') = L_{lm}(G)$ by taking $Q := \{ [A] \mid A \in V \} \cup \{[?]\}$, $Q^\forall := \{ [A] \mid A \in U \}$, and by defining P' as follows:

$$\begin{aligned}([?], A) &\rightarrow ([A], A) \quad \text{for all } A \in V, \\ ([A], xA\alpha) &\rightarrow ([?], x\beta) \quad \text{for all } (xA\alpha \rightarrow x\beta) \in P.\end{aligned}$$

Let $wA\gamma$ be a sentential form of G, where $w \in \Sigma^*$, $A \in V$, and $\gamma \in (V \cup \Sigma)^*$. Assume further that G contains the production $(xA\alpha \rightarrow x\beta)$, and that $w = w_0 x$ and $\gamma = \alpha\delta$ hold. Then with respect to G, we have the leftmost derivation step $wA\gamma = w_0 xA\alpha\delta \Rightarrow_G w_0 x\beta\delta$. From the definition above we see that G' can execute the leftmost derivation $([?], wA\gamma) \Rightarrow_{G'} ([A], wA\gamma) = ([A], w_0 xA\alpha\delta) \Rightarrow_{G'} ([?], w_0 x\beta\delta)$. Here the first step is existential, and the second step is existential if and only if the above step of G is existential. Thus, it follows immediately that the leftmost derivations mod G are in one-to-one correspondence to the leftmost derivations mod G'. Thus, $L_{lm}(G') = L_{lm}(G)$.   □

An analogous result holds for alternating phrase-structure grammars.

Lemma 2. $\mathcal{L}_{lm}(\mathsf{APSG}) \subseteq \mathcal{L}_{lm}(\mathsf{sAPSG})$.

However, for APSGs we even have the following result.

Lemma 3. $\mathcal{L}_{lm}(\mathsf{APSG}) \subseteq \mathcal{L}_{lm}(\mathsf{sACFG})$.

Proof. Let $G = (V, U, \Sigma, P, S)$ be an APSG, let $k := \max\{ |\ell| \mid (\ell \rightarrow r) \in P \}$, and let $\overline{\Sigma} := \{ \bar{a} \mid a \in \Sigma \}$ be a set of new variables that are in one-to-one correspondence to Σ. By $^-: (V \cup \Sigma)^* \rightarrow (V \cup \overline{\Sigma})^*$ we denote the morphism that is defined through $A \mapsto A$ for all $A \in V$ and $a \mapsto \bar{a}$ for all $a \in \Sigma$. We define an sACFG $G' := (Q', U', V', \Sigma, P', S', [\varepsilon], \{[\varepsilon]\})$ as follows, where S', $\$$, and $\cent$ are new variables, and $\hat{V} := \{\, yx\alpha \mid x, y \in \Sigma^*, |yx| \leq k, \alpha \in (V \cup \Sigma)^*, |\alpha| \leq k \,\}$:

$$Q' := \{\, [yx\alpha], [y\#x\alpha] \mid yx\alpha \in \hat{V} \,\},$$
$$U' := \{\, [y\#x\alpha] \mid yx\alpha \in \hat{V},\ \alpha \in U \cdot (V \cup \Sigma)^* \,\},$$
$$V' := V \cup \overline{\Sigma} \cup \{S', \$, \cent\}.$$

The set P' contains the following productions, where $a \in \Sigma$, $X \in V$, $A \in V \cup \overline{\Sigma}$, and $\alpha \in (V \cup \Sigma)^*$:

(0) $([\varepsilon], S') \to ([\varepsilon], S\cent)$.
(1) (a) $([\alpha], X) \to ([\alpha X], \varepsilon)$,   (b) $([\alpha], X) \to ([\alpha X], \$)$ for all $|\alpha| < 2k$.
      (c) $([\alpha], \bar{a}) \to ([\alpha a], \varepsilon)$,   (d) $([\alpha], \bar{a}) \to ([\alpha a], \$)$ for all $|\alpha| < 2k$.
(2) $([a\alpha], A) \to ([\alpha], aA)$ for all $|\alpha| < 2k$.
(3) $([y\#xA\alpha], \$) \to ([yx], \bar{\beta})$ for all $yx \in \Sigma^*$, $|yx| \leq k$, if $(xA\alpha \to \beta) \in P$.
(4) $([\varepsilon], \bar{a}) \to ([\varepsilon], a)$.
(5) $([\varepsilon], \cent) \to ([\varepsilon], \varepsilon)$.
(6) $([yx\alpha], \$) \to ([y\#x\alpha], \$)$ for all $yx\alpha \in \hat{V}$.

The states of G' of the form $[yx\alpha]$ and $[y\#x\alpha]$ are introduced to hold information on the portion $xA\alpha$ of a G-sentential form $yxA\alpha\gamma$ such that $xA\alpha$ could be the lefthand side of a G-production. The first step of a G'-derivation is $([\varepsilon],\ S') \Rightarrow ([\varepsilon],\ S\cent)$, and the $\cent$-symbol will remain unchanged until the sentential form contains this symbol as the only variable left.

Let $xA\alpha \to x\beta$ be a production in P, and consider a leftmost G-derivation step $zyxA\alpha\gamma \Rightarrow zyx\beta\gamma$, where $zy \in \Sigma^*$, $|y| \leq k - |x|$, and $\gamma \in (V \cup \Sigma)^*$. Assume that in G' we have already derived the sentential form $([yx], zA\bar{\alpha}\bar{\gamma}\cent)$. Then we can execute the leftmost G'-derivation

$$([yx], zA\bar{\alpha}\bar{\gamma}\cent) \Rightarrow ([yxA], z\bar{\alpha}\bar{\gamma}\cent) \quad \Rightarrow^* ([yxA\alpha], z\$\bar{\gamma}\cent)$$
$$\Rightarrow ([y\#xA\alpha], z\$\bar{\gamma}\cent) \Rightarrow\ ([yx], z\bar{\beta}\bar{\gamma}\cent).$$

Productions of type (1) choose nondeterministically either to further expand the substring α of the current sentential form $x\alpha\gamma$ obtained so far as a candidate for the lefthand side of a production to be applied to the sentential form or to terminate this process.

Note that $xA\alpha$ is a universal (existential, resp.) string for G iff $A \in U$ ($A \in V \setminus U$, resp.) iff $[y\#xA\alpha]$ is a universal (existential, resp.) state of G' for all $y \in \Sigma^*$ satisfying $|y| \leq k - |x|$, and that by the definition of the productions of group (3), exactly one production $xA\alpha \to \beta$ of G corresponds to the productions of the form $([y\#xA\alpha], \$) \to ([yx], \bar{\beta})$ of G'. Thus, the above simulation of a G-derivation step by G' remains valid also if the G-derivation step is universal.

Next the derivation can proceed according to one of the following two cases.

Case 1 : If $yx\beta\gamma = uvB\delta$ with $uv \in \Sigma^*$, $B \in V$, and $\delta \in (V \cup \Sigma)^*$, then there is a G'-derivation $([yx], z\bar{\beta}\bar{\gamma}\cent) \Rightarrow^* ([v], zuB\bar{\delta}\cent)$ by using productions of type (1), (2), and (4), which enables the application of a production for the next derivation step, whose lefthand side is of the form $v'B\delta_1$ for some suffix v' of v and a prefix δ_1 of δ.

Case 2 : If $\beta\gamma$ is a terminal string, G' can complete the derivation by using productions of type (2) and (4) followed by production (5) to terminate the derivation: $([yx],\ z\bar\beta\bar\gamma\mathcourt) \Rightarrow^* ([\varepsilon],\ zyx\bar\beta\bar\gamma\mathourt) \Rightarrow^* ([\varepsilon],\ zyx\beta\gamma\mathourt) \Rightarrow ([\varepsilon],\ zyx\beta\gamma)$.

It is easily seen that G' generates all the strings in $L_{lm}(G)$ with respect to the leftmost derivation mode, and the converse inclusion can be proved similarly. □

Next we see that also the converse of Lemma 1 holds.

Lemma 4. $\mathcal{L}_{lm}(\mathsf{sACSG}) \subseteq \mathcal{L}_{lm}(\mathsf{ACSG})$.

Proof. Let $G = (Q, U, V, \Sigma, P, q_0, S, F)$ be an sACSG. We construct an ACSG $G' = (V', U', \Sigma, P', [q_0, S])$ that generates the same language as G in leftmode mode. Let $\overline{\Sigma} := \{\, \bar a \mid a \in \Sigma \,\}$ be a new set of variables in one-to-one correspondence to Σ, and let $^- : (V \cup \Sigma)^* \to (V \cup \overline{\Sigma})^*$ be the corresponding morphism. We choose

$$V' := \{\, [q, A]_e, [q, A] \mid q \in Q,\ A \in V \cup \overline{\Sigma} \,\} \cup V \cup \overline{\Sigma},$$
$$U' := \{\, [q, A] \mid q \in U,\ A \in V \,\},$$

and we let P' consist of the following productions, where $p, q \in Q$, $A, B \in V$, $a, b \in \Sigma$, $x, y \in \Sigma^*$, and $\alpha, \beta \in (V \cup \Sigma)^*$:

(1) $[q, A]_e \to [q, A]$ for all $q \in Q$ and all $A \in V$,
(2) $x[p, A]\bar\alpha \to xy[q, B]_e\bar\beta$ for all $((p, xA\alpha) \to (q, xyB\beta)) \in P$,
(3) $x[p, A]\bar\alpha \to x[q, \bar a]_e\bar y$ for all $((p, xA\alpha) \to (q, xay)) \in P$,
(4) $[q, \bar a]_e\bar b \to a[q, \bar b]_e$ for all $a, b \in \Sigma$,
(5) $[q, \bar a]_e B \to a[q, B]_e$ for all $a \in \Sigma$ and all $B \in V$,
(6) $[q, \bar a]_e \to a$ for all $q \in F$ and all $a \in \Sigma$.

As G is a context-sensitive grammar, G' is context-sensitive as well. Further, as G does not contain any ε-rules, the productions of G' of type (2) and (3) are in one-to-one correspondence to the productions of G. Actually, this correspondence respects the type of the productions of being existential or universal.

The basic idea of the simulation of G by G' is the following. The actual state of G is combined with the leftmost variable in the current sentential form. Thus, a sentential form $(p, uxA\alpha\gamma)$ of G, where $p \in Q$, $u, x \in \Sigma^*$, $A \in V$, and $\alpha, \gamma \in (V \cup \Sigma)^*$, is encoded by the sentential form $ux[p, A]_e\bar\alpha\bar\gamma$ of G'. The production $(p, xA\alpha) \to (q, xyB\beta)$ then yields the leftmost derivation step $(p, uxA\alpha\gamma) \Rightarrow_G (q, uxyB\beta\gamma)$. In G' this is simulated by $ux[p, A]_e\bar\alpha\bar\gamma \Rightarrow_{G'} ux[p, A]\bar\alpha\bar\gamma \Rightarrow_{G'} uxy[q, B]_e\bar\beta\bar\gamma$. If a production of the form $(p, xA\alpha) \to (q, xay)$ is used, then we obtain the leftmost derivation step $(p, uxA\alpha\gamma) \Rightarrow_G (q, uxay\gamma)$, which is simulated in G' by $ux[p, A]_e\bar\alpha\bar\gamma \Rightarrow_{G'} ux[p, A]\bar\alpha\bar\gamma \Rightarrow_{G'} ux[q, \bar a]_e\bar y\bar\gamma \Rightarrow^*_{G'} uxayz[q, B]_e\delta$, provided that $\gamma = zB\delta$ for some $z \in \Sigma^*$ and $B \in V$.

A successful leftmost derivation of G ends with a sentential form (q, w) for some $q \in F$ and $w \in \Sigma^+$, where in the last step the last occurring variable is replaced by a terminal string. In G' the corresponding leftmost derivation will yield a sentential form $ux[q,\bar{a}]_e \bar{y}\bar{b}$, where $w = uxayb$. However, using productions of type (4) and (6) we can complete the corresponding leftmost derivation of G' by $ux[q,\bar{a}]_e \bar{y}\bar{b} \Rightarrow^*_{G'} uxay[q,\bar{b}]_e \Rightarrow_{G'} uxayb = w$. It follows that $L_{\mathsf{lm}}(G') = L_{\mathsf{lm}}(G)$ holds.   □

Combining Lemmas 1 and 4 we obtain the following equivalence.

Theorem 1. $\mathcal{L}_{\mathsf{lm}}(\mathsf{ACSG}) = \mathcal{L}_{\mathsf{lm}}(\mathsf{sACSG})$.

The proof above can also be adapted to the case of alternating phrase-structure grammars, which yields the following result.

Lemma 5. $\mathcal{L}_{\mathsf{lm}}(\mathsf{sAPSG}) \subseteq \mathcal{L}_{\mathsf{lm}}(\mathsf{APSG})$.

From Lemmas 3 and 5 and the facts that $\mathcal{L}_{\mathsf{lm}}(\mathsf{sACFG}) \subseteq \mathcal{L}_{\mathsf{lm}}(\mathsf{sAPSG})$ and that $\mathcal{L}_{\mathsf{lm}}(\mathsf{sACFG}) = \mathcal{L}(\mathsf{APDA})$ [14] we obtain the following equivalence.

Theorem 2. $\mathcal{L}_{\mathsf{lm}}(\mathsf{APSG}) = \mathcal{L}_{\mathsf{lm}}(\mathsf{sAPSG}) = \mathcal{L}_{\mathsf{lm}}(\mathsf{sACFG}) = \mathcal{L}(\mathsf{APDA})$.

As $\mathcal{L}_{\mathsf{lm}}(\mathsf{ACSG}) \subseteq \mathcal{L}_{\mathsf{lm}}(\mathsf{APSG})$ holds, Theorem 1 and Theorem 2 yield the following consequence.

Corollary 1. $\mathcal{L}_{\mathsf{lm}}(\mathsf{ACSG}) = \mathcal{L}_{\mathsf{lm}}(\mathsf{sACSG}) \subseteq \mathcal{L}_{\mathsf{lm}}(\mathsf{sACFG}) = \mathcal{L}(\mathsf{APDA})$.

Note that the grammar G' in the proof of Lemma 3 is not context-sensitive. It is still open whether the converse inclusion of Corollary 1 holds.

Now we turn to the unrestricted derivation mode. By using essentially the same proof ideas the above results carry over to this derivation mode. Thus, we have the following equalities, where RE denotes the class of recursively enumerable languages.

Corollary 2. (a) $\mathcal{L}(\mathsf{ACSG}) = \mathcal{L}(\mathsf{sACSG})$.
(b) $\mathcal{L}(\mathsf{APSG}) = \mathcal{L}(\mathsf{sAPSG}) = \mathsf{RE}$.

4   ACSGs and Alternating Linear Bounded Automata

An *alternating linear bounded automaton*, ALBA for short, M is a linear bounded automaton for which some of its states are distinguished as *universal states*. A *configuration* of M is given through a string of the form $\mathvarphi uqav\$$, where q is the current state, uav is the current tape inscription with $a \in \Gamma$, and $u, v \in \Gamma^*$, and the head of M is currently scanning the tape cell containing the distinguished occurrence of the letter a. Here Γ is the tape alphabet of M, and the symbols $\mathcent$ and $\$$ are used as delimiters for the work space.

If state q is existential, then an applicable transition of M is chosen non-deterministically, and M executes the corresponding transformation. If state q is universal, then M applies all applicable transformation steps simultaneously,

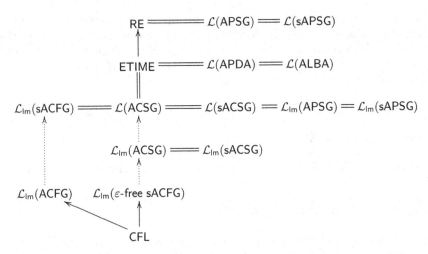

Fig. 1. Inclusion relations among language classes defined by various types of alternating grammars. An arrow denotes a proper inclusion, while a dotted arrow denotes an inclusion that is not known to be proper.

which yields a finite number of successor configurations. In this way, a computation of M can be seen as a tree the nodes of which are labelled by configurations.

The initial configuration for an input $w \in \Sigma^*$ has the form $q_0 \mathord{\text{¢}} w \$$, where q_0 is the initial state of M. The word w is *accepted* by M, if there exists a finite computation tree of M the root of which is labelled with the initial configuration $q_0 \mathord{\text{¢}} w \$$, and all leaves are labelled with accepting configurations, that is, with configurations in which M is in an accepting state. By $L(M)$ we denote the language consisting of all words that are accepted by M.

It is known that $\mathcal{L}(\text{ALBA}) = \mathcal{L}(\text{APDA})$ [3,10]. The next lemma shows that ACSGs are of sufficient expressive power to generate all languages that are accepted by ALBAs. It can be proved by an adaptation of the standard proof that linear bounded automata only accept context-sensitive languages (see, e.g., [6] Theorem 9.6). However, the proof must be modified in such a way that universal steps of the ALBA considered are being simulated faithfully by universal steps of the sACSG. The details can be found in [17].

Lemma 6. $\mathcal{L}(\text{ALBA}) \subseteq \mathcal{L}(\text{sACSG})$.

Also we have the converse of Lemma 6, which can also be proved by an appropriate modification of the standard construction of a linear bounded automaton from a monotone grammar (see, e.g., [6]).

Lemma 7. $\mathcal{L}(\text{sACSG}) \subseteq \mathcal{L}(\text{ALBA})$.

Thus, we obtain the following theorem.

Theorem 3. $\mathcal{L}(\text{sACSG}) = \mathcal{L}(\text{ALBA})$.

By Corollary 2 (a) this yields the following consequence.

Corollary 3. $\mathcal{L}(\mathsf{ACSG}) = \mathcal{L}(\mathsf{sACSG}) = \mathcal{L}(\mathsf{ALBA}) = \mathcal{L}(\mathsf{APDA}) = \mathcal{L}_{\mathsf{lm}}(\mathsf{sACFG})$.

From Corollaries 1 and 3, the following inclusion follows.

Corollary 4. $\mathcal{L}_{\mathsf{lm}}(\mathsf{ACSG}) \subseteq \mathcal{L}(\mathsf{ACSG})$.

It remains to consider the converse of the above inclusion. By Corollary 3 this is equivalent to the question of whether the inclusion $\mathcal{L}_{\mathsf{lm}}(\mathsf{sACFG}) \subseteq \mathcal{L}_{\mathsf{lm}}(\mathsf{ACSG})$ holds. As each ε-free sACFG is context-sensitive, at least the following special case holds.

Corollary 5. $\mathcal{L}_{\mathsf{lm}}(\varepsilon\text{-free sACFG}) \subseteq \mathcal{L}_{\mathsf{lm}}(\mathsf{ACSG})$.

The diagram in Figure 1 depicts the inclusion relations among the classes of languages we have discussed here.

5   Concluding Remarks

We have generalized the notion of alternation from context-free grammars to general phrase-structure grammars. Our main result shows that with respect to the leftmost derivation mode alternating phrase-structure grammars are just as expressive as state-alternating context-free grammars, and that alternating context-sensitive grammars working in the unrestricted derivation mode have the same expressive power, too. In this way we have obtained new grammar-based characterizations for the class of languages that are accepted by alternating pushdown automata. However, an important relation concerning alternating context-sensitive grammars remains open.

Problem 1. Does $\mathcal{L}_{\mathsf{lm}}(\mathsf{ACSG}) = \mathcal{L}(\mathsf{ACSG})$ hold?

The corresponding problem is also still open for alternating context-free grammars. In the light of our results this problem can be expressed as follows.

Problem 2. Does $\mathcal{L}_{\mathsf{lm}}(\mathsf{ACFG}) = \mathcal{L}_{\mathsf{lm}}(\mathsf{APSG})$ hold?

This can be seen as the counterpart for alternating grammars to Matthews' result that in leftmost mode phrase-structure grammars only generate context-free languages [12]. Another obvious problem is the following.

Problem 3. Is $\mathcal{L}_{\mathsf{lm}}(\mathsf{ACFG})$ contained in $\mathcal{L}_{\mathsf{lm}}(\mathsf{ACSG})$?

Finally, there are other derivation modes like the *leftish mode* that we have not considered in this paper. Also it remains to study alternating versions of growing context-sensitive (that is, strictly monotone) grammars [1,5] and their relationship to the shrinking alternating two-pushdown automaton studied by the authors in [19].

Acknowledgement. The first author was supported in part by Waseda University Grant for Special Research Projects #2006B-073, which he gratefully acknowledges. The authors thank Hartmut Messerschmidt from Universität Kassel for fruitful discussions on the notions and results presented in this paper.

References

1. Buntrock, G., Otto, F.: Growing context-sensitive languages and Church-Rosser languages. Information and Computation 141, 1–36 (1998)
2. Chandra, A.K., Stockmeyer, L.J.: Alternation. In: Proc. 17th FOCS, pp. 98–108. IEEE Computer Society Press, Los Alamitos (1976)
3. Chandra, A.K., Kozen, D.C., Stockmeyer, L.J.: Alternation. Journal of the Association for Computing Machinery 28, 114–133 (1981)
4. Chen, Z.Z., Toda, S.: Grammatical characterizations of P and PSPACE. IEICIE Transactions on Information and Systems E 73, 1540–1548 (1990)
5. Dahlhaus, E., Warmuth, M.: Membership for growing context-sensitive grammars is polynomial. Journal of Computer and System Sciences 33, 456–472 (1986)
6. Hopcroft, J.E., Ullman, J.D.: Introduction to Automata Theory, Languages, and Computation. Addison-Wesley, Reading (1979)
7. Ibarra, O.H., Jiang, T., Wang, H.: A characterization of exponential-time languages by alternating context-free grammars. Theoretical Computer Science 99, 301–313 (1992)
8. Kasai, T.: An infinite hierarchy between context-free and context-sensitive languages. Journal of Computer and Systems Sciences 4, 492–508 (1970)
9. Ladner, R.E., Lipton, R.J., Stockmeyer, L.J.: Alternating pushdown automata. In: Proc. 19th FOCS. IEEE Computer Society Press, Los Alamitos (1978)
10. Ladner, R.E., Lipton, R.J., Stockmeyer, L.J.: Alternating pushdown and stack automata. SIAM Journal on Computing 13, 135–155 (1984)
11. Mateescu, A., Salomaa, A.: Aspects of classical language theory. In: Rozenberg, G., Salomaa, A. (eds.) Handbook of Formal Languages. Word, Language, Grammar, vol. 1, pp. 175–251. Springer, Berlin (1997)
12. Matthews, G.: A note on symmetry in phrase structure grammars. Information and Control 7, 360–365 (1964)
13. Moriya, E.: A grammatical characterization of alternating pushdown automata. Theoretical Computer Science 67, 75–85 (1989)
14. Moriya, E., Hofbauer, D., Huber, M., Otto, F.: On state-alternating context-free grammars. Theoretical Computer Science 337, 183–216 (2005)
15. Moriya, E., Nakayama, S.: Grammatical characterizations of alternating pushdown automata and linear bounded automata. Gakujutsu Kenkyu, Series of Math., vol. 45, pp. 13–24. School of Education, Waseda Univ. (1997) (in Japanese)
16. Moriya, E., Otto, F.: Two ways of introducing alternation into context-free grammars and pushdown automata. IEICIE Transactions on Information and Systems E 90- D, 889–894 (2007)
17. Moriya, E., Otto, F.: On alternating non-context-free gammars. Kasseler Informatik Schriften 2007, 6. Fachbereich Elektrotechnik/Informatik, Universität Kassel (2007), https://kobra.bibliothek.uni-kassel.de/handle/urn:nbn:de:hebis:34-2007110719587
18. Okhotin, A.: Conjunctive grammars. Journal of Automata, Languages and Combinatorics 6, 519–535 (2001)
19. Otto, F., Moriya, E.: Shrinking alternating two-pushdown automata. IEICIE Transactionns on Information and Systems E 87- D, 959–966 (2004)

A Two-Dimensional Taxonomy of Proper Languages of Lexicalized FRR-Automata

Friedrich Otto[1] and Martin Plátek[2]

[1] Fachbereich Elektrotechnik/Informatik, Universität Kassel, D-34109 Kassel
otto@theory.informatik.uni-kassel.de
[2] Charles University, Faculty of Mathematics and Physics,
Department of Computer Science, CZ-118 00 Praha 1
Martin.Platek@mff.cuni.cz

Abstract. We study proper languages of (strongly) lexicalized FRR-automata, which are a theoretical model for the analysis by reduction that is used in structural analysis of (natural) languages. We obtain two variants of a two-dimensional hierarchy of language classes based on two types of constraints: (1) the number of rewrite operations per cycle, and (2) the number of occurrences of auxiliary symbols (categories) in the sentences (words) of the corresponding characteristic language. The former type of constraints models non-local valences (dependencies), and the latter type models the use of categories during syntactic disambiguation of the sentence being analyzed.

1 Introduction

Automata with a restart operation were introduced originally to describe a method of grammar-checking for the Czech language (see, e.g., [5]). These automata, which work in a fashion similar to the automata used in this paper, started the investigation of restarting automata as a suitable tool for modeling the so-called *analysis by reduction*. Analysis by reduction in general facilitates the development and testing of categories for syntactic and semantic disambiguation of sentences of natural languages. It is often used (implicitly) for developing formal descriptions of natural languages based on the notion of *dependency* [6,7,14]. In particular, the Functional Generative Description (FGD) for the Czech language developed in Prague (see, e.g., [8]) is based on this method.

Analysis by reduction consists in stepwise simplifications (reductions) of a given extended sentence (enriched by syntactical and semantical categories) until a correct simple sentence is obtained. Each simplification replaces a small part of the sentence by an even shorter phrase. Here we formalize analysis by reduction by using deterministic restarting automata for proper languages. These automata work on so-called *characteristic languages*, that is, on languages that include auxiliary symbols (categories) in addition to the input symbols. The proper language is obtained from a characteristic language by removing all auxiliary symbols from its words (sentences). By requiring that the automata considered are *lexicalized* we

C. Martín-Vide, F. Otto, and H. Fernau (Eds.): LATA 2008, LNCS 5196, pp. 409–420, 2008.

restrict the lengths of the blocks of auxiliary symbols that are allowed on the tape by a constant. This restriction is quite natural from a linguistic point of view, as these blocks of auxiliary symbols model the meta-language categories from individual linguistic layers with which an input string is being enriched when its disambiguated form is being produced (see, e.g., [8]). We use deterministic restarting automata in order to ensure the *correctness preserving property* for the analysis. In fact, we mainly consider *strongly lexicalized* restarting automata. This additional restriction requires that all rewrite operations must be deletions. For example, this type of automaton can be used for modelling the surface (syntactic) level(s) of the Functional Generative Description.

We need a type of automaton that allows us to handle non-local dependencies (valences). Therefore, we choose the *freely rewriting restarting automaton*, FRR-automaton for short, from [10] as our basic model, since it can in general perform an unlimited number of rewrite operations per cycle. However, here we use it in a different way in order to obtain a suitable model for the analysis by reduction. Instead of input (and characteristic) languages as in [10], which correspond to the modelling of *syntactic analysis*, we consider the proper languages of these automata. We use this model in order to study the combination of two types of restrictions that influence the degree of complexity (of analysis by reduction).

The first type restricts the number of rewrite operations per cycle. In linguistic terms this number measures the degree of non-local dependencies (valences) in a sentence. The second type restricts the *word-expansion factor*, that is, the number of auxiliary symbols that may appear concurrently on the tape while a sentence from the characteristic language is being processed. In linguistic terms this corresponds to the number of categories which may be used during a deterministic analysis by reduction. It serves as a measure for the degree of ambiguity (of a certain type) of individual sentences of the language considered. The latter type of restriction was introduced in [12] (see also [11]) for the simpler type of RRWW-automata. For a (formal) language L, the minimal word-expansion factor for any lexicalized (deterministic) restarting automaton with proper language L can also be seen as a measure for the degree of nondeterminism of L. From a language-theoretic point of view this is quite natural, as the auxiliary symbols inserted in an input sentence can be interpreted as information that is used to single out a particular computation of an otherwise nondeterministic restarting automaton. Corresponding notions have been investigated before for finite-state automata and some other devices [2,3]. Our main results establish two variants of a two-dimensional hierarchy of language classes based on the two types of constraints mentioned above.

This paper is structured as follows. In Section 2 we define the deterministic FRR-automaton, and we restate some basic results on this model. We will see in particular that the class of proper languages of deterministic FRR-automata is almost universal. In Section 3 we introduce (strongly) lexicalized FRR-automata, and we define the two types of restrictions we are interested in. Then in Section 4 we present the announced hierarchy results. The paper closes with a short summary in Section 5.

2   FRR-Automata

Here we describe in short the type of restarting automaton we will be dealing with. It is a variant of a model that was introduced in [10].

A *freely rewriting deterministic restarting automaton*, det-FRR-automaton for short, consists of a finite-state control, a flexible tape, and a read/write window of a fixed size $k \geq 1$. It is described as $M = (Q, \Sigma, \Gamma, \mathrm{c}, \$, q_0, k, \delta)$. Here Q denotes a finite set of (internal) states that contains the initial state q_0, Σ is a finite input alphabet, and Γ is a finite tape alphabet that contains Σ. The elements of $\Gamma \setminus \Sigma$ are called *auxiliary symbols*. The additional symbols $\mathrm{c}, \$ \notin \Gamma$ are used as markers for the left and the right end of the workspace, respectively. They cannot be removed from the tape. The behaviour of M is described by a transition function δ that associates transition steps to certain pairs of the form (q, u) consisting of a state q and a possible content u of the read/write window. There are four types of transition steps: *move-right steps*, *rewrite steps*, *restart steps*, and *accept steps*. A *move-right step* simply shifts the read/write window one position to the right and changes the internal state. A *rewrite step* causes M to replace a non-empty prefix u of the content of the read/write window by a shorter word v, thereby shortening the length of the tape, and to change the state. Further, the read/write window is placed immediately to the right of the string v. A *restart step* causes M to place its read/write window over the left end of the tape, so that the first symbol it sees is the left sentinel c, and to reenter the initial state q_0. Finally, an *accept step* simply causes M to halt and accept.

Observe that the det-FRR-automaton is obtained from the FRR-automaton studied in [10] by two essential restrictions: it is deterministic, while an FRR-automaton is in general nondeterministic, and it only has length-reducing rewrite steps, while an FRR-automaton has rewrite steps that are only required to be weight-reducing with respect to some weight function.

A *configuration* of M is described by a string $\alpha q \beta$, where $q \in Q$, and either $\alpha = \lambda$ (the empty word) and $\beta \in \{\mathrm{c}\} \cdot \Gamma^* \cdot \{\$\}$ or $\alpha \in \{\mathrm{c}\} \cdot \Gamma^*$ and $\beta \in \Gamma^* \cdot \{\$\}$; here q represents the current state, $\alpha\beta$ is the current content of the tape, and it is understood that the window contains the first k symbols of β or all of β when $|\beta| \leq k$. A *restarting configuration* is of the form $q_0 \mathrm{c} w \$$, where $w \in \Gamma^*$.

Any computation of M consists of certain phases. A phase, called a *cycle*, starts in a restarting configuration. The window is shifted along the tape by move-right and rewrite operations until a restart operation is performed and thus a new restarting configuration is reached. If no further restart operation is performed, the computation necessarily finishes in a halting configuration – such a phase is called a *tail*. It is required that in each cycle M performs at least one rewrite step. As each rewrite step shortens the tape, we see that each cycle reduces the length of the tape. We use the notation $u \vdash_M^c v$ to denote a cycle of M that begins with the restarting configuration $q_0 \mathrm{c} u \$$ and ends with the restarting configuration $q_0 \mathrm{c} v \$$; the relation $\vdash_M^{c*}$ is the reflexive and transitive closure of $\vdash_M^c$.

A word $w \in \Gamma^*$ is *accepted* by M, if there is a computation which starts from the restarting configuration $q_0 \mathrm{c} w \$$, and which ends with an application of an accept step. By $L_C(M)$ we denote the language consisting of all words accepted by M. It is the *characteristic language* of M.

By Pr^Σ we denote the projection from Γ^* onto Σ^*, that is, Pr^Σ is the morphism defined by $a \mapsto a$ $(a \in \Sigma)$ and $A \mapsto \lambda$ $(A \in \Gamma \setminus \Sigma)$. If $v := \mathsf{Pr}^\Sigma(w)$, then v is the Σ-*projection* of w, and w is an *expanded version* of v. For a language $L \subseteq \Gamma^*$, $\mathsf{Pr}^\Sigma(L) := \{\, \mathsf{Pr}^\Sigma(w) \mid w \in L \,\}$.

In recent papers (see, e.g., [13]) restarting automata were mainly used as acceptors. The (*input*) *language* accepted by a restarting automaton M is the set $L(M) := L_C(M) \cap \Sigma^*$. Here, motivated by linguistic considerations to model the processing of sentences that are enriched by syntactic and semantic categories, we are rather interested in the so-called *proper language of M*, which is the set of words $L_P(M) := \mathsf{Pr}^\Sigma(L_C(M))$. Hence, a word $v \in \Sigma^*$ belongs to $L_P(M)$ if and only if there exists an expanded version u of v such that $u \in L_C(M)$.

For each type X of restarting automata, we use $\mathcal{L}_C(\mathsf{X})$ and $\mathcal{L}_P(\mathsf{X})$ to denote the class of all characteristic languages and the class of all proper languages of automata of this type. As a det-FRR-automaton can easily be simulated by a two-tape Turing machine in quadratic time, we have the following result.

Proposition 1. *If M is a deterministic FRR-automaton, then the membership problems for the languages $L_C(M)$ and $L(M)$ are solvable in quadratic time.*

The deterministic RRWW-automaton (see, e.g., [13]) is essentially a det-FRR-automaton that only performs a single rewrite step in each cycle. In [11] it is shown that the class $\mathcal{L}_P(\text{det-RRWW})$ of proper languages of deterministic RRWW-automata is 'almost' universal. Accordingly we have the following result that is in stark contrast to Proposition 1.

Proposition 2. *There exists a deterministic FRR-automaton M such that the language $L_P(M)$ is non-recursive.*

We close this section with two basic properties of det-FRR-automata that are used repeatedly in proofs (see, e.g., [4] and [13]).

Proposition 3 (Correctness Preserving Property)
Each det-FRR-automaton M is correctness preserving, *that is, if $u \in L_C(M)$ and $u \vdash_M^{c*} v$, then $v \in L_C(M)$, too.*

Proposition 4 (Pumping Lemma)
For any det-FRR-automaton M, there exists a constant p such that the following property holds. Assume that $uxvyz \vdash_M^c ux'vy'z$ is a cycle of M, where $u = u_1 u_2 \cdots u_n$ for some non-empty words $u_1, \ldots, u_n$ and an integer $n > p$. Then there exist $r, s \in \mathbb{N}_+$, $1 \le r < s \le n$, such that

$$u_1 \cdots u_{r-1}(u_r \cdots u_{s-1})^i u_s \cdots u_n xvyz \vdash_M^c u_1 \cdots u_{r-1}(u_r \cdots u_{s-1})^i u_s \cdots u_n x'vy'z$$

holds for all $i \ge 0$, that is, $u_r \cdots u_{s-1}$ is a 'pumping factor' in the above cycle. Similarly, such a pumping factor can be found in any factorization of length greater than p of v or z. Such a pumping factor can also be found in any factorization of length greater than p of a word accepted in a tail computation.

3   Strongly Lexicalized FRR-Automata

From Propositions 1 and 2 we know that proper languages of deterministic FRR-automata are in general far more complex than the corresponding input and characteristic languages. Therefore we restrict our attention to deterministic FRR-automata for which the use of auxiliary symbols is restricted as in [11,12].

Definition 1. *Let* $M = (Q, \Sigma, \Gamma, \mathс, \$, q_0, k, \delta)$ *be a* det-FRR-*automaton.*

(a) *A word* $w \in \Gamma^*$ *is not immediately rejected by* M *if, starting from the restarting configuration* $q_0 \mathс w \$$, M *either performs a cycle of the form* $w \vdash^c_M z$ *for some word* $z \in \Gamma^*$, *or* M *accepts* w *in a tail computation. By* NIR(M) *we denote the set of all words that are not immediately rejected by* M.
(b) *The* det-FRR-*automaton* M *is called* lexicalized *if there exists a constant* $j \in \mathbb{N}_+$ *such that, whenever* $v \in (\Gamma \setminus \Sigma)^*$ *is a factor of a word* $w \in$ NIR(M), *then* $|v| \leq j$.
(c) M *is called* strongly lexicalized *if it is lexicalized, and if each of its rewrite operations just deletes some symbols.*

Strong lexicalization is a technique that is used in dependency based formal descriptions of natural languages [8]. If M is a lexicalized FRR-automaton, and if $w \in \Gamma^*$ is an extended version of an input word $v = \mathrm{Pr}^\Sigma(w)$ such that w is not immediately rejected by M, then $|w| \leq (j+1) \cdot |v| + j$ for some constant $j > 0$. Accordingly we have the following result.

Corollary 1. *If* M *is a lexicalized* FRR-*automaton, then the proper language* $L_P(M)$ *is context-sensitive.*

In what follows we are mainly interested in (strongly) lexicalized FRR-automata and their proper languages. By LRR (SLRR) we denote the class of (strongly) lexicalized FRR-automata, and by t-LRR (t-SLRR) we denote the class of (strongly) lexicalized FRR-automata which execute at most t rewrite steps in any cycle. Recall from the definition that lexicalized FRR-automata are deterministic. We now introduce a static complexity measure for LRR-automata.

Definition 2. *Let* $M = (Q, \Sigma, \Gamma, \mathс, \$, q_0, k, \delta)$ *be an* LRR-*automaton, and let* $m \in \mathbb{N}$. *The automaton* M *has* word-expansion m, *denoted by* W$(M) = m$, *if each word from* NIR(M) *contains at most* m *occurrences of auxiliary symbols, that is, if* $w \in \Gamma^*$ *is not immediately rejected by* M, *then* $|\mathrm{Pr}^{\Gamma \setminus \Sigma}(w)| \leq m$.

We use the prefix W(m)- to denote classes of deterministic FRR-automata that have word-expansion m. The following result is a generalization of a result for lexicalized RRWW-automata given in [11,12].

Theorem 1. *If* M *is a* W(m)-LRR-*automaton for some* $m \in \mathbb{N}$, *then the membership problem for* $L_P(M)$ *is solvable deterministically in time* $O(n^{m+2})$.

As the 1-(S)LRR-automaton is essentially identical to the (strongly) lexicalized RRWW-automaton considered in [11,12], we have the following results, where the symbol $\subset$ denotes the proper inclusion relation.

Theorem 2. [11,12]

(a) $\qquad\qquad$ DCFL $\subset \mathcal{L}_P(W(0)\text{-}1\text{-SLRR}) \not\subseteq$ CFL.

(b) $\qquad\qquad$ CFL $\subset \mathcal{L}_P(1\text{-SLRR})$.

(c) $\qquad\qquad$ CFL $\not\subseteq \bigcup_{i \geq 0} \mathcal{L}_P(W(i)\text{-}1\text{-LRR})$.

(d) $\quad \mathcal{L}_P(W(i)\text{-}1\text{-(S)LRR}) \subset \mathcal{L}_P(W(i+1)\text{-}1\text{-(S)LRR})$ *for all* $i \in \mathbb{N}$.

(e) $\bigcup_{i \geq 0} \mathcal{L}_P(W(i)\text{-}1\text{-(S)LRR}) \subset \mathcal{L}_P(1\text{-(S)LRR})$.

Further we obtain the following result from the proof of Corollary 2 of [10].

Theorem 3. $\mathcal{L}_P(W(0)\text{-}t\text{-(S)LRR}) \subset \mathcal{L}_P(W(0)\text{-}(t+1)\text{-(S)LRR})$ *for all* $t \in \mathbb{N}_+$.

The transition relation of a t-FRR-automaton $M = (Q, \Sigma, \Gamma, \math{c}, \$, q_0, k, \delta)$ can be described more transparently through a finite sequence of so-called *meta-instructions* of the form $(E_1, u_1 \to v_1, E_2, u_2 \to v_2, E_3, \ldots, E_i, u_i \to v_i, E_{i+1})$, where $i \leq t$, $E_1, \ldots, E_{i+1}$ are regular expressions, and for each $j = 1, \ldots, i$, $u_j, v_j \in \Gamma^*$ satisfying the condition $k \geq |u_j| > |v_j|$. Here each rule $u_j \to v_j$ ($1 \leq j \leq i$) embodies a rewrite step of M that replaces the factor u_j by v_j. Starting from a restarting configuration $q_0 \math{c} w \$$, M can execute this meta-instruction only if w admits a factorization of the form $w = w_1 u_1 w_2 u_2 \cdots w_i u_i w_{i+1}$ such that $\math{c} w_1 \in L(E_1)$, $w_2 \in L(E_2)$, $\ldots$, $w_{i+1} \$ \in L(E_{i+1})$, where $L(E_l)$ denotes the language described by the regular expression E_l. In this case the leftmost of these factorizations is chosen, and $q_0 \math{c} w \$$ is transformed into the restarting configuration $q_0 \math{c} w_1 v_1 w_2 v_2 \cdots w_i v_i w_{i+1} \$$. In order to describe tails of accepting computations of M (during which M cannot apply any rewrite operations at all), we use meta-instructions of the form $(\math{c} \cdot E \cdot \$, \mathsf{Accept})$, which accepts the sentences from the regular language $L(E)$.

We will use meta-instructions for describing individual examples of t-LRR-automata in the next section. Observe, however, that meta-instructions are inherently nondeterministic. Therefore we will only use them to describe t-LRR-automata in a more readable way. To obtain an exact definition of the automaton presented, one needs to derive an explicit transition function.

4   Two-Dimensional Hierarchies

We study a number of example languages in order to establish a two-dimensional hierarchy of classes of proper languages of (strongly) lexicalized FRR-automata.

Definition 3. *Let* $\Sigma_0 := \{a, b\}$, *let* c, d *be two additional letters, and let* L_{rp}, $L_{\mathrm{d}}(j)$, *and* $L_{\mathrm{dp}}(i, j)$ *be the following languages, where* w^R *denotes the* reversal *of a word* w:

(1) $L_{\mathrm{rp}} \quad := \{\, wwc \mid w \in \Sigma_0^* \,\}$,

(2) $L_{\mathrm{d}}(j) \quad := \{\, (wc)^j dw^R \mid w \in \Sigma_0^* \,\}$ *for all* $j \in \mathbb{N}_+$,

(3) $L_{\mathrm{dp}}(i, j) := (L_{\mathrm{rp}})^i \cdot \{d\} \cdot L_{\mathrm{d}}(j) \qquad$ *for all* $i \geq 1$ *and* $j \geq 2$.

For processing the language $L_{\mathrm{d}}(j)$ no auxiliary symbols are needed; however, at least j rewrite steps per cycle are required.

Proposition 5. *For all* $j \in \mathbb{N}_+$, $L_d(j) \in \mathcal{L}_P(W(0)\text{-}j\text{-SLRR})$, *while for* $j > 1$, $L_d(j) \notin \mathcal{L}_P((j-1)\text{-LRR})$.

Proof. Let $j \in \mathbb{N}_+$, and let $M_d^{(j)}$ be the strongly lexicalized j-FRR-automaton that is given by the following meta-instructions, where $x \in \Sigma_0$:

(1) $(\mathfrak{c} \cdot \Sigma_0^*, xc \to c, \Sigma_0^*, xc \to c, \Sigma_0^*, \ldots, xc \to c, \Sigma_0^*, xcdx \to cd, \Sigma_0^* \cdot \$)$,
(2) $(\mathfrak{c} \cdot c^j d \cdot \$, \mathsf{Accept})$.

The automaton $M_d^{(j)}$ executes exactly j rewrite steps per cycle, it does not use any auxiliary symbols, and it is easily seen that $L_P(M_d^{(j)}) = L(M_d^{(j)}) = L_d(j)$ holds. Thus, it follows that $L_d(j) \in \mathcal{L}_P(W(0)\text{-}j\text{-SLRR})$.

Now let $j > 1$, and assume that M is a $(j-1)$-LRR-automaton on Γ such that $L_P(M) = L_d(j)$. Let $A(m,n,j) := ((a^m b^m)^n c)^j d(b^m a^m)^n$, where $m, n \in \mathbb{N}_+$ are sufficiently large. Obviously, $A(m,n,j) \in L_d(j)$. Hence, there exists an expanded version $w \in \Gamma^*$ of $A(m,n,j)$ such that $w \in L_C(M)$. Assume that w is a shortest expanded version of $A(m,n,j)$ in $L_C(M)$. The computation of M on input w is accepting, but based on the Pumping Lemma (Prop. 4) it is easily seen that this computation cannot just consist of an accepting tail. Thus, it begins with a cycle of the form $w \vdash_M^c x$. From the Correctness Preserving Property it follows that $x \in L_C(M)$, which in turn implies that $\mathrm{Pr}^\Sigma(x) \in L_d(j)$. As all rewrite steps of M are length-reducing, $|x| < |w|$ follows. Thus, our choice of $w \in L_C(M)$ as a shortest expanded version of $A(m,n,j)$ implies that x is not an expanded version of $A(m,n,j)$. Since M executes at most $j-1$ rewrite steps in the above cycle, it follows that $\mathrm{Pr}^\Sigma(x) \notin L_d(j)$. Hence, $L_P(M) \neq L_d(j)$, which implies that $L_d(j) \notin \mathcal{L}_P((j-1)\text{-LRR})$. $\qquad\square$

In contrast to the situation for $L_d(j)$, the language $L_{rp}^i := (L_{rp})^i$ only requires two rewrite steps per cycle, but in that case it needs word expansion i.

Proposition 6. *For all* $i \geq 1$, *the following results hold:*

(a) $L_{rp}^i \in \mathcal{L}_P(W(i)\text{-}2\text{-SLRR})$.
(b) $L_{rp}^i \notin \mathcal{L}_P(1\text{-LRR})$.
(c) $L_{rp}^i \notin \mathcal{L}_P(W(i-1)\text{-}j\text{-LRR})$ *for any* $j \in \mathbb{N}_+$.

Proof. (a) Let $i \geq 1$, and let M be the 2-SLRR-automaton with input alphabet $\Sigma := \Sigma_0 \cup \{c\}$ and tape alphabet $\Gamma := \Sigma \cup \{D\}$ that is given through the following meta-instructions, where $x \in \Sigma_0$:

(0) $(\mathfrak{c} \cdot (Dc)^i \cdot \$, \mathsf{Accept})$,
(1) $(\mathfrak{c}, x \to \lambda, \Sigma_0^*, Dx \to D, \Sigma_0^* \cdot c \cdot (\Sigma_0^* \cdot D \cdot \Sigma_0^* \cdot c)^{i-1} \cdot \$)$,
(2) $(\mathfrak{c} \cdot Dc, x \to \lambda, \Sigma_0^*, Dx \to D, \Sigma_0^* \cdot c \cdot (\Sigma_0^* \cdot D \cdot \Sigma_0^* \cdot c)^{i-2} \cdot \$)$,
$\cdots \quad \cdots \qquad\qquad \cdots$
$(i-1)$ $(\mathfrak{c} \cdot (Dc)^{i-2}, x \to \lambda, \Sigma_0^*, Dx \to D, \Sigma_0^* \cdot c \cdot \Sigma_0^* \cdot D \cdot \Sigma_0^* \cdot c \cdot \$)$,
(i) $(\mathfrak{c} \cdot (Dc)^{i-1}, x \to \lambda, \Sigma_0^*, Dx \to D, \Sigma_0^* \cdot c \cdot \$)$.

Then $L_C(M) = \{ w_1 D w_1 c w_2 D w_2 c \cdots w_i D w_i c \mid w_1, w_2, \ldots, w_i \in \Sigma_0^* \}$, which implies that $L_P(M) = L_{rp}^i$. Obviously, M has word expansion i. This proves part (a).

(b) We proceed as in the proof of the Proposition 5. Let $i \geq 1$, and assume that M is a 1-LRR-automaton on Γ such that $L_P(M) = L_{rp}^i$. Let $B(m, n, i)$ denote the string $B(m, n, i) := ((a^m b^m)^{2n} c)^i$, where $m, n \in \mathbb{N}_+$ are sufficiently large. Obviously, $B(m, n, i) \in L_{rp}^i$. Hence, there exists an expanded version $w \in \Gamma^*$ of $B(m, n, i)$ such that $w \in L_C(M)$. Assume that w is a shortest expanded version of $B(m, n, i)$ in $L_C(M)$. The computation of M on input w is accepting, but based on the Pumping Lemma (Prop. 4) it is easily seen that this computation cannot just consist of an accepting tail. Thus, it begins with a cycle of the form $w \vdash_M^c x$, where $|x| < |w|$. From the Correctness Preserving Property we see that $x \in L_C(M)$. Because of our choice of w as a shortest expanded version of $B(m, n, i)$ in $L_C(M)$, it follows that $\mathsf{Pr}^\Sigma(x) \neq B(m, n, j)$. Recall, however, that M only executes a single rewrite step in the above cycle. This implies that $\mathsf{Pr}^\Sigma(x)$ cannot possibly be an element of L_{rp}^i. Hence, we conclude that $L_P(M) \neq L_{rp}^i$, which yields that $L_{rp}^i \notin \mathcal{L}_P(\text{1-LRR})$.

(c) First we consider the case $i = 1$. Assume that $j \geq 2$ is a minimal integer such that there exists a j-LRR-automaton M' with word expansion 0 satisfying $L_{rp} = L_P(M')$. As each word from the language L_{rp} contains just a single occurrence of the symbol c as the very last letter, we can assume without loss of generality that in each cycle of each accepting computation M' never executes more than a single rewrite step with the symbol c in its read/write window.

Let $x_0 = (a^m b^m)^n$, where $n, m \in \mathbb{N}_+$ are sufficiently large integers that depend on the constant for M' from the Pumping Lemma, the size of the read/write window of M', and the number of internal states of M'. Then $x_0 x_0 x_0 x_0 c \in L_{rp}$. Starting from the restarting configuration corresponding to $x_0 x_0 x_0 x_0 c$, M' will execute an accepting computation, but clearly this computation cannot just consist of an accepting tail because of the Pumping Lemma. Thus, M' executes a cycle of the form $x_0 x_0 x_0 x_0 c \vdash_{M'}^c v$ for some word $v \in (\Sigma_0 \cup \{c\})^*$.

Let us state some observations about this cycle. As M' is deterministic, and as it executes at most j rewrite operations in the cycle above, we can conclude from the Pumping Lemma that all these rewrite operations (with at most a single exception) are applied while the read/write window is still inside the prefix $c x_0$. In fact, if there is a rewrite step which is not executed on this prefix, then it must be applied to the very end of $x_0 x_0 x_0 x_0 c$ (that is, with the symbol c already inside the window). Recall that m and n are (very) large integers, which means that M' will not execute any more rewrite operations before encountering the symbol c once it has moved across a sufficiently large number of blocks of the form $a^m b^m$. It now follows that $v \notin L_{rp}$, since in the above cycle the prefix $c x_0$ and possibly the suffix $x_0 c$ are changed, while the infix (the middle blocks) $x_0 x_0$ of $x_0 x_0 x_0 x_0 c$ remain unchanged. This contradicts the Correctness Preserving Property for M'. Hence, the language L_{rp} is not the proper language of any j-LRR-automaton with word expansion 0.

Finally, assume that $j \geq 2$, and that M' is a j-LRR-automaton with word expansion $i - 1$ such that $L_{rp}^i = L_P(M')$. Let $w := (x_0 x_0 x_0 x_0 c)^i$. Then $w \in L_{rp}^i$, and hence, there exists an expanded version W of w such that $W \in L_C(M')$. Thus, the computation of M' on input W is accepting, and clearly it cannot just consist of an accepting tail. Hence, M' executes a cycle of the form $W \vdash^c_{M'} W'$, and $W' \in L_C(M')$ and $|W'| < |W|$ follow. Thus, we have $w' := \mathrm{Pr}^{\Sigma}(W') \in L_P(M') = L^i_{\mathrm{rp}}$.

As M' has word expansion $i - 1$, we see that at least one factor of the form $x_0x_0x_0x_0c$ of w does not contain an occurrence of an auxiliary symbol in W. Thus, the processing of this particular factor by M' starts without any auxiliary symbols. Now to the processing of this factor the arguments from the proof for $i = 1$ apply. Recall that we consider the proper language of M', which means that the ability of M' to rewrite up to j factors of this particular factor $x_0x_0x_0x_0c$ of W by shorter words possibly containing (up to) $i - 1$ occurrences of auxiliary symbols does not interfere with the arguments from the proof for the case $i = 1$. Thus, we obtain the same contradiction as above. This completes the proof of part (c). □

By combining Proposition 5 and Proposition 6 we obtain the following results.

Proposition 7. *For all $i \geq 1$ and all $j \geq 2$, the following results hold:*

(a) $L_{\mathrm{dp}}(i,j) \in \mathcal{L}_P(\mathsf{W}(i)\text{-}j\text{-SLRR})$.
(b) $L_{\mathrm{dp}}(i,j) \notin \mathcal{L}_P((j-1)\text{-LRR})$.
(c) $L_{\mathrm{dp}}(i,j) \notin \mathcal{L}_P(\mathsf{W}(i-1)\text{-}m\text{-LRR})$ *for any* $m \in \mathbb{N}_+$.

Thus, we have the following hierarchy results.

Theorem 4. *For all $\mathsf{X} \in \{\mathsf{LRR}, \mathsf{SLRR}\}$, for all $i \geq 0$, and for all $j \geq 1$, we have the following proper inclusions:*

(a) $\mathcal{L}_P(\mathsf{W}(i)\text{-}j\text{-}\mathsf{X}) \subset \mathcal{L}_P(\mathsf{W}(i+1)\text{-}j\text{-}\mathsf{X})$.
(b) $\mathcal{L}_P(\mathsf{W}(i)\text{-}j\text{-}\mathsf{X}) \subset \mathcal{L}_P(\mathsf{W}(i)\text{-}(j+1)\text{-}\mathsf{X})$.
(c) $\quad\quad \mathcal{L}_P(j\text{-}\mathsf{X}) \subset \mathcal{L}_P((j+1)\text{-}\mathsf{X})$.

Proof. The inclusions in (a) contain Theorem 2 (d) as the special case $j = 1$, and the inclusions in (b) contain Theorem 3 as the special case $i = 0$. All other results follow from Proposition 7. □

As 1-LRR-automata coincide with lexicalized RRWW-automata, we see from [11] Propositions 3.2 and 3.3 that the class of proper languages of 1-LRR-automata is a proper subclass of the class of growing context-sensitive languages. On the other hand, there exists a W(0)-2-SLRR-automaton M such that the proper language of M coincides with the language $L_{\mathrm{rep}} := \{\, wcw \mid w \in \Sigma_0^* \,\}$, which is not growing context-sensitive (see, e.g., [1]). Thus, while all classes $\mathcal{L}_P(\mathsf{W}(i)\text{-}1\text{-}(\mathsf{S})\mathsf{LRR})$ only consist of growing context-sensitive languages, we see that each class $\mathcal{L}_P(\mathsf{W}(i)\text{-}j\text{-}(\mathsf{S})\mathsf{LRR})$, $j \geq 2$, contains languages that are not growing context-sensitive.

We also want to separate the classes of proper languages of (strongly) lexicalized FRR-automata with unbounded degree of word expansion or unbounded number of rewrites per cycle from those with bounded degree of word expansion or bounded number of rewrites per cycle, respectively. For that purpose we consider the following example languages.

Definition 4. Let $L_{d+} := \bigcup_{j\geq 1} L_d(j)$ and $L_{rp+} := \bigcup_{j\geq 1}(L_{rp})^j$.

From the proofs of Proposition 5 and 6 we obtain the following results.

Proposition 8. (a) $L_{d+} \in \mathcal{L}_P(W(0)\text{-SLRR})$.
    (b) $L_{d+} \notin \mathcal{L}_P(j\text{-LRR})$ for any $j \in \mathbb{N}_+$.

Proposition 9. (a) $L_{rp+} \in \mathcal{L}_P(2\text{-SLRR})$.
    (b) $L_{rp+} \notin \mathcal{L}_P(W(i)\text{-}j\text{-LRR})$ for any $i \in \mathbb{N}$ and $j \in \mathbb{N}_+$.

This yields the following proper inclusions, where the result in (a) contains Theorem 2 (e) as the special case $j = 1$.

Corollary 2. For all $X \in \{\text{LRR}, \text{SLRR}\}$,

(a) $\bigcup_{i\geq 0} \mathcal{L}_P(W(i)\text{-}j\text{-X}) \subset \mathcal{L}_P(j\text{-X})$ for all $j \geq 1$.
(b) $\bigcup_{j\geq 1} \mathcal{L}_P(j\text{-X}) \subset \mathcal{L}_P(X)$.

Finally, we want to separate the hierarchy of proper languages of strongly lexicalized FRR-automata from the corresponding hierarchy for lexicalized FRR-automata. To this end we consider the example language

$$L_{\text{expo}} := \{ a^{i_0}ba^{i_1}b\cdots a^{i_{n-1}}ba^{i_n} \mid n \geq 0, i_0,\ldots,i_n \geq 0, \text{ and }$$
$$\exists m \geq 0 : \sum_{j=0}^{n} 2^j \cdot i_j = 2^m \} \cup b^*,$$

for which we have the following result.

Proposition 10
$L_{\text{expo}} \in \mathcal{L}_P(W(0)\text{-}1\text{-LRR})$, but $L_{\text{expo}} \notin \mathcal{L}_P(j\text{-SLRR})$ for any $j \in \mathbb{N}_+$.

Proof. Let M_{expo} be the deterministic 1-FRR-automaton that is given through the following meta-instructions:

    (1) $(\text{¢} \cdot a^*, aab \to ba, \Sigma_0^* \cdot \$)$,   (3) $(\text{¢} \cdot a^*, a^4 \to baa, \$)$,
    (2) $(\text{¢}, b \to \lambda, \Sigma_0^* \cdot \$)$,   (4) $(\text{¢} \cdot \{\lambda, a, aa\} \cdot \$, \text{Accept})$.

In [11] it is shown that this automaton, which is actually a deterministic RRW-automaton, accepts the language L_{expo}. Thus, $L_{\text{expo}} \in \mathcal{L}_P(W(0)\text{-}1\text{-LRR})$.

Assume now that there exists a strongly lexicalized j-FRR-automaton M with input alphabet $\Sigma_0 := \{a, b\}$ and tape alphabet Γ such that $L_{\text{expo}} = L_P(M)$ holds, and let $z := a^{2^n} \in L_{\text{expo}}$, where n is a large integer. Then there exists an expanded version $w \in \Gamma^*$ of z such that $w \in L_C(M)$. Assume that w is a shortest expanded version of z in $L_C(M)$. The computation of M on input w is accepting, and based on the Pumping Lemma (Prop. 4) it is easily seen that it cannot just consist of an accepting tail. Thus, it begins with a cycle of the form $w \vdash_M^c w'$ for some word $w' \in \Gamma^*$ satisfying $|w'| < |w|$. From the Correctness Preserving Property it follows that $w' \in L_C(M)$, which in turn implies that $\text{Pr}^{\Sigma_0}(w') \in L_{\text{expo}}$. However, w' is not an expanded version of z due to our assumption on w. Thus, $\text{Pr}^{\Sigma_0}(w') = a^m$ for some integer $m < 2^n$. In the above cycle M executes at most j rewrite (that is, delete) operations, and so we see that $m \geq 2^n - j \cdot k$, where k is the size of the read/write window of M. This contradicts the fact that a^m must be a power of 2. It follows that L_{expo} differs from the language $L_{\mathrm{P}}(M)$, which implies that $L_{\mathrm{expo}} \notin \bigcup_{j \geq 1} \mathcal{L}_{\mathrm{P}}(j\text{-SLRR})$. $\square$

Using the same proof idea it can be shown that the Church-Rosser language $L_{\mathrm{expo}} \cap a^+ = \{ a^{2^n} \mid n \geq 0 \}$ is not contained in $\mathcal{L}_{\mathrm{P}}(j\text{-LRR})$ for any $j \in \mathbb{N}_+$, which yields the following consequence.

Corollary 3. $\mathsf{GCSL} \not\subset \bigcup_{j>0} \mathcal{L}_{\mathrm{P}}(j\text{-LRR})$.

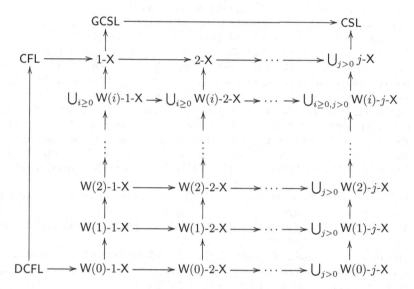

Fig. 1. Inclusion relations between language classes defined by various types of lexicalized FRR-automata. Here X denotes either LRR or SLRR, a node labeled by an automata type $\mathcal{A}$ denotes the class $\mathcal{L}_{\mathrm{P}}(\mathcal{A})$, and an arrow denotes a proper inclusion.

5   Conclusion

We have studied the classes of proper languages of (strongly) lexicalized restarting automata with multiple rewrites. We have investigated the influence of two parameters on the expressive power of these automata: the number of rewrites per cycle, and the number of auxiliary symbols that may appear on the tape at the same time. The resulting two-dimensional hierarchies are shown in Figure 1. Language classes which are not connected (by an oriented path) in that diagram are incomparable under inclusion. The only possible exception concerns the inclusion $\mathsf{CFL} \subset \bigcup_{i \geq 0, j > 0} \mathcal{L}_{\mathrm{P}}(\mathsf{W}(i)\text{-}j\text{-LRR})$, which remains currently open. However, we conjecture that the language $L_{\mathrm{pal}}^+ = \{ w_1 w_1^R \cdots w_n w_n^R \mid n \geq 1, w_1, \ldots, w_n \in \Sigma_0^* \}$, which is context-free, is not the proper language of any $\mathsf{W}(i)\text{-}j\text{-LRR}$-automaton.

Acknowledgement. Martin Plátek was supported by the program 'Information Society' under project 1ET100300517. The authors also want to thank Dana Pardubská from Comenius University, Bratislava, and František Mráz from Charles University, Prague, for numerous discussions on the notions and results of this paper.

References

1. Buntrock, G., Otto, F.: Growing context-sensitive languages and Church-Rosser languages. Information and Computation 141, 1–36 (1998)
2. Goldstine, J., Kintala, C., Wotschke, D.: On measuring nondeterminism in regular languages. Information and Computation 86, 179–194 (1990)
3. Goldstine, J., Leung, H., Wotschke, D.: Measuring nondeterminism in pushdown automata. Journal of Computer and System Sciences 71, 440–466 (2005); An extended abstract appeared in Reischuk, R., Morvan, M. (eds.) STACS 1997, LNCS, Vol. 1200, pp. 295–306. Springer, Heidelberg (1997)
4. Jančar, P., Mráz, F., Plátek, M., Vogel, J.: On monotonic automata with a restart operation. Journal of Automata, Languages and Combinatorics 4, 287–311 (1999)
5. Kuboň, V., Plátek, M.: A grammar based aproach to a grammar checking of free word order languages. In: COLING 1994, Kyoto, Japan, vol. II, pp. 906–910 (1994)
6. Kunze, J.: Abhängigkeitsgrammatik. Studia Grammatica XII. Akademie-Verlag, Berlin (1975)
7. Lopatková, M., Plátek, M., Kuboň, V.: Modeling syntax of free word-order languages: Dependency analysis by reduction. In: Matoušek, V., Mautner, P., Pavelka, T. (eds.) TSD 2005. LNCS (LNAI), vol. 3658, pp. 140–147. Springer, Heidelberg (2005)
8. Lopatková, M., Plátek, M., Sgall, P.: Towards a formal model for functional generative description: Analysis by reduction and restarting automata. The Prague Bulletin of Mathematical Linguistics 87, 7–26 (2007)
9. Lothaire, M.: Combinatorics on Words. Addison-Wesley, Mass., Reading (1982)
10. Mráz, F., Otto, F., Plátek, M.: Free word order and restarting automata. In: Loos, R., Fazekas, S.Z., Martin-Vide, C. (eds.) LATA 2007, Preproc., Research Group on Math. Linguistics, Universitat Rovira i Virguli, Tarragona, pp. 425–436 (2007)
11. Mráz, F., Otto, F., Plátek, M.: The Degree of Word-Expansion of Lexicalized RRWW-Automata – A New Measure for The Degree of Nondeterminism of (Context-Free) Languages. Kasseler Informatikschriften (2007)
12. Mráz, F., Plátek, M., Otto, F.: A measure for the degree of nondeterminism of context-free languages. In: Holub, J., Žďárek, J. (eds.) CIAA 2007. LNCS, vol. 4783, pp. 192–202. Springer, Heidelberg (2007)
13. Otto, F.: Restarting automata. In: Ésik, Z., Martin-Vide, C., Mitrana, V. (eds.) Recent Advances in Formal Languages and Applications, Studies in Computational Intelligence, vol. 25, pp. 269–303. Springer, Berlin (2006)
14. Sgall, P., Hajičová, E., Panevová, J.: The Meaning of the Sentence in Its Semantic and Pragmatic Aspects. Reidel Publishing Company, Dordrecht (1986)

Minimalist Grammars with Unbounded Scrambling and Nondiscriminating Barriers Are NP-Hard*

Alexander Perekrestenko

Rovira i Virgili University
Research Group on Mathematical Linguistics
International PhD School in Formal Languages and Applications
Pl. Imperial Tarraco 1; 43005 Tarragona, Spain
alexander.perekrestenko@estudiants.urv.cat
http://www.grlmc.com
http://www.urv.cat

Abstract. Minimalist Grammars were proposed in [15] as a formalization of the basic structure-building component of the Minimalism Program, a syntactic framework introduced in [2] and [3]. In the present paper we investigate the effects of extending this formalism with an unrestricted scrambling operator together with nondiscriminating barriers. We show that the recognition problem for the resulting formalism NP-hard. The result presented here is a generalization of the result shown by the author in [14] for Minimalist Grammars with unrestricted scrambling and category-sensitive barriers.

1 Introduction

Minimalist Grammars (MGs) were proposed in [15] as a formal tool for modeling some fundamental structure-building operations of the Minimalist Program, an approach adopted within the Chomskyan branch of syntactic theory [2,3]. Unrestricted Minimalist Grammars originally introduced in [15] belong to the class of mildly context-sensitive grammar formalisms and are weakly equivalent to *linear context-free rewriting systems* (LCFRS) which was shown in [10], [7] and [11].

The generative power of MGs is crucially affected by the presence or absence of so-called *locality constraints* (LCs). The two best investigated LCs in terms of their effect on the weak generative capacity of MGs are the *shortest-move constraint* (SMC) and the *specifier island constraint* (SPIC). The SMC prohibits competitive movement of constituents, while the SPIC bars movement from within constituents that have already moved. It was shown that adding these LCs has a non-monotonic effect on the generative power of MGs so that

---

* This research work has been supported by the Russian Foundation for the Humanities as a part of the project "The typology of free word order languages" (grant RGNF 06-04-00203a).

C. Martín-Vide, F. Otto, and H. Fernau (Eds.): LATA 2008, LNCS 5196, pp. 421–432, 2008.

the following inclusion results hold for the corresponding language classes ('+' and '−' stand here for presence or absence of the mentioned LCs):[1]

- $L(\mathrm{MG}^{SMC+}_{SPIC+}) \subset L(\mathrm{MG}^{SMC+}_{SPIC-}) = L(\mathrm{LCFRS})$,
- $L(\mathrm{MG}^{SMC-}_{SPIC+}) = \mathrm{Type0}$.

In order to model some specific syntactic operations not incorporated into the original version of MG, the formalism was extended with the operations of *scrambling* and *(cyclic) adjunction* in [5] and *countercyclic adjunction* in [13]. However, the scrambling operator introduced in [5] was restricted by SMC which reduced it to an operation similar to non-obligatory movement making the generalized description of this syntactic phenomenon impossible. Formally, the SMC does not make much sense for scrambling as an SMC-restricted scrambling can be simulated by movement. What on the contrary is important for scrambling from the linguistic point of view is *barriers*, i.e., constituents that prevent other constituents from being displaced.

In [14] it was shown that extending MGs with unbounded scrambling and barriers makes the recognition problem for the resulting formalism NP-hard.[2] It was proved using two category-sensitive barriers, i.e., barriers prohibiting scrambling of constituents with a given base category and not affecting scrambling of other constituents. In this paper we prove a stronger result showing that even nondiscriminating barriers blocking scrambling of *any* constituent are enough to make the recognition problem for the scrambling-extended MG NP-hard.

2   MGs with Unbounded Scrambling and Barriers

Below we give a definition of the Minimalist Grammars with unbounded scrambling and nondiscriminating barriers which is closely based on the definition of MG proposed in [6].

Definition 1 (MG$^{scr}_{B0}$). *A Minimalist Grammar with unbounded scrambling and nondiscriminating barriers, MG^{scr}_{B0}, is a tuple $G = \langle \neg Syn, Syn, c, \#, Lex, \Omega \rangle$, such that*[3]

- *$\neg Syn$ is a finite set of non-syntactic features partitioned into the sets of phonetic (Phon) and semantic (Sem) features;*
- *Syn is a finite set of syntactic features disjoint from $\neg Syn$ and partitioned into the following sets:*
  - *base (syntactic) categories, Base, partitioned into*
    the set of categories without barrier $B = \{ x_1, x_2, \ldots, x_n \}$ and
    the set of categories with barrier $\bar{B} = \{ \bar{x}_1, \bar{x}_2, \ldots, \bar{x}_n \}$,

---

[1] The result on the proper inclusion of $L(\mathrm{MG}^{SMC+}_{SPIC+})$ into $L(\mathrm{MG}^{SMC+}_{SPIC-})$ can be found in [12]. The Turing equivalence of $\mathrm{MG}^{SMC-}_{SPIC+}$ was proved in [9].

[2] The term *unbounded scrambling* will further be used to denote scrambling that is not restricted by SMC.

[3] This particular use of $\neg Syn$ and Syn in the definition is motivated by the tradition to expressly separate syntactic and nonsyntactic features.

- $m(erge)$-selectors $M = \{ =x \mid x \in B \}$,
- $m(ove)$-licensees $E = \{ -x \mid x \in B \}$,
- $m(ove)$-licensors $R = \{ +x \mid x \in B \}$,
- $s(cramble)$-licensees, $S = \{ \sim x \mid x \in B \}$, and
- '$\#$', a special symbol;

- c is a distinguished element of Base, the completeness category;
- Lex is a lexicon, defined further on;
- Ω is the set of the structure-building operators 'merge', 'move' and 'scramble' specified later.

We will denote by Feat the union of the syntactic and non-syntactic features: $Feat = Syn \cup \neg Syn$.

Definition 2 (Expression). *An expression over the set of features Feat, also called a minimalist tree, is a five-tuple* $\tau = \langle N_\tau, \vartriangleleft_\tau^*, \prec_\tau, <_\tau, label_\tau \rangle$, *obeying the following conditions:*

- $\langle N_\tau, \vartriangleleft_\tau^*, \prec_\tau \rangle$ *is a finite binary ordered tree, where* N_τ *is a non-empty finite set of nodes,* $\vartriangleleft_\tau$ *is the binary relation of immediate dominance on* N_τ, $\vartriangleleft_\tau^*$ *is its reflexive transitive closure, and* $\prec_\tau$ *is the binary relation of precedence on* N_τ;
- $<_\tau \subseteq N_\tau \times N_\tau$ *is the asymmetric relation of immediate projection that holds for any two sibling nodes, so that for each* $x \in N_\tau$ *which is not the root of* τ *either* $x <_\tau sibling(x)$ *or* $sibling(x) <_\tau x$; *in the case* $x <_\tau y$, *we say that* x *immediately projects over* y;
- $label_\tau$ *is a leaf-labeling function assigning to each leaf of* $\langle N_\tau, \vartriangleleft_\tau^*, \prec_\tau \rangle$ *an element from* $Syn^* \{\#\} Syn^* Phon^* Sem^*$, *where Syn, Phon, Sem and '$\#$' are the same as in the definition of* MG_{B0}^{scr}.

We will denote by $Exp(Feat)$ the set of all expressions over the features $Feat$. Let $\tau = \langle N_\tau, \vartriangleleft_\tau^*, \prec_\tau, <_\tau, label_\tau \rangle \in Exp(Feat)$ be an expression.

Leaf $z \in N_\tau$ is the *head* of a given node $x \in N_\tau$ if either z and x are the same node or $x \vartriangleleft_\tau^+ z$ and for each $y \in N_\tau$, such that $x \vartriangleleft_\tau^+ y \vartriangleleft_\tau^* z$, the following holds: $y <_\tau sibling_\tau(y)$.[4] Expression τ is said to be a *head*, or a *simple* expression, if N_τ contains exactly one node. Otherwise τ is said to be a *non-head*, or a *complex* expression. The head of a tree is the head of its root. The root of the tree τ is denoted as r_τ.

Expression corresponding to a given subtree ϕ of τ is referred to as a *subexpression* of τ. Such subexpression ϕ with root $x \in N_\tau$ is said to be a *maximal projection* in τ if either x is the root of τ (and, as a consequence, does not have any siblings with respect to τ) or ϕ is a proper subexpression of τ and $sibling_\tau(x) <_\tau x$. The set of all maximal projections of τ is denoted as $MaxProj(\tau)$.

---

[4] Analogously to the notation $\vartriangleleft^*$, we use the shorthand $\vartriangleleft^+$ to denote the non-reflexive transitive closure of the dominance relation.

Expression $\phi \in MaxProj(\tau)$ over *Feat* is said to *display* feature $f \in Feat$ if its label is in $\alpha \# f \beta$ where $\alpha, \beta \in Feat^*$; expression $\phi \in MaxProj(\tau)$ is said to *contain* feature $f \in Feat$ if its label is in $\alpha f \alpha' \# \beta \beta'$ or in $\alpha \alpha' \# \beta f \beta'$ where $\alpha, \alpha', \beta, \beta' \in Feat^*$.

Expression τ is *complete* if its head label is in $Syn^* \{\#\} \{c\} S^? Phon^* Sem^*$ and the labels of all of its leaves are in $Syn^* \{\#\} S^? Phon^* Sem^*$.[5]

Maximal projection τ is *licensed for scrambling* (to x) if the head label of τ displays feature $\sim x$ for some $x \in B$ or $\bar{x} \in \bar{B}$.

Subexpression $\phi \in MaxProj(\tau)$ is *barred for scrambling to* τ if there exists a $\chi \in MaxProj(\tau)$ such that $\phi \in MaxProj(\chi)$, $r_\phi \neq r_\chi$, $r_\chi \neq r_\tau$ and the label of χ contains feature $\bar{x} \in \bar{B}$. The head of the subtree χ will be called a *barrier*. Sometimes we will use the word *barrier* to refer to a full projection whose head is a barrier.

Subexpression $\phi \in MaxProj(\tau)$ is said to be a *candidate for scrambling to expression* τ if τ is a maximal projection, for some $x \in B$ (or $\bar{x} \in \bar{B}$) the head label of τ displays category x (or $\bar{x}$), the subexpression ϕ is licensed for scrambling to x and it is not barred for scrambling to τ. For a given maximal projection τ, the set of all candidates for scrambling to τ will be denoted as $ScrS(\tau)$.

For two expressions $\phi, \chi \in Exp(Feat)$, $[_< \phi, \chi]$ (respectively, $[_> \phi, \chi]$) denotes the complex expression $\psi = \langle N_\psi, \lhd^*_\psi, \prec_\psi, <_\psi, label_\psi \rangle \in Exp(Feat)$ with $r_\psi \lhd_\psi r_\phi$, $r_\psi \lhd_\psi r_\chi$, $r_\phi \prec_\psi r_\chi$, and $r_\phi <_\psi r_\chi$ (respectively, $r_\chi <_\psi r_\phi$).

The phonetic yield of the (complex) expression τ, $Y_{Phon}(\tau)$, is defined as the concatenation of the *Phon* string of the leaf labels in the order in which they appear in the tree τ.

The *lexicon* of MG^{scr}_{B0}, *Lex*, is a finite set of simple expressions over *Feat*, each of which is of the form $\phi = \langle N_\phi, \lhd^*_\phi, \prec_\phi, <_\phi, label_\phi \rangle$ with $N_\phi = \{\epsilon\}$ and the leaf-labeling function $label_\phi$ assigns to the only node of ϕ an element from $\{\#\} M^* R^* Base (E \cup S)^? Phon^* Sem^*$.

An expression thus defined directly translates into a binary tree whose leaves are elements of the lexicon (lexical entries) and the nonleaf nodes are marked with a symbol of the immediate projection relation ('<' or '>').

The structure-building operators Ω of MG^{scr}_{B0} are defined below in terms of their mapping type, domain and operation.

Operator *merge*

Type: partial mapping $Exp(Feat) \times Exp(Feat) \to Exp(Feat)$.

Domain: for any $\phi, \chi \in Exp(Feat)$, the tuple $\langle \phi, \chi \rangle$ is in $Dom(merge)$ iff for some $x \in B$ the head label of ϕ displays m-selector $=x$ and the head label of χ displays category x or $\bar{x}$.

Operation:

$$merge(\phi, \chi) = \begin{cases} [_< \phi', \chi'] & | \ \phi \text{ is simple} \\ [_> \chi', \phi'] & | \ \phi \text{ is complex} \end{cases}$$

---

[5] Here it is not required that all scrambling licensees be eliminated. In this way the optionality of scrambling is guaranteed.

where ϕ' and χ' result from the corresponding ϕ and χ by swapping the # symbol with the feature immediately following it to the right.

Operator *move*

Type: partial mapping $Exp(Feat) \rightarrow Exp(Feat)$.

Domain: for any $\phi \in Exp(Feat)$, the expression ϕ is in $Dom(move)$ iff for some $x \in B$ the head label of ϕ displays m-licensor $+x$ and there is a unique $\chi \in MaxProj(\phi)$ displaying m-licensee $-x$.[6]

Operation: $move(\phi) = [_> \chi', \phi']$, where ϕ' and χ' result from the corresponding ϕ and χ by swapping the # symbol with the feature immediately following it to the right and replacing the subtree χ in ϕ with a single empty node labeled ϵ.

Operator *scramble*

Type: partial mapping $Exp(Feat) \rightarrow 2^{Exp(Feat)}$.

Domain: for any $\phi \in Exp(Feat)$, the expression ϕ is in $Dom(scramble)$ iff $ScrS(\phi) \neq \emptyset$.

For $\Phi \subseteq Exp(Feat)$, let $S'(\Phi) = \{ [_> \chi', \phi'] \mid \phi \in \Phi,\ \phi \in Dom(scramble),\ \chi \in ScrS(\phi),\ \phi'$ results from ϕ by replacing subtree χ by a single empty node labeled ϵ, and χ' results from χ by swapping the # symbol with the feature immediately following it to the right $\}$.

Let $S^0(\Phi) = \Phi$ and $S^{k+1}(\Phi) = S^k(\Phi) \cup S'(S^k(\Phi))$ for $k \geq 0$, $k \in I\!N$.

Operation: $scramble(\phi) = \bigcup_{k \in I\!N} S^k(\{\phi\})$.

Let $G = \langle \neg Syn, Syn, c, \#, Lex, \Omega \rangle$ be an MG^{scr}_{B0}. Let $CL^0(G) = Lex$. For $k > 0$, $k \in I\!N$, $CL^k(G)$ will be defined as follows:

$$CL^{k+1}(G) = CL^k(G) \cup \{ merge(\phi, \chi) \mid \phi, \chi \in CL^k(G),\ \langle \phi, \chi \rangle \in Dom(merge) \} \cup$$
$$\{ move(\phi) \mid \phi \in CL^k(G),\ \phi \in Dom(move) \} \cup$$
$$\{ scramble(\phi) \mid \phi \in CL^k(G),\ \phi \in Dom(scramble) \}.$$

Let $CL(G) = \bigcup_{k \in I\!N} CL^k(G)$. The tree language of G, $MT^{scr}_{B0}(G)$, is defined in the following way:

$$MT^{scr}_{B0}(G) = \{ \tau \mid \tau \in CL(G) \text{ and } \tau \text{ is complete } \}.$$

The string language of G, $ML^{scr}_{B0}(G)$, is defined as the yields of its tree language:

$$ML^{scr}_{B0}(G) = \{ Y_{Phon}(\tau) \mid \tau \in MT^{scr}_{B0}(G) \}.$$

---

[6] The uniqueness of χ prohibiting the occurrence of two or more competing movement candidates is the way the SMC is implemented for the *move* operator.

3 MG_{B0}^{scr} is NP-Hard

3.1   Preliminaries: NP-Hardness

A problem X is NP-hard if and only if an NP-complete problem N can be transformed ("reduced") to X in polynomial time in such a way that a (hypothetical) polynomial-time algorithm solving X could also be used to solve N in polynomial time.

Let $L()$ be the word recognition problem for the language L. Let L, L_1 and L_2 be languages over an alphabet Σ^* such that $L = L_1 \cup L_2$ and $L_1 \cap L_2 = \emptyset$. Let $p(w)$ be a polynomial-time computable function with domain $Dom(p) = \Sigma^*$ such that for any $w \in L$ it returns *true* if $w \in L_1$ and *false* otherwise. (For a $w \notin L$, it can return either *true* or *false*.) We will need the following proposition:

Proposition 1. *If $L_1()$ is NP-hard, then $L()$ is also NP-hard.*

3.2   The Idea of the Proof

The NP-hardness of the word recognition problem for MG_{B0}^{scr} will be proved by constructing a grammar $G \in MG_{B0}^{scr}$ that generates a language $L = L_1 \cup L_2$, $L_1 \cap L_2 = \emptyset$, such that a known NP-complete problem can be reduced to $L_1()$ in polynomial time, i.e., $L_1()$ is NP-hard, and the question whether a word $w \in L$ belongs to L_1 or to L_2 can be solved in deterministic polynomial time. In the proof we will use the 3-Partition Problem which is known to be strongly NP-complete:

> Given a multiset of $3k$ natural numbers $\{n_1, n_2, \ldots, n_{3k}\}$ and a constant m, decide whether this multiset can be partitioned into k subsets consisting each of three elements whose sum is m.

This problem can be described as a language

$$L_{3P} = \{ ax^{n_1} ax^{n_2} \ldots ax^{n_{3k}} b^m \mid a, b, x \in \Sigma \}$$

such that it consists of *all* the words for which $\langle n_1, n_2, \ldots, n_{3k}, m \rangle$ represents an instance of the 3-Partition Problem as described above. The word recognition for this language is NP-hard.[7]

3.3   Proving NP-Hardness

Let $G = \langle \neg Syn, Syn, s, \#, Lex, \Omega \rangle$ be an MG_B^{scr} where

- $Phon = \{a, b, c, d\}$, $Sem = \emptyset$, and
- $Base = \{ a_1, a_2, a_3, a_1', a_2', a_3', a_1'', a_2'', a_3'', b, b', b^0, c_1, c_2, c_3, c_1', c_2', c_3', c_1'',$ $c_2'', c_3'', d_1, d_2, d_3, d_1', d_2', d_3', d_1'', d_2'', d_3'', g, e, p, s, \bar{t}, \bar{t}', u, u', u'', v, v', w,$ $w' \}$.

---

[7] Without loss of generality we will only consider positive natural numbers and assume $k \geq 1$.

The lexicon of the grammar, Lex, consists of the following entries:[8]

1. (a) $\#. = c_3''. a_3''. \sim s. a;$ $\#. = d_3''. = b^0. c_3''. c;$
   $\#. = c_3''. = b. d_3''. d;$ $\#. = e. = b. d_3''. d;$ $\#. e;$

   (b) $\#. = c_2''. a_2''. \sim s. a;$ $\#. = d_2''. = b^0. c_2''. c;$
   $\#. = c_2''. = b. d_2''. d;$ $\#. = a_3''. = b. d_2''. d;$

   (c) $\#. = c_1''. a_1''. \sim s. a;$ $\#. = d_1''. = b^0. c_1''. c;$
   $\#. = c_1''. = b. d_1''. d;$ $\#. = a_2''. = b. d_1''. d;$

   (d) $\#. = u''. w. -w;$ $\#. = a_1''. u''. -u';$

2. (a) $\#. = c_3. a_3. \sim s. a;$ $\#. = d_3. c_3. c;$
   $\#. = c_3. = b'. d_3. d;$ $\#. = w. = b'. d_3. d;$

   (b) $\#. = c_2. a_2. \sim s. a;$ $\#. = d_2. c_2. c;$
   $\#. = c_2. = b'. d_2. d;$ $\#. = a_3. = b'. d_2. d;$

   (c) $\#. = c_1. a_1. \sim s. a;$ $\#. = d_1. c_1. c;$
   $\#. = c_1. = b'. d_1. d;$ $\#. = a_2. = b'. d_1. d;$

   (d) $\#. = t. +v. w'. -w';$ $\#. = u. \bar{t};$
   $\#. = v. +u'. +w. u. -u;$ $\#. = a_1. v. -v;$

3. (a) $\#. = c_3'. a_3'. \sim s. a;$ $\#. = d_3'. c_3'. c;$
   $\#. = c_3'. = b. d_3'. d;$ $\#. = w'. = b. d_3'. d;$

   (b) $\#. = c_2'. a_2'. \sim s. a;$ $\#. = d_2'. c_2'. c;$
   $\#. = c_2'. = b. d_2'. d;$ $\#. = a_3'. = b. d_2'. d;$

   (c) $\#. = c_1'. a_1'. \sim s. a;$ $\#. = d_1'. c_1'. c;$
   $\#. = c_1'. = b. d_1'. d;$ $\#. = a_2'. = b. d_1'. d;$

   (d) $\#. = t'. +v'. w. -w;$ $\#. = u'. \bar{t}';$
   $\#. = v'. +u. +w'. u'. -u';$ $\#. = a_1'. v'. -v';$

4. $\#. = w. g;$ $\#. = w'. g;$

5. $\#. = g. +u'. +w. s;$ $\#. = g. +u. +w'. s;$

6. $\#. b. \sim c_1. b;$ $\#. b. \sim c_2. b;$ $\#. b. \sim c_3. b;$ $\#. b. \sim g. b;$
   $\#. b'. \sim c_1'. b;$ $\#. b'. \sim c_2'. b;$ $\#. b'. \sim c_3'. b;$ $\#. b'. \sim g. b;$ $\#. b^0. b$

Proposition 2. *The language L generated by the grammar G is a union of two disjoint languages, $L = L_{3p} \cup L'$, $L_{3p} \cap L' = \emptyset$, such that L_{3p} consists of all the words*

$$a(bcd)^{n_1} a(bcd)^{n_2} \ldots a(bcd)^{n_{3k}} b^m$$

with $a, b, c, d \in \Sigma$, where $\langle n_1, n_2, \ldots, n_{3k}, m \rangle$ is an instance of the 3-Partition Problem, as described above, and there exists a polynomial-time computable function $p(w)$ such that for any word $w \in L$ it returns true if $w \in L_{3p}$ and false otherwise; for $w \notin L$ it returns either true or false.

---

[8] For better legibility we separate features with dots.

We will prove the proposition 2 by following the derivation of the language L steps whose numbering corresponds to that of the lexical entries standing for these derivation steps.[9]

Subwords in $a(bcbd)^+a(bcbd)^+a(bcbd)^+$ or $a(cbd)^+a(cbd)^+a(cbd)^+$ generated at the step 1, 2 or 3 will be referred to as *triples*. The $a(bcbd)^+$ or $a(cbd)^+$ subwords of a triple generated during the (a), (b) or (c) substep of the corresponding step will be called *a-blocks*.

The derivation starts at step 1.

Step 1. The derivation starts with the lexical entries in (1a) generating the following (sub)tree:

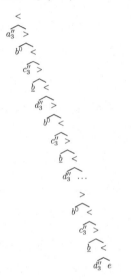

The yield of this tree is $a(bcbd)^+$. Each b located between a c and a d (the corresponding base category is underlined) is licensed for scrambling to c_1, c_2, c_3 or g to be introduced at a later point in the derivation, as every such b has s-licensee $\sim c_1$, $\sim c_2$, $\sim c_3$ or $\sim g$. The whole a_3''-headed subtree is licensed for scrambling to s since its head a_3'' has s-licensee $\sim s$. After this, subtrees headed by a_2'' and a_1'' are generated by the entries in (1b) and (1c) respectively. The generation proceeds in the same way as in the case of the a_3''-headed subtree; the b nodes are licensed for scrambling to c_1, c_2, c_3 or g, and the a_2'' and a_1'' subtrees are themselves licensed for scrambling to s.

Applying the lexical entries in (1d), the resulting a_1''-headed tree merges to u'' and the u''-headed tree merges to w:

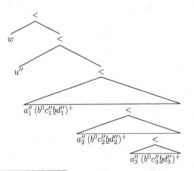

The phonetic yield generated at this point is a triple

$$a(bcbd)^+a(bcbd)^+(bcbd)^+$$

The derivation continues further to the step 2 or 4.

---

[9] For legibility and in order to avoid cumbersome notation, we will only use base category symbols in the illustrations below. In the grammar G, the lexical entries are made in such a way that the phonetic symbol—if it is present—can be obtained by stripping the base category symbol of its indices and bars.

Step 2. First, subtrees headed by a_3, a_2 and a_1 are generated by the entries in (2a), (2b) and (2c) respectively, similarly to the previously performed step. All of them are licensed for scrambling to s. The b'_1, b'_2 and b'_3 nodes inside these subtrees are licensed for scrambling to c'_1, c'_2, c'_3 or g. Some of the b nodes *introduced at the previous step* scramble to some of the c_1, c_2 and c_3 nodes introduced at the present step. The resulting a_1-headed subtree merges to v:

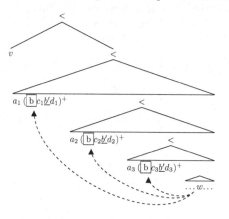

If the previous step was 3, the only b nodes that can scramble here are those generated at that previous step. This is because the v'-headed subtree that contains these nodes moved out of the barrier subtree $\bar{t}'$ at the previous step, but prior to this movement the subtrees containing the rest of the b (and b') nodes moved to the specifiers of u' which is itself located inside $\bar{t}'$ (this will become clear further on as the step 3 is similar to the step 2 modulo bars on the base category symbols).

Further, the v-headed subtree merges to u, the u'-headed (or u''-headed) subtree generated at the previous step moves from within w to u and then w itself moves to u. After that, the u-headed subtree merges to $\bar{t}$:

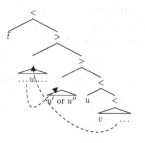

Thus we get a v-headed tree where only the b' nodes *generated at the present step* remain, and the rest of the b and b' nodes are contained within the specifiers of u. This configuration makes it possible to displace the b' nodes generated at the present step beyound the barrier in one go leaving the rest of the b and b' nodes inside the barrier.

Further, the $\bar{t}$-headed subtree merges to w' and the v-headed subtree moves to w':

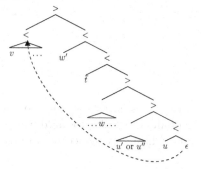

As a consequence of this movement, the b nodes *generated at the present step* get displaced out of the barrier $\bar{t}$ and become scrambling candidates, while the rest of the b (and b') nodes remain within $\bar{t}$.

After this step the derivation continues to the step 3 or 4.

Step 3. This derivation step is performed by the entries in (3). It is similar to the step 2 with the only difference that instead of the categories with a bar of step 2, the corresponding categories without bars are used here (e.g., b' for b, etc), and vice versa. It can only be preceded by step 2. After this step the derivation goes further to the step 2 or 4.

Step 4. A subtree headed by g is generated by the entries in (4). The g head merges w or w' depending on which was the previous step. Some of the b' or b nodes *introduced in the previously performed step* scramble to g:

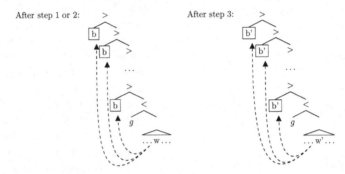

The derivation continues further to the step 5.

Step 5. A subtree headed by s is generated by one of the entries in (5) merging the g-headed subtree generated in step 4. The s head triggers the movement of w and u' (or w' and u respectively) getting the corresponding subtrees out of the barrier $\bar{t}$ (or $\bar{t}'$). Further, (some of) the a-, a'- or a''-headed subtrees generated at previous steps scramble from these two subtrees to s in an arbitrary order:

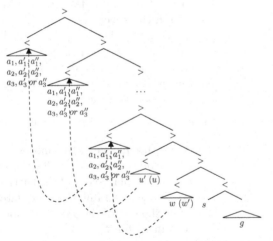

In this way, the a-blocks are ordered arbitrarily on the left of the yield of the g-headed subtree which itself stores the "counter". No further b and b' scrambling is possible here because of the lack of an appropriate scrambling domain.

The language L generated by this grammar can be seen as the union of two languages, $L = L_1 \cup L_2$, such that L_1 consists of all the words produced with all b and b' nodes having scrambled and each c and c' head having accepted *exactly one* scrambling b or b' node, and L_2 contains the rest of the words. The language L_1 consists of all the words

$$a(bcd)^{n_1} a(bcd)^{n_2} \ldots a(bcd)^{n_{3k}} b^m$$

such that for all positive natural numbers k and m, the multiset $\{n_1, n_2, \ldots, n_{3k}\}$ can be partitioned into k multisets of cardinality 3, each of which sums to m. This can be seen following the generation of the words of the language.

On the yield level,[10] each $a(\boxed{b}\,cbd)^+ a(\boxed{b}\,cbd)^+ a(\boxed{b}\,cbd)^+$ triple generated at the step 2 or 3 receives the scrambling symbols b from the neighboring triple on the right generated during the previous step and later "gives away" through scrambling its symbols b located between c and d to the triple generated during the next step. Barriers in combination with movement guarantee that the symbols b generated during one step can only scramble to the triples generated during the text step. The symbols b scrambling from the last generated triple are stored as a "counter" at the step 4. After that, at the step 5, the a-blocks move to the left of the counter and scramble arbitrarily.

In case all the b and b' nodes have scrambled and each c_1, c_2, c_3, c'_1, c'_2 and c'_3 head have received through scrambling exactly one b or b', all the triples will contain an equal number of bcd subwords while the number of these subwords in different a-blocks of one and the same triple may vary. The "counter" will consist of as many symbols b as there are bcd subwords in each triple.

Each word in L_2 will contain at least one following subword in positions to the left from the rightmost occurrence of d: bb (more than one b have scrambled to the same c head), ac, dc (omission of scrambling to a particular c head), cb (b has not scrambled), while no word in L_1 will follow this pattern. This means that $L_1 \cap L_2 = \emptyset$, and there exists a *polynomial-time computable* function $p(w)$ such that for any $w \in L$, $p(w) = true$ if $w \in L_1$ and $p(w) = false$ otherwise. For a $w \notin L$, it will return *true* or *false*.

The language L_1 constitutes the unary encoding of the 3-Partition Problem whereby we have proved the proposition 2 which together with the proposition 1 gives us the following result:

Proposition 3. *The word recognition problem for* MG^{scr}_{B0} *is NP-hard.*

4 Conclusions

In this paper we have proved that the fixed recognition problem for MG^{scr}_{B0} is NP-hard, which means that the generalized description of scrambling is probably impossible in MG even if we operate with one single barrier blocking all scrambling. This result shows that non-locality has similar consequences in the case of MGs as it does for TAG-based formalisms [1,16]. A further step in this research could be investigation of other properties of the proposed formalism and determining more exactly the complexity of the recognition problem for it.

---

[10] Here we write in squares those symbols b that have scrambled and the underlined symbols are those that have not.

References

1. Champollion, L.: Lexicalized non-local MCTAG with dominance links is NP-complete. In: Penn, G., Stabler, E. (eds.) Proceedings of Mathematics of Language 10. CSLI On-Line Publications UCLA (2007)
2. Chomsky, N.: The Minimalist Program. MIT Press, Cambridge (1995)
3. Chomsky, N.: Derivation by phase. In: Kenstowicz, M. (ed.) Ken Hale: A Life in Language, pp. 1–52. MIT Press, Cambridge (2001)
4. de Groote, P., Morrill, G., Retoré, C. (eds.): LACL 2001: Proceedings of the 4th International Conference on Logical Aspects of Computational Linguistics, London, UK. LNCS, vol. 2099. Springer, Heidelberg (2001)
5. Frey, W., Gärtner, H.: On the treatment of scrambling and adjunction in Minimalist Grammars. In: Jäger, G., Monachesi, P., Penn, G., Wintner, S. (eds.) Proceedings of the 7th Conference on Formal Grammar, pp. 41–52 (2002)
6. Gärtner, H., Michaelis, J.: Some remarks on locality conditions and Minimalist Grammars. In: Gärtner, H., Sauerland, U. (eds.) Interfaces + Recursion = Language? Chomsky's Minimalism and the View from Syntax and Semantics, pp. 161–195. Walter de Gruyter, Berlin (2007)
7. Harkema, H.: A characterization of Minimalist languages. In: de Groote, P., Morrill, G., Retoré, C. (eds.) LACL 2001. LNCS (LNAI), vol. 2099, pp. 193–211. Springer, Heidelberg (2001)
8. Jäger, G., Monachesi, P., Penn, G., Wintner, S. (eds.): FG-MOL 2005: Proceedings of the 10th conference on Formal Grammar and the 9th Meeting on Mathematics of Language, Edinburgh, Scotland (2005)
9. Kobele, G., Michaelis, J.: Two type 0-variants of Minimalist Grammars. In: Jäger, et al. (eds.) [8]
10. Michaelis, J.: Derivational Minimalism is mildly context-sensitive. In: Moortgat, M. (ed.) LACL 1998. LNCS (LNAI), vol. 2014, pp. 179–198. Springer, Heidelberg (2001)
11. Michaelis, J.: Transforming linear context-free rewriting systems into Minimalist Grammars. In: de Groote, P., Morrill, G., Retoré, C. (eds.) LACL 2001. LNCS (LNAI), vol. 2099, pp. 228–244. Springer, Heidelberg (2001)
12. Michaelis, J.: An additional observation on strict derivational Minimalism. In: Jäger, et al. (eds.) [8]
13. Michaelis, J., Gärtner, H.: A note on countercyclicity and Minimalist Grammars. In: Jäger, G., Monachesi, P., Penn, G., Wintner, S. (eds.) Proceedings of the 8th Conference on Formal Grammar, pp. 103–114 (2003)
14. Perekrestenko, A.: A note on the complexity of the recognition problem for the Minimalist Grammars with unbounded scrambling and barriers. In: Madrigal, V.D., de Ros Salamanca, F.E. (eds.) Actas del XXIII Congreso de la Sociedad Española para el Procesamiento del Lenguaje Natural, Seville, Spain. Revista de la SEPLN, vol. 39, pp. 27–34 (2007)
15. Stabler, E.: Derivational Minimalism. In: Retoré, C. (ed.) LACL 1996. LNCS (LNAI), vol. 1328, pp. 68–95. Springer, Heidelberg (1997)
16. Søgaard, A., Lichte, T., Maier, W.: On the complexity of linguistically motivated extensions of tree-adjoining grammar. In: RANLP 2007: Proceedings of the Conference on Recent Advances in Natural Language Processing, Borovets, Bulgaria (2007)

Sorting and Element Distinctness on One-Way Turing Machines

Holger Petersen

Univ. Stuttgart, FMI
Universitätsstraße 38
D-70569 Stuttgart
petersen@informatik.uni-stuttgart.de

Abstract. For nondeterministic Turing machines with one work-tape and one-way input a lower time bound $\Omega(m^2\ell)$ on sorting m strings of length ℓ each is shown, which matches the upper bound. For the related Element Distinctness Problem with input of the same format we prove the upper bound $O(m^2)$ if $\ell = O(m/\log m)$, showing this problem to be easier than sorting. The bound $O(m^2)$ also holds for deterministic machines if $\ell = c + \log m$ with constant c. For this problem Szepietowski has shown the bound $\Theta(m^2 \log m)$ on deterministic Turing machines without input tape.

1 Introduction

Sorting is among the most widely studied tasks in computer science. Many algorithms have been developed for computational models that treat the keys of items to be sorted as atomic units. There are however interesting solutions like radix-sort that access portions of the keys. The investigation of such algorithms clearly requires a model that processes its input on a bit-level. The traditional model in this field is the Turing machine. As discussed in Section 2, multi-tape Turing machines are quite efficient at solving the sorting problem.

In the present work we consider the sorting problem for binary strings or equivalently natural numbers in binary encoding. This is no severe restriction, since pointers to data can be stored in the least significant digits. We denote the problem of sorting m strings of length ℓ by SORT(m, ℓ), where we assume that m and ℓ are related by some easily computable function. When considering nondeterministic sorting procedures we require that within the given time bound at least one computation writes the sorted input on a work-tape and all other cells contain blanks. No terminating computation produces an incorrect output.

A decision problem which is closely connected to sorting is the Element Distinctness Problem (EDP). The EDP asks whether m elements given as strings of ℓ bits each are pairwise different. We denote this problem by EDP(m, ℓ). As a language recognition problem EDP can be defined in the following way:

$$\text{EDP}(m, \ell) = \{x_1 \# x_2 \# \cdots \# x_m \mid x_i \in \{0, 1\}^\ell, x_i \neq x_j \text{ for } i \neq j\}.$$

Since on many computational models EDP can be efficiently reduced to sorting, lower bounds on the decision problem EDP carry over to the sorting problem.

C. Martín-Vide, F. Otto, and H. Fernau (Eds.): LATA 2008, LNCS 5196, pp. 433–439, 2008.

2 Previous Work

It is well-known that deterministic Turing machines with three work-tapes can sort much faster than machines with one such tape. Using merge-sort and radix-sort machines of the former type can solve $\text{SORT}(m, \ell)$ in time $O(m(\log m)\ell)$ and $O(m\ell^2)$ resp. [Rei90, p. 152].

Radix-sort can be implemented on machines with two work-tapes by making two passes over the sequence of strings when sorting according to a bit, but for large ℓ (e.g., $\ell = m^2$) the resulting bound $O(m\ell^2)$ is worse than sorting by selection. One of the factors ℓ in this bound can be replaced by $\log m$ with the help of a merge-sort strategy that first determines the final position of each string by counting.

If the input-tape of a two-tape Turing machine is restricted to be read-only we obtain the off-line Turing machine investigated in [Wie92] by Wiedermann. He gave a nondeterministic solution running in time $O(m^{3/2}\ell)$. A deterministic algorithm for Turing machines with a separate two-way input-tape having the same complexity was presented in [Pet08]. This bound cannot be improved in general by the lower bound from [DMS91].

The most restricted model we mention here is the single-tape machine that receives its input on the work-tape. For this model Wiedermann could show the bound $\Theta(m\ell \log \binom{m+2^\ell - 1}{m})$ in [Wie92], which is better than the naive quadratic algorithm for small ℓ.

The investigation of the EDP on Turing machines with a single work-tape was initiated by López-Ortiz [LO94]. He proved the lower bound $\Omega(m^2 \log m)$ on the time necessary for a nondeterministic Turing machine without separate input-tape to accept the EDP with m elements of $\ell = k \log m$ bits each for $k \geq 2$. There is a straight-forward deterministic solution in time $O(m^2 \log^2 m)$ and López-Ortiz conjectured that the latter bound was optimal. Szepietowski [Sze96] showed that the lower bound $\Omega(m^2 \log m)$ also holds for elements of size $c + \log m$ with constant c. In this case there is a matching upper bound. The nondeterministic version of the conjecture of López-Ortiz was disproved in [Pet02] with a solution of complexity $O(m^2 (\log m)^{3/2} (\log \log m)^{1/2})$. An even better algorithm in [BABP03] showed the lower bound to be tight in the case of nondeterministic machines, while the deterministic version of the conjecture turned out to be true.

In [Pet08] the EDP was solved by Turing machines with a single work-tape and a separate two-way input-tape. For the EDP with m elements of size $\ell = O(m/ \log^2 m)$ a nondeterministic solution of time complexity $O(m^{3/2}\ell^{1/2})$ was presented, which is slightly faster than the sorting procedures mentioned above.

In the present work we investigate computational models between the machines discussed above, namely Turing machines with one work-tape and a separate one-way input-tape. Probabilistic Turing machines of this kind have been investigated by Kalyanasundaram and Schnitger [KS92], who could show the lower bound $\Omega(m^2/ \log m)$ on EDP with elements of size $2 \log m$. We prove here that in the case of nondeterministic machines the EDP is easier than sorting by showing an optimal lower bound on sorting and a faster algorithm for EDP.

3   Sorting with One Work-Tape and One-Way Input

First we outline the easy upper bound for nondeterministic machines.

Proposition 1. *Sorting m binary strings of length ℓ each (SORT(m, ℓ)) for $\ell = O(m)$ can be done by a nondeterministic Turing machine with one-way input in time $O(m^2\ell)$.*

Proof. The Turing machine guesses the correct position for each string, leaving space for the remaining elements. This position can be reached in $O(m\ell)$ steps per string, since this is the size of the input. Then in time $O(m\ell^2)$ the machine can check that the strings are of the same length and appear in sorted order. By the assumption $\ell = O(m)$ the claimed bound follows.   □

The next result shows that this bound is optimal for $\ell \geq 2\log m$. Notice that this is not a severe restriction, since $\ell < \log m$ is impossible for positive instances of EDP.

Theorem 1. *The problem SORT(m, ℓ) requires $\Omega(m^2\ell)$ steps on nondeterministic Turing machines with one-way input if $\ell \geq 2\log m$.*

Proof. Let M be a Turing machine with q states solving SORT(m, ℓ) and of the kind mentioned in the theorem. Let m be sufficiently large, divisible by 10 and let $n = m(\ell + 1) - 1$.

The proof uses the concept of Kolmogorov Complexity, see [LV97] for definitions and references to other papers applying this technique. Let w be an incompressible binary string of length $(m/5) \cdot (\ell - \log m)$. Form $m/5$ strings $w_0, \ldots, w_{m/5-1}$ of length $\ell - \log m$ each, such that $w = w_0 \cdots w_{m/5-1}$. Denote by $b(i)$ the binary encoding of i in $\log m$ bits (with leading 0's if necessary). Form the following input x for M:

$$b(0)w_0 \# b(m-1)w_1 \# \cdots \# b(m/10-1)w_{m/5-2} \#$$
$$b(m-m/10)w_{m/5-1} \# b(m/10)0^\ell \# \cdots \# b(m-m/10-1)0^\ell$$

The input is formed by concatenating binary encodings of small numbers with portions of w having an even index and by concatenating encodings of large numbers with portions having an odd index. These strings alternate. The trailing elements consist of encodings of numbers of intermediate size followed by zeroes.

Fix a terminating computation of minimum length of M on x. The length of the computation can be bounded from above by $n^2 = m^2\ell^2$, since otherwise the lower bound $\Omega(m^2\ell)$ holds (and we already know that the upper bound is $O(m^2\ell)$ from Proposition 1). Therefore also the number of tape-squares used is bounded by n^2. We assign numbers to the work-tape squares eventually containing the sorted strings starting with 0 (squares to the left of the sequence are assigned negative numbers). If for i with $0 \leq i \leq m/10 - 1$ the position of the work-tape head is between square $-(m/5)\ell$ and $(2m/5)(\ell + 1)$ when some bit of $b(i)w_{2i}$ is scanned by the input head we say that w_{2i} is *directly copied*. Similarly w_{2i+1} is directly copied, if some bit of $b(m-i)w_{2i+1}$ is scanned by the input head while the work-tape head is between square $(3m/5)(\ell+1)$ and $(6m/5)(\ell+1)$. If for at least half of the i both w_{2i} and w_{2i+1} are directly copied, then M makes at least $(m/20)(m/5)(\ell+1) \geq (1/100)m^2\ell$ steps.

Now assume that for at least half of the i one of the strings w_{2i} and w_{2i+1} is not directly copied. Then at least one fourth of the strings with even or odd index is not directly copied. We just consider the former case, the other being similar.

We will show that not both from square $-(m/5)\ell$ to 0 and from square $(m/5)(\ell+1)$ to $(2m/5)(\ell+1)$ a short crossing sequence of M is possible. Let c_1 be a crossing sequence in the first interval and c_2 be a a crossing sequence in the second interval. In addition to the states of M the crossing sequences include information about the position of the input head when the work-tape head crosses the border.

The string w can be reconstructed from the following information:

- A formalized version of the decoding algorithm outlined below ($O(1)$ bits).
- An encoding of machine M ($O(1)$ bits).
- A self-delimiting binary encoding of m and ℓ ($O(\log n)$ bits).
- The distance of c_1 and c_2 to the initial position of the work-tape head in self-delimiting binary encoding ($O(\log n)$ bits).
- The length of c_1 and c_2 in self-delimiting binary encoding ($O(\log n)$ bits).
- The crossing sequences c_1 and c_2 including the positions of the input head $((|c_1|+|c_2|)(\log n + \log q)$ bits).
- A sequence f of $m/10$ bits indicating those w_{2i} which are directly copied.
- A binary string v consisting of all w_{2i} (concatenated without separators) which are directly copied and all w_{2i+1} (at most $(7m/40) \cdot (\ell - \log m)$ bits).

The above information can be extracted from a terminating computation of M.

The reconstruction algorithm first sets up an input-string u for M by generating the address fields and markers, copying the w_{2i} which are indicated by f, all w_{2i+1}, and filling the remaining positions with 0's. It can do so by using the encodings of m and ℓ and the above information. Notice that u may not be identical to the original input of M, but in the course of the simulation none of the differing bits will be queried.

Then a blank segment of the work-tape between c_1 and c_2 is initialized. Now the simulator checks whether the initial position of the work-tape head is to the left of c_1, between c_1 and c_2 or to the right of c_2. In the first and last case M is started in the state according to the first entry of the respective crossing sequence and from its position. Otherwise M's head is placed inside the segment. Now a simulation of M is started, storing each of the simulated steps and the intermediate work-tape contents in order to backtrack when a branch of the simulation fails. Whenever a new symbol of the input is read by M, the simulator looks up the corresponding information in u.

If the work-tape head of M is about to cross the position of c_1 or c_2, then the simulator checks that the number of input symbols read during the simulation and the current state is consistent with the crossing sequence. If it isn't, backtracking starts. Otherwise the simulation is resumed at the next entry of the crossing sequence.

Eventually the simulation terminates, since there is at least one terminating computation. Then the work-tape contains the missing strings of w, which can be combined with u in order to recover w.

Now assume that $|c_1| + |c_2| \leq m/100$. Then $(|c_1| + |c_2|)(\log n + \log q) \leq (m/50)\log n$ for n sufficiently large and w has a description of size at most $(m/50 + 7m/40) \cdot (\ell - \log m) + o(n) = 39m/200 \cdot (\ell - \log m) + o(n)$. Thus w can be compressed, contrary to our assumption. By choosing disjoint pairs of positions for c_1 and c_2 we obtain the desired bound $\Omega(m^2\ell)$. $\square$

The restriction to "large" ℓ in Theorem 1 cannot be omitted, since for $\ell \leq \log m$ Wiedermann proved the upper bound $O(m\ell 2^\ell(\log m - \ell + 1))$ even for the weaker model without any input-tape [Wie92, Theorem 1].

4   EDP with One Work-Tape and One-Way Input

Szepietowski [Sze96] investigated the problem of distinctness for m short numbers of size $c + \log m$, where c is constant. He could prove that on a deterministic one-tape Turing machine without separate input this problem has complexity $\Theta(m^2 \log m)$. Here we show that the problem can be solved faster with an additional one-way input-tape, thus separating the two computational models with a natural problem.

Theorem 2. *The problem $EDP(m, c + \log m)$ with c constant can be accepted by a deterministic Turing machine with one work-tape and a one-way input-tape in time $O(m^2)$.*

Proof. After recording the first element x_1 on the work-tape the Turing machine M solving the EDP allocates $m/\log m$ buckets. Each bucket contains a unique address of $\log m - \log \log m$ bits and an array of $2^c \cdot \log m$ flags. The buckets appear in increasing order of their addresses. All flags are initially false.

Element x_1 is now shifted along the tape on a separate track and for every flag passed its numerical value is decreased by one. This continues until the value zero is reached and the corresponding flag is marked as true. The work-head is then moved back to the first bucket.

For each remaining element x_i machine M does the following. It repeatedly reads one bit and shifts the prefix of x_i read so far along the separate track until the prefix matches a prefix of a bucket address. This continues until the first $\log m - \log \log m$ bits of x_i have been read. The following $c + \log \log m$ bits are then copied onto the special track and are used as a binary counter in order to locate the flag within the bucket belonging to x_i. If the flag is already set M rejects, otherwise it sets the flag.

Initializing the buckets and shifting x_1 can be done in $O(m \log m)$ steps. Each of the other elements requires $O(m)$ steps, since a prefix of i bits is moved along at most $2^{\log m - \log \log m - i} = m/(2^i \log m)$ buckets of size $O(\log m)$. Locating the flag can be done in time $O(\log m \log \log m)$. $\square$

The following result shows that the same bound holds for longer strings if the machines considered are nondeterministic. Note that the EDP is thus easier than sorting on this model of computation.

Theorem 3. *The problem $EDP(m, \ell)$ with $\ell = O(m/\log m)$ can be accepted by a nondeterministic Turing machine with one work-tape and a one-way input-tape in time $O(m^2)$.*

Proof. We describe a computation of a nondeterministic Turing machine M on a positive instance of the problem respecting the time bound. In the case of an instance that contains duplicates the computation can be aborted by maintaining a counter.

First M marks off $m/\log m$ blocks with space $O(\log m)$ each, for a total $O(m)$ space usage. In block i an upper bound u_i of length ℓ is guessed, such that the number of elements from the input in the range from $u_{i-1} + 1$ to u_i is $\log m$ (we let $u_0 = -1$). We call them the elements belonging to block i

In each block M stores information for hash functions perfect for the elements belonging to this block. The techniques of Fredman, Komlós, and Szemerédi [FKS84] combined with the space efficient rehashing of Slot and van Emde Boas [Sv88] allow an implementation in space $O(\log m)$ per block. More details can be found in the proof of Theorem 1 of [Pet08]. Computations and access to the information in the block can be done in poly-logarithmic time.

For each element encountered in the input the Turing machine M guesses the correct block and verifies that it belongs to the block. Then M computes the hash function in time $O(m)$ and finds the relevant information in time $O(\log^2 m)$ by shifting counters along the tape. Thus time $O(m)$ is sufficient for each element, leading to the claimed bound.   □

5   Conclusion and Open Problems

For Turing machines with one-way input the result of Theorem 1 shows the straight-forward sorting algorithm to be optimal for nondeterministic machines. We don't have a deterministic algorithm that matches the lower bound and possibly the lower bound can be improved for deterministic machines.

In the case of the EDP it seems reasonable to conjecture that the upper bound from Theorem 3 is optimal.

Acknowledgments. The author is grateful to the referees for suggesting improvements of the presentation.

References

[BABP03]   Ben-Amram, A.M., Berkman, O., Petersen, H.: Element distinctness on one-tape Turing machines. Acta Informatica 40, 81–94 (2003)

[DMS91]   Dietzfelbinger, M., Maass, W., Schnitger, G.: The complexity of matrix transposition on one-tape off-line Turing machines. Theoretical Computer Science 82, 113–129 (1991)

[FKS84] Fredman, M.L., Komlós, J., Szemerédi, E.: Storing a sparse table with $O(1)$ worst case access time. Journal of the Association for Computing Machinery 31, 538–544 (1984)

[KS92]   Kalyanasundaram, B., Schnitger, G.: Communication complexity and lower bounds for sequential machines. In: Buchmann, J., Ganzinger, H., Paul, W.J. (eds.) Informatik – Eine Festschrift zum 60. Geburtstag von Günter Hotz, Teubner, Stuttgart. Teubner-Texte zur Informatik, vol. 1, pp. 253–268 (1992)

[LO94]   López-Ortiz, A.: New lower bounds for element distinctness on a one-tape Turing machine. Information Processing Letters 51, 311–314 (1994)

[LV97]   Li, M., Vitányi, P. (eds.): An Introduction to Kolmogorov Complexity and its Applications, 2nd edn. Graduate Texts in Computer Science. Springer, Heidelberg (1997)

[Pet02] Petersen, H.: Bounds for the element distinctness problem on one-tape Turing machines. Information Processing Letters 81, 75–79 (2002)

[Pet08] Petersen, H.: Element distinctness and sorting on one-tape off-line Turing machines. In: Geffert, V., Karhumäki, J., Bertoni, A., Preneel, B., Návrat, P., Bieliková, M. (eds.) SOFSEM 2008. LNCS, vol. 4910, pp. 406–417. Springer, Heidelberg (2008)

[Rei90] Reischuk, K.R.: Einführung in die Komplexitätstheorie. B. G. Teubner (1990)

[Sv88]   Slot, C., van Emde Boas, P.: The problem of space invariance for sequential machines. Information and Computation 77, 93–122 (1988)

[Sze96] Szepietowski, A.: The element distinctness problem on one-tape Turing machines. Information Processing Letters 59, 203–206 (1996)

[Wie92] Wiedermann, J.: Optimal algorithms for sorting on single-tape Turing machines. In: van Leeuwen, J. (ed.) Algorithms, Software, Architecture, Proceedings of the IFIP 12th World Computer Congress, Madrid, Spain, vol. I, pp. 306–314. Elsevier Science Publishers, Amsterdam (1992)

On Periodicity of Generalized Two-Dimensional Words

Svetlana Puzynina[*]

Sobolev Institute of Mathematics,
pr. Koptyuga 4, Novosibirsk 630090, Russia
Novosibisk State University,
Pirogova street, 2, Novosibirsk 630090, Russia
puzynina@math.nsc.ru

Abstract. A generalized two-dimensional word is a function on $\mathbb{Z}^2$ with finite number of values. The main problem we are interested in is periodicity of two-dimensional words satisfying some local conditions. Let $A = (a_{ij})_{i,j=1}^{n}$ be an integer matrix. The function $\varphi : \mathbb{Z}^2 \to \mathbb{Z}^n$ is a *generalized centered function of radius r with the matrix A* if

$$\sum_{\substack{(y_1, y_2) : \\ 0 < |x_1 - y_1| + |x_2 - y_2| \le r}} \varphi(y_1, y_2) = \varphi(x_1, x_2) A$$

for every integers x_1, x_2. We prove that every bounded generalized centered function of radius $r > 1$ is periodic. For $r = 1$ periodicity depends on spectrum of the matrix A. Similar results are obtained for the infinite triangular and hexagonal grids.

1   Introduction

We consider some special types of two-dimensional words as functions on vertices of the graphs of the infinite rectangular, triangular and hexagonal grids.

We begin with definitions for an arbitrary graph. Let $G = (V, E)$ be a graph. The distance between two vertices $\mathbf{x}$ and $\mathbf{y}$, denoted by $d(\mathbf{x}, \mathbf{y})$, is the usual graph metric. A ball $B_r(\mathbf{x})$ of radius r with the center at the vertex $\mathbf{x}$ is defined in the following way:

$$B_r(\mathbf{x}) = \{\mathbf{y} \in V | \ d(\mathbf{x}, \mathbf{y}) \le r\}.$$

Let $A = (a_{ij})_{i,j=1}^{n}$ be an integer matrix. A function $\varphi : V \to \mathbb{Z}^n$ is called *generalized centered of radius r with the matrix A* if

$$\sum_{\mathbf{y} \in B_r(\mathbf{x}), \ \mathbf{y} \ne \mathbf{x}} \varphi(\mathbf{y}) = \varphi(\mathbf{x}) A$$

for every $\mathbf{x} \in V$.

[*] This work was supported in part by Russian Foundation of Basic Research (grant 07-01-00248) and by Russian Science Support Foundation.

C. Martín-Vide, F. Otto, and H. Fernau (Eds.): LATA 2008, LNCS 5196, pp. 440–451, 2008.

In fact, the notion of generalized centered function is a generalization of the notion of perfect coloring and ordinary centered function. Let G be a graph, $A = (a_{ij})_{i,j=1}^n$ be an integer nonnegative matrix, r be an integer, $r \geq 1$. Consider a coloring of vertices of the graph G into n colors and an arbitrary vertex x of a color i. If the number of vertices of a color j (distinct from x) at distance at most r from the vertex x does not depend on the vertex x and is equal to a_{ij}, then the coloring is called *perfect of radius* r with the matrix A. A perfect coloring into n colors with matrix A can be considered as generalized centered function with the value area $\{e_1, \cdots, e_n\}$, where e_i is a unit vector with 1 in its i-th coordinate. The other name used for perfect colorings is equitable partitions. This notion naturally arises in different fields of mathematics, such as algebraic combinatorics, graph theory and coding theory. A function $f : V(G) \to \mathbb{Z}$ is called *centered of radius* r if the sum of its values in every ball of radius r is equal to 0. An ordinary centered function can be considered as a generalized centered function for $n = 1$, $A = -1$.

The notion of centered function was introduced as a generalization of the notion of perfect code in the hypercube $\mathbb{H}^n$ [1]. The notion of perfect coloring also generalizes the notion of perfect code and several other well-known codes, such as Preparata code, completely regular code, uniformly packed code. Namely, these codes can be interpreted as perfect colorings into two or more colors. This means that generalized centered functions can be used as an instrument for studying perfect colorings and different codes.

We prove, that every bounded generalized centered function of radius $r > 1$ on the infinite rectangular grid is periodic. For $r = 1$, A such that $\det A \neq 0$ generalized centered function with the matrix A is also periodic. If $r = 1$ and $\det A = 0$, then there exist non-periodic and periodic generalized centered functions. Similar results are obtained for the infinite triangular and hexagonal grids. These results are obtained using method of R-prolongable words, which was earlier used for obtaining some results about periodicity of perfect colorings and ordinary centered functions [2], [6], [7]. In this paper we develop this method of proving periodicity of words with local conditions.

Periodicity of two-dimensional words have been studied before. There exist many methods of proving periodicity and many theorems about periodicity of different types of words. The following hypothesis about connection of local complexity and periodicity is known as Nivat's conjecture [5]: if there exist a pair (n, m) such that the complexity function $p_w(n, m)$ of a two-dimensional word w satisfies the condition $p_w(n, m) \leq mn$, then w has at least a periodicity vector. Weak forms of the conjecture for $p_w(n, m) \leq mn/144$ and for $p_w(n, m) \leq mn/16$ were proved by C. Epifanio, M. Koskas, F. Mignosi in [4] and by A. Quas, L. Zamboni in [9], respectively. In [3] V. Berthe, L. Vuillon explore the notion of minimal complexity for two-dimensional sequences, in particular, they give an example of two-dimensional sequence of complexity $p_w(n, m) = mn + m$, for every (m, n), which is uniformly recurrent and which has no rational periodic direction.

2   The Infinite Rectangular Grid

The graph of the infinite rectangular grid is 4-regular, its vertices are all possible ordered pairs of integers. Two vertices $\mathbf{x} = (x_1, x_2)$ and $\mathbf{y} = (y_1, y_2)$ are adjacent if $|x_1 - y_1| + |x_2 - y_2| = 1$.

Let φ be a function from $\mathbb{Z}^2$ to $\mathbb{Z}^n$, i. e. $\varphi(\mathbf{x})$ is a vector of length n. Denote i-th coordinate of the vector $\varphi(\mathbf{x})$ by $\varphi_i(\mathbf{x})$: $\varphi(\mathbf{x}) = (\varphi_1(\mathbf{x}), \cdots, \varphi_n(\mathbf{x}))$. We study M-bounded functions on $\mathbb{Z}^2$, i. e. $\varphi : \mathbb{Z}^2 \to \mathbb{Z}^n$, such that $|\varphi_i(\mathbf{x})| \leq M$ for $i = 1, \cdots, n$, where M is a positive integer. These functions can be considered as two-dimensional words over the finite alphabet of $(2M+1)^n$ vectors of length n.

A two-dimensional word ω is $\mathbf{v}$-*periodic* (or $\mathbf{v}$ is a vector of periodicity of the word ω) if $\omega(\mathbf{x} + \mathbf{v}) = \omega(\mathbf{x})$ for all $\mathbf{x} \in \mathbb{Z}^2$. A perfect coloring that is $\mathbf{v_1}$- and $\mathbf{v_2}$-periodic for some noncollinear $\mathbf{v_1}$ and $\mathbf{v_2}$, is called *periodic*. It is easy to show that we can take $\mathbf{v_1} = (p, p)$, $\mathbf{v_2} = (q, -q)$.

We say that a two-dimensional word ω is R-*prolongable* if for any $\mathbf{x}, \mathbf{y} \in \mathbb{Z}^2$ an equality $\omega|_{B_R(\mathbf{x})} = \omega|_{B_R(\mathbf{y})}$ implies $\omega|_{B_{R+1}(\mathbf{x})} = \omega|_{B_{R+1}(\mathbf{y})}$. Notation $\omega|_{B_R(\mathbf{x})} = \omega|_{B_R(\mathbf{y})}$ means that $\omega(\mathbf{x} + \mathbf{z}) = \omega(\mathbf{y} + \mathbf{z})$ for $|\mathbf{z}| \leq R$.

Lemma 1. *Let ω be a two-dimensional word on a finite alphabet. If ω is R-prolongable for some $R \geq 0$, then ω is periodic.*

The proof of this lemma is simple and can be found in [2].

Example. Now we will give an example of generalized centered function of radius 1, where the matrix A is a degenerate matrix. Let $\boldsymbol{v}$ be its left eigenvector corresponding to the eigenvalue $\lambda = 0$: $A\boldsymbol{v} = \boldsymbol{0}$. In the further text we omit the word 'left' but always mean left eigenvectors.

The function $\theta^v : \mathbb{Z}^2 \to \mathbb{Z}^n$, given by the following formula:

$$\theta^v(x_1, x_2) = \begin{cases} \boldsymbol{0}, & \text{if } x_1 \neq x_2, \\ \boldsymbol{v}, & \text{if } x_1 = x_2 \text{ even}, \\ -\boldsymbol{v}, & \text{if } x_1 = x_2 \text{ odd} \end{cases}$$

is a non-periodic generalized centered function with the matrix A of radius 1.

Denote by $\chi_{\mathbf{y}}^v$ a function, which is obtained from θ^v by translation by a vector $\mathbf{y}$: $\chi_{\mathbf{y}}^v(\mathbf{x}) = \theta^v(\mathbf{x} + \mathbf{y})$. Denote by $\chi_{\mathbf{y}}^{*v}$ a function, which is obtained from $\chi_{\mathbf{y}}^v$ by rotation by $\pi/2$: $\chi_{\mathbf{y}}^{*v}(x_1, x_2) = \chi_{\mathbf{y}}^v(-x_2, x_1)$. Functions $\chi_{\mathbf{y}}^v$ and $\chi_{\mathbf{y}}^{*v}$ are called *rectangular alternating functions*.

Notice that the sum of generalized centered functions is generalized centered function with the same matrix.

Theorem 1. *1. Let $\varphi : \mathbb{Z}^2 \to \mathbb{Z}^n$ be a bounded generalized centered function of radius $r > 1$. Then φ is periodic.*

2. If φ is generalized centered function of radius 1 with a matrix A and $\det A \neq 0$, then φ is periodic. If $\det A = 0$, then there exist non-periodic and periodic generalized centered functions. A periodic function can be obtained from a non-periodic one by adding rectangular alternating functions.

Proof. 1. The first part of this theorem is a generalization of results, obtained in [2] and [8].

We need some notation to prove the theorem.

A sphere $S_\rho(\mathbf{x})$ of radius ρ with the center at the vertex $\mathbf{x}$ is defined in the following way:

$$S_\rho(\mathbf{x}) = \{\mathbf{y} \in V \mid d(\mathbf{x}, \mathbf{y}) = \rho\}.$$

Consider an arbitrary sphere $S_\rho(\mathbf{x})$. It consists of five sets of vertices: $S_\rho(\mathbf{x}) = \bigcup_{i=1}^{5} S_\rho^i(\mathbf{x})$, where

$$S_\rho^1(\mathbf{x}) = \{(x_1, x_2) + (j, j - \rho) \mid j = 1, 2, ..., \rho\},$$
$$S_\rho^2(\mathbf{x}) = \{(x_1, x_2) + (-j, j - \rho) \mid j = 1, 2, ..., \rho\},$$
$$S_\rho^3(\mathbf{x}) = \{(x_1, x_2) + (j, \rho - j) \mid j = 1, 2, ..., \rho\},$$
$$S_\rho^4(\mathbf{x}) = \{(x_1, x_2) + (-j, \rho - j) \mid j = 1, ..., \rho\},$$
$$S_\rho^5(\mathbf{x}) = \{(x_1, x_2) + (0, -\rho), (x_1, x_2) + (0, \rho), (x_1, x_2) + (\rho, 0), (x_1, x_2) + (-\rho, 0)\}.$$

In Fig.1 one can see the ball $B_5(\mathbf{x})$ of radius 5, its boundary is marked by bold. Vertices from each set $S_6^i(\mathbf{x})$ of the sphere $S_6(\mathbf{x})$ are marked by i.

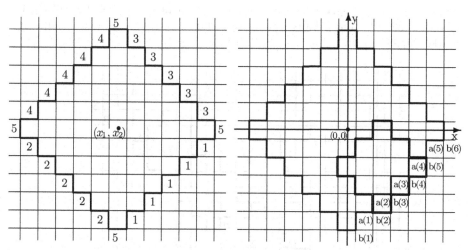

Fig. 1. The ball $B_5(\mathbf{x})$ of radius 5   **Fig. 2.** The illustration for the proof of Theorem 1 for $R = 5$, $r = 2$

Due to Lemma 1 it is sufficient to prove that φ is R-prolongable for some $R > r$. We will prove it for $R > 2r^2 + 5r + 1$.

Consider two arbitrary balls $B_R(\mathbf{x})$ and $B_R(\mathbf{y})$ such that $\varphi|_{B_R(\mathbf{x})} = \varphi|_{B_R(\mathbf{y})}$. It suffices to prove that $\varphi|_{S_{R+1}(\mathbf{x})} = \varphi|_{S_{R+1}(\mathbf{y})}$. Consider the function

$$\psi(\mathbf{t}) = \varphi(\mathbf{x} + \mathbf{t}) - \varphi(\mathbf{y} + \mathbf{t}).$$

We have $\psi|_{B_R(\mathbf{0})} = \mathbf{0}$ by definition of the function ψ. To prove that φ is R-prolongable it suffices to prove that $\psi|_{S_{R+1}(\mathbf{0})} = \mathbf{0}$.

First we will prove that $\psi(\mathbf{x}) = \mathbf{0}$ for $\mathbf{x} \in S_{R+1}^1$. Denote $a(i) = \psi(i, i - 1 - R)$, $i = 1, ..., R$. Fig. 2 illustrates our reasoning for the case $R = 5$, $r = 2$.

By definition of generalized centered function we have $\psi(j, j-1+r-R)A =$
$$= \sum_{\substack{(x,y) \in B_r(j, j-1+r-R), \\ (x,y) \neq (j, j-1+r-R)}} \psi(x,y).$$ It holds $\psi|_{B_R(0)} = 0$, so $\psi(j, j-1+r-R) = 0$, thus

we get $\sum_{i=j}^{j+r} a(i) = 0$, $j = 1, ..., R-r$. Therefore $a(j+r+1) = a(j)$ for every
$1 \leq j \leq R-r-1$, i. e., the sequence $a(i)$ is periodic with period $(r+1)$.

Suppose that there exists $\mathbf{x} \in S_{R+1}^1$ such that $\psi(\mathbf{x}) \neq 0$. This means that
there exists $i : 1 \leq i \leq R$ such that $a(i) \neq 0$. Denote $a(i) = d$. Therefore
$a(i + k(r+1)) = d$,

$$\sum_{j=i+1+k(r+1)}^{i+r+k(r+1)} a(j) = -d \tag{1}$$

for every $1 \leq i + k(r+1) \leq R - r$.

Now we consider elements of the sphere $S_{R+2}(0)$, more precisely, the set
$S_{R+2}^1(0)$. Denote the values of the function ψ in the vertices of this set in the
following way: $b(i) = \psi(i, i-2-R)$, $i = 1, ..., R+1$ (see Fig. 3). Consider the
balls $B_r(\mathbf{v_k})$, $\mathbf{v_k} = (i+1+k(r+1), i-1-R+r+k(r+1))$, where $k \in \mathbb{Z}$,
$1 \leq i+1+k(r+1) \leq R-r+2$. In the Fig. 3 one can see part of the ball $B_R(0)$
of radius $R = 17$ and 5 balls $B_r(\mathbf{v_k})$ of radius $r = 2$. Boundaries of balls are
marked by bold line.

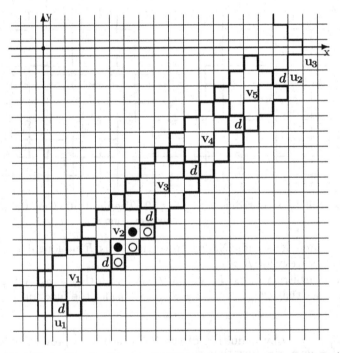

Fig. 3. The illustration for the proof of Theorem 1: part of the ball $B_R(0)$ and balls
$B_r(\mathbf{v_i})$, $R = 17$, $r = 2$. Black circles mark $B_r(\mathbf{v_2}) \cap S_{R+1}(0)$, white circles mark
$B_r(\mathbf{v_2}) \cap S_{R+2}(0)$.

Now we are going to apply the definition of generalized centered function to the vertex $\mathbf{v_k}$. The ball $B_r(\mathbf{v_k})$ consists of vertices with zero values in the ball $B_R(\mathbf{0})$, vertices from $B_r(\mathbf{v_k}) \cap S_{R+1}(\mathbf{0})$, which are marked by black circles in the Fig. 3 (vertices with the values $a(i+1+k(r+1))$, ... , $a(i+r+k(r+1))$), and vertices from $A_k = B_r(\mathbf{v_k}) \cap S_{R+2}(\mathbf{0})$, which are marked by white circles in the Fig. 3 (vertices with the values $b(i+1+k(r+1))$, ... , $b(i+r+1+k(r+1))$). By definition of generalized centered function we have that $\psi(\mathbf{v_k})A = \sum\limits_{\substack{\mathbf{x} \in B_r(\mathbf{v_k}), \\ \mathbf{x} \neq \mathbf{v_k}}} \psi(\mathbf{x})$.

Combining it with (1), we get

$$\sum_{j=i+1+k(r+1)}^{i+r+1+k(r+1)} b(j) = \mathbf{d}. \tag{2}$$

The set $S_{R+2}^1(\mathbf{0})$ can be represented as a union of disjoint sets A_k and the set D of vertices in $S_{R+2}^1(\mathbf{0})$ that do not belong to one of the sets A_k (boundary effects):

$$S_{R+2}^1(\mathbf{0}) = \bigcup_k A_k \cup D, \ 1 \leq i + k(r+1) \leq R - r + 2.$$

In Fig. 3 the set D consists of 3 vertices $\mathbf{u_1}$, $\mathbf{u_2}$, $\mathbf{u_3}$. The number of elements in the set D is not more than $2r$: $|D| \leq 2r$, denote $\sum\limits_{\mathbf{x} \in D} \psi(\mathbf{x}) = \mathbf{c}$.

Let us calculate the sum of the values in the set $S_{R+2}^1(\mathbf{0})$:

$$\sum_{\mathbf{x} \in S_{R+2}^1(\mathbf{0})} \psi(\mathbf{x}) = \sum_{j=1}^R b(j) = \sum_k \sum_{\mathbf{x} \in A_k} \psi(\mathbf{x}) + \sum_{\mathbf{x} \in D} \psi(\mathbf{x}) = k\mathbf{d} + \mathbf{c}.$$

Consider a nonzero coordinate of $\mathbf{d} = (d_1, ..., d_n)$, let it be l-th coordinate. Without loss of generality $d_l \geq 1$. So, if we take $k \geq 2r + 1$ (therefore, $R \geq (2r+1)(r+1) + 2r = 2r^2 + 5r + 1$), then the vector $\sum\limits_{j=1}^R b_j$ is greater than 0 in its l-th coordinate.

We have that $\sum\limits_{i=j}^{j+r} a(i) = \mathbf{0}$, so there exists m such that $a(m) = \mathbf{f} = (f_1, ..., f_n)$ and $f_l \leq 0$. Arguing as above we get that the l-th coordinate of the vector $\sum\limits_{j=1}^R b(j)$ is less than 0. A contradiction.

Thus $a(i) = \mathbf{0}$ for $i = 1, ..., R$.

So, we proved that $\varphi|_{S_{R+1}^1} = \mathbf{0}$. The proof is similar for the sets S_{R+1}^2, S_{R+1}^3, S_{R+1}^4. Now, $\psi(0, -R-1) = \mathbf{0}$, because ψ is equal to $\mathbf{0}$ in all other vertices of the ball $B_r(0, r - R - 1)$. Similarly for other elements of the set S_{R+1}^5.

Thus, we have $\varphi|_{S_{R+1}(\mathbf{x})} = \varphi|_{S_{R+1}(\mathbf{y})}$, therefore, φ is R-prolongable for $R \geq 2r^2 + 5r + 2$. Now, by Lemma 1, φ is periodic. The first part of the theorem is proved.

2. We will need the following auxiliary proposition.

Proposition 1. *Let φ be a generalized centered function of radius 1 with a matrix A. Then there exist R_0, such that for $R \geq R_0$ the condition $\varphi|_{B_R(\mathbf{x})} = \varphi|_{B_R(\mathbf{y})}$ implies either*

$$\varphi|_{S^1_{R+1}(\mathbf{x})} = \varphi|_{S^1_{R+1}(\mathbf{y})}$$

or

$$\varphi|_{S^1_{R+1}(\mathbf{x})} = (\varphi + \chi^{\mathbf{v}}_{(y_1, y_2 - R - 1)})|_{S^1_{R+1}(\mathbf{y})},$$

where $\mathbf{v}$ is an eigenvector of A, corresponding to the eigenvalue 0: $\mathbf{v}A = \mathbf{0}$.

Proof. Consider a function $\psi(\mathbf{t}) = \varphi(\mathbf{x} + \mathbf{t}) - \varphi(\mathbf{y} + \mathbf{t})$. We have $\psi|_{B_R(\mathbf{0})} = \mathbf{0}$ by definition of function ψ.

Consider the values of ψ on the set $S^1_{R+1}(\mathbf{0})$. Denote $a(i) = \psi(i, i - 1 - R)$, $i = 1, ..., R$. Fig. 2 illustrates our reasoning for $R = 5$. Denote $a(1) = \mathbf{u}$, consider a ball $B_1(1, 1 - R)$. The value of ψ in its center is $\mathbf{0}$, the sum of values of ψ in this ball is equal to $\mathbf{0}$. So $a(2) = -\mathbf{u}$. Considering sums of values of ψ in the balls $B_1(j, j - R)$, we get $a(j + 1) = -a(j) = (-1)^j \mathbf{u}$.

Now we consider the values of ψ on the set $S^1_{R+2}(\mathbf{0})$. Denote $b(i) = \psi(i, i - 2 - R)$, $i = 1, ..., R + 1$ (see Fig. 2). Denote $b(1) = \mathbf{w}$, consider a ball $B_1(1, -R)$. The value of ψ in its center is $\mathbf{u}$, the sum of values of ψ in this ball is equal to $\mathbf{u}A$: $b(1) + b(2) = a(1)A$. So $b(2) = \mathbf{u}A - \mathbf{w}$. Considering sums of values of ψ in the balls $B_1(j, j - 1 - R)$ (one of these balls is marked by bold line in Fig. 2), we get $b(j) + b(j + 1) = a(j)A$ for $1 \leq j \leq R$, whence $b(j + 1) = (-1)^{j+1}(j\mathbf{u}A - \mathbf{w})$. The values $b(j)$ are growing in modulus in nonzero coordinates of the vector $\mathbf{u}A$. The function φ is M-bounded, so ψ is $2M$-bounded and if $R > 4M + 1$, then $\mathbf{u}A = \mathbf{0}$. If $\mathbf{u} = \mathbf{0}$, then $\psi|_{S^1_{R+1}(\mathbf{0})} = \mathbf{0}$, i. e. $\varphi|_{S^1_{R+1}(\mathbf{x})} = \varphi|_{S^1_{R+1}(\mathbf{y})}$. If $\mathbf{u}A = \mathbf{0}$ and $\mathbf{u} \neq \mathbf{0}$, then $\mathbf{u}$ is an eigenvector of A, corresponding to the eigenvalue $\lambda = 0$. In this case we have that $\psi|_{S^1_{R+1}(\mathbf{0})} = \chi^{\mathbf{u}}_{(0, -R-1)}|_{S^1_{R+1}(\mathbf{0})}$, i. e. $\varphi|_{S^1_{R+1}(\mathbf{x})} = (\varphi + \chi^{\mathbf{u}}_{(y_1, y_2 - R - 1)})|_{S^1_{R+1}(\mathbf{y})}$. The proposition is proved. $\quad\square$

Notice that $\chi^{\mathbf{v}}_{(y_1, y_2 - R - 1)}$ in this lemma is bounded by $2M$.

It is easy to obtain similar assertions for the values of φ on the sets $S^i_{R+1}(\mathbf{x})$ and $S^i_{R+1}(\mathbf{y})$, $i = 2, 3, 4$.

Proposition 2. *Let φ be a generalized centered function with the matrix A of radius 1. If $\varphi|_{B_R(\mathbf{x})} = \varphi|_{B_R(\mathbf{y})}$ and $\varphi|_{S^i_{R+1}(\mathbf{x})} = \varphi|_{S^i_{R+1}(\mathbf{y})}$ for $i = 1, 2, 3, 4$, then*

$$\varphi|_{S^5_{R+1}(\mathbf{x})} = \varphi|_{S^5_{R+1}(\mathbf{y})}.$$

Proof. To prove that $\varphi(x_1 - R - 1, x_2) = \varphi(y_1 - R - 1, y_2)$ it is sufficient to compare the values of the function φ in the balls $B_1(x_1 - R, x_2)$ and $B_1(y_1 - R, y_2)$. Similarly for other vertices of the sets $S^5_{R+1}(\mathbf{x})$ and $S^5_{R+1}(\mathbf{y})$. $\quad\square$

Now we proceed to proof of the theorem. If $\det A \neq 0$, then $\mathbf{v}A = \mathbf{0}$ implies $\mathbf{v} = \mathbf{0}$. In this case Propositions 1 and 2 imply that φ is R-prolongable. By Lemma 1 it is periodic.

Let φ be non-periodic, $\det A = 0$. Consider the set of balls $\{B_R(x, -x) | x \in \mathbb{Z}\}$, where $R > R_0 + 1$. Vertices of these balls lie in a region $(z_1, z_2) : |z_1 + z_2| \leq R$ (see Fig. 4). This set of balls is infinite, the alphabet is finite, so we have finite number of subwords on B_R, this number is less or equal $|\Sigma|^{|B_R|} = ((2M+1)^n)^{(2R^2+2R+1)}$, where Σ is the alphabet of values of φ, $|B_R|$ is the number of vertices in a ball. This means that we have two balls $B_R(x, -x)$ and $B_R(y, -y)$, such that

$$\varphi|_{B_R(x,-x)} = \varphi|_{B_R(y,-y)}$$

and $|x - y| \leq (2M + 1)^{n(2R^2+2R+1)}$. To be definite, assume that $x \leq y$. In Fig. 4 boundaries of these balls are marked by bold line.

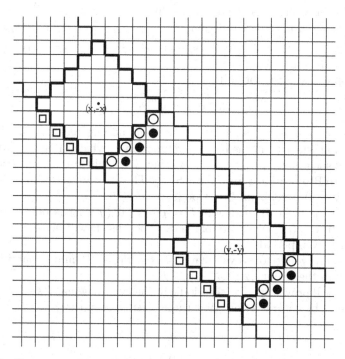

Fig. 4. The illustration for the proof of Proposition 2: the region $(z_1, z_2) : |z_1 + z_2| \leq R$ and two balls $B_R(x, -x)$ and $B_R(y, -y)$. White circles mark $S^1_{R+1}(x, -x)$ and $S^1_{R+1}(y, -y)$, black circles mark $S^1_R(x + 1, -x - 1)$ and $S^1_R(y + 1, -y - 1)$, squares mark the sets $S^2_{R+1}(x, -x)$ and $S^2_{R+1}(y, -y)$.

Consider the sets $S^1_{R+1}(x, -x)$ and $S^1_{R+1}(y, -y)$. In Fig. 4 they are marked by white circles. By Proposition 1 we have either

$$\varphi|_{S^1_{R+1}(x,-x)} = \varphi|_{S^1_{R+1}(y,-y)}$$

or

$$\varphi|_{S^1_{R+1}(x,-x)} = (\varphi + \chi^v_{(y,-y-R-1)})|_{S^1_{R+1}(y,-y)},$$

where v is such that $vA = 0$. In the first case we define $\varphi' = \varphi$. In the second case we subtract the function $\chi^v_{(y,-y-R-1)}$ from the function φ: $\varphi' = \varphi - \chi^v_{(y,-y-R-1)}$. Notice that if φ is M-bounded, then φ' is $3M$-bounded. Therefore,

$$\varphi'|_{S^1_{R+1}(x,-x)} = \varphi'|_{S^1_{R+1}(y,-y)}.$$

We have that $\varphi'|_{B_{R-1}(x+1,-x-1)} = \varphi'|_{B_{R-1}(y+1,-y-1)}$. Now consider the sets $S_R^1(x+1, -x-1)$ and $S_R^1(y+1, -y-1)$. In Fig. 4 they are marked by black circles. By Proposition 1 we have either

$$\varphi'|_{S_R^1(x+1,-x-1)} = \varphi'|_{S_R^1(y+1,-y-1)}$$

or

$$\varphi'|_{S_R^1(x+1,-x-1)} = (\varphi' + \chi_{(y+1,-y-1-R)}^u)|_{S_R^1(y+1,-y-1)},$$

where u is such that $uA = 0$. In the first case we define $\varphi'' = \varphi'$. In the second case we subtract the function $\chi_{(y+1,-y-1-R)}^u$ from the function φ': $\varphi'' = \varphi' - \chi_{(y+1,-y-1-R)}^u$. Therefore, $\varphi''|_{S_R^1(x+1,-x-1)} = \varphi''|_{S_R^1(y+1,-y-1)}$. Notice that φ'' is also $3M$-bounded.

By Proposition 2 we obtain that

$$\varphi''|_{B_R(x+1,-x-1)} = \varphi''|_{B_R(y+1,-y-1)}.$$

Arguing as above we proceed unit by unit adding rectangular alternating functions if necessary and then obtain a function $\tilde{\varphi}$, which satisfies $\tilde{\varphi}(\mathbf{z}) = \tilde{\varphi}(\mathbf{z} + \mathbf{y} - \mathbf{x})$ for $\mathbf{z}$ such that $|z_1 + z_2| \leq R$. Note that we should also use assertion that is analogous to Proposition 1 for the set S_{R+1}^3. The function $\tilde{\varphi}$ is also $3M$-bounded and it is $(\mathbf{y} - \mathbf{x})$-periodic in the stipe $|z_1 + z_2| \leq R$.

If $(y - x)$ is even, then denote $t = y - x$, if $(y - x)$ is odd, then denote $t = 2(y - x)$, $\mathbf{t} = (t, -t)$ (we double period, because it will be convenient for us to deal with even period).

Now we are going to prove, that $\tilde{\varphi}$ is $\mathbf{t}$-periodic. Remind that we have $\mathbf{t}$-periodicity in the region $\{\mathbf{z} : |z_2 + z_1| \leq R\}$.

First we will prove that $\tilde{\varphi}$ is $\mathbf{t}$-periodic in the next diagonal $\{(s-R, -s+1)|s \in \mathbb{Z}\}$. Suppose, by contradiction, that there exists q such that

$$\tilde{\varphi}(q - R, -q+1) \neq \tilde{\varphi}(q - R + t, -q + 1 - t).$$

Denote $w = \tilde{\varphi}(q - R, -q+1) - \tilde{\varphi}(q - R + t, -q + 1 - t)$.

The function $\tilde{\varphi}$ is $\mathbf{t}$-periodic in the region $|z_1 + z_2| \leq R$, so we have that

$$\tilde{\varphi}|_{B_R(q,-q)} = \tilde{\varphi}|_{B_R(q+t,-q-t)}.$$

So by assertion that is analogous to Proposition 1 for the set S_{R+1}^2 we have that

$$\tilde{\varphi}|_{S_{R+1}^2(q,-q)} = (\tilde{\varphi} + \chi_{(q-R,-q+1)}^w)|_{S_{R+1}^2(q+t,-q-t)},$$

where w is such that $wA = 0$. In Fig. 4 the sets $S_{R+1}^2(q, -q)$ and $S_{R+1}^2(q + t, -q - t)$ (for $q = x$, $(y - x)$ even) are marked by squares. In particular, this means that

$$w = \tilde{\varphi}((q + 2) - R, -(q + 2) + 1) - \tilde{\varphi}((q + 2) - R + t, -(q + 2) + 1 - t).$$

Now by induction we get that

$$w = \tilde{\varphi}((q + k) - R, -(q + k) + 1) - \tilde{\varphi}((q + k) - R + t, -(q + k) + 1 - t)$$

for every even integer k. Using this equality for $k = t, 2t, \ldots, mt$ we obtain

$$\tilde{\varphi}(q-R, -q+1) = \tilde{\varphi}(q+t-R, -q-t+1)+\boldsymbol{w} = \tilde{\varphi}(q+2t-R, -q-2t+1)+2\boldsymbol{w} = \cdots =$$

$$= \tilde{\varphi}(q + mt - R, -q - mt + 1) + m\boldsymbol{w}.$$

This implies that $\tilde{\varphi}$ in not bounded. A contradiction.

Thus we obtain $\mathbf{t}$-periodicity in extended region (in a stripe with additional diagonal). Continuing this line of reasoning, we obtain $\mathbf{t}$-periodicity for all $\mathbb{Z}^2$.

Therefore we obtained $(t, -t)$-periodic function $\tilde{\varphi}$ by adding rectangular alternating functions χ and we did not use functions $\chi*$. (t, t)-periodicity can be organized for the function $\tilde{\varphi}$ in the same way by adding functions $\chi*$. Note that adding functions $\chi*$ does not break $(t, -t)$-periodicity. The theorem is proved.   □

Perfect coloring and centered function are partial cases of generalized centered function, so Theorem 1 implies the results, obtained in [8] and [2]:

Corollary 1. *Every perfect coloring of radius $r \geq 2$ on the infinite rectangular grid is periodic.*

Corollary 2. *Every centered function of radius $r \geq 1$ on the infinite rectangular grid is periodic.*

3   The Infinite Triangular and Hexagonal Grids

In this section we consider periodicity of generalized centered functions on the infinite triangular and hexagonal grids (see Fig. 5 and Fig. 6). All necessary definitions can be given in the same way as for the infinite rectangular grid. The sets of vertices of the infinite hexagonal and triangular grids we denote by H and T correspondingly.

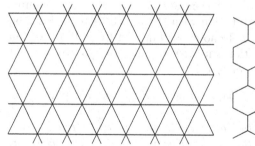 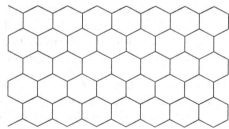

Fig. 5. The infinite triangular grid   **Fig. 6.** The infinite hexagonal grid

These graphs are dual. A function on vertices of a flat graph can be considered as a function on faces of dual graph. The pictures illustrate functions on faces of dual graphs.

Example. Let A be an integer matrix, such that $\lambda = -2$ is an eigenvalue of A, v be corresponding eigenvector: $vA = -2v$. The example of non-periodic generalized centered function with the matrix A on the infinite triangular grid (on faces of hexagonal grid) is in Fig. 8. Denote this function by φ. If a function ψ can be obtained from φ by translation and/or rotation by $\pm\pi/3$, we say that ψ is a *triangular alternating function*.

Theorem 2. *1. Let $\varphi : T \rightarrow \mathbb{Z}^n$ be a bounded generalized centered function of radius $r > 1$. Then φ is periodic.*

2. If φ is generalized centered function of radius 1 with a matrix A and $\lambda = -2$ is not an eigenvalue of the matrix A, then φ is periodic. If $\lambda = -2$ is eigenvalue of A, then there exist non-periodic and periodic generalized centered functions with the matrix A. A periodic function can be obtained from a non-periodic one by adding triangular alternating functions.

Perfect coloring and centered function are partial cases of generalized centered function, so Theorem 1 implies the following results (Corollary 4 was obtained in [2]):

Corollary 3. *Every perfect coloring of radius $r \geq 2$ on the infinite triangular grid is periodic.*

Corollary 4. *Every centered function of radius $r \geq 1$ on the infinite triangular grid is periodic.*

Example. Let A be an integer matrix, such that $\lambda = 1$ is an eigenvalue of A^2, v be corresponding eigenvector: $vA^2 = v$. Denote $vA = u$. The example of non-periodic generalized centered function with the matrix A of radius 1 on the infinite hexagonal grid (on faces on triangular grid) is in Fig. 7. Denote this function by φ. If a function ψ can be obtained from φ by translation and/or rotation by $\pm\pi/3$, we say that ψ is a *hexagonal alternating function of type 1*.

Let A be an integer matrix, such that $\lambda = 1$ is an eigenvalue of $(A+2E)^2$, v be corresponding eigenvector: $v(A+2E)^2 = v$. Denote $v(A+2E) = u$. The example of non-periodic generalized centered function with the matrix A of radius 2 on the infinite hexagonal grid is in Fig. 7. Denote this function by φ. If a function ψ can be obtained from φ by translation and/or rotation by $\pm\pi/3$, we say that ψ is a *hexagonal alternating function of type 2*.

Theorem 3. *1. Let $\varphi : H \rightarrow \mathbb{Z}^n$ be a bounded generalized centered function of radius $r > 2$. Then φ is periodic.*

2. If φ is generalized centered function of radius 1 with a matrix A and $\lambda = 1$ is not an eigenvalue of the matrix A^2, then φ is periodic. If $\lambda = 1$ is an eigenvalue of A^2, then there exist non-periodic and periodic generalized centered functions with the matrix A. A periodic function can be obtained from a non-periodic one by adding hexagonal alternating functions of type 1.

3. If φ is generalized centered function of radius 2 with a matrix A and $\lambda = 1$ is not an eigenvalue of the matrix $(A + 2E)^2$, then φ is periodic. If $\lambda = 1$ is an eigenvalue of $(A + 2E)^2$, then there exist non-periodic and periodic generalized *centered functions with the matrix A. A periodic function can be obtained from a non-periodic one by adding hexagonal alternating function of type 2.*

Perfect coloring and centered function are partial cases of generalized centered function, so Theorem 1 implies the following results (Corollary 6 was obtained in [2]):

Corollary 5. *Every perfect coloring of radius $r \geq 3$ on the infinite hexagonal grid is periodic.*

Corollary 6. *Every centered function of radius $r \geq 3$ on the infinite hexagonal grid is periodic.*

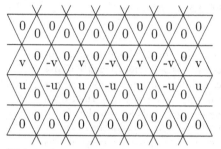

Fig. 7. An example of non-periodic generalized centered function on faces of the infinite triangular grid

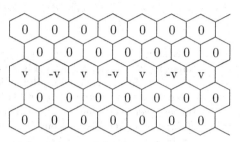

Fig. 8. An example of non-periodic generalized centered function on faces of the infinite hexagonal grid

References

1. Avgustinovich, S.V., Vasil'eva, A.Y.: Reconstruction of centered function by its values at two middle levels of the hypercube. Discrete analysis and operations research 10(2), 3–16 (2003) (in Russian)
2. Puzynina, S.A., Avgustinovich, S.V.: On periodicity of two-dimensional words. Special issue of Theoretical Computer Science (accepted)
3. Berthe, V., Vuillon, L.: Tilings and rotations: a two-dimensional generalization of Sturmian sequences. Discrete Math. 223, 27–53 (2000)
4. Epifanio, C., Koskas, M., Mignosi, F.: On a conjecture on bidimensional words. Theoretical Computer Science 299(1-3), 123–150 (2003)
5. Nivat, M.: Invited talk at ICALP 1997 (1997)
6. Puzynina, S.A.: Periodicity of perfect colorings of the infinite rectangular grid. Discrete analysis and operations research 11(1), 79–92 (2004)
7. Puzynina, S.A.: Perfect colorings of the infinite rectangular grid. Bayreuther Mathematischen Schriften, Heft 74, 317–331 (2005)
8. Puzynina, S.A.: Perfect colorings of radius $r > 1$ of the infinite rectangular grid. Siberian Electronic Mathematical Reports (Submitted)
9. Quas, A., Zamboni, L.: Periodicity and local complexity. Theoretical Computer Science 319(1-3), 229–240 (2004)

On the Analysis of "Simple" 2D Stochastic Cellular Automata

Damien Regnault[1], Nicolas Schabanel[1,2], and Éric Thierry[1,3]

[1] IXXI – LIP, Université de Lyon, École normale supérieure de Lyon, 46 allée d'Italie, 69364 Lyon Cedex 07, France
http://perso.ens-lyon.fr/{damien.regnault,eric.thierry}
[2] CNRS, Centro de Modelamiento Matemático, Universidad de Chile, Blanco Encalada 2120 Piso 7, Santiago de Chile, Chile
http://www.cmm.uchile.cl/~schabanel
[3] CNRS, Laboratoire d'Informatique Algorithmique: Fondements et Applications, Université Paris 7, 75205 Paris Cedex 13

Abstract. We analyze the dynamics of a two-dimensional cellular automaton, 2D Minority, for the Moore neighborhood (eight neighbors per cell) under fully asynchronous dynamics (where only one random cell updates at each time step). Even if 2D Minority seems a simple rule, from the experience of Ising models and Hopfield nets, it is known that models with negative feedback are hard to study. This automaton actually presents a rich variety of behaviors, even more complex than what was observed and analyzed in a previous work on 2D Minority for the von Neumann neighborhood (four neighbors per cell) [1], including particles and a wider range of stable configurations. Nevertheless our work suggests that predicting the behavior of this automaton although difficult is possible, opening the way to analyze the class of totalistic automata.

1 Introduction

Cellular automata are attractive models for complex systems in various fields, like physics, biology or social sciences. Their relevance is supported by many observations of natural phenomena which closely match the dynamics of some cellular automaton, as illustrated by Fig. 1. An example of challenging issue in biology is to predict the expression of genes in a set of cells which share the same gene regulatory network. Cellular automata can be used to model such systems [2,3]. For example consider the simple gene regulatory networks where a gene exerts a feedback inhibition of its expression. The state of a cell is whether it expresses this gene or not. Assuming that each cell starts expressing the gene when less than half of its neighbors (including itself) express it, and that otherwise it stops expressing it, leads to the Minority rule [4]. If cells are assembled into a two-dimensional grid, it yields 2D Minority. Such a model is of course an extreme simplification of any real phenomena but understanding this "simple" model is already an indispensable step towards the study of more involved interaction networks. Surprisingly, it already exhibits an astonishingly rich behavior which is investigated in this paper.

C. Martín-Vide, F. Otto, and H. Fernau (Eds.): LATA 2008, LNCS 5196, pp. 452–463, 2008.
© Springer-Verlag Berlin Heidelberg 2008

 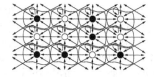

1.a – The pattern growth of the shell of the widespread species *Conus textile* is governed by a mathematical function presenting similarities with the Rule 30 CA above.

1.b – 2D minority induced by a set of cells expressing (in black) or not (in white) a gene which tends to inhibit its expression in neighboring cells.

Fig. 1. Cellular automata as models in biology

The 2D Minority automaton belongs to the class of threshold cellular automata which have been intensively studied under synchronous dynamics (at each time step, all the cells update simultaneously) [5]. However models for natural phenomena rather update asynchronously. Empirical studies [6,7,8] have shown how the behavior can change drastically when introducing asynchronism. However only few mathematical analyses are available and they mainly concern one-dimensional stochastic cellular automata [9,10,11,12]. Providing analyses of 2D rules remains a real challenge. For instance the mean-field approach does not succeed in approximating tightly such stochastic dynamics [13]. Fig. 2 illustrates for three 2D cellular automata the differences between the synchronous dynamics and the *fully asynchronous dynamics* where only one random cell updates at each time step. Some related stochastic models like Ising models or Hopfield nets have been studied under asynchronous dynamics (our model of asynchronism corresponds to the limit when temperature goes to 0 in the Ising model). These models are acknowledged to be harder to analyze when it comes to two-dimensional topologies [14] or negative feedbacks [15].

For all these reasons 2D Minority under fully asynchronous dynamics turns out to be an interesting and challenging candidate for mathematical analyses. This paper focuses on the Moore neighborhood: at each time step, the fired cell updates to the minority state among its eight closest neighbors and itself. It carries on a work initiated in [1] where 2D Minority was analyzed for the von

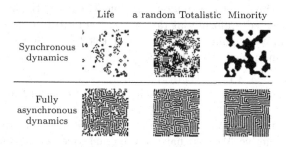

Fig. 2. Typical configurations observed during the evolution of some 2D cellular automata (Moore neighborhood). Similar stripes emerge in asynchronous regime even if their synchronous behavior differ drastically.

Neumann neighborhood (four neighbors instead of eight). One might have hoped
for minors adjustments to deal with the Moore neighborhood, however the results
do not come out so easily. Experiments discussed in Section 3 show that new
patterns (wider variety of striped patterns) and new phenomena occur (particle-
like behaviors). Several key ideas presented in [1] (energy, borders and regions)
apply, but their use requires some innovations. We show that the initial stage
of the dynamics is characterized by a fast energy drop (Theorem 3). We exhibit
borders that separate striped regions competing with one another and we manage
to prove how final stable (horizontally or vertically) stripes configurations are
reached almost surely from typical configurations occuring at the end of the
process. Furthermore, we prove that this convergence occurs in polynomial time
(Theorem 5). In the proof, we show that as the regions crumble, inflate or retract,
the overall structure admits a recursive description which persists over time. The
proofs of the study of such dynamical systems, know as complex systems, involve
unavoidable tedious case studies and one of the important contribution of this
paper is to set up a compact, possibly elegant, and thus safe framework to deal
with these enumerations of cases. Note that in the course of the paper, we present
an interesting characterization of the stable configurations for 2D Minority for
the Moore neighborhood (Theorem 2). As far as we know, it had only been
solved for the von Neumann neighborhood [5].

2   Definitions

We consider in this paper the 2D 2-states cellular automaton Minority under
fully asynchronous dynamics over finite configurations with periodic boundary
conditions. We are given two positive integers n and m, let $N = nm$. We denote
by $\mathbb{T} = \mathbb{Z}_n \times \mathbb{Z}_m$ the set of *cells* and $Q = \{0, 1\}$ the set of *states* (0 stands for
white and 1 for black in the figures). We consider the *Moore neighborhood*: two
cells (i, j) and (k, l) are *neighbors* if $\max(|i - k|_n, |j - l|_m) \leqslant 1$ (where $|i - j|_p$
denotes the distance in $\mathbb{Z}_p$). A $n \times m$-*configuration* c is a function $c : \mathbb{T} \to Q$; c_{ij}
is the *state* of the cell (i, j) in configuration c.

We consider the *fully asynchronous dynamics of 2D Minority*. Time is discrete
and let c^t denote the configuration at time t; c^0 is the *initial configuration*. The
configuration at time $t + 1$ is a random variable defined by the following process:
a cell (i, j) is selected uniformly at random in $\mathbb{T}$ and its state is updated to the
minority state in its neighborhood (we say that cell (i, j) *fires* at time t), all the
other cells remain in their current state: $c_{ij}^{t+1} = 1$ if $(c_{ij}^t + c_{i-1,j}^t + c_{i+1,j}^t + c_{i,j-1}^t +
c_{i,j+1}^t + c_{i-1,j+1}^t + c_{i-1,j-1}^t + c_{i+1,j-1}^t + c_{i+1,j+1}^t) \leqslant 4$, and $c_{ij}^{t+1} = 0$ otherwise;
and $c_{kl}^{t+1} = c_{kl}^t$ for all $(k, l) \neq (i, j)$. A cell is said *active* if its state would change
if fired.

A configuration c is *stable* if it remains unchanged under the dynamics, *i.e.*, if
all its cells are inactive. We say that the random sequence (c^t) *converges almost
surely* from an *initial configuration* $c^0 = c$ if the random variable $T = \min\{t :
c^t$ is stable$\}$ is finite with probability 1. We say that the convergence occurs in
polynomial time on expectation if $\mathbb{E}[T] \leqslant p(N)$ for some polynomial p.

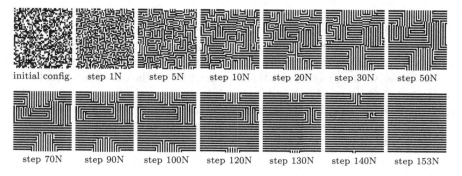

initial config.   step 1N   step 5N   step 10N   step 20N   step 30N   step 50N

step 70N   step 90N   step 100N   step 120N   step 130N   step 140N   step 153N

Fig. 3. A typical execution of stochastic 2D Moore minority with $N = 50 \times 50$ cells

3   Experiments

Typical behavior. Like other 2D automata (such as Game of Life [6])the asynchronous behavior of 2D Minority differs radically from its synchronous dynamics. In particular, [5] proved that the synchronous dynamics eventually leads to stable configurations or cycles of two opposite configurations. The latter case is the typical behavior in synchronous simulations where one can observe big flashing islands (Fig. 2). On the contrary, as can be observed in Fig. 3, the states of most of the cells are very stable over time in fully asynchronous regime and present typically very rapidly striped patterns (horizontal or vertical) that tend to extend and merge with each other until one gets over the others and covers the whole configuration (when at least one of the dimensions n or m is even). A goal of this paper is to explore how such stripes arise and end up covering the whole configuration. Note also that such stripes arise as well in many other asynchronous automata such as the totalistic cellular automata (see Section 1). Very rarely a random initial configuration may converge to more exotic stable configurations. Fig. 6 gives some examples of more or less exotic stable configurations under 2D minority dynamics.

Borders and Particles. Part of the richness of 2D Minority under fully asynchronous behavior is due to some specific configurations where "particles" can be observed. Several patterns can be identified as particles although for now we do not have a formal definition. We say that there is a border between two diagonally neighboring cells if they are in the same state (more details in the next section). Active cells are always located near the borders. When the borders draw a pattern (where red spots indicate active cells), then, depending of which of the two active cell fires the pattern will move in different directions: forward or backward. Such patterns which "move" along borders can be called particles. In some configurations, the set of all the borders form a network of "rails" carrying several particules. These particules follow random walks along the rails and vanish when they collide. Note that the dynamics is a lot more intricate than 2D random walks because the rails are modified by the passage of the particules and if two rails become too close, a whole part of the rail network collapses. Fig. 4 illustrates this kind of phenomena. Configurations with particles and rails are rarely reached from a random initial configuration. Nevertheless, we have to consider them when we study the convergence and these phenomena are extremely difficult to analyze mathematically. Such a system of particles is not observed in asynchronous 2D minority with von Neumann neighborhood [1] or in related models like the ferromagnetic Ising model or Hopfield networks with positive feedback.

4   Energy, Borders, Diamonds and Stable Configurations

4.1   Borders, Diamonds and Stripes

The following definitions allow to highlight the underlying structure of a configuration with respect to the dynamics. These tools turn out to be a key step to prove the convergence.

If n and m are even, a set of cells $\mathcal{R}$ is said to be tiled with *even horizontal stripes* (resp. *odd horizontal, even* and *odd vertical stripes*) if $c_{ij} = i \bmod 2$ (resp. $i + 1$, j, $j + 1 \bmod 2$), for all cell $(i, j) \in \mathcal{R}$. Note that cells whose whole neighborhood is striped are inactive. We say that there is a *border* between two diagonally neighboring cells (i, j) and $(i + \epsilon, j + \eta)$, with $\epsilon, \eta \in \{-1, 1\}$, if they are in the same state, *i.e.*, if $c_{ij} = c_{i+\epsilon, j+\eta}$. If n and m are even, we say that a cell (i, j) is *even* (resp. *odd*) if $i + j$ is even (resp. odd). We say that the border between two cells is *blue* if the cells are even, and *green* otherwise; furthermore, we say that there is a *diamond* over cell (i, j) if its state coincides with even horizontal stripes, *i.e.*, if $c_{ij} = i \bmod 2$; the diamond is *blue* if the cell is even and *green* otherwise.

Proposition 1 (Borders are boundaries). *The borders are the exact boundary of regions tiled with stripes patterns (odd/even horizontal/vertical). Moreover, when n and m are even, the blue (resp. green) borders are the exact boundary of the regions covered by the blue (resp. green) diamonds.*

4.a – A sequence of updates in a configuration starting with 4 particles where two of them move along rails and ultimately vanish after colliding with each other

4.b – A sequence of updates where the rails cannot sustain the pertubations due to the movements of the particles: at some point, rails get to close with each other, new active cells appear, and part of the rail network collapses

Fig. 4. Some examples of the complex behavior of particles in a 20×20 configuration

Since cells whose neighborhood is striped are inactive, the only active cells in a configuration may be found along the borders.

4.2   Energy

As in Ising model [14] or Hopfield networks [15], we define a natural global parameter that one can consider to be the energy of the system since it counts the number of interactions between neighboring cells in the same state. This parameter will provide key insights on the evolution of the system.

The *potential* v_{ij} of cell (i, j) is the number of its neighboring cells in the same state as itself minus 2.[1] By definition, if $v_{ij} \leqslant 1$, then the cell is in the minority state in its own neighborhood and is thus *inactive* (its state will not change if fired); whereas, if $v_{ij} \geqslant 2$ then the cell is *active* and its state will change if fired. Note that a configuration c is *stable* iff for all cell $(i, j) \in \mathbb{T}$, $v_{ij} \leqslant 1$. Let say that a subset of cells $\mathcal{R}$ is *fat* if for each cell $(i, j) \in \mathcal{R}$, there exists a square $Q = \{(i, j), (i + \epsilon, j), (i + \epsilon, j + \eta), (i, j + \eta)\}$, for some $\epsilon, \eta \in \{1, -1\}$, such that $Q \subset \mathcal{R}$. The *energy* $E_{\mathcal{R}}$ of set $\mathcal{R}$ in a given configuration is defined as: $E_{\mathcal{R}} = \sum_{(i,j) \in \mathcal{R}} v_{ij}$. We denote by E the energy of the whole configuration c.

The next proposition shows that the energy is non-negative for almost every subset of cells of a configuration. This means that there cannot be too many cells with negative potential. This implies that the decrease of energy over time (Proposition 3 and Theorem 3) is not due to the increase of the number of cells with negative potential, but to the decrease of the potentials of the cells with positive potential, which explains intuitively why the striped patterns which have minimum energy (Proposition 4) arise naturally very rapidly.

Proposition 2 (Energy is non-negative). *For any fat subset of cells $\mathcal{R}$ of size q:* $0 \leqslant E_{\mathcal{R}} \leqslant 6q$.

The following easy fact will be very handy in order to prove the convergence of the dynamics.

Fact 1. *When an active cell (i, j) is fired, its new potential is $v_{ij} := 4 - v_{ij}$ and the total energy varies by $8 - 2v_{ij}$. Note that if $v_{ij} = 2$, both remain unchanged.*

Proposition 3 (Energy is non-increasing). *Under fully asynchronous dynamics, the energy is a non-increasing function of time and decreases each time a cell with potential $\geqslant 3$ fires.*

Proposition 4 (Minimum energy configurations). *The energy of a configuration c is 0 iff c is a striped configuration.*

---

[1] The offset -2 is convenient since its ensures that the minimum energy of a configuration is 0 (see Proposition 2 below).

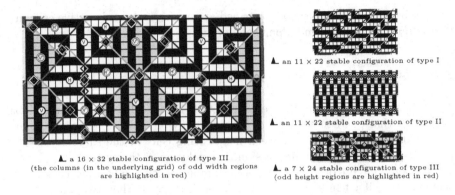

▲ a 16 × 32 stable configuration of type III
(the columns (in the underlying grid) of odd width regions
are highlighted in red)

▲ an 11 × 22 stable configuration of type I

▲ an 11 × 22 stable configuration of type II

▲ a 7 × 24 stable configuration of type III
(odd height regions are highlighted in red)

Fig. 5. Examples of stable configurations illustrating most of the possibilities

4.3   Stable Configurations

As opposed to the von Neumann fully asynchronous dynamics in [1], stable configurations under the Moore neighborhood exhibit rather complex structures as shown on Fig. 6. Although there is a great variety of stable configurations, a general structure can be extracted and they can be characterized thanks to the borders. We first describe the stable configurations when n and m are even and deduce from there the structure of the stable configurations in the general case by doubling the odd dimension. Fig. 5 gives examples of each type of stable configurations.

Theorem 2 (Stable configurations). *When at least one of n and m is even, there are exactly three types of stable configurations:*

- *Type I: the borders are parallel straight (diagonal) lines such that: two lines of the same color are at (ℓ_1-)distance $\geqslant 2$; if two lines blue and green are at distance 1, there is no other line at distance $\leqslant 4$ from each of them; the number of lines of each color along each row (resp., column) of the configuration has the same parity as m (resp., n).*
- *Type II: all the blue and green borders are all pairwise interlaced either horizontally according to the pattern* ▨ *or vertically according to* ▨ *the pairs of interlaced borders are at distance $\geqslant 2$ from each other; and the number of interlaced pairs has the parity of n if interlaced horizontally, and of m otherwise.*
- *Type III: the borders define a bicolor (horizontal/vertical stripes) underlying toric grid s.t.:*

    - *the segments of borders between two intersections are straight lines at distance at least 2 from each other;*
    - *two borders of the same color cannot intersect;*
    - *the number of borders of each color crossed by every row (resp. column) in the configuration has the same parity as m (resp. n);*

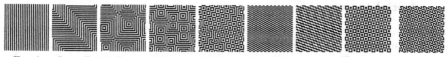

$E = 0$ Low E Higher E Med E High E The four highest energy config. ($E = N$)

Fig. 6. Examples of stable configurations for 2D Minority at various levels of energy

- the borders of opposite colors intersect at the corners of the cells only, and according to the following (possibly overlapping) patterns:

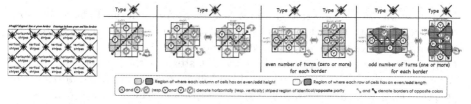

Furthermore, no stable configuration can have both crossings of types C and D and if a region has a crossing of type C (resp., D), all the crossings at the same vertical (resp., horizontal) level in the underlying grid are of type C (resp., D); moreover, the partity of the number of such horizontal (resp., vertical) levels of C-crossings (resp., D-) equals the parity of m (resp. n).

And any $n \times m$-configuration of type I, II or III is a stable configuration.

Corollary 1. *If n and m are odd, no stable configuration exists, and the dynamics never converges. If only one of n and m is odd, stable configurations of type I, II, and III exist with the parity restrictions mentioned in Theorem 2.*

Proposition 5 (Stable configurations energy). *The energy of a stable configuration c satisfies: $0 \leqslant E \leqslant N$. The only configurations with minimum energy (zero) are tiled with a striped 1×2-pattern. And the only stable configuration with maximum energy N are of four types: either tiled with the 2×4-pattern ▄▄, or the 8×8-patterns ▤▦, ▦▦, or ▨▨ (see Fig. 6).*

5   Analysis of the Convergence

In this section, we give our results on the existence and speed of the convergence of the dynamics towards a stable configuration from an arbitrary initial configuration. As opposed to the von Neuman dynamics where we were able to analyse the whole convergence, because of the existence of particles following sophisticated guided random walks (see Section 3), we are only able to describe the first steps and the last steps of the convergence. These results rely on the study of the energy function which is combined with an other parameter to obtain a variant. This variant allows to reduce the study of the randomly evolving 2D shape to an one dimensional random walk. The section ends with challenging conjectures on the overall convergence of the process.

5.1 Initial Energy Drop

According to experiments, the energy of a configuration drops very fast during the first steps until it converges, most of the time to a striped configuration of minimal energy. The following theorem provides a bound on the speed of this initial energy drop.

Theorem 3. *The energy of any configuration of size N is at most $N + 2N/3$ after $O(N^2)$ fully asynchronous minority updates on expectation.*

5.2 The Last Steps of Convergence

From now on, we assume that n and m are even. As mentioned above, in most of the experiments striped regions arise quickly, then they extend, compete with each other, merge until only one covers the whole configuration. In this section, we provide an analysis of the very last steps of the convergence to this stable configuration: the case where there remains only one *single horizontally striped region* within a *vertically striped background*, which we will call a *standard configuration*. We then show that the background ends up covering the whole configuration in polynomial time on expectation as expected according to the experiments (Fig. 3 when $t \geqslant 100N$). This involves studying the randomly evolving shape defined by the horizontally striped regions.

Note that every configuration is completely determined by its set of diamonds. When starting from a standard configuration, we show that the configuration exhibits a recursive structure that is conserved over time. We then show by studying a combination of the energy with the area of the random shape, that the random shape of the set of diamonds tends to vanish. Interestingly enough, we show that the horizontally striped region can flip the parity of its stripes but cannot extend beyond its initial surrounding rectangle. Let us now start with some definitions.

A *blue rectangle* (resp. *green rectangle*) is a rectangle such that its sides are parallel to the diagonals and its corners are located at the centers of odd (resp. even) cells. A blue or green rectangle is *enclosing* a set of diamonds D if all the diamonds are contained in the rectangle, and it is *surrounding* D if it is the smallest enclosing rectangle of that color for D. We say that a configuration is *standard* if it consists in a finite set of diamonds of the same color forming a rectangle (i.e., a set of diamonds of the same color whose borders match its surrounding rectangle). Two diamonds are *neighbors* if they have a side in common (and are thus of the same color). A set D of diamonds is: *connected* if D is connected for the neighborhood relationship; *convex* if for all $\epsilon \in \{1, -1\}$ and for any pair of diamonds centered on cells (i, j) and $(i + k, j + \epsilon k)$ in D, the diamonds centered on cells $(i + \ell, j + \epsilon\ell)$ for $0 \leqslant \ell \leqslant k$ belong to D; an *island* if it is connected and convex.

Definition 1 (Valid configurations). *A* valid configuration *(or* valid diamonds set*) is defined recursively by a tree structure of interlocked blue or green rectangles where each subtree describes the diamond set enclosed within the corresponding rectangle. Precisely:*

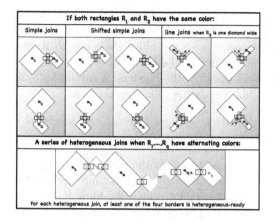 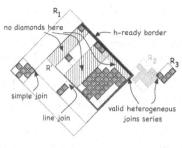

Fig. 7. To the left: valid combinations of valid configurations (the underlying cells of the automaton are shown at the junction of the rectangles). To the right: a valid configuration and its diamond set with a valid decomposition.

- *A set of diamonds consisting of an island is a valid configuration.*
- *The composition of two valid diamonds sets D_1 and D_2 enclosed by two rectangles R_1 and R_2 of the same color laying next to each other according to the patterns given in Fig. 7, is valid.*
- *The juxtaposition of q valid diamonds sets $D_1, \ldots, D_q$ enclosed in q rectangles $R_1, \ldots, R_q$ of alternating colors as shown in Fig. 7, is valid if at each junction, either both a blue and a green diamonds are located at the corresponding corners of the surrounding blue and green rectangles, or at least one of the four borders of these rectangles is h-ready; we say that the northeast border of an enclosing rectangle R of a valid configuration is h-ready if, within the smallest rectangle R' corresponding to the node enclosing all the diamonds laying along this border in the construction tree of R, the diamonds are located as follows: no diamond may lay in R' to the south-west of the diamonds along the north-east border of R nor one row below (see Fig. 7) (the definition extends naturally to NW, SW, and SE borders by rotation).*

A configuration is valid if its corresponding set of diamonds is valid. Each valid configuration is recursively described by a construction tree: a binary tree where each leaf is an island and each internal node stands for a join operation whose two edges pointing downwards are labeled by the two, blue or green, joint rectangles enclosing the two valid diamond sets described by the left and right subtrees.

Fig. 7 gives an example of a valid configuration starting with a blue island composed with several line joins, followed by a simple join with another blue island, and ending with a series of two heterogenous joins with the two islands to its right. A valid configuration can be represented by several construction trees. Rearranging construction trees according to certain rules is one of the keys to the following results (like Theorem 4).

The set of valid configurations is closed under the minority dynamics. In the fully asynchronous dynamics, only one cell fires at each time step, thus only one diamond is added or removed at each time step. Since there are horizontal stripes inside an island, the cells which are not at the borders are not active. All the deletions and additions of diamonds occur at the borders. A careful analysis of the actives cells in valid configuration yields the following theorem.

Theorem 4 (Closure and Reachability). *The set of valid configurations is the set of all reachable configurations from standard configurations.*

The energy function is not sufficiently precise to follow the evolution of the valid configurations since it may remain constant for long period of time whereas the configuration evolves towards a stable configuration. But combining it with the *area* A of the configuration, defined as the number of its diamonds, yields a variant from which we will deduce a polynomial bound on the expected convergence time.

Proposition 6. *The energy of a valid configuration is equal to twice the number of its blue and green borders minus twice the number of intersections of blue and green borders. Thus, $E \leqslant 8A$.*

The variant. Let $\Phi = A + E/4$, for any given configuration. Let us denote by $\mathbb{E}[\Delta\Phi]$ the expected variation of Φ for this configuration after one fully asynchronous minority update.

Proposition 7. *For any valid configuration constructed from k islands with ℓ joins: $\mathbb{E}[\Delta\Phi] \leqslant \dfrac{3\ell - 3k}{N}$.*

Proof. (Sketch) The proof proceeds by induction on the construction tree of the valid configuration. By following clockwise the borders of the island and counting the active cells, we can show that the expected variation of Φ for a configuration with only one island is at most $-\frac{3}{N}$. If the configuration is obtained by joining two valid configurations then whatever the join is, it can be checked that an active cell in one of the two configurations remains active with the same characterization in the joined configuration. A cell which is inactive in both configurations is inactive in the joined configuration, except around the join where at most three cells may have their activity changed. Then the expected variation of Φ is the sum of the expected variation of the two configurations plus the effect of these three cells which is in every case at most $+\frac{3}{N}$. $\Box$

Theorem 5. *Every valid configuration of area A converges to the background configuration in finite time with probability 1. The expected convergence time is $O(AN)$, which is thus $O(N^2)$.*

Proof. The construction of a valid configuration can be expressed as a binary tree where the leaves are the islands and the internal nodes are the joins (an heterogenous series is encoded as a series of two-by-two joins). Thus, $\ell = k - 1$ if the tree is not empty, and by Proposition 6 and 7, as long as the configuration is not stable, $\mathbb{E}[\Delta\Phi] \leqslant -\frac{3}{N}$. For the initial configuration $\Phi \leqslant 3A$ and the stable configuration with vertical stripes is the only configuration where $\Phi = 0$. Thus it converges in finite time with probability 1 to this stable configuration and the expected convergence time is $O(AN)$ (see e.g. Lemma 5 in [9]). □

Conclusion. The behavior of 2D Minority with the Moore neighborhood under fully asynchronous dynamics is surprisingly rich and difficult to analyze. The approach outlined in [1] for the von Neumann neighborhood is useful. The analysis of the energy and of the competing regions requires however a very accurate comprehension of the combinatorics of the automaton, which turned out to be more complex for the Moore neighborhood. A key to complete the analysis seems to find the most appropriate definitions for particles and rails and explain precisely how they evolve.

References

1. Regnault, D., Schabanel, N., Thierry, É.: Progresses in the analysis of stochastic 2D cellular automata: a study of asynchronous 2D Minority. In: Kučera, L., Kučera, A. (eds.) MFCS 2007. LNCS, vol. 4708, pp. 320–332. Springer, Heidelberg (2007)
2. Ermentrout, G.B., Edlestein-Keshet, L.: Cellular automata approaches to biological modelling. Journal of Theoretical Biology 160, 97–133 (1993)
3. Silva, H.S., Martins, M.L.: A cellular automata model for cell differentiation. Physica A: Statistical Mechanics and its Applications 322, 555–566 (2003)
4. Demongeot, J., Aracena, J., Thuderoz, F., Baum, T.P., Cohen, O.: Genetic regulation networks: circuits, regulons and attractors. C.R. Biologies 326, 171–188 (2003)
5. Goles, E., Martinez, S.: Neural and automata networks, dynamical behavior and applications. Maths and Applications, vol. 58. Kluwer Academic Publishers, Dordrecht (1990)
6. Bersini, H., Detours, V.: Asynchrony induces stability in cellular automata based models. In: Proceedings of Artificial Life IV, pp. 382–387. MIT Press, Cambridge (1994)
7. Buvel, R., Ingerson, T.: Structure in asynchronous cellular automata. Physica D 1, 59–68 (1984)
8. Schönfisch, B., de Roos, A.: Synchronous and asynchronous updating in cellular automata. BioSystems 51, 123–143 (1999)
9. Fatès, N., Morvan, M., Schabanel, N., Thierry, E.: Fully asynchronous behaviour of double-quiescent elementary cellular automata. TCS 362, 1–16 (2006)
10. Fatès, N., Regnault, D., Schabanel, N., Thierry, E.: Asynchronous behaviour of double-quiescent elementary cellular automata. In: Correa, J.R., Hevia, A., Kiwi, M. (eds.) LATIN 2006. LNCS, vol. 3887. Springer, Heidelberg (2006)
11. Fukś, H.: Non-deterministic density classification with diffusive probabilistic cellular automata. Phys. Rev. E 66(2) (2002)
12. Fukś, H.: Probabilistic cellular automata with conserved quantities. Nonlinearity 17(1), 159–173 (2004)
13. Balister, P., Bollobás, B., Kozma, R.: Large deviations for mean fields models of probabilistic cellular automata. Random Struct. & Alg. 29, 399–415 (2006)
14. McCoy, B., Wu, T.T.: The Two-Dimensional Ising Model. Harvard University Press (1974)
15. Rojas, R.: The Hopfield Model. In: Neural Networks: A Systematic Introduction, ch. 13. Springer, Heidelberg (1996)

Polycyclic and Bicyclic Valence Automata

Elaine Render and Mark Kambites*

School of Mathematics, University of Manchester,
Manchester M13 9PL, England
E.Render@maths.manchester.ac.uk
Mark.Kambites@manchester.ac.uk

Abstract. We study the classes of languages defined by valence automata with rational target sets (or equivalently, regular valence grammars with rational target sets), where the valence monoid is drawn from the important class of polycyclic monoids. We show that for polycyclic monoids of rank 2 or more, such automata accept exactly the context-free languages. For the polycyclic monoid of rank 1 (that is, the bicyclic monoid), they accept a class of languages strictly including the partially blind one-counter languages. Key to the proofs is a description of the rational subsets of polycyclic and bicyclic monoids, other consequences of which include the decidability of the rational subset membership problem for these monoids, and the closure of the class of rational subsets under intersection and complement.

1 Introduction

Both mathematicians and computer scientists have found applications for finite automata augmented with registers which store values from a given group or monoid, and are modified by multiplication. These automata, variously known as *extended finite automata*, *valence automata* or *M-automata*, provide an algebraic method to characterize important language classes such as the context-free, recursively enumerable and blind counter languages (see [6,8,10]). Their study provides insight into computational problems in algebra (see, for example, [9]). These automata are also closely related to *regulated rewriting systems*, and in particular the *valence grammars* introduced by Paun [11]: the languages accepted by *M*-automata are exactly the languages generated by regular *M*-valence grammars [5].

Traditionally, automata are considered in which the monoid registers are initialised to the identity element, and a word is accepted only if it can be read by a successful computation which results in the register being returned to the identity. Several authors have observed that the power of these automata to describe language classes may be increased by allowing a more general set of accepting values in the register. Fernau and Stiebe [4] began the systematic study of the resulting *valence automata with target sets*, along with the corresponding class of regulated grammars. In particular they considered the natural restriction that the target set be a *rational subset* of the register monoid.

---

* The research of the second author was supported by an RCUK Academic Fellowship.

C. Martín-Vide, F. Otto, and H. Fernau (Eds.): LATA 2008, LNCS 5196, pp. 464–475, 2008.

Of particular interest, when considering semigroups and monoids in relation to automata theory, is the class of *polycyclic monoids*. The polycyclic monoid of rank n is the natural algebraic model of a pushdown store on an n-letter alphabet. For M a polycyclic monoid of rank 2 or more, it is well-known that M-automata are equivalent to pushdown automata, and hence accept exactly the context-free languages. The polycyclic monoid of rank 1 is called the *bicyclic monoid* and usually denoted B; we shall see below that B-automata accept exactly the *partially blind one-counter languages* defined by Greibach [7].

One of the main objectives of this paper is to consider the class of languages accepted by polycyclic monoid valence automata with rational target sets. It transpires that, for M a polycyclic monoid of rank 2 or more, every language accepted by an M-automaton with rational target set is context-free, and hence is accepted by an M-automaton with target set $\{1\}$. In the rank 1 case the situation is rather different; a language accepted by a B-automaton with rational target set need not be a partially blind one-counter language, but it is always a finite union of languages, each of which is the concatenation of two partially blind one-counter languages.

A key element of the proofs is a simple but extremely useful characterisation of the rational subsets of polycyclic monoids (Corollary 6 below). From this we are easily able to derive a number of other consequences which may be of independent interest. These include the facts that the rational subsets of a finitely generated polycyclic monoid form a boolean algebra (with operations effectively computable), and that membership is uniformly decidable for rational subsets of polycyclic monoids.

2 Preliminaries

Firstly, we recall some basic ideas from formal language theory. Let Σ be a finite alphabet. Then we denote by Σ^* the set of all words over Σ and by ϵ the empty word. Under the operation of concatenation and with the neutral element ϵ, Σ^* forms a free monoid. A *finite automaton* over Σ^* is a finite directed graph with each edge labelled with an element of Σ^*, and with a distinguished initial vertex and a set of distinguished terminal vertices. A word $w \in \Sigma^*$ is accepted by the automaton if there exists some path connecting the initial vertex with some terminal vertex, the product of whose edge labels in order is w. The set of all words accepted by the automaton is denoted L or for an automaton A sometimes $L(A)$, and is called the *language* accepted by A. A language accepted by a finite automaton is called *rational* or *regular*.

More generally, if M is a monoid then a *finite automaton over* M is a finite directed graph with each edge labelled with an element of M, and with a distinguished initial vertex and a set of distinguished terminal vertices. An element $m \in M$ is accepted by the automaton if there exists some path connecting the initial vertex with some terminal vertex, the product in order of whose edge labels is m. The *subset accepted* is the set of all elements accepted; a subset of M which is accepted by some finite automaton is called a *rational subset*. The rational subsets of M are exactly the homomorphic images in M of regular languages.

We now recall the definition of a finite valence automaton, or M-automaton. Let M be a monoid with identity 1 and let Σ be an alphabet. An *M-valence automaton* (or M-automaton for short) over Σ is a finite automaton over the direct product $M \times \Sigma^*$. We say that it accepts a word $w \in \Sigma^*$ if it accepts $(1, w)$, that is if there exists a path connecting the initial vertex to some terminal vertex labelled $(1, w)$.

Intuitively, we visualize an M-automaton as a finite automaton augmented with a memory register which can store an element of M; the register is initialized to the identity element, is modified by right multiplication by elements of M, and for a word to be accepted the element present in the memory register on completion must be the identity element. We write $F_1(M)$ for the class of all languages accepted by M-automata, or equivalently for the class of languages generated by regular M-valence grammars [5]. More generally, an *M-automaton with (rational) target set* is an M-valence automaton together with a (rational) subset $X \subseteq M$. A word $w \in \Sigma^*$ is accepted by such an automaton if it accepts (x, w) for some $x \in X$. We denote by $F_{Rat}(M)$ the family of languages accepted by M-automata with rational target sets. We recall the following result from [4].

Theorem 1 (Fernau and Stiebe). *Let G be a group. Then $F_{Rat}(G) = F_1(G)$.*

3   Automata, Transductions and Closure Properties

In this section we study the relationship between rational transductions and M-automata with target sets. Consider a finite automaton over the direct product $\Omega^* \times \Sigma^*$. We call an automaton of this type a *rational transducer* from Ω to Σ; it recognises a relation $R \subseteq \Omega^* \times \Sigma^*$ called a *rational transduction*. The image of a language $L \subseteq \Omega^*$ under the relation R is the set of $y \in \Sigma^*$ such that $(x, y) \in R$ for some $x \in L$. We say that a language K is a *rational transduction of* a language L if K is the image of L under some rational transduction. The following is a straightforward generalisation of a well-known observation concerning M-automata (see for example [8, Proposition 2]); the proof, which proceeds similarly, can be found in [12].

Proposition 1. *Let X be a subset of a monoid M, and let $L \subseteq \Sigma^*$ be a regular language. Then the following are equivalent:*

(i) L is accepted by an M-automaton with target set X;

(ii) there exists a finite alphabet Ω and a morphism $\omega : \Omega^* \to M$ such that L is a rational transduction of $X\omega^{-1}$.

If M is finitely generated then the following condition is also equivalent to those above.

(iii) for every finite choice of generators $\omega : \Omega^* \to M$ for M, L is a rational transduction of $X\omega^{-1}$.

In particular, Proposition 1 gives a characterisation in terms of rational subsets and transductions of each class of languages accepted by M-automata with rational target sets.

Proposition 2. *Let M be a monoid and $L \subseteq \Sigma^*$ a language. Then the following are equivalent.*

(i) $L \in F_{Rat}(M)$;
(ii) there exists a finite alphabet Ω, a morphism $\omega : \Omega^* \to M$ and a rational subset $X \subseteq M$ such that L is a rational transduction of $X\omega^{-1}$.

If M is finitely generated then the following condition is also equivalent to those above.

(iii) there exists a rational subset $X \subseteq M$ such that for every finite choice of generators $\omega : \Omega^* \to M$ for M, L is a rational transduction of $X\omega^{-1}$.

Recall that a *rational cone* (also known as a *full trio*) is a family of languages closed under rational transduction, or equivalently under morphism, inverse morphism, and intersection with regular languages [1, Section V.2]. Since rational transductions are closed under composition [1, Theorem III.4.4] we have the following immediate corollary.

Corollary 1. *$F_{Rat}(M)$ is a rational cone. In particular, it is closed under morphism, inverse morphism, intersection with regular languages, and (since it contains a non-empty language) union with regular languages.*

Our next objective is to show that adjoining a zero to a monoid M makes no difference to the families of languages accepted either by M-automata or by M-automata with rational target sets. Recall that if M is a monoid, the result of *adjoining a zero* to M is the monoid M^0 with set of elements $M \cup \{0\}$ where 0 is a new symbol not in M, and multiplication is such that if $s, t \in M$ then st is equal to the usual product st over M, and zero otherwise.

We begin with the M-automaton case, where the required result is a very simple observation.

Proposition 3. *Let M be a monoid. Then $F_1(M^0) = F_1(M)$.*

Proof. That $F_1(M) \subseteq F_1(M^0)$ is immediate, so we need only prove the converse. Suppose $L \in F_1(M^0)$, and let A be an M^0-automaton accepting L. Clearly any path in A containing an edge with first label component 0 will itself have first label component 0; thus, no accepting path in A can contain such an edge. It follows that by removing all edges whose label has first component 0, we obtain a new M^0-automaton B accepting the language L. But now since M is a submonoid of M^0, B can be interpreted as an M-automaton accepting L, so that $L \in F_1(M)$ as required. □

Next we establish the corresponding result for M-automata with rational target sets, which is a little more involved.

Theorem 2. *Let M be a monoid. Then $F_{Rat}(M^0) = F_{Rat}(M)$.*

Proof. That $F_{Rat}(M) \subseteq F_{Rat}(M^0)$ is immediate. For the converse, suppose $L \in F_{Rat}(M^0)$. Then we may choose an M^0-automaton A accepting L with rational target set $X \subseteq M$.

Let L_0 be the language of words $w \subseteq \Sigma^*$ such that $(0, w)$ labels a path from the initial vertex to a terminal vertex. Let L_1 be the set of words w such that (m, w) labels a path from the initial vertex to a terminal vertex for some $m \in X \setminus \{0\}$. Clearly either $L = L_0 \cup L_1$ (in the case that $0 \in X$) or $L = L_1$ (if $0 \notin X$). We claim that L_0 is regular and $L_1 \in F_{Rat}(M)$. By Corollary 1 this will suffice to complete the proof.

The argument to show that $L_1 \in F_{Rat}(M)$ is very similar to the proof of Proposition 3. We construct from the M^0-automaton A a new M-automaton B by simply removing each edge with label of the form $(0, m)$. The new automaton B has target set $X \setminus \{0\}$. It is straightforward to show that B accepts exactly the language L_1.

It remains to show that L_0 is regular. Let Q be the vertex set of the automaton A, and let $Q_0 = \{q_0 \mid q \in Q\}$ and $Q_1 = \{q_1 \mid q \in Q\}$ be disjoint copies of Q. We define from A a finite automaton C with

- vertex set $Q_0 \cup Q_1$;
- for each edge in A from p to q with label of the form (m, x)
  - an edge from p_0 to q_0 labelled x and
  - an edge from p_1 to q_1 labelled x;
- for each edge in A from p to q with label of the form $(0, x)$
  - an edge from p_0 to q_1 labelled x and
  - an edge from p_1 to q_1 labelled x;
- initial vertex q_0 where q is the initial vertex of A; and
- terminal vertices q_1 whenever q is a terminal vertex of A.

A simple argument (see [12]) shows that C accepts exactly the language L_0. $\square$

Combining Theorem 2 with the result of Fernau and Stiebe [4] mentioned above (Theorem 1) gives us the following immediate corollary.

Corollary 2. *Let G be a group. Then*

$$F_{Rat}(G^0) = F_{Rat}(G) = F_1(G) = F_1(G^0).$$

4   Polycyclic Monoids

In this section we study the language classes $F_1(M)$ and $F_{Rat}(M)$, where M is drawn from the class of *polycyclic monoids*, which form the natural algebraic models of pushdown stores. In the process, we obtain a number of results about rational subsets of these monoids which may be of independent interest.

Let X be a set. Recall that the *polycyclic monoid* on X is the monoid $P(X)$ generated, under the operation of relational composition, by the partial bijections of the form

$$p_x : X^* \to X^*, \quad w \mapsto wx$$

and

$$q_x : X^*x \to X^*, \quad wx \mapsto w.$$

The monoid $P(X)$ models a pushdown store on the alphabet X, with p_x and q_x corresponding to the operations of *pushing* x and *popping* x (where defined) respectively, and composition to performing these operations in sequence. For a more detailed introduction see [8].

Clearly for any $x \in X$, the composition $p_x q_x$ is the identity map, while if $x, y \in X$ with $x \neq y$ then $p_x q_y$ is the *empty map* which constitutes a zero element in $P(X)$. In the case $|X| = 1$, say $X = \{x\}$, $P(X)$ is called the *bicyclic monoid*, and denoted B. The partial bijections p_x and q_x alone (which we shall often denote just p and q) do not generate the empty map, and so the bicyclic monoid has no zero; to avoid treating it as a special case we write $P^0(X)$ for the union of $P(X)$ with the empty map; thus $P^0(X) = P(X)$ if $|X| \geq 2$ but $P^0(X)$ is isomorphic to B^0 if $|X| = 1$.

Let $P_X = \{p_x \mid x \in X\}$, $Q_X = \{q_x \mid x \in X\}$, and z be a new symbol not in $P_X \cup Q_X$ which will represent the zero element. Let $\Sigma_X = P_X \cup Q_X \cup \{z\}$. Then there is an obvious morphism $\sigma : \Sigma_X^* \to P^0(X)$ and indeed $P^0(X)$ admits the monoid presentation

$$P^0(X) = \langle \Sigma_X \mid p_x q_x = 1, p_x q_y = z,$$
$$z p_x = z q_x = p_x z = q_x z = z z = z \text{ for all } x, y \in X, \ x \neq y \rangle.$$

For $|X| \geq 2$, a $P(X)$-automaton is equivalent to a pushdown automaton with stack alphabet X, so that the language class $F_1(P(X))$ is exactly the class of context-free languages [6,8]. In the one-symbol case, a B-automaton is easily seen to be equivalent to a *partially blind one-counter automaton* of the type introduced and studied by Greibach [7], so that $F_1(B)$ is the class of partially blind one-counter languages. Indeed more generally, $F_1(B^n)$ is the class of partially blind n-counter languages; see [12] for a more detailed explanation.

We now turn our attention to the classes $F_{Rat}(P(X))$ of languages accepted by polycyclic monoid automata with rational target sets. For $|X| \geq 2$, it transpires that every language accepted by a $P(X)$-automaton with rational target set is accepted by a $P(X)$-automaton, and hence that $F_{Rat}(P(X))$ is the class of context-free languages. In order to prove this, we shall need some results about rational subsets of polycyclic monoids, which we establish using techniques from string rewriting theory.

Recall that a *monadic rewriting system* Λ over an alphabet Σ is a subset of $\Sigma^* \times \{\Sigma \cup \{\epsilon\}\}$. We normally write an element $(w, x) \in \Lambda$ as $w \to x$. Then we write $u \Rightarrow v$ if $u = rws \in \Sigma^*$ and $v = rxs \in \Sigma^*$ with $w \to x$. Denote by $\Rightarrow^*$ the transitive, reflexive closure of the relation $\Rightarrow$. If $u \Rightarrow^* v$ we say that u is an *ancestor* of v under Λ and v is a *descendant* of u under Λ; we write $L\Lambda$ for the set of all descendants of words in L. It is well-known that if L is regular then $L\Lambda$ is again a regular language; if moreover the rewriting system Λ is finite, a finite automaton recognising $L\Lambda$ can be effectively computed from a finite automaton recognising L. For more information on such systems see [2,3].

Theorem 3. *Let X be a finite alphabet and R a rational subset of $P^0(X)$. Then there exists a regular language $L \subseteq Q_X^* P_X^* \cup \{z\}$ such that $L\sigma = R$. Moreover,* *there is an algorithm which, given an automaton recognizing a regular language $G \subseteq \Sigma_X^*$, constructs an automaton recognising a language $L \subseteq Q_X^* P_X^* \cup \{z\}$ with $L\sigma = G\sigma$.*

Proof. Since R is rational, there exists a regular language $K \subseteq \Sigma_X^*$ such that $K\sigma = R$. We define a confluent monadic rewriting system Λ on Σ_X^* with the following rules:

$$p_x q_x \to \epsilon, \quad p_x q_y \to z, \qquad z q_x \to z, \quad p_x z \to z,$$
$$z p_x \to z, \quad q_x z \to z, \quad zz \to z$$

for all $x, y \in X$ with $x \neq y$. Notice that the language of Λ-irreducible words is exactly $Q_X^* P_X^* \cup \{z\}$. With this in mind, we define $L = K\Lambda \cap (Q_X^* P_X^* \cup \{z\})$. Certainly L is regular, and moreover an automaton for L can be effectively computed from an automaton for K. Thus, it will suffice to show that $L\sigma = R$.

By definition $L\sigma \subseteq (K\Lambda)\sigma$, and since the rewriting rules are all relations satisfied in $P^0(X)$, $(K\Lambda)\sigma \subseteq K\sigma = R$. Conversely, if $s \in R$ then $s = w\sigma$ for some $w \in K$. Now the rules of Λ are all length-reducing, so w must clearly have an irreducible descendant, say w'. But now $w' \in L$ and $w'\sigma = w\sigma = s$ so that $s \in L\sigma$. Thus, $L\sigma = R$ as required.   □

As a corollary, we obtain a corresponding result for bicyclic monoids.

Corollary 3. *Let R be a rational subset of a bicyclic monoid B, and $\sigma : \{p, q\}^* \to B$ be the natural morphism. Then there exists a regular language $L \subseteq q^* p^*$ such that $L\sigma = R$. Moreover, there is an algorithm which, given an automaton recognizing a regular language $G \subseteq \{p, q\}^*$, constructs an automaton recognising a language $L \subseteq q^* p^*$ with $L\sigma = G\sigma$.*

Before proceeding to apply the theorem to polycyclic monoid automata with target sets, we note some general consequences of Theorem 3 for rational subsets of polycyclic monoids. Recall that a collection of subsets of a given *base set* is a *boolean algebra* if it is closed under union, intersection and complement within the base set.

Corollary 4. *The rational subsets of any finitely generated polycyclic monoid form a boolean algebra. Moreover, the operations of union, intersection and complement are effectively computable.*

Proof. The set of rational subsets of a monoid is always (effectively) closed under union, as a simple consequence of non-determinism. Since intersection can be described in terms of union and complement, it suffices to show that the rational subsets of polycyclic monoids are closed (effectively) under complement. To this end, suppose first that R is a rational subset of a finitely generated polycyclic monoid $P(X)$ with $|X| \geq 2$. Then by Theorem 3, there is a regular language $L \subseteq (Q_X^* P_X^* \cup \{z\})$ such that $L\sigma = R$. Let $K = (Q_X^* P_X^* \cup \{z\}) \setminus L$. Then K is regular and, since $Q_X^* P_X^* \cup \{z\}$ contains a unique representative for every element of $P(X)$, it is readily verified that $K\sigma = P(X) \setminus (L\sigma)$. Thus, $P(X) \setminus (L\sigma)$ is a rational subset of $P(X)$, as required.

For effective computation of complements, observe that given an automaton recognizing a language $R = \Sigma_X^*$, we can by Theorem 3 construct an automaton recognizing a regular language $L \subseteq (Q_X^* P_X^* \cup \{z\})$ with $L\sigma = R\sigma$. Clearly we can then compute the complement $K = (Q_X^* P_X^* \cup \{z\}) \setminus L$ of L in $(Q_X^* P_X^* \cup \{z\})$, and since $K\sigma = P(X) \setminus (L\sigma)$, this suffices.

In the case that $|X| = 1$, the statement can be proved in a similar way but using Corollary 3 in place of Theorem 3. $\square$

Recall that the *rational subset membership problem* for a monoid M is the algorithmic problem of deciding, given a rational subset of M (specified using an automaton over a fixed generating alphabet) and an element of M (specified as a word over the same generating alphabet), whether the given element belongs to the given subset. The decidability of this problem is well-known to be independent of the chosen generating set [9, Corollary 3.4]. As another corollary, we obtain the decidability of this problem for finitely generated polycyclic monoids.

Corollary 5. *Finitely generated polycyclic monoids have decidable rational subset membership problem.*

Proof. Let $|X| \geq 2$ [respectively, $|X| = 1$]. Suppose we are given a rational subset R of $P(X)$ (specified as an automaton over Σ_X^* [respectively $\{p,q\}^*$]) and an element w (specified as a word in the appropriate alphabet). Clearly, we can compute $\{w\}$ as a regular language. Now by Corollary 4 we can compute a regular language $K \subseteq \Sigma_X^*$ [respectively, $\{p,q\}^*$] such that $K\sigma = R \cap \{w\}\sigma$. But $w\sigma \in R$ if and only if $R \cap \{w\sigma\}$ is non-empty, that is, if and only if K is non-empty. Since emptiness of regular languages is decidable, this completes the proof. $\square$

Corollary 6. *Let R be a rational subset of $P^0(X)$ and suppose that $0 \notin R$. Then there exists an integer n and regular languages $Q_1, \ldots, Q_n \subseteq Q_X^*$ and $P_1, \ldots, P_n \subseteq P_X^*$ such that*

$$R = \bigcup_{i=1}^{n} (Q_i P_i)\,\sigma.$$

Proof. By Theorem 3, there is a regular language $L \subseteq Q_X^* P_X^*$ such that $L\sigma = R$. Let A be a finite automaton accepting L, with vertices numbered $1, \ldots, n$. Suppose without loss of generality that the edges in A are labelled by single letters from $Q_X \cup P_X$. For each i let Q_i be the set of all words in Q_X^* which label paths from the initial vertex to vertex i. Similarly, let P_i be the set of all words in P_X^* which label words from vertex i to a terminal vertex.

Now if $w \in Q_i P_i$ then $w = uv$ where $u \in Q_X^*$ labels a path from the initial vertex to vertex i, and $v \in P_X^*$ labels a path from vertex i to a terminal vertex. Hence $uv = w$ labels a path from the initial vertex to a terminal vertex, and so $w \in L$. Conversely, if $w \in L \subseteq Q_X^* P_X^*$ then w admits a factorisation $w = uv$ where $u \in Q_X^*$ and $v \in P_X^*$. Since the edge labels in A are single letters, an accepting path for w must consist of a path from the initial vertex to some vertex i labelled u, followed by a path from i to a terminal vertex labelled v. It follows that $u \in Q_i$ and $v \in P_i$, so that $w \in Q_i P_i$. Thus we have

$$L = \bigcup_{i=1}^{n} Q_i P_i \quad \text{and so} \quad R = L\sigma = \left(\bigcup_{i=1}^{n} Q_i P_i \right) \sigma = \bigcup_{i=1}^{n} (Q_i P_i)\sigma$$

as required.   □

For the next proposition, we shall need some notation. For a word $q = q_{x_1} q_{x_2} \cdots q_{x_n} \in Q_X^*$, we let $q' = p_{x_n} \cdots p_{x_2} p_{x_1} \in P_X^*$. Similarly for a word $p = p_{x_1} p_{x_2} \cdots p_{x_n} \in Q_X^*$, we let $p' = q_{x_n} \cdots q_{x_2} q_{x_1} \in Q_X^*$. Note that $p'' = p$ and $q'' = q$. Note also that $p'\sigma$ is the unique right inverse of $p\sigma$, and $q'\sigma$ is the unique left inverse of $q\sigma$.

Proposition 4. *Let $u \in \Sigma_X^*$, and let $q \in Q_X^*$ and $p \in P_X^*$. Then $u\sigma = (qp)\sigma$ if and only if there exists a factorisation $u = u_1 u_2$ such that $(q'u_1)\sigma = 1 = (u_2 p')\sigma$.*

Proof. Suppose first that $u\sigma = (qp)\sigma$. Let Λ be the monadic rewriting system defined in the proof of Theorem 3. Then u is reduced by Λ to qp. Notice that the only rules in Λ which can be applied to words not representing zero remove factors representing the identity; it follows easily that u admits a factorisation $u = u_1 u_2$ where $u_1 \sigma = q\sigma$ and $u_2 \sigma = p\sigma$. Now we have

$$(q'u_1)\sigma = (q'\sigma)(u_1\sigma) = (q'\sigma)(q\sigma) = 1$$

and symmetrically $(u_2 p')\sigma = 1$ as required.

Conversely, $q\sigma$ is the unique right inverse of $q'\sigma$, so if $(q'u_1)\sigma = (q'\sigma)(u_1\sigma) = 1$ then we must have $u_1 \sigma = q\sigma$. Similarly, if $(u_2 p')\sigma = 1$ then $u_2 \sigma = p\sigma$, and so we deduce that $u\sigma = (u_1 u_2)\sigma = (qp)\sigma$ as required.   □

We are now ready to prove our main theorem about M-automata with rational target sets where M is a polycyclic monoid.

Theorem 4. *Suppose $L \in F_{Rat}(P^0(X))$. Then L is a finite union of languages, each of which is in $F_1(P^0(X))$ or $F_1(P^0(X))^2$.*

Proof. Let $M = P^0(X)$ and let C be an M-automaton with rational target set R accepting the language L. By Corollary 6 there exists an integer n and regular languages $Q_1, \ldots, Q_n \subseteq Q_X^*$ and $P_1, \ldots, P_n \subseteq P_X^*$ such that

$$R = R_0 \cup \bigcup_{i=1}^{n} (Q_i P_i)\sigma.$$

where either $R_0 = \emptyset$ or $R_0 = \{0\}$ depending on whether $0 \in R$. For $1 \leq i \leq n$, we let $R_i = (Q_i P_i)\sigma$. It follows easily that we can write $L = L_0 \cup L_1 \cup \cdots \cup L_n$ where each L_i is accepted by an M-automaton with target set R_i. Clearly it suffices to show that each L_i is a finite union of languages, each of which is the concatenation of at most two languages in $F_1(M)$.

We begin with L_0. Let $Z = \{u \in \Sigma_X^* \mid u\sigma = 0\}$ and $W = \{w \in \Sigma_X^* \mid w\sigma = 1\}$. It is easily seen (for example, by considering the rewriting system Λ from the proof of Theorem 3) that $u \in Z$ if and only if either u contains the letter z, or u factorizes as $u_1 p_x u_2 q_y u_3$ where $x, y \in X$, $x \neq y$ and $u_1, u_2, u_3 \in \Sigma_X^*$ are such that u_2 represents the identity, that is, such that $u_2 \in W$. Thus,

$$Z = \Sigma_X^* \{z\} \Sigma_X^* \cup \bigcup_{x,y \in X, x \neq y} \Sigma_X^* \{p_x\} W \{q_y\} \Sigma_X^*.$$

From this expression it is a routine matter to show that Z is a rational transduction of W. By Proposition 1, L_0 is a rational transduction of the language Z. Since the class of rational transductions is closed under composition, it follows that L_0 is a rational transduction of W, and hence by Proposition 1 that $L_0 \in F_1(M)$, as required.

We now turn our attention to the languages L_i for $i \geq 1$. Recall that L_i is accepted by an M-automaton with target set $R_i = (Q_i P_i)\sigma$. Let

$$P_i' = \{(p', \epsilon) \mid p \in P_i\} \subseteq Q_X^* \times \Sigma^*$$

and similarly

$$Q_i' = \{(q', \epsilon) \mid q \in Q_i\} \subseteq P_X^* \times \Sigma^*.$$

It is readily verified that P_i' and Q_i' are rational subsets of $\Sigma_X^* \times \Sigma^*$; let A_P and A_Q be finite automata accepting P_i' and Q_i' respectively, and assume without loss of generality that the first component of every edge label is either a single letter in Σ_X or the empty word ϵ.

By Proposition 1 there is a rational transduction $\rho \subseteq \Sigma_X^* \times \Sigma^*$ such that $w \in L_i$ if and only if $(u, w) \in \rho$ for some $u \in \Sigma_X^*$ such that $u\sigma \in R_i$. Let A be an automaton recognizing ρ, again with the property that the first component of every edge label is either a single letter in Σ_X or the empty word ϵ. We construct a new M-automaton B with

- vertex set the disjoint union of the state sets of A_Q, A, and A_P;
- all the edges of A_Q, A and A_P;
- initial vertex the initial vertex of A_Q;
- terminal vertices the terminal vertices of A_P;
- an extra edge, labelled (ϵ, ϵ), from each terminal vertex of A_Q to the initial vertex of A; and
- an extra edge labelled (ϵ, ϵ), from each terminal vertex of A to the initial vertex of A_P.

It is immediate that B recognizes the relation

$$\tau = Q_i' \rho P_i' = \{(q' x p', w) \mid q \in Q_i, p \in P_i, (x, w) \in \rho\} \subseteq \Sigma_X^* \times \Sigma^*$$

and again has the property that the first component of every edge label is either a single letter or the empty word.

Let Q be the vertex set of A, viewed as a subset of the vertex set of B. For each vertex $y \in Q$, we let K_y be the language of all words w such that (u, w)

labels a path in B from the initial vertex of B to y for some u with $u\sigma = 1$. By considering B as an transducer but with terminal vertex y, we see that K_y is a rational transduction of the word problem of $P(X)$, and hence by Proposition 1 lies in the class $F_1(P(X))$.

Dually, we let L_y be the language of all words w such that (u, w) labels a path in B from y to a terminal vertex for some u with $u\sigma = 1$. This time by considering B as a transducer but with initial vertex y, we see that L_y is also a rational transduction of the word problem of $P(X)$, and hence also lies in $F_1(P(X))$.

We claim that

$$L_i = \bigcup_{y \in Q} K_y L_y,$$

which will clearly suffice to complete the proof.

Suppose first that $w \in L_i$. Then there exists a word $u \in \Sigma_X^*$ such that $u\sigma \in R_i$ and that $(u, w) \in \rho$. Since $R_i = (Q_i P_i)\sigma$ we have $u\sigma = (qp)\sigma$ for some $q \in Q_i$ and $p \in P_i$. Note that $(q'up', w) \in \tau$ is accepted by B. By Proposition 4, u admits a factorization $u = u_1 u_2$ such that $(q'u_1)\sigma = 1$ and $(u_2p')\sigma = 1$. Now in view of our assumption on the edge labels of B, w must admit a factorization $w = w_1 w_2$ such that B has a path from the initial vertex to some vertex y labelled $(q'u_1, w_1)$ and a path from y to a terminal vertex labelled (u_2p', w_2); moreover, the vertex y can clearly be assumed to lie in Q. Since $(q'u_1)\sigma = 1 = (u_2p')\sigma$, it follows that $w_1 \in K_y$ and $w_2 \in L_y$ so that $w = w_1 w_2 \in K_y L_y$, as required.

Conversely, suppose $y \in Q$ and that $w = w_1 w_2$ where $w_1 \in K_y$ and $w_2 \in L_y$. Then B has a path from the initial vertex to vertex y labelled (u_1, w_1) and a path from the vertex y to a terminal vertex labelled (u_2, w_2) for some u_1 and u_2 with $u_1\sigma = u_2\sigma = 1$. Since $y \in Q$, it follows from the definition of B that $u_1 = q'v_1$ and $u_2 = v_2p'$ for some $q \in Q_i$ and $p \in P_i$ and v_1 and v_2 such that $(v_1v_2, w) \in \rho$. But now $(q'v_1)\sigma = u_1\sigma = 1$ and $(v_2p')\sigma = u_2\sigma = 1$, so by definition of q' and p' we deduce that $v_1\sigma = q\sigma$ and $v_2\sigma = p\sigma$. But then $(v_1v_2)\sigma = (qp)\sigma \in R_i \subseteq R$ and $(v_1v_2, w) \in \rho$, from which it follows that $w \in L_i$ as required.

Thus, we have written L as a finite union of languages L_i where each L_i either lies in $F_1(M)$ (in the case $i = 0$) or is a finite union of concatenations of two languages in $F_1(M)$. This completes the proof. $\square$

In the case that $|X| \geq 2$, we have $P^0(X) = P(X)$ and $F_1(P(X))$ is the class of context-free languages, which is closed under both finite union and concatenation. Hence, we obtain the following easy consequence.

Theorem 5. *If $|X| \geq 2$ then $F_{Rat}(P(X))$ is the class of context-free languages.*

In the case $|X| = 1$, we have that $P^0(X)$ is isomorphic to the bicyclic monoid $B = P(X)$ with a zero adjoined. Combining Theorem 4 with Proposition 3 and Theorem 2 we thus obtain.

Corollary 7. *Every language in $F_{Rat}(B)$ is a finite union of languages, each of which is in either $F_1(B)$ or $F_1(B)^2$.*

Since the class $F_1(B)$ of partially blind one-counter languages is not closed under concatenation, however, we cannot here conclude that $F_{Rat}(B) = F_1(B)$. Indeed, the following result, a full proof of which can be found in [12], shows that this is not the case.

Theorem 6. *The language $\{a^i b^i a^j b^j \mid i, j \geq 0\}$ lies in $F_{Rat}(B)$ but not in $F_1(B)$.*

It is possible, however, to describe concatenations of partially blind one-counter languages using partially blind two-counter automata. Indeed more generally we have the following simple proposition, a proof of which can again be found in [12].

Proposition 5. *Let M_1 and M_2 be monoids and L_1 and L_2 languages over the same alphabet. If $L_1 \in F_1(M_1)$ and $L_2 \in F_1(M_2)$ then $L_1 L_2 \in F_1(M_1 \times M_2)$.*

Since classes of the form $F_1(M)$ are closed under union, Theorem 4 and Proposition 5 combine to give the following inclusion.

Corollary 8. $F_{Rat}(B) \subseteq F_1(B^2)$.

References

1. Berstel, J.: Transductions and Context-Free Languages. Teubner Studienbucher, Stuttgart (1979)
2. Book, R.V., Jantzen, M., Wrathall, C.: Monadic Thue systems. Theoretical Computer Science 19, 231–251 (1982)
3. Book, R.V., Otto, F.: String rewriting systems. Springer, New York (1993)
4. Fernau, H., Stiebe, R.: Valence grammars with target sets. In: Yu, S., Ito, M. (eds.) Words, Semigroups and Transductions, pp. 129–140. World Scientific, Singapore (2001)
5. Fernau, H., Stiebe, R.: Sequential grammars and automata with valences. Theoretical Computer Science 276, 377–405 (2002)
6. Gilman, R.H.: Formal languages and infinite groups. In: Geometric and Computational Perspectives on Infinite Groups, Minneapolis, MN and New Brunswick, NJ, 1994. DIMACS Series Discrete Mathematics and Theoretical Computer Science, Providence R, vol. 25. American Mathematical Society (1996)
7. Greibach, S.A.: Remarks on blind and partially blind one-way multicounter machines. Theoretical Computer Science 7(3), 311–324 (1978)
8. Kambites, M.: Formal languages and groups as memory. Communications in Algebra (to appear)
9. Kambites, M., Silva, P.V., Steinberg, B.: On the rational subset problem for groups. J. Algebra 309(2), 622–639 (2007)
10. Mitrana, V., Stiebe, R.: Extended finite automata over groups. Discrete Applied Mathematics 108(3), 247–260 (2001)
11. Paun, G.: A new generative device: valence grammars. Rev. Roumaine Math. Pures Appl. XXV(6), 911–924 (1980)
12. Render, E., Kambites, M.: Rational subsets of polycyclic monoids and valence automata. arXiv:0710.3711v1 [math.RA]

Length Codes, Products of Languages and Primality

Arto Salomaa[1], Kai Salomaa[2], and Sheng Yu[3]

[1] Turku Centre for Computer Science, Joukahaisenkatu 3-5 B, 20520 Turku, Finland
asalomaa@utu.fi
[2] School of Computing, Queen's University, Kingston, Ontario, Canada K7L 3N6
ksalomaa@cs.queensu.ca
[3] Department of Computer Science, University of Western Ontario, London, Ontario,
Canada N6A 5B7
syu@csd.uwo.ca

Abstract. We continue the investigation, [9,11,5,1,2], of representing a
language as a catenation of languages, each of which cannot be further
decomposed in a nontrivial fashion. We study such prime decompositions,
both finite and infinite ones. The notion of a length code, an extension of
the notion of a code leads to general results concerning decompositions
of star languages. Special emphasis is on the decomposition of regular
languages. Also some open problems are mentioned.

Keywords: Catenation of languages, language decomposition, prime
decomposition, length code, star language.

1 Introduction

Products or catenations of languages, viewed as subsets of the free monoid,
are needed in many applications. However, because of the noncommutativity,
many phenomena are not yet properly understood. For instance, although the
condition for the commutation of two words can be explicitly stated, the equation
$L_1 L_2 = L_2 L_1$ for languages presents various difficulties. (See [11,8] and their
references.)

This paper investigates products of languages where the individual factors L
cannot be decomposed further in a nontrivial way, that is, presented in the form
$L = L_1 L_2$, where neither of the languages L_1 and L_2 consists of the empty word
alone. The initial work on such "prime decompositions", [11,9], concentrated
mainly on finite languages. Then a prime decomposition can always be found,
although it is not necessarily unique. It is decidable whether or not a regular
language is prime but the complexity of this problem is not known.

Various cases were considered in [5] where a language has a prime decomposition consisting of infinitely many factors but none with a finite number of
factors. This paper continues the investigation of such finitary and infinitary
prime decompositions. A generalization of the notion of a code, a *length code*,
will be a useful tool in this investigation. In this paper we are mainly concerned

C. Martín-Vide, F. Otto, and H. Fernau (Eds.): LATA 2008, LNCS 5196, pp. 476–486, 2008.

with regular languages. As will be seen below, it makes a big difference whether or not the prime factors of a regular language are also required to be regular.

There has been much work recently concerning orthogonal (unambiguous) catenation. (See, for instance, [3].) Studies similar to those in this paper could be carried out also with respect to orthogonal catenation, instead of ordinary catenation.

A brief outline about the contents of this paper follows. The next section discusses various possibilities of defining a prime decomposition and presents examples of the different cases. More explicit comparisons are made in Section 3. The notion of a *length code* is introduced and its basic properties are discussed in Section 4, and its connection with prime decompositions of star languages is shown. Techniques for obtaining prime decompositions, beyond those using length codes, are presented in Section 5. In case of regular languages, the techniques lead to nonregular factors. Possibilities of actually getting regular factors are discussed in Section 6.

2 Different Types of Prime Factorizations

We assume that the reader is familiar with the basics of formal languages. Whenever necessary, [10] may be consulted. As customary, we use small letters from the beginning of the English alphabet a, b, c, d, possibly with indices, to denote letters of our formal alphabet Σ. Words are usually denoted by small letters from the end of the English alphabet. The empty word is denoted by ε. Following the regular expression notation, we sometimes denote the union by "+" and singleton sets $\{\alpha\}$ simply by α. Thus, $\varepsilon + ab$ stands for the set $\{\varepsilon\} \cup \{ab\}$. The following definition, [9], contains the basic notions of this paper.

Definition 1. *A nonempty language L has a nontrivial decomposition if, for some $L_1 \neq \{\varepsilon\}$ and $L_2 \neq \{\varepsilon\}$, we have $L = L_1 L_2$. A nonempty language $L \neq \{\varepsilon\}$ having no nontrivial decomposition is* prime. *A language L has a* prime factorization *(or a* prime decomposition*) if*

$$L = L_1 \dots L_m, \ m \geq 1,$$

where each of the languages L_i, $1 \leq i \leq m$, is prime.

Observe that by a "prime factorization" without further specifications we mean a *finite* factorization. Another definition is given below for infinitary prime factorizations.

Products of subsets of the free monoid are not yet very well understood. One can also visualize different ways of defining a "prime" language. For instance, in [4] a language L is termed *indecomposable* if the equation $L = L_1 L_2$ implies that either $L = L_1$ or $L = L_2$. It is clear that if a language is prime in the sense of Definition 1, then it is also indecomposable. On the other hand, languages ϕ and $\{\varepsilon\}$ are indecomposable but not prime. We do not know other indecomposable languages that are not prime.

It is obvious that every finite language has a prime factorization. It is not necessarily unique, for instance,

$$(\varepsilon + a^2 + a^3)(\varepsilon + a^3 + a^4) = (\varepsilon + a^2)(\varepsilon + a^3 + a^4 + a^5),$$

where it is easy to verify that all four factors listed are indeed prime. A prefix-free regular language has a *unique* prime factorization if it is additionally required that the factors are regular prefix-free languages, [2,7], but infix-free regular languages do not possess the analogous property, [6]. Decompositions of factorial languages, that is, languages closed under the subword operation are investigated in [1,4].

The notion of a *strongly prime decomposable* language was introduced in [5]. A language L is strongly prime decomposable if, for some integer t, any decomposition of L contains at most t nontrivial factors. When L is strongly prime decomposable, any way of iteratively decomposing L has to stop after a finite number of steps, i.e., the refinement of any decomposition results in a prime decomposition in a finite number of steps. In this case we say also that L has a *strong finitary prime decomposition*.

The language Σ^* possesses many nontrivial decompositions. Some of them are prime factorizations, as shown in [5]. Indeed, consider any nonempty language $H \subseteq \Sigma^n$, $n \geq 1$. (Thus, possibly $H = \Sigma$.) Then

$$H^* = (\varepsilon + H)(\varepsilon + H(H^2)^*),$$

where both factors on the right side are prime.

It is also shown in [5] that the language

$$H_d = \varepsilon + \{a^{i_1}b^{i_1}a^{i_2}b^{i_2}\ldots a^{i_k}b^{i_k} | k \geq 1,\ 1 \leq i_1 < i_2 < \ldots < i_k\}$$

does not have a prime factorization. (The language H_d is not context-free but its complement is context-free.) This reflects the fact that Definition 1 requires the prime factorization to be finitary. If this requirement is relaxed, we can write

$$H_d = \prod_{i=1}^{\infty}(\varepsilon + a^i b^i),$$

where each factor is prime. Moreover, this infinitary prime factorization of H_d is *unique* in the sense of Definition 2 given below. On the other hand, the language H^* considered above has many infinitary prime factorizations. For instance, any infinite product of factors $(\varepsilon + w)$, where w runs through all nonempty words of H^*, constitutes such a factorization. Thus, the language H^* possesses both a prime factorization and (nondenumerably) many infinitary prime factorizations.

Some clarifying remarks are in order. When we consider infinite products $\prod_{i=1}^{\infty} L_i$, where each L_i is a language, we consider only finite words defined by the product. Then we also assume that each L_i contains the empty word. Indeed, an infinite product of languages defines finite words only if all of these languages, with at most finitely many exceptions, contain the empty word. In this case there is a language K and an integer $m \geq 1$ such that the original product can be written as

$$\prod_{i=1}^{\infty} L_i = K \prod_{i=m}^{\infty} L_i,$$

where each language in the product on the right side contains the empty word. If every language in the product $\prod_{i=1}^{\infty} L_i$ contains the empty word, then each word in the product belongs to a finite prefix of the product.

In the following definition we assume that each of the languages L_i and K_i *properly* contains the empty word.

Definition 2. *A language L has a* unique infinitary prime factorization *if $L = \prod_{i=1}^{\infty} L_i$, where each L_i is prime and, whenever $L = \prod_{i=1}^{\infty} K_i$, where each K_i is prime, then $L_i = K_i$, for all i. If L is over a one-letter alphabet, it is only required that the languages K_i are the languages L_i in some order.*

Since languages over one letter are commutative, the relaxation of uniqueness given in the definition is very natural. It is not difficult to see that the infinitary prime factorization given above for the language H_d is unique. A language can have both a prime factorization in the sense of Definition 1 and an infinitary prime factorization. For instance, as seen above, Σ^* has a prime factorization and also an infinitary prime factorization

$$\Sigma^* = \prod_{w} (\varepsilon + w),$$

where w runs through all nonempty words over Σ.

3 Comparisons between Different Notions

We now summarize the many notions of prime factorizations discussed above. Thus, a prime factorization (or decomposition) of a language can be

1. finitary,
2. unique finitary,
3. strong finitary,
4. infinitary,
5. unique infinitary.

Altogether this gives numerous possibilities for a language L: each of the properties 1)-5) can be present or absent. Many combinations are excluded by definition. For instance, if L has property 2) (resp. 5)), it surely possesses also property 1) (resp. 4)). For most of the 32 subsets of the properties 1)-5), it is easy to give an example of a language possessing each of the properties in that subset but none of the others, or show that no such language exists. We now present a few examples, some of them from [5].

The languages L_1, L_2, and L_3 defined below are over the one-letter alphabet $\{a\}$. Thus, for them uniqueness is up to the order of the factors.

The language L_1 consists of all words a^n such that every number 1 in the binary representation of n occurs in an even position, counted from the right. Hence, the six shortest words in L are

$$\varepsilon, \ a^2, \ a^8, \ a^{10}, \ a^{32}, \ a^{34}.$$

The sets $\{\varepsilon, a^{2^{2n+1}}\}$, $n \geq 0$, constitute the collection of prime languages appearing in any decomposition of L_1. This means that L_1 has no finitary but a unique infinitary prime factorization. These properties are also shared by the language L_2 defined as follows. It consists of all words a^i such that the binary representation of i is in the regular language

$$(00000 + 00001 + 00010 + 00100 + 01000)^*.$$

But now the factors in the prime factorization are of cardinality 5, instead of the cardinality 2 in L_1.

The language L_3 consists of all words a^i such that the binary representation of i is in the regular language

$$(000 + 010 + 100 + 110 + 011 + 101 + 111)^*.$$

Then L_3 has no finitary prime factorization but nondenumerably many infinitary ones.

Clearly, the language $L_i b$, $1 \leq i \leq 3$ has no prime factorization at all in any of our five senses.

We already noticed that Σ^* has both a finitary and an infinitary prime factorization, in fact, infinitely many of both. The following is a basic open question.

Open problem. Can a language have both a finitary prime factorization and a *unique* infinitary one?

4   Length Codes

We now introduce a notion useful in considerations about prime factorizations of finite languages. We believe that the notion is also important on its own right.

Definition 3. *A language L is a* length code *if every equation*

$$u_1 \ldots u_k = v_1 \ldots v_l, \ u_i, v_j \in L, \ 1 \leq i \leq k, \ 1 \leq j \leq l,$$

implies that $k = l$.

Clearly, every *code* (see [10]) is also a length code. The converse does not hold true. For instance, the language $\{a, ab, ba\}$ is not a code but it is a length code. This can be seen as follows.

Consider an equation as in Definition 3. We may assume that

$$u_1 = a, \ v_1 = ab$$

because, otherwise, the equation can be shortened or is not valid. Consequently, $u_2 = ba$, which implies that $v_2 = a$ (leads to a shortening) or $v_2 = ab$ (leads to a loop). Therefore, the only possibility is the equation

$$(ab)^i a = a(ba)^i, \ i \geq 0.$$

The notions of a code and a length code can be defined for morphisms as well. Then one can also speak of a *Parikh code*. By definition, a morphism $h : \Sigma^* \to \Delta^*$ is a code (resp. Parikh code, length code) if the equation $h(x) = h(y)$ always implies the equation $x = y$ (resp. $\Psi(x) = \Psi(y)$, $|x| = |y|$). (Here $\Psi(x)$ denotes the Parikh vector of x.) It follows that the set of codes is included in the set of Parikh codes which, in turn, is included in the set of length codes. It is also easy to see that both inclusions are proper.

Sets $H \subseteq \Sigma^n$ considered in Section 2 are codes and, hence, also length codes. We are now ready to generalize the prime decomposition of H^* (considered above) to concern all finite length codes.

Theorem 1. *If L is finite and a length code, then L^* has a prime decomposition consisting of regular factors.*

Proof. It follows by the assumption that the empty word is not in L. Clearly,

$$L^* = (\varepsilon + L)(\varepsilon + L(L^2)^*).$$

The first factor on the right side is finite and, hence, has a prime decomposition. The second factor is itself prime. This follows because its decompositions must have the form

$$\varepsilon + L(L^2)^* = (\varepsilon + H_1)(\varepsilon + H_2), \ H_1, H_2 \subseteq L(L^2)^*,$$

and, consequently, each word in $H_1 \cup H_2$ is a product of an odd number of words in L. Since L is a length code, any catenation of a word in H_1 and a word in H_2 is a product of an even number of words in L, which is impossible. Hence, one of the sets H_1 and H_2 is empty and, consequently, the left side is prime. This concludes the proof. Observe that, in the proof, the finiteness of L is *not* required in showing that the language $\varepsilon + L(L^2)^*$ is prime.   □

Languages over the unary alphabet of cardinality at least two are not length codes. The language

$$K = \{ab, aba, bab\} \tag{1}$$

constitutes a simple example over the binary alphabet. Indeed,

$$ababab = (ab)(ab)(ab) = (aba)(bab)$$

and, thus, the same word equals the product of both two and three words of K.

Observe that a language L is not a length code exactly in case, for some i and j, $i \neq j$, we have $L^i \cap L^j \neq \emptyset$. For instance, $K^2 \cap K^3 \neq \emptyset$. In general, the minimal difference between i and j can be arbitrarily large. For any $t \geq 2$, there is a finite language F_t that is not a length code but $F_t^i \cap F_t^j = \emptyset$ wherever $1 \leq |i - j| < t$. For instance, this holds for

$$F_t = \{(ab)^{3t}, aba, bab\}.$$

We now consider the problem of how long words we have to test in order to find out that a finite language F is a length code. We already noticed that there is no upper bound, independent of F, for the minimal difference between the number of factors i and j in two decompositions. On the other hand, there is an upper bound for both i and j depending on F (in the sense made precise below).

By a *proper suffix* of a word w we mean a suffix of w different from w and the empty word. By m_F we denote the length of the longest word(s) in F, by S_F the set of all proper suffixes of the words in F, and s_F the cardinality of S_F. (Thus, a suffix appearing in several words of F is counted only once.) Finally, we denote $c_F = m_F s_F$.

Theorem 2. *Assume that F is a finite language (not containing ε) such that the inequality $F^i \cap F^j \neq \emptyset$ holds for some i and j, $i \neq j$. Then, for some i_1 and j_1, $i_1 \neq j_1$,*

$$F^{i_1} \cap F^{j_1} \neq \emptyset, \quad i_1 \leq 2c_F, \quad j_1 \leq 2c_F.$$

Proof. By the assumption, we have

$$u_1 \cdots u_i = v_1 \cdots v_j, \quad i \neq j, \qquad (*)$$

where each of the u-words and v-words is in F. We assume that this equation is *minimal* with respect to i and j, that is, there are no i' and j', $i' \neq j'$, $i' < i$, $j' < j$, such that $x_1 \cdots x_{i'} = y_1 \cdots y_{j'}$, where each of the x-words and y-words is in F. To complete the proof we have to show that, under this minimality assumption, neither one of the indices i and j exceeds $2c_F$.

If $u_1 = v_1$ then $u_2 \cdots u_i = v_2 \cdots v_j$, which contradicts the minimality assumption. Thus, one of the words is a proper prefix of the other, say, $v_1 = u_1 s_1$, $s_1 \neq \varepsilon$. We say that the suffix $s_1 \in S_F$ *appears* and that u_1 and v_1 are the u-words and v-words *so far listed*. We now list u-words until we obtain the first one, say u_α, such that $|u_1 \cdots u_\alpha| > |v_1|$. Then we list v-words until we obtain the first one, say v_β, such that $|v_1 \cdots v_\beta| > |u_1 \cdots u_\alpha|$. Consequently,

$$v_1 \cdots v_\beta = u_1 \cdots u_\alpha s_2, \quad s_2 \in S_F,$$

and we say that the suffix s_2 *appears*. The process is continued: list u-words (respectively v-words) until the listed u-part exceeds in length the so far listed v-part (respectively u-part). In this way we get also a sequence $s_1, s_2, \ldots$ of appearing suffixes.

Assume now that one of original indices i and j exceeds $2c_F$. We will show that this contradicts minimality.

When we have listed more than c_F u-words or v-words, some suffix appears at least twice. Similarly, when we have listed more than $2c_F$ of them, some suffix s' appears at least three times. Consequently, there are indices

$$\alpha_1 < \alpha_2 < \alpha_3 \text{ and } \beta_1 < \beta_2 < \beta_3$$

such that

$$u_1 \cdots u_{\alpha_t} s' = v_1 \cdots v_{\beta_t}, \quad 1 \le t \le 3.$$

This means that we can remove the factor $u_{\alpha_1+1} \cdots u_{\alpha_2}$ from the left and the factor $v_{\beta_1+1} \cdots v_{\beta_2}$ from the right side of the equation $(*)$. This contradicts minimality, unless the number of factors is the same on both sides of the new equation. (This happens if $\alpha_2 - \alpha_1 - \beta_2 + \beta_1 = i - j$.) In this case we either remove also the factors $u_{\alpha_2+1} \cdots u_{\alpha_3}$ and $v_{\beta_2+1} \cdots v_{\beta_3}$, or else replace the former two factors by the latter two factors and remove the latter two factors from their original positions. In every case the new equation contradicts minimality and has a different number of factors on its two sides.   □

Theorem 2 gives an algorithm for deciding whether or not a given finite language is a length code. It is an open problem to construct an efficient algorithm.

The argument is simpler and the upper bound smaller if there is no comparison between the number of factors on the two sides of the equation. By an *F-factorization* of a word w we mean an equation $w = u_1 \cdots u_i$, where each of the words on the right side is in F. The proof of the following result follows the lines of the preceding proof.

Theorem 3. *Assume that a finite language F not containing the empty word is not a code. Then some word in F^{c_F} has two different F-factorizations.*

In the sequel we will be quite much concerned with languages such as K^*. We will see that K^* possesses a prime decomposition of two factors but a technique very different from the one in Theorem 1 is needed.

5   Techniques for Star Languages

We now consider some special techniques for obtaining prime decompositions for star languages. It turns out that the resulting factors are sometimes "strange" in comparison with the original language. We begin with the following simple result.

Theorem 4. *If a language L is prime then, for every nonempty word $w \in L$, there is a word $w' \in L$ (resp. $w'' \in L$) such that $ww' \notin L$ (resp. $w''w \notin L$).*

Proof. Assume the contrary: no such word w' exists for a nonempty word $w \in L$. This means that $ww' \in L$, for all words $w' \in L$. Consequently, $(\varepsilon+w)L = L$, which shows that L is not prime. If no w'' exists, we obtain similarly $L(\varepsilon + w) = L$.   □

The *converse* of Theorem 4 is not valid. For instance, the language

$$L = \varepsilon + \{ab^{2i+2}a, ab^{2j+1}a, ab^{2i+2}a^2b^{2j+1}a | i, j \ge 0\}$$

satisfies the condition of Theorem 4. Indeed, for every nonempty $w \in L$, we have $w^2 \notin L$. However, L is not prime because

$$L = (\varepsilon + \{ab^{2i+2}a | i \ge 0\})(\varepsilon + \{ab^{2j+1}a | j \ge 0\}).$$

We will now establish a general result about prime decompositions of star languages. The language K^* considered above will have a prime decomposition consisting of two languages.

Let F be a finite language not containing the empty word. We say that a word $x \in F$ is *independent* if no other word of F appears as a prefix of x, and x itself is not a prefix of any other word of F. Clearly, if a longest (resp. shortest) word in F has no proper prefix in F (resp. is not a proper prefix of any word in F), then it is independent.

Theorem 5. *If F is a finite language containing an independent word, then F^* has a prime decomposition consisting of two factors.*

The proof is omitted due to the page limit.

The language K considered above contains the independent word bab. Thus, K^* has a prime decomposition into two factors.

It seems likely that Theorem 5 remains valid without the assumption concerning the independent word. The simultaneous parsing of a word into 2α and α words in F seems always to lead into a contradiction.

If we consider arbitrary (finite) prime decompositions, instead of ones with only two factors, we obtain the result without the assumption concerning the independent word.

Theorem 6. *If F is finite, then the language F^* has a prime decomposition.*

The proof is omitted due to the page limit.

We *conjecture* that every regular language has a prime decomposition. This result was established in [5] for regular languages over a one-letter alphabet. Theorem 6 constitutes a step for a possible proof in the general case. As such our constructions do not work for arbitrary star languages because then the length arguments fail.

6   On Regular Prime Decompositions of K^* and Related Languages

We consider, finally, possible prime decompositions of K^* and related languages, where K is as in (1). That is, we consider the star of a language that is not a length code. It seems very likely that, although K^* has a prime decomposition of two factors, it still does not have any *regular* prime decomposition (i.e., one where the factors are regular), not even an infinitary one. Thus, there would be regular languages having a prime decomposition but having no prime decomposition consisting of regular languages.

Any prime decomposition of K^* is of the form

$$K^* = (\varepsilon + H_1)(\varepsilon + H_2)\ldots(\varepsilon + H_n),$$

where the languages H_i, $1 \le i \le n$, are contained in K^* and do not contain the empty word. Moreover, at least one of them has to be infinite.

The most immediate decompositions for star languages (also in use in Theorem 1) are of the form

$$L^* = (\varepsilon + L + \ldots L^{m-1})(\varepsilon + L^m(L^m)^*), \ m \geq 2.$$

However, the following result shows that no decomposition of this type can lead to a prime decomposition of K^*.

Lemma 1. *An infinite union*

$$K_1 = \varepsilon + \bigcup_{j=1}^{\infty} K^{i_j}, \ 1 \leq i_1 < i_2 < \ldots,$$

is not prime, provided there is a bound B such that $i_{j+1} - i_j \leq B$, for all $j \geq 1$.

Proof. Choose the smallest r such that $i_r \geq 3B$, and consider the word

$$w = (ab)^{i_r} = (ab)^{3B}(ab)^{\nu},$$

where $\nu \geq 0$. Let $w' \in K_1$ be arbitrary. Clearly,

$$ww' \in K^{3B+\nu+i_s},$$

for some i_s. (If $w' = \varepsilon$, then $i_s = 0$.) Each of the B factors $(ab)(ab)(ab)$ constituting the prefix of w can be parsed also $(aba)(bab)$, which shows that

$$ww' \in K^{3B+\nu+i_s-\mu}, \text{ for all } \mu, \ 0 \leq \mu \leq B.$$

By the assumption concerning B, there is a μ, $0 \leq \mu \leq B$, such that $3B+\nu+i_s-\mu$ is one of the exponents i_j. This shows that $ww' \in K_1$. Since $w' \in K_1$ was arbitrary, we conclude that

$$(\varepsilon + w)K_1 = K_1.$$

This shows that K_1 is not prime.   □

As regards regular languages, the result of Lemma 1 can be presented in the following form.

Lemma 2. *No regular language $R \subseteq K^*$, containing the empty word and infinitely many powers of K, is prime.*

Proof. Since R is regular, there is a bound B between the exponents of consecutive powers of K in R, as in Lemma 1. In addition, R may contain "loose" words that do not belong to any full power of K contained in R. The argument in the proof of Lemma 1 remains valid, with K_1 replaced by R.   □

According to Lemma 2, in any regular prime decomposition of K^*, finitary or infinitary, every factor contains only finitely many full powers of K.

Instead of K, we can start with any finite language L that is not a length code. (Since L does not necessarily contain an independent word, we do not get a prime decomposition of two factors but have to use Theorem 6.) We obtain, thus, the following summarizing result.

Theorem 7. *If L is finite and a length code, then L^* has a prime decomposition consisting of two regular factors. If L is finite but not a length code, then in any regular prime decomposition of L^*, finitary or infinitary, every factor contains only finitely many full powers of L. However, L^* has a (finitary) prime decomposition.*

7   Conclusion

The notion of a length code is interesting and seems to be applicable in various contexts. We hope to return to a further study of it. Some of the basic problems concerning products and primality of languages are challenging. As we have seen, one of such problems deals with the prime decompositions of regular languages.

References

1. Avgustinovich, S.V., Frid, A.: A unique decomposition theorem for factorial languages. Internat. J. Alg. Comp. 15, 149–160 (2005)
2. Czyzowicz, J., Fraczak, W., Pelc, A., Rytter, W.: Linear-time prime decompositions of regular prefix codes. Internat. J. Found. Comp. Sci. 14, 1019–1031 (2003)
3. Daley, M., Domaratzki, M., Salomaa, K.: On the operational orthogonality of languages. In: Proc. Workshop on Language Equations, vol. 44, pp. 43–53. TUCS General Publications (2007)
4. Frid, A.: Commutation in binary factorial languages. In: Harju, T., Karhumäki, J., Lepistö, A. (eds.) DLT 2007. LNCS, vol. 4588, pp. 193–204. Springer, Heidelberg (2007)
5. Han, Y.-S., Salomaa, A., Salomaa, K., Wood, D., Yu, S.: Prime decompositions of regular languages. Theor. Comp. Sci. 376, 60–69 (2007)
6. Han, Y.-S., Wang, Y., Wood, D.: Infix-free regular expressions and languages. Internat. J. Found. Comp. Sci. (to appear)
7. Han, Y.-S., Wood, D.: The generalization of generalized automata: Expression automata. In: Domaratzki, M., Okhotin, A., Salomaa, K., Yu, S. (eds.) CIAA 2004. LNCS, vol. 3317, pp. 156–166. Springer, Heidelberg (2005)
8. Kunc, M.: The power of commuting with finite sets of words. Theory Comput. Syst. 40, 521–551 (2007)
9. Mateescu, A., Salomaa, A., Yu, S.: Factorizations of languages and commutativity conditions. Acta Cybernetica 15, 339–351 (2002)
10. Rozenberg, G., Salomaa, A. (eds.): Handbook of Formal Languages, pp. 1–3. Springer, Berlin (1997)
11. Salomaa, A., Yu, S.: On the decomposition of finite languages. Developments in Language Theory. In: DLT 1999, pp. 22–31. World Scientific Publ.Co., Singapore (2000)

An Efficient Algorithm for the Inclusion Problem of a Subclass of DPDAs*

Ryo Yoshinaka

Hokkaido University
ry@ist.hokudai.ac.jp

Abstract. This paper presents an efficient algorithm solving the inclusion problem of a new subclass of context-free languages. The languages are accepted by the special kind of real-time deterministic pushdown automata, called *strongly forward-deterministic pushdown automata*, that go to the same state and push the same sequence of stack symbols whenever transition is allowed on the same input symbol. Our algorithm can be applied to efficient identification in the limit of that class from positive data.

1 Introduction

Decidability and/or complexity of inclusion problems, which are generally harder than equivalence problems or non-emptiness problems, is a fundamental issue in formal language theory. While the inclusion of two regular languages is trivially decidable, not many nontrivial classes of context-free languages are known to admit an inclusion checking algorithm and rather many negative results have been obtained on this issue (cf. e.g. Asveld and Nijholt[1]). For instance, it is undecidable even for simple deterministic context-free languages[2] or unambiguous sequential linear languages[1]. On the other hand, Linna[5] has shown that inclusion is decidable for languages that are accepted by final state and empty stack by *stack-uniform* DPDAs. Greibach and Friedman [4] have presented an algorithm which decides whether $L \subseteq L'$ for an arbitrary context-free language L and the language L' accepted by final state and empty stack by a *superdeterministic* PDA. The language class defined by Greibach and Friedman is a superclass of the one by Linna. While both algorithms run in time doubly exponential in the description size of the automata, some subclasses of them are known to admit a more efficient inclusion checking algorithm. Wakatsuki and Tomita[8] have given an inclusion checking algorithm for *very simple languages*. Based on their algorithm, Yoshinaka [11] has presented an algorithm that decides the inclusion between an arbitrary context-free language and a *length-uniform simple language*, which is accepted by empty stack by a stateless stack-uniform DPDA. Those algorithms

---

* This research was partially supported by a grant from the Global COE Program, "Center for Next-Generation Information Technology based on Knowledge Discovery and Knowledge Federation", from the Ministry of Education, Culture, Sports, Science and Technology of Japan.

C. Martín-Vide, F. Otto, and H. Fernau (Eds.): LATA 2008, LNCS 5196, pp. 487–498, 2008.

run in single exponential time. Actually the latter work is motivated for using the algorithm as a subroutine of an efficient algorithm learning a restricted kind of length-uniform simple languages, called *right-unique simple languages*, from positive data. A most standard strategy for identifying languages in the limit from positive data is to output an automaton that accepts a minimal language among automata consistent with the given data. For this purpose, an efficient inclusion checking algorithm would be useful, if one gets a finite number of reasonable consistent automata to be considered from the given data. Actually when the inclusion checking algorithm for length-uniform simple languages is used as a subroutine of the learning algorithm for right-unique simple languages [11], it runs in polynomial time in the size of the data given to the learning algorithm.

Along this line, this paper introduces another subclass of stack-uniform DPDAs, called *strongly forward-deterministic PDAs* (SFDPDAs) and presents an algorithm deciding the inclusion between a context-free language and the language accepted by an SFDPDA. Our algorithm is as efficient as the inclusion checking algorithm for length-uniform simple languages. Although no polynomial-time upper bound in the description size of the automata is given, it runs in polynomial time when it is used as a subroutine of a reasonable learning algorithm for the class of SFD-PDAs. Therefore our algorithm provides a means for efficient learning of languages accepted by SFDPDAs. While the class of languages accepted by SFDPDAs is incomparable with that of length-uniform simple languages, it contains very simple languages, right-unique simple languages and languages accepted by *Szilard strict restricted one-counter automata*, all of which are known to be efficiently identifiable in the limit from positive data[9,10,11,7]. We however do not go into the detail of learning in this paper; we concentrate on the inclusion problem. Apart from learning, our algorithm runs in single exponential time, which gives an improvement to the result by Greibach and Friedman [4] for the special cases.

After preliminaries in Section 2, Section 3 gives a formal definition and some important properties of SFDPDAs, which are the target of this paper. Our algorithm for the inclusion problem of SFDPDAs is given in Section 4. Then we conclude the paper with Section 5.

2   Preliminaries

$\mathbb{Z}$ and $\mathbb{N}$ denote the set of integers and the set of nonnegative integers, respectively. ε is the empty sequence and $\varnothing$ is the empty set. $|X|$ denotes the length of the sequence X or the cardinality of the set X depending on the context. For a set X, X^* denotes the set of finite (possibly empty) sequences of elements of X, $X^+ = X^* - \{\varepsilon\}$, $X^k = \{ x \in X^* \mid |x| = k \}$ and $\mathcal{P}(X)$ denotes the power set of the set X. For a sequence X of length at least k, $\mathrm{Pref}_k(X)$ denotes the prefix of X of length k.

This paper extends some relations and operations on sets to sequences of sets as follows:

- $x_1 \ldots x_m \in X_1 \ldots X_m$ iff $x_i \in X_i$ for all $i \in \{1, \ldots, m\}$,
- $X_1 \ldots X_m \subseteq Y_1 \ldots Y_m$ iff $X_i \subseteq Y_i$ for all $i \in \{1, \ldots, m\}$,

- $X_1 \ldots X_m \cap Y_1 \ldots Y_m$ denotes the sequence $(X_1 \cap Y_1) \ldots (X_m \cap Y_m)$,
- $X_1 \ldots X_m \cup Y_1 \ldots Y_m$ denotes the sequence $(X_1 \cup Y_1) \ldots (X_m \cup Y_m)$.

Note that a sequence of sets cannot be identified with the Cartesian product of the sets if some of the sets are empty.

A *context-free grammar* (CFG) is denoted by a quadruple $G = \langle N, \Sigma, P, S \rangle$, where N is the finite set of *nonterminal symbols*, Σ is the finite set of *terminal symbols*, P is the finite set of *production rules* and $S \in N$ is the *start symbol*. A production rule in P has the form $A \to \beta$ for some $A \in N$ and $\beta \in (N \cup \Sigma)^*$. If $A \to \beta \in P$, we write $\alpha A \gamma \Rightarrow_G \alpha \beta \gamma$ for any $\alpha, \gamma \in (N \cup \Sigma)^*$. $\Rightarrow_G^*$ is the reflexive and transitive closure of $\Rightarrow_G$. The subscript G of $\Rightarrow_G$ is omitted if it is understood from the context. The language $L(G)$ defined by G is the set $L(G, S)$, where $L(G, \alpha) = \{ w \in \Sigma^* \mid \alpha \overset{*}{\Rightarrow} w \}$ for $\alpha \in (N \cup \Sigma)^*$. The description size of G is defined as $\|G\| = \sum_{A \to \alpha \in P} (1 + |\alpha|)$.

A *pushdown automaton* (PDA) is denoted by $M = \langle Q, \Gamma, \Sigma, \Delta, q_0, Z_0, F \rangle$ where Γ is the set of *stack symbols*, Q is the set of *states*, Σ is the set of *input symbols*, Δ is the set of *transition rules*, $q_0 \in Q$ is the *initial state*, $Z_0 \in \Gamma$ is the *initial stack* and $F \subseteq Q$ is the set of *final states*. Each transition rule in Δ has the form $\langle p, X \rangle \overset{a}{\to} \langle q, \zeta \rangle$ for $p, q \in Q$, $X \in \Gamma$, $a \in \Sigma \cup \{\varepsilon\}$, $\zeta \in \Gamma^*$. If $\langle p, X \rangle \overset{a}{\to} \langle q, \zeta \rangle$ is a rule of M, we have $\langle p, X\eta \rangle \overset{a}{\underset{M}{\to}} \langle q, \zeta\eta \rangle$ for any $\eta \in \Gamma^*$. If the automaton is clear from the context, we omit the subscript M. We also write $\langle p_0, \zeta_0 \rangle \xrightarrow{a_1 \ldots a_n} \langle p_n, \zeta_n \rangle$ if $\langle p_{i-1}, \zeta_{i-1} \rangle \overset{a_i}{\to} \langle p_i, \zeta_i \rangle$ for all $i \in \{1, \ldots, n\}$. There are several slightly different manners to define the language of a PDA. This paper defines it as $L(M) = \{ w \in \Sigma^* \mid \langle q_0, Z_0 \rangle \overset{w}{\to} \langle p, \varepsilon \rangle$ with $p \in F \}$ (accept by final state and empty stack). The description size of M is defined as $\|M\| = \sum_{\langle p, X \rangle \overset{a}{\to} \langle q, \zeta \rangle \in \Delta} (4 + |\zeta|)$.

A *real-time deterministic pushdown automaton* (real-time DPDA) is a PDA such that whenever $\langle p, X \rangle \overset{a}{\to} \langle q_1, \zeta_1 \rangle$ and $\langle p, X \rangle \overset{a}{\to} \langle q_2, \zeta_2 \rangle$ are rules, it holds that $a \neq \varepsilon$, $q_1 = q_2$ and $\zeta_1 = \zeta_2$. Regarding Δ as a partial function from $Q \times \Gamma \times \Sigma$ to $Q \times \Gamma^*$, we write $\Delta(p, X, a) = \langle q, \zeta \rangle$ when the real-time DPDA has the rule $\langle p, X \rangle \overset{a}{\to} \langle q, \zeta \rangle$. We also write $\Delta(p, X, a) = \bot$ if there are no q, ζ such that $\Delta(p, X, a) = \langle q, \zeta \rangle$.

3   Strongly Forward-Deterministic Pushdown Automata

Definition 1. A *stack-uniform* DPDA (SUDPDA) (Linna [5]) is a real-time DPDA such that if $\langle p_1, X_1 \rangle \overset{a}{\to} \langle q_1, \zeta_1 \rangle$ and $\langle p_2, X_2 \rangle \overset{a}{\to} \langle q_2, \zeta_2 \rangle$ are rules, then $|\zeta_1| = |\zeta_2|$ for any $a \in \Sigma$.

A *strongly forward-deterministic pushdown automaton* (SFDPDA) is an SUD-PDA such that if $\langle p_1, X_1 \rangle \overset{a}{\to} \langle q_1, \zeta_1 \rangle$ and $\langle p_2, X_2 \rangle \overset{a}{\to} \langle q_2, \zeta_2 \rangle$ are rules, then $q_1 = q_2$ and $\zeta_1 = \zeta_2$ for any $a \in \Sigma$.

A *simple* DPDA is a real-time DPDA such that $Q = F = \{q_0\}$.

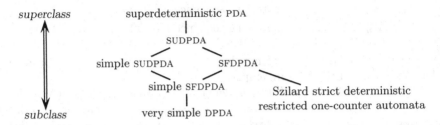

Fig. 1. Related classes of DPDAs to SFDPDAs

Example 1. Let an SFDPDA M consist of the following transition rules:

$$\Delta(q_0, Z_0, a) = \Delta(q_1, Z_0, a) = \langle q_1, Z_0 X \rangle, \ \Delta(q_0, Z_0, b) = \Delta(q_2, Z_0, b) = \langle q_2, Z_0 Y \rangle,$$
$$\Delta(q_1, Z_0, c) = \Delta(q_2, Z_0, c) = \langle q_1, \varepsilon \rangle, \quad \Delta(q_1, Z_0, d) = \langle q_2, \varepsilon \rangle,$$
$$\Delta(q_1, X, e) = \Delta(q_1, Y, e) = \Delta(q_2, Y, e) = \langle q_1, \varepsilon \rangle,$$
$$\Delta(q_2, X, f) = \Delta(q_1, Y, f) = \Delta(q_2, Y, f) = \langle q_2, \varepsilon \rangle,$$

and $F = \{q_1, q_2\}$. We have

$$L(M) = \{\, a^n c e^n \mid n \geq 1 \,\} \cup \{\, a^n d f^n \mid n \geq 1 \,\} \cup \{\, b^n c x \mid x \in \{e, f\}^n, n \geq 1 \,\}.$$

Yoshinaka[11] presented an efficient algorithm that identifies simple SFDPDAs in the limit from positive data, which has a sub-algorithm that decides inclusion of simple SUDPDAs[1]. There are other subclasses of SUDPDAs known to be polynomial-time identifiable in the limit from positive data: very simple DPDAs[9,10] and Szilard strict deterministic restricted one-counter automata[7]. The largest superclass of those classes that is known to admit an algorithm deciding the inclusion problem is of *superdeterministic* PDAs[4]. The relation between the classes of languages defined by those automata are shown in Figure 1.[2]

The following definition plays a very important role throughout this paper.

Definition 2. Let Σ be the set of input symbols. A function $\sharp$ from Σ^* to $\mathbb{Z}$ is called a *shape* if

- $\sharp(xy) = \sharp(x) + \sharp(y)$ for all $x, y \in \Sigma^*$ (homomorphism),
- $\sharp(a) \geq -1$ for all $a \in \Sigma$.

For an SUDPDA $M = \langle Q, \Gamma, \Sigma, \Delta, q_0, Z_0, F \rangle$, *the shape of M* denoted by $\sharp_M$ is the shape satisfying that

$$\sharp_M(a) = |\zeta| - 1 \quad \text{if } \Delta(p, X, a) = \langle q, \zeta \rangle.$$

---

[1] In[11], simple SFDPDAs and simple SUDPDAs are represented by the form of CFGs and called right-unique simple grammars and length-uniform simple grammars, respectively.

[2] Not every SUDPDA is a superdeterministic PDA, but any language accepted by an SUDPDA is also accepted by some superdeterministic PDA[3,4].

A function $\flat_M$ mapping Σ^* to $\mathbb{N}$ is defined as follows:

$$\flat_M(x) = \begin{cases} 0 & \text{if } x = \varepsilon, \\ \max\{\, 1 - \sharp_M(x') \mid x' \text{ is a proper prefix of } x \,\} & \text{if } x \in \Sigma^+. \end{cases}$$

Lemma 1. *Let M be an* SUDPDA. *If $\langle p, \zeta \rangle \xrightarrow{x} \langle q, \eta \rangle$, then*

(1) $|\zeta| \geq \flat_M(x)$ and $|\eta| = |\zeta| + \sharp_M(x)$,
(2) $\langle p, \operatorname{Pref}_{\flat_M(x)}(\zeta) \rangle \xrightarrow{x} \langle q, \operatorname{Pref}_{\flat_M(x) + \sharp_M(x)}(\eta) \rangle$.

In particular for $w \in L(M)$, $\sharp_M(w) = -1$ and $\flat_M(w) = 1$.

We say that a shape $\sharp$ is *compatible with a language L* if $\sharp(w) = -1$ and $\flat(w) = 1$ for all $w \in L$ where $\flat$ is defined from $\sharp$ as $\flat_M$ is defined from $\sharp_M$. Therefore, it is necessary that $\sharp_M$ is compatible with L for $L \subseteq L(M)$.

SFDPDAs have the following strong property. When $|x| = 1$ in the following lemma, it turns to the definition of an SFDPDA.

Lemma 2 (strong forward determinacy). *Let M be an* SFDPDA. *For any $x \in \Sigma^+$, if $\langle p_1, \zeta_1 \rangle \xrightarrow{x} \langle q_1, \eta_1 \rangle$ and $\langle p_2, \zeta_2 \rangle \xrightarrow{x} \langle q_2, \eta_2 \rangle$ with $|\zeta_1| = |\zeta_2| = \flat_M(x)$, then we have $q_1 = q_2$ and $\eta_1 = \eta_2$.*

4   Inclusion Problem

In this section, we present an algorithm for the inclusion problem of SFDPDAs. More precisely, our algorithm decides whether $L(G) \subseteq L(M)$ for an SFDPDA M and an arbitrary CFG G. We note that the opposite relation, i.e., whether $L(M) \subseteq L(G)$, is undecidable by the standard result on CFGs. Our algorithm consists of two parts. The first part decides the compatibility of $\sharp_M$ with $L(G)$. Although the compatibility is no more than a necessary condition, the procedure computes some values defined when $\sharp_M$ is compatible with $L(G)$ that are necessary for the second part of the algorithm. In the second part, we check whether M has enough rules for including $L(G)$. Throughout this section, we fix a CFG $G = \langle N, \Sigma, P, S \rangle$ and an SFDPDA $M = \langle Q, \Gamma, \Sigma, \Delta, q_0, Z_0, F \rangle$. Recall that no SFDPDA accepts the empty string ε. Thus if $\varepsilon \in L(G)$, one can immediately conclude that $L(G) \not\subseteq L(M)$. Hence we may assume that G has no useless rules or ε-rules without loss of generality.

4.1   Checking the Compatibility of the Shape[3]

Suppose that $\sharp$ is compatible with $L(G)$. It is not hard to see that for every $A \in N$, there are integers n_A and m_A such that $\sharp(y) = n_A$ and $\flat(y) \leq m_A$ for all $y \in L(G, A)$. This fact entails that the following extensions $\tilde{\sharp}$ and $\tilde{\flat}$ of $\sharp$ and $\flat$, respectively, are well-defined.

---

[3] One may regard Tozawa and Minamide's algorithm for checking "shape-balancedness" [6] as a special case of our compatibility checking.

Definition 3. Let $\sharp$ be a shape compatible with $L(G)$. Functions $\tilde{\sharp}$ from $(N \cup \Sigma)^*$ to $\mathbb{Z}$ and $\tilde{\flat}$ from $(N \cup \Sigma)^*$ to $\mathbb{N}$ are defined as follows:

$$\tilde{\sharp}(\alpha) = \sharp(x) \qquad \text{for } x \in L(G, \alpha),$$

$$\tilde{\flat}(\alpha) = \max\{\,\flat(x) \mid x \in L(G, \alpha)\,\}.$$

In particular, we have $\tilde{\sharp}_M(S) = -1$ and $\tilde{\flat}_M(S) = 1$ if $\sharp_M$ is compatible with $L(G)$. The algorithm in Figure 2 decides whether $\sharp_M$ is compatible with $L(G)$ by trying to compute $\tilde{\sharp}_M(A)$ and $\tilde{\flat}_M(A)$ for all $A \in N$ at Stage 1.

Lemma 3. *The algorithm in Figure 2 goes into Stage 2 if and only if $\sharp_M$ is compatible with $L(G)$. In this case, $\tilde{\sharp}(A) = \tilde{\sharp}_M(A)$ and $\tilde{\flat}(A) = \tilde{\flat}_M(A)$ hold for all $A \in N$.*

If $\sharp_M$ is not compatible with $L(G)$, it is immediately concluded that $L(G) \not\subseteq L(M)$.

4.2   Comparison Forest

Hereafter we assume that $\sharp_M$ is compatible with $L(G)$ and simply write $\sharp$ and $\flat$ for $\tilde{\sharp}_M$ and $\tilde{\flat}_M$, respectively. Our algorithm tries to find the corresponding transition of M for each leftmost derivation of G. For this purpose, the algorithm constructs a collection of trees, called *the comparison forest* $\mathcal{F}$, which consists of trees T_A for all $A \in N$. T_A is the prefix tree (trie) of the set $\{\,\alpha \mid A \to \alpha \in P\,\}$, which represents leftmost derivations starting from A. To each node of T_A, $\mathcal{F}$ gives a label, which is related to the corresponding configurations of M. In other words, the comparison forest is a function $\mathcal{F}$ whose domain is

$$\mathrm{dom}(\mathcal{F}) = \{\,[A : \alpha] \mid A \to \alpha\beta \in P \text{ and } \alpha, \beta \in (N \cup \Sigma)^*\,\}.$$

Each element of $\mathrm{dom}(\mathcal{F})$ is called a *node*. The node $[A : \varepsilon]$ is *the root node* of T_A, and a node $[A : \alpha]$ is called a *final node* of T_A if $A \to \alpha \in P$. The value $\mathcal{F}([A : \alpha])$ is called the *label* of a node $[A : \alpha]$. For simplicity we write $\mathcal{F}(A : \alpha)$ instead of $\mathcal{F}([A : \alpha])$. The node $[A : \alpha]$ concerns leftmost derivations of G of the following form and the corresponding transition of M:

$$\left.\begin{array}{c} S \overset{*}{\underset{G}{\Rightarrow}} xA\gamma \Rightarrow x\alpha\beta\gamma \overset{*}{\Rightarrow} xy\beta\gamma \\[4pt] \langle q_0, Z_0 \rangle \xrightarrow[M]{x} \langle p, \zeta \rangle \xrightarrow{y} \langle q, \eta \rangle \end{array}\right\} \tag{$*$}$$

where $y \in L(G, \alpha)$. By Lemma 1 and Definition 3, here we must have

$$\langle p, \mathrm{Pref}_{\flat(A)}(\zeta) \rangle \xrightarrow[M]{y} \langle q, \mathrm{Pref}_{\flat(A)+\sharp(\alpha)}(\eta) \rangle.$$

The basic idea behind our algorithm is to find the corresponding transition of M to each derivation of G as in $(*)$ by putting $\langle p, \mathrm{Pref}_{\flat(A)}(\zeta) \rangle$ on the label of the root node of T_A and putting $\langle q, \mathrm{Pref}_{\flat(A)+\sharp(\alpha)}(\eta) \rangle$ on the label of the node

---

Input: a CFG $G = \langle N, \Sigma, P, S \rangle$ and an SFDPDA $M = \langle Q, \Gamma, \Sigma, \Delta, q_0, Z_0, F \rangle$;
Output: whether $L(G) \subseteq L(M)$?
Begin Algorithm
— *Stage 1. Compatibility checking* —
— *Stage 1.1. Compute* $\bar{\#}_M$ —
let $\bar{\#}(a) := \#_M(a)$ for all $a \in \Sigma$;
let $\bar{\#}(A)$ be undefined for all $A \in N$;
let every production rule in G be *unmarked*;
while there remains an *unmarked* rule **do**
  take some *unmarked* rule $A \to \alpha \in P$ such that $\bar{\#}(\alpha)$ is defined,
    where $\bar{\#}(B_1 \ldots B_m)$ is defined to be $\bar{\#}(B_1) + \cdots + \bar{\#}(B_m)$;
  if $\bar{\#}(A)$ is not defined yet **then** define $\bar{\#}(A) := \bar{\#}(\alpha)$;
   elseif $\bar{\#}(A) \neq \bar{\#}(\alpha)$ **then** output "No" and **halt**;
  fi
  mark the rule $A \to \alpha$;
  if $\bar{\#}(S)$ is defined and $\bar{\#}(S) \neq -1$ **then** output "No" and **halt**; **fi**
od
— *Stage 1.2. Compute* $\tilde{b}_M$ —
let $\bar{b}_0(A) := 1$ for all $A \in N \cup \Sigma$;
for $n = 0$ to $2|N|$ **do**
  let $\bar{b}_{n+1}(a) := 1$ for each $a \in \Sigma$;
  let $\bar{b}_{n+1}(A) := \max\{ -\bar{\#}(B_1 \ldots B_{k-1}) + \bar{b}_n(B_k) \mid A \to B_1 \ldots B_m \in P, 1 \leq k \leq m \}$
for each $A \in N$;
od
if $\bar{b}_{2|N|}(S) \neq 1$ or $\bar{b}_{2|N|+1} \neq \bar{b}_{2|N|}$ **then** output "No" and **halt**; **fi**
let $\bar{b} = \bar{b}_{2|N|}$;
— *Stage 2. Compute* $\mathcal{F}$ —
define $\mathcal{F}(A : \alpha) := \varnothing^{\max\{1, \bar{b}(A) + \bar{\#}(\alpha)\}}$ for all $[A : \alpha] \in \text{dom}(\mathcal{F})$;
define $\text{Start}(S) := \{\langle q_0, \overline{Z}_0 \rangle\}$;
until none of the following if-clauses is satisfied **do**
 if $\text{Final}(S) \not\subseteq F$ **then** output "No" and **halt**; **fi**
 if there is $[A : \alpha a] \in \text{dom}(\mathcal{F})$ with $a \in \Sigma$ such that
    $\langle q, X \rangle \in I_1$ or $\langle q, \overline{X} \rangle \in I_1$ for $\mathcal{F}(A : \alpha) = I_1 \ldots I_{\bar{b}(A) + \bar{\#}(\alpha)}$ and $\Delta(q, X, a) = \bot$
  then output "No" and **halt**;
 fi
 if there is $[A : \alpha B] \in \text{dom}(\mathcal{F})$ with $B \in N$ such that
   $\text{Pref}_{\bar{b}(B)}(\text{Bar}(\mathcal{F}(A : \alpha))) \not\subseteq \text{Start}(B)$
  then redefine $\text{Start}(B) := \text{Start}(B) \cup \text{Pref}_{\bar{b}(B)}(\text{Bar}(\mathcal{F}(A : \alpha)))$;
 fi
 if there is $[A : \alpha B] \in \text{dom}(\mathcal{F})$ with $B \in N \cup \Sigma$ such that
   $\mathcal{F}(A : \alpha B) \neq \text{Trans}(\mathcal{F}(A : \alpha), B)$
  then redefine $\mathcal{F}(A : \alpha B) := \text{Trans}(\mathcal{F}(A : \alpha), B)$;
 fi
od
output "YES" and **halt**;
End Algorithm

Fig. 2. Algorithm for the inclusion problem of SFDPDAS

$[A : \alpha]$ in a special format. Our actual algorithm defines the label $\mathcal{F}(A : \alpha)$ of a node $[A : \alpha]$ to be a sequence of sets such that

$$\mathcal{F}(A : \alpha) \in \begin{cases} \mathcal{P}(Q) & \text{if } \flat(A) + \sharp(\alpha) = 0, \\ \mathcal{P}(Q \times (\Gamma \cup \overline{\Gamma}))\mathcal{P}(\Gamma \cup \overline{\Gamma})^{\flat(A)+\sharp(\alpha)-1} & \text{otherwise,} \end{cases}$$

where $\overline{\Gamma} = \{ \overline{X} \mid X \in \Gamma \}$. Then necessary memory space for storing the label of a node $[A : \alpha]$ is bounded by $O((|Q| + \flat(A) + \sharp(\alpha))|\Gamma|)$. The label of the root node of T_A is always an element of $\mathcal{P}(Q \times \overline{\Gamma})\mathcal{P}(\overline{\Gamma})^{\flat(A)-1}$ and occurrences of elements of $\overline{\Gamma}$ on a node $[A : \alpha]$ are carried from the root node and represent stack symbols that have not been popped by reading a string in $L(G, \alpha)$, while elements of Γ represent stack symbols that are newly pushed on the stack by reading a string in $L(G, \alpha)$.

At the beginning, the algorithm initializes the value $\mathcal{F}(A : \alpha)$ to $\varnothing^{\max\{1, \flat(A) + \sharp(\alpha)\}}$ for every $[A : \alpha] \in \mathrm{dom}(\mathcal{F})$ other than $\mathcal{F}(S : \varepsilon) = \{\langle q_0, \overline{Z_0}\rangle\}$. The algorithm monotonically adds new elements to the labels on nodes so that the following statements are satisfied:

I. for $\mathcal{F}(A : \alpha) = I_1 \ldots I_{\flat(A)+\sharp(\alpha)}$ and $\flat(A) + \sharp(\alpha) > 0$,
   i. $\langle q, X \rangle \in I_1$ with $X \in \Gamma$ iff $\exists x \in \Sigma^*$, $\exists y \in L(G, \alpha)$ such that $(*)$ holds where $\flat(y) + \sharp(y) \geq 1$ and the first element of η is X.
   ii. $X \in I_k$ with $k \geq 2$ and $X \in \Gamma$ iff $\exists x \in \Sigma^*$, $\exists y \in L(G, \alpha)$ such that $(*)$ holds where $\flat(y) + \sharp(y) \geq k$ and the k-th element of η is X.
   iii. $\langle q, \overline{X} \rangle \in I_1$ with $\overline{X} \in \overline{\Gamma}$ iff $\exists x \in \Sigma^*$, $\exists y \in L(G, \alpha)$ such that $(*)$ holds where $\flat(y) + \sharp(y) = 0$ and the first element of η is X.
   iv. $\overline{X} \in I_k$ with $k \geq 2$ and $\overline{X} \in \overline{\Gamma}$ iff $\exists x \in \Sigma^*$, $\exists y \in L(G, \alpha)$ such that $(*)$ holds where $\flat(y) + \sharp(y) < k$ and the k-th element of η is X.
II. for $\mathcal{F}(A : \alpha) \in \mathcal{P}(Q)$ $(\flat(A) + \sharp(\alpha) = 0)$,
   i. $q \in \mathcal{F}(A : \alpha)$ iff $\exists x \in \Sigma^*$, $\exists y \in L(G, \alpha)$ such that $(*)$ holds and $\flat(A) + \sharp(\alpha) = 0$.

Example 2. Let G consist of the production rules

$$S \to aAf \mid bAg, \quad A \to c \mid de$$

and M (simple SFDPDA) consist of the transition rules

$$\langle q_0, Z_0 \rangle \xrightarrow{a} \langle q_0, Z_0 X \rangle, \; \langle q_0, Z_0 \rangle \xrightarrow{b} \langle q_0, Z_0 Y \rangle, \; \langle q_0, Z_0 \rangle \xrightarrow{c} \langle q_0, \varepsilon \rangle,$$

$$\langle q_0, Z_0 \rangle \xrightarrow{d} \langle q_0, \varepsilon \rangle, \; \langle q_0, X \rangle \xrightarrow{e} \langle q_0, W \rangle, \; \langle q_0, Y \rangle \xrightarrow{e} \langle q_0, W \rangle,$$

$$\langle q_0, X \rangle \xrightarrow{f} \langle q_0, \varepsilon \rangle, \; \langle q_0, Y \rangle \xrightarrow{g} \langle q_0, \varepsilon \rangle, \; \langle q_0, W \rangle \xrightarrow{g} \langle q_0, \varepsilon \rangle,$$

and $F = \{q_0\}$. $\sharp_M$ is compatible with $L(G)$ and $\tilde{\sharp}_M(S) = -1$, $\tilde{\sharp}_M(A) = -1$, $\flat_M(S) = 1$, $\flat_M(A) = 2$. The algorithm constructs the forest as follows.

First the algorithm lets all the node labels be sequences of the empty sets except for $\mathcal{F}(S : \varepsilon) = \{\langle q_0, \overline{Z_0}\rangle\}$. By $\langle q_0, Z_0 \rangle \xrightarrow[M]{a} \langle q_0, Z_0 X \rangle$, we let $\mathcal{F}(S : a) =$

$\{\langle q_0, Z_0 \rangle\}\{X\}$. The edge going out from $[S : a]$ labeled with A refers to the tree T_A. $\mathcal{F}(S : a) = \{\langle q_0, Z_0 \rangle\}\{X\}$ is copied to $\mathcal{F}(A : \varepsilon)$ with adding bars. By $\langle q_0, Z_0 X \rangle \xrightarrow{c} \langle q_0, X \rangle$, we put $\{\langle q_0, \overline{X} \rangle\}$ into $\mathcal{F}(A : c)$, where $\overline{X}$ is carried down from the label of the parent node $[A : \varepsilon]$. This stack symbol plays no role for reading c. Similarly we get $\mathcal{F}(A : d) = \{\langle q_0, \overline{X} \rangle\}$. By $\langle q_0, X \rangle \xrightarrow{e} \langle q_0, W \rangle$, we put $\{\langle q_0, W \rangle\}$ into $\mathcal{F}(A : de)$, where W is newly pushed on to the stack when de is read. At this moment, we get the following trees:

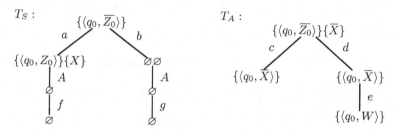

On the other hand, concerning the rule $S \to bAg$, we get $\mathcal{F}(S : b) = \{\langle q_0, Z_0 \rangle\}\{Y\}$. Then copying this with adding bars to the root node of T_A, we get $\mathcal{F}(A : \varepsilon) = \{\langle q_0, \overline{Z_0} \rangle\}\{\overline{X}, \overline{Y}\}$. Similarly $\langle q_0, \overline{Y} \rangle$ is added to $\mathcal{F}(A : c)$ and $\mathcal{F}(A : d)$. Then we get $\mathcal{F}(A : c) = \mathcal{F}(A : d) = \{\langle q_0, \overline{X} \rangle, \langle q_0, \overline{Y} \rangle\}$.

Recall that the node $[A : c]$ is a final node of T_A. $\langle q_0, \overline{X} \rangle \in \mathcal{F}(A : c)$ means that we have $(*)$ where $q = q_0$ and the first element of η is X that is just carried over from the second element of ζ due to the fact $\sharp(A) = -1$. Therefore, we put $\langle q_0, X \rangle$ to $\mathcal{F}(S : aA)$ because X occurs in the second element of $\mathcal{F}(S : a) = \{\langle q_0, Z_0 \rangle\}\{X\}$, while we do not put $\langle q_0, X \rangle$ to $\mathcal{F}(S : bA)$.

On the other hand, the fact $\langle q_0, W \rangle \in \mathcal{F}(A : de)$, where $[A : de]$ is a final node, means that there is $y \in L(G, A)$ (actually $y = de$ here) such that $\langle q, \zeta \rangle \xrightarrow{y} \langle q_0, W \rangle$ and $\flat_M(y) + \sharp_M(y) = 1$. Here W always appears on the stack after the automaton successfully reads $y \in L(G, A)$ independently of the starting configuration $\langle q, \zeta \rangle$ of this computation (strong forward determinacy). Therefore, we put $\langle q_0, W \rangle$ into both $\mathcal{F}(S : aA)$ and $\mathcal{F}(S : bA)$. In this way, we get the following forest:

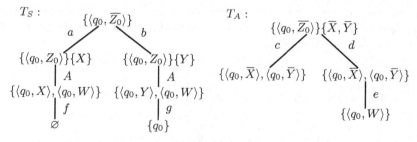

The fact that $\langle q_0, W \rangle \in \mathcal{F}(S : aA)$ implies that there is $x \in L(G, aA)$ such that

$$S \underset{G}{\Rightarrow} aAf \overset{*}{\Rightarrow} xf,$$

$$\langle q_0, Z_0 \rangle \xrightarrow[M]{x} \langle q_0, W \rangle.$$

The node $[S : aA]$ has a child node $[S : aAf]$, however $\Delta(q_0, W, f) = \bot$. The algorithm concludes $L(G) \not\subseteq L(M)$. In fact $adef \in L(G) - L(M)$.

Now we give the definition of functions appearing in the algorithm.

- $\mathrm{Start}(A) = \mathcal{F}(A : \varepsilon)$.
- $\mathrm{Final}(A) = \bigcup\{ \mathcal{F}(A : \alpha) \mid A \to \alpha \in P \}$.
- $\mathrm{Bar}(I) = \begin{cases} \{ \langle q, \overline{X} \rangle \mid \langle q, X \rangle \in I \text{ or } \langle q, \overline{X} \rangle \in I \} & \text{for } I \in \mathcal{P}(Q \times (\Gamma \cup \overline{\Gamma})), \\ \{ \overline{X} \mid X \in I \text{ or } \overline{X} \in I \} & \text{for } I \in \mathcal{P}(\Gamma \cup \overline{\Gamma}). \end{cases}$
  $\mathrm{Bar}(I_1 I_2 \ldots I_m) = \mathrm{Bar}(I_1)\mathrm{Bar}(I_2)\ldots\mathrm{Bar}(I_m)$.
- For $a \in \Sigma$, $\mathrm{Trans}(I_1 \ldots I_m, a) =$
  $\begin{cases} \{q\} & \text{if } m = 1 \text{ and } \sharp(a) = -1, \\ (\{q\} \times I_2)I_3 \ldots I_m & \text{if } m \geq 2 \text{ and } \sharp(a) = -1, \\ \{\langle q, X_0 \rangle\}\{X_1\} \ldots \{X_{\sharp(a)}\}I_2 \ldots I_m & \text{if } \sharp(a) \geq 0, \end{cases}$
  where $\Delta(p, X, a) = \langle q, X_0 \ldots X_{\sharp(a)} \rangle$ for some $p \in Q$ and $X \in \Gamma$.
- For $A \in N$, $\mathrm{Trans}(I_1 \ldots I_m, A) =$
  $\begin{cases} \mathrm{Final}(A) & \text{if } m + \sharp(A) = 0, \\ (\mathrm{Final}(A) \times I_{\flat(A)+1})I_{\flat(A)+2} \ldots I_m & \text{if } m + \sharp(A) > \flat(A) + \sharp(A) = 0, \\ (\vec{J} \cup \varnothing^n \vec{K})I_{\flat(A)+1} \ldots I_m & \text{otherwise, where} \end{cases}$

$$n = \max\{ 0, \sharp(A) \},$$
$$\vec{J} = (Q \times \Gamma)\Gamma^{\flat(A)+\sharp(A)-1} \cap \mathrm{Final}(A),$$
$$\vec{K} = K_{n+1} \ldots K_{\flat(A)+\sharp(A)} \text{ where for } \mathrm{Final}(A) = J_1 \ldots J_{\flat(A)+\sharp(A)},$$

$$K_i = \begin{cases} \{ \langle q, X \rangle \mid X \in I_{1-\sharp(A)} \text{ and } \langle q, \overline{X} \rangle \in J_1 \} \\ \quad \cup \{ \langle q, \overline{X} \rangle \mid \overline{X} \in I_{1-\sharp(A)} \text{ and } \langle q, \overline{X} \rangle \in J_1 \} & \text{if } i = 1, \\ \{ X \mid X \in I_{i-\sharp(A)} \text{ and } \overline{X} \in J_i \} \\ \quad \cup \{ \overline{X} \mid \overline{X} \in I_{i-\sharp(A)} \text{ and } \overline{X} \in J_i \} & \text{otherwise.} \end{cases}$$

The termination of the algorithm is easily seen, because the algorithm monotonically expands the sets constituting node labels and those sets have the upper bound $Q \times (\Gamma \cup \overline{\Gamma})$, $\Gamma \cup \overline{\Gamma}$ or Q. Lemma 4 shows the soundness of the algorithm and Lemma 5 shows the completeness.

Lemma 4. *If the algorithm in Figure 2 outputs "*YES*", then $L(G) \subseteq L(M)$.*

Proof. Let us write $\langle q, \zeta\overline{\eta} \rangle \in I_1 \ldots I_m$ if one of the following conditions holds:

- $\zeta = \eta = \varepsilon$, $q \in I_1$ and $m = 1$, or
- $\zeta = \varepsilon$, $\eta = Y_1 \ldots Y_m$, $\langle q, \overline{Y_1} \rangle \in I_1$ and $\overline{Y_i} \in I_i$ for $2 \leq i \leq m$, or
- $\zeta = X_1 \ldots X_k$ with $k \geq 1$, $\eta = Y_{k+1} \ldots Y_m$, $\langle q, X_1 \rangle \in I_1$, $X_i \in I_i$ for $2 \leq i \leq k$ and $\overline{Y_i} \in I_i$ for $k < i \leq m$.

One can prove the following claim by induction on the number of derivation steps of $\alpha \overset{*}{\Rightarrow} y$. Applying this claim to $w \in L(G, S)$, we get the desired property.

Suppose that the algorithm outputs "YES". For any rule $A \to \alpha\beta \in P$ and $y \in L(G, \alpha)$, if $\langle p, \overline{\zeta_1 \zeta_2} \rangle \in \text{Start}(A)$ with $|\zeta_1| = \flat(y)$, then for some $\eta_1 \in \Gamma^{\flat(y)+\sharp(y)}$

$$\langle p, \zeta_1\zeta_2 \rangle \xrightarrow[M]{y} \langle q, \eta_1\zeta_2 \rangle \text{ and } \langle q, \eta_1\overline{\zeta_2} \rangle \in \mathcal{F}(A : \alpha). \qquad \square$$

Lemma 5. *If $L(G) \subseteq L(M)$, then the algorithm in Figure 2 outputs "YES".*

Proof. One can prove by induction on the number of steps of computation of the algorithm that the "only if" directions of conditions I–II (3 pages before) always hold if $L(G) \subseteq L(M)$. This claim ensures that the algorithm never outputs "NO". $\qquad \square$

4.3 Time Complexity

We evaluate the efficiency of our algorithm with the parameter τ_G defined as

$$\tau_G = \max\{ \tau(G, A) \mid A \in N \} \text{ where } \tau(G, A) = \min\{ |y| \mid y \in L(G, A) \}.$$

It is easy to see that $\tau_G \leq \rho_G^{|N|}$ where ρ_G is the length of a longest rule of G and that τ_G cannot be polynomially bounded in $\|G\|$. Nevertheless if an algorithm taking a CFG G as input runs in polynomial time in $\|G\|\tau_G$, it deserves to be called efficient for some application. When from a finite language L one constructs a CFG G whose all rules are useful for deriving some element of L, τ_G is not bigger than the length of a longest string in L. For instance, Yoshinaka's algorithm[11] for deciding whether $L(G) \subseteq L(M)$ for a CFG G and a simple SUDPDA M runs in polynomial time in $\tau_G\|G\|\|M\|$, and his learning algorithm using that algorithm as a subroutine runs in polynomial time in the size of the given data.

Lemma 6. *If $\sharp_M$ is compatible with $L(G)$, then $-\tau_G \leq \tilde{\sharp}_M(A) < \|G\|\tau_G$ and $1 \leq \tilde{\flat}_M(A) \leq \|G\|^2\tau_G$ for every $A \in N \cup \Sigma$. Moreover, Stage 1 of the algorithm in Figure 2 decides the compatibility in $O(\|M\| + \|G\|^2)$ time.*

Concerning the comparison forest, $|\text{dom}(\mathcal{F})| \leq \|G\|$, $|\mathcal{F}(A : \alpha)| = \max\{1, \tilde{\flat}(A) + \sharp(\alpha)\}$ and each element of $\mathcal{F}(A : \alpha)$ has the upper bound $Q \times (\Gamma \cup \tilde{\Gamma})$, $\Gamma \cup \tilde{\Gamma}$ or Q. Since the labels are expanded monotonically, Stage 2 runs in polynomial time in $\|G\|$, $|Q|$, $|\Gamma|$ and the maximums of $\tilde{\flat}(A)$, $\tilde{\sharp}(A)$ for $A \in N \cup \Sigma$. Lemma 6 entails the following theorem.

Theorem 1. *Let G be a CFG and M be an SFDPDA. One can decide whether $L(G) \subseteq L(M)$ in time polynomial in $\|M\|$, $\|G\|$ and τ_G.*

5   Concluding Remarks

This paper has presented an efficient algorithm that decides whether $L(G) \subseteq L(M)$ for a CFG G and an SFDPDA M. The algorithm resembles the author's algorithm for deciding whether $L(G) \subseteq L(M)$ for a CFG G and a simple SUDPDA M[11]. Both run in polynomial time in τ_G, $\|G\|$ and $\|M\|$. When comparing a PDA M' instead of a CFG G with M, (slight modifications of) those algorithms run in single exponential time in $\|M'\|$ and $\|M\|$. The comparison forests constructed by those two algorithms have the same form and the difference is only the labels on the forest. In fact for SUDPDAs, which form a common superclass of SFDPDAs and simple SUDPDAs, the basic idea still works with giving appropriate node labels. However no inclusion checking algorithm for SUDPDAs is known to run in single exponential time. Thus we now have two different kinds of restriction on SUDPDAs that allow crucially more efficient algorithms for the inclusion problem. On the other hand, if the language of a DPDA is defined by accept by final state (the stack may be nonempty), the inclusion problem for SUDPDAs turns to be undecidable[3]. It is not yet clear if it is the case for SFDPDAs too.

Our result is interesting not only from the point of view of computational complexity theory but also from the view point of grammatical inference. Our algorithm can be utilized for constructing an efficient algorithm learning SFD-PDAs from positive data as Yoshinaka's learning algorithm for simple SFDPDAs has a subroutine that solves the inclusion problem.

References

1. Asveld, P.R.J., Nijholt, A.: The inclusion problem for some subclasses of context-free languages. Theoretical Computer Science 230(1-2), 247–256 (2000)
2. Friedman, E.P.: The inclusion problem for simple languages. Theoretical Computer Science 1(4), 297–316 (1976)
3. Friedman, E.P., Greibach, S.A.: Superdeterministic dpdas: The method for accepting does affect decision problems. Journal of Computer and System Science 19(1), 79–117 (1979)
4. Greibach, S.A., Friedman, E.P.: Superdeterministic PDAs: A subcase with a decidable inclusion problem. Journal of the Association for Computing Machinery 27(4), 675–700 (1980)
5. Linna, M.: Two decidability results for deterministic pushdown automata. Journal of Computer and System Science 18, 92–107 (1979)
6. Tozawa, A., Minamide, Y.: Complexity results on balanced context-free languages. In: Seidl, H. (ed.) FOSSACS 2007. LNCS, vol. 4423, pp. 346–360. Springer, Heidelberg (2007)
7. Wakatsuki, M., Teraguchi, K., Tomita, E.: Polynomial time identification of strict deterministic restricted one-counter automata in some class from positive data. In: Paliouras, G., Sakakibara, Y. (eds.) ICGI 2004. LNCS (LNAI), vol. 3264, pp. 260–272. Springer, Heidelberg (2004)
8. Wakatsuki, M., Tomita, E.: A fast algorithm for checking the inclusion for very simple deterministic pushdown automata. IEICE transactions on information and systems E76-D(10), 1224–1233 (1993)
9. Yokomori, T.: Polynomial-time identification of very simple grammars from positive data. Theoretical Computer Science 298, 179–206 (2003)
10. Yokomori, T.: Polynomial-time identification of very simple grammars from positive data. Theoretical Computer Science 377(1–3), 282–283 (2003); Theoret. Comput. Sci. 298, 179–206 (2003)
11. Yoshinaka, R.: Polynomial-time identification of an extension of very simple grammars from positive data. In: Sakakibara, Y., Kobayashi, S., Sato, K., Nishino, T., Tomita, E. (eds.) ICGI 2006. LNCS (LNAI), vol. 4201, pp. 45–58. Springer, Heidelberg (2006)

Author Index